STRUCTURE REPORTS

Volume 36

CUMULATIVE INDEX

for 1961-1970

Structure Reports is prepared under the guidance of a Commission of the International Union of Crystallography.

STRUCTURE REPORTS

Volume 36

CUMULATIVE INDEX

for 1961-1970

Editor

W. B. Pearson

Compilers

N. E. Sennema

W. R. Sennema

SPRINGER-SCIENCE+BUSINESS MEDIA, B.V.

ISBN 978-94-017-3162-1 ISBN 978-94-017-3160-7 (eBook)
DOI 10.1007/978-94-017-3160-7

First published in 1983

Originally published by D. Reidel Publishing Company in 1983
Softcover reprint of the hardcover 1st edition 1983

TABLE OF CONTENTS

Introduction, VI

Cumulative Subject Index, 1

Cumulative Formula Index, 148

Cumulative Index of Carbon Compounds, 213

Cumulative Author Index, 241

INTRODUCTION

This Cumulative Index of *Structure Reports* is for the years 1961 to 1970 (Vols. 26 to 35). The Subject Index is arranged in strict alphabetical succession regardless of the construction of words, although in the listing of organic compounds certain prefixes such as *mono, o-, m-, p-, D* and *L* are disregarded. Nevertheless, some inconsistencies in the rendering of these prefixes and others such as *trans, cyclo* and *iso* remain, and where a name is sought which contains these, it should be searched for both with and without regard for the prefix. The Formula Index which lists Metals and Inorganic substances, is arranged in alphabetical order of chemical symbols. Organic compounds are listed in the Index of Carbon Compounds, which is indexed first by C, then H with other elements following in alphabetical order of chemical symbols. A carbon compound not appearing in this Index should also be sought in the Formula Index.

The scheme usually employed for the transliteration of Russian is given below.

Waterloo
20 October 1982

W. B. PEARSON

TRANSLITERATION OF RUSSIAN

а	a	и	i	р	r	ш	š
б	b	и	j	с	s	щ	šč
в	v	к	k	т	t	ы	y
г	g	л	l	у	u	ъ	"
д	d	м	m	ф	f	ь	'
е	e	н	n	х	kh	э	è
ж	ž	о	o	ц	c	ю	ju
з	z	п	p	ч	č	я	ja

CUMULATIVE SUBJECT INDEX

This index contains (i) the names of the substances as printed at the heads of the reports, (ii) certain alternative names, which may or may not appear in the reports, and (iii) names of substances represented only as formulae in the reports.

Numerical prefixes, *mono* (not di, tri, bis, ...), *o*-, *m*-, D-, L-, *etc.* are ordinarily disregarded in fixing the alphabetical order. Occasionally double entries are made.

In double salts *etc.* the metal names are alphabetical: aluminum sodium fluoride, not sodium aluminum fluoride.

Abernathyite, 29, 377
ABO_4 crystals, 27, 520
Acenaphthalenepentacarbonyldi-iron, 33B, 550
Acenaphthene derivatives, 27, 933
cis-1,2-Acenaphthenediol, 28, 435, 674
trans-1,2-Acenaphthenediol, 28, 674
cis-1,2-Acenaphthenediol dinitrate, 28, 674; 29, 681
Acenaphthenequinone, 28, 434, 674
Acenaphthylene, 35B, 908
Acenaphthylenediiron pentacarbonyl, 35B, 572
Acepleiadylene, S-trinitrobenzene complex, 31B, 279
Acetamide, 30B, 12
Acetamide hemibromohydrate, 34B, 3
Acetamide hemihydrobromide, 32B, 9
Acetamide-sodium bromide, 27, 761
Acetamide-sodium bromide complex, 31B, 271
Acetamide-sodium iodide, 27, 761
2-Acetamido-2-deoxy-D-glucose, 31B, 171
Acetanilide, 31B, 70
Acetate acidium hydrogensulphate, 33B, 38
Acetato-*C*-*rac*-(5,7,7,12,14,14-hexamethyl-1,4,8,11-tetraaza-cyclotetradecane)nickel(II) perchlorate, 35B, 772
Acetatopentaamminecobalt(III) chloride perchlorate, 30B, 344
Acetic acid, 35B, 3
Acetic acid-sulphuric acid, 33B, 38
Acetohydroxamic acid hemihydrate, 35B, 20
Acetoinenediol cyclophosphate, 32B, 220
Acetone hydrate, 26, 344, 551
Acetone-sodium iodide, 27, 761
Acetonitrile adduct of icosaborane-16, 31B, 292
Acetonitrile: antimony pentachloride complex, 32B, 266
Acetonitrile-bromine, 33B, 238
Acetonitrile dihydrobromide, 32B, 28
1-Acetonyl-1-thionia-5-thiacyclo-octane perchlorate, 35B, 223
5α-Acetoxy-6β-bromohexahydrophysalin A, 35B, 400
5α-Acetoxy-6β-bromo-hexahydrophysalin A methanol solvate, 34B, 344
(-)-3β-Acetoxy-4α-t-butylcarbamoyl-3α-methyl-7β-(*p*-bromophenacyl)-acetamidocepham 1α-oxide, 35B, 915
(+)-*trans*-2-Acetoxycyclopropyl-trimethylammonium iodide, 35B, 922
2β-Acetoxy-7α, 11α-dibromolanostane-8α, 9α-epoxide, 31B, 208
3β-Acetoxy-17α-hydroxy-16β-bromo-5α-pregnan-11,20-dione, 33B, 194
5-Acetoxy-6-methoxy-8-nitro-quinoline, 32B, 160
π-(*exo*-2-Acetoxynorbornenyl)-*exo*-tricarbonylchromium, 35B, 917
4-Acetylacenaphthene, 27, 933
Acetylacetonato-bis(carbonyl) rhodium (I), 32B, 495
Acetylacetonatobis-(*N*-phenylamino-troponiminato)manganese(III), 35B, 913
Acetylacetonato(cyclo-octa-2,4-dienyl)-palladium, 30B, 393; 31B, 370
Acetylacetonato-π-duroquinone-rhodium, 35B, 574
Acetylacetonato(π-pentamethyl-bicyclo[2.2,0]hexa-2,5-dienyl-methyl)palladium(II), 35B, 576
Acetylacetone-*mono*-(*o*-hydroxyanil)-copper(II), 30B, 361

Acetylacetone metal chelates, 31B, 400
bis-Acetylacetone nickel (II), 26, 673
bis-Acetylacetone zinc (II), 26, 673
Acetylacetoniminotris (acetylacetonato)ytterbium(III), 35B, 612, 614
3'-O-Acetyladenosine, 35B, 297, 911
Acetylbenchrotrene[acetophenonetricarbonylchromium], 35B, 918
1-Acetyl-1'-benzoylferrocene, 34B, 454
Acetyl-bromoacetyldihydroenmein, 31B, 227
Acetyl-bromoacetyldroenmein, 29, 771
N-Acetylbromoanisomycin, 33B, 188
4-Acetyl-2'-bromobiphenyl, 30B, 428
4-Acetyl-3'-bromobiphenyl, 34B, 81
5-Acetyl-9-bromo-1,3,7,8,10-pentamethyl-1,5-dihydroisoalloxazine, 35B, 199
2-Acetyl-5-bromothiophene, 35B, 218
Acetyl chloride, 26, 546
4-Acetyl-2'-chlorobiphenyl, 30B, 428; 33B, 72
Acetylcholine chloride, 27, 991; 35B, 321
3-Acetyl-2,6-dioxo-4-hydroxy-5-hydroxyethylidenepyran, 35B, 210
N(a)-Acetyl-7-ethyl-5-desethylaspidospermidine N(b)-methiodide, 30B, 229
N-Acetylethylenimine, 33B, 531
cis-9-Acetyl-4a-ethyl-1,2,3,4,4*a*,9*a*-hexahydrocarbazole, 35B, 923
4-Acetyl-2'-fluorobiphenyl, 30B, 427; 33B, 73
N-Acetylglucosamine, 29, 766
N-Acetyl-α-D-glucosamine, 31B, 171
N-Acetylglycine, 27, 764
Acetyl hexachloroantimonate, 32B, 318
2-Acetyl-3-hydroxythiophene, 34B, 180
3-Acetylimino-5-acentamido-1,2-dithiole, 28, 670
2-Acetyl-3-indazolinone, 34B, 151
4-Acetyl-2'-iodobiphenyl, 30B, 428
DL-Acetylleucine *N*-methylamide, 34B, 272
(±)-1-Acetyl-3-methylaspidospermidine 9-methiodide, 35B, 389
(+-)-1-Acetyl-16-methylaspidospermidine-14-methiodide, 32B, 491
4-Acetyl-2'-nitrobiphenyl, 35B, 910
N-Acetyl-N'-phenyl-selenourea, 34B, 33
α-Acetyl β-phenyl selenourea, 28, 675
Acetyl-2-pyridine chlorohydrate, 30B, 397
2-Acetylpyridine hydrochloride, 31B, 123
Acetylsalicylic acid (aspirin), 29, 653
7-Acetyl-2,5,8-trihydroxy-3-methoxy-6-methyl-1,4-naphthoquinone, 30B, 104
Acetyltriphenylgermane, 33B, 306
Acetyltriphenylsilane, 34B, 373
Acid chloromaleates, 26, 554
Acid hydrogen potassium fumarate, 26, 554; 35B, 16
1:1 Acid potassium myristate, 30B, 430
Acid salts of nitrobenzoic acid, 26, 705-709
Acid with formamide, 34B, 349
Acorenone-B4-iodo-2-nitrophenylhydrazone, 33B, 549
Acorone, 30, 397
Acorone *p*-bromophenylsulphonylhydrazone, 31B, 216
Acridine III, 26, 780
Acridine-cytosine-water complex, 33B, 554
Acrolein, 33B, 530; 35B, 892
Acrylamide, 31B, 3
Acrylates, 35B, 909
Acrylic acid, 28, 365
Acrylonitrile, 35B, 892
Actinide(IV) *N*,*N*-diethyldithiocarbamates, 33B, 543
Actinide(IV) fluorides, 31A, 249
Actinium, 26, 3
Actinium hydride, 26, 3
Actinomycin C_3, 29, 771
Acutimine, 33B, 223
Acutimine acetate, 33B, 223
Adamantane, 30B, 115; 32B, 148
Addition compound dithiane-iodoform, 26, 773
1:1 Addition compound of formic acid with formamide, 34B, 349
1:1 Addition compound of iodoform-1,4-dioxan, 34B, 348
Addition compound iodoform-sulphur, 27, 747
Addition compound pyridine and iodine, 26, 778
Addition compounds of ammonia with ammonium halides, 26, 298
1:0.5:1 Adduct of tribromobis(phenyldimethylphosphine)nickel(III), dibromobis(phenyldimethylphosphine)nickel(II), and benzene, 35B, 836
Adelpholite, 30A, 455
$(AdenineH)_2Cu_3Cl_8.4H_2O$, 35B, 917
Adenosine-5-bromouridine, 28, 685
Adenosine, 5'-bromouridine monohydrate (1:1 hydrogen-bonded complex), 30B, 196
Adenosine 3',5'-cyclic phosphate, 33B, 554
Adenosine-5'phosphate, 28, 521
Adenosine 3'-phosphate dihydrate, 31B, 185
Adenosine triphosphate, 35B, 922
β-Adenosine-2'-β-uridine-5'-phosphoric acid, 29, 771
β-Adenosine-2'-β-uridine-5'-phosphoric acid tetrahydrate, 34B, 265

Adenylic acid b, 31B, 185
Adiantol B bromoacetate, 35B, 403, 910
Adipamide, 30B, 396, 413; 31B, 9
Adipic acid, 26, 559; 30B, 4
Adipic acid-urea, 26, 572
Aegirine-augite, 31A, 242
Aenigmatite, 29, 428; 30A, 458
Aerugite, 30A, 457
Aeschynite, 27, 535
Aflatoxin B1, 35B, 374
Aflatoxin B2, 35B, 374
Aflatoxins, 35B, 374
Aikinite, 30A, 458; 35A, 30
Aksaite, 27, 558
Akuammidine methodide monohydrate, 28, 557
Alamosite, 33A, 454
β-Alanine, 30B, 207
D-Alanine, 27, 763
L-Alanine, 27, 763; 31B, 187
DL-α-Alanine hydrochloride, 27, 764
L-Alanyl-L-alanine hydrochloride, 34B, 270
L-Alanylglycine, 35B, 308
(β-Alanyl-L-histidinato)copper(II) dihydrate, 32B, 421
β-Alanyl-L-histidine copper(II) complex, 30B, 395
Albite, 27, 676; 29, 398; 34A, 368
Alchoreine bromomethylate, 35B, 918
Aldanite, 27, 673
Aldehyde 2,4-dinitrophenylhydrazone, 26, 701
Aldrin, 30B, 425
Alfalfa mosaic virus, 27, 1061
Alginic acid, 27, 1056
Alginic acids, 35B, 922
Aliphatic carboxylic acids, 26, 831
Alizarine, 26, 751
Alkali-alkaline earth arsenates, 26, 469-470
Alkali-alkaline earth phosphates, 26, 469-470
Alkali-alkaline earth vanadates, 26, 470
Alkali azides, 30A, 256
Alkali fluoroberyllates, 27, 552
Alkali halides, 26, 296; 30A, 260
Alkali halide high pressure forms, 27, 424
Alkali hexahalogentellurates, 27, 463
Alkali imidodisulphates, 30A, 367
Alkali iron(III) sulphates, 31A, 246
Alkali metals, 26, 3
Alkali metal acetylides, 33B, 296
Alkali-metal amides, 27, 420
Alkali-metal aurates, 35A, 214
Alkali metal t-butylates, 33B, 297
Alkali-metal compounds, 29, 3
Alkali metal cuprates, 34A, 247
Alkali-metal germanates, 31A, 252
Alkali-metal hexahalogomolybdates, 27, 464
Alkali-metal-iron-sulphur, 33A, 3
Alkali-metal mesouranates, 27, 544
Alkali metal methylates, 35B, 472
Alkali-metal molybdates, 35A, 232
Alkali-metal nickelates, 34A, 293
Alkali-metal noble metal oxides, 33A, 280
Alkali-metal phosphides, 27, 420
Alkali metal propynylides, 33B, 296
Alkali-metal selenates, 27, 605
Alkali-metal silicates, 26, 505-509
Alkali-metal sulphates, 27, 604
Alkali-metal trimethylsilanolates, 35B, 440
Alkali metal uranium fluorides, 31A, 249
Alkali-metal wolframates, 35A, 232
Alkaline amphiboles, 28, 258
Alkaline-earth cuprates, 35A, 246
Alkaline-earth fluoborates, 27, 459
Alkaline-earth germanates, 27, 541
Alkaline-earth metaborates, 34A, 354
Alkaline-earth metal uranates, 34A, 289
Akaline-earth rare-earth oxides, 35A, 230
Alkaline-earth rare-earth sulphides, 27, 327
Alkali rhenium bronzes, 31A, 260
Alkali titanate bronze, 35A, 276
Alkali titanates, 29, 318; 31A, 242
Alkali titanoniobates, 29, 327
Alkaloid II methiodide (from daphniphyllum macropodum), 31B, 258
Alkaloids of calabash curare, 26, 830
p-Alkoxyphenylsuccinimides, 35B, 906
n-Alkylamine-water complexes, 33B, 528
Alkylammonium-vermiculite complexes, 35B, 905
n-Alkylarsinic acids, 35B, 909
Alkyl chloride, 26, 575
Allactite, 33A, 422
Allanite, 29, 439; 31A, 255
DL-Allantoin, 30B, 125
All-*trans*-cyclododecatriene-1,5,9-nickel(0), 30B, 394
All-*cis*-1,6-dichlorocyclodeca-1,3,6,8-tetraene, 32B, 486
Alleghanyite, 35A, 467
Allenetriphenylphosphinediiron hexacarbonyl, 33B, 546
Allenic ketones, 35B, 916
Allevardite, 31A, 242
Allitol, 35B, 918
Allophane, 30A, 458
Alloxan, 29, 715; 30B, 159
Alloxantin dihydrate, 30B, 168

Alloys and compounds of yttrium, 26, 276, 277
Alluaudite, 31A, 244
Allyl alcohol, 29, 453
N-Allyl-*o*-(carboxymethoxy)benzamide, 29, 763
Allylchloroplatinum, 35B, 914
Allylcyclopentadienyl palladium, 30B, 421
π-Allyl-π-cyclopentadienyl palladium, 31B, 343; 33B, 331
2-*trans*-Allyl-3,5-dichloro-1-hydroxy-4-oxo-(1s,5s)-2-cyclopentene-1-carboxylic acid methyl ester, 34B, 339
π-Allyldi(thiourea)nickel(II) chloride, 35B, 515
2-Allyl-2'-hydroxy-5,9-dimethyl-6,7-benzomorphan hydrobromide monohydrate, 35B, 379
π-Allyl-palladium acetate, 29, 768
Allylpalladium chloride complex, 30B, 299
Allyl-palladium complexes, 33B, 554
Allylthiourea, 30B, 42
Allyltricarbonyl iron iodide, 30B, 418
π-Allyl-tricarbonyl-iron iodide, 31B, 343; 33B, 330
Allyltricarbonyl iron nitrate, 30B, 419
π-Allyltriphenylphosphinedicarbonyl-iron iodide, 33B, 340
π-Allyl(triphenylphosphine)palladium-trichlorotin, 33B, 551
Almandine, 29, 440
Alnico, 27, 9; 30A, 186
Alpha brass, 26, 139
dl-Alphaprodine hydrochloride, 26, 782
Alstonia alkaloid C, 28, 677
Alumina, 26, 346-348
α-Alumina, 26, 347
β-Alumina, 33A, 275
β'-Alumina, 33A, 276
β"-Alumina, 33A, 277
κ-Alumina, 26, 347
χ-Alumina, 26, 347
Alumina-silica, 27, 707
Aluminates, 26, 324, 370, 394; 27, 518
Alumino-aeschynite, 29, 439
Aluminosilicates, 26, 505, 517
Aluminosiloxanes, 33B, 268
Aluminum, 26, 3, 279; 27, 381, 385
Aluminum antimony bromide, 34A, 213
Aluminum arsenide, 27, 3
Aluminum barium hydrates, 35A, 292
Aluminum-beryllium-boron, 27, 3
Aluminum-beryllium-iron, 27, 4
Aluminum borohydride-trimethylamine, 30B, 393
Aluminum-boron, 26, 4; 29, 4; 31A, 3; 32A, 3
Aluminum-boron-carbon, 29, 4; 34A, 3; 35A, 6
Aluminum-boron-cobalt, 27, 4, 92
Aluminum-boron-copper, 35A, 3
Aluminum-boron-iron, 34A, 6
Aluminum-boron-magnesium, 35A, 4
Aluminum-boron-manganese, 31A, 3
Aluminum-boron-molybdenum, 31A, 4
Aluminum-boron-nickel, 27, 4, 92
Aluminum bromide-benzene complex, 26, 653
Aluminum bronze, 27, 12, 383
Aluminum cadmium oxides, 31A, 247
Aluminum calcium borate, 33A, 444
Aluminum calcium hydrogarnets, 32A, 426
Aluminum calcium hydroxide, 28, 151
Aluminum calcium oxides, 26, 369; 33A, 274
Aluminum-calcium-silicon, 32A, 5
Aluminum-carbon-chromium, 28, 3
Aluminum carbonitrides, 31A, 5
Aluminum-carbon-managanese, 27, 272
Aluminum-carbon-nitrogen, 28, 3; 31A, 5
Aluminum-carbon-platinum, 26, 225
Aluminum-carbon-silicon, 26, 5
Aluminum-cerium, 26, 174; 30A, 3
Aluminum-cerium-cobalt, 27, 5
Aluminum-cerium-copper, 26, 175; 27, 5
Aluminum-cerium-iron, 26, 175; 27, 5
Aluminum-cerium-manganese, 26, 175; 27, 5
Aluminum-cerium-nickel, 27, 5
Aluminum-cerium-sulphur, 33A, 4
Aluminum-cerium-thorium, 26, 5
Aluminum chloride, diethyl ether (1:1 complex), 30B, 415
Aluminum chloride hexahydrate, 33A, 202
Aluminum chloride-nitrobenzenes, 26, 704
Aluminum chloride-nitromethane, 26, 703
Aluminum-chromium-molybdenum-silicon, 26, 21
Aluminum-chromium-silicon, 26, 21
Aluminum-cobalt, 26, 6; 27, 8
Aluminum-cobalt-dysprosium, 27, 264
Aluminum-cobalt-iron-nickel, 27, 9
Aluminum-cobalt-manganese, 27, 10
Aluminum-cobalt-silicon, 27, 11
Aluminum-cobalt-titanium, 27, 11
Aluminum-cobalt-zirconium, 35A, 7
Aluminum-copper, 26, 6; 27, 12, 185; 28, 6; 30A, 3; 33A, 72
Aluminum-copper-iron, 26, 7
Aluminum-copper-magnesium, 27, 12
Aluminum-copper-manganese, 26, 8; 27, 15; 33A, 95
Aluminum-copper-nickel, 31A, 39

Aluminum-copper-oxygen, 26, 396
Aluminum-copper-titanium, 27, 296
Aluminum-copper-zirconium, 27, 296
Aluminum-deuterium-thorium, 26, 27
Aluminum-dysprosium, 26, 236, 237; 29, 5
Aluminum dysprosium borate, 27, 562
Aluminum-dysprosium-iron, 27, 264
Aluminum-dysprosium-nickel, 27, 264
Aluminum-erbium, 26, 174
Aluminum erbium borate, 27, 562
Aluminum ethoxide, 28, 671
Aluminum europium borate, 27, 562
α-Aluminum fluoride, 26, 300
Aluminum-gadolinium, 26, 236, 237; 29, 4
Aluminum gadolinium borate, 27, 562
Aluminum gadolinium garnet, 26, 414
Aluminum gallium fluoride, 26, 300
Aluminum gallium lithium silicate, 26, 523
Aluminum gallium oxide, 31A, 244
Aluminum-germanium-manganese, 26, 9; 34A, 9
Aluminum-gold, 35A, 7
Aluminum-hafnium, 26, 9, 10, 237; 27, 16, 378
Aluminum-hafnium-silicon, 26, 10; 27, 17
Aluminum-hafnium-silicon-zirconium, 26, 11
Aluminum holmium borate, 27, 562
Aluminum hydride, 34A, 153
Aluminum hydride-N,N,N',N'-tetramethylethylenediamine, 29, 536
Aluminum-hydrogen-thorium, 28, 6
Aluminum hydroxide, 29, 298
Aluminum-iron, 26, 11-16; 27, 18; 30A, 4
Aluminum-iron-lithium-oxygen, 26, 394
Aluminum iron oxide, 31A, 244
Aluminum-iron-uranium, 27, 20
Aluminum-iron-zirconium, 35A, 7
Aluminum isopropoxide, 28, 671
Aluminum-lanthanum, 32A, 6
Aluminum-lithium, 27, 21; 33A, 6
Aluminum lithium fluoride, 33A, 181
Aluminum-lithium-magnesium-zinc, 27, 21; 29, 6
Aluminum lithium molybdate(VI), 35A, 247
Aluminum-lithium-oxygen, 26, 413
Aluminum lithium silicate, 26, 506, 523
Aluminum lithium silicate, high-quartz, 35A, 453
Aluminum lithium tetraethyl, 29, 535
Aluminum-magnesium, 27, 33; 30A, 5; 33A, 8
Aluminum magnesium hydroxide, 27, 423
Aluminum-magnesium hydroxycarbonate, 32A, 421
Aluminum magnesium oxide, 27, 487, 545
Aluminum-magnesium-oxygen, 26, 396, 413
Aluminum magnesium sodium oxide, 34A, 250
Aluminum-magnesium-zinc, 27, 22; 35A, 8
Aluminum-manganese, 26, 15; 28, 7
Aluminum-manganese-nickel, 26, 8
Aluminum-manganese-palladium, 33A, 95
Aluminum-manganese-silicon, 31A, 8
Aluminum-manganese-silicon-zinc, 26, 16
Aluminum-manganese-silver, 27, 22
Aluminum-manganese-zinc, 26, 17
Aluminum mellitate hexahydrate, 30B, 425
Aluminum methyl chloride dimer, 28, 596
Aluminum-molybdenum, 27, 23, 377; 29, 7, 9
Aluminum-molybdenum-silicon, 26, 18; 27, 25
Aluminum-neodymium, 26, 236, 238
Aluminum neodymium borate, 27, 562
Aluminum nickel borate, 35A, 288
Aluminum-nickel-cerium, 33A, 13
Aluminum-nickel-holmium, 33A, 13
Aluminum-nickel-niobium, 27, 26
Aluminum-nickel-nitrogen, 26, 219
Aluminum-nickel-oxygen, 26, 413
Aluminum-nickel-silicon, 27, 27
Aluminum-nickel-titanium, 27, 296
Aluminum-nickel-zirconium, 27, 296; 35A, 7
Aluminum-niobium, 26, 19
Aluminum niobium oxide, 27, 523
Aluminum-niobium-silicon, 26, 21
Aluminum nitrate nonahydrate, 30A, 402
Aluminum nitride, 26, 348
Aluminum-nitrogen cage compound, 35B, 480
Aluminum-osmium, 29, 10; 30A, 10
Aluminum oxide, 27, 499
β-Aluminum oxide, 32A, 265
Aluminum oxide-chromium oxide, 27, 500
Aluminum oxycarbide, 26, 295
Aluminum-palladium, 28, 7; 29, 12
Aluminum periodate, 26, 500
Aluminum phosphate, 29, 266, 367; 31A, 183, 245
Aluminum-platinum, 26, 225; 27, 321; 29, 12; 31A, 10
Aluminum-plutonium, 26, 231; 27, 28; 30A, 11
Aluminum potassium molybdate, 35A, 254
Aluminum-rare earths, 31A, 12; 32A, 8
Aluminum rare earth oxides, 28, 136
Aluminum-rhenium, 26, 23; 27, 29; 29, 9; 32A, 11
Aluminum-rhodium, 32A, 12; 33A, 15
Aluminum-ruthenium, 30A, 12; 31A, 16;

33A, 16
Aluminum samarium borate, 27, 562
Aluminum-selenium, 31A, 17
Aluminum-silicon-tantalum, 26, 25
Aluminum-silicon-titanium, 26, 18; 27, 30
Aluminum-silicon-vanadium, 26, 21
Aluminum-silicon-wolfram, 26, 25
Aluminum-silicon-zirconium, 26, 10; 27, 31; 30A, 14
Aluminum-silver, 26, 187; 27, 33
Aluminum sodium germanate, 32A, 492
Aluminum sodium selenate, basic, 28, 217
Aluminum solid solutions, 27, 32
Aluminum sulphate, basic, 28, 218
Aluminum sulphate guanidinium hexahydrate, 27, 828
Aluminum-sulphur, 28, 8
Aluminum-sulphur-zinc, 32A, 13
Aluminum tantalate, 29, 420
Aluminum-technetium, 27, 30; 29, 9; 32A, 11
Aluminum tellurium oxide, 27, 526
Aluminum terbium borate, 27, 562
Aluminum ternary alloys, 27, 3
Aluminum tetrahydroborate-trimethylamine, 33B, 303
Aluminum tetroxycarbide, 28, 66
Aluminum-thorium, 26, 256; 33A, 17
Aluminum-titanium, 26, 28; 27, 35, 378
Aluminum-titanium-vanadium, 27, 35
Aluminum triethyl-potassium fluoride, 26, 611
Aluminum trifluoride, 27, 438
Aluminum trifluoride monohydrate, 29, 253
Aluminum-uranium, 26, 30; 27, 36
Aluminum-uranium-zirconium, 26, 30
Aluminum vanadium oxides, 31A, 246
Aluminum wolframate, 33A, 352
Aluminum ytterbium borate, 27, 562
Aluminum-yttrium, 26, 238; 32A, 14
Aluminum yttrium borate, 27, 562
Aluminum-yttrium garnet, 26, 414; 30A, 452
Aluminum yttrium oxide, 27, 520
Aluminum-zinc, 27, 33; 31A, 17
Aluminum-zirconium, 26, 10, 31; 27, 36, 378
Alumobritholite, 26, 536
Alums, 26, 446; 32A, 339
Alunite, 27, 634; 30A, 376
Alvanite, 28, 334
Amarantite, 32A, 510; 33A, 377
Americium, 26, 32; 27, 39
Americium(III) chloride, 31A, 251
Americium trichloride, 35A, 151
Amesite, 27, 690
Amidinium carboxylates, 26, 732-734
2-Amido-4-ethoxy-3,4-dihydropteridin-1-ium bromide, 34B, 146
DL-μ-Amido-μ-hydroperoxy-bis(bis (ethylenediamine) cobalt(III) tetranitrate, 32B, 494
DL-μ-Amido-μ-hydroperoxo-bis(bis (ethylenediamine) cobalt(III) tetranitrate dihydrate, 32B, 495
μ-Amido-μ-nitro-bis tetraammine cobalt(III) tetrachloride tetrahydrate, 35A, 409
trans-μ-Amido-μ-nitrosyl-bis(h^5-cyclopentadienyl-nitrosylchromium), 35B, 543
6-Amido-3-pyridazone, 28, 458
DL-μ-Amido-μ-superoxo-bis(bis (ethylenediamine) cobalt(III) tetranitrate monohydrate, 32B, 494, 495
μ-Amido-μ-superoxo-bis(tetrammine cobalt(III)]tetranitrate, 34A, 345
Amino acid copper salts, 26, 661-663
Amino acid derivatives, 35B, 326
2-(4'-Amino-5'-aminopyrimidyl)-2-pentene-4-one, 29, 770; 30B, 171
2-(4'-Amino-5'-azamethenylpyrimidyl) -3-pentene-4-ol, 29, 771
p-Aminobenzamide, 30B, 392
m-Aminobenzenearsonic acid, 27, 905
p-Aminobenzenearsonic acid, 26, 700
p-Aminobenzene sulphonamide, 32B, 93
o-Aminobenzoic acid, 32B, 77; 33B, 552
p-Aminobenzoic acid, 30B, 55; 31B, 62; 32B, 79
α-*p*-Aminobenzoic acid, 33B, 42
Aminoboranes, 35B, 435
*mono*Aminoborazine, 34A, 153
2-Amino-4'-bromobenzophenone, 29, 763; 32B, 99
bis-(β-Aminobutyrato)-copper(II) dihydrate, 26, 665
DL-α-Amino-*n*-butyric acid, 33B, 180
ε-Aminocaproic acid, 32B, 15
2-Amino-4'-chlorobenzophenone, 29, 763; 32B, 99
4-Amino-3-chloro-4'-nitrobiphenyl, 35B, 910
1-Aminocyclopentane carboxylic acid hydrobromide, 33B, 552
α-D-2'-Amino-3'-deoxyadenosine monohydrate, 35B, 296
12-Amino-11,13-diazo-9β,14β-11,12-dehydroestrone methyl ether hydrobromide, 35B, 331
α-Amino-3,5-dichloro-4-hydroxyphenylacetic acid, 33B, 551
2-Amino-1,6-dihydro-1-methylpyrimidinium-6-sulphonate, 35B, 164
2-Amino-ethanol phosphate, 26, 591
1-(2-Aminoethyl)biguanidecyanoguanidinecopper(II) sulphate mono-

hydrate, 33B, 550
2-Aminoethylphosphonic acid, 31B, 301
2-Aminoethylsulphonic acid, 31B, 187
Aminoffite, 32A, 466
2-Amino-4'-halogeno-benzophenones, 32B, 98
4-Amino-3-hydrazino-5-mercapto-1,2,4-triazole, 35B, 914
2-Amino-3-hydroxy-4-hydroxymethyl-7-methyl-Δ^2-furo[2,3-*c*]pyridine hydrobromide, 35B, 229
6-Amino-2-hydroxypurine sulphate tetrahydrate, 32B, 489
1,2-*o*-Aminoisopropylidene-α-D-glucopyranoside hydroiodide, 31B, 172
Aminomalonic acid, 31B, 3
2-Amino-3-methylbenzoic acid, 28, 414
trans-1,4-Aminomethylcyclohexane-carboxylic acid, 33B, 82
cis-1,4-Aminomethylcyclohexane-carboxylic acid hydrobromide, 30B, 90
trans-4-Aminomethylcyclohexane-1-carboxylic acid hydrobromide, 31B, 88
cis-4-Aminomethylcyclohexane-1-carboxylic acid hydrochloride, 31B, 88
1-Aminonaphthalenetricarbonylchromium, 33B, 361
3-Amino-1,4-naphthoquinone, 31B, 98; 34B, 101
4-Amino-1,2-naphthoquinone, 35B, 118
4-Amino-1,2-naphthoquinone hemihydrate, 33B, 551
o-Aminophenol hydrochloride, 30B, 70
2-Amino-5-propyl-7-methyl-*s*-triazolo[2,3-*c*] pyrimidine hydrochloride, 27, 978
4-Aminopyridine, 28, 670
3-Aminoquinazolines, 35B, 923
2-Amino resorcinol, 35B, 924
7-Amino-3-(α-D-ribofuranosyl)pyrazolo [4,3-*d*] pyrimidine hydroxbromide monohydrate, 31B, 194
2-Amino-1,3-selenazol-2-inium bromide, 33B, 545
5-Aminotetrazole monohydrate, 32B, 152
α-(Amminechlorotriethylenetetramine)-cobalt(III) nitrate (racemate), 35B, 679
Ammonia, 26, 298; 29, 233
Ammonia-ammonium bromide, 26, 298
Ammonia-ammonium chloride, 26, 298
Ammonia-ammonium iodide, 26, 298
Ammonium acetate, 32B, 22
Ammonium acetate-acetic acid complex, 34B, 16
Ammonium acetate *monammine*, 33B, 13
Ammonium acetate diammine, 33B, 15
Ammonium analcite, 26, 517
Ammonium antimonyl tartrate, 29, 762
Ammonium antimonyl D-tartrate trihydrate, 32B, 320
Ammonium barbiturate, 29, 714
Ammonium bisoxalato cuprate dihydrate, 27, 817
Ammonium boride, 30A, 284
Ammonium bromide, 26, 298; 29, 284; 30A, 457
Ammonium bromochloroiodate(I), 32A, 219
Ammonium cadmium trimetaphosphate, 32A, 379
Ammonium calcium trimetaphosphate, 32A, 379
Ammonium catena-di-μ-chromato-diamminezincate(II), 34A, 276
Ammonium ceric nitrate, 33A, 426
Ammonium chloranilate monohydrate, 32B, 48
Ammonium chlorate, 27, 641
Ammonium chloride, 27, 426
Ammonium chloride thionate, 34A, 224
Ammonium chromate, 34A, 274; 35A, 384
Ammonium chromium iodate, 26, 500
Ammonium cobalt phosphate monohydrate, 33A, 395
Ammonium cobalt(II) sulphate hexahydrate, 32A, 343
Ammonium cobalt trimetaphosphate, 32A, 379-381
Ammonium copper(II) chloride, 28, 85
Ammonium copper heptafluoro silicate tetrahydrate, 31A, 245
Ammonium copper heptafluoro stannate tetrahydrate, 31A, 245
Ammonium copper(II) heptafluorotitanate(IV) tetrahydrate, 31A, 258
Ammonium copper oxalate dihydrate, 27, 818
Ammonium copper pentafluoro-dioxowolframate, 31A, 245
Ammonium copper sulphate hexahydrate, 30A, 374, 458; 34A, 305
Ammonium copper(I) sulphite, 33A, 384
Ammonium cuprous thiomolybdate, 35A, 144
Ammonium diamidothiophosphate, 33A, 406
Ammonium difluorophosphate, 34A, 324
Ammonium dihydrogen nitrate, 35A, 322
Ammonium dihydrogen phosphate, 28, 291
Ammonium μ-dihydroxy hexacarbonatodizirconium(IV), 35A, 417
Ammonium diisoselenocyanatobis(dimethylglyoximato)cobaltate(III) tri-

hydrate, 35B, 919
Ammonium diisothiocyanatobis(dimethylglyoximato)cobaltate(III) trihydrate, 35B, 919
Ammonium dimolybdomalate, 33B, 410
Ammonium dimolybdomalate hydrate, 31B, 407
Ammonium dioxovanadium (V) bisoxalate dihydrate, 30B, 407
Ammonium dioxovanadium (IV) trisoxalate hexahydrate, 30B, 407
Ammonium diselenocyanatobis dimethylglyoxime cobaltate (III) trihydrate, 27, 842
Ammonium dithiocarbamate, 32B, 486; 33B, 543
Ammonium dithiocyanatobis dimethylglyoxime cobaltate (III) trihydrate, 27, 842
Ammonium diuranyl trioxalate, 33B, 550
Ammonium fluoride, 34A, 172
Ammonium fluoride oxoperoxotetrafluoromolybdate(VI), 35A, 421
Ammonium fluoroberyllates, 26, 302, 303; 27, 453
Ammonium fluoromanganate, 26, 309
Ammonium fluorosulphate, 35A, 191
Ammonium formate, 33B, 11
Ammonium halides, 35A, 145
Ammonium heptafluorozirconate, 35A, 169
Ammonium heptamolybdate tetrahydrate, 32A, 296, 510
Ammonium hexabromoantimonate, 31A, 108
Ammonium hexabromotellurate(IV), 31A, 109
Ammonium hexachloropalladate(IV), 31A, 102
Ammonium 1,1,2,6,7,7-hexacyanoheptatrienide, 35B, 31
Ammonium hexafluorocerate(IV), 34A, 193
Ammonium hexafluorogallate, 29, 266
Ammonium hexafluorosilicate, 31A, 88
Ammonium hexaiodotellurate, 27, 463
Ammonium hydrogen *o*-carboxybenzenesulphonate, 32B, 81
Ammonium hydrogen croconate, 31B, 38
Ammonium hydrogen di-*p*-chloro-benzoate, 28, 416
Ammonium hydrogen dicinnamate, 28, 387
Ammonium hydrogen peroxide, 31A, 128
Ammonium hydroxide, 26, 298
Ammonium iodide, 27, 425; 31A, 243
Ammonium iron(II) sulphate hexahydrate, 32A, 343
Ammonium lithium sulphate, 34A, 305
Ammonium magnesium phosphate hexahydrate, 35A, 329
Ammonium magnesium sulphate hexahydrate, 27, 619
Ammonium magnesium trimetaphosphate, 32A, 379; 33A, 416
Ammonium manganese pentafluoride, 34A, 201
Ammonium manganese polyphosphate, 35A, 347
Ammonium manganese trimetaphosphate, 32A, 379
Ammonium methyl alum, 27, 813
Ammonium methyl periodate, 27, 640
Ammonium nickel fluoride, 27, 454
Ammonium nickel phosphate hexahydrate, 31A, 243
Ammonium nickel sulphate hexahydrate, 28, 213
Ammonium nitranilate, 29, 648; 32B, 47
Ammonium nitrate, 27, 566
Ammonium nitrate sulphate, 27, 569
Ammonium octachlorodimolybdate chloride hydrate, 35A, 418
Ammonium octachloroditechnicate dihydrate, 35A, 426
Ammonium octafluorouranate, 35A, 170
Ammonium osmium chloride, 27, 423
Ammonium oxalate monohydrate, 30B, 21
Ammonium oxamate, 28, 382
Ammonium oxovanadium (IV) bisoxalate dihydrate, 30B, 407
Ammonium oxypentabromomolybdenum(V), 35A, 423
Ammonium parawolframates, 26, 393; 29, 422; 34A, 283
β-Ammonium pentaborate tetrahydrate, 34A, 353
Ammonium pentafluorodioxyuranate(VI), 31A, 258
Ammonium pentafluoroxyuranate, 34A, 218
Ammonium perchlorate, 27, 639
Ammonium perchlorate-copper pentammino-perchlorate, 26, 497
Ammonium pertechnetate, 27, 671
Ammonium picrate, 34B, 68
Ammonium salts, 27, 421
Ammonium scandium sulphates, 31A, 255
Ammonium selenate, 27, 606
Ammonium sodium trioxo-oxalato molybdate (VI) dihydrate, 27, 820
Ammonium sulphate, 28, 206; 31A, 197
Ammonium sulphite monohydrate, 30A, 378
Ammonium tetrabis [4,4,4-trifluoro-1-(2-thienyl)-1,3-butanedione]

praesodymate monohydrate, 32B, 391
Ammonium tetrachloridocuprate, 29, 278; 34A, 204
Ammonium tetrachloridocuprate(II) dihydrate, 34A, 204
Ammonium tetrachlorocuprate, 29, 278; 34A, 204
Ammonium tetrachloropalladate(II), 31A, 102
Ammonium tetrafluoroborate, 34A, 189
Ammonium tetrametaphosphate, 29, 361; 35A, 348
Ammonium tetraselenowolframate, 33A, 387
Ammonium tetrathioarsenate, 28, 153
Ammonium tetroxalate dihydrate, 32B, 23
Ammonium thiowolframate, 28, 227
Ammonium thorium(IV) octafluoride, 34A, 194
Ammonium tricyanomethide, 30A, 291; 30B, 39
Ammonium trifluoroacetate, 29, 456
Ammonium tri-iodide, 35A, 147
Ammonium tris(pentasulphido)platinum (IV) dihydrate, 34A, 296
Ammonium Tutton's salts, 32A, 343
Ammonium uranyl arsenate, 30A, 458
Ammonium uranyl cupferrate, 29, 769
Ammonium uranyl peroxyoxalate, 29, 762
Ammonium uranyl sulphates, 30A, 453
Ammonium vanadium(II) sulphate hexahydrate, 32A, 343
Ammonium vanadyl (+)-tartrate monohydrate, 32B, 416
Ammonium yttrium(II) oxalate hydrate, 32B, 415
Ammonium zinc fluoroberyllate, 33A, 180
Ammonium zinc molybdate hydrate, 32A, 301
Ammonium (and sodium and potassium) zinc (and cobalt and nickel) molybdates, 31A, 246
Ammonium zinc trimetaphosphate, 32A, 379
Amobarbital I, 34B, 220
Amobarbital II, 34B, 220
Amosite, 30A, 455
Amphibole, 30A, 455; 31A, 244
Amphiboles, 26, 526-528; 28, 258; 29, 439, 31A, 243
Ampicillin I, 33B, 554
Amylose-*n*-butanol, 33B, 526
Amylose V complexes, 29, 759
Amylose hydrate-iodine complex, 26, 810
Amylose-potassium bromide, 33B, 527
Amyrolin, 35B, 369
Analcime, 28, 241; 30A, 424
Analcite, 29, 439; 31A, 219
Anandite, 32A, 477
Anaphe moloneyi, 26, 827
Andalusite, 26, 502, 505; 27, 710
Andersonite, 28, 177
Andradite, 30A, 441, 455; 31A, 219
1,4-Androstadien-17β-ol-3-one, 30B, 437
5β-Androstan-4β-bromo-3,17-dione, 30B, 439
1,4-Androstandien-17β-ol-3-one, *p*-bromophenol (1:1), 30B, 438
5α-Androstan-3,17-dione, 30B, 437
Androstanes, 27, 1038
5α-Androstan-17β-ol-3-one, 30B, 437
5α-Androstan-17β-ol-3-one 17-benzoate, 30B, 438
5α-Androstan-3α-ol-17-one, *p*-bromophenol (1:1), 30B, 438
5α-Androstan-17β-ol-3-one, *p*-bromophenol (1:1), 30B, 438
5β-Androstan-3α-ol-17-one, *p*-bromophenol (1:1), 30B, 438
5β-Androstan-17β-ol-3-one, *p*-bromophenol (1:1), 30B, 438
4-Androsten-6-bromo-17β-ol-3-one 17-acetate, 30B, 439
5-Androsten-3β, 17-diol, 30B, 437
4-Androsten-3,17-dione, 30B, 437
4-Androsten-17β-ol-3-one, 30B, 437
4-Androsten-17β-ol-3-one 17-acetate, 30B, 438
4-Androsten-17β-ol-3-one 17-bromoacetate, 30B, 439
Δ^4-Androsten-17β-ol-3-one, *p*-bromophenol (1:1), 30B, 438
Δ^5-Androsten-3β-ol-17-one, *p*-bromophenol (3:1), 30B, 438
4-Androsten-3,11,17-trione, 30B, 437
Androsterone, 28, 686; 31B, 201
5α-Androsterone-3α-01-17-one, 31B, 201
Anemonin, 30B, 397; 31B, 142
Anglesite, 28, 78, 208
Anhydrite, 27, 607; 28, 207
1,5-Anhydro-4-deoxy-D-*arabo*-hexitol, 29, 762
2,6-Anhydro-3-deoxy-D-*gluco*heptitol, 29, 766
Δ-8,14-Anhydrodigitoxigenin, 35B, 338
3,6-Anhydro-α-D-glucosyl-1,4:3,6-dianhydro-β-D-fructoside, 35B, 913
Anhydrohirundigenin *p*-bromobenzoate, 33B, 555
Anhydro-*N*-hydroxymethyldeoxyangustifoline perchlorate, 32B, 264
Anhydroignavinol methiodide, 35B, 923
Anhydro-iodic acid, 31A, 212
Anhydro-isocalebassin methyl ether di-iodide monohydrate, 34B, 333
Anhydro-5-mercapto-2,3-diphenyltetrazolium hydroxide, 35B, 162

Anhydro-2-methylthio-3-amino-4-hydroxy-5-phenyl-thiazolium hydroxide, 32B, 194
Anhydrous copper(II)formate, 26, 547
Anhydrous lithium acetate, 26, 546
Anhydrous manganic acetate, 34B, 531
Anhydrous stannous sulphate, 27, 608
Aniline hydrobromide, 26, 698
Aniline *p*-thiocyanate, 30B, 64
2-Anilino-3-bromo-tetrahydro-exo-dicyclopentadiene, 32B, 149
1-Anilino-4-chloro-pyromellitic acid-*di*-phenylimide, 26, 782
5-Anilino-3-oxo-2-phenyl-2,3-dihydro-1H-pyrazolo(3,4-D)-thiazole, 34B, 212
Anilite, 35A, 57
Anisaldehyde azine, 33B, 75
Anisic acid, 32B, 84
[16]Annulene, 33B, 553; 35B, 115
[18]Annulene, 30B, 94
Anorthite, 26, 518; 27, 679; 31A, 252
Anorthoclase, 31A, 242
Antarticite, 30A, 449
Antheraea mylitta, 26, 827
Anthophyllite (Fe, Al), 28, 334
Anthracene, 26, 739; 27, 925; 29, 682
Anthracenechromium tricarbonyl, 33B, 365
Anthracene derivatives, 26, 748
Anthracene, pyromellitic dianhydride (1:1) complex), 30B, 254
Anthracene - tetracyanoethylene complex, 35B, 421
Anthracene-*s*-trinitrobenze, 29, 686
9,10-Anthrahydroquinone dibenzoate, 27, 936
9-Anthraldehyde dimer, 33B, 106
Anthranilic acid, 32B, 77
Anthraquinone, 32B, 128
1,4-Anthraquinone(form 1), 26, 750
1,4-Anthraquinone(form 2), 26, 750
9,10-Anthraquinone, 26, 751
1,4-Anthraquinone and derivatives, 26, 750
Anthraquinones, 35B, 908, 922
Anthrone, 26, 750; 27, 932; 29, 683; 35B, 121
Anti-9-benzonorbornenyl *p*-bromobenzenesulphonate, 35B, 910
Antibiotic X-537A, 35B, 406
Antibiotic X-537A (barium salt), 35B, 912
Antigorite, 26, 511; 30A, 458
Antimonate spinels, 26, 410
Antimonious xanthate, 26, 610
Antimonpearceite, 28, 288
Antimony, 27, 76, 385; 28, 8; 35A, 9
Antimony-arsenic-manganese, 27, 57
Antimony-barium-lithium, 35A, 9
Antimony-bismuth, 27, 76
Antimony-bismuth-indium, 26, 42
Antimony-bismuth-nickel, 27, 287
Antimony-bismuth-palladium, 27, 287
Antimony-bismuth-platinum, 27, 287
Antimony-bismuth-selenium-tellurium, 26, 33
Antimony-bismuth-tellurium, 27, 40
Antimony-bismuth-tellurium-thallium, 27, 362
Antimony-cadmium-zinc, 26, 34; 27, 41
Antimony-caesium, 26, 35, 54
Antimony-cerium, 27, 335
Antimony-cerium-hydrogen, 27, 335
Antimony-cerium-sulphur, 27, 335
Antimony-chromium, 35A, 10
Antimony-cobalt, 26, 35, 47
Antimony cobalt oxide, 27, 527
Antimony-cobalt-oxygen, 26, 410, 412
Antimony-copper, 27, 185
Antimony-copper-lead-sulphur, 35A, 11
Antimony-copper-manganese, 27, 41; 33A, 95
Antimony-copper-selenium, 29, 14
Antimony-copper-sulphur, 26, 37; 29, 15; 32A, 25
Antimony-erbium, 27, 335
Antimony-germanium-tellurium, 26, 38; 31A, 18; 33A, 19
Antimony-gold, 26, 42
Antimony-hafnium-zinc, 33A, 21
Antimony-holmium, 26, 164; 27, 335
Antimony hydrogen bis(thioglycollate), 33B, 287
Antimony-indium, 26, 42; 27, 42
Antimony-indium-selenium, 26, 42
Antimony-indium-tellurium, 26, 43
Antimony(III) iodide sulphide, 32A, 223
Antimony-iridium, 26, 35, 48, 176
Antimony-iron, 35A, 10
Antimony iron oxide, 31A, 241
Antimony-lanthanum, 35A, 12
Antimony-lead-oxygen-tin, 27, 534
Antimony-lithium-strontium, 35A, 9
Antimony magnesium oxide, 29, 316
Antimony-molybdenum, 31A, 18
Antimony niobate, 31A, 241
Antimony-niobium, 29, 18; 30A, 14
Antimony-niobium-tellurium, 32A, 17
Antimony-osmium-tellurium, 26, 43
Antimony oxide, 26, 362
Antimony-oxygen-zinc, 26, 410, 412
Antimony-palladium, 34A, 10; 35A, 14
Antimony pentachloride, *o*-chlorophenyl diazonium chloride (double salt), 30B, 418
Antimony pentachloride complexes, 33B, 293
Antimony pentachloride-phosphorus oxychloride, 28, 95
Antimony pentachloride-trimethyl-phosphine oxide, 26, 595; 28, 95
Antimony phosphate, 35A, 328
Antimony-platinum, 34A, 11

Antimony-potassium, 26, 44
Antimony potassium tartrate, 30B, 395
Antimony-rare-earth, 32A, 18; 33A, 21
Antimony-rhodium, 26, 35, 48; 27, 61
Antimony-rubidium, 26, 45, 54
Antimony rubidium bromide, 33A, 515
Antimony-scandium, 27, 42, 88
Antimony-selenium, 27, 200
Antimony-silver, 26, 249
Antimony-silver-sulphur, 29, 21; 31A, 19; 33A, 22; 34A, 12; 35A, 15
Antimony-silver-tellurium, 26, 46
Antimony sulphide iodide, 31A, 112
Antimony tantalate, 27, 522
Antimony-tellurium, 27, 42, 200; 32A, 30
Antimony-tellurium-thallium, 26, 46
Antimony-tellurium-thallium-tin, 27, 363
Antimony-tellurium-tin, 26, 38, 47
Antimony-terbium, 26, 255; 27, 334
Antimony(V) tetrachloride fluoride, 31A, 85
Antimony thiooxinate, 35B, 908
Antimony thorium nitride, 35A, 113
Antimony-titanium, 27, 43
Antimony tribromide, 27, 439
α-Antimony tribromide, 29, 285
Antimony trichloride-aniline, 33B, 288
Antimony trichloride-benzene, 33B, 528
Antimony trifluoride, 35A, 147
Antimony triiodide, 31A, 111
Antimony triiodide-sulphur, 28, 103
Antimony trioxide, 27, 719
Antimony-uranium, 32A, 21
Antimony uranium nitride, 35A, 113
Antimony-vanadium, 27, 46; 33A, 24
Antimony-ytterbium, 31A, 20; 32A, 22; 35A, 17
Antimony-yttrium, 27, 46
Antimony-zinc, 29, 22; 30A, 16
Antimony zinc oxide, 27, 490, 527
Antimony-zirconium, 27, 379; 28, 27; 33A, 24
Anti-LaF_3 structures, 30A, 268
Antiviral epidithiapiperazinedione derivative, LL-*S*88α, 33B, 553
Antlerite, 26, 439, 440; 28, 211
Apatite, 28, 188; 31A, 242
Apatite (Sr), 28, 334
Apatites, 26, 468; 29, 369, 425; 30A, 455
Aplowite, 30A, 458
Apophyllite, 30A, 455
Apoquinamine, 27, 1009
Aquacreptite, 30A, 442
*mono*Aquobis(acetylacetonato)zinc, 28, 618
Aquo-*cis*-bis(histamino)perchlorato-nickel(II) perchlorate, 35B, 655
Aquobis(DL-lactato)copper(II) hemi-hydrate, 33B, 403
Aquo bis(L-leucinato)copper(II), 34B, 545
Aquo(dihydrogen-ethylenediamine-tetra-acetato) copper(II), 34B, 492
Aquo-N,N'-ethylene bis(acetylacetone-iminato)copper(II), 34B, 500
Aquomagnesium tetraphenylporphyrin, 34B, 278
Aquozinc(II) tetraphenyl porphine, 32B, 237
D,L-Arabinitol, 33B, 157
β-Arabinose, 26, 622
β-DL-Arabinose, 32B, 211
Arabinose *p*-bromophenylhydrazone, 27, 790
Aragnonite, 30A, 458; 31A, 244
Arborinol, 30B, 396
Ardennite, 30A, 455; 33A, 491
Arfvedsonite, 28, 258
Argentous silicate, 27, 706
L-Arginine dihydrate, 29, 479
L-Arginine dihydriodide, 32B, 489
L-Arginine hydrobromide monohydrate, 30B, 420; 31B, 189
Arginine hydrochloride *mono*hydrate, 32B, 489
L-Arginine hydrochloride *mono*hydrate, 30B, 420; 35B, 319
L-Arginine hydrohalides, 29, 765
Argon, 26, 279; 27, 386, 720; 29, 414; 30A, 449, 457; 31A, 254
Argon-cordierite, 27, 694
Argon-oxygen, 31A, 250
Aristeromycin hydrobromide, 32B, 490
Armepavine, 27, 1003
Arsenic, 31A, 239; 34A, 13
Arsenic acid hydrate, 33A, 418
Arsenic-boron, 26, 67
Arsenic-cadmium, 33A, 25; 34A, 15; 35A, 18
Arsenic-cadmium-indium-tin, 27, 47
Arsenic-cadmium-silicon, 35A, 19
Arsenic-cadmium-zinc, 27, 47
Arsenic-caesium, 26, 54
Arsenic-carbon-vanadium, 32A, 24
Arsenic chalcogenide glasses, 28, 293
Arsenic-chromium, 30A, 17
Arsenic-chromium-nitrogen, 33A, 28
Arsenic-cobalt, 26, 47; 27, 61; 31A, 21
Arsenic-cobalt-iron-nickel, 27, 48
Arsenic-cobalt-sulphur, 28, 9; 30A, 17
Arsenic-copper, 27, 185
Arsenic-copper-germanium-selenium, 26, 48
Arsenic-copper-selenium, 33A, 30
Arsenic-copper-silver, 26, 48
Arsenic-copper-sulphur, 29, 23; 31A, 18; 32A, 25; 35A, 19
Arsenic(III) *NN*-diethyldithiocarba-mate, 33B, 546

Arsenic-gallium, 27, 50
Arsenic-gallium-phosphorus, 27, 51
Arsenic-gallium-sulphur, 27, 51
Arsenic-germanium, 27, 52; 35A, 20
Arsenic-germanium-indium-zinc, 26, 49
Arsenic-germanium-selenium, 26, 38
Arsenic-germanium-tellurium, 26, 38
Arsenic-hafnium, 27, 54
Arsenic-holmium, 26, 164
Arsenic-indium, 27, 50
Arsenic-indium-selenium, 26, 42; 27, 247
Arsenic-indium-tellurium, 27, 248
Arsenic-indium-tin-zinc, 26, 49
Arsenic-iridium, 26, 48, 55, 176; 27, 61
Arsenic-iron, 34A, 16
Arsenic-iron-sulphur, 26, 50; 27, 55
Arsenic-lead-silver-sulphur, 34A, 17
Arsenic lead sulphides, 26, 428; 34A, 23
Arsenic lead sulphur glasses, 26, 431
Arsenic-lithium-manganese, 26, 187-189; 33A, 31
Arsenic-lithium-oxygen-zinc, 26, 417
Arsenic-manganese, 26, 53; 27, 57; 29, 25
Arsenic-manganese-phosphorus, 33A, 33; 35A, 21
Arsenic-molybdenum, 29, 18; 30A, 19; 31A, 22
Arsenic-nickel, 27, 58; 33A, 34; 35A, 14
Arsenic-nickel-sulphur, 27, 58; 32A, 26
Arsenic-niobium, 28, 9; 29, 18; 30A, 14
Arsenic-nitrogen-oxygen-thorium, 34A, 135
Arsenic-nitrogen-uranium, 34A, 135
Arsenic-osmium, 26, 55
Arsenic(V) oxide hydrate, 31A, 121
Arsenic-palladium, 26, 55; 34A, 25
Arsenic-palladium-thallium, 35A, 22
Arsenic-platinum, 26, 55; 33A, 34
Arsenic-rhenium, 27, 27
Arsenic-rhodium, 26, 48, 55; 27, 60
Arsenic-rubidium, 26, 53, 54, 55
Arsenic-selenium, 30A, 22
Arsenic-silicon, 30A, 22; 33A, 36
Arsenic-silver, 26, 249
Arsenic-silver-sulphur, 29, 25; 31A, 19; 33A, 37; 34A, 28
Arsenic spodiosite, 35A, 193
Arsenic-sulphur, 27, 61
Arsenic-sulphur glass, 26, 431
Arsenic-tantalum, 29, 18
Arsenic-tellurium, 28, 9
Arsenic-terbium, 26, 255; 27, 335
Arsenic-tin, 33A, 146
Arsenic-tin-zinc, 27, 47
Arsenic-titanium, 29, 27
Arsenic tribromide, 30A, 285; 32A, 206
Arsenic tricyanide, 26, 339; 28, 105
Arsenic triiodide, 30A, 287; 31A, 239
Arsenic trisulphide-iodine glass, 27, 401
Arsenic-wolfram, 30A, 19
Arsenides of transition metals, 26, 55
Arsenobenzene, 26, 736
Arsenomethane polymer, 35B, 466
Arsenopolybasite, 28, 288
Arsenopyrite, 26, 50
Arthurite, 29, 426
Artinite, 30A, 408
Asbecasite, 31A, 254; 34A, 381
Ascharite, 28, 158; 30A, 458
Ascochlorin *p*-bromobenzenesulphonate, 34B, 286
Ascorbic acid, 26, 808
L-Ascorbic acid (vitamin C), 29, 771; 33B, 158
Asparagine monohydrate, 26, 602
Aspartic acid, 26, 826
DL-Aspartic acid, 30B, 415; 33B, 177
L-Aspartic acid, 33B, 176
(-)-Aspidospermine N(b)-methiodide, 33B, 552
Aspirin, 29, 653
Astrophyllite, 28, 322; 32A, 473
Asym-αβ-naphthazine, 27, 988
Atacamite, 29, 250
Atactic polyvinyl fluoride, 26, 633
Atrovenetin organge trimethyl ether ferrichloride, 30B, 242
Attakolite, 29, 439; 30A, 455
Augelite, 33A, 403
Auramine, 26, 828
Auramine perchlorate, 32B, 100
Aureomycin hydrochloride, 28, 534
Aurichalcite, 30A, 454
Aurous chloride trichlorophosphine, 27, 471
Austenite, 26, 99; 27, 155, 251
Austenitic steels, 31A, 31
Authigenic maximum microcline, 27, 678
Autunite, 26, 468
1-Azabicyclo[3.2.0]heptane-1-methyl chloride, 33B, 127
3-Azabicyclo[3.3.1]nonane hydrobromide, 29, 507
3-Azabicyclo[3.3.1]nonane hydrochloride, 29, 762
Azacyclododecane hydrochloride, 29, 513
8-Azaestrone hydrobromide, 34B, 289
8-Azaguanine hydrochloride monohydrate, 29, 764
8-Azaguanine monohydrate, 30B, 138; 33B, 150
Azelaic acid, 26, 561, 575; 29, 765; 32B, 18
Azelamide, 31B, 15
Azepine adduct, 30B, 399
1H-Azepine derivatives, 33B, 553
L-Azetidine-2-carboxylic acid, 34B, 124

Azide methaemoglobin, 31B, 269
6-Azido-5,6-dideoxy-5-iodo-1,2-*O*-isopropylidene-β-L-idofuranose, 33B, 550; 35B, 262
Azidodinitrogenbis(ethylenediamine) ruthenium(II) hexafluorophosphate, 35B, 592
Azidopentamminecobalt(III) azide, 29, 246
*mono*Azidopentamminecobalt(III) chloride, 32A, 193
Azidopurine, 33B, 151
Aziridine borane, 34B, 359
Azobenzene, 31B, 47
cis-Azobenzene dioxide, 35B, 917
Azobenzene-palladium(II) complex, 33B, 548
Azobenzene-2-sulphenyl bromide, 28, 675
Azobenzene-2-sulphenyl cyanide, 28, 675; 33B, 63
Azodicarbonamide, 26, 589
4,4'-Azodiphenetole, 35B, 70
Azomethane, 35B, 890, 891
p-Azotoluene, 31B, 48
p-Azoxyanisole, 33B, 529; 35B, 69
Azulene, 27, 926
Azulene diiron pentacarbonyl, 32B, 376
trans-Azulenedimanganese hexacarbonyl, 33B, 364
Azulene dimolybdenum hexacarbonyl, 32B, 494
Azulene-1,3-dipropionic acid, 31B, 107
Azulenetricarbonylmethylmolybdenum dimer, 33B, 382
Azulene, *s*-trinitrobenzene (1:1 complex), 30B, 250

Babefphite, 31A, 248; 32A, 368
Baccatin-V, 35B, 916
Bacterial viruses, 26, 827
Baddeleyite, 30A, 307
Bafertisite, 28, 268
Bakerite, 27, 706
Baotite, 26, 536; 27, 716; 28, 274; 34A, 382
Barbital I, 34B, 216
Barbital II, 34B, 216
Barbital-caffeine (2:1), 35B, 922
Barbituric acid, 28, 501
Barbituric acid dihydrate, 26, 764
Barene, 30B, 399
Barite, 28, 78, 208; 32A, 327
Barium acetate trihydrate mixed crystals with various dyes, 26, 828
Barium aluminate, 30A, 316
Barium aluminate hydrates, 34A, 252
Barium azide, 34A, 169
Barium bis-dioxomolybdenum(V) oxalate pentahydrate, 30B, 346
Barium-bismuth, 26, 62
Barium bismuth niobium oxide, 27, 538
Barium-bismuth-niobium-oxygen-titanium, 26, 402
Barium-bismuth-oxygen-titanium, 26, 402
Barium bismuth tantalum oxide, 27, 538
Barium bismuth titanium oxide, 27, 539
Barium borate, 30A, 415; 31A, 171; 33A, 445
Barium borate glasses, 27, 559
Barium borosilicate, 28, 292
Barium bromate monohydrate, 27, 643
Barium bromide, 26, 335; 28, 77
Barium bromine phosphate, 26, 469
Barium cadmium oxides, 27, 517
Barium-calcium-oxygen-tantalum, 26, 420
Barium calcium silicate, 26, 524
Barium-cerium-sulphur, 27, 63, 327
Barium chlorate monohydrate, 33A, 448
Barium chloride, 28, 77, 79
Barium chloride dihydrate, 28, 79
Barium chlorine phosphate, 26, 469
Barium chromate, 27, 609
Barium cobalt fluoride, 35A, 165
Barium cobalt(II) wolfram(VI) oxide, 32A, 272
Barium diborate tetrahydrate, 31A, 249
Barium diethyl phosphate, 31B, 177
Barium dimetasilicates, 27, 699
Barium dithionate dihydrate, 31A, 208
Barium dysprosium niobate, 27, 509
Barium-dysprosium-sulphur, 27, 327
Barium erbium niobate, 27, 509
Barium-erbium-sulphur, 27, 327
Barium europium niobate, 27, 509
Barium *mono*ferrite, 26, 395
m-Barium ferrite, 32A, 290
Barium fluoberyllates, 26, 302
Barium fluoride, 31A, 243
Barium fluorine phosphate, 26, 469
Barium fluorometallates(II), 32A, 168
Barium gadolinium niobate, 27, 509
Barium-gadolinium-sulphur, 27, 327
Barium germanate, 26, 397, 398; 27, 502, 541; 31A, 134
Barium-germanium-magnesium, 35A, 24
Barium halides, 28, 77
Barium hexaboride, 26, 69
Barium hexafluoroferrate(II), 31A, 245
Barium holmium niobate, 27, 509
Barium hydrogen oxalate dihydrate, 35B, 15
Barium hydroxide *mono*hydrate, 31A, 256
Barium hydroxide octahydrate, 28, 119; 29, 297
Barium indium niobate, 27, 509
Barium inosine-5'-phosphate, 33B, 545
Barium iodide, 28, 77
Barium iridium oxide, 31A, 137

Barium iron niobate, 26, 422; 27, 509
Barium-iron-oxygen, 26, 420
Barium iron silicate, 26, 523
Barium iron tantalate, 26, 422
Barium iron uranium oxide, 31A, 161; 33A, 356
Barium lanthanum niobate, 26, 423; 27, 509
Barium-lanthanum-sulphur, 27, 327
Barium lanthanum tantalate, 26, 423
Barium lanthanum titanate, 26, 372
Barium-lead, 29, 29
Barium lead niobium titanium oxide, 27, 506
Barium-lead-thallium, 35A, 23
Barium lead titanate, 26, 372
Barium-lithium, 30A, 23
Barium lithium hydride, 33A, 160
Barium-lithium-silicon, 32A, 27; 33A, 39
Barium lutecium niobate, 27, 509
Barium-magnesium, 26, 175, 176; 27, 94
Barium-magnesium-silicon, 35A, 24
Barium manganate, 27, 662
α-Barium manganate, 27, 662
β-Barium manganate, 27, 663
Barium manganese fluoride, 35A, 164
Barium manganese niobate, 26, 423
Barium manganese silicate, 26, 523
Barium manganese tantalate, 26, 423
Barium manganite, 26, 394
Barium metaborate pentahydrate, 32A, 504
Barium metaphosphate, 32A, 380
Barium metatantalate, 33A, 342
Barium minerals, 30A, 455
Barium molybdate, 35A, 226
Barium neodymium niobate, 26, 423; 27, 509
Barium-neodymium-sulphur, 27, 327
Barium neodymium tantalate, 26, 423
Barium nickel arsenate, 34A, 338
Barium-nickel-sulphur, 35A, 25
Barium nickel wolframate, 27, 509
Barium nickel(II) wolfram(VI) oxide, 32A, 272
Barium niobium silicon oxide, 35A, 269
Barium nitrate, 32A, 351
Barium-nitrogen, 35A, 26
Barium orthoferrite, 31A, 249
Barium orthotitanate, 26, 371
Barium orthovanadate, 35A, 363
Barium oxalate, 30B, 543
Barium oxychloride, 35A, 186
Barium-oxygen-strontium-tantalum, 26, 420
Barium-oxygen-tantalum, 26, 421
Barium-palladium, 35A, 27
Barium permanganate, 27, 662
Barium-phosphorus, 33A, 40; 35A, 28
Barium potassium arsenate, 26, 470
Barium potassium phosphate, 26, 469; 27, 582
Barium potassium vanadate, 26, 470
Barium praseodymium niobate, 26, 423
Barium-praseodymium-sulphur, 27, 327
Barium praseodymium tantalate, 26, 423
Barium rare-earth oxide, 33A, 341
Barium ribose-5-phosphate, 27, 792
Barium ruthenium oxide, 30A, 329; 31A, 137
Barium samarium niobate, 26, 423; 27, 509
Barium-samarium-sulphur, 27, 327
Barium samarium tantalate, 26, 423
Barium selenate, 27, 609
Barium silicate, 26, 513, 537; 30A, 421
Barium silicide, 28, 11; 31A, 25
Barium silicon tantalum oxide, 35A, 267
Barium sodium arsenate, 26, 469
Barium sodium niobate, 34A, 269
Barium sodium phosphate, 26, 469
Barium sodium vanadate, 26, 470
Barium strontium silicate, 31A, 254
Barium sulphate, 28, 208, 209
Barium-sulphur-uranium, 35A, 28
Barium-sulphur-ytterbium, 27, 327
Barium-sulphur-yttrium, 27, 327
Barium-sulphur-zinc, 26, 424
Barium tantalum bronze, 32A, 313
Barium tantalum oxide, 35A, 261
Barium tetraborate, 30A, 413; 34A, 356
Barium tetrafluoroferrate(II), 31A, 245
Barium tetragermanate, 26, 398; 33A, 398
Bariumtetrahydroxoborate, 34A, 358
Barium tetra sulphide monohydrate, 34A, 294
Barium-thallium, 29, 3
Barium thiosulphate monohydrate, 27, 630
Barium thulium fluoride, 35A, 172
Barium thulium niobate, 27, 509
Barium titanate, 27, 507; 34A, 255
Barium titanate ceramics, 26, 372
Barium titanates, 26, 370-373
Barium titanium germanate, 26, 398
Barium titanium oxide, 35A, 271
α-Barium-titanium-phosphate phosphor, 26, 467
β-Barium-titanium-phosphate phosphor, 26, 467
Barium titanium sulphide, 28, 151
Barium uranate, 30A, 447; 31A, 160
D(+)-Barium uridine-5'-phosphate heptahydrate, 30B, 206
Barium-vanadium muscovite, 31A, 243
Barium vanadium oxide, 30A, 319

Barium vanadium sulphide, 34A, 295
Barium violurate, 30B, 398
Barium wolframate, 35A, 226
Barium wolframate ferrate, 34A, 286
Barium wolframates, 26, 391
Barium ytterbium niobate, 27, 509
Barium yttrium niobate, 26, 423
Barium yttrium tantalate, 26, 423
Barium zirconium sulphide, 28, 151
Barnesite, 28, 352
Barringtonite, 30A, 454
Barsanovite, 28, 322; 30A, 455
Barylite, 27, 704; 29, 398; 31A 243
Barysilite, 29, 399, 428; 31A, 220
Basic aluminum selenate, 27, 627
Basic aluminum sulphate, 27, 625
Basic beryllium benzoate, 29, 763
Basic beryllium nitrate, 26, 474
Basic bismuth chromate, 29, 320
Basic indium sulphate, 26, 454
Basic lead azides, 26, 293
Basic lead carbonate, 29, 383
Basic lead(II) perchlorate, 34A, 360
Basic tellurium nitrate, 31A, 175
Basic tin(II) nitrate, 26, 475
Basic zinc *OO*-di-*n*-butyl phosphorodithioate, 30B, 436
Basic zinc *OO*-di-iospropyl phosphorodithioate, 30B, 436
Bastnasite, 29, 428; 30A, 456
Batisite, 27, 705; 28, 334
Batrachotoxinin A *p*-bromobenzoate, 33B, 552
Baumhauerite, 26, 428; 27, 402; 34A, 17
Bavenite, 29, 428; 31A, 221
Bayerite, 26, 518; 29, 299; 32A, 243
Bazzite, 31A, 248
Becquerelite, 27, 673
Belovite, 29, 374
Bemenite, 28, 334
Bence-Jones protein, 35B, 907
Benitoit, 34A, 380
Benstonite, 26, 483; 27, 563
Benzalazine, 26, 731; 35B, 63
Benzaldehyde phenylhydrazone, 30B, 427
Benzamide-hydrogen triiodide, 29, 656
Benzamidine benzoate, 26, 732
Benzene, 27, 1064; 29, 636; 33B, 528
Benzene II, 34B, 79
bis(Benzene azo-β-naphthol) copper(II), 26, 663
p-Benzenebisdiazonium tetrachlorozincate, 33B, 52
Benzene *mono*bromide chloride, 26, 293
Benzenechromium tricarbonyl, 30B, 316
Benzene copper aluminum tetrachloride complex, 31B, 365
Benzene diazonium chloride, 28, 402
Benzenediazonium chloride-acetic acid, 33B, 243
Benzenediazonium copper(I) bromide complex, 30B, 357
Benzenediazonium tribromide, 27, 889
Benzene-dinitrogen tetroxide, 33B, 243
Benzene-1,2-dithiolatodi(π-cyclopentadienyl)molybdenum(IV), 35B,553
Benzene *mono*fluoride, 26, 693
Benzene hexacarboxylic acid, 26, 713
Benzene *mono*iodide, 26, 693
Benzene-metal complexes, 26, 650
Benzene sandwich complexes, 35B, 567
Benzene silver (I) aluminum tetrachloride complex, 31B, 364
Benzenesulphonamides, 34B, 620
Benzene tellurenyl bromide thiourea complex, 31B, 337
Benzene tellurenyl chloride thiourea complex, 31B, 337
Benzene-tricarbonylchromium, 26, 685
Benzene-1,3,5-tricarboxylic acid, 34B, 62
Benzil, 30B, 81
Benzil α-monoxime *p*-bromobenzoate, 32B, 114
Benzocaine, 33B, 542
Benzo[*c*]cinnolinebis(tricarbonyliron, 35B, 699
Benzocyclobutadienequinone, 29, 763
cis-Benzocyclobutene-1,2-diol dinitrate, 29, 763; 35B, 123
Benzocyclobutene-1,2-dione, 35B, 121
Benzo-1,2,4,5-dicyclobutene, 30B, 397; 34B, 110
1,2-Benzodithiol-3-one oxime, 34B, 189
Benzofuroxane, 26, 701
1,2-Benzoisothiazolin-3-one, 35B, 235
Benzo[*d*]naphtho[1,2-*d*']benzo[1,2-*b*:-4,3-*b*']dithiophene, 35B, 921
Benzo-norbornen-anti-9-yl-*p*-bromobenzene sulphonate, 34B, 118
syn-9-Benzonorbornenyl *p*-bromobenzenesulphonate, 33B, 205
Benzo[*c*]phenanthrene, 28, 444
Benzophenone, 32B, 113; 33B, 71
p-Benzoquinone-*p*-bromophenol complex, 32B, 275
p-Benzoquinone-*p*-chlorophenol complex, 32B, 275
p-Benzoquinone-Di(*p*-bromophenol), 32B, 280
p-Benzoquinone-Di(*p*-chlorophenol) complex, 32B, 280
p-Benzoquinone-dihalogenophenol complexes, 32B, 279
p-Benzoquinone-*p*-halogenophenol complexes, 32B, 275
[1]Benzothieno[2,3-*b*][1]benzothiophene, 35B, 923
Benzotrifurazan triethyl phosphate, 33B, 547
Benzotrifuroxan, 30B, 417; 31B, 152

Benzotrifuroxan, anthracene (1:1), 30B, 418
Benzotrifuroxan, benzene (1:1), 30B, 417
Benzotrifuroxan, bis-8-hydroxyquinolinato copper(II) (1:1 complex), 30B, 263
Benzotrifuroxan, diphenyl (1:1), 30B, 417
Benzotrifuroxan, 13,14-dithiatricyclo-[8,2,1,1^{4,7}]tetradeca-4,6,10,12-tetraene (1:1) complex), 30B, 267
Benzotrifuroxan, naphthalene (1:1), 30B, 418
Benzotrifuroxan, perylene (1:1), 30B, 417
Benzotrifuroxan, pyrene (1:1), 30B, 416
3-Benzoylanthranil, 27, 975
O-Benzoylbenzoic acid, 34B, 58
9-Benzoyl-3α-bromo-2β-hydroxy-9-azabicyclo[3.3.1]nonane, 33B, 129
3-Benzoyl-5-*p*-bromophenyl-2-methylthio-6A-thiathiophthen, 32B, 488; 34B, 193
Benzoyl chloride-aluminum chloride complex, 30B, 398; 31B, 326
6-Benzoyl-3-hydroxyl-1-methyl-5-phenyl-thiabenzene-1-oxide, 33B, 555
2-Benzoylimino-3-methylthiazolid-5-one, 26, 717
*mono*Benzoylosmocene, 30B, 431
β-Benzoyl β-phenyl selenourea, 28, 675
Benzoyl(triphenylphosphoranylidene)-methyl bromide, 30B, 275
Benzoyl(triphenylphosphoranylidene)-methyl chloride, 30B, 275
Benzoyl(triphenylphosphoranylidene)-methyl iodide, 30B, 275
3,4-Benzpyrene 1,3,7,9-tetramethyluric acid complex, 32B, 287
D- or L-3-Benzylamino-4-hydroxypentanoic acid lactone hydrobromide, 35B, 86
±3-Benzylamino-4-hydroxypentanoic acid lactone hydrobromide and hydrochloride, 33B, 546
DL-3-Benzylamino-4-hydroxypentanoic acid lactone hydrochloride, 35B, 86
2-Benzyl-5-*p*-bromobenzylidene cyclopentanone, 35B, 918
N-Benzyl-1,4-dihydronicotinamide, 26, 787
N-Benzyldithiocyanopyrrole, 27, 957
6-(N-Benzylformamido)penicillanic acid, 29, 772
(+)-2-Benzylglutamic acid hydrobromide dihydrate, 32B, 232
Benzylideneacetophenone, 35B, 55
Benzylideneaniline, 33B, 545; 35B, 61
Benzylideneaniline-*p*-carboxylic acid, 35B, 61
Benzylidene-*p*-carboxyaniline, 33B, 545
Benzylidene-cetazine, 35B, 918
Benzylidene-glucoside oxidation derivatives, 26, 627-628
Benzyllithium triethylenediamine, 35B, 474
3-(*N*-Benzyl-*N*-methylaminoethyl)-2-norbornanol, 32B, 247
(+)-N-((2-Benzylmethylamino) propyl) propionanilide hydrobromide, 34B, 75
N-Benzyl-N-methylcamphidinium iodide, 28, 677
N-Benzyl-4-methylthiazolium bromide, 35B, 239
Benzyloxycarbonyl-glycine, 26, 809
Benzyloxycarbonyl-glycyl-L-proline, 26, 809
Benzyloxycarbonyl-glycyl-L-prolyl-L-leucine, 26, 809
Benzyloxycarbonyl-glycyl-L-prolyl-L-leucyl-glycine, 26, 809
4-Benzyloxy-2,2,6-trimethyl piperidine hydrochloride, 28, 672
1-Benzylphosphole, 35B, 921
1-Benzyl-1,2,2-trimethylazetidinium bromide, 33B, 109
Beraunite, 32A, 365
Berryite, 30A, 453; 31A, 257
Bertrandite, 26, 509
Beryl, 30A, 427, 31A, 221; 33A, 480
Beryllia, 27, 473
β-Beryllia, 30A, 304
Beryllium, 26, 56; 27, 65
Beryllium borates, 32A, 511; 33A, 447
Beryllium-boron, 26, 57; 27, 65
Beryllium-boron-hafnium, 26, 64
Beryllium-boron-zirconium, 26, 64
Beryllium caesium fluorides, 33A, 176
Beryllium-calcium, 27, 68
Beryllium calcium oxide, 27, 546
Beryllium-cerium, 27, 71
Beryllium-cobalt-manganese, 26, 57, 174
Beryllium di-*n*-butylphosphinate polymer, 33B, 525
Beryllium fluoride, 27, 430
Beryllium-hafnium, 26, 58
Beryllium-hafnium-silicon, 26, 58
Beryllium-iron, 27, 4
Beryllium-lanthanum, 27, 71
Beryllium lanthanum oxide, 33A, 271
Beryllium-magnesium, 27, 68
Beryllium-molybdenum, 27, 70
Beryllium-nitrogen, 34A, 29
Beryllium-nitrogen-silicon, 26, 61; 32A, 28
Beryllium-osmium, 27, 70
Beryllium oxide, 29, 296; 34A, 238
Beryllium-platinum, 27, 332

Beryllium rare-earth compounds, 27, 70
Beryllium-rhenium, 27, 71
Beryllium-rhodium, 35A, 29
Beryllium-ruthenium, 27, 74
Beryllium-scandium, 27, 70
Beryllium-silicon-zirconium, 26, 63
Beryllium strontium oxide, 27, 547; 34A, 249
Beryllium sulphate tetrahydrate, 34A, 298
Beryllium-tantalum, 26, 58
Beryllium-titanium, 26, 58, 175
Beryllium-wolfram, 27, 70
Beryllium-yttrium, 27, 71
Beryllium yttrium oxide, 27, 547; 32A, 264
Beryllonite, 26, 456; 27, 580
Beta alumina, 33A, 275
Betafite, 26, 387
Betaine hydrobromide, 29, 771
Betaine hydrochloride, 29, 771; 35B, 313
dl-Betaprodine derivatives, 28, 462
dl-Betaprodine hydrobromide, 28, 463
dl-Betaprodine hydrochloride, 28, 464
Beudanite, 31A, 255
Beyerol monoethylidene iodoacetate, 31B, 229
Biacetylbis(mercaptoethylimine)nickel-(II), 30B, 380
1-Biadamantane, 33B, 206
Bi(anthracene-9,10-dimethylene) photo-isomer, 31B, 114
1-Biapocamphane, 33B, 206
*octahapto*Bicyclo[4.2.2]deca-2,4,7,9-tetraenediiron hexacarbonyl, 35B, 919
Bicyclo[2.2.1]hepta-2,5-diene, 33B, 537
Bicyclo[2.2.1]heptane, 33B, 537
Bicyclo-[2.2.1]-heptane derivatives, 27, 784-786
Δ^6-Bicyclo[3.2.0]heptene, 31B, 93
Bicyclo[2.2.2]hexa-2,5-diene, 35B, 902
Bicyclohexylidene, 29, 766; 30B, 91
Bicyclo[3.3.1]nonane, 30B, 425
Bicyclo[2,2,2]octa-2,5-diene-2,3-dicarboxylic acid, 35B, 137
Bicyclo[3.2.1]-octadienyliron tricarbonyltetrafluoroborate, 30B, 397
Bicyclo[2.2.2]octane, 35B, 900
Bicyclo[2,2,2]octane-1,4-dicarboxylic acid, 33B, 547
Bicyclo(2.2.2)octene-2,3-endodicarboxylic anhydride, 34B, 113
trans-Bicyclo[4.2.0]oct-1-yl-3,5-dinitrobenzoate, 35B, 142
Bicyclopentadienylberyllium, 29, 524
Bicyclo[1.1.1]pentane, 35B, 896
Bicyclo[2.1.0]pentane, 35B, 896
Bicyclo[2.1.0]pent-2-ene, 35B, 905
Bicyclopropyl, 32B, 116
Biferrocenyl, 29, 549
Bigeranyl tetrahydrochloride, 33B, 28
Biguanide acid sulphate monohydrate, 28, 669
3,3'-Biisoxazole, 32B, 190
3,4'-Biisoxazole, 34B, 195
5,5'-Biisoxazole, 28, 670; 33B, 145
3,3'-Bi-2-isoxazoline, 31B, 153
Billietite, 27, 673
1,1'-Binaphthyl, 26, 747
Bindheimite, 27, 534
Binnite, 31A, 164
1-Binorbornane, 33B, 206
Binuclear molybdenum(VI) oxalate complex, 29, 621
Binuclear nickel(II) complex of 1,4-dihydrazinophthalazine, 34B, 561
Biotite, 31A, 243
Biotite-phlogopite, 27, 697
Biotites, 28, 288
Biphenyl, 26, 726; 27, 907
2,2'-Biphenyl bis-(2-iminomethylene phenolato)copper(II), 28, 645
NN'-(2,2'-Biphenyl)bis(salicylald-iminato) copper(II), 31B, 398
Biphenylene, 26, 830; 27, 935; 31B, 109
C-Biphenylene-N^{α}-(4-iodophenyl)-N^{β}-cyanoazomethinimine, 32B, 138
2-Biphenylylferrocene, 32B, 361
4-Biphenylylferrocene, 35B, 558
2,2'-Bipyridine, 29, 763
Bipridyl-(1-acetylacetonyl)trimethyl-platinum(IV), 26, 831
2,2'-Bipyridinium tetrabromo-cobaltate(II), 35B, 911
Bipyridyl$(CO)_3ClMoSnCH_3Cl_2$, 33B, 552
Bis(acetamidine) diammineplatinum(II) chloride, 27, 834
Bisacetatobis(ethylenethiourea) cobalt(II) 35B, 632
Bis(acetonitrile) niobium(IV) bromide, 32B, 496
Bis (acetonitrile) zinc chloride, 27, 829
Bis(acetylacetonato)manganese(II) dihydrate, 35B, 606
Bis(acetylacetonato)nickel(II), 30A, 338
Bis(acetylacetonato) quinoline-copper(II), 34B, 498
Bis-acetylacetone nickel(II), 26, 673
Bis-acetylacetone zinc(II), 26, 673
Bis-acetylferrocenyl, 30B, 396
Bis-1-(1'-acetylferrocenyl), 27, 861
cis-Bis-(D-alaninato)copper(II), 34B, 537
trans-Bis-(L-α-alaninato)copper(II), 31B, 413

Bis(allyl) chromium(II), <u>32B</u>, 493
Bis-π-allyl rhodium chloride, <u>32B</u>, 493
Bis-(β-aminobutyrato)-copper(II) dihydrate, <u>26</u>, 665
Bis-1-aminocyclopentanecarboxylato copper(II), <u>28</u>, 642
Bis-(2-aminoethylsalicylideneiminato)chromium(III) iodide, <u>33B</u>, 547
Bis(2-amino-2-methyl-3-butanone oximato)nickel(II) chloride monohydrate, <u>33B</u>, 424
Bis(2-amino-2-methyl-propionato) nickel (II) tetrahydrate, <u>27</u>, 839
Bis-(4-amino-3-penten-2-ono) copper (II), <u>29</u>, 768
Bis-(4-amino-3-penten-2-ono) nickel (II), <u>29</u>, 768
Bis-(4-amino-3-penten-2-ono) palladium (II), <u>29</u>, 768
Bis-(3-amino-1-phenyl-2-buten-1-ono) copper(II), <u>29</u>, 768; <u>33B</u>, 446
Bis-(3-amino-1-phenyl-2-buten-1-ono) nickel(II), <u>29</u>, 768
Bis-(6-aminopurine) copper(II) tetrahydrate, <u>34B</u>, 559
Bis-(4-aminopyrimidine-2-one) copper(II) chloride, <u>33B</u>, 548
Bis-arene-metal complex, <u>30B</u>, 393
Bis-(arsenic-diphenyl)-acetylene, <u>26</u>, 736
Bis-(π-azulene)iron, <u>33B</u>, 549
Bis(benzene azo-β-naphthol) copper (II), <u>26</u>, 663
Bis(benzoylacetonato) zinc monoethanolate, <u>34B</u>, 504
Bis(biguanide)copper(II) perchlorate, <u>35B</u>, 909
Bis(2,2'-bipyridyl) copper(II) iodide, <u>26</u>, 657
Bisbipyridyl-μ-dihydroxo-dicopper(II) nitrate, <u>35B</u>, 728
Bis(bipyridyl)lanthanum nitrate, <u>33B</u>, 190
Bis[(+)-3-bromocamphor-9-sulphonate], <u>33B</u>, 190
Bis-[1,2-bis(diphenylphosphino)ethane] thorium(I) perchlorate, <u>35B</u>, 864
Bis(*cis*-1,2-bis(trifluoromethyl)ethylene-1,2-dithiolate)cobalt dimer, <u>30B</u>, 372
Bis-biuret-copper(II) dichloride, <u>31B</u>, 437
Bis-biuret zinc dichloride, <u>28</u>, 645
Bis(*m*-bromobenzoyl)methane, <u>27</u>, 913
trans-Bis(α-bromoisopropyl)-1,4-cyclohexane, <u>33B</u>, 545
1,2-Bis(bromomethyl)carborane, <u>29</u>, 531
2,4-Bis(4-bromophenyl)benz[*d*][1,3,6] oxadiazepine, <u>33B</u>, 146
Bis-(4-bromophenylsulphonyl)methane, <u>34B</u>, 45
Bis-(4-bromophenylthio)methane, <u>35B</u>, 918
Bis-(5-bromosalicylaldiminato)-nickel(II), <u>26</u>, 672
Bis(butadiene)rhodium(I) chloride, <u>34B</u>, 441
Bis(*t*-butyl isocyanide)(tetracyanoethylene)nickel(0), <u>35B</u>, 523
Bis(N-*n*-butylsalicylaldiminato) copper(II), <u>34B</u>, 518
Bis(N-t-butylsalicylaldiminato)-copper(II), <u>30B</u>, 393; <u>31B</u>, 395
Bis(N-t-butylsalicylaldiminato)-palladium(II), <u>33B</u>, 399
Bis-N-butylsalicylaldimine-cobalt, <u>26</u>, 679
Bis(N-*n*-butylsalicylaldimine) copper, <u>27</u>, 872
Bis-(N-*n*-butylsalicylaldimine)nickel, <u>27</u>, 872
Bis-*N*-butylsalicylaldimine palladium, <u>29</u>, 582
Bis-(N-*n*-butylsalicylaldimine) palladium, <u>27</u>, 872
Bis-N-butylsalicylaldimine-zinc, <u>26</u>, 679
Bis(caesium) tri-μ-bromo-octabromo-trirhenate(III), <u>31A</u>, 110
Bis-calcium strontium hexapropionate, <u>32B</u>, 492
Bis-[1-1'-carbomethoxyferrocenyl)], <u>27</u>, 861; <u>33B</u>, 355
Bis-(*m*-chlorobenzoyl)-methane, <u>29</u>, 655
Bis-chloroferrocenyl, <u>30B</u>, 396
Bis-1-(1'-chloroferrocenyl), <u>27</u>, 861
Bis-[1(2'-chlorferrocenyl)], <u>30B</u>, 310; <u>31B</u>, 354
trans-Bis(α-chloroisopropyl)-1,4-cyclohexane, <u>33B</u>, 545
2,3-Bis-(*cis*-4-chloro-1-methylcyclohexyl)-*trans*-2-butene, <u>33B</u>, 82
Bis(*p*-chlorophenyl)dichlorosulphide, <u>34B</u>, 42
Bis(chloropyridylmercury)tetracarbonyl-iron, <u>35B</u>, 914
Bis-(5-chlorosalicylaldoximato) copper(II), <u>29</u>, 598
Bis(chlorosulphur)nitrogen-tetrachloroborate, <u>34A</u>, 206
1,4-Bis(cyanomethylene)cyclohexane, <u>33B</u>, 545
Bis(cyclohexylbenzene)silver(I) perchlorate, <u>33B</u>, 547
Bis(*N*-cyclohexylsalicylaldiminato) copper(II), <u>35B</u>, 907
Bis(*N*-cyclohexylthiopicolinamido-copper(II), <u>35B</u>, 919
Bis(cyclo-octa-1,5-diene) iridium trichlorotin, <u>32B</u>, 376
Bis-cycloocta-1,5-dienenickel(0), <u>30B</u>, 321
Bis(cyclooctatetraene)iron, <u>33B</u>, 553

Bis(cyclo-octatetraene)titanium, 34B, 469
Bis(cyclooctatetraene)triruthenium tetracarbonyl, 33B, 378
Bis-π-cyclopentadienyl(2-aminoethane-thiolate)-molybdenum iodide, 34B, 458
Bis(π-cyclopentadienyl cobalt) bis(t-butylimine) carbonyl, 32B, 493
2,4-Bis-(π-cyclopentadienylcobalt-cyclopentadien)ylcyclopentadiene, 30B, 2, 4, 302
Bis-N,N'-(π-cyclopentadienyl-cobalt)-di-t-butyl urea, 34B, 455
Bis(π-cyclopentadienyldicarbomethoxy-acetylenecarbonylniobium), 35B, 562
trans-1,4-Bis(π-cyclopentadienyldicar-bonyl-iron)buta-1,3-diene, 34B, 425
Bis(π-cyclopentadienyldicarbonyl-iron)-diphenylsulphonyltin, 33B, 549
Bis(cyclopentadienyldicarbonyl-ruthenium), 32B, 357
Bis(π-cyclopentadienyl niobium tolane carbonyl), 34B, 448
Bis(cyclopentadienyl oxomolybdenum sulphide), 32B, 350
Bis(π-cyclopentadienyl)titanium pentasulphide, 35B, 913
Bis(cyclopentamethylene)diphosphine disulphide, 35B, 451
Bis(cyclotetramethylene) diphosphine disulphide, 34B, 382
Bis(diethyl-(2-mercaptoethyl)-phosphine)platinum(II)-silver(I) nitrate complex, 34B, 596
1,8-Bis dehydro[14]-annulene, 31B, 94
Bisdethiodi(methylthio)-acetylaranotin, 33B, 553
Bis(1,3-diaminobutane)copper(II) perchlorate, 33B, 544
Bis[di(2-aminoethyl)amine]nickel(II) chloride monohydrate, 35B, 702
Bis(1,3-diaminopropane)copper(II) chloride monohydrate, 30B, 420
Bis(1,3-diamino propane)copper(II) nitrate, 34B, 555
Bis(1,3-diaminopropane)copper(II) nitrite, 35B, 676
Bis(1,3-diaminopropane)nickel(II) nitrate dihydrate, 33B, 544
(-)-*cis,trans,cis*-Bis(diamino-propionato)-cobalt(III) bromide, 34B, 539
Bis[di-(3-aminopropyl)amine]nickel(II) perchlorate, 35B, 702
Bis-(1,4-diazacycloheptane) copper(II) nitrate hemihydrate, 34B, 562
3,6-Bisdiazocyclohexanetetraone, 34B, 89
Bis(*N,N*-di-*n*-butyldithiocarbamato)-gold(III) bromide, 35B, 806
Bis(*N,N*-di-*n*-butyldithiocarbamato) gold(III) dibromoaurate(I), 33B, 493
Bis(*N,N*-di-*n*-butylthiocarbamato)-gold(III) dibromocuprate(I), 35B, 810
Bis-(3-1,7-dicarbollylnickelate(II) dianion, 35B, 431
trans-1,4-Bis(dicarbonyl-π-cyclo-pentadienyliron)buta-1,3-diene, 33B, 549
Bis(dicarbonyl-π-cyclopentadienyliron)-dichlorogermane, 32B, 355
Bis-(dicarbonyl-π-cyclopentadienyl-iron)-dicyclopentadienyl tin, 33B, 549; 34B, 414
Bis(dicarbonyl-π-cyclopentadienyliron) dimethyllead, 33B, 310
Bis(dicarbonyl-π-cyclopentadienyliron) dimethyltin, 33B, 310
Bis(dicarbonyl-π-cyclopentadienyliron) tin dinitrite, 33B, 310
Bis(dichlorophosphate) bis(phosphoryl chloride) magnesium, 35A, 186
Bis(dichlorophosphato)-bis(ethyl ace-tate)manganese(II), 28, 649
3,6-Bis(dicyanomethylene)cyclohexa-1,4-diene-hexamethylbenzene, 35B, 415
1,4-Bis(dicyanomethylene)cyclo-hexane, 35B, 108
Bis-(1,2-diethoxycarbonyl-ethyl)tin dibromide, 33B, 313
Bis(DL-N,N-diethyl-α-alaninato) copper(II), 34B, 543
NN-Bis-(2-diethylaminoethyl)-2-oxyethylaminonickel(II) perchlorate, 35B, 777
Bis(diethyldithiocarbamate)dioxo-(triphenylarsine oxide)-uranium(VI), 35B, 819
Bis(diethyldithiocarbamate)dioxo-triphenylphosphine oxide)-uranium(VI), 35B, 819
Bis-diethyldithiocarbamate of phenyl-arsine, 30B, 393
Bis-(N,N-diethyldithiocarbamato)-copper(II), 26, 668
Bis-(N,N-diethyldithiocarbamato)-nickel(II), 26, 672
Bis(O,O'-diethyldithiophosphato) bispyridinenickel(II), 32B, 457
Bis(diethyldithiophosphato) nickel(II), 32B, 454; 34B, 592
Bis(diethylenetriamine) copper(II) bromide monohydrate, 34B, 489
Bis(diethylenetriamine) copper(II) nitrate, 34B, 486
Bis(*N,N'*-diethylethylenediamine)-copper(II)nitrate, nitrate hydrate, and bromide hydrate, 33B, 543

Bis(N,N'-diethylethylenediamine) diaquo copper(II) nitrate, 34B, 494
Bis(N,N-diethylthiocarbamato) nickel (II), 28, 673
Bis(diethylthiophosphoryl) diselenide, 31B, 301
Bis-[diiron hexacarbonyl methyl mercaptan] sulphide, 32B, 455
Bis(*O*,*O*'-diisopropylphosphorothionyl) disulphide, 35B, 452
trans-Bis(dimethylamine)diammine platinum(II) chloride, 35B, 666
Bis(dimethylamino)beryllium, 34B, 401
Bis(dimethylamino) beryllium trimer, 32B, 492
Bis(dimethylamino)boronium chloride, 33B, 264
4,4'-Bis(dimethylamino)diphenylamine dihydroiodide, 33B, 542
4,4'-Bis(dimethylamino)diphenylamine perchlorate, 32B, 71
Bis(2-dimethylaminoethyl(methyl) amino)di(methyl-magnesium), 34B, 399
Bis(*p*-*N*,*N*-dimethylaminophenyl diazonium) tetrachlorozincate(II), 31B, 46
Bis(*N*-γ-diethylaminopropylsalicylaldiminato) nickel(II), 32B, 406
1,1-Bis(dimethylamino)-2,4,6-triphenylphosphorin, 35B, 911
Bis-(*o*-dimethylarsinophenyl)methylarsine-copper-manganese pentacarbonyl, 30B, 396
1,2-Bis(dimethylarsino)-3,3,4,4-tetrafluorocyclobutene-bis (irontricarbonyl), 32B, 374
1,2-Bis(dimethylarsino)-tetrafluorocyclobutene-di-irontricarbonyl complex, 31B, 466
1,2-Bis(dimethylarsino)tetrafluorocyclobutene nonacarbonyl triiron, 34B, 603
Bis-(μ-4,4'-dimethylbenzophenone hydrazonato tricarbonyl iron), 32B, 497
Bis(*NN*-dimethyldithiocarbamato)-nitrosyliron, 35B, 792
Bis-(*N*,*N*-dimethyldithiocarbamato) pyridinezinc(II), 32B, 412
Bis(dimethyldithiophosphato)nickel (II), 34B, 586
Bis(dimethyldithiophosphinato)-nickel(II), 33B, 546; 34B, 587
Bis(N,N'-dimethylethylenediamine) copper(II) nitrate, 34B, 488
Bis(*N*,*N*'-dimethylethylenediamine)-copper(II) sulphate tetrahydrate, 35B, 909
Bis(*N*,*N*'-dimethylethylenediamine)-copper(II) thiocyanate, 35B, 678
Bis(N,N'-dimethylethylenediamine) nickel(II) bromide, 34B, 489
p-Bis(dimethylhydroxysilyl) benzene, 32B, 300
Bis(*N*,*N*-dimethyl-β-mercaptoethylamine nickel(II), 32B, 453
Bis-(diphenylarsenic) oxide, 28, 592
Bis(diphenyldiazomethane) tris(tricarbonyl iron), 32B, 497
Bis(diphenylglyoximato) nickel(II) iodine complex, 32B, 491
4,4'-Bis(diphenylmethyl)biphenyl dicarbonium antimony(V) hexachloride, 32B, 486
Bisdiphenylmethyl diselenide, 34B, 52
Bis-μ-diphenylphosphido-bis-(dicarbonylnickel), 35B, 848
Bis(diphenylphosphino)acetylene, 34B, 390
Bis(diphenylphosphino)ethylamine ethyl iodide, 30B, 278
Bis(diphenylphosphino)ethylamine-molybdenum tetracarbonyl, 30B, 387
Bis-(2-diphenylphosphinoethyl)-aminenickel(II) dibromide, 33B, 550
Bis-(diphenyl-phosphorus)-acetylene, 26, 736
Bis(1,3-diphenyl-1,3-propanedionato) copper(II), 31B, 387
trans-Bis(diphenyl-*o*-selenolatophenylphosphine)nickel(II), 35B, 832
Bis(dipivaloyimethanido)zinc(II), 29, 616
Bis(N,N-di-*n*-propyldithiocarbamato) copper(II), 27, 844
Bis[di(pyridine-*N*-oxide)copper-dibromide], 35B, 912
Bis(2,2'-dipyridyliminato)palladium(II), 29, 768; 30B, 354
Bis(dithiobenzil)nickel, 32B, 460
Bis(dithiobenzoato)palladium, 33B, 547
Bis(dithiobiureto)palladium(II), 33B, 550
Bis[di(trimethylsilyl)amino]-beryllium, 35B, 901
Bis(*o*-dodecacarborane), 30B, 272
Bis-*l*-ephedrine copper(II) benzene, 29, 609
Bis(ethylacetato) calcium difluorophosphate, 34B, 397
Bis(*N*-ethyl-2-benzothiazolyl)phosphamethinecyanine perchlorate, 31B, 302
1,4-Bis-(*N*-ethyl-1,2-dihydrobenzthiazol-2-ylidene)tetrazene, 32B, 198
Bisethylene-1,2-bis(methylphenylphosphine)palladium(II) chloride, 35B, 855

trans-Bis(ethylenediamine)-bis(isothiocyanato)nickel(II), 26, 670; 28, 616
cis-(+)D-Bis(ethylenediamine)cobalt (III) dicyanide chloride monohydrate, 32B, 494
Bis(ethylenediamine)copper(II) fluoroborate, 33B, 383
Bisethylenediaminecopper(II) nitrate, 29, 593
Bis(ethylenediamine)copper(II) perchlorate, 32B, 379
Bisethylenediaminecopper(II) thiocyanate, 29, 595
Bis-ethylenediamine copper(II) thiocyanate complex, 26, 670
trans(Bis(ethylenediamine)-dichlorocobalt(III) chloride hydrochloride dihydrate, 32B, 494
Bis(ethylenediamine) lithium bromide, 32B, 267
Bisethylenediamine lithium chloride, 31B, 378; 32B, 268
Bis(ethylenediamine) lithium halides, 32B, 267
Bis(ethylenediamine)nickel(II) dibromoargentate(I), 33B, 544
Bis(ethylenediamine)nickel(II) diiodoargentate(I), 33B, 544
Bis-ethylenediamine nickel(II) thiocyanate complex, 26, 670
Bis(ethylenediamine)palladium(II) chloride, 31B, 375
Bisethylenediaminepalladium(II) *cis*-dithiosulphate-ethylenediaminepalladite, 35B, 598
Bis(ethylenediamine)platinum(II) chloride, 28, 670
Bis(ethylenediammonium) dibromide tetrabromocuprate, 32B, 485
Bis-[1'-(1-ethylferrocenyl)], 27, 861; 30B, 308; 31B, 361
α-*trans*-Bis(N-ethylidene-DL-threoninato)copper(II)dihydrate, 26, 660
β-*trans*-Bis(N-ethylidene-DL-threoninato)copper(II)dihydrate, 26, 660
Bis(μ-ethylmercaptide μ-ethylthioxanthate)di(iron ethylthioxanthate) 34B, 597
μ,μ-Bis(ethylmercapto)hexacarbonyldiiron(I), 28, 654
Bis-(1-ethyl-2-quinolyl)-phosphamethinecyanine perchlorate, 33B, 545
Bis-(*N*-ethylsalicylaldiminato)copper-(II), 32B, 397
Bis-(*N*-ethylsalicylaldiminato)-nickel, 32B, 495
Bis-*N*-ethylsalicylaldimine palladium, 29, 582
Bis(ethyl thioxanthato)-μ-bis-(ethyl thioxanthato)-μ'-bis-(ethylthio)-diiron(III), 35B, 802
Bis(fulvalene)di-iron, 34B, 456
trans-Bisglycinatoplatinum(II), 34B, 538
Bisglycino-copper(II) monohydrate, 29, 605
Bis(glycollato)copper(II), 33B, 403
Bis(glyoximato)nickel(II), 32B, 428
Bis(glyoximato) platinum(II), 34B, 550
Bis(hexafluoroacetylacetonato) 2,2'-bipyridine-copper(II), 34B, 503
Bis-(1,1,1,5,5,5-hexafluorpentane-2,4-dionato)bis-(*NN*-dimethylethylenediamine)copper(II), 35B, 916
Bis-hexamethylenetetramine calcium(II) bromide decahydrate, 32B, 276
Bis(*N*-*n*-hexyl-*o*-hydroxyacetophenoniminato)copper(II), 35B, 907
Bis(histamino)-bis(tetrafluoroborato)copper(II), 35B, 918
Bis(L-histidinato)zinc(II) dihydrate, 28, 636
Bis(histidinato)zinc(II) pentahydrate, 28, 636
Bis(L-histidine)copper(II) dinitrate dihydrate, 34B, 544
Bis-histidino-cadmium(II), 30B, 395
Bis(histidino)cadmium dihydrate, 32B, 422
Bis-histidino-cobalt(II), 30B, 395
Bis-(L-histidino)cobalt(II) monohydrate, 33B, 421
Bis-histidino-nickel(II), 30B, 395
Bis(histidino)nickel(II) monohydrate, 32B, 424
1,1'-Bishomocubane $C_{10}H_{10}$Isomer, 35B, 923
Bis(hydrazinecarboxylato-*N'*,*O*) cadmium, 33B, 436
Bis(hydrazinecarboxylato-N',O) cadmium monohydrate, 34B, 547
Bis(hydrazinecarboxylato-*N'*,*O*)-manganese(II) dihydrate, 32B, 425
Bis(hydrazinecarboxylato-*N'*,*O*)-nickel dihydrate, 35B, 720
Bis(hydrazinecarboxylato-*N'O*)-zinc, 35B, 912
Bis-(hydrazine)manganese(II) chloride, 28, 89
Bis(hydrazine)zinc bis(hydrazincarboxylate-N',O), 30A, 409
Bishydrazine zinc chloride, 28, 92
Bis(hydrazine)zinc isothiocyanate, 30A, 298
Bis(hydrogen *o*-phthalato) diaquocopper(II), 34B, 535
Bis[hydrotris(1-pyrazolyl)borato]-cobalt(II), 35B, 434
Bis-(2-hydroxyacetophenone)trimethylenediiminocopper(II), 35B, 727

Bis(*N*-(2-hydroxyethyl)ethylenediamine)copper(II) nitrate, 33B, 544
Bis(*N*-(2-hydroxyethyl)-ethylenediamine)copper(II) perchlorate, 33B, 544
Bis(2-hydroxyethylimino-pyruvato)copper(II)tetrahydrate, 34B, 529
Bis(2-hydroxyethyliminosalicylaldehydato)copper(II), 35B, 622
Bis-(*N*-2-hydroxyethylsalicylaldiminato)copper(II), 31B, 393
Bis-α-hydroxy-α-phenylbutyramidine-copper dihydrate, 32B, 438
Bis(2-hydroxypropane-1,3-diamine)-copper(II) chloride, bromide, and nitrate, 35B, 909
Bis-(6-hydroxypurine)copper(II) chloride trihydrate, 35B, 694
Bis(8-hydroxyquinolato)copper(II), 28, 687; 32B, 435
Bis-8-hydroxyquinolinatocopper(II)-1,2,4,5-tetracyanobenzene, 33B, 451
Bis-8-hydroxyquinolinatopalladium(II), chloranil (1:1 complex), 30B, 262
Bis-8-hydroxyquinolinatopalladium(II)-1,2,4,5-tetracyanobenzene, 1:1 complex, 31B, 280
Bis(8-hydroxyquinoline)silver(I) (pyridine solvate), 33B, 443
Bis(imidazole)-α,β,γ,δ-tetraphenylporphinatoiron(III) chloride, 34B, 279
Bis(imidazoline-2-thione)cadmium chloride, 33B, 549
Bis(imidotetramethyldithiodiphosphino-*SS*)-iron(II), 35B, 914
Bis(2-imino-4-amino-4-methylpentane)-copper(II) nitrate, 29, 600
Bis(2-imino-4-amino-4-methylpentane) nickel(II) nitrate, 34B, 571
Bis[2,2'-iminobis(acetamidoxime)]-copper(II) chloride, 35B, 687
Bis(2,2'-imino-bis(acetamidoxime)) nickel(II) chloride dihydrate, 32B, 496; 35B, 687
Bis-indenyl ruthenium, 27, 863; 32B, 359
1,4-Bis(iodo-diphenyl-tin)butane, 34B, 414
3,7-Bis(*p*-iodophenyl)-4,5,6-triphenyl-4H-1,2-diazepine, 35B, 184
Bis-(*N*-isopropyl-3-methylsalicylaldiminato)nickel, 31B, 397
Bis-(*N*-isopropyl-3-ethylsalicylaldiminato)nickel(II), 32B, 403
Bis-(*N*-isopropyl-3-ethylsalicylaldiminato)palladium(II), 32B, 400, 404
Bis-(*N*-isopropylsalicylaldiminato) copper(II), 31B, 393
Bis-(*N*-isopropylsalicylaldiminato) nickel(II), 28, 686; 29, 568
1,2-Bis(isopropylseleno)ethane-palladium(II) dichloride, 35B, 831
Bis(maleonitriledithiolato) nickelate (II) ion, 28, 685
Bis(meso-stilbenediamine)nickel(II) dichloroacetate, 29, 571
Bis(meta-chlorobenzoyl)methane, 28, 684
Bis-methallylnickel, 28, 685; 30B, 298
Bis-methanol, mercuric chloride, 30B, 368
Bis(methinyltricobaltenneacarbonyl)-acetylene, 35B, 504
2,3-Bis-(methoxycarbonyl)-2π,5-norbornadien-7-yl-(π-cyclopentadienyl)nickel(II), 28 606
Bis-*N*-methyl-*N*'-diazabicyclo[2.2.2]-octonium di-μ-chloro-hexachlorodinickelate(II), 35B, 878
Bis(*N*-methylenediamine)copper(II) chloride, 35B, 906
Bis(*N*-methylethylenediamine) copper (II) bromide, 35B, 909
Bis(*N*-methylethylenediamine) copper (II) iodide, 35B, 909
Bis(*N*-methylethylenediamine)copper-(II) nitrate, 33B, 544
Bis(*N*-methylethylenediamine) copper(II) perchlorate, 35B, 677
Bis(*N*-methylethylenediamine)copper (II) wolframate(VI) dihydrate, 35B, 909
Bis(N-methyl-2-hydroxy-1-naphthal-diminato)copper(II), 29, 768
Bis(1-methyl-5-iodouracil)-9-ethyl-adenine complex, 34B, 231
Bis(1-methyl-5-iodouracil):9-ethyl-2,6-diamino-purine complex, 34B, 233
Bis-(3-methylpentane-2,4-dionato) copper(II), 32B, 384
Bis(2-methylpentane-2,4-dioxydimethyl-titanium), 35B, 747
Bis(2-methylpropane-1,2-diamine)-copper(II) bromide, 35B, 909
Bis(2-methylpropane-1,2-diamine) copper(II) nitrate, 35B, 909
1,2-Bis(*N*-methyl-4-pyridinium)-ethylenetetracyanonickelate(II), 35B, 177
Bis-(1-methyl-2-quinoline)hydrogen hexafluoroarsenate(V), 34B, 150
Bis(2-methyl-8-quinolinolato)-aluminum(III)-μ-oxobis(2-methyl-8-quinolinolato)aluminum(III), 35B, 732
Bis-(N-methylsalicylaldiminato)-copper, 26, 667

Bis-(*N*-methylsalicylaldiminato) copper(II) (γ-form), 33B, 397
Bis-(4-methylsalicylaldiminato)-nickel (II), 26, 672; 32B, 395
Bis methylthiomercury, 27, 835
Bis(1-methyl thymine):9-ethyl-2,6-diamino-purine complex monohydrate, 34B, 232
Bismuth, 26, 62, 280; 27, 75, 385; 35A, 9
Bismuth-II, 32A, 29
Bismuth-caesium, 26, 54, 62
Bismuth caesium iodide, 33A, 241
Bismuth calcium niobium oxide, 27, 538
Bismuth calcium tantalum oxide, 27, 538
Bismuth calcium titanium oxide, 27, 538
Bismuth-cerium, 27, 335
Bismuth *mono*chloride, 27, 450; 28, 81
Bismuth(I)chloro-aluminate, 26, 324
Bismuth-copper, 26, 63
Bismuth-copper-gadolinium, 34A, 34
Bismuth-copper-lead-sulphur, 35A, 30
Bismuth-copper-sulphur, 33A, 41
Bismuth double oxides, 33A, 321
Bismuth ferrate, 27, 506
Bismuth ferrate-lead titanate, 27, 506
Bismuth ferrite, 28, 135; 29, 323; 31A, 248, 249; 32A, 269
Bismuth fluoride, 27, 724
Bismuth(III) formate, 34B, 395
Bismuth-gadolinium, 34A, 34
Bismuth gallium titanium oxides, 27, 539
Bismuth germanium oxide, 32A, 292
Bismuth-germanium-tellurium, 30A, 28; 33A, 42; 35A, 31
Bismuth halide complex salts, 32B, 157
Bismuth-holmium, 26, 164
Bismuth hydroxychromate, 27, 657
Bismuth-indium, 26, 42; 27, 80; 34A, 31
Bismuth-indium-tellurium, 26, 43
Bismuth iodide, 31A, 111
Bismuth-lead, 27, 76, 81
Bismuth lead niobium oxide, 26, 402; 27, 538
Bismuth-lead-niobium-oxygen, 26, 402, 27, 538
Bismuth-lead-niobium-oxygen-titanium, 26, 402
Bismuth-lead-oxygen-tantalum, 26, 402; 27, 538
Bismuth-lead-oxygen-titanium, 26, 402; 27, 538
Bismuth-lead-platinum, 27, 287
Bismuth-lead-selenium, 26, 63; 31A, 25; 33A, 43
Bismuth-lead-sulphur, 27, 81, 405
Bismuth lead tantalum oxide, 26, 402; 27, 538
Bismuth-lead-tellurium, 26, 63; 27, 83; 34A, 32; 35A, 32
Bismuth-lead-tellurium-thallium, 27, 363
Bismuth lead titanium oxide, 26, 402; 27, 538
Bismuth lithium oxide, 29, 316
Bismuth magnesium titanate, 33A, 305
Bismuth minerals, 27, 599
Bismuth molybdate, 31A, 242, 261
Bismuth molybdenum oxide, 31A, 250
Bismuth-nickel-rhodium, 27, 287
Bismuth niobium strontium oxide, 27, 538
Bismuth niobium titanium oxide, 27, 538
Bismuth nitrate pentahydrate, 30A, 401
Bismuth oxide, 26, 402; 29, 311; 30A, 446
Bismuth oxide-wolfram oxide system, 31A, 257
Bismuth oxyfluoride, 29, 271
Bismuth-oxygen-titanium, 26, 402
Bismuth-palladium-platinum, 27, 287
Bismuth phosphate, 27, 588
Bismuth-platinum, 27, 85
Bismuth-potassium, 26, 53
Bismuth potassium titanate, 27, 508
Bismuth potassium titanum oxide, 27, 539
Bismuth-rare-earth, 34A, 34
Bismuth-rhodium, 26, 35, 65; 27, 87
Bismuth-rubidium, 26, 54
Bismuth-scandium, 27, 88
Bismuth-selenium, 27, 88; 32A, 30; 33A, 44
Bismuth-selenium-silver-sulphur, 33A, 44
Bismuth-selenium-tellurium, 26, 65; 28, 12
Bismuth-selenium-thallium, 27, 89
Bis(selenourea)metal(II) thiocyanates, 34A, 234
Bismuth-silver-tellurium, 29, 30
Bismuth sodium titanate, 27, 508
Bismuth sodium titanium oxide, 27, 508, 539; 32A, 268
Bismuth solid solutions, 27, 75
Bismuth-strontium, 26, 62
Bismuth strontium tantalum oxide, 27, 538
Bismuth strontium titanates, 28, 289
Bismuth strontium titanium oxide, 27, 539; 28, 289
Bismuth subchloride, 27, 450; 28, 81
Bismuth-sulphur-tellurium, 27, 90
Bismuth tantalum titanum oxide, 27, 538
Bismuth-tellurium, 27, 76, 90; 32A, 30; 33A, 45
Bismuth-tellurium-thallium, 26, 46
Bismuth-tellurium-thallium-tin, 27, 363

Bismuth-terbium, 26, 255
Bismuth thorium nitride, 35A, 113
Bismuth-tin, 27, 76
Bismuth titanium oxide, 27, 538
Bismuth trifluoride, 27, 724
Bismuth triiodide, 31A, 111
α-Bismuth trioxide, 35A, 210
Bismuth-uranium, 32A, 21
Bismuth uranium nitride, 35A, 113
Bismuth-yttrium, 27, 92
Bismuth-zirconium, 33A, 24
Bis-nitrilotriacetatozirconate(IV) ion, 30B, 396
Bisnitrito-2,2'-bipyridylcopper(II), 34B , 552
Bis(ω-nitroacetophenonato)copper(II), 35B, 916
Bis-(*o*-nitrophenyl)disulphide, 35B, 58
Bis(*p*-nitrophenyl) disulphide, 34B, 44
Bis(nitrosocyclohexane), 35B, 111
cis-Bis(nitrosomethane), 28, 404
22,23-Bisnor-5-cholenic acid-3β-01, 30B, 441
Bisnorquassin *m*-bromobenzoate, 35B, 391
cis-1,6-8,13-Bis-ocido-[14]-annulene, 32B, 178
Bis(octamethylcyclotetraphosphonitrilium) tetrachlorocobaltate(II), 35B, 461
3,6-Bis-2-oxobut-3-ylidene-1,2,4,5-tetrathiin, 31B, 148
trans-Bis(pentacarbonylmanganate)-bispyridineplatinum(II), 33B, 551
Bis(pentacarbonylmanganese)diphenyl-tin, 29, 769
Bis(pentadienyl)dinickel, 34B, 438
Bis(pentafluorobenzenethiolato)bis-tributylphosphine)platinum(II), 35B, 824
1,1'-Bis(pentamethyldisilanyl)-ferrocene, 33B, 353
Bis(2,4-pentanedionato)bis(pyridine)-nickel(II), 30B, 432
Bis(2,4-pentanedionato)bis(pyridine N-oxide)nickel(II), 33B, 394
trans-Bis(2,4-pentanedionato) dipyridinecobalt(II), 33B, 392
trans-Bis(2,4-pentanedionato) dipyridinenickel(II), 33B, 393
Bis(perchlorato)bis(octamethyl-pyrophosphoramide)copper(II), 35B, 744
Bisperchloratotetrakis-(3,5-dimethylpyridine)nickel(II), 33B, 447
Bis[*cis*-(1,2-perfluoromethylene-1,2-dithiolato)]nickel, 35B, 804
2,4-Bis(phenacylmethylidene)-1,3-dithiete, 32B, 488
Bis(phenanthroline) cobalt chloride trihydrate, 31B, 427
Bis(phenoxyacetato)triaquocopper(II), 35B, 914
Bis-(3-phenylacetylacetonato)copper, 27, 844
Bis-(N-phenylacetylacetonimino)-nickel(II), 27, 843
Bis(2-phenyl-π-allyl-palladium chloride), 34B, 444
Bis(1-phenyl-1,3-butanedionato) copper, 31B, 383
Bis(1-phenyl-1,3-butanedionato) palladium(II), 32B, 387
Bis(1-phenyl-1,3-butanedionato)vanadyl (*cis* form), 30B, 340
Bis(phenyldithiophosphinato)nickel(II), 33B, 495
Bis(*o*-phenylenediamino)nickel, 33B, 425
Bis-*o*-phenylene disulphide, 30B, 425
trans-Bis(phenylethynyl)bis(tri-ethylphosphine)nickel(II), 32B, 474
Bis-(μ-phenyloxycarbene-tricarbonyl-iron), 34B, 427
Bis(3-phenyl-2,4-pentanedionato) copper(II), 30B, 337
Bis-(N-phenylsalicylaldiminato) copper(II), 29, 597
Bis-(N-phenylsalicylaldiminato) copper(II) pyridine adduct, 30B, 395; 33B, 401
Bis-(N-phenylsalicylaldimino)-nickel (II), 27, 843
Bis(phosphorodibromidato)-bis(ethyl acetate)calcium, 30B, 422
Bispicolinatocopper(II), 33B, 543
trans-Bis(picolinato)copper(II) dihydrate, 35B, 765
α-Bis(picoline)copper dinitrate, 35B, 914
Bispiperidinium pentabromobismuth-ate(III), 33B, 294
Bis(1,3-propanediamine)copper(II) benzoate, 35B, 909
Bis(1,2-propanediamine)copper(II) bromide, 33B, 544
Bis(1,2-propanediamine)copper(II) dichromate, 33B, 543; 35B, 675
Bis(1,2-propanediamine)copper(II) iodide, 33B, 544
Bis(1,2-propanediamine)copper(II) nitrate, 35B, 674
Bis(1,3-propanediamine)copper(II) selenate monohydrate, 35B, 681
Bis(1,2-propanediamine)copper(II) sulphate hydrate, 33B, 544
Bis-(1,3-propanediamine)copper(II) sulphate monohydrate, 35B, 681
Bis(propylenethiourea)tellurium dibromide, 26, 583
Bis(propylenethiourea)tellurium dichloride, 26, 583

Bis(propylenethiourea)tellurium dimethylthiosulphonate, 26, 584
Bis(propylenethiourea)tellurium dithiocyanate, 26, 584
Bis-N-*n*-propylsalicylaldiminato)copper-(II), 34B, 517
Bis-(+)-pseudoephedrine complex of Cu(II), 33B, 547
Bis(pyrazinamide)cobalt(II) nitrate monohydrate, 35B, 909
Bis(pyrazine)cobalt(II) bromide, 35B, 909
Bispyridinebisacetatocopper(II) pyridine, 32B, 419
Bis(pyridine-2-carboxamido)nickel(II) dihydrate, 32B, 496; 33B, 427
Bis(pyridine-2-carboxylato)-copper(II) potassium thiocyanate, 35B, 412
Bis(pyridine)copper(II) azide, 34B, 554
Bis-pyridinelithium(I) chloride hydrate, 32B, 271
Bis(pyridine)nickel(II) monothiobenzoate, 35B, 907
Bis(pyridine-N-oxide) copper(II) nitrate, 34B, 580
Bispyridinesilver(I) nitrate monohydrate, 35B, 762
Bis(8-quinolinethiolato)palladium, 30B, 422
Bis-salicylaldehydatocopper(II), 28, 687; 29, 599, 763, 768; 30B, 369
Bis(salicylaldehyde)ethylenediimine cobalt(II) monochloroformate, 34B, 514
Bis-salicylaldiminatocopper(II), 31B, 389
Bis(salicylaldoximato)copper(II), 29, 596
Bis-salicylaldoximatonickel(II), 32B, 392
Bis(salicylaldoximato)palladium(II), 35B, 619
N,N'-Bis(salicylal)ethylenediiminenickel, 35B, 620
Δ-2,2'-Bis(salicylideaminato)-+D-(R)-6,6'-dimethylbiphenyl-cobalt (II), 34B, 516
2,2'-Bis-(salicylidenamino)-diphenyl-copper(II), 28, 679
Δ-2,2'-Bis(salicylideneaminato)-$(+)_D$-(*R*)-6,6'-dimethylbiphenyl-cobalt(II), 33B, 550
NN'-Bis(salicylideneiminato)-iron(III) chloride nitromethane solvate, 32B, 396
Bis-selenoureacadmium dichloride, 33B, 556
Bis-semicarbazidecopper(II) chloride, 30B, 360
Bis-semicarbazide-zinc chloride, 30B, 360
Bis(L-serinato)copper(II), 34B, 541
Bis(L-serinato) zinc, 32B, 496; 35B, 651
Bis(*meso*-stilbenediamine)nickel(II) dichloroacetate, 29, 571
Bis(succinodinitrile) copper(I) perchlorate, 34B, 556
trans-Bis(tetracarbonylcobaltate)-bispyridineplatinum(II), 33B, 551
Bis(tetracarbonylcobalt) mercury, 33A, 164
Bis(tetracarbonyliron-diethylgermanium), 33B, 551
Bis[tetracarbonyltriethylphosphine-manganese(O)], 33B, 509
Bis(3,3',5,5'-tetramethyldipyromethenato)nickel(II), 35B, 715
Bis-tetramethyldisilylene dioxide, 28, 583
Bis(2,2,6,6-tetramethylheptane-3,5-dionato)nickel(II), 31B, 386
Bis(tetramethylthiourea)bis(thiourea) tellurium dibromide, 26, 582
Bis(tetramethylthiourea)bis(thiourea) tellurium dichloride, 26, 581
Bis(tetramethylthiourea) tellurium dibromide, 26, 581
Bis(tetramethylthiourea) tellurium dichloride, 26, 581
Bis(tetramethylthiourea) tellurium diiodide, 26, 581
Bis(tetramethylurea)dichloro-oxovanadium(IV), 35B, 737
Bis(tetraphenylarsonium) tris-(1,2-dithiolato)molybdate, 35B, 914
Bis*peri*-tetraphenylnaphthalene, 35B, 922
Bis(thioacetamide)nickel(II) thiocyanate, 33B, 483
Bis(thioacethydrazidato)nickel(II), 35B, 788
cis-Bis(thioacetohydroxamato)nickel (II), 33B, 480
trans-Bis(thioacetohydroxamato) nickel(II), 34B, 574
Bis-(2-thioimidazolidine)-nickel(II) thiocyanate, 31B, 402
Bis (thiosemicarbazidato) nickel (II), 27, 837
Bisthiourea-cadmium formate, 27, 835; 30B, 344
Bis(thiourea)copper(I) chloride, 33B, 547; 35B, 784
Bis(thiourea)iodine(I) iodide, 35B, 912
Bis-thiourea mercury(II) iodide, 33B, 467
Bis thiourea nickel(II) thiocyanate, 31B, 402
Bis(thiourea)silver(I) chloride, 33B, 469
Bis-thiourea zinc acetate, 32B, 408

Bis(*p*-toluidinium) hexachloro-
rhenate(IV), 32B, 98; 33B, 552
Bis-(1-*m*-tolylazo-2-naphtolato)
nickel(II), 33B, 450
(+)5461-Bis(tribenzo[*b*,*f*,*j*][1,5,9]-
triazocycloduodecine)cobalt(III)
iodide, 35B, 716
Bis[tri-*n*-butylphosphine)tricarbonyl-
cobalt], 33B, 513
α-Bis-tricarbonylchromium diphenyl,
26, 650
β-Bis-tricarbonylchromium diphenyl,
26, 650
Bis(tricarbonyltributylphosphine-
cobalt), 33B, 550
Bis(tricarbonyltriethylphosphine-
cobalt)mercury, 33B, 550
Bis(tricobalt enneacarbonyl)acetone,
34B, 430
Bis(triethanolamine)lead(II)acetate,
28, 769
trans-Bis-triethylphosphinedihalo-
platinum, 31B, 457
4,4-Bis(trifluoromethyl)-2,2,2-
triphenyl-3-(triphenylphosphor-
anylidene)-1,2-oxaphosphetane,
33B, 284
Bis(trimethylaluminum)dioxanate, 32B,
328
Bis-trimethylamine alane, 28, 598
Bis-trimethylamine chromium trichloride,
32B, 496
Bis(trimethylamine)oxovanadium(IV)
dichloride, 33B, 437
Bis(trimethylbenzylammonium)tetra-
chlorocuprate(II), 32B, 35
Bis-(trimethylphosphineoxide)-cobalt(II)
dinitrate, 28, 647
Bis(trimethylphosphine)tetra-
chlorosilicon(IV), 35B, 919
1-Bis(trimethylsilyl)methylsilyl-
3α, 4β, 5α, 6α,-tetrakis
(trimethylsilyl)-1-cyclohexane,
32B, 491
Bistrimethylsulphonium mercuritetra-
iodide, 31B, 57
N,*N*'-Bis-(β,β,β-trinitroethyl)urea,
35B, 36
Bis-(triphenylarsine oxide) mercury(II)
dichloride, 28, 652
Bis(triphenylmethylarsonium)tetrachloro-
nickel(II), 31B, 314
Bis(triphenylphosphine)(carbon di-
sulphide) palladium, 33B, 500
Bis(triphenylphosphine) diphenylacety-
lene platinum, 32B, 493
Bis(triphenylphosphine) ethylene
nickel(O), 32B, 493; 33B, 343
trans-Bis(triphenylphosphine)-
methyldiiodorhodium(III), 33B, 547
Bis(triphenylphosphine)methyl tetra-
chloroferrate, 32B, 486
Bistriphenylphosphineplatinum(O)-
oxygen-acetone, 33B, 550
Bis(triphenylsilyl)chromate, 35B,
445
Bis-[μ-(tri-1,2,4-triazolo-*N*[1],*N*[2])
-triaquonickel]nickel, 33B, 546
NN'-Bisuccinimidyl, 31B, 10
3,3'-*commo*-Bis[undecahydro-1,2-
dicarba-3-nickela-*closo*-
dodecaborane], 35B, 429
Bis-(C-vinylbarenyl) mercury, 30B,
411
Bis-(*o*-xylene)silver perchlorate
dimer, 35B, 917
p,*p*' Bitolyl, 34B, 82
Biuret hydrate, 26, 570
9,9'-Bixanthenylidine, 28, 491
Blandfordite, 28, 334
Boehmite, 26, 349, 517; 29, 299
Bombyx mori, 26, 827
Bonattite, 27, 634
Boracite, 27, 551
Boracites, 35A, 288
Borate glasses, 27, 559
Borazine, 34A, 152
Borcarite, 30A, 456
Borine hydrazine, 32A, 150
Borkarite, 33A, 440
Bornite, 26, 137; 27, 182
Borohydridobis(triphenylphosphine)
copper(I), 32B, 478
Boron, 26, 539; 27, 394; 30A, 251,
449; 35A, 127
α-Boron, 26, 67
Boron, amorphous, 31A, 240
Boron-carbon, 26, 67
Boron-carbon-chromium, 31A, 26
Boron-carbon-hafnium, 26, 67
Boron-carbon-molybdenum, 28, 12;
34A, 36
Boron-carbon-scandium, 30A, 28
Boron-carbon-thorium, 26, 68
Boron-carbon-titanium, 26, 70
Boron-carbon-uranium, 26, 70
Boron-carbon-zirconium, 26, 67
Boron-chromium, 26, 71; 33A, 46;
35A, 34
Boron-chromium-iron, 28, 31
Boron-chromium-phosphorus, 27, 101
Boron-chromium-yttrium, 35A, 36
Boron-cobalt-gallium, 27, 92
Boron-cobalt-hafnium, 34A, 38
Boron-cobalt-molybdenum, 33A, 47
Boron-cobalt-phosphorus, 27, 102, 321
Boron-cobalt-tantalum, 26, 81
Boron-cobalt-tin, 27, 92, 95
Boron-cobalt-wolfram, 31A, 27; 34A,
39
Boron-cobalt-zirconium, 34A, 38
Boron-gallium, 26, 72
Boron-gallium-nickel, 27, 92
Boron-germanium-nickel, 27, 92

Boron-gold, 26, 73
Boron-hafnium-nitrogen, 26, 73
Boron-hafnium-silicon, 26, 64
Boron hydride derivative, 26, 619
Boron hydrides, 26, 290; 28, 60; 29, 230; 31A, 77
Boron-indium-nickel, 27, 92
Boron-iridium, 28, 13
Boron-iron-phosphorus, 27, 95
Boron-iron-tantalum, 26, 81
Boron-iron-zirconium, 27, 102
Boron-lanthanum, 26, 74
Boron-magnesium-nickel, 27, 92
Boron-manganese, 35A, 38
Boron-manganese-phosphorus, 27, 102
Boron-molybdenum, 26, 75
Boron-nickel, 32A, 32
Boron-nickel-phosphorus, 27, 102
Boron-nickel-silicon, 35A, 38
Boron-nickel-tantalum, 26, 81
Boron-nickel-tin, 27, 92, 103
Boron-nickel-zinc, 27, 92, 103
Boron nitride, 26, 291; 27, 408; 31A, 250
Boron-nitrogen-titanium, 26, 70
Boron-nitrogen-zirconium, 26, 73
Boron-osmium, 26, 76; 28, 13
Boron oxide, 27, 718; 33A, 258; 35A, 202
Boron oxide, high pressure form, 33A, 259
Boron-oxygen, 26, 67
Boron-palladium, 26, 76-79
Boron-phosphorus, 26, 67, 79
Boron-rhenium, 27, 103
Boron-ruthenium, 26, 76; 28, 13; 32A, 34; 34A, 41
Boron sesquioxide, 35A, 202
Boron-silicon, 26, 67
Boron-silicon-zirconium, 26, 64
Boron-silver, 26, 73
Boron-sodium, 35A, 39
Boron-sulphur, 26, 79
Boron-tantalum, 26, 80; 27, 25
Boron tetrathiophosphate, 28, 153
Boron-thorium-uranium, 26, 81
Boron-thulium, 26, 81
Boron-titanium, 27, 92
Boron trichloride, 31A, 239
Boron trifluoride, 31A, 239
Boron trifluroide dihydrate, 29, 252
Boron trifluoride-hydrazine, 26, 539
Boron trifluloride-sulphur nitride, 32A, 161
Boron triiodide, 27, 438
Boron-uranium-zirconium, 27, 105
Boron-wolfram, 26, 75; 31A, 29; 32A, 34; 34A, 41
Boron-yttrium, 30A, 29; 34A, 43
Boroxine, 34A, 153
Bostrycin *p*-bromobenzoate, 33B, 556
Botallackite, 29, 250
Botryogen, 32A, 338
Botryogens, 33A, 379
Boulangerite, 27, 405
Bournonite, 35A, 11
Bovine pancreatic ribonuclease, 27, 1059
Bovine serum albumen, 26, 819
Brandisite, 26, 521
Brannerite, 31A, 139
α-Brass, 27, 383
Brassylic acid, 33B, 7
Brassylique acid, 30B, 393
Bredigite, 31A, 243
Breunnerite, 29, 439
Brewsterite, 29, 399
Brilliant yellow, 26, 828
Britholite, 26, 536; 29, 428, 429
Brockite, 27, 598; 31A, 244
Bromanil, 26, 714
Brominated graphite, 28, 65
Bromine, 27, 720
Bromine calcium phosphate, 26, 468
Bromine-methanol (1:2), 29, 451
Bromine strontium phosphate, 26, 469
Bromine trifluoride-antimony pentafluoride, 32A, 508
p-Bromoacetanilide, 33B, 59
α-Bromoacetophenone, 35B, 924
17β-Bromoacetoxy-9β, 10α-androst-4-ene-3-one, 32B, 243
O-(Bromoacetyl)tetrahydrodouglanine, 35B, 380
3-Bromoambrosin, 31B, 213
1-Bromo-2-aminonaphthalene-picric acid, 33B, 245
2-Bromo-3-amino-1,4-naphthoquinone, 30B, 395; 31B, 98
3-Bromo-4-amino-1,2-naphthoquinone-methanol complex, 35B, 918
4α-Bromo-5α-androst-2-ene-1,17-dione, 35B, 333
3-Bromoanhydrodehydrodihydropulchellin, 35B, 372
Bromoanhydrotetrodoic lactone hydrobromide, 28, 673; 35B, 367
α-Bromoanthraquinone, 34B, 103
1-Bromoanthraquinone, 28, 675
2α-Bromoarborinone, 32B, 252
Bromoazido-1,1,7,7-tetraethyldiethylenetriaminecopper(II), 33B, 548
B-Bromobarene, 30B, 410
1-Bromo-2-barenyl-ethane, 30B, 411
p-Bromobenzeneazotribenzoylmethane, 31B, 50
2-*o*-(*p*-Bromobenzenesulphonyl)-1,4:3,6-dianhydro-D-glucitol-5-nitrate, 30B, 184
1-*p*-Bromobenzenesulphonyloxymethylbicyclo[2,2,2]octane, 33B, 546
1-*p*-Bromobenzenesulphonyloxymethyl-5-methylbicyclo[3,3,1]nonan-9-OL, 30B, 116

17aβ-*p*-Bromobenzensulphonyloxy-17α-methyl-19-nor-9β, 10α-D-homo-androst-4-ene-3-one, 32B, 245
1-*o*-(*p*-Bromobenzenesulphonyl)-4,5,7-tri-*o*-acetyl-2,6-anhydro-3-deoxy-D-glucoheptitol, 30B, 183
O-*p*-Bromobenzoate of batrachotoxinin A, 34B, 334
16β-O-*p*-Bromobenzoate of 6-keto-leucotylin, 33B, 555
m-Bromobenzoic acid, 32B, 76
o-Bromobenzoic acid, 27, 895
p-Bromobenzoic anhydride, 29, 655
α-2-Bromo-*p*-benzoquinone-4-oxime acetate, 26, 709
6-(*p*-Bromobenzoyl)-6-azabicyclo[3.1.0] hexane, 33B, 128
p-Bromobenzoyldimethylamine, 32B, 34; 33B, 61
p-Bromobenzoylethylenimine, 32B, 154
2-*p*-Bromobenzoyl-5-isopropyl-7-methyl-8H-azuleno[1,8-*bc*]thiophene, 33B, 552
p-Bromobenzoyl-laserol, 32B, 251
p-Bromobenzoyl-norbormide, 31B, 210
3β-*p*-Bromobenzoyloxy-13α-androst-5-en-17-one, 35B, 341
syn-3-*exo*-*p*-Bromobenzoyloxybicyclo-[3,2,2]non-6-ene-8,9-*endo*-*cis*-dicarboxylic acid anhydride, 35B, 147
α-3-*p*-Bromobenzoyloxy(cyclopentadiene dimer), 33B, 547
2-*p*-Bromobenzoyl-1,5,5-trimethyl-bicyclo-[2,2,2]-octa-6,8-dione, 34B, 120
p-Bromobenzylideneacetone, 35B, 84
3-(β-*p*-Bromobenzylthiocrotonoyl)-1-methyl-2-oxo-imidazoline, 33B, 548
2'-Bromobiphenyl-4-carboxylic acid, 30B, 427
Bromo-bis(diethyldithiocarbamato)-arsenic, 35B, 918
2α-Bromo-5β-bromomethyl-5α-methyl-2β-oxo-11,3,2-dioxaphosphorinane, 34B, 380
Bromobruceol, 28, 563
p-Bromocarbobenzoxy-glycyl-propyl-leucyl-glycine, 34B, 276
5-Bromo-6-chloroacenaphthene, 29, 679
Bromochloroacetylene, 35B, 903
10-Bromo-2-chloro-2-nitrosocamphane, 26, 649
α'-Bromo-*cis*-cinnamic acid, 29, 769
α-Bromo-*trans*-cinnamic acid, 29, 769
Bromocyanoacetylene, 33B, 29
5-Bromo-2'-deoxyuridine, 31B, 183
2-Bromo-(-)-β-desmotroposantonin, 32B, 249
Bromodiacetylene, 35B, 893
p-Bromodiazoaniline, 32B, 61
2-Bromodiazofluorene, 35B, 126
4-Bromo-2,6-di-t-butylphenol, 29, 769
Bromodihydroacronycine, 35B, 357
7-Bromo-2,3-dihydro-1,4-benzoxazepin-5(4*H*)-one, 33B, 544
2-Bromodihydroisophoto-α-santonic lactone acetate, 30B, 227
10-Bromo-5,10-dihydrophenarsazine, 31B, 312
(-)-Bromodihydroumbellulone, 34B, 305
3-Bromo-4,4-dihydroxypimelic acid dilactone, 35B, 5
2-Bromo-9-(ω-dimethylaminopropylidene)-thioxanthene, 29, 764
(-)-(*S*)-3-Bromo-1,4-dimethyl-1-isopropyl-3-cyclo-penten-2-one, 34B, 305
3-Bromo-1,8-dimethylnaphthalene, 30B, 97
N-1-Bromo-3,5-dimethylphenylbenzene-sulphonamide, 30B, 398; 34B, 624
Bromo(2,12-dimethyl-3,7,11,17-tetra-azabicyclo[11.3.1]heptadeca-1(17), 2,11,13,15-pentaene)nickel(II) bromide monohydrate, 33B, 428
3-Bromo-2,7-dinitrothionaphthen-5-diazo-4-oxide, 27, 982
Bromodiphenylarsine, 27, 867
2-Bromo-1,1-diphenyl-prop-1-ene, 31B, 34; 32B, 41
2-Bromo-1,1-di-*p*-tolylethylene, 32B, 42
*mono*Bromoduclauxin, 30B, 433; 33B, 235
4-Bromoduphaston, 31B, 202
Bromodurene, 29, 763
Bromoemimycin, 28, 529
1-Bromo-2,4-epidithio-1-nitro-4-phenylbutadiene, 35B, 913
6β-Bromo-7β,7αβ-epoxy-4α,6α-dihydroxy-7α-methoxy carbonyl-3β,4β-dimethyl-3αβ,7αβ-octahydrobenzo(*c*) furan-1-one, 34B, 171
Bromoepoxy*nor*cafestanone, 27, 1031
4-Bromoestradiol, 29, 750
Bromoestrone, 28, 536
5-(6'-Bromo-3'-ethyl-2'-methylbenzimidazolium) barbiturate monohydrate, 30B, 161
9β-Bromofraxinellone, 35B, 371
Bromogeigerin acetate, 27, 1026
3β-Bromogorgostene, 35B, 921
5-Bromogriseofulvin, 28, 533
8-Bromoguanosine dihydrate, 34B, 259
Bromohelenalin, 29, 771
17-Bromoheptadecanoic acid, 29, 762
3-Bromo-4-hydroxycoumarin monohydrate, 30B, 146
1-(5'-Bromo-2'-hydroxy-4'-methoxyphenyl)-3-oxoisoindolo-1-en, 26, 722
9α-Bromo-17β-hydroxy-17α-methylandrost-4-ene-3,11-dione, 33B, 192

6α-Bromo-17β-hydroxy-17α-methyl-4-oxa-5α-androstan-3-one, 35B, 334
12α-Bromo-11β-hydroxyprogesterone, 33B, 193
2-Bromo-3-hydroxypyrazine-N-oxide-1, 28, 529
7-Bromo-4-hydroxytetracycloxide, 30B, 399
Bromoindole derivative of 3β-methoxy-21-keto-Δ^{13}-serratene, 35B, 404
Bromo-isofenchone, 29, 771
6-Bromoisofenchone, 34B, 306
Bromoisotenulin, 28, 676
α-Bromoisotutin, 29, 745
α-Bromo-isotutinone, 28, 685; 30B, 245
2-Bromoketofluorene, 35B, 126
(-)-*trans*-3-Bromo-khellactone methyl ether, 34B, 338
Bromolactone, 26, 831
2-Bromolumisantonin, 33B, 209
Bromomalonic dialdehyde, 29, 454
5-Bromo-2-methoxy-*cis*-cinnamic acid, 29, 769
α-Bromo-*p*-methoxy *cis*-cinnamic acid, 29, 769
trans-Bromo-(5-methoxynortricycl-3-enyl)dipyridinepalladium, 35B, 916
3-Bromo-2-methoxytropone, 29, 675
7-Bromo-2-methoxytropone, 29, 675
9-Bromo-10-methylanthracene, 26, 749
erythro-α-Bromo-α-methylbenzyl-α-methylbenzyl sulphone, 35B, 920
2-(*cis*-4-Bromo-1-methylcyclohexyl)-3-(4-hydroxyimino-1-methylcyclohexyl)-*trans*-2-butene, 34B, 93
2-Bromomethyl-3,4-diacetyl-5-methylfuran, 33B, 133
2-Bromoethyl-2,3-dihydrofuro[2,3,b] tropone, 32B, 176
5-(Bromomethylene)-10,11-dihydro-5H-dibenzo[a,d]-cycloheptene, 35B, 127
8-Bromo-2,2-methylenediquinoline, 28, 476
Bromo-2-methyl-3-naphthoquinone-1,4, 30B, 393
N-1-Bromo-3-methylphenyl-benzenesulphonamide, 34B, 623
Bromomexicanin-E, 32B, 248
Bromomiroestrol, 29, 751
16-Bromomyricanol, 35B, 916
8-Bromo-2-naphthol, 33B, 88
2-Bromo-1,4-naphthoquinone, 26, 742
3-Bromo-1,2-naphthoquinone, 35B, 918
5-Bromo-2-nitrobenzaldehyde, 29, 652
m-Bromonitrobenzene, 28, 418
(--)-2-Bromo-2-nitrocamphane, 26, 830; 27, 1000
3-Bromo-2'-nitrochalcone, 35B, 54
4-Bromo-2'-nitrochalcone (Forms A and B), 35B, 54
2-Bromo-2-nitrofenchane, 33B, 551
p-Bromo-*m*-nitro-*n*-methylaniline, 32B, 486
2-Bromo-2-nitro-1,3,3-trimethyl(2,2,1-bicycloheptane), 32B, 487
3-Bromonopinone, 29, 771
Bromonoranisatinone, 30B, 398
D-L-Bromo-octadecanoic acid, 31B, 22
5α-Bromo-6β,19-oxido-pregnan-3β-ol-20-one, 34B, 294
(+)-*N*-(3-*endo*-Bromo-2-oxo-9-bornanesulphonyl)methyl-*p*-tolylysulphoximide, 35B, 921
9-Bromo-1,3,7,8,10-pentamethyl-1,5-dihydro-alloxazine, 34B, 154
p-Bromophenacyl dichloroacetate, 35B, 17
p-Bromophenacyl difluoroacetate, 35B, 17
p-Bromophenacyl ester of thujic acid, 27, 1025
p-Bromophenacyl labdanolate, 33B, 216
p-Bromophenol, 33B, 543
(-)-*m*-Bromophenoxypropionic acid, 31B, 16
p-Bromophenylacyl isamate, 35B, 913
C-*p*-Bromophenylbarene, 30B, 411
N-*p*-Bromophenylbenzene sulphonamide, 32B, 69; 34B, 622
p-Bromophenyl *N*-carbamoylthiamine anhydride, 35B, 910
N-*p*-Bromophenyl-*p*-chlorobenzenesulphonamide, 29, 770; 34B, 620
N-*p*-Bromophenyl-2,2-diphenylvinylideneamine, 35B, 64
α-Bromo-γ-phenyl-γ-ethylglutaconimide, 27, 967
p-Bromophenylhydrazones, 28, 684
3-(4-Bromophenyl)-2-hydroxyimino-6,7-dihydro-4(5*h*)-benzofuranone, 35B, 917
p-Bromophenyl*isothiocyanate*, 27, 893
2-(α-*p*-Bromophenyl-β-nitroethyl)-cyclohexanone, 31B, 89
p-Bromophenylosazones, 28, 684
3-*p*-Bromophenyl-10-phenyl-3,10-diazatricyclo[4,2,1,$1^{2,5}$[decan-4-one], 32B, 170
α-1-(*p*-Bromophenyl)phenylmethylene-3-oxo-1,2-diazetidinium inner salt, 33B, 548
3-(*p*-Bromophenyl)phthalide, 34B, 173
N-(*p*-Bromophenyl)-α-D-ribopyranosylamine 2,4-diphenyl-boronate, 34B, 253
p-Bromophenyl-N-succimide, 26, 720
N-*p*-Bromophenyl succinimide, 28, 366
N-*p*-Bromophenylsydnone, 28, 494
3-(*p*-Bromophenyl)thiazolo(2,3,a) isoquinolinium-2-thione betaine, 32B, 196

2-*o*-Bromophenyl-6α,7,9-trimethoxy-trans-*m*-dioxano[5,4-e][1,4]-dioxepan, 26, 627
N-*p*-Bromophenyltriphenylphosphine imide, dimethylacetylenedicarboxylate adduct, 30B, 278
4-Bromophenylurethane of a proposed hinesol intermediate, 35B, 348
5-Bromo-2-phthalimidobenzoic acid monohydrate, 28, 474
α_1-Bromopicrotoxinin, 27, 1017
β-Bromopicrotoxinin, 33B, 233
4-Bromo-9β-10α-pregn-4,6-dione-3,20-dione, 30B, 398
4-Bromo-9β,10α-pregna-4,6-diene-3,20-dione, 31B, 202
6β-Bromo-Δ(4)-pregnene-3,20-dione, 34B, 293
6β-Bromo-progesterone, 34B, 293
cis-β-Bromo-β-(2-pyridyl)styrene, 35B, 171
2-Bromo-N-salicylidene aniline, 34B, 73
2-Bromo-α-santonin, 30B, 226
2-Bromo-β-santonin, 34B, 309
Bromo-*meso*-sarcosinuric acid, 35B, 252
N-Bromo-succinimide, 26, 600
6-Bromo-8,8,9,9-tetracyano-2-methoxycarbonyl-2-azabicyclo[3,2,2]nona-3,6-diene toluene solvate, 32B, 167
1-Bromo-2,3,5,6-tetramethylbenzene, 30B, 79
2-Bromo-3,3,5,5-tetramethylcyclohexanone, 29, 493
9-Bromo-3,7,8,10-tetramethylisoalloxazine monohydrate, 35B, 910
(4-Bromo-1,2,3,4-tetraphenyl-*cis*,*cis*-1,3-butadienyl)dimethyltin bromide, 35B, 491
1-Bromothiophanium bromide, 35B, 217
2-Bromothiophene, 35B, 893
2-Bromo-2-*p*-tolylethyldimethylammonium bromide, 33B, 545
Bromo(tribromogermyl)-2,2'-bipyridinetricarbonylwolfram, 35B, 870
3-Bromo-2,4,6-triphenylphenoxyl dimer, 33B, 66
1-Bromotriptycene, 33B, 101
α-Bromotutin, 28, 675
α_2-Bromotutin, 28, 675
α-Bromotutinone, 28, 675
Bromouliginosin B, 33B, 234
11-Bromoundecanoic acid, 27, 740; 28, 673
11-Bromoundecanoic acid(D-form), 28, 367
11-Bromoundecanoic acid(E-form), 28, 369
2-*mono*-11-Bromoundecanoin, 30B, 429
2-11-Bromoundecanoyl-1,1'-dicaprin, 33B, 236
5-Bromouridine, 31B, 183
5-Bromouridine-dimethylsulphone 1:1 complex, 32B, 272
5-Bromoveratric acid, 33B, 552
Brookite, 26, 351
6-*O*-Brosylgynocardin, 35B, 913
N-Brosylmitomycin-A benzene solvate, 32B, 240
N-Brosylpolyoxin C monohydrate, 33B, 545
Brownmillerite, 29, 418
Brucite, 26, 348; 30A, 445; 32A, 242
Brushite, 27, 574
Buddingtonite, 29, 429
Buergerite, 31A, 242; 34A, 391
Bulgarene dihydrobromide, 33B, 555
Bullvalene, 32B, 487; 33B, 96
Bullvalene-silver(I) fluoroborate complex, 32B, 494
Bullvalene-silver tetrafluoroborate, 33B, 368
Bullvalene-silver tetrafluoroborate monohydrate, 33B, 546
Bultfonteinite, 28, 256
Bundlin B *p*-bromophenylhydrazone, 34B, 287
Burbankite, 32A, 510; 33A, 437
Bustamite, 27, 701; 28, 246
Butadiene, 33B, 530
μ-Butadiene bis(cyclopentadienylmanganesedicarbonyl), 32B, 347
Butadiene cobalt dicarbonyl, 31B, 346
π-Butadiene-π-cyclopentadienyl dichloromethylgermyliron, 35B, 916
Butadieneiron tricarbonyl, 28, 605; 35B, 895
Butadieneiron tricarbonyl-sulphur dioxide-boron trifluoride adduct, 35B, 516
Butane-2,3-dione-bis(2'-pyridylhydrazonato) nickel, 34B, 558
Butatriene, 26, 553
cis-2-Butene, 35B, 894
trans-2-Butene, 35B, 894
trans-2-Butene-1,4-dicarboxylic acid 31B, 8
cis-2-Butene episulphone, 32B, 66
tert-Butyl acetylene, 27, 735
sec-Butyl alcohol, 26, 554
t-Butylamine hydrate, 32B, 289
n-Butyl bromide, 28, 666; 29, 765
n-Butyl chloride, 27, 733
sec-Butyl chloride, 26, 554; 27, 733
t-Butyl chloride, 28, 666; 29, 765; 33B, 28
t-Butyl cyanide, 27, 735
sec-Butyl derivatives, 26, 554
t-Butyl-N,N-dimethyltrithiopercarbamate, 34B, 45
n-Butyllithium, tertlithium butoxide (addition compound), 30B, 415

2-*t*-Butyl-4-(4-methoxyphenyl)-5,6-dihydro-naphtho[1,2-*b*]-phosphorin, 33B, 556
5-*t*-Butyl-2-methyl-2-oxo-1,3,2-dioxaphosphorinan, 35B, 449
bis-N-Butylsalicylaldimine-cobalt, 26, 679
bis-N-Butylsalicylaldimine-zinc, 26, 679
(2-Butyne)dicyclopentadienylnickel, 30B, 428
But-2-yne with iron hydrocarbonyl complex, 26, 682
Butyric acid, 27, 737
Buxus alkaloid A,N,N'-dimethyl derivative, 28, 680
Buxus alkaloid A monomethobromide of N,N'-dimethyl derivative, 28, 680
Byssochlamic acid bis-*p*-bromo-phenyl hydrazide, 28, 564
Bytownite, 31A, 222

Cacodyl disulphide, 29, 539
Cacodylic acid, 30B, 282
Cacoxenite, 26, 462
(±)-Cadinene dihydrochloride, 28, 539
Cadmium, 26, 279
Cadmium acetate trihydrate mixed crystals with various dyes, 26, 828
Cadmium ammonium sulphate hexahydrate, 31A, 203
Cadmium borate glass, 28, 292
Cadmium bromide, 27, 431
Cadmium *n*-butyl xanthate, 30B, 371
Cadmium caesium chlorides, 29, 280
Cadmium calcium acetate hexahydrate, 32B, 496
Cadmium carbonate, 26, 481
Cadmium chloride bis(pyridine complex), 34B, 553
Cadmium chloride, 2,6-dimethyl-γ-pyrone (3:2 complex)dihydrate, 30B, 429
Cadmium-copper, 27, 106; 32A, 36; 33A, 72
Cadmium-copper-gallium, 27, 261
Cadmium-copper-germanium, 27, 261
Cadmium-copper-germanium-sulphur, 34A, 47
Cadmium-copper-indium, 27, 261
Cadmium diaquo sulphate monothiourea complex, 32B, 407
Cadmium diborate, 27, 549; 31A, 166
Cadmium(II) *NN*-diethyldithiocarbamate, 33B, 486
Cadmium dimetaphosphate, 33A, 417
Cadmium fluoroborate hexahydrate, 26, 304
Cadmium formate, 28, 669
Cadmium formate dihydrate, 28, 669
Cadmium-gallium-oxygen, 26, 413
Cadmium-gallium-sulphur, 27, 202
Cadmium germanate, 34A, 260
Cadmium-germanium-phosphorus, 33A, 49
Cadmium-gold, 26, 82; 27, 107; 30A, 32
Cadmium-gold-silver, 27, 107
Cadmium-gold-tin, 27, 108
Cadmium hexacyanochromate, 26, 343
Cadmium hexacyanochromate hexahydrate, 27, 653
Cadmium hydraxinecarboxylate, 30B, 409
Cadmium hydrazine chloride, 27, 471
Cadmium hydroxide, 31A, 130, 256
Cadmium hydroxychloride, 26, 327
Cadmium-indium, 27, 108
Cadmium-indium-selenium, 32A, 41
Cadmium iodide, 27, 431; 29, 287; 30A, 288; 33A, 517
Cadmium-iron-manganese-sulphur-zinc, 26, 252
Cadmium-iron-sulphur-zinc, 26, 252
Cadmium-magnesium, 27, 109
Cadmium-magnesium carbonate, 26, 481
Cadmium manganese oxide, 32A, 285
β-Cadmium metavanadate, 34A, 265
Cadmium molybdate, 31A, 156
Cadmium-nickel, 30A, 33; 35A, 40
Cadmium nitrate-tetradeuterate, 34A, 342
Cadmium nitrate tetrahydrate, 28, 179; 31A, 178
Cadmium orthoarsenate, 35A, 352
Cadmium oxalate trihydrate, 30B, 407
Cadmium palladium hexafluoride, 33A, 188
Cadmium phosphide, 35A, 41
Cadmium-phosphorus, 34A, 49, 50; 35A, 41
Cadmium-platinum-zinc, 35A, 41
Cadmium pyroborate, 27, 549
Cadmium pyrochlore fluoantimonate, 28, 200
Cadmium pyrophosphate, 34A, 327
Cadmium pyrovanadate, 32A, 274
Cadmium rhenium oxide, 30A, 446
Cadmium rhodite, 26, 399
Cadmium ruthenite, 26, 399
Cadmium salicylalethylenediiminate, 28, 676
Cadmium-scandium, 27, 109; 29, 32
Cadmium-selenium, 27, 109, 382
Cadmium-selenium-sulphur-zinc, 27, 110
Cadmium silicates, 29, 397
Cadmium-silver, 26, 248; 27, 111, 190
Cadmium-silver-zinc, 26, 83
Cadmium-sodium, 27, 113; 30A, 33
Cadmium sodium chloride, 26, 316

Cadmium sodium silicate, 32A, 454; 33A, 469
Cadmium sulphate, 26, 443; 28, 210
Cadmium sulphate monohydrate, 35A, 372
Cadmium-sulphur, 27, 109, 115, 382
Cadmium-sulphur-tellurium-zinc, 27, 111
Cadmium-sulphur-thallium, 34A, 52
Cadmium-sulphur-yttrium, 26, 277
Cadmium-sulphur-zinc, 26, 251
Cadmium-tellurium, 27, 382
Cadmium tetraisoselenocyanato-mercurate, 31A, 255
Cadmium tetrakisthiourea tetraisothiocyanatozincate(II), 32A, 240
Cadmium tetrathiourea:zinc tetrathiocyanate, 32B, 409
Cadmium-thorium, 26, 256
Cadmium tin zinc spinel, 33A, 286
Cadmium-titanium, 27, 116
Cadmium trihydrogenhexaoxoiodate, 35A, 397
Cadmium uranate, 27, 543
β-Cadmium uranate, 26, 405
Cadmium wolframate, 26, 393; 31A, 158
Cadmium-yttrium, 33A, 50
Cadmium zinc phosphates, 33A, 399
Cadrela odorata substance B, 30B, 392
Caesium, 26, 3
Caesium acetylenediolate, 29, 467
Caesium acid chloromaleate, 26, 555
Caesium alum, 31A, 202
Caesium aluminate, 29, 318
Caesium aluminum germanate, 31A, 249
Caesium acquopentachlororuthenate(III), 31A, 101
Caesium azide, 27, 414; 30A, 256
Caesium beryllium fluorides, 33A, 176
Caesium bis(1,2-dicarbollyl)cobaltate, 32B, 294
Caesium bismuth iodide, 33A, 241
Caesium bis(oxo-μ-trioxo-chromium(VI) chromium(III), 32A, 281
Caesium bromide, 26, 297
Caesium bromoiodoiodate(III), 31A, 112
Caesium carbonyl cobaltate, 33A, 165
Caesium cerium fluoride, 26, 312
Caesium chloride, 26, 297
Caesium chloride. 1/3 hydronium bichloride, 31A, 259
Caesium chloride-thiourea hydrate, 33B, 239
Caesium chromium chlorides, 27, 457
Caesium chromium iodate, 26, 500
Caesium chromium oxychloride, 29, 281
Caesium copper(II) chloride hydrate, 31A, 244
Caesium(I) di-*n*-butyldithiocarbamate, 34B, 401
Caesium dicarbollylrhenium tricarbonyl, 31B, 294
Caesium difluorophosphate, 32A, 369
Caesium dimedone-sulphonate, 30B, 421
Caesium dioxopentathiocyanato-uranium(VI), 28, 114
Caesium dodecachlorotrirhenate(III), 28, 87
Caesium enneaborate, 32A, 499
Caesium erbium sodium hexafluoride, 33A, 186
Caesium ferric oxide, 26, 395
Caesium fluoride-thiourea hydrate, 33B, 239
Caesium fluoroberyllate. 27, 552
Caesium fluoromanganate, 26, 309
Caesium-germanium, 26, 84, 157
Caesium germanium(II) chloride, 30A, 278
Caesium-gold, 26, 86
Caesium halide-hydronium bihalides, 33A, 201
Caesium heptafluoroterbate, 26, 313
Caesium hexabromomolybdate, 27, 464
Caesium hexabromotellurate(IV), 31A, 109
Caesium hexachlorocerate(IV), 31A, 104
Caesium hexachloromolybdate, 27, 464
Caesium hexaiodotellurate, 27, 463
Caesium hydrogen di-trifluoroacetate, 30B, 13
Caesium iodide, 26, 297
Caesium iron sulphide, 33A, 3
Caesium lithium fluoride, 27, 451; 30A, 269
Caesium magnesium chloride, 35A, 160
Caesium magnesium fluoride, 34A, 188
Caesium manganese chloride dihydrate, 27, 456
Caesium manganese fluoride, 27, 455
Caesium methylate, 35B, 472
Caesium methylsulphonate, 32B, 64
Caesium molybdenum bronze, 35A, 277
Caesium molybdenum oxychloride, 29, 281
Caesium niobium oxychloride, 29, 281
Caesium octabromodirhenate(III), 35A, 171
Caesium oxalate monohydrate, 30B, 407
Caesium oxomercurate, 29, 338
Caesium ozonide, 31A, 257
Caesium pentachloro titanium(III). acetic acid, 27, 816
Caesium perchlorate, 27, 725
Caesium permanganate, 31A, 259
Caesium pertechnetate, 27, 671
Caesium praseodymium fluoride, 26, 312
Caesium rhenium bromide, 34A, 213
Caesium rhenium chlorobromide hydrate, 34A, 213
Caesium selenate, 27, 605

Caesium-silicon, 26, 84, 156
Caesium sulphate, 27, 605
Caesium-sulphur, 33A, 60
Caesium *trans*-tetrachloroaquocarbonyl-ruthenate(II), 35A, 429
Caesium tetrachloromanganate(II) dihydrate, 29, 275
Caesium tetracyanoquinodimethanide, 29, 650; 31B, 43
Caesium tetrafluoroborate, 34A, 189
Caesium tetrakis (hexafluoro-acetyl-acetonato) americate, 33B, 554; 34B, 500
Caesium tetrakis (hexafluoro-acetyl-acetonato) europate, 33B, 554; 34B, 500
Caesium tetrakis(hexafluoroacetyl-acetonato) yttrate, 31B, 383; 33B, 391
Caesium tetramethylammonium tri-decahydroundecaborate, 32B, 296
Caesium titanate, 29, 218
Caesium tribromide, 34A, 180
Caesium tribromoplumbate(II), 31A, 110
Caesium trichlorocobaltate, 26, 326; 33A, 214
Caesium trichlorocuprate(II), 31A, 103
Caesium trichlorostannate(II), 35A, 159
Caesium trichromate(III, VI), 31A, 150
Caesium trifluoroferrate(II), 31A, 249
Caesium trivanadate, 31A, 141
Caesium uranate, 26, 404
Caesium uranium(VI) hexafluoride, 32A, 178
Caesium uranyl nitrate, 26, 477
Caesium uranyl oxychloride, 29, 282
Caesium vanadium sulphate hydrate, 27, 723
Caesium wolframaluminate, 27, 672
Caesium wolfram bronze, 26, 356
Caesium wolfram ferrate, 27, 672
Caesium wolfram manganate, 27, 672
Caesium wolfram oxychloride, 29, 281
Cafarsite, 31A, 254
Cafestol, 29, 772
Caffeine-5-chlorosalicylic acid, 33B, 245
Caffeine-pyrogallol, 33B, 244
Cahnite, 26, 471
Calciferyl 4-iodo-3-nitrobenzoate, 28, 537
Calciostrontianite, 26, 443
Calcirtite 26, 379
Calcite, 27, 561; 30A, 408
Calcium alkyl phosphoric acid esters, 26, 552
Calcium(+)-allo-hydroxycitrate lactone tetrahydrate, 34B, 18
Calcium aluminate hydrates, 34A, 251
Calcium D-arabonate, 27, 788
Calcium D-arabonate pentahydrate, 27, 788
Calcium beryllate, 31A, 130
Calcium borate, 26, 491; 29, 391; 34A, 356
Calcium *mono*borate, 29, 387
Calcium borate hydrate, 27, 554
Calcium borite, 35A, 283
Calcium bromide, 28, 97
Calcium carbide III, 26, 87
Calcium-carbon, 26, 87, 91; 27, 117
Calcium carbonate, 26, 481
Calcium carbonate hexahydrate, 35A, 296
Calcium-cerium-sulphur, 27, 327
Calcium-chabazite, 28, 250
Calcium-chabazite-chlorine, 27, 692
Calcium chloride glycylglycylglycine trihydrate, 32B, 489
Calcium chlorine phosphate, 26, 468
Calcium chromate(V) chloride, 32A, 202
Calcium copper acetate hexahydrate, 32B, 496
Calcium copper(II) oxide, 34A, 248
Calcium cyanamide, 27, 419
Calcium dialuminate, 35A, 290
Calcium *trans*-1,2-diaminocyclohexane-*N,N'*-tetraacetatoaquoferrate(III) octahydrate, 31B, 429
Calcium diferrite, 27, 548
Calcium dioxalatoplatinate(II) hydrate, 33B, 556
Calcium 1,3-diphosphorylimidazole hexahydrate, 33B, 107
Calcium dipicolinate, 29, 771
Calcium dipicolinate trihydrate, 33B, 113
Calcium dithionate monohydrate, 27, 643
Calcium-dysprosium-sulphur, 27, 327
Calcium-erbium-sulphur, 27, 327
Calcium ethylenediaminetetracetate, 33B, 554
Calcium *mono*ethyl phosphate, 27, 751
Calcium ferrate(IV), 35A, 243
Calcium ferrite, 31A, 241, 246, 249
Calcium fluoride aluminate, 27, 519
Calcium fluorine phosphate, 26, 468
Calcium fluoroborates, 27, 459
Calcium fluorosilicate, 29, 439
Calcium formate, 28, 669
Calcium-gadolinium-sulphur, 27, 327
Calcium gallates, 28, 144
Calcium germanate, 27, 501, 541
Calcium germanate, high pressure, 35A, 313

Calcium-germanium, 28, 14
Calcium-germanium-lithium, 35A, 43
Calcium germanium-magnesium, 34A, 53
Calcium α-D-glucoisosaccharate, 33B, 168
Calcium hexa-antipyrine perchlorate, 33B, 465
Calcium hexaboride, 26, 69
Calcium hexacyanoisobutene hexahydrate, 32B, 323
Calcium hexa-hydroxystannate, 32A, 247
Calcium hydrazinecarboxylates, 35B, 913
Calcium hydride, 27, 118
Calcium hydrogen potassium phosphate, 27, 577
Calcium hydroxide, 27, 422
α'-Calcium-hydroxyborate, 32A, 502
Calcium hydroxychloride, 26, 327
Calcium(-)-hydroxycitrate lactone tetrahydrate, 34B, 18
Calcium hypochlorite, dibasic, 33A, 217
Calcium indate, 32A, 509
Calcium-indium, 29, 3
Calcium iridium(IV) oxide, 30A, 333
Calcium iron vanadium garnet, 35A, 259
Calcium 5-keto-D-gluconate dihydrate, 30B, 182
Calcium 2-keto-D-gluconate trihydrate, 30B, 427
Calcium-lanthanum-sulphur, 27, 327
Calcium-lead, 26, 87; 28, 14
Calcium luteciate, 35A, 231
Calcium magnesium carbonate, 26, 481
Calcium-magnesium-silicon, 34A, 53
Calcium-magnesium-tin, 34A, 53
Calcium-magnesium-zinc, 26, 90
Calcium malate dihydrate, 31B, 26
Calcium manganate, 31A, 136
Calcium-manganese carbonate, 26, 481
Calcium manganese(III) silicate, 31A, 217
Calcium manganite, 30A, 322
Calcium metaborate, 28, 162; 34A, 354
Calcium *mono*methyl phosphate, 27, 751
Calcium 1-naphthyl phosphate trihydrate, 29, 770; 30B, 188
Calcium-neodymium-sulphur, 27, 327
Calcium nickel tetracyanide pentacyanide, 33A, 517
Calcium niobate, 27, 537; 35A, 240
Calcium niobium oxides, 26, 382
Calcium nitride, 33A, 163
Calcium-nitrogen, 31A, 29; 33A, 50, 163; 34A, 54
Calcium orthophosphate, 26, 458
Calcium orthophosphate monohydrate, 26, 458
Calcium oxalate hydrate, 27, 819
Calcium oxalate *mono*hydrate, 30B, 408
Calcium oxalate polyhydrate, 30B, 22
Calcium palladium hexafluoride, 33A, 188
Calcium palladium oxide, 29, 334
Calcium palladium tetracyanide pentahydrate, 33A, 517
Calcium phosphate chloride, 32A, 202
Calcium-platinum, 27, 322
Calcium polygermanosilicate, 32A, 463
Calcium potassium arsenate, 26, 470
Calcium potassium phosphate, 26, 469
Calcium potassium vanadate, 26, 470
Calcium-praseodymium-sulphur, 27, 327
Calcium *mono-n*-propylphosphoric acid ester, 26, 553
Calcium pyrochlore fluoantimonate, 28, 200
Calcium pyrophosphate, 31A, 191; 33A, 407
β-Calcium pyrophosphate, 27, 591
Calcium-samarium-sulphur, 27, 327
Calcium scandate, 30A, 316; 31A, 136
Calcium silicate, 28, 394; 31A, 252
Calcium-silicon, 33A, 51
Calcium-silicon-strontium, 32A, 43
Calcium-silver, 27, 119; 29, 33
Calcium sodium arsenate, 26, 469
Calcium sodium hypophosphite, 34A, 334
Calcium-sodium metahewettite, 27, 589
Calcium sodium phosphate, 26, 469
Calcium sodium vanadate, 26, 381, 470; 33A, 310
Calcium stannate, 32A, 305
Calcium, strontium, and barium fluoride solid solutions, 31A, 248
Calcium strontium carbide, 35A, 135
Calcium sulphate, 27, 604, 607
Calcium-sulphur-ytterbium, 27, 327
Calcium-sulphur-yttrium, 26, 88; 27, 327
Calcium tantalate, 28, 201; 35A, 258
Calcium tartrate tetrahydrate, 33B, 16
Calcium tetrahydroxoborate dihydrate, 28, 166
Calcium thymidylate, 26, 803
Calcium-tin, 26, 87
Calcium tripotassium hydrogen phosphate, 34A, 329
Calcium uranate, 31A, 160
Calcium uranium hexafluoride, 33A, 193
Calcium wolframate, 27, 542; 29, 331
Calcium ytterbate, 32A, 310; 35A, 231

Calcium-zinc, 26, 90
Calcium zirconium sulphide, 28, 151
Calcium zirconium titanate, 29, 319
Caldariomycin, 33B, 190
C-Calebassin alkaloid, 30B, 394
Caledonite, 35A, 376
Californium oxyfluoride, 33A, 175
Californium sesquioxide, 31A, 247
Californium trichloride, 31A, 247
Calycanthine hydrobromide, 27, 1013
Camptothecin iodoacetate, 33B, 226
Cancentrine, 35B, 921
Cancrinite, 30A, 428; 35A, 458
Cancrinite, basic, 35A, 458
Cancrinite, chromate, 35A, 460
Cancrinite, molybdate, 35A, 460
Cancrinite, nitrate, 35A, 459
Canthaxanthin, 33B, 84
Capaurimine *p*-bromobenzoate, 35B, 916
Capaurine hydrobromide, 33B, 555; 34B, 332
ε-Caprolactam, 28, 765
Caracolite, 32A, 333; 34A, 311
Caracurine II, 26, 830
Caracurine-II dimethiodide, 30B, 230
Carbamide phosphate, 32B, 485
Carbamolycholine bromide, 35B, 918
O-Carbamoylhydroxylamine, 33B, 35
Carbazole, 33B, 556
N-Carbazole, 34B, 157
χ-Carbide, 31A, 30
Carbobenzoxy-glycyl-L-alanine ethyl ester, 27, 1050
Carbobenzoxy-glycyl-L-prolyl-L-leucyl-glycine, 27, 1051
Carbobenzoxy-glycyl-L-prolyl-leucyl-glycyl-L-proline, 26, 723; 27, 1051
Carbobenzoxy-L-leucyl-L-tyrosyl-L-leucine-methyl ester, 27, 1050
Carboborite, 29, 427
Carbocernite, 26, 483; 31A, 261; 32A, 417
3-Carbomethoxy-5-anisyl-1,2,4-trioxacyclopentane, 35B, 212
Carbomethoxybenzocyclobutadiene-iron tricarbonyl, 35B, 920
Carbomethoxymethyl-bis(dimethylglyoximato) pyridinato-cobalt, 32B, 497
Carbon, 27, 395; 28, 289; 32A, 147; 34A, 149
1-Carbonamido-2-methyl-4-(1-hydroxy-ethyl-idene-5-iminoyl-cyclopent-1-ene bromide, 27, 945
Carbonatodiaquobisimidazolecobalt(II) 35B, 759
Carbonatopentaamminecobalt(III) bromine hydrate, 30A, 286
Carbonatopentamminecobalt(III) iodide monohydrate, 32A, 218
Carbonatotetra-amine cobalt(III) bromide, 27, 654
Carbon blacks, 27, 396
Carbon-cerium, 26, 91, 92
Carbon-chromium, 26, 96; 27, 122; 34A, 56
Carbon-chromium-iron-niobium-vanadium, 26, 95
Carbon-chromium-nitrogen, 31A, 26
Carbon-chromium-vanadium, 31A, 26
Carbon-cobalt, 26, 95
Carbon-cobalt-titanium-wolfram, 26, 97
Carbon-cobalt-wolfram, 26, 97
Carbon-cobalt-zinc, 26, 97
Carbon-deuterium-zirconium, 32A, 44
Carbon disulphide, 27, 719; 33A, 161
(Carbon disulphide)bis(triphenylphosphine)platinum(O), 35B, 825
Carbon disulphide hydride dicarborane cobaltate, 34B, 360
Carbon-dysprosium, 33A, 52
Carbon-gallium-platinum, 26, 225
Carbon-hafnium, 27, 121
Carbon-hafnium-iron-sulphur, 30A, 35
Carbon-hafnium-uranium, 26, 93
Carbon-hydrogen-zirconium, 30A, 35; 32A, 44
Carbon-indium-palladium, 26, 225
Carbon-iron, 26, 99; 27, 122; 28, 15, 29, 34; 30A, 36; 31A, 30
Carbon-iron-manganese, 26, 99; 30A, 37; 31A, 31
Carbon-iron-silicon, 26, 100
Carbon-iron-tin, 26, 100
Carbon-iron-uranium, 27, 125
Carbon-iron-wolfram, 29, 35; 30A, 38; 32A, 45
Carbon-lanthanum, 26, 91, 92; 33A, 53
Carbon-lithium, 27, 126; 30A, 39
Carbon-lutecium, 26, 91, 92
Carbon-manganese, 30A, 36
Carbon-manganese-silicon, 32A, 102
Carbon-molybdenum, 26, 101; 27, 126; 28, 15
Carbon-molybdenum-niobium, 26, 101
Carbon-molybdenum-silicon, 26, 103; 30A, 39
Carbon-molybdenum-uranium, 29, 36; 31A, 32
Carbon-molybdenum-vanadium, 27, 126
Carbon-molybdenum-zirconium, 27, 129
Carbon*monoxy*-erythrocruorin, 35B, 888
Carbon-nickel-zinc, 26, 101
Carbon-niobium, 26, 96, 102; 27, 121; 34A, 58
Carbon-niobium-uranium, 26, 93
Carbon-palladium-tin, 26, 225
Carbon-phosphorus-vanadium, 33A, 130
Carbon-plutonium, 26, 233; 27, 132
Carbon-plutonium-uranium, 27, 132
Carbon-praseodymium, 26, 92
Carbon rare earths, 31A, 32
Carbon-rhenium-wolfram, 26, 102
Carbon-scandium, 26, 103; 34A, 59
Carbon-scandium-titanium, 26, 103

Carbon-silicon-titanium, 32A, 46
Carbon-silicon-uranium, 30A, 40; 31A, 33; 33A, 54
Carbon-silicon-wolfram, 26, 103
Carbon suboxide, 29, 296
Carbon-tantalum, 26, 104, 253; 27, 121; 30A, 40, 41; 34A, 58
Carbon-tantalum-uranium, 26, 93
Carbon-tantalum-vanadium, 35A, 43
Carbon-tantalum-wolfram, 27, 133
Carbon-technetium, 26, 104
Carbon-terbium, 26, 91, 92; 34A, 60
Carbon tetrabromide-*p*-xylene complex, 27, 891
Carbon-thorium, 27, 134; 33A, 56
Carbon-thorium-uranium, 26, 104
Carbon-thorium-wolfram, 27, 135
Carbon-thorium-zirconium, 27, 136
Carbon-titanium, 26, 264; 27, 121, 136
Carbon-uranium, 26, 91, 92; 27, 137; 31A, 34
Carbon-uranium-wolfram, 27, 135
Carbon-uranium-zirconium, 26, 93
Carbon-vanadium, 26, 96, 105; 27, 121, 137; 30A, 40; 31A, 35; 33A, 57; 35A, 114
Carbon-vanadium-wolfram, 27, 138
Carbon-wolfram, 26, 105; 27, 140
Carbonyl diurea, 30A, 41; 31A, 45
Carbonyl fluoride, 27, 729
Carbonylsalicylamide, 29, 763
Carbon-ytterbium, 26, 91, 92
Carbon-yttrium, 26, 91, 92; 34A, 60
Carbon-zirconium, 27, 121
m-Carborane, 35B, 908
o-Carborane, 35B, 908
O-Carborane, brominated derivatives, 31B, 283
Carboranes, 29, 526-532
N(1')-Carboxybiotin-DI-(*p*-bromoanilide), 30B, 242
Carboxylic acids and carboxyamides of picoline and pyrazine derivatives, 26, 762
2-Carboxyquinoline-N-oxide, 29, 763
Carminite, 27, 602, 678; 28, 197
Carnosine copper(II) complex, 30B, 395
Carnotite, 30A, 348
β-Carotene, 29, 496
Carpesterol *p*-iodobenzenesulphonate, 35B, 921
Carpholite, 31A, 251
Carvone hydrosulphide, 27, 1002
α-Caryophyllenealcohol(*p*-bromobenzene sulphonate), 29, 771; 35B, 386
Caryophyllene chlorohydrin, 28, 675
Caryophyllene iodonitrosite, 33B, 210
Casein, 27, 823
Casimidine dihydrochloride, 27, 993; 28, 545
Cassipourine, 34B, 321
Catalase, 33B, 327
Catapleiite, 30A, 456
Catechol, 31B, 71
Catena-DI-μ-hydrazine-zinc diacetate, 30B, 365
Catena-μ-trichloroacetato-acetylacetonatocopper(II), 35B, 913
Catoptrite, 31A, 223
Cedrelone, 26, 830; 27, 1032
Cedrelone iodoacetate, 28, 542
Celadonite, 29, 429
Celebixanthone, 27, 1006; 28, 562
Celestite, 26, 443; 30A, 365
Cellobiose, 26, 625; 33B, 161
β-Cellobiose, 31B, 174
Cellodextrins, 27, 798
Cellopentaose, 27, 798
Cellotetraose, 27, 798
Cellotriose, 27, 798
Cellulose, 26, 628; 27, 798; 29, 759
Cellulose II, 27, 798
Cellulose triacetate, 26, 628
Cembrene, 26, 799, 34B, 315
Cementite, 26, 99; 27, 122; 28, 15; 29, 34
Cement phase Y, 35A, 470
Cenosite, 28, 275
Cephalexin, 35B, 908
Cephaloglycin, 35B, 908
Cephaloglycine acetic acid hydrate, 35B, 384
Cephaloridine hydrochloride monohydrate, 35B, 384
Cephalothin sodium, 28, 676
Ceric fergusonite, 30A, 456
Cerium, 26, 106, 233; 27, 141, 334; 34A, 62
Cerium(IV)acetylacetonate, 28, 622
Cerium(III) chloride, 27, 724; 31A, 251
Cerium-cobalt, 26, 239; 27, 7, 142
Cerium-cobalt-plutonium, 29, 37
Cerium-cobalt-silicon, 34A, 63
Cerium complex fluorides, 26, 312
Cerium-copper, 26, 107
Cerium dioxide, 27, 480
Cerium-dysprosium, 27, 148
Cerium-erbium, 27, 148
Cerium formate, 27, 815
Cerium-gadolinium, 27, 148
Cerium-gallium, 26, 146
Cerium-germanium, 27, 341
Cerium hexaboride, 26, 67
Cerium-holmium, 27, 148
Cerium-hydrogen-oxygen, 27, 332
Cerium-hydrogen-tellurium, 27, 332
Cerium-iron, 27, 6
Cerium-iron-silicon, 35A, 44
Cerium-lanthanum, 27, 145, 148
Cerium lanthanum sulphide, 29, 341
Cerium-lanthanum-thorium, 27, 145
Cerium-lutecium, 27, 148

Cerium-magnesium, 26, 189; 32A, 47
Cerium magnesium nitrate hydrate, 27, 571; 28, 181
Cerium-magnesium-sulphur, 27, 327
Cerium-manganese-sulphur, 27, 327
Cerium-neodymium, 27, 148
Cerium-nickel, 26, 108, 239
Cerium-nickel-silicon, 31A, 35; 34A, 63, 64
Cerium niobate, 26, 386; 27, 521
Cerium-nitrogen, 27, 304
Cerium orthovanadate, 27, 589
Cerium oxytelluride, 27, 424
Cerium-platinum, 26, 239
Cerium-plutonium, 26, 231
Cerium potassium fluoride, 26, 312
Cerium-praseodymium, 27, 148
Cerium-rare-earth alloys, 27, 148
Cerium rubidium fluoride, 26, 312
Cerium-ruthenium, 27, 149
Cerium-scandium, 27, 148
Cerium-selenium, 27, 335; 33A, 58
Cerium sodium fluoride, 26, 312
Cerium-strontium-sulphur, 27, 327
Cerium sulphate enneahydrate, 33A, 374
Cerium sulphide halides, 31A, 247
Cerium-sulphur, 33A, 60
Cerium sulphur iodide, 34A, 214
Cerium-terbium, 27, 148
Cerium-thorium, 27, 145
Cerium thorium oxides, 31A, 250
Cerium-thulium, 27, 148
Cerium triacetate hydrate, 32B, 496
Cerium vanadate, 27, 724
Cerium-yttrium, 27, 148
Cerium-zinc, 26, 175
Cerolite, 28, 322
Ceroplastol 1 *p*-bromobenzoate, 33B, 552; 34B, 320
Cerusite, 29, 426
Cervantite, 27, 478
Chabazite, 29, 401; 31A, 242
Chalcocite, 28, 19
Chalcokyanite, 26, 438
Chalcone, 35B, 55
Chalcosiderite, 30A, 454
Chamaecyenol 4-bromo-3-nitrobenzoate, 34B, 311
Chambersite, 27, 551
Chamosite, 29, 429
Chelated dinuclear peroxytitanium(IV), 35B, 740
Chenevixite, 29, 429
Cheralite, 30A, 454; 32A, 359
Chervetite, 28, 354; 30A, 456; 32A, 275
Chevkinite, 26, 523; 31A, 242, 253
Chillagite, 29, 420
Chimonanthine didydrobromide, 30B, 232
Chitan, 33B, 522
β-Chitin, 26, 821
Chloral hydrate, 27, 731; 28, 400; 29, 766
Chloranil, 26, 714; 27, 899
Chloranil acid dihydrate, 32B, 106
Chloranil-hexamethylbenzene, 27, 909
Chloranil-8-hydroxyquinoline complex, 32B, 283
Chloranilic acid, 32B, 105
Chlorine, 27, 716; 30A, 252
Chlorine dioxide, 26, 346; 34B, 243
Chlorine heptoxide, 27, 719; 30A, 445
Chlorine strontium phosphate, 26, 469
Chlorite, 26, 505, 529; 27, 696; 28, 270; 30A, 428
Chlorites, 32A, 474
Chloroacetaldehyde polymers, 29, 759
p-Chloroacetanilide, 31B, 70
*mono*Chloroacetic anyhydride, 34B, 5
Chloroacetiminium chloride, 33B, 23
mono-Chloroacetonitrile, 26, 588
α-Chloroacetophenone, 31B, 78
N-Chloroacetyl-alanine, 32B, 489
DL-*N*-Chloroacetylalanine, 35B, 305
α-Chloroacrolein, 26, 550
μ-Chloro-μ-amido-octaamminedicobalt tetrachloride tetrahydrate, 35A, 406
2-Chloro-3-amino-1,4-naphthoquinone, 29, 770
p-Chloroaniline, 31B, 69
α-Chloroanthraquinone, 34B, 105
1-Chloroanthraquinone, 28, 675
2-Chloro-1,4-anthraquinone, 26, 750
Chloroaquobis(ethylenediamine)copper (II) chloride, 32B, 379
β-(Chloroaquo-triethylene-tetramine) cobalt(III) perchlorate, 33B, 547
DL-*cis*-β-(Chloro-aquo-triethylene-tetramine) cobalt(III) perchlorate, 34B, 482
anti-*p*-Chlorobenzaldoxime, 29, 657; 35B, 79
syn-*p*-Chlorobenzaldoxime, 29, 657
p-Chlorobenzene-*anti*-diazoimidoglyoxy-nitrile, 33B, 57
o-Chlorobenzoic acid, 26, 694
p-Chlorobenzoic acid anhydride, 31B, 65
7-Chloro-1,2-benzoisothiazolin-3-one, 33B, 554; 34B, 206
Chloro-1,4-benzoquinone, 35B, 71
α-2-Chloro-*p*-benzoquinone-4-oxime acetate, 26, 709
β-2-Chloro-*p*-benzoquinone-4-oxime acetate, 26, 709
o-Chlorobenzoylacetylene, 31B, 79
3,4-(*p*-Chlorobenzoyl)-1,2,5-oxa-diazole *N*-oxide, 35B, 233

σ-*p*-Chlorobenzoyl-rhenium pentacarbonyl, 34B, 420
5-(3-Chlorobenzyldimethylammonium) tetrazolate, 35B, 914
1-(4-chlorobenzyl)-1-nitroso-1(4,5-dihydro-2-imidazolyl)hydrazine monohydrate, 30B, 124
7-*endo*-Chlorobicyclo[3.2.0]hept-2-en-6-*endo*-01 *p*-bromobenzoate, 35B, 914
2-Chlorobicyclo[3,3,1]nonan-9-one, 32B, 144
2-Chlorobiphenyl-4-carboxylic acid, 34B, 63
2'-Chlorobiphenyl-4-carboxylic acid, 30B, 427; 34B, 64
Chloro-bis(acetonethiosemicarbazone)-nickel(II) chloride monohydrate, 34B, 585
N-Chloro-bis(chlorotetra(benzimidazole) nickel(II)) chloride-4-acetone, 33B, 435
Chlorobis(π-cyclopentadienyl dicarbonyl iron)-(π-cyclopentadienyl tricarbonyl molybdenum) tin(IV), 32B, 494
*mono*Chlorobis(diethyldithiocarbamato)iron(III), 35B, 798
Chloro-1,2-bis(diphenylphosphino) ethanemonocarbonyl-π-cyclopentadienylmolybdenum(II), 35B, 854
Chlorobispentane-2,4-dianatoiron(III), 35B, 916
Chlorobis-(*N*-*n*-propylsalicylaldiminato)iron(III), 35B, 916
Chloro-NN'-bis-salicylidene-ethylenediamine)iron(III), 32B, 394
Chlorobis(triphenylstibine)tetrakis-(trifluoromethyl)rhodiacyclopentadienedichloromethane solvate, 35B, 512
4-Chloro-4'-bromoacenaphthene, 26, 747
4-Chloro-5-bromoacenaphthene, 28, 685
Chloro-9-bromo 10-anthracene, 26, 748
2β,3α[*a*,*a*] Chloro-bromo-cholestane, 27, 1037
Chlorocarbonyl bis(triphenyl phosphine)-iridium, oxygen (1:1 complex), 30B, 386
Chlorocarbonylnitrosylbis(triphenylphosphine)iridium tetrafluoroborate, 33B, 430
Chlorocarbonyl(sulphur dioxide)bis (triphenylphosphine) iridium, 31B, 460
Chlorocarbonyl-(sulphur dioxide)-bis(triphenylphosphine)rhodium, 34B, 613
α-Chloro-α-chlorosulphenyl-4-nitro-2,5-dimethoxyphenylacetyl chloride, 32B, 103
3β-Chloro-5-cholestene, 30B, 439
m-Chloro-*trans*-cinnamic acid, 29, 769
cis-β-Chlorocrotonic acid, 33B, 3
Chlorocuprates, 26, 326
Chlorocyanoacetylene, 33B, 29
Chlorocyclohexane, 26, 642
Chloro-π-cyclopentadienylbis-8-quinolinolatotitanium(IV), 35B, 559
3-Chloro-2-decalone, 34B, 109
p-Chlorodiazoaminobenzene, 33B, 63
μ-Chloro-(dichloro-methyl-tin)-(2,2'-bipyridyl-tricarbonyl molybdenum), 34B, 409
μ-Chloro(dichloromethyltin)-2,5-dithiahexane-tricarbonyl-wolfram, 34B, 585
7-Chloro-5-(2,4-dichlorophenyl)-4,5-dihydro-1,4-dimethyl-3H-1,4-benzodiazepin-2-one, 32B, 164
Chloro(diethylphosphonato)mercury(II), 35B, 834
Chlorodifluormethane, 27, 729
8-Chloro-6,7-dihydro-3-methyldipyridazino-[2,3-*a*:4,3-*d*]pyrrole, 33B, 124
10-Chloro-5,10-dihydrophenarsazine, 30B, 283; 31B, 312
2-Chloro-9(ω-dimethylaminopropylidene)-thioxanthene, 29, 735
3-Chloro-10-(3'-dimethylamino-*n*-propyl)phenothiazine, 34B, 213
*mono*Chloro-dimethyl ether, 30B, 402
Chlorodinitro(ethylenediamine)ammine cobalt(III), 31B, 380
Chlorodiphenylarsin, 27, 867
2-Chloro-1,3-dithia-2-stibacyclopentane, 32B, 319
Chlorodurene, 29, 763
α-5-(2'-chloroethoxy)-O-quinone-2-oxime, 31B, 76
9-β-Chloroethyl-7,8-dihydro-9H-imidazo (2,1-*i*)purine methiodide, 28, 512
Chloro(ethylenediamine) (diethylenetriamine)cobalt(III) chloride, 35B, 919
ω-Chloro(ethylenediamine)diethylenetriaminecobalt(III) chloride, 35B, 924
ω-Chloro(ethylenediamine)diethylenetriaminecobalt(III) iodide, 35B, 910
ω-Chloro(ethylenediamine)diethylenetriamine tetrachlorozincate(II), 35B, 910
β-Chloro(ethylenediamine)dipropylenetriamine)cobalt(III) tetrachlorozincate(II), 35B, 599
β-Chloroethyl triptycene, 34B, 121
Chlorofluoromethane hydrates, 29, 762
cis-1-Chloro-2-fluoropropene, 33B, 540
Chloroform, 31B, 33
Chloroglycyl-L-methionatoplatinum(II) monohydrate, 35B, 917
α-Chlorohemin, 30B, 217

2-Chloro-3-hydroxy-1,4-naphthoquinone, 29, 770
Chloro-*N*-(2-hydroxypropyl)-salicylaldiminocopper(II), 35B, 617
6-Chloro-2-hydroxypyridine, 34B, 128
7-Chloro-4-hydroxytetracycloxide, 30B, 399
Chloroiron(II) derivative of tetraphenylporphine, 32B, 234
Chloroisothiocyanodiethylenediaminenickel, 33B, 385
6-Chloro-2-methoxy-9-thiolacridine, 29, 764
7-Chloro-4-methyl coumarin, 30B, 423
6-Chloromethyl-2-hydroxy-3,5-dimethyl-1,4-benzoquinone, 35B, 73
2-Chloro-3-methyl-1,4-naphthoquinone, 29, 770
4-Chloro-2-methyl-3-phenylcyclobut-2-enone, 34B, 67
Chloromethyl silane, 27, 732
β-Chloronaphthalene, 28, 672; 30A, 423
2-Chloro-1,4-naphthoquinone, 26, 742
3-Chloro-1,2-naphthoquinone, 35B, 918
2-Chloro-4-nitroaniline, 30B, 68
6-Chloro-2-nitrobenzaldehyde, 29, 652
m-Chloronitrobenzene, 27, 893; 30B, 63
p-Chloronitrobenzene, 27, 892
2-Chloro-5-nitrobenzoic acid, 27, 897
3-Chloro-2'-nitrochalcone, 35B, 54
1-Chloro-8-nitronaphthalene, 28, 674
trans-Chloronitrosylbis(ethylenediamine)cobalt(III) perchlorate, 35B, 588
3-Chloronopinone, 29, 771
7-Chloropentacyclo(7.5.0.0(2,7).0(5,13).0(6,12))tetradeca-3,10-diene-8-one, 34B, 115
Chloropentamethylbenzene, 33B, 70
1-Chloro-2,2,3,4,4-pentamethylphosphetan 1-oxide, 35B, 448
Chloropentamminecobalt(III) dichloride, 28, 90; 33A, 208
Chloropentaamminecobalt(III) hexafluorosilicate, 32A, 167
p-Chlorophenol, 33B, 543, 556
p-Chlorophenoxyacetic acid, 30B, 420
1A-(*p*-Chlorophenoxy)-1A,7B-dihydrobenzo(D)cyclopropa(B) pyran-3(1H)one, 34B, 174
trans-Chloro-2-(phenylazo)phenylbis-(triethylphosphine) palladium(II), 35B, 842
2-*p*-Chlorophenyl-1,3-dioxane, 35B, 211
1-(*p*-Chlorophenyl)-5-isopropylbiguanide hydrochloride, 32B, 59
N-(*p*-Chlorophenyl)-α-iospropyl-β-phenyl-β-lactam, 35B, 149
p-Chlorophenylmethylsulphone, 26, 695; 29, 771
N-(4-Chlorophenyl)phthalamic acid, 35B, 48
N-(4-Chlorophenyl)phthalimide, 33B, 551
Chlorophoenicite, 33A, 423
2-Chloro-1,8-phthaloylnaphthalene, 32B, 141
Chlorophyll, 33B, 544
Chloroplasts, 27, 1061
Chloroplatinate of an alkaloid derived from senecio kirkii, 31B, 243
Chloropromazine, 34B, 213
Chloroquine diphosphate monohydrate, 35B, 915
2-Chloroquinoline, 28, 470; 30B, 422; 33B, 123
6-Chloroquinoline, 33B, 124
Chlororhenium(III)-*N*-tetrabenzoatochlororhenium(III)-chloroform, 33B, 414
5-Chlorosalicylaldoxime, 26, 711
2-Chloro-N-salicylideneaniline, 29, 660
N-5-Chlorosalicylideneaniline, 29, 660
N-Chloro-succinimide, 26, 599
7-Chlorotetracycline hydrochloride, 28, 534
αα-Chlorotetraethylenepentaminecobalt(III) chloride perchlorate, 35B, 686
1-Chloro-2,3,5,6-tetramethylbenzene, 33B, 69
Chlorotetramine(sulphur dioxide)-ruthenium(II) chloride, 30A, 444
3-Chloro-1,2,3,4-tetraphenylcylobutenium pentachlorostannate, 28, 770
Chlorothiazine, 35B, 910
Chlorothiodiazyl chloride, 31A, 79
2-Chlorothiophene, 35B, 893
*mono*Chloro*mono*thiosemicarbazide-silver, 30B, 397
Chlorotrifluoroethylene, 26, 587
Chloro(trimethyl)pyridinetin(IV), 28, 603
Chlorotriphenylphosphonium bis(*cis*-1,2-bis(trifluoromethyl)ethene-1,2-dithiolato)gold, 33B, 496
Chlorotris(*o*-methylthiophenyl) phosphinenickel(II) perchlorate, 34B, 607
Chlorotris(pentacarbonylmanganese) tin, 32A, 510
α-Chlorotropane monohydrate, 35B, 135
2-Chloroundecane complex, 26, 575
6-Chloroundecane heptadecanol-9 complexes, 26, 575
α-Chloro-δ-valerolactam, 32B, 11
trans-β-Chlorovinyl mercury chloride, 31B, 341
Chloropromazine base, 30B, 431
Chloropromazine hydrochloride, 30B, 431

5β-Cholanic acid-3α,6α-diol, 30B, 441
5β-Cholanic acid-3α,7α-diol, 30B, 441
5β-Cholanic acid-3α,12α-diol.ethanol, 30B, 441
5β-Cholanic acid-3α,7α-diol-12-one diacetate, 30B, 442
5β-Cholanic acid-3α-ol-acetate methyl ester, 30B, 442
5β-Cholanic acid-3α-ol methyl ester, 30B, 442
5β-Cholanic acid-3α-ol-12-one acetate methyl ester, 30B, 442
5β-Cholanic acid-3α,7α,12α-triol 3,7-diacetate methyl ester hemihydrate, 30B, 442
5β-Cholanic acid-3α,7α,12α-triol methyl ester, ethanol, 30B, 442
5β-Cholanic acid-3α,7α,12α-triol sodium salt, 30B, 441
5β-Cholanic acid-3α,7α,12α-triol tetrahydrate, 30B, 441
5β-Cholan-24-ol, 30B, 441
Cholestan-4-one-3-spiro(2,5-oxathiolane), 33B, 201
Cholesterol monohydrate, 28, 680
Choline chloride, 26, 603
Choline-O-sulphate, 29, 765
Chondrodite, 35A, 469
Chromate-apatites, 30A, 321
Chromatite, 28, 328
Chrome apatites, 30A, 455
Chrome-diopsides, 26, 512
Chrome pyrope garnet, 35A, 258
Chrome-spinels, 26, 410
Chrome spodiosites, 30A, 455
Chrome-urania, 26, 407
Chromian muscovite, 30A, 456
Chromites, 27, 492, 493; 30A, 457
Chromium, 26, 109; 27, 150; 32A, 51
Chromium alums, 33A, 372
Chromium carboxyl acetophenone complex, 30B, 294
Chromium(II)chloride, 26, 317
Chromium(III)chloride, 26, 318; 29, 272
Chromium chlorite, 32A, 475
Chromium(II) chromium(III) pentafluoride, 29, 255
Chromium cobalt oxide, 27, 493
Chromium-cobalt-phosphorus, 27, 320
Chromium-cobalt-silicon, 27, 152, 376
Chromium-cobalt-sulphur, 32A, 53
Chromium complexes, 33B, 452; 35B, 507
Chromium-copper-oxygen, 26, 396
Chromium-copper-selenium, 32A, 53; 33A, 61
Chromium-copper-tellurium, 32A, 53
Chromium dioxide, 27, 478
Chromium disilicide-titanium disilicide, 31A, 257
Chromium disilicide-vanadium disilicide, 31A, 257
Chromium fluoride, 26, 300
Chromium gadolinium borate, 27, 562
Chromium-gallium-sulphur, 26, 117
Chromium-germanium, 32A, 70
Chromium-germanium-thorium, 29, 91
Chromium-gold, 27, 226
Chromium-hafnium, 26, 162
Chromium hexacarbonyl, 32A, 393
Chromium hexacyanoiridate, 26, 343
Chromium hexamine bromide, 27, 657
Chromium hexamine chloride, 27, 656, 657
Chromium hexamine halides, 27, 656
Chromium hexaminehydrate chloride, 27, 657
Chromium hexamine iodide, 27, 657
Chromium(III) hydrogen ethylenediaminetetraacetic acid monohydrate, 28, 673
Chromium-indium-sulphur, 26, 117
Chromium(VI)-iodine(V) compounds, 26, 500
Chromium-iron, 27, 155
Chromium-iron-nickel, 27, 155
Chromium-iron-nickel-titanium, 26, 110
Chromium-iron-oxygen, 29, 39, 326
Chromium-iron-selenium, 32A, 54
Chromium-iron-silicon, 27, 156
Chromium-lead-molybdenum, 26, 391
Chromium-lithium-nitrogen, 26, 187
Chromium lithium oxide, 26, 390
Chromium-lithium-sulphur, 35A, 44
Chromium-magnesium-oxygen, 26, 396
Chromium-manganese-nickel-silicon, 26, 175
Chromium-manganese-oxygen-zinc, 26, 409
Chromium-manganese-silicon, 27, 375
Chromium-molybdenum-nickel, 30A, 40
Chromium-molybdenum-niobium, 26, 110
Chromium molybdenum oxide, 31A, 249
Chromium-molybdenum-wolfram, 26, 111; 27, 156
Chromium neodymium titanium pentoxide, 35A, 237
Chromium-nickel, 26, 111; 27, 157
Chromium-nickel-phosphorus, 27, 157
Chromium-nickel-silicon, 27, 152, 374
Chromium-nickel-silicon-tantalum, 26, 175
Chromium-nickel-silicon-vanadium, 26, 175
Chromium-nickel-sulphur, 31A, 36
Chromium-niobium, 26, 112
Chromium-niobium-silicon, 26, 113; 33A, 64
Chromium octahydrotriborotetracarbonyl, 32A, 510
Chromium orthovanadate, 27, 525, 526
Chromium oxides, 26, 390; 27, 499; 30A, 304
Chromium(IV) oxide, 26, 354
Chromium oxide chloride, 27, 469

Chromium oxide hydroxide, 31A, 129
Chromium oxide-iron oxide, 27, 500
Chromium(II) oxybromide, 31A, 107
Chromium(III) oxychloride, 26, 329
Chromium-oxygen-vanadium, 26, 356
Chromium peroxy compounds, 27, 659
Chromium-phosphorus, 27, 319
Chromium-platinum, 27, 254, 323
Chromium pnictides, 33A, 63
Chromium potassium fluoride, 26, 305
Chromium potassium iodate, 26, 500
Chromium rhodite, 26, 399
Chromium rubidium iodate, 26, 500
Chromium ruthenite, 26, 399
Chromium ruthenium, 26, 114
Chromium-selenium, 26, 115; 27, 159; 29, 40; 35A, 45
Chromium-silicon-vanadium, 27, 375
Chromium-silicon-wolfram, 26, 103; 27, 159
Chromium(III) strontium tetrafluoride, 35A, 163
Chromium sulphate, 31A, 246
Chromium(III) sulphide bromide, 31A, 253
Chromium-sulphur, 32A, 55
Chromium sulphur spinels, 26, 117
Chromium-sulphur-yttrium, 26, 277
Chromium-tantalum, 26, 118
Chromium-technetium, 27, 361
Chromium-tellurium, 35A, 47
Chromium tellurium oxide, 27, 526
Chromium-tellurium-titanium, 31A, 37
Chromium-tin, 31A, 37
Chromium-titanium-selenium, 33A, 66
Chromium-titanium sulphur, 33A, 66
Chromium tribromide, 27, 439
Chromium tricarbonyl complex, 26, 684
Chromium trioxide, 35A, 206
Chromium wolfram oxide, 27, 527; 31A, 259
Chromium xanthate, 27, 844
Chromous acetate, 35B, 920
Chromous acid, 28, 223
Chromous bromide, 27, 433
Chromous halides, 27, 432
Chromous iodide, 27, 432
Chrysazin, 30B, 104
Chrysene, 27, 925
Chrysoberyl, 28, 141; 29, 317
Chrysocolla, 33A, 472
Chrysotile, 29, 429; 31A, 255
Chymotrypsin, 35B, 887, 920
γ-Chymotrypsin, 29, 758
Chymotrypsinogen, 27, 1060
β-Ciliatine, 31B, 301
Cinchomeronic acid, 28, 671
Cinchomeronic acid hydrochloride, 28, 671
cis-Cinnamic acids, 29, 763
trans-Cinnamic acids, 29, 763, 769
trans-Cinnamide, 29, 769
Citric acid, 34B, 7
Clathrate hydrates, 28, 678, 679; 29, 765
Clay minerals, 26, 529, 530
Cleavamine methiodide, 29, 746
Clerodin, 26, 830
Clerodin bromolactone, 27, 1029
Clinochlore, 29, 429
Clinoclase, 30A, 397
Clinoenstatite, 30A, 456; 31A, 252
Clino-ferrosilite, 30A, 455, 456
Clinohedrite, 28, 261
Clino-holmquistite, 30A, 458
Clinopyroxenes, 31A, 242
Clinozoisite, 26, 521; 33A, 475
Closo-tetrakis(methyltrimethyl silanolato beryllium), 34B, 368
Coalinite, 30A, 456
Cobalt, 35A, 49
Cobalt(II) acetylacetonate, 27, 821, 29, 767; 30B, 334
Cobalt aerugite, 35A, 359
Cobalt arsenate, 35A, 350
Cobalt(III) bis dimethyl glyoximino diammine chloride, 26, 677
Cobalt(III) bis dimethyl glyoximino diammine nitrate, 26, 676
Cobalt(III) bis dimethyl glyoximino diammine perchlorate, 26, 676
Cobalt(III) bis dimethyl glyoximino diammine thiocyanate, 26, 675
$(-)_{589}$-[Cobaltbis(isothiocyanate)-bis(trimethylenediamine)][antimony-((+)-tartrate)]dihydrate, 35B, 911
Cobalt(III) bis-(3-methoxysalicylaldehyde)ethylenediimine complex, 35B, 915
Cobalt biuret, 30A, 409
Cobalt boracite, 27, 551
Cobalt(II) bromide dihydrate, 32A, 204
Cobalt(II) bromide hexahydrate, 26, 334
Cobalt(III) bromide tris(L-propylene-1,2-diamine), 31B, 378
Cobalt carbonate, 26, 481
Cobalt carbonyl complexes, 35B, 505
Cobalt carbonyls, 35B, 502
Cobalt carbonyl sulphur complex, 32A, 404
Cobalt(II) chloride, 34A, 178
Cobalt(II) chloride dihydrate, 27, 435; 28, 76; 31A, 95
Cobalt chloride hexahydrate, 27, 435; 35A, 402
Cobalt(II) chloride-methanol, 33B, 239
Cobalt complexes, 26, 479; 27, 842; 33B, 266; 35B, 716
Cobalt compound, 35A, 394
Cobalt coordination compounds, 26, 675
Cobalt-copper, 27, 160
Cobalt-copper-germanium, 26, 129
Cobalt-copper-nickel-oxygen, 26, 412
Cobalt(II) diethoxyphosphonylacetylmethane, 33B, 464

Cobalt(II)diphenylphosphinothionate, 35B, 810
Cobalt(II) di(tris(2-aminoethoxido)-cobalt(III) acetate, 34B, 556
Cobalt-dysprosium, 26, 236, 239
Cobalt-erbium, 26, 239
Cobalt fluoride, 26, 300
Cobalt fluoroborate hexahydrate, 26, 304
Cobalt formate dihydrate, 32B, 414
Cobalt-gadolinium, 26, 118, 236, 239; 27, 161; 30A, 42; 35A, 49
Cobalt-gallium-magnesium-oxygen, 26, 412
Cobalt-gallium-nitrogen, 27, 306
Cobalt-gallium-oxygen, 26, 412
Cobalt-gallium-titanium, 27, 161, 206
Cobalt-gallium-vanadium, 27, 161
Cobalt garnet, 27, 514
Cobalt-germanium, 27, 161
Cobalt-germanium-hafnium, 27, 162
Cobalt germanium lithium oxide, 27, 486
Cobalt-germanium-molybdenum, 26, 174
Cobalt-germanium-nickel, 27, 162
Cobalt-germanium-niobium, 27, 162
Cobalt-germanium-nitrogen, 27, 306
Cobalt-germanium-oxygen, 26, 413
Cobalt-germanium-tantalum, 27, 162
Cobalt-germanium-thorium, 29, 91
Cobalt-germanium-zirconium, 27, 162; 34A, 66
Cobalt-gold, 27, 160
Cobalt-hafniumm, 26, 162
Cobalt-hafnium-phosphorus, 33A, 67
Cobalt-hafnium-silicon, 27, 162, 377
Cobalt hexacyanochromate, 26, 343
Cobalt hexacyanochromate hexahydrate, 27, 653
Cobalt hexacyanoiridate, 26, 343
Cobalt hexamine bromide, 27, 657
Cobalt hexamine chloride, 27, 657
Cobalt hexamine halides, 27, 656
Cobalt-holmium, 27, 163; 30A, 42; 34A, 68
Cobalt(II)hydrazinecarboxylate, 30B, 413
Cobalt hydroxybromide, 26, 327; 27, 470
Cobalt hydroxychloride, 26, 327
Cobaltic acid, 34A, 237
Cobaltic oxide, 29, 325
Cobalt-indium-nitrogen, 27, 306
Cobalt(II) iodide hexahydrate, 28, 102
Cobalt-iron, 26, 120; 27, 163
Cobalt iron nickel arsenides, 26, 120
Cobalt-iron-nickel-sulphur, 26, 121; 27, 164
Cobalt iron oxide, 33A, 324
Cobalt-iron-phosphorus, 27, 164, 320
Cobalt-iron-silicon, 26, 124
Cobalt iron zinc sulphide, 26, 125
Cobalt isoselenocyanato dimethylforna-mide complex, 29, 764
Cobalt isothiocyanato dimethylforma-mide complex, 29, 764
Cobaltite, 28, 50
Cobalt-lanthanum, 26, 127, 239
Cobalt lithium vanadate, 26, 409
Cobalt-ludwigite, 31A, 253
Cobalt-magnesium-nickel, 26, 174; 27, 165
Cobalt maleonitriledithiolate, 29, 767
Cobalt-manganese-nickel-oxygen, 26, 412
Cobalt-manganese-oxygen, 26, 408
Cobalt-manganese-oxygen-tin, 26, 410
Cobalt-manganese-phosphorus, 27, 320
Cobalt-manganese-silicon, 27, 165, 376; 35A, 50
Cobalt manganese spinels, 30A, 451
Cobalt manganite, 27, 493
Cobalt mercury carbonyl, 32A, 510
Cobalt metagermanate, 33A, 504
Cobalt(II) N-*p*-methoxyphenylsalicyl-iminate, 30B, 433
Cobalt molybdate, 27, 522; 30A, 326
Cobalt-molybdenum, 27, 166; 28, 17; 30A, 43
Cobalt-molybdenym-silicon, 26, 269; 27, 376; 32A, 57
Cobalt molybdenum sodium oxychloride, 35A, 262
Cobalt-molybdenum-sulphur, 34A, 70
Cobalt-nickel-silicon, 26, 124
Cobalt-nickel-sulphur, 26, 121
Cobalt nickel vanadate, 27, 491
Cobalt-nickel-zinc, 26, 219
Cobalt niobates, 26, 389
Cobalt-niobium-silicon, 27, 162, 377; 30A, 44; 33A, 70
Cobalt-nitrogen-tin, 27, 306
Cobalt-olivine, 28, 350
Cobalt orthosilicate, 29, 396
Cobaltous chloride-methanol system, 33B, 542
Cobaltous chloroaluminate, 27, 459
Cobaltous fluoride, 28, 290
Cobaltous oxide, 27, 474; 33A, 268
Cobaltous sulphate monohydrate, 26, 437
Cobalt(II) oxide, 31A, 251
Cobalt-oxygen-titanium, 26, 413
Cobalt-palladium, 26, 127
Cobaltpentadiene complex, 30B, 394
Cobalt(II)perchlorate-*o*-phenylene-bis-(dimethylarsine) complex, 32B, 468
Cobalt-phosphorus, 26, 47; 27, 319; 34A, 117
Cobalt-phosphorus-silicon, 27, 321
Cobalt phthalocyanine, 30B, 435
Cobalt-platinum, 29, 42
Cobalt-plutonium, 26, 232; 28, 17
Cobalt potassium fluoride, 26, 305
Cobalt potassium molybdate hydrate, 32A, 301
Cobalt-praseodymium, 26, 239
Cobalt-rare-earth, 32A, 59; 34A, 72
Cobalt-rhenium-silicon, 27, 376
Cobalt rhodium oxide, 26, 390

Cobalt(II) rubidium pentakischloride, 32A, 190
Cobalt(II) rubidium trischloride, 32A, 190
Cobalt salicylalbutyliminate, 26, 678
Cobalt-scandium, 26, 236; 27, 342
Cobalt selenate, 29, 357
Cobalt-selenium, 34A, 73
Cobalt-selenium-sulphur, 27, 168
Cobalt-silicon, 27, 375
Cobalt-silicon-tantalum, 27, 162, 377
Cobalt-silicon-titanium, 27, 162, 377
Cobalt-silicon-vanadium, 26, 174, 269, 270; 27, 376
Cobalt-silicon-wolfram, 26, 269; 27, 376
Cobalt-silicon-zirconium, 27, 162, 377
Cobalt sodium molybdate, 29, 329
Cobalt squarate, 33B, 542
Cobalt stannate, 27, 490
Cobalt strontium molybdate, 27, 509
Cobalt strontium wolframate, 27, 509
Cobalt sulphate, 26, 436, 437, 438; 27, 526; 30A, 370
Cobalt sulphate hexahydrate, 27, 616
Cobalt-sulphur, 26, 121, 208-210; 27, 169
Cobalt-talc, 28, 350
Cobalt-tantalum, 26, 389; 27, 170
Cobalt-tellurium, 35A, 75
Cobalt tetra-*iso*selenocyanatomercurate, 26, 343
Cobalt-thorium, 26, 239; 34A, 75; 35A, 53
Cobalt-tin, 27, 258
Cobalt titanate, 27, 507; 29, 318
Cobalt trideuterium hexacyanide, 35A, 404
Cobalt-triol bromide, 32A, 208
Cobalt-triol iodide, 32A, 208
Cobalt(III) tris(*N*,*N*-diethyldithiocarbamate), 33B, 491
Cobalt(III)tris(O-ethylxanthate, 34B, 524
Cobalt(III) tris glycinate dihydrate, 26, 675
Cobalt(II) tris(pyridine-2-carboxaldehyde)-1,1,1-tris(aminomethyl)-ethane diperchlorate, 35B, 921
Cobalttris(triphenylphosphine) nitrogen hydride, 34B, 618
Cobalt uranate, 26, 406; 27, 523
Cobalt uranium oxide, 33A, 356
Cobalt-vanadium-zirconium, 28, 19
Cobalt-yttrium, 26, 236, 239; 27, 163
Cobalt-zinc, 27, 173
Cobalt zinc stannates, 27, 490
Cobalt zinc sulphide, 26, 125
Cobalt-zirconium, 26, 128; 35A, 53
l-Cocaine derivatives, 28, 548
l-Cocaine hydrobromide, 28, 548
l-Cocaine hydrochloride, 28, 548
Coclaurine hydrobromide monohydrate, 33B, 223
Cocrystallised cyclo adducts from 3 and 4-bromo-N-carbomethoxyazepine and tetracyanoethylene, 31B, 138
Codeine, 29, 749
Codeine hydrobromide, 27, 1023; 29, 749
Coesite, 34A, 363
Coffinite, 26, 515
Colemanite, 31A, 172
Collagen, 26, 823; 27, 1054; 29, 758
Collidine-mercuric chloride, 30B, 397
Coloradorite, 28, 320
Columbite, 31A, 244
π-Complexes, 35B, 513
Complex molybdates, 30A, 349
1:1 Complex of 5-bromouridine and dimethylsulphoxide, 32B, 272
Complex oxides, AB_2O_6 type, 31A, 250
Complex oxychlorides, 29, 281
Compounds with α-$NaFeO_2$ structure, 27, 518
Compreignacite, 29, 422
Condensed ring hydrocarbons, 26, 739; 27, 925
Congressane, 30B, 119
Conichalcite, 28, 198
Coordination compounds of nickel(II), 27, 843
Copolyamides, 29, 759
Copolymers of ethylene-*cis*-butene-2 and ethylene cyclopentene, 27, 801
Copper, 26, 3, 129
Copper acetate-quinoline adduct, 33B, 412
Copper acetylacetonate, 27, 1065; 34B, 497
Copper(II) 6-aminohexanoic acid complexes, 35B, 917
Copper ammonium heptafluorotitanate tetrahydrate, 32A, 171
Copper ammonium sulphate dihydrate, 26, 449
Copper ammonium sulphate hexahydrate, 26, 449; 31A, 203
Copper arsenate, 33A, 419
Copper azide, 32A, 232; 33A, 510
Copper(II) benzoate trihydrate, 31A, 634
Copper biguanide chloride, 29, 768
Copper biguanide chloride dihydrate, 30B, 415
Copper biguanide nitrate dihydrate, 30B, 416
Copper(II) bis-2,5-dimethyloxazolidine-4-carboxylate dihydrate, 35B, 916
Copper 1,7-bis-(2-pyridyl)-2,6-diazeheptane dithiocyanate, 35B, 915
Copper-N-*p*-bromophenylsalicylaliminate, 30B, 433
Copper(II) carbonate diammine, 27, 660
Copper(II) carboxylates, 33B, 403

Copper(I) chloride-acetonitrile complex, 34B, 546
Copper(II) chloride-bis-(*N,N*-dimethylacetamido)thioether, 35B, 795
Copper(II) chloride bis(dimethylsulphoxide, 35B, 734
Copper(II) chloride complexes, 29, 590
Copper(I) chloride diethylsulphide complex, 32B, 450
Copper(II) chloride dihydrate, 35A, 197
Copper(I) chloride hydrate, 31A, 248
Copper *o*-chloro-mandelamidine, 29, 763
Copper croconate, 29, 669
Copper(II) cupferronate, 27, 1065; 28, 648
Copper(I) cyanide-ammonia, 30A, 290
Copper(I) cyanide hydrazine, 31A, 116
Copper decaborane, 27, 399
Copper di-β-alanine hexahydrate, 26, 662
Copper(I) diazoaminobenzene, 26, 655
Copper dibenzoylmethanate, 34B, 505
Copper(I) diethyldithiocarbamate, 28, 625; 30B, 373; 31B, 404
Copper 8,12-diethyl-2,3,7,13,17,18,21-heptamethylcorrole, 35B, 912
Copper diglycine monohydrate, 26, 661
Copper(II) *N,N*-dimethylthioformohydrazide, 33B, 543
Copper(I) dipropylthiocarbamate hexamer, 35B, 828
Copper(II) dirhodanodiammine, 32A, 239
Copper(II) NN'-disalicylidene-*o*-phenylenediamine, 26, 670
Copper(II) distrontium hexafluoride, 35A, 163
Copper-dysprosium, 26, 237
Copper elutamate, 30B, 395
Copper-erbium-sulphur, 30A, 44
Copper(II) ethylacetoacetate, 30B, 336
Copper ethylenediaminetetraacetate tetrahydrate, 35B, 924
Copper ethylsulphate tetrahydrate, 34A, 306
Copper ferrite, 26, 411
Copper ferrite-chromite, 26, 411
Copper(II) fluorides and oxyflourides, 31A, 245
Copper fluorotitanate tetrahydrate, 32A, 170
Copper(II) formate dihydrate, 30B, 346; 33B, 402
Copper(II) formate dioxan complex, 31B, 406
Copper formate, hemidioxane, 30B, 415
Copper(II) formate monourea, 35B, 633
Copper-gadolinium, 26, 237, 239
Copper-gadolinium-germanium, 35A, 54
Copper-gallium, 27, 175, 185
Copper-gallium-magnesium, 26, 174
Copper-gallium-manganese, 26, 174
Copper gallium spinels, 31A, 251
Copper-germanium, 27, 185
Copper-germanium-iron, 26, 129
Copper-germanium-magnesium, 26, 131, 174
Copper-germanium-manganese, 26, 129, 174
Copper-germanium-nickel, 26, 129
Copper-germanium-phosphorus, 26, 227
Copper-germanium-selenium, 26, 134
Copper-germanium-silicon, 26, 132
Copper-germanium-sulphur, 26, 133, 134
Copper-germanium-tellurium, 26, 134
Copper-germanium-thorium, 29, 91
Copper glutamate dihydrate, 31B, 411
Copper(II) *mono*glycylglycine trihydrate, 26, 657
Copper glycyl-*l*-leucyl-*l*-tyrosine hydrate etherate, 33B, 553
Copper-gold, 26, 134, 135; 27, 176, 185; 31A, 37; 33A, 75
Copper-gold-palladium, 31A, 38
Copper-gold-zinc, 31A, 39
Copper-hafnium, 27, 234
Copper hexacyanochromate, 26, 343
Copper hexacyanochromate hexahydrate, 27, 653
Copper(II) hexafluorotitanate(IV) tetrahydrate, 31A, 258
Copper(II) hexamethylenedithiocarbamate, 33B, 488
Copper-holmium, 26, 239
Copper hydrogen phosphite, 34A, 333
Copper(II) hydrogen *o*-phthalate dihydrate, 33B, 554
Copper hydroxybromide, 26, 327
Copper hydroxychloride, 26, 327
Copper hydroxyiodide, 26, 327
Copper(II)-2-(O-hydroxyphenyl)-benzoxazole, 34B, 572
Copper α-hydroxy-α-phenyl-butyramidine, 29, 768
Copper 8-hydroxyquinolinate, 28, 685; 29, 612
Copper-indium, 27, 185
Copper-indium-magnesium, 26, 136, 174
Copper-indium-manganese, 26, 174
Copper indium oxide, 33A, 284
Copper-indium-silver-tellurium, 27, 181
Copper-indium-tellurium, 27, 181
Copper-indium-titanium, 27, 11, 296
Copper N-*p*-iodophenylsalicylaliminate, 30B, 433
Copper-iron-nitrogen, 27, 306
Copper-iron-oxygen, 26, 396
Copper-iron-sulphur, 26, 137; 27, 182; 29, 43
Copper isoserine trihydrate, 26, 663
Copper(I) lanthanide, oxides, 34A, 252
Copper-lanthanum, 26, 239
Copper lithium oxide 35A, 216
Copper-lithium-tin, 35A, 55
Copper-magnesium-oxygen, 26, 396

Copper maleonitriledithiolate, 29, 767
Copper-manganese, 27, 182, 186
Copper-manganese-nickel-oxygen, 26, 412
Copper-manganese-tin, 26, 174
Copper-manganese-zinc, 27, 261
Copper manganite, 27, 494
Copper manganite-chromite, 26, 394
Copper-mercury, 34A, 77
Copper mercury oxynitrate trihydrate, 34A, 343
Copper(II) mercury(II) thiocyanate, 28, 112
Copper metagermanate, 32A, 494
Copper molybdate, 33A, 335
Copper-nickel, 27, 186
Copper nickel stannate, 27, 518
Copper nickel strontium oxide, 35A, 245
Copper nickel titanate, 27, 518
Copper-niobium-selenium, 26, 138
Copper-niobium-selenium-vanadium, 26, 138
Copper-niobium-silicon, 33A, 77
Copper-niobium-sulphur, 26, 138
Copper-niobium-tantalum-tellurium, 26, 138
Copper(II) nitrate, 30A, 400
Copper(II) nitrate hydrate, 34A, 340
Copper(II) nitrate-methyl cyanide, 33B, 453
1:1 Copper(II) nitrate-nitromethane complex, 31B, 435
Copper(II) nitrate-pyrazine, 35B, 721
Copper(II) nitrate trihydrate, 31A, 177
Copper(II) nitrosophenylhydroxylaminate, 28, 648
Copper(II) oxalate hydrates, 33B, 402
Copper oxides, 26, 397
Copper oxinate, 29, 763
Copper-oxygen-titanium, 28, 20
Copper-palladium, 27, 186
Copper perchlorate hexahydrate, 26, 496
Copper N-phenyl-5-bromosalicylaliminate, 30B, 433
Copper N-phenyl-4-methylsalicylaliminate, 30B, 433
Copper N-phenylsalicylaliminate, 30B, 433
Copper-phosphorus, 30A, 45
Copper-phosphorus-silicon, 26, 227
Copper phthalocyanine, 30B, 435; 33B, 431
Copper-platinum, 27, 186
Copper-platinum-zinc, 35A, 56
Copper potassium chloride dihydrate, 35A, 197
Copper potassium fluoride, 26, 305
Copper potassium polyphosphate, 35A, 346
Copper-praseodymium, 26, 239
Copper propionate, 30B, 419; 35B, 640
Copper propionate monohydrate, 30B, 419
Copper(II) propionate-*p*-toluidine, 35B, 921
Copper pyridoinate, 27, 1065
Copper(II) pyridoxylidene (+-)-valine complex, 32B, 496
Copper(II) 1-(2-pyridylazo)-2-naphtholato perchlorate monohydrate, 32B, 497
Copper pyrophosphate, 26, 466; 31A, 192
α-Copper pyrophosphate, 32A, 376
Copper rhodite, 26, 399
Copper rhodium oxide, 26, 390
Copper rubidium fluoride, 27, 454
Copper ruthenite, 26, 399
Copper salicylalbutyliminate 26, 678
Copper(II) salicylates, 33B, 552
Copper-scandium, 27, 342
Copper selenate, 29, 357
Copper-selenium, 27, 183; 30A, 46; 31A, 40
Copper-selenium-sulphur-tantalum-tellurium, 26, 138
Copper-selenium-sulphur-vanadium, 26, 138
Copper-selenium-tantalum, 26, 138
Copper-selenium-tin, 26, 134
Copper-selenium-vanadium, 26, 138
Copper sodium fluoride, 27, 454
Copper solid solutions, 27, 185
Copper strontium oxide, 35A, 217
Copper(II) succinate dihydrate, 31B, 408
Copper sulphate, 26, 436, 438
Copper(II) sulphate monohydrate, 26, 439; 31A, 246
Copper sulphate pentahydrate, 27, 614
Copper sulphate trihydrate, 33A, 368
Copper(I,II) sulphite dihydrate, 30A, 380
Copper-sulphur, 28, 19; 29, 46; 33A, 79; 35A, 56
Copper-sulphur-tantalum, 26, 138; 29, 46
Copper-sulphur-tin, 26, 134
Copper-sulphur-vanadium, 26, 137, 138, 31A, 41
Copper tartrate trihydrate, 30B, 414
Copper(II) tellurites, 31A, 253
Copper-tellurium, 26, 138; 29, 48; 32A, 61; 33A, 80; 34A, 79
Copper tellurium oxide, 33A, 392
Copper-tellurium-tin, 26, 134
Copper tetrammine complexes, 34A, 307
Copper tetraphenylporphine, 28, 686; 29, 617
Copper tetrathiocyanatomercurate, 27, 660

Copper thallium fluoride, 27, 454
Copper thiocyanate derivatives, 28, 109
Copper-thorium, 26, 256
Copper tin, 27, 185; 31A, 41
Copper-tin-titanium, 27, 296
Copper-titanium, 27, 186; 28, 20; 30A, 47
Copper titanium oxide, 28, 140
Copper-titanium-zinc, 27, 187, 296
Copper-titanium-zirconium, 26, 138
Copper-tri-ethylenediammine sulphate, 26, 605
Copper trilonate, 27, 1065
Copper uranium tetroxide, 33A, 357
Copper violurate tetrahydrate, 33B, 551
Copper wolframate, 35A, 226
Copper wolfram oxide, 34A, 284; 35A, 226
Copper-yttrium, 26, 239, 277; 28, 21
Copper yunainate trihydrate, 35B, 763
Copper-zinc, 26, 139; 27, 185, 188; 33A, 72
Copper-zinc-zirconium, 27, 297
Copper-zirconium, 27, 234
Coquimbite, 35A, 437
Corannulene, 32B, 486
Cordeauxia-quinone, 30B, 104
Cordierite, 26, 505; 27, 724; 29, 440
Coriose, 35B, 259
Cornetite, 29, 371
Coronene, 26, 739; 30B, 111
Cortisone alcohol, 27, 1037
Corundophilite, 30A, 456
Corundum, 26, 347, 348, 517
Corundum structures, 34A, 239
Corymine hydrobromide monohydrate, 30B, 233
Cosalite, 26, 432; 30A, 456
Cotton, 26, 629
Cotunnite, 26, 324; 28, 78
Coulsonite, 27, 491
Coulteropine hydrobromide, 33B, 555
p-Coumaric acid, 32B, 497; 35B, 44
Covellite, 33A, 79
Cranadallite, 31A, 254
Creedite, 30A, 378
p-Cresol, 31B, 74; 35B, 65
Cristobalite, 26, 517; 31A, 244
β-Cristobalite, 31A, 242
Cristobalite, low, 30A, 420
Crocidolite, 26, 528
Crocoite, 27, 609; 30A, 320
Croconate ions, 28, 684
Cronstedtite, 27, 697; 30A, 458
Crotepoxide iodohydrin, 34B, 342
Crotocine, 28, 675
Crotocol, 28, 675
Cryophyllite, 27, 685
Cryptomelane, 28, 334
Cryptoperthite, 30A, 456
Cryptoperthites, 26, 525
Crytopine, 33B, 226
Crystal violet, 26, 828
Cubane, 29, 766
Cucurbitine perchlorate, 30B, 212
Cumene, 33B, 536
Cummingtonite, 26, 526; 30A, 456, 458; 31A, 224
Cummingtonite-grunerite, 29, 429
Cupric chloride-1,2,4-triazole, 27, 859
Cupric dithizonate, 26, 663
Cupric fluoride dihydrate, 27, 436
Cupric hydroxide, 26, 358
Cupric nitrate, 35A, 322
Cupric oxide, 27, 474; 35A, 207
Cupric pyrophosphate, 33A, 410
Cupric tropolone, 30B, 370
Cuprous iodide, 26, 335
Curite, 28, 150
Curium, 34A, 80
Curium(III) phosphate, 30A, 458
Cuspidine, 26, 513
Custonilide silver nitrate, 35B, 923
Cyanamide, 26, 586
Cyanamides, 29, 239
Cyanates, 27, 653
Cyanein, 28, 676
Cyanoacetamide, 35B, 21
8-Cyano-2,3-benzobicyclo[4,2,0]-octa-2,4-diene, 33B, 549
2-Cyano-5-bromobenz[*f*][1,3]oxazepine, 35B, 232
α-Cyanocinnamonitrile, 34B, 28
Cyano complexes, 27, 653
Cyano 9-dihydro 9-10-anthracene, 26, 748
Cyanodimethylarsine, 28, 586
Cyanogen, 28, 52, 401
4-Cyano-N-(2-(*p*-methoxyphenyl) ethyl)pyridinium iodide, 34B, 135
(*Z*)-2-Cyano-2-phenylhydrazono-acetonitrile, 35B, 59
Cyano-triphenylphosphine gold, 34B, 604
Cyano tris(3-dimethylarsinopropyl) phosphine nickel(II) perchlorate, 32B, 465
Cyanuric acid, 29, 771
Cycl(3,2,2)azine, 26, 789
DL-Cyclazocine, 34B, 326
L-Cyclazocine hydrobromide mono-hydrate, 34B, 326
Cyclic cytidine 2,3'-phosphate, 35B, 922
Cyclic hexameric phosphonitrilic molecule, 30B, 399
Cyclic polypeptides, 35B, 328
Cyclic urethane oligomers, 26, 637
Cyclic uridine-3',5'-phosphate, 33B, 554

Cyclo-D-alanyl-L-alanyl, 35B, 318
Cyclo-L-alanyl-L-alanyl, 35B, 318
Cycloalliin hydrochloride monohydrate, 30B, 420; 31B, 157
Cyclobarbital, 35B, 918
Cyclobutadieneiron tricarbonyl, 35B, 895
Cyclo-butadiene-tricarbonyl-triphenyl-phosphine-molybdenum(O), 32B, 494
Cyclobutane, 26, 641; 27, 733
cis-1,2-Cyclobutanedicarboxylic acid, 32B, 486
cis-1,3-Cyclobutanedicarboxylic acid, 33B, 552; 34B, 86
trans-1,2-Cyclobutanedicarboxylic acid, 35B, 97
trans-1,3-Cyclobutanedicarboxylic acid, 32B, 115, 486
trans-1,3-Cyclobutanedicarboxylic acid(disodium acid salt), 33B, 78
Cyclobutane-1,5-spiro-2,4,6-triketo-hexahydropyrimidine, 30B, 169
Cyclobutene-1,2-dicarboxylic acid dimethyl ester, 28, 671
cis, cis-Cyclodeca-3,8-diene-1,6-dione, 33B, 553
Cyclodecasulphur, 31A, 258
Cyclodecylamine hydrochloride, 26, 830
Cyclodecylamine hydrochloride sesquihy-drate, 29, 511
α-Cyclodextrin complexes, 35B, 906
γ-Cyclodextrin complexes, 35B, 906
1,5,9-*trans-trans-trans*-Cyclododecatriene, 27, 787; 32B, 119
Cyclododecatriene-silver nitrate, 29, 769
Cycloheptaamylose complexes, 33B, 542
1,3-Cycloheptadiene, 31B, 93
1,3,5-Cycloheptatriene, 29, 503
π-Cycloheptatrienyldicarbonyl-molybdenum σ-pentafluorophenyl, 33B, 552
Cycloheptatrienyl-tricarbonylchromium, 26, 685
5-(1'-Cyclohepten-1'-yl)-5-ethyl-barbituric acid, 34B, 222
Cyclohexaamylose potassium acetate complex, 30B, 186
1,3-Cyclohexadiene, 33B, 534
DL-1,4-Cyclohexadiene-1-glycine, 33B, 552
π-Cyclohexadienylmanganese tricarbonyl, 34A, 162
Cyclohexadione diguanides, 29, 762
Cyclohexadione diureides, 29, 762
Cyclo(hexaglycyl)hemihydrate, 28, 525
Cyclohexakis[bis-(μ-1-hydroxyethane-2-thiolato)-nickel(II), 35B, 788
Cyclohexane, 27, 1064; 28, 666; 35B, 907
Cyclohexane-1,1'-diacetic acid, 35B, 107
cis-1,2-Cyclohexanedicarboxylic acid, 35B, 106
trans-1,2-Cyclohexane-dicarboxylic acid, 33B, 551
(+)-*trans*-1,2-Cyclohexane dicarboxylic acid, 34B, 90
(±)-*trans*-1,2-Cyclohexane dicarboxylic acid, 34B, 91
trans-1,4-Cyclohexane-dicarboxylic acid, 31B, 87
Cyclohexane-1,4-dione, 28, 686; 29, 491
Cyclohexane-1,4-dione, diiodoacetylene (1:1 complex), 30B, 248
Cyclohexane-1,4-dione-mercuric chloride, 29, 492
Cyclohexane-1,4-dioxime, 33B, 80
Cyclohexanone, 33B, 540
Cyclohexene, 35B, 897
5-(1-Cyclohexene-1-yl)-5-ethyl-barbituric acid [cyclobarbital], 35B, 918
1-Cyclohexenyl-1-cyclobutenedione, 29, 667
Cyclohexylammonium chloride, 34B, 92
Cyclohexylammonium phenyl phosphate, 32B, 486
Cyclohexyl chloride, 27, 784
Cyclohexyl *o*-cresyl ketone, 33B, 552
Cyclo-(d-α-Hydroxy-isovaleryl)-L-methylisoleucyl-D-α-hydroxy-isovaleryl-L-methyl-leucyl), 34B, 270
Cycloneosamandione hydroiodide, 29, 771; 30B, 432
Cycloneosamandions, 28, 686
Cyclononanone, 33B, 543
cis,cis,cis-1,4,7-Cyclononatriene, 29, 766
cis, cis, cis-1,4,7-Cyclononatriene silver nitrate complex, 32B, 373
1,3-Cyclooctadiene, 35B, 899
1,5-Cyclooctadiene, 29, 766
1,5-Cyclooctadienebis(iron tetra-carbonyl), 35B, 575
1,5-Cyclooctadiene-copper(I) chloride, 27, 850
1,5-Cyclooctadiene-duroquinone-nickel, 30B, 323
Cyclooctane-*cis*-1,2-dicarboxylic acid, 33B, 85
Cyclooctane-*trans*-1,2-dicarboxylic acid, 31B, 92
Cyclooctanone, 33B, 543
1,3,5,7-Cyclooctatetraenecarboxylic acid, 30B, 92
Cyclooctatetraene copper(I) chloride, 29, 587
μ-Cyclooctatetraene-di-(bismethylplati-num(II)), 26, 690

Cyclo-octatetraene di(cyclopentadienyl cobalt), 32B, 494
Cyclooctatetraene di-(irontricarbonyl), 26, 681
Cyclooctatetraenediruthenium hexacarbonyl, 33B, 369
Cyclooctatetraene-iron pentacarbonyl complex, 31B, 369
Cyclooctatetraene iron tricarbonyls, 26, 681; 27, 854-857
Cyclooctatetraenemolybdenum tricarbonyl, 31B, 368
Cyclo-octatetraene palladium (II) chloride, 32B, 494
(1,3,5-Cyclo-octatriene)di-iron hexacarbonyl, 34B, 475
(-)-*trans*-Cyclooctene, 35B, 921
Cyclopentadiene, 30B, 87
Cyclopentadiene niobium tetracarbonyl, 28, 672
π-Cyclopentadienyl bis(acetylacetonato)chlorozirconium(IV), 34B, 464
π-Cyclopentadienylbis(tribluoroacetato)vanadium, 35B, 540
π-Cyclopentadienyl-*cis*-1,2-bis(trifluoromethyl) ethanedithione cobalt, 31B, 348
π-Cyclopentadienyl-bis(triphenylphosphite)iron iodide, 33B, 358
π-Cyclopentadienyl-π-cycloheptatrienylvanadium, 28, 614
Cyclopentadienylcyclooctatetraenetitanium, 35B, 575
Cyclopentadienyl[1-(cyclopentadienyl)-1,2,3,4-tetramethyl-π-cyclobutenyl]-nickel, 30B, 322
π-Cyclopentadienyl-π-(1)-2,3-dicarbollyl-iron(III), 30B, 305
π-Cyclopentadienyl-dicarbonyl(4-carboethoxy-5-hydroxy-1,2,3-molybdenodiazacyclopenta-3,5-diene), 34B, 426
π-Cyclopentadienyl(dicarbonyl) carboxymethyl-iron, 34B, 418
Cyclopentadienyldicarbonyliron triphenyltin, 32B, 335
Cyclopentadienyl dicarbonyl molybdenum methylenemethylsulphide complex, 34B, 438
π-Cyclopentadienyl(1,2-dicyanoethene-1,2-dithiolato)cobalt, 33B, 344
Cyclopentadienyldinitrosylchromium chloride, 31B, 349
π-Cyclopentadienyl-π-duroquinonerhodium, 35B, 573
π-Cyclopentadienyl-hexakis (trifluoromethyl) benzene rhodium, 28, 684; 31B, 366
Cyclopentadienylindium(I), 28, 598
π-Cyclopentadienyliron dicarbonyl σ-cyclopentadiene, 31B, 353
Cyclopentadienyl manganese tricarbonyl, 28, 608
π-Cyclopentadienyl molybdenum tricarbonyl chloride, 33B, 343
π-Cyclopentadienyl niobium di(tolane)-carbonyl, 34B, 447
π-Cyclopentadienyl niobium tolane carbonyl π-(tetraphenyl cyclobutadiene), 34B, 450
π-Cyclopentadienyl-σ-pentafluorophenyl(triphenylphosphine) nickel, 33B, 357
π-Cyclopentadienyl 1-phenylcyclopentadiene cobalt, 28, 684; 29, 558
π-Cyclopentadienyl-rhodium-π-(2,6-di-*t*-butyl-1,4-benzoquinone), 34B, 467
Cyclopentadienyl-tetramethylcyclopentadienone cobalt(II), 26, 652
π-Cyclopentadienyl-π-tetraphenylcyclobutadiene-vanadium dicarbonyl, 34B, 479
Cyclopentadienylthallium(I), 28, 598
Cyclopentadienyl titanium(IV) trichloride, 28, 609
π-Cyclopentadienyl tolane complexes of niobium and tantalum, 33B, 556
π-Cyclopentadienyl(tricarbonyl) carboxymethyl-molybdenum, 34B, 418
π-Cyclopentadienyltricarbonylmolybdenum-manganese pentacarbonyl, 33B, 347
π-Cyclopentadienyltricarbonyl-σ-phenylwolfram, 33B, 349, 547
π-Cyclopentadienyl(triethylphosphine)-copper(I), 35B, 541
π-Cyclopentadienyl-σ-trifluoromethyl (triphenylphosphine)-nickel, 35B, 530
Cyclopentadienyltrimethylgermane, 35B, 900
Cyclopentadienyl trimethyl platinum, 34B, 418
Cyclopentadienyltrimethylsilane, 35B, 900
π-Cyclopentadienyl triphenylphosphine-σ-benzoyl-iron-carbonyl, 34B, 435

π-Cyclopentadienyl-triphenylphosphine-*mono*carbonyl-σ-benzoyliron, 33B, 556
π-Cyclopentadienyltriphenylphosphine-*mono*carbonyl-σ-phenyliron, 30B, 434; 31B, 361; 34B, 434
π-Cyclopentadienyl(triphenylphosphine)-phenyl nickel, 34B, 463
π-Cyclopentadienyl vanadium tetra-carbonyl, 32B, 350
4H-Cyclopenta[2,1-*b*:3,4-*b*']dithiophene, 35B, 221
Cyclopentane, 26, 641; 35B, 897
Cyclopentathio-1,3,5-tri-imine, 31A, 81
Cyclopentene, 35B, 896
Cyclopentenone photodimer, 29, 766
1,2-Cyclopentenophenanthrene, 26, 752
Cyclopropane, 29, 490
Cyclopropanecarboxamide, 34B, 84
Cyclopropanecarboxylic acid chloride, 30B, 403
Cyclopropanone, 33B, 552
Cyclopropene, 35B, 892
Cyclopropyl carboxaldehyde, 30B, 402
Cyclopropyl chloride, 27, 732
Cyclopropyl methyl ketone, 30B, 403
Cyclo sulphur nitrogen phosphorus compounds, 34A, 177
Cyclotetradecaheptaene, 27, 787
Cyclotetramethylene-tetranitramine, HMX, 28, 467; 32B, 487
β-Cyclotetramethylene tetranitramine, 35B, 179
Cyclotetra[μ-oxo-chloro-π-cyclo-pentadienyltitanium(IV)], 33B, 550; 35B, 554
Cyclotetrasarcosyl, 35B, 181
Cyclo-2,4,6,8-tetrathiotetrimine, 32A, 154
Cyclotetratriacontane, 33B, 87
Cyclo-tris(bis(trimethylsilyl) amidolithium), 34B, 372
μ-Cycloundeca-allyl-heptacarbonyldi-iron, 35B, 578
Cycloveratryl, acetone (clathrate), 30B, 437
Cycloveratryl, benzene (clathrate), 30B, 436
Cycloveratryl, butyric acid (clathrate), 30B, 437
Cycloveratryl, carbon disulphide (clathrate), 30B, 437
Cycloveratryl, chlorobenzene (clathrate), 30B, 436
Cycloveratryl, chloroform (clathrate), 30B, 436
Cycloveratryl, toluene (clathrate), 30B, 436
Cymrite, 29, 429
L-Cysteic acid, 33B, 174
L-Cysteine, 33B, 172
L-Cysteine ethyl ester hydrochloride-urea (1:1 complex), 30B, 208
L(+)Cysteine hydrochloride monohydrate, 30B, 393; 33B, 174
Cystine, 26, 826
L-Cystine dihydrobromide, 29, 484
L-Cystinediamide dihydrochloride, 32B, 489; 33B, 184
L-Cystinocopper(II), 27, 838
Cytidine, 30B, 198
Cytidine-3'-phosphate, 30B, 202; 32B, 223
Cytidylic acid B, 29, 771; 30B, 202
Cytosine, 29, 703
Cytosine-5-acetic acid, 27, 965
Cytosine-5-fluorouracil complex monohydrate, 34B, 226
Cytosine *mono*hydrate, 28, 503

Dachiardite, 28, 252; 30A, 458
Dalyite, 30A, 430
Damascenine acid, 33B, 544
Dammarane-type triterpenes, 33B, 555
Dannemorite, 27, 689
D'Ansite, 26, 453
Daphmacrine methiodide, 33B, 547
Daphmacrine methiodide acetone solvate, 34B, 333
Daphnetoxin bisbromoacetate, 35B, 920
Daphniphylline hydrobromide, 31B, 257
Darapskite, 32A, 332
Datolite, 31A, 251; 32A, 425
Daucyl-D,L-alaninate hydrobromide, 32B, 490
Deca-*trans*-3, *trans*-7-dienedioic acid 32B, 20
Davallol iodoacetate, 31B, 232
Davidite, 26, 378; 28, 334
Dawsonite, 32A, 419
Deacetyl-aspidospermine N(a)-hydroiodide, 34B, 328
Deacetylated triacetate, 26, 628
Deacetylcascarillin acetal iodoacetate, 31B, 225
Debye temperature copper, 26, 129
Decaborane, 34A, 151
B-Decachloroneobarene, 30B, 413
Decachloropentaphosphonitrile, 31A, 259
Deca-*trans*-3-*trans*-7-diene-1,10-dioic acid, 34B, 13
Decafluorobiphenyl, 33B, 537, 556
Decahydro-1,2,4:5,6,8-dimetheno-*S*-indacenedione, 35B, 921
trans-Decalin derivatives, 35B, 907
Decalins, 35B, 907
trans-2-Decalone, 33B, 537
Decamethylenebis[trimethylammonium bromide]dihydrate, 30B, 34
Deca(methylisonitrile)dicobalt(II) perchlorate, 29, 565
Decammine-μ-amidodicobalt(III) pen-tanitrate, 28, 183; 33A, 428

Decammine-μ-peroxo-dicobalt disulphate tetrahydrate, 33A, 375
Decammine-μ-peroxo-dicobalt pentanitrate, 28, 184; 33A, 427
Decammine-μ-peroxo-dicobalt *mono*-sulphate tris(bisulphate), 31A, 206
Decammine-μ-peroxodicobalt(III) tetrathiocyanate, 30A, 297
Decane dicarboxylic acid, 26, 575
Decaoxodiodate dihydrogen tetrapotassium octahydrate, 27, 641
n-Decylammonium chloride, 33B, 545
Deerite, 30A, 459
Dehydrated Ca-chabazite, 27, 691
Dehydrated natrolite, 27, 690
Dehydroacetic acid, 28, 671
Dehydroascrobic acid, 35B, 910
15,15'-Dehydrocanthaxanthin, 33B, 84
15,15'-Dehydro-β-carotene, 29, 499
Dehydromucate of polyethylene, 33B, 527
1,5-Dehydro-*vic*-triazolo[2,1a]-*vic*-triazole derivatives, 28, 456
Delhayelite, 26, 535
Demesmaekerite, 30A, 453
(+)-Demethanolaconinone hydroiodide trihydrate, 26, 793; 29, 749
4-Demethylhasubanonine brosylate, 33B, 553
Demycarosyl leucomycin A3 hydrobromide, 35B, 402
Demycarosyl leucomycin A3 hydrobromide ethanol solvate, 32B, 490
Denningite, 26, 455; 30A, 382
Denudatine methiodide, 35B, 913
Deoxyadenosine monohydrate, 30B, 197
Deoxyanisoin, 27, 911
7-Deoxy-7(S)-chlorolincomycin hydrochloride monohydrate, 32B, 490
3-Deoxy-3,4-*C*-(dichloromethylene)-1,2:5,6-di-*O*-isopropylidene-α-D-galactofuranose, 35B, 170
Deoxycholic acid, 27, 1039
2'-Deoxycytidine hydrochloride, 35B, 279
Deoxycytidylic acid(5') monohydrate, 35B, 910
Deoxy-erythrocruorin, 35B, 888
Deoxyguanosine-5-bromodeoxycytidine, 29, 771; 30B, 200
Deoxyguanosine, 5-bromodeoxycytidine (1:1 hydrogen-bonded complex), 30B, 200
2-Deoxy-*cis*-inosa-1,3-diamine dihydrochloride, 32B, 117
Deoxynupharidine hydrobromide, 35B, 373
5-(1(2'-Deoxy-α-D-ribofuranosyl) uracilyl) disulphide, 32B, 224
Deoxyribonucleic acid, 26, 817
1,2-Desacetyl-ε-caesalpin 2-*p*-bromobenzoate, 34B, 317
(+)-Des-(oxymethylene)-lycoctonine hydriodide monohydrate, 26, 795
Despujolsite, 33A, 381
Destigloylswietenine *p*-iodobenzoate, 29, 771
2-Desylidene-1,3-dithiolane, 32B, 488
Detigloylswietenine *p*-iodobenzoate, 31B, 236
Deuterated oxalic acid dihydrate, 29, 459
α-*mono*Deuterioglycolate, 30B, 396
Deuterium, 26, 285; 27, 393; 31A, 240
Deuterium chloride, 32A, 180
Deuterium-gold-palladium, 33A, 120
Deuterium-holmium, 29, 48
Deuterium-magnesium, 28, 22
Deuterium-niobium, 33A, 81
Deuterium-tantalum, 26, 139; 29, 50
Deuterium-uranium, 27, 194
Deuterium-zirconium, 28, 23
Deuteroammonia, 26, 290, 298
Deuteroethane, 30B, 401
Deuteroethylene, 30B, 402
Deutero fulminic acid, 34A, 233
Deuteromethane, 26, 543
Deuteromethylamine, 30B, 400
Develline, 30A, 453
Deweylite, 26, 538
Diacetamide-calcium bromide, 35B, 746
3,5-Diacetamido-1,2-dithiolium bromide, 28, 669; 31B, 146
3,5-Diacetamido-1,2-dithiolium chloride, 28, 669
3,5-Diacetamido-1,2-dithiolium iodide, 28, 670; 30B, 396
trans-Di-μ-acetato-bis(dioxotriphenylphosphine oxide acetate)-diuranium(VI), 34B, 533
Diacetatodiaquotetrachlorodirhenium(II), 35B, 637
Di-μ-acetato-hydroxo-aquoamminerhodium (II)-acetatoaquoamminerhodium(II), 35B, 908
cis-1,2-Diacetonyl-1,2,3,3-tetrachlorocyclopropane, 35B, 99
2,7-Diacetoxy-*trans*-15,16-dimethyl-15,16-dihydropyrene, 30B, 110
6,11-Diacetylanhydrotetrodotoxin hydroiodide, 31B, 241
Diacetylferrocene, 35B, 550
1,1'-Diacetylferrocene, 32B, 493
Diacetylruthenocene, 28, 612
Diacidobis(ethylenediaminenickel, 30B, 399
Diacidobis(ethylenediamine)nickel compounds, 30B, 399
Di-O-acylglycerol dihydrogen phosphates, 27, 783
Dialkalimetal hydroxy bis chromate(VI) chromium(III), 35A, 389

Di-*n*-alkylphosphinate copolymers, 33B, 525
Dialuric acid monohydrate, 30B, 167
Di-μ-amido-bis(tricarbonyliron), 33A, 163
1,6-*trans*-Diaminocyclodecane dihydrochloride, 26, 645; 31B, 93
1,6-*cis*-Diaminocyclodecanedihydrochloride dihydrate, 26, 645
1,3-*trans*-Diaminocyclohexane dihydrochloride, 30B, 89
1,4-*trans*-Diaminocyclohexane dihydrochloride, 30B, 394; 31B, 87
4,4'-Diamino-3,3'-dichlorobiphenyl, 33B, 64
4,4'-Diamino-3,3'-dimethylbiphenyl (*o*-tolidine), 33B, 64
2,2'-Diaminodiphenyl disulphide, 32B, 70; 35B, 60
Diaminodiphenylsulphone, 30B, 392
4,4'-Diaminodiphenyl sulphone, 35B, 915
3,5-Diamino-1,2-dithiolium bromide monohydrate, 28, 669
3,5-Diamino-1,2-dithiolium chloride monohydrate, 28, 669
3,5-Diamino-1,2-dithiolium iodide, 28, 669; 30B, 148
Diaminofumaronitrile, 33B, 555
Diaminomaleonitrile, 26, 600
2,7-Diamino-9-phenyl-10-ethylphenanthridinium bromide methanol solvate, 34B, 160
p-Diaminotetrachlorobenzene, 33B, 556
2,4-Diamino-5-(3,4,5-trimethoxybenzyl) pyrimidine-1-oxide dihydrate, 35B, 172
1,3-Diamino-2,4,6-trinitrobenzene, 32B, 91
Diamminebis(isothiocyanato)copper(II), 31A, 118
Diamminecopper(II) carbonate, 28, 176
Diamminecopper(II) thiocyanate, 28, 108
Diamminecopper(II) trisisothiocyanatoargentate(I), 31A, 118
Diammine diacetato copper(II), 28, 631
cis-Diamminedichloropalladium, 31A, 250
trans-Diamminedichloropalladium, 31A, 250
Diammine-(*o*-phthalato)copper(II), 35B, 725
Di(2-ammonioethyl)ammonium monochloride tetrachlorocuprate(II), 32B, 485
Diammonium amidothiophosphate, 34A, 325
Diammonium croconate, 29, 668
Diammonium dihydrogen hypophosphate, 29, 361
Diammonium hexachlorotellurate(IV), 31A, 100
Diammonium hydrogen arsenate, 35A, 353
Diammonium oxotetrakis(isothiocyanato)-vanadate pentahydrate, 28, 114; 29, 294
Diammonium vanadyl oxalate dihydrate, 33B, 543
Diamond, 27, 395
Dianin's compound, 35B, 213
Dianin's compound, chloroform adduct, 35B, 214
10,10'-Dianthronyl, 32B, 139
Diaquobis(acetylacetonato)magnesium(II), 32B, 324
Diaquobis(acetylacetonato)manganese(II), 33B, 388
Diaquobis(acetylacetonato)nickel(II), 29, 567
Diaquobis-(2,2'-biimidazole)nickel(II) dinitrate, 34B, 569
Diaquobis(glycinato)nickel(II), 33B, 416
Diaquobis(2-hydroxy-2-methylpropionato) copper(II), 33B, 403
Diaquobis(methoxyacetato)copper(II), 33B, 403
Diaquobis(phenoxyacetato)copper(II), 33B, 403
trans-Diaquo-bis(picolinato)nickel(II) dihydrate, 35B, 768
trans-Diaquo-bis(picolinato)zinc(II) dihydrate, 35B, 768
trans-Diaquo-bis(pyridine-2-carboximide) nickel(II) chloride, 32B, 431
Diaquobis(salicylaldehydato)nickel, 26, 674
Diaquobis(*L*-serinato)nickel(II), 34B, 542
Diaquodihydrazinecopper sulphate, 35B, 924
Di-μ-aquo-di-μ_3-hydroxo-deca(1,1,1-trifluoropentane-2,4-dionato)-hexanickel(II) toluene solvate, 34B, 511
Diaquodithiosemicarbazidenickel(II) dinitrate, 33B, 472
Diaquo-europium(III)-tris acetylacetonate monohydrate, 34B, 497
Diaquo-oxonium salicylic acid 5-sulphonate, 35B, 43
Diaquotris(acetylacetonato)lanthanum-(III), 33B, 390
Diaquo zinc glycylglycylglycinato hemisulphate dihydrate, 35B, 648
Diarylfuroxans, 35B, 912
Diarylureas, 29, 762; 33B, 542
4-(1,5-Diazabicyclo[3.2.1]oct-8-yl) pyridine, 31B, 137
(1,5-Diazacyclooctane-*N*,*N'*-diacetate)-aquonickel dihydrate, 33B, 553
1,8-Diazacyclotetradecane dihydrobromide, 30B, 136

1,8-Diazacyclotetradecane-2,7-dione, 33B, 120
1,3-Diaza-2,4-diborolidine derivative, 33B, 555
8,13-Diazo-18-noroestrone methyl ether, 33B, 549
cis-Diazidobisethylenediaminecobalt (III) nitrate, 33B, 384
Diazidodiamminecopper(II), 31A, 77
Diazidodipyridinecadmium, 35B, 691
Diazidodipyridinezinc, 35B, 693
μ-Diazido-tetrakis(triphenylphosphine)-dicopper(I), 35B, 920
meso-1,4-Diaziridinyl-2,3-butanediol, 26, 784
Diazoaminobenzene, 33B, 553
Diazofluorene, 32B, 486
p-Diazoniobenzenesulphonate, 29, 770
2-Diazonium-4-phenosulphonate monohydrate, 34B, 36
Diazonium salts, 29, 763
2-Diazo-1-oxide-4-benzene sulphonic acid monohydrate, 27, 981
4,4'-Diazulene tetra-iron decacarbonyl dichloroethane solvate, 34B, 477
1:2,5:6-Dibenzanthraquinone, 34B, 111
Dibenzanthrone(violanthrone), 29, 698
Dibenzene chromium, 28, 613; 33B, 556
Dibenzene molybdenum, 26, 650
Dibenzenesulphonamide, 35B, 84
Dibenzene thiosulphonato bis(ethylene thiourea) tellurium II, 26, 582
Dibenzene thiosulphonato bis(tetramethylthiourea) tellurium II, 26, 583
Dibenzene vanadium, 28, 674
Dibenzene wolfram, 26, 650
1,2:7,8-Dibenzocoronene, 26, 757
Dibenzodimethoxy polycyclic, 33B, 105
Dibenzoequinene, 33B, 105
2,3,11,12-Dibenzo-1,4,7,10,13,16-hexaoxocyclo-octadeca-2,11-diene-('dibenzo-18-crown-6')-rubidium sodium isothiocyanate complex, 35B, 476
1,2,7,8-Dibenzophenazine, 28, 677
Dibenzoquinolizine derivative, 35B, 199
Dibenzoselenophene, 35B, 226
Dibenzothiophen, 35B, 225
Dibenzothiophene sulphone, 33B, 144
3,4:7,8-Dibenzotricyclo[4.2.0.0^{2,5}] octa-3,7-diene, 35B, 144
Dibenzo[*b,f*]tropone, 33B, 555
cis-1,2-Dibenzoylethylene, 35B, 57
1,2-Dibenzoylethyleneiron tetracarbonyl (*cis* and *trans*), 33B, 556
Dibenzoylmethane, 30B, 430; 31B, 80
Dibenzoyl peroxide, 32B, 85
1,1'-Dibenzoylruthenocene, 30B, 433
Dibenzyl, 35B, 908
1,1-Dibenzyl-3,3-dimethylazetidinium bromide, 34B, 127
Dibenzyl disulphide, 34B, 49
Di-μ-bis(diphenylarsino)methane-bis(*trans*-chlorocarbonyl rhodium(I)), 34B, 616
Di(bis(triphenylarsine oxide)hydronium dimercury hexabromide, 32B, 492
Diborane, 31A, 259
β-Diborane, 30A, 252
Diboron tetrachloridetrimethylamine, 35B, 921
Dibortetrachloride, 34A, 173
Dibrombis-(2-diphenylphosphinoethyl)-aminenickel(II), 35B, 850
Dibromethane deoxycholic acid adduct, 27, 1039
2,4-Dibromoacenaphthene, 27, 933
4,4'-Dibromoacenaphthene, 26, 747
3,5-Dibromo-*p*-aminobenzoic acid, 27, 898; 30B, 57
Di(*p*-bromoanilide) of CO_2-biotin, 29, 771
1,5-Dibromoanthraquinone, 26, 751; 28, 431; 32B, 124
2,3-Dibromo-1,4-anthraquinone, 26, 751; 32B, 486
10,10-Dibromoanthrone, 28, 433
trans-p,p'-Dibromoazobenzene, 31B, 46
3,5-Dibromobenzamidine benzoate, 26, 733
3,5-Dibromobenzamidine 3,5-dibromobenzoate, 26, 733
Dibromobis-(benzyldiphenylphosphine)-nickel(II), 35B, 858
Di-μ-bromobis-π-cycloheptenyldipalladium, 33B, 550
Dibrombis(1,1-dimethylurea)cobalt hexahydrate, 30B, 419
Dibrombis(1,1-dimethylurea(nickel hexahydrate, 30B, 419
Dibrombis(1,1-dimethylurea)zinc, 30B, 419
Dibrombis(ethylenethiourea) tellurium(II), 26, 577
Dibromo-bis-triphenylphosphine nickel-(II), 28, 680; 33B, 515
Dibromocannabicyclol, 35B, 917
2α,3β-Di-bromo-5α-cholestane, 30B, 396; 31B, 203
2α,3β[*e,e*] Dibromo-cholestane, 27, 1037
2β,3α[*a,a*] Dibromo-cholestane, 27, 1037
4,4'-Dibromocinnamaldazine, 32B, 63
1,4-Dibromocycl(3,2,2)azine, 26, 790
trans-3,10-Dibromocyclodeca-1,2-dione, 35B, 114
1,6-*trans*-Dibromocyclodecane, 29, 509
2,12-Dibromo-cyclododecanone, 34B, 96
Dibromodeca-*O*-methylhopeaphenol benzene solvate, 35B, 405
Dibromo dehydrobispulegone, 34B, 345

(+)-Dibromodehydrotetrahydro-rugulosin, 33B, 556; 35B, 398
p-Dibromodiazoaminobenzene, 26, 731
2,4-Dibromodiazoaminobenzene, 30B, 49
mm'-Dibromodibenzoylmethane, 30B, 430
pp'-Dibromodibenzoylmethane, 30B, 430
4,4'-Dibromo-dibenzoyl peroxide, 33B, 43
Di-p-Bromo-dibromo-trans-bis(dimethyl sulphide)-dipalladium(II), 33B, 481
Dibromo-dibutoxytitanium(IV) butanol adduct, 28, 674
cis-2,4-Dibromo-2,4-di-t-butyl cyclobutanone, 31B, 84
N,N-Dibromo-di-n-butyldithio-carbamatocopper(III), 33B, 485
9,10-Dibromo-1,7-dicarbadodecarborane (12), 32B, 292
cis-1,2-Dibromo-1,2-dicarbometh-oxycyclobutane, 31B, 85
trans-1,2-Dibromo-1,2-dicarbo-methoxycyclobutane, 31B, 85
2,6-Dibromo-4,4-dichlorocyclo-hexa-2,5-diene-1-one, 33B, 79
1,5-Dibromo-4,8-dichloronaphthalene, 30B, 99
p,p'-Dibromo-α, α'-difluorostilbene, 32B, 41
6,6-Dibromo-2,3:4,5-dimethano-2,4-dinitrocyclohexanone, 30B, 106
2,3-Dibromo-2,3-dimethylbutane, 35B, 41
Dibromo-[6,6'-dimethyldi-(2-pyridyl-methyl)amine-NNN]-nickel(II), 35B, 702
1,5-Dibromo-4,8-diiodonaphthalene, 27, 923
Dibromo-2,5-dimethylpyrazinenickel (II), 29, 570
trans-2,3-Dibromo-1,4-dioxane, 28, 485
2,3-Dibromo-1,4-dithiane, 30B, 414
trans-2,5-Dibromo-1,4-dithiane, 31B, 143
Dibromodithioureatellurium(II), 31B, 335
22,23-Dibromo-9β-ergost-4-en-3-one, 34B, 303
Dibromoeriostoic acid, 32B, 265
3,5-Dibromofuschone, 34B, 80
trans-6,8-Dibromo-1,2,3,4,4a,9a-hexahydro-4a,9-dimethylcarbazole, 35B, 913
3,5-Dibromo-hydroxytriphenyl-methane carbinol, 31B, 75; 33B, 551
Dibromoleucodrin, 29, 771
Dibromo-[1-(o-methoxyphenyl)-2,6-diazaoctane]nickel(II), 33B, 441
B-Dibromo-C-methylbarene, 30B, 411
8,8'-Dibromo-2,2'-methylene diquinoline, 31B, 134
4,12-Di(bromomethyl)[2,2]meta-cyclophane, 33B, 100
1,8-Di-(bromomethyl)-naphthalene, 28, 674
α,4-Dibromo-α-(4-methyl-2-nitrophenyl-azo)acetanilide, 32B, 62
22,23-Dibromo-12-methyl-18-norergosta-8,11,13-trien-3β-yl acetate, 29, 751
1,4-Dibromonaphthalene, 26, 741
2,3-Dibromo-1,4-naphthoquinone, 31B, 97; 32B, 124
meso-3,3'-Di-(p-bromophenyl) bis-3-phthalidyl, 32B, 86
2,4-Dibromophenyldiazocyanide, 32B, 60
2,6-Di-p-bromophenyl-1,1-dimethyl-1-sila-2,4,6-tirazacyclohexane-3,5-dione, 29, 538
Di(p-bromophenyl)thiosulphonate, 32B, 68
11,13-Dibromopulchellin, 35B, 923
Di-2-bromopyridine-decaborane, 27, 883
3,5-Dibromopyridine oxide-mercury(II) chloride complex, 31B, 439; 34B, 576
Dibromo-1-(2-pyridyl)-2,5-diaza-5-methyl-hexa-1-enezinc(II), 35B, 696
11β,12α-Dibromo-3α,9-oxidocholanic acid methyl ester, 34B, 300
11β,12β-Dibromo-3α,9-oxidocholanic acid methyl ester, 34B, 301
Dibromoradicine(2,3-dihydro-3-hydroxy-2-methyl-7-(1,2-dibromo-propyl)-4H,5H-pyrano[4,3-b] pyran-4,5-dione), 35B, 213
Dibromotetrachloro-benzenes, 33B, 67
1,2-Dibromo-1,1,2,2-tetrachloro-ethane, 33B, 545
2,6-Dibromo-3,3,5,5-tetramethylcyclo-hexanone, 29, 493
Dibromotetrapyrazolenickel(II), 34B, 568
Dibromotrimethylsiloxyaluminum, 32B, 326
trans-8,8-Dibromo-1,4,4-trimethyl-tricyclo[5.10.$0^{3,5}$]octane, 35B, 125
Dibutadienerhodium(I) chloride, 30B, 398
Di-t-butylacetylene p-toluidine plat-inum(II) chloride, 32B, 493
Di-t-butylberyllium, 33B, 535
o-Di-t-butylquinoxaline, 33B, 552
Dicaesium di-μ-carbonyl-bis[π-(3)-1,2-dicarbollylcarbonyliron, 35B, 432
Dicaesium indium sodium hexafluoride, 35A, 158
Dicaesium lithium hexacyanidocobaltate (III), 34A, 231

Dicaesium potassium hexafluoro-argentate(III), 31A, 260
Dicaesium tetrachlorodioxouranium(VI), 31A, 105
Dicalcium ferrite, 27, 547
Dicalcium iridium(IV) oxide, 31A, 138
Dicalcium nitride, 33A, 163
Dicalcium silicate, 26, 511
α-Dicalcium silicate, 30A, 423
γ-Dicalcium silicate, 31A, 243
γ-Dicalcium silicate hydrate, 31A, 252
Dicalcium silicate hydrates, 26, 511
closo-2,3-Dicarbahexaborane(6), 35B, 904
2,3-Dicarbahexaborane(8), 29, 526
1,2-Dicarboclovododecaborane, 29, 529, 531
Dicarbonylbiscyclopentadienylbis-μ-diphenylacetylenediniobium, 33B, 550
1,2':1',2-Di-μ-carbonyl-bis[1,2-dicarba-*closo*-dodecaborane(12], 35B, 427
Di-carbonyl-1-carbomethoxy-2-phenyl-2[π-2',4'-dicarbomethoxy-3',5'-diphenyl-1'-cyclopentadienyloxy] σ-vinyl iron, 31B, 362
trans-Di-μ-carbonyl-bis(π-2,3-*cis*-dimethylbutadiene carbonylcobalt), 35B, 525
Dicarbonyl-3-[π-(2-cyclohexadienyl)]-σ-propenoyliron, 35B, 915
Dicarbonyl-π-cyclopentadienyl-bicyclo[2,2.1]-hepta-2π,5-dienemanganese(I), 33B, 350
Dicarbonyl cyclopentadienyl cobalt-mercury(II) chloride complex, 32B, 493
Dicarbonyl-π-cyclopentadienyl[dichloro-(phenyl)stannyl]iron, 35B, 545
Dicarbonylcyclopentadienyl(tri-bromostannyl)iron, 35B, 539
Dicarbonylcyclopentadienyl(tri-chlorostannyl)iron, 35B, 538
trans-Dicarbonyl-π-cyclopentadienyl-(triphenylphosphine)molybdenum acetyl, 33B, 356
cis-Di-μ-carbonyl-dicarbonyldi-π-cyclopentadienyldi-iron(Fe-Fe), 35B, 546
trans-Di-μ-carbonyl-dicarbonyldi-π-cyclopentadienyldi-iron(Fe-Fe), 35B, 546
Di-μ-carbonyl-μ-(1,3-di-π-cyclopentadienyl-*NNN'N'*-tetramethylethylenediamine)-bis-(carbonyliron), 35B, 557
Dicarbonylrhodium chloride, 26, 332
2,2'-Dicarboxydiethyl sulphide, 27, 773
Dicarboxyldimethyl sulphide, 27, 769
Dicarboxylic acid complexes, 26, 575
Dicarboxylic acid-urea adducts, 26, 571
4-Di-(β-chlorethyl)amino phenyl isothiocyanate, 28, 673
trans-Dichloridodiaquoethylenediamine chromium(III) chloride, 34B, 480
4,4'-Dichloroacenaphthene, 26, 746
Di-μ-chloro allyl palladium, 27, 820
2,5-Dichloroaniline, 28, 420
3,5-Dichloroanthranilic acid, 34B, 57
1,5-Dichloroanthraquinone, 26, 751
2,3-Dichloro-1,4-anthraquinone, 26, 750
Dichloro aquo-(2,9-dimethyl-1,10-phenanthroline) copper(II), 32B, 497
4,4-Dichloro-2a-aza-a-homocholestan-3-one, 35B, 342
trans,p,p'-Dichloroazobenzene, 34B, 38
B-Dichlorobarene, 30B, 410
p-Dichlorobenzene, 28, 424
2,3-Dichloro-1,4-benzoquinone, 30B, 398; 35B, 71
2,5-Dichloro-1,4-benzoquinone, 35B, 71
2,6-Dichloro-1,4-benzoquinone, 35B, 71
3,6-Dichloro-11,12-benzo-tetracyclo-$[5.3.2.0.^{2,6}.0^{3,8}]$dodeca-9-one, 35B, 145
cis-1,2-Di-*p*-chlorobenzoylethylene, 35B, 57
Dichloro(benzyclacetophenone)bis-(dimethylsulphoxide)iridium(III), 35B, 811
N(1)-(2,6-Dichlorobenzyl)-3-carbamido-pyridinium iodide monohydrate, 34B, 134
1-(2,6-Dichlorobenzyl)-6-hydroxy-1,4,5,6-tetrahydronicotinamide dihydrate, 34B, 140
7,7-Dichlorobicyclo[4.1.0]heptane, 35B, 898
trans-[Dichloro-{*NN'*-bis(*S*-2-amino-3-phenylpropyl)trans-*R*-cyclohexane-1,2-diamine δλδ}cobalt(III)] perchlorate, 35B, 914
μ,μ-Dichloro-bis(cyclooctadiene-1.5-copper(I)), 28, 614
μ-Dichloro-bis(π-cycloocta-1,5-diene)-diiridium, 35B, 908
Dichlorobis(dicarbonyl-π-cyclopentadienyliron)tin(IV), 32B, 356
trans-Dichloro-bis(diethylselenide)-palladium(II), 35B, 798
Di-μ-chloro bis(π-1,3-dimethylallyl), dipalladium(II), 32B, 493
Dichlorobis(2,6-dimethyl-γ-pyrone)-copper, 30B, 429

Dichlorobis(2,6-dimethyl-γ-pyrone)-
zinc, 30B, 429
cis-Dichloro-*cis*-bis(dimethylsulphoxide)-
trans-dimethyltin(IV), 33B, 548; 35B, 735
trans-Dichlorobis(dimethylsulphoxide) palladium(II), 32B, 450
Dichlorobis(1,1-dimethylurea)cobalt hexahydrate, 30B, 419
Dichlorobis(1,1-dimethylurea)nickel hexahydrate, 30B, 419
Dichlorobis(1,1-dimethylurea)zinc, 30B, 420
Di-μ-chloro-bis-(μ-diphenylphosphidotetracarbonylironpalladium), 33B, 550
Dichloro-bis(diphenylphosphino)ethylaminepalladium(II), 30B, 387
Di-μ-chloro-bis(π-1-ethoxycarbonyl-2-hydroxyallyl)dipalladium(II), 33B, 548; 35B, 521
Dichlorobis(ethylenediamine)cobalt-(III) chloride monohydrate, 33B, 554
cis-(+)$_{589}$-Dichlorobis(ethylenediamine)-cobalt(III) chloride monohydrate, 35B, 591
trans-Dichlorobis(ethylenediamine)-cobalt(III) hexathionate monohydrate, 30B, 330
trans-Dichlorobis(ethylenediamine)-cobalt(III) nitrate, 28, 615
Dichlorobis(ethylenediamine)iridium-(III) chloride, 35B, 922
(+)$_{450}$-*cis*-Dichlorobis(ethylenediamine)-platinum(IV) chloride, 35B, 587
Dichlorobisformamidecadmium, 32B, 444
Dichlorobis(glycinato)diaquomanganese (II), 27, 838
Dichlorobis(2,6-lutidine-*N*-oxide) copper(II), 34B, 578
Dichlorobis(2,6-lutidine *N*-oxide) zinc(II), 33B, 458
Dichlorobis-(2-methylpyridine) copper(II), 34B, 566
Dichloro-bis(phenoxathiin) mercury, 34B, 589
trans-Dichlorobis-[o-phenylene(bis-dimethylarsine)]cobalt(III) perchlorate, 35B, 869
trans-Dichlorobis(*l*-propylenediamine) cobalt(III) chloride hydrochloride monohydrate, 27, 833
Dichlorobis-8-quinolinolato-titanium(IV), 33B, 445
Dichlorobis(2,3,5,6-tetrahapto-norbornadienedicarbonylcobalt)tin, 35B, 921
Dichlorobis(tetramethylthiourea)-copper(II), 35B, 913
Dichloro-bis(thiourea)cobalt(II), 34B, 521
Dichlorobis-(1,4-thioxan)mercury(II), 32B, 452
cis-1,2-Dichloro-1,2-bis-(*p*-tolylthio)-ethylene, 33B, 550
Dichloro-bis-(tricyclohexylphosphine)-nickel(II), 28, 659
trans-Dichlorobistrimethylene-diaminecobalt(III) chloride hydrate, 35B, 911
cis-Dichlorobis(trimethylphosphine) platinum(II), 32B, 464
Dichloro-bis-(triphenylphosphine)-nickel(II), 28, 658
Di-μ-chloro-bis(triphenylphosphite)-(cyclo-octa-1,5-diene)dirhodium(I), 35B, 583
trans-Dichlorobis(tripropylphosphine)-μ,μ'-dichloro-diplatinum, 34B, 606
Dichlorobis-(4-vinylpyridine)-cobalt(II), 33B, 549
(-)-*cis*-Dichloro(*trans*-2-butene) ((S)-α-phenylethylamine) platinum (II), 34B, 439
trans-Dichloro-(*cis*-2-butene)-[(S)-α-phenethylamine]platinum(II) complex, 30B, 395
Dichloro(chloromercury)(nitrosyl)bis-(triphenylphosphine)osmium(II), 35B, 915
2α,3β[*e,e*] Dichloro-cholestane, 27, 1037
2α,3β-Dichloro-5α-cholestane, 31B, 203
2β,3α[*a,a*] Dichloro-cholestane, 27, 1037
2α,3β-Dichloro-5α-cholestane (modification β), 32B, 206
cis-1,6-Dichlorocyclodeca-1,3,6,8-tetraene, 35B, 113
trans-1,4-Dichlorocyclohexane, 29, 490
trans-1,4-Dichloro-cyclo-octane, 34B, 95
3,6-Dichloro-2,5-diamino-1,4-benzoquinone, 34B, 78
p-Dichlorodiazoaminobenzene, 26, 731
2,6-Dichloro-4-diazo-2,5-cyclohexadiene-1-one, 35B, 103
mm'-Dichlorodibenzoylmethane, 30B, 430
4,4'-Dichloro-dibenzoyl peroxide, 33B, 43
1,5-Dichloro-4,8-dibromonaphthalene, 27, 923
Dichloro-dibutoxytitanium(IV) butanol adduct, 28, 674
trans-Dichloro-(π-di-*t*-butylacetylene)-*p*-toluidineplatinum(II), 35B, 526
Di-μ-chloro-[dichloro-*trans*-bis(trimethylarsine)platinum(II), 33B, 547; 35B, 865

Di-μ-chloro-dichlorodimethyl tetracarbonyl di-iridium(III), 32B, 493
Di-μ-chlorodichlorohexacarbonyldiruthenium, 34A, 165
Dichloro-diethoxytitanium(IV) ethanol adduct, 28, 674
4,8-Dichloro-2,6-diethylbenzo-(1-2,4-5)bisoxazole, 33B, 542
1,2-Dichlorodiethylenediaminecobalt-(III) chloride monohydrate (*cis* form), 30B, 331
4,4'-Dichloro-α,β-diethylstilbene, 30B, 393; 32B, 43
cis-11,12-Dichloro-9,10-dihydro-9,10-ethanoanthracene, 28, 676
trans-11,12-Dichloro-9,10-dihydro-9, 10-ethanoanthracene, 28, 676
Dichlorodimethyl-bis(pyridine N-oxide)tin(IV), 34B, 410
trans-4,5-Dichloro-1,2-dimethylcyclohexene, 28, 671
Dichloro(2,7-dimethylocta-2,6-diene-1,8-diyl)ruthenium(IV), 30B, 398
Dichloro-(2,9-dimethyl-1,10-phenanthroline)nickel(II) dimer, 33B, 548
Dichloro-(2,9-dimethyl-1,10-phenanthroline)zinc(II), 34B, 557
cis-2,3-Dichloro-1,4-dioxane, 28, 482
trans-2,3-Dichloro-1,4-dioxane, 28, 483
trans-2,5-Dichloro-1,4-dioxane, 28, 487
Dichloro-di-μ-(pentamethylbicyclo-[2.2.0]hexa-2,5-dienylmethyl)-diplatinum(II), 35B, 579
p,p'-Dichlorodiphenoxy-1,2-ethane, 32B, 104
Dichlorodiphenoxytitanium (IV), 31B, 445
p,p'-Dichlorodiphenylamine, 26, 727
meso-1,1'-Dichloro-1,1'-diphenyl-1,1'-azopropane, 35B, 33
Dichlorodi-(*o*-phenylenebisdimethylarsine) platinum(II), 29, 578
5,6-Dichloro-11,12-diphenylnaphthacene, 27, 940; 28, 685; 29, 697
5,11-Dichloro-6,12-diphenylnaphthacene, 27, 940
5,12-Dichloro-6,11-diphenylnaphthacene, 27, 940
4,4'-Dichlorodiphenylsulphone, 26, 728
Di-*p*-chlorodiphenyltellurium diiodide, 27, 864
Dichlorodipyridine-[*o*-(di-*o*-tolylphosphino)benzyl]-rhodium(III)-0.61 chloroform, 35B, 852
Dichlorodipyridine-zinc, 31B, 424
2,2-Dichloro-1,3-dithiacyclobutanone-4, 33B, 556
trans-2,3-Dichloro-1,4-dithiane, 30B, 150
α-and-β-Dichloro-di-μ-thiocyanato-bis(tri-*n*-propylphosphine) platinum(II), 35B, 814
Dichlorodithioureatellurium (II), 31B, 335
trans-9,10-Dichloro-1,2,3,4,5,6,7,8,9,10,11,12-dodecahydrophenanthrene, 28, 675
Dichloro(dodeca-2,6,10-triene-1,12-diyl)ruthenium(IV), 29, 769; 33B, 335
1,2-Dichloroethane, 26, 545
cis-Dichloroethylene, 30B, 402
4,4'-Dichloro-3,3'-ethylenebis-(sydnone), 34B, 194
Dichlorofluoromethane, 29, 449
Dichloro-hexane complex, 26, 575
Dichloromalonamide, 35B, 919
Dichloro-DL-methioninepalladium(II), 35B, 646
Dichloro-DL-methionineplatinum(II), 35B, 917
Dichloro-L-methionineplatinum(II), 35B, 917
Dichloro-π-methylallyl-bis-(triphenylarsine)rhodium(III), 33B, 550; 35B, 534
1,8-Dichloro-9-methylanthracene, 34B, 107
1,8-Dichloro-10-methylanthracene, 35B, 918
B-Chloro-C-methylbarene, 30B, 411
1,2-Dichloronaphthalene, 29, 763
1,3-Dichloronaphthalene, 26, 740
2,6-Dichloronaphthalene, 29, 678
2,3-Dichloro-1,4-naphthoquinone, 26, 742
2,5-Dichloronitrobenzene, 30B, 418
1,4-Dichloronorbornane, 33B, 534
Dichloro(oxo)-2,2'-bipyridylethyoxyniobium(V), 35B, 766
Dichloro[oxydiethylene-bis(diphenylphosphine)]nickel(II), 35B, 849
Dichloropalladium II-norbornadiene, 26, 687
Dichloro(1,10-phenanthroline) zinc, 31B, 427
meso-3,3'-Di-(*p*-chlorophenyl) bi-3-phthalidyl, 34B, 65
2,2-Dichloro-3-phenylcyclobutenone, 29, 770; 30B, 86
1,6-Di-*o*-chlorophenyl-3,4-dimethylhexatriene, 31B, 35
1,6-Di-*p*-chlorophenyl-3,4-dimethylhexatriene, 31B, 35
syn-8,8-Dichloro-4-phenyl-3,5-dioxabicyclo[5,1,0]octane, 35B, 917
Di-*p*-chlorophenyl hydrogen phosphate, 29, 663
1,3-Di-*p*-chlorophenyl-2-triethylcarbinyl-4-ethyl-5,5-diethyl-1,3-diaza-2,4-diborolidine, 35B, 435
Di-*m*-chlorophenylurea, 33B, 545
2,2-Dichloropropane, 35B, 908

Dichloro(pyridine-2-aldehyde 2-pyridylhydrazone) cobalt(II), 31B, 425
Dichloroquinaldine, 30B, 423
Dichloro-2,2',2"-terpyridylcobalt(II), 35B, 703
trans-Dichlorotetraaquochromium(III) chloride dihydrate, 30A, 280
Dichloro-1,4,8,11-tetra-azacyclotetradecanenickel(II), 30B, 393
Dichloro(tetrahydrogenethylenediaminetetra-acetato)palladium(II) pentahydrate, 35B, 697
trans-Dichlorotetrakis(ethylenethiourea)nickel(II), 32B, 410
trans-Dichlorotetrakisthioureacobalt(II), 33B, 548
Dichlorotetrakisthioureanickel(II), 28, 623
Dichlorotetrakis(trimethylenethiourea)nickel(II), 33B, 492
Dichloro(1,3,5,7-tetramethyl-4,8,9-trioxabicyclo(3.3.1)nona-2,6-diene)platinum(II), 34B, 468
Dichloro(1,7,10,16-tetraoxa-4,13-diazacycloctadecane)copper(II), 35B, 919
2,2-Dichloro-4,4,6,6-tetraphenylcyclotriphosphazatriene, 31B, 307
trans-Dichlorotetrapyridine rhodium(III) hydrogen dinitrate, 32B, 497
trans-2,3-Dichloro-1,4-thioxane, 32B, 199
2,2'-Dichlorotrimethylene sulphite, 33B, 147
1,3-Dichloro-2,4,6-trinitrobenzene, 27, 902; 32B, 90
Di-μ-chlorotris(*trans*-cyclooctene)-dicopper(I), 35B, 581
Dichlorotris(tetrahydrofuran)-*p*-tolylchromium(III), 31B, 444; 32B, 448
Dichlorotris(triphenylphosphine)-dicopper(I)-benzene, 35B, 920
Dichlorotris(triphenylphosphine)-ruthenium(II), 30B, 390
Dichloro-(1,3,5-trithian) mercury (II) and derivatives, 31B, 447
2,5-Dichloro-1,4-xylene, 33B, 542
4,5-Dichloro-1,2-xylene, 33B, 542
Dichotine hydrobromide, 35B, 920
Dicinnamyl disulphide, 29, 663
Dickite, 26, 504
Dicobalt octacarbonyl, 29, 241
Dicopper octafluorozirconate dodecahydrate, 32A, 508
Dicoumarol, 33B, 552
Dictyoploca japonica, 26, 827
Dicyandiamide, 29, 765; 31B, 41
trans-Dicyano-bis(ethylenediamine)-cobalt(III) chloride, 35B, 592, 921
β,β'-Dicyanodiethyl sulphide, 28, 671
Dicyano-dispiro[2.4.2.0.]dec-5-ene, 35B, 907
8,8-Dicyanoheptafulvene, 30B, 399; 31B, 91
Dicyano-5,7,7,12,12,14-hexamethyl-1,4,8,11-tetra-azacyclotetradeca-4,14-dienecobalt(III) perchlorate, 35B, 712
10-Dicyanomethyleneanthrone, 32B, 131
9-Dicyanomethylene-2,7-dinitrofluorene, 33B, 91
9-Dicyanomethylene-2,4,7-trinitrofluorene, 32B, 137
2,4-Dicyanopentane, 31B, 42
trans-αβ-Dicyanostilbene, 26, 730
cis-Dicyanotetra(methyl isocyanide) iron(II)-chloroform, 33B, 326
Dicyanotris(phenyldiethoxyphosphine)nickel(II), 34B, 610
Dicyanotris(phenyldimethylphosphine)nickel(II), 34B, 609
Dicyanotrisulphane, 29, 291
3,6-*spiro*-Dicycloheptylidene-1,2,4,5-tetraoxacyclohexane, 32B, 180
Di[cyclohexane-1,2-dioximato(1-)]diimidazoleiron(II), 29, 554
Dicyclohexylthiophosphinic acid, 29, 762
Di-π-cyclooctatetraene-uranium, 34B, 470
3,6-*spiro*-Dicyclooctylidene-1,2,4,5-tetraoxacyclohexane, 32B, 180
Dicyclopentadiene-platinum chloride, 26, 689
Dicyclopentadienylberyllium, 28, 672; 33B, 536
Dicyclopentadienyl(ethynylbenzene)-dinickel, 30B, 431, 433
Dicyclopentadienyl(1-hexyne)dinickel, 30B, 431
Dicyclopentadienyl-lead, 31B, 332
Dicyclopentadienyl mercury, 30B, 424
Dicyclopentadienylnickel, 35B, 902
Di-cyclopentadienyl nickel dimethyl acetylene dicarboxylate, 28, 606
Dicyclopentadienyltitanium diethylaluminum dimer, 32B, 362
Dicyclopentadienyltricarbonylrhodium, 32B, 352
Dicyclopentadienylzirconium dichloride, 35B, 901
Dicyclopentamethylenethiuram monosulphide, 35B, 170
4,6-Dideoxy-4-(*N*,*N*-dimethylamino)-α-D-talopyranoside methiodide, 35B, 264
3,7-Dideoxy-7-iodoheptulosonate methyl ester, 30B, 433
2,3-Di(3,4-dihydroxybenzyl) butane, 34B, 343
Di-μ-dimethylamido-bis(cyclopentadienylnitrosylchromium, 35B, 548

Di-μ-(4,4'-dimethylbenzophenoniminato) bis(tricarbonyl iron), 32B, 497
Di-(ω,ω'-(*N*,*N*'-dimethyl)ethylene-diamino)-*o*-tolymercury, 33B, 328
Di-μ-dimethylphosphidobis(tricarbonyl) triethylphosphinemolybdenum, 32B, 469
Di-μ-diphenylphosphidobis(tricarbonyl nickel), 32B, 475
Di-μ-Diphenylphosphinatoacetylacetonatochromium(III), 30B, 333
1,3':1,3'-Diepoxy-3,3'-diphenyl-2,2'-bi-indolinyl, 33B, 547
1,8-Diethoxy-3α,4,7,7α-tetrahydro-4,7-phosphinidene-phosphindole-1,8-dioxide, 34B, 385
Diethoxytitanium dichloride, 33B, 454
Diethylacetylene metal carbonyls, 33B, 552
N-β-Diethylamineethyl-5-chlorosalicylaldimine cobalt(II) complex, 30B, 398
N-β-Diethylamineethyl-t-chlorosalicylaldimine nickel(II) complex, 30B, 398
Diethylamine hydrate, 32B, 288
Diethylaminoborondifluoride, 35B, 424
(2-Diethylaminoethanolato)copper(II) chloride, 35B, 909
N-β-Diethylaminoethyl-5-chlorosalicylaldimine (nickel(II) complex), 31B, 400
2-Diethylaminoethyl *p*-methoxybenzoate hydrochloride, 35B, 50
Diethylaminoethyl succinate diiodomethylate, 35B, 906
Diethylammonium and piperidinium tetrakis(benzoylacetonato)-europate(III) 35B, 909
Diethylammonium tetracyanopalladate(II), 33B, 546
5,5'-Diethylbarbituric acid, 34B, 216
4-Diethylcarbamoyl-1-cyclohexene-5-carboxylic acid, 35B, 111
trans-15,16-Diethyldihydropyrene, 32B, 144
Diethyldithiocarbamate of copper, 28, 687
Diethyldithiocarbamate of zinc, 28, 687
cis-Diethylenediaminediacetatocobalt(III) perchlorate, 35B, 906
trans-Diethylenediaminediacetatocobalt(III) perchlorate, 35B, 906
Di-(ω,ω'-ethylenedioxy)-*o*-tolylmercury, 33B, 328
(Diethylenetriamine)chromium tricarbonyl, 31B, 376
cis-(Diethylenetriamine)molybdenum tricarbonyl, 30B, 355
Diethylether-monobromodichloro-methane, 29, 450
Diethylmagnesium, 30B, 284
Diethyl(salicylaldehydato)thallium (III), 32B, 329
Diethylstilbestrol, 35B, 332
Diethylstilbesterol monohydrate, 34B, 77
Diethylstilbestrol monohydrate methanol solvate, 34B, 76
1,3-Diethyl-2-thiobarbituric acid, 33B, 551
Di-μ-ethylthiolatobis(ethyltrithiocarbonato)dinickel(II), 35B, 916
1,1'-Diferrocenylferrocene, 27, 862
Diferrocenyl ketone, 31B, 359
Diferrocenyls, 27, 860
1,5-Difluoroanthraquinone, 26, 751; 32B, 126
cis-Difluoroethylene, 26, 544; 28, 664
Difluoro[1,3-bis(*p*-bromophenyl)-1,3-propanedionato]-boron, 30B, 430
Difluorobromine(III)hexafluoroantimonate(V), 34A, 198
Difluorochlorine(III) hexafluoroantimonate(V), 35A, 178
α,α'-Difluorostilbenes, 29, 763, 770
3,5-Difluoro-1,3,5-trioxo-1-phenylcyclotriazathiane, 35B, 234
cis-1,2-Difluorovinylpentacarbonylmanganese, 32B, 339
Digadolinium cobalt manganese hexaoxide, 34A, 292
Digenite, 28, 20
Digitoxogenin, 34B, 298
Diglycine barium chloride monohydrate, 28, 670
Diglycine nitrate, 28, 687; 33B, 183
Dihaloanthraquinones, 26, 751; 28, 431
1,5-Dihaloanthraquinones, 28, 431
cis-1,2-Dihalobenzocyclobutenes, 33B, 90
Dihalogeno-cholestanes, 27, 1037
2,2'-Dihalogenodibenzoyl peroxides, 35B, 91
trans-Dihalogenotetrammineplatinum salts, 35B, 910
Dihistaminocopper(II) perchlorate, 35B, 662
Dihydrazinesilver nitrate, 35B, 924
Dihydrazinium sulphate, 30A, 366
Dihydridodi-π-cyclopentadienylmolybdenum, 27, 849; 30B, 311; 31B, 352
Di-μ-hydrido-diphenylsiliconbis-(tetracarbonylrhenium), 35B, 871
Dihydroaflatoxin B_1, 29, 772
7,7'-Dihydro-β-carotene, 29, 497
5,10-Dihydro-5,10-dimethylarsanthren, 33B, 546
Dihydro-β-erythroidine hydrobromide, 28, 546
Dihydrofomannosin *p*-bromobenzoylurethane, 33B, 213
Dihydrogedun-3β-yl iodoacetate, 27, 1033

Dihydroheliangine monochloroacetate, 31B, 264
μ-(π,π-1,6-Dihydroheptalene)-bis(tricarbonyl molybdenum), 34B, 471
1,3-Dihydro-1-hydroxy-3-oxo-1,3-benziodoxole, 30B, 157
9,10-Dihyro-9-hydroxy-9-phosphaphenanthrene-9-oxide, 27, 919
Dihydrolycorine hydrobromide, 33B, 220
Dihydromalvalic acid, 26, 831; 28, 375
trans-3,5-Dihydro-3-methyl-4,1-benzothiazepin-2(1H)-one-4-oxide, 34B, 211
2,3-Dihydro-2,3-methylene-1,4-naphthaquinone, 27, 906
9,10-Dihydrophenanthrenechromium tricarbonyl, 33B, 366
1,4-Dihydropyridine dimers, 35B, 909
5,10-Dihydrosilanthrene, 30B, 426
Dihydrositsirikine *p*-bromobenzoate, 29, 764
Dihydroteleocidin B monobromoacetate, 31B, 194
2,5-Dihydrothiophene-1,1-dioxide, 32B, 65
Dihydrothymidine, 35B, 290
Dihydrothymine, 33B, 154
3,4-Dihydro-2,4,6-triphenyl-5-tetrazin-1(2H)-yl free radical, 34B, 56
Dihydrouracil, 33B, 548; 35B, 249
Di-μ-hydroxobis (dimethylamine) copper (II) sulphate monohydrate, 31B, 421
Di-μ-hydroxo-bis-(*N*,*N*,*N'*,*N'*-tetramethylethylenediamine)dicopper(II) bromide, 35B, 700
Di-μ-hydroxo-tetrakis-μ-hydroxobis (tetramminechromium(III)bis(diamminechromium(III)), 33A, 206
Di-μ-hydroxotetrakis-μ-hydroxobis-(tetraamminechromium(III)) bis (diamminechromium(III)) bromine dihydrate, 35A, 399
Di-μ-hydroxobistetramminedicobalt(III) chloride tetrahydrate, 27, 655
1,4-Dihydroxyanthraquinone, 32B, 486
1-5-Dihydroxyanthraquinone, 28, 675; 29, 770
1,8-Dihydroxyanthraquinone, 27, 931
1,2-Dihydroxy-9,10-anthraquinone, 32B, 129
1,5-Dihydroxy-9,10-anthraquinone, 31B, 103; 32B, 129
5,5-Dihydroxy barbituric acid, 30B, 159
5,5-Dihydroxybarbituric acid trihydrate, 30B, 160
2,5-Dihydroxy-1,4-benzoquinone dipotassium salt, 33B, 555
(+)-2,2'-Dihydroxy-1,1'-binaphthalene-3,3'-dicarboxylic acid dimethyl ester bromobenzene solvate, 34B, 100
(+)-2,2'-Dihydroxy-1,1'-binaphthyl-3,3'-dicarboxylic acid dimethyl ester (bromobenzene solvate), 33B, 555
Di-μ-hydroxybis(bipyridyl)dicopper(II) sulphate pentahydrate, 35B, 915
Di-μ-hydroxy-bis{*N*,*N*,*N'*,*N'*-tetramethylethylenediaminecopper(II)} nitrate, 35B, 909
3β,17α-Dihydroxy-16β-bromo-5α-pregnan-11,20-dione, 34B, 296
3β,17α-Dihydroxy-21-bromo-5α-pregnan-11,20-dione, 33B, 194
o,*o'*-Dihydroxydiaryl-*trans*-azo-dyemolecule 2:1-chromium complex, 30B, 395
1,8-Dihydroxy-1,8-diazacyclotetradecane, 31B, 129
Dihydroxyfumaric acid dihydrate, 33B, 3
2,8-Dihydroxy-5,6,11,12,4*b*,10*b*-hexahydrochrysene, 28, 442
7,9-Dihydroxy-6α-methoxy-2-phenyl-*trans*-*m*-dioxano[5,4-e][1,4]-dioxepan, 26, 627
3,4-Dihydroxyphenylalanine(L-DOPA), 35B, 910
2,3-*cis*-3,4-*trans*-3,4-Dihydroxy-L-proline, 34B, 278; 35B, 307
4,4'-Dihydroxythiobenzophenone monohydrate, 27, 915; 29, 770; 30B, 82
Di-imidazole-zinc(II) dichloride, 31B, 421
3,5-Diiodoacenaphthene, 28, 438
Di-iodacetylene-1,4-diselenane complex, 31B, 270
Di-iodoacetylene-1,4-dithiane complex, 31B, 270
1,5-Diiodoanthraquinone, 26, 751; 28, 432; 32B, 124
B-Diiodobarene, 30B, 410
2,5-Di-(*p*-iodobenzal-)cyclopentanone, 26, 734
Diiodobis(dimethyl-*o*-methylthiophenylarsine) palladium(II), 35B, 839
trans-Di-iodobis(dimethylphenylphosphine)palladium(II), 30B, 393; 33B, 507
Diiodobis(1,1-dimethylurea)zinc, 30B, 420
Diiodobis(ethylenethiourea) tellurium (II), 26, 578
Di-iodo(bis(*o*-methylthiophenyl) phenylphosphine)nickel(II), 34B, 588
Di-iodo-bis(*o*-phenylene-bis(dimethylarsine))gold(III) iodide, 34B, 606
Diiodo-bis-triphenylphosphine nickel(II), 28, 680

Di-iodocarbomethoxycarbonyl(2,2'-bipyridyl)iridium, 34B, 430
p,p-Diiododiphenylamine, 26, 728
Diiododi-)o-phenylenebisdimethylarsine) nickel(II), 27, 869; 29, 576
Di-iododi-(o-phenylenebisdimethylarsine) palladium (II), 27, 869
Di-iododi-(o-phenylenebisdimethylarsine) platinum (II), 27, 869
Diiodo-dithiourea-tellurium complex, 26, 576
Diiodomethylarsine, 28, 585
B-Diiodoneobarene, 30B, 412
B,B'-Diiodoneocarborane, 33B, 530
trans-Diiodotetra(ethyleneimine) rhodium(III) iodide, 34B, 550
Di-μ-iodotetramethyl-μ-(dimethylsulphide)-bis(dimethylsulphide)-dirhodium(III), 33B, 553
Di-iodo-L-tyrosine dihydrate, 32B, 230
Di-iron hexacarbonyl diphenylacetylene complex, 32B, 348
Di-iron-sulphonyl-octacarbonyl, 32A, 400
2,6-Di(iron tricarbonyl)-3,5-dimethylhepta-2,5-diene-4-one, 34B, 422
1,5-Di(iron tricarbonyl)-3-methylene-penta-1,4-diene (dark-red isomer), 30B, 303
1,5-Di(iron tricarbonyl)-3-methylene-penta-1,4-diene (red-orange isomer), 30B, 296
3-Di-isopropylamino-benzisothiazol-1,1-dioxide, 34B, 204
1,2:3,4-Di-O-isopropylidene-5-O-chloroacetyl-α-D-glucoseptanose, 35B, 912
Di-isopropylphosphoryl trypsin, 28, 680
Diisothiocyanatobis(aniline)zinc, 30B, 428
cis-Diisothiocyanato-bis(2,2'-bipyridyl)iron(II), 35B, 917
Diisothiocyanatobis(ethylenediamine)-cadmium(II), 35B, 922
4,4'-Diketo-β-carotene, 33B, 84
Diketo-cyclobutenediol, 32B, 486
Diketopiperazine, 26, 767
1,1'-[α,α-Diketo-γ-(2-thienyl)-pentamethylene]ferrocene, 33B, 543
Dilactic acid, 35B, 922
Dilactic acid amide, 35B, 922
Dilauroyl, 27, 783
Dilithium diferrous molybdate, 35A, 251
Dilithium metagermanate, 33A, 503
Dilituric acid, 28, 499
Dilituric acid trihydrate, 29, 719
Dimagnesium nitride fluoride, 35A, 151
Dimanganese decacarbonyl, 34A, 162
Dimanganesedivanadate, 32A, 272
Dimedone-β-nitrostyrene condensation product, 35B, 923
2,2'-Dimercaptodiethylsulphide-nickel(II), 34B, 591
2,2'-Dimercaptodiethylsulphide palladium(II) trimer, 34B, 594
Dimeric acid product of 1-methyl-1,4-dihydronicotinamide, 32B, 168
Dimeric bis(acetylacetonato) aquocobalt (II), 31B, 384
Dimeric bis(N-methylsalicylaldiminato) zinc(II), 31B, 389
Dimeric 2-carboxyethyl-π-allylnickel bromide, 32B, 345
Dimeric cycloctanone peroxide, 30B, 395
Dimeric phosphinobovine derivatives, 26, 72
trans-Dimer of nitroisobutane, 26, 608
trans-Dimesityl-bis-(diethylphenyl-phosphine)cobalt(II), 28, 660
Dimethane thiosulphonato bis(ethylene-thiourea) tellurium II, 26, 582
Dimethane thiosulphonato bis(tetra-methyl thiourea) tellurium II, 26, 583
Dimethane thiosulphonato-dithiourea tellurium II, 26, 582
p,p'-Dimethoxybenzophenone, 27, 909
3β,6β-Dimethoxy-5β,19-cyclo-androstan-17-one N-acetyl-p-bromobenzenesulphonylhydrazone, 33B, 199
2,6-Dimethoxy-3,5-dichloro-4-methyl-benzoic acid, 27, 899
4,4'-Dimethoxy-α,β-diethylstilbene, 30B, 393
Dimethoxyfuranochrome, kellin, 27, 1008
1,6-Dimethoxy phenazine, 32B, 166
Dimethoxytriphenylantimony, 33B, 290
1,1-Dimethoxy-2,4,6-triphenylphosphorin, 34B, 392
erythro(±)-α,β-Dimethylacetyl-choline iodide, 35B, 322
threo(±)-α,β-Dimethylacetylcholine iodide, 35B, 322
meso-β-β'-Dimethyladipic acid, 31B, 12; 32B, 18
1,2-Dimethylallyl-dicyclopentadienyl titanium(III), 32B, 346
π-1,3-Dimethylallyl palladium chloride dimer, 32B, 493
Dimethylaluminumoxybenzylidene-aniline dimer, 33B, 550
Dimethylamine, 33B, 530, 539
Dimethylamine copper trichloride, 31B, 29
Dimethylamine-vanadium trichloride, 27, 828
N-(ω-Dimethylaminoalkyl)-phthalimide methiodides, 26, 717
6-Dimethylamino-5-aza-azulene, 34B, 149

Dimethylaminoborine dimer, 27, 771
Dimethylaminoborine trimer, 26, 618
Dimethylaminoboron dichloride dimer, 28, 580
Dimethylaminoboron difluoride, 31B, 289
6-Dimethylamino-5,7-diaza-azulene, 35B, 187
Dimethylaminodichloroborane, 35B, 890
Dimethylamino-difluorophosphine, 34B, 376
Dimethylaminodifluorophosphine-tetraborane, 34B, 377
Dimethylamino-dimethylaluminum dimer, 35B, 479
(Dimethylamino)dimethylborane, 35B, 425
1-*p*-(2-Dimethylaminoethoxyphenyl)-1,2-*cis*-diphenylbut-1-ene hydrobromide, 33B, 546; 35B, 97
2-[(2-Dimethylaminoethyl)-2-thenylamino]-pyridine hydrochloride, 35B, 920
Dimethylamino(methyl)carbene-pentacarbonyl chromium, 34B, 423
10-(2-Dimethylaminopropyl) phenothiazine, 33B, 551
10-(2-Dimethylaminopropyl)-phenothiazine hydrobromide, 33B, 551
Bis-dimethylamino sulphone, 28, 399
Dimethylammonium chloride, 33B, 24
N-Dimethylaniline, 30B, 403
2-*p*-Dimethylanilino-4-phenylthio-thiophthene, 35B, 910
Dimethylarsino dimethyldithioarsinate ("cacodyl disulphide"), 29, 539
8,8-Dimethyl-8-azoniabicyclo[5.1.0] octane iodide, 29, 506
9,9-Dimethyl-9-azonia-bicyclo-[6.1.0] nonane iodide, 28, 471
cis-13, 13-Dimethyl-13-azoniabicyclo [10,1,0]tridecane iodide, 32B, 163
trans-13,13-Dimethyl-13-azoniabicyclo [10.1.0]tridecane iodide, 28, 472
2,3-Dimethyl-4,6-benzal-methyl-β-idoside, 30B, 429
9,10-Dimethyl-1,2-benzanthracene, 29, 696
5,6-Dimethylbenzimidazolylcobamide coenzyme, 26, 807
2,6-Dimethylbenzoic acid, 28, 672; 32B, 84
1,12-Dimethylbenzo[*c*]phenanthrene, 28, 444
2,3-Dimethyl-1,4-benzoquinone, 32B, 108
2,5-Dimethyl-1,4-benzoquinone, 29, 645; 32B, 109
2,6-Dimethyl-1,4-benzoquinone, 32B, 110
N,N'-Dimethyl-4,4'-bipyridinium tetrachlorocuprate, 34B, 138
Dimethylbis(8-hydroxyquinolate) tin(IV), 32B, 333
Dimethylbis(quinuclidine)magnesium, 35B, 709
N-3-Dimethyl-4-bromo-2,6-dinitroaniline, 32B, 96
2,3-Dimethylbutadiene, 33B, 534
[π-(2,3-Dimethylbutadiene)Co(CO)$_2$]$_2$, 35B, 913
2,3-Dimethyl-2,3-butanediol hexahydrate, 35B, 29
6,9-Dimethylcarborane, 31B, 291
anti-2,6-Dimethyl-4-chloro-N-methyl-benzaldoxime, 34B, 40
4,4'-Dimethyl-α,α'-cinnamaldazine, 35B, 918
2,6-Dimethyl-4-(3'-cyano-cyclo-pentadienylidene) thiopyran, 34B, 186
7,7-Dimethylcycloheptatriene-3-carboxylic acid *p*-bromophenacyl ester, 31B, 262
Dimethyl 1,4-cyclohexadien-2,5-diol-1,4-dicarboxylate, 28, 427
5,5-Dimethyl-1,3-cyclohexanedione, 28, 672
1,5-Dimethyl-5-cyclohexenyl-barbituric acid, 35B, 918
Dimethyl-π-cyclopentadienyl(methyl-cyclopentadiene)rhenium, 30B, 392; 32B, 354
4a,8a-Dimethyl-*trans*-decal-1-yl *p*-bromobenzoate, 35B, 920
N,N'-Dimethyl-13,14-diazatricyclo(6,4,1.1(2,7)tetradeca-3,5,9,11-tetraene dihydrobromide, 32B, 488
p-Dimethyldiazoaminobenzene, 29, 662
2,5-Dimethyl-3,6-dibromo-1,3a,4,6a-tetraazapentalene, 28, 457
Dimethyl-1,6-dicarba-*closo*-decaborane(10), 35B, 426
Dimethyl-1,6-dicarbaclovonona-borane(9), 33B, 263
1,7-Dimethyl-1,7-dicarbaclovoocta-borane(8), 33B, 262
2,3-Dimethyl-2,3-dicarbahexaborane, 29, 526
6,8-Dimethyl-6,8-dicarbanonaborane, 32B, 293
1,2-Dimethyl-1,2-dicarbollide complex of nickel, 35B, 430
1,6-Dimethyl-8,8-dicyanohepta-fulvene, 33B, 555
2,5-Dimethyl-7,7-dicyanonorcaradiene, 31B, 104
p,p-Dimethyl-α,α'-difluorostilbene, 33B, 74
Dimethyl 3,6-dihydro-2,5-dihydroxy terephthalate, 28, 427
N,N'-Dimethyl-4,4'-dipyridylium tetrachlorocobaltate, 34B, 131
N,N'-Dimethyl-4,4'-dipyridylium tetrachlorodicuprate, 34B, 131
N,N'-Dimethyl-4,4'-dipyridylium tetrachloropalladite, 34B, 131

N,N'-Dimethyl-4,4'-dipyridinium tetracyanonickelate(II), 34B, 138
3,7-Dimethyl-1,5-dithia-3,7-diaza-cyclo-octane, 32B, 488
NN-Dimethyldithiocarbamatotri-methylstannane, 35B, 793
2,5-Dimethyl-dithio-furophthene, 26, 761
Dimethyldodecylamine oxide-deuterium oxide system, 33B, 529
3,4-Dimethylenecyclobutene, 33B, 533
trans-(1,4),(5,8)-Dimethylene-*cis*, *anti*,*cis*-perhydroanthraquinone, 30B, 119
Dimethylene triurea, 33B, 544
Dimethyl ester of anthraquinone-2,6-di-carboxylic acid, 28, 675
Dimethyl ether, 28, 664
DL-4β,10β-Dimethyl-6-ethylamino-4-hydroxycarbonyl-2,3,5β,10-tetra-hydrophenanthr-1-one lactam, 35B, 351
1,12-Dimethyl(1.1)ferrocenophane, 32B, 494
Dimethylformamide, 27, 759
Dimethylformamide-antimony(V) chloride complex, 31B, 324
Dimethylformamide-lithium bromide, 27, 762
Dimethylformamide-lithium chloride, 27, 762
Dimethylformamide-lithium iodide, 27, 762
Dimethylformamide NaI complex, 27, 760
Dimethylformamide-sodium iodide, 27, 762
Dimethylfulvene, 35B, 899
α,α'-Dimethylglutaric acid, 29, 762, 765
meso-α,α'-Dimethylglutaric acid, 32B, 16
Dimethylglyoximatocopper(II), 35B, 760
Dimethylglyoxime, 26, 596
Dimethylgold(III) hydroxide, 33B, 321
3,4-Dimethyl-2,4-hexadienes, 35B, 900
1,4-Dimethylhexahydro-*S*-tetrazine, 35B, 913
5,5-Dimethylhomochroman, 27, 988
5,5-Dimethylhomochroman carboxylic acid, 27, 988
1,1-Dimethyl hydrazine fluoborate, 26, 594
O,O-Dimethyl-(1-hydroxy-2,2,2-trichloroethyl) phosphonate, 34B, 378
1,3-Dimethyl-2(3H)-imidizolethione, 35B, 912
Di-μ_3-methylimido-tris(tricarbonyl-iron), 34B, 564
N,*N*-Dimethylisopropylideniminium perchlorate, 31B, 30
1-*a*-Dimethyl-7-isopropyl-perhydro-phenanthrene, 27, 1036
Dimethylmalonic acid, 31B, 6
4,12-Dimethyl(2,2)metacyclophane, 27, 918
4,5-Dimethyl-2-methoxy-2-oxo-1,3,2-dioxaphospholene, 32B, 220
Dimethyl(methylthiol) aluminum, 34B, 403
Dimethyl(methylamido)aluminum trimers, 35B, 917
2,3-Dimethyl-2-(3-methylbenzothiazolin-2-ylidene)methyl benzothiazoline, 32B, 488
2,2-Dimethyl-3-methyl-3-methoxy butyl mercury iodide, 27, 824
2,2-Dimethyl-3-methyl-3-methoxy butyl mercury thiocyanate, 27, 824
cis-*NN*-Dimethyl-9-(3-(4-methyl-1-piperazinyl) propylidene) thioxan-threne-2-sulphonamide, 32B, 488
S,S-Dimethyl-N-(methylsulphonyl)sul-philimine, 32B, 38, 485
Dimethyl-*trans*-*trans*-muconate, 32B, 17
1,5-Dimethylnaphthalene, 30B, 96
1,8-Dimethylnaphthalene, 28, 674
2,3-Dimethyl-1,4-naphthoquinone, 31B, 102; 32B, 123
N,N-Dimethyl-*p*-nitroaniline, 30B, 68
Dimethylnitrosoamine, 33B, 529
Dimethylnitrosaminecopper(II) chloride, 33B, 545; 34B, 573
trans-2,7-Dimethylocta-2,4,6-triene-1:4,5:8-diolide, 31B, 140
cis-Dimethylpentaborane, 31B, 287
N,*N*'-Dimethyl-*trans*-perhydro-cyclo-penta-[1,2-*c*:3,4-*c*']dipyrrol-bis-methiodide, 33B, 125
2,7-Dimethylperhydropyrene, 29, 502
2,9-Dimethyl-1,10-phenanthroline-MCl_2 complexes, 33B, 543
2,9-Dimethyl-1,10-phenanthroline-zinc chloride complex, 32B, 496
2,3-Dimethylphenol, 30B, 421; 31B, 73; 32B, 102
2,5-Dimethylphenol, 30B, 74, 421
2,6-Dimethylphenol, 30B, 421
3,4-Dimethylphenol, 30B, 421
3,5-Dimethylphenol, 30B, 421
Dimethylphenylaluminum dimer, 35B, 913
5,6-Dimethyl-4-phenylcarbam-oylcholine bromide, 35B, 918
5,7-Dimethyl-2-phenylcyclopent-[*c*,*d*]azulene, 35B, 130
N,*N*-Dimethyl-*p*-phenylenediamine bromide, 33B, 60
2,6-Dimethyl-4-phenylphosphobenzene, 33B, 545
2,6-Dimethyl-4-phenylphosphorin, 35B, 454

dl-B-β-1,3-Dimethyl-4-phenyl-4-propionoxypiperidine derivatives, 28, 462
1,1-Dimethyl-3-phenylpyrazolium-5-oxide, 35B, 161
Dimethylphenylsulphonium perchlorate, 29, 485
Di-(3 methyl-1-phenyl-5-*p*-tolylformazyl) nickel(II), 32B, 443
Di-μ-dimethylphosphido-bis(iodotricarbonyliron), 33B, 501
μ-(Dimethylphosphido)hydridodi-π-cyclopentadienyltetracarbonyldimolybdenum, 30B, 312
Dimethylphosphinic acid, 32B, 304
Dimethylphosphinoborine tetramer, 27, 989; 29, 766
N,N'-Dimethylpiperazine, 28, 666
trans-3,6-Dimethylpiperazine-2,5-dione, 34B, 129; 35B, 318
1-[N-Dimethylpiperidinium-(4')-]-4-benzyl-5-phenyl-pyrazole-3-one iodide, 27, 961
2,2-Dimethyl-1,3-propanediol, 26, 559
2,6-Dimethyl-γ-pyrone hydrobromide monohydrate, 30B, 145
Dimethylsilylamine pentamer, 32B, 301
Di-μ-dimethylstannylene-bis(tricarbonyl(trimethyl-stannyl))-ruthenium, 34B, 411
α,α'-Dimethylsuccinic acid, 29, 762, 765; 32B, 14
Dimethylsulphide, 26, 586
Dimethylsulphide-dodecahydrodecaborane, 27, 772
Dimethyl sulphone, 28, 406, 35B, 890, 917
Dimethylsulphone-5-bromouridine 1:1 complex, 32B, 272
Dimethylsulphone di-imine, 32B, 485
Dimethyl sulphoxide, 29, 766; 31B, 55, 56
4,4-Diphenylcyclohexanone, 32A, 93
μ-Diphenylamino-μ-methyl-tetramethyl-dialuminum, 34B, 406
Dipentene platinum(II) chloride,30B,327
Dioxygenyl hexafluoroplatinate(V), 31A, 92
4,5-Dioxo-2-thioxo-1,3-dithiolane, 34B, 179
Dioxodi-8-quinolinolato-8-quinolinol-uranium(VI)-chloroform, 29, 769
2-Dimethylsulphuranylidenemalononitrile, 34B, 27
β-Dimethyltellurium di-iodide, 32B, 337
Dimethyltetrathionitrosyl-nickel(II), 28, 661
1,4-Dimethyltetrazadieneiron tricarbonyl, 33B, 549
α,γ-Dimethyltetronic acid, 34B, 172
Dimethyl thallium perchlorate 1,10-phenanthroline complex, 32B, 492
3,5-Dimethyl-1-thiaindenesulphone, 35B, 907
2,5-Dimethyl-2,5-*endo*-thio-1,4-dithiane, 32B, 183
1,3-Dimethyl-2-thio-4-oxo-1,2,3,4,6,7,8-heptahydro-pyrimido (1,2-A) (1,3,5)triazine, 32B, 487
2,5-Dimethyl-thio-thiophthene, 27, 473
1,3-Dimethylthymine photodimer A, 35B, 257
Dimethyltin chloride *N,N*-dimethyldithiocarbamate, 35B, 488
Dimethyltin dichloride, 35B, 485
Dimethyltin difluoride, 31B, 328
Dimethyltin diisothiocyanate, 35B, 486
1,1,-Dimethyl-2,4,6-triphenylphosphorin, 35B, 465
μ-Dimethylurenylene-bis(tricarbonyliron), 33B, 423
α-Dimorphite, 35A, 140
Dimyristoyl, 27, 783
1,4-Dinaphthoquinone, 26, 748
2,2'-Di(1,4-naphthaquinone), 34B, 103
N-Diniobium pentoxide, 32A, 257
trans-Dinitratobis(trimethylenediamine)-cobalt(III) nitrate, 35B, 678
Dinitrato-(*NNN'N'*-tetramethyl-*o*-phenylenediamine)nickel(II), 35B, 762
Dinitritobis-(NN-dimethylethylene-diamine) nickel(II), 29, 768
Dinitroalkane alkali salts, 29, 765
m-Dinitrobenzene, 26, 695; 31B, 67
2-(2',4'-Dinitrobenzyl)pyridine, 33B, 119
2,2'-Dinitrobiphenyl, 27, 908
Dinitro-bis(ethylenediamine)nickel-(II), 29, 768; 34B, 481
(+)$_{589}$-*cis*-Dinitrobis[(-)$_{589}$-1,2-propylenediamine]cobalt(III), 35B, 672
1,5-Dinitro-4,8-dihydroxyanthraquinone, 32B, 127
Dinitro(2,9-dimethyl-1,10-phenanthroline)palladium(II), 35B, 920
(-)$_{546}$-*cis*-β-Dinitro-(L-3,8-dimethyltriethylenetetramine)cobalt(III) perchlorate, 35B, 684
4,4'-Dinitrodiphenyl, 26, 830; 28, 419
N,N'-Dinitroethylenediamine, 33B, 33
Dinitrogen tetroxide, 28, 53; 31A, 258
Dinitrogen tetroxide-1,4-dioxane, 27, 963
Dinitrogen tetroxide, 1,4-dioxan (1:1 complex), 30B, 246
Dinitrogen trioxide, 34A, 240
Dinitromethane salts, 33B, 556
3,5-Dinitro-4-methylbenzoic acid, 34B, 60

1,8-Dinitronaphthalene, 28, 763, 770; 30B, 98
1-(2',4'-Dinitrophenyl)-4-bromopyrazole, 34B, 126
1-(2',4'-Dinitrophenyl)-4-chloropyrazole, 35B, 158
2,4-Dinitrophenylhydrazone ketone, 26, 701
o-Dinitrosobenzene, 27, 889
2,2'-Dinitrosodiphenyl, 27, 889
Dinitrosylbis(triphenylphosphine)-iridium perchlorate, 35B, 856
Dinitrosylcobalt chloride, 32A, 195
Dinitrosyl metal iodides, 34A, 183
p-Dinitrotetrachlorobenzene, 33B, 556
Dinitro-(NNN'N'-tetramethylenediamine) nickel(II), 30B, 394
trans-Dinitrotetraminecobalt(III) acetate, 33B, 541
trans-Dinitrotetramminocopper, 26, 478
Dinuclear metal carbonyl anions, 35B, 506
Dioctadecyl adipate, 29, 762
Dioctahedral chlorite, 32A, 474
Diolefin of [2.2] paracyclophane, 28, 452
Diol-oligo-urethanes, 26, 637
Diopside, 27, 700
Di-L-ornithatocopper(II) dichloride dihydrate, 35B, 656
Diosgenin iodoacetate, 28, 685; 32B, 246
1,4-Dioxan, 28, 666
1,4-Dioxan germanium dichloride, 35B, 407
1,4-Dioxan-lithium chloride complex, 27, 761; 31B, 273
1,4-Dioxan lithium chloride monohydrate, 33B, 298
Dioxan-mercuric bromide, 35B, 918
1,4-Dioxan, oxalyl bromide (1:1 complex), 30B, 246
1,4-Dioxan, oxalyl chloride (1:1 complex), 30B, 246
Dioxocarbonyliodobis(triphenylphosphine)iridium dichloromethane, 32B, 479
Dioxo-dibromo-2,2'-bipyridylmolybdenum(VI), 34B, 551
Dioxodichlorobis(*N*,*N*-dimethylformamido)molybdenum(VI), 33B, 456
Dioxodi-8-quinolinolato-8-quinolinoluranium(VI) chloroform, 29, 769; 32B, 442
Dioxomolybdenum(VI) complexes, 35B, 919
Dioxomolybdenum(VI) diethyldithiocarbamate, 34B, 524
Dioxonium ethan-1,2-disulphonate, 35B, 40
4,5-Dioxo-2-thioxo-1,3-dithiolane, 33B, 556; 34B, 179
5,9-Dioxo-1,7,13-trioxa-4,10-diazacyclopentadecane, 35B, 917
cis-N-N'-Dioxyazomethane, 28, 404
Dioxygenyl hexafluoroplatinate, 27, 666; 31A, 92
Dipalmitoyl, 27, 783
Dipara-anthracene, 31B, 112
Di-μ-pentafluorobenzenethiolatobis-(pentafluorobenzenethiolato)-bis(triphenylphosphine)dipalladium (II), 35B, 830
Dipentene platinum chloride, 29, 768; 30B, 327
Diperoxoaquoethylenediaminechromium(IV) monohydrate, 30B, 328
Diperoxotriammine chromium(IV), 29, 302
Diphenantridylmethane, 30B, 398
1,2-Diphenoxyethane, 32B, 104
N,*N*-Diphenylacetamide, 33B, 9
(Diphenylacetylene)bis(cyclopentadienylnickel), 33B, 341
Diphenylacetylene iron carbonyl complex, 30B, 297
Diphenylacetylene-tri(iron)tricarbonyl complex, 31B, 347
Diphenyl aluminum-phenyl-*p*-bromophenylketimine dimer benzene solvate, 34B, 406
2-Diphenylaminoethyl-trimethylammonium iodide, 32B, 485
μ-Diphenylamino-μ-methyltetramethyldialuminum, 33B, 552; 34B, 406
Diphenyl arsenical bromide, 26, 728
Diphenyl arsenical chloride, 26, 728
Diphenyl arsenical iodide, 26, 728
Diphenyl arsenical oxide, 26, 728
Diphenyl-arsenonitrile tetramer, 32B, 492
Di(*o*-phenyl azophenyl)disulphide, 29, 770
p-Diphenyl benzene, 26, 734
2,6-Diphenyl-3-benzyl-2H-thiopyran-5-carboxaldehyde, 32B, 189
Diphenylbis(2,3,5,6-*tetrahapto*-norbornadienedicarbonylcobalt)tin, 35B, 921
2,6-Diphenyl-4-(4-bromophenyl)-*N*-(*p*-oxy-*m*,*m*'-diphenyl)-phenyl pyridinium betaine monoethanol solvate, 34B, 143
4,4-Diphenylcyclohexanone, 33B, 548; 34B, 93
N,*N*'-Diphenyl-1,5-diaminoanthraquinone, 32B, 132
N,*N*'-Diphenyl-1,8-diaminoanthraquinone, 32B, 133
Diphenyl disulphide, 34B, 48
2,5-Diphenyl-1,4-dithiin 1-oxide, 35B, 916
2,4-Diphenyl-dithiofurophthene, 34B, 189

2,4-Diphenyl-1,3-dithiolium perchlorate, 28, 670
Di-μ-*NN'*-*m*-phenylene-tetrakis(salicyclideneiminato)-dicopper(II) chloroform solvate, 35B, 624
Di(phenylethyl) mercury, 27, 864
Diphenylfluoro-*N*-methylphosphine imide, 35B, 454
Diphenylmagnesium dietherate, 29, 763
Diphenyl mercury, 27, 863; 29, 628; 33B, 537
Diphenyl methane, 26, 727
N,*N'*-Diphenyl-*p*-phenylenediamine iodide, 32B, 72
2,4-Diphenyl-5-(3'-phenyl-2'-thiobuten-4'-yl)isothiazolium iodide, 35B, 914
2-(Diphenylphosphine oxide)ethyl-diethyl-ammonium dichlorocuprate (I), 35B, 916
Diphenylphosphinetetracarbonyl-iron, 34B, 600
Diphenylphosphinic acid, 30B, 195
Diphenylphosphino-cyclopentadienyl-cobalt dimer, 32B, 364
Diphenylphosphino-cyclopentadienyl-nickel dimer, 32B, 364
Diphenyl-phosphonitrile tetramer, 32B, 492
1,1-Diphenylphosphonitrilic fluoride trimer, 32B, 491; 34B, 383
2,2-Diphenyl-1-picrylhydrazyl, 30B, 431
2,2-Diphenyl-1-picrylhydrazyl benzene solvate, 32B, 73
Diphenyl-Δ^2-pyrazoline, 33B, 110
Diphenylsilanediol, 33B, 543
2,2-Diphenyl-4,4,6,6-tetrachlorocyclotriphosphazatriene, 30B, 280
3,6-Diphenyl-1,2,4,5-tetraoxacyclohexane, 32B, 179
Di-μ-phenylthio-bis(cyclopentadienylcarbonyliron), 33B, 549
2,5-Diphenylthiodiazole, 30B, 156
trans-Di-μ-phenylthio-dinitrosylbis-(π-cyclopentadienyl)dichromium(I), 33B, 354
trihapto-1,2-Diphenylthioketocarbene-hexacarbonyldiiron, 35B, 920
2,4-Diphenyl-thiothiophthene, 31B, 151; 34B, 191
2,5-Diphenyl-thiothiophthene, 33B, 544
sym-Diphenylthiourea, 29, 770
Diphenyltin-*m*-xylene, 29, 769
Diphenyltrichlorostibine, 26, 729
Diphenyltrichlorostibine monohydrate, 32B, 321
1,5-Diphenyltricyclo[2,1,0,0^{2,5}]pent-3-yl *p*-bromobenzoate, 32B, 151
sym-Diphenylurea, 29, 762
μ-Diphenylureylene-bis(tricarbonyl iron), 32B, 437, 497
Diphosphorus hexathiodibromide, 30A, 355
Di(phosphoryl trichloride)-hexachloroditin(IV) di-μ-dichlorophosphate, 33B, 546; 34, 222
Dipotassium bis(glycylglycinato) cuprate(II) hexahydrate, 33B, 418
Dipotassium bis(trimethylenedinitramine) nickelate(II) tetrahydrate, 30B, 353
Dipotassium carbonate trihydrate, 32A, 414
Dipotassium 2,6-dinitro-benzene-1,4-disulphonate monohydrate, 32B, 486
Dipotassium dipyraphene, 30B, 434
Dipotassium enneahydridotechnate(VII), 29, 236
Dipotassium ethylenediamine tetra-acetate monohydrate, 34B, 20
Dipotassium ethylenetetracarboxylate, 30B, 31
Dipotassium fumarate dihydrate, 26, 563; 35B, 14
Dipotassium glucose-1-phosphate dihydrate, 30B, 191
Dipotassium heptafluoroprotactinate(V), 32A, 159
Dipotassium hexachlorotechnetate(IV), 32A, 192
Dipotassium hydrogen di-iodate chloride, 32A, 509
Dipotassium hydroxy bis chromate(VI) chromium(III), 35A, 389
Dipotassium lithium trimetaphosphate monohydrate, 27, 594
Dipotassium octachlorodirhenate(III) dihydrate, 30A, 283
Dipotassium pentachloromonohydroxy-technetate(IV), 32A, 192
Dipotassium phenyl phosphate sesquihydrate, 32B, 220
Dipotassium 1,1,2,2-tetranitro-ethanide, 32B, 485
Dipotassium tetranitroethide, 33A, 432, 33B, 34
Dipotassium trimolybdate, 31A, 154; 32A, 509; 33A, 333
Dipterocarpol, 27, 1036
Dipyridinecopper(II) sulphate dihydrate, 29, 608
Dipyridine copper(I) thiocyanate, 28, 672
Di-μ-(pyridine-N-oxide)-bis(dichloro-copper(II)), 32B, 446; 34B, 575
Dipyridine tetraacetatocopper(II), 26, 669
Dipyridinium di-μ-hydroperoxo-tetra-peroxodioxodimolybdate(IV), 33B, 549
Dipyridinium μ-oxo-tetraperoxodioxo-bis(aquo)dimolybdate(VI), 33B, 549
Dipyridinium(II) oxytetrachloro-selenate(IV), 35B, 493
o-Dipyridyldioxide perchlorate, 29, 770
1,2-Di-2-Pyridylethendiol-1,2, 30B, 132

Dipyrrolidyltrimethincyanin perchlorate, 35B, 924
Dirhenium heptoxide dihydrate, 33A, 269; 35A, 279
Dirubidium ethylenediaminetetra-acetate dihydrate, 35B, 26
Dirubidium salt of 2,5-dimethyl-1,3a,4,6a-tetraazapentalene-3,6-dicarboxylic acid dihydrate, 28, 456
Dirubidium 1,3a,4,6a-tetra-azapentalene-2,5-dicarboxylate dihydrate, 28, 671
N,N'-Disalicylidenbenzidine copper, 27, 873
NN'-Disalicylidene-ethylenediamine nickel (II), 26, 671
NN'-Disalicylidene-ethylenediaminezinc (II) monohydrate, 31B, 392
NN'-Disalicylidene-o-phenylenediamine nickel (II), 26, 670
dl-1,2-Diselenane,3,6-dicarboxylic acid, 29, 520
1,4-Diselenane iodine complex, 26, 771
1-4-Diselenane tetrachloride, 28, 686; 29, 521
1,4-Diselenane-tetraiodoethylene, 30B, 249
Diselenium tri-iron nonacarbonyl, 28, 70
α,α'-Diselenobisformamidinium dichloride, 35B, 38
1,4-Diselenocyanatobenzene, 35B, 81
1,2-Disilacenaphthene, 30B, 424
1,5-Disilapyracene, 30B, 427
1,6-Disilapyracene, 30B, 426
Disilazane, 34A, 155
Disilazanes, 33B, 548
Disilylmethane, 35B, 888
Disodium adenosine triphosphate, 33B, 541
Disodium dihydrogen silicate octahydrate, 31A, 215
Disodium dihydrogen silicate tetrahydrate, 31A, 216
Disodium dimolybdate, 32A, 298
Disodium di-μ-oxo bis(oxocysteinato molybdenum(V)) pentahydrate, 34B, 540
Disodium dipyraphene, 30B, 435
Disodium β-glycerolphosphate pentahydrate, 31B, 176
Disodium glycylglycylglycylglycino-cuprate(II) decahydrate, 29, 768; 30B, 349
Disodium hydrogen metaphosphate, 27, 596
Disodium hydrogen orthoarsenate heptahydrate, 35A, 335
Disodium hydrogen orthophosphate heptahydrate, 35A, 335
Disodium hydroxy bis chromate(VI) chromium(III), 35A, 390
Disodium inosine-5'-phosphate, 33B, 545
Disodium 4-oxyprimidine-2-sulphinate hexahydrate, 34B, 224
Disodium *N*-phosphorylcreatine hydrate, 33B, 275
Disodium salt of 2,4,6,8-tetrahydroxy-pyrimido[5,4-*d*] pyrimidine (tetrahydrate), 31B, 131
Disodium tetraglycylglycinatocuprate (II) 4.5-hydrate, 35B, 653
Disodium tetranitritonitrosohydroxy-ruthenate(III), 30A, 407
Distearoyl, 27, 782
1,3-Disubstituted phthalans, 35B, 923
Disulphitodiamminepalladate, 34A, 317
Disulphur dichloride, 34A, 178
Ditechnetium decacarbonyl, 30A, 443; 31A, 259
Ditelluryl sulphate, 31A, 199
Diterpenes, 35B, 923
2,5-Di-tert-butylthiophensulphone, 31B, 150
Di(tetra-*n*-butylammonium)cobalt(II) bis(maleonitrile dithiolate), 29, 560
Di-tetraethylammonium tetrachlorooxo-titanate(IV), 33B, 460
Di-tetraethylammonium tetrachlorooxo-vanadate(IV), 33B, 460
Di(tetramethylammonium) bis(maleo-nitriledithiolato)cuprate(II), 33B, 543
Di(tetramethylammonium)bis(maleo-nitrile dithiolato)nickelate(II), 30B, 377
Di(tetramethylammonium) vanadium tris(maleodinitrile dithiolate), 32B, 485
1,3-Dithiacyclobutanethione-(4), 35B, 909
1,3-Dithiacyclobutanone-(4), 35B, 909
1,6-Dithiacyclodeca-*cis*-3,*cis*-8-dienebis(mercuric chloride), 30B, 376
3,6-Dithiacyclohexane-1,2-dicarboxylic acid anhydride, 33B, 555
3,6-Dithiacyclohexene-1,2-dicarboxylic acid anhydride II, 34B, 181
1,4-Dithiane-antimony tri-iodide complex, 31B, 273
dl-1,2-Dithiane-3,6-dicarboxylic acid, 29, 520
15,18-Dithia-1,5,8,12-tetraaza-3,4:9,10:13,14:19,20-tetra-benzocycloeicosane-1,11-diene-iron(II) perchlorate methanolate, 35B, 823
2,2'-Dithienyl, 33B, 140
2,3'-Dithienyl, 33B, 140
3,3'-Dithienyl, 33B, 140
Dithiocyanatobis(ethylenethiourea)tel-lurium(II), 26, 578
Dithiocyanato-S-bis(tri-*n*-propyl phos-phine) palladium(II), 28, 677

trans-Dithiocyanatodiamminecopper, 30B, 289
trans-Dithiocyanato-tetrakis-(N,N'-diethylthiourea)nickel(II), 34B, 522
μ-Dithiocyanatotetrakis(methyldiphenylphosphine)dicopper(I), 35B, 829
1,2-Dithiolane-4-carboxylic acid, 31B, 143
1,2-Dithiole-3-thione, 28, 570
Dithionitrosylamminolead, 31A, 81
Dithio-oxamide, 30B, 36
2,5-Dithio-1-phenyl-1-thiophosphorus(V)-cyclopentane, 35B, 922
Dithiosemicarbazidenickel(II) sulphate (β-form), 33B, 471
2,4-Dithiouracil, 32B, 204
2,4-Dithiouridine, 35B, 919
Dithioxohexacarbonyldiiron, 30B, 383
Dithioxohexacarbonyldiiron, dithioxononacarbonyltriiron (1:1 complex), 30B, 384
Ditoluenechromium iodide, 26, 679
1,1-Di-*p*-tolyl-2-bromoethylene, 30B, 397
Di-*p*-tolyldisulphide, 32B, 71
1,1-Di-*p*-tolylethylene, 35B, 93
p,p'-Ditolyliminosulphur, 35B, 913
Di-*p*-tolylmercury, 29, 769; 35B, 511
Di-μ-trimethylaluminumcarbonyl-bis(iron-cyclopentadienyl carbonyl), 34B, 459
1,4-Di(trimethyl tin) butane, 27, 1064
Divalent metal chromates, 27, 609
Divalent metal germanates, 27, 500
Divalent metal selenates, 27, 609
Divalent tellurium-propylenethiourea complexes, 26, 583
Divalent tellurium-tetramethyl-thiourea complex, 26, 581
Divalent tellurium thiosulphonate-thiourea complexes, 26, 582
m-Divinylbenzene-diiron hexacarbonyl, 35B, 920
p-Divinylbenzene-diiron hexacarbonyl, 35B, 920
Di-(4-vinylpyridine)copper dichloride, 33B, 548
Diyttrium silicon beryllate, 34A, 377
Di-*p*-xylylene, 28, 769
DNA, 29, 758; 31B, 269; 33B, 527
DNA-proflavine complex, 26, 818
Dodecadimethylaminocyclohexaphosphazahexaene, 33B, 282
Dodecafluoro-cyclohexane, 33B, 532
Dodecafluoro-tricyclo(3,3,0,0(2,6)) octane, 30B, 112
Dodecahydro-1,4,7,9b-tetraazaphenalene trihydrochloride hemihydrate, 30B, 142
Dodecanedioic acid, 29, 765; 31B, 20
Dodecaphenylcyclohexastannane *m*-xylene, solvate, 28, 604
Dodecasodium *myo*-inositol hexaphosphate (phytic acid) hydrates, 35B, 273
Dodecawolframates, 31A, 247
Dolerophanite, 26, 440; 27, 621; 28, 216
Dopamine hydrochloride, 33B, 231
Dothistromin, 35B, 917
Double metaphosphates, 31A, 246
Double selenates, 31A, 241
Doverite, 27, 563
Drimenol, 26, 800
Dufrenite, 35A, 337
Dufrenoysite, 26, 427; 31A, 261; 32A, 227
Dumontite, 27, 583
Durylmethylsulphone, 29, 770
Durylsulphonyl chloride, 29, 770
Dussertite, 31A, 255
Dysprosium, 26, 141, 234; 27, 333
Dysprosium borate, 26, 491
Dysprosium chromite, 33A, 316
Dysprosium dihydride, 27, 237
Dysprosium dimanganate, 32A, 287
Dysprosium-gallium, 26, 146, 235
Dysprosium-germanium, 26, 235; 29, 5
Dysprosium-gold, 34A, 81
Dysprosium-indium, 26, 237, 238
Dysprosium-iron, 26, 236
Dysprosium lutecium oxide, 27, 516
Dysprosium-magnesium, 27, 195
Dysprosium-magnesium-sulphur, 27, 327
Dysprosium-manganese, 26, 236
Dysprosium manganese oxide, 31A, 249
Dysprosium-manganese-sulphur, 27, 327
Dysprosium-manganese-thulium, 27, 264
Dysprosium-nickel, 26, 235, 236, 239
Dysprosium niobate, 27, 521; 29, 329
Dysprosium-nitrogen, 27, 335
Dysprosium oxide, 26, 365
Dysprosium-oxygen-zirconium, 26, 365; 27, 535
Dysprosium oxytelluride, 27, 424
Dysprosium-platinum, 26, 235, 236, 238
Dysprosium-rhodium, 26, 236
Dysprosium-silver, 26, 237; 34A, 81
Dysprosium sodium sulphide, 29, 342
Dysprosium stannate, 26, 399
Dysprosium-strontium-sulphur, 27, 327
Dysprosium tantalate, 27, 521; 29, 329
Dysprosium-thorium, 26, 237, 238
Dysprosium trihydride, 27, 237
Dysprosium zirconium oxide, 26, 365, 27, 535
Dzhalindite, 28, 320

Eastonite, 29, 429
Ecdyson, 28, 686; 30B, 222
Echitamine, 26, 830
Echitamine bromide, 27, 1010
Echitamine bromide dihydrate, 26, 792

Echitamine bromide dihydrate, 26, 792
Echitamine bromide methanol, 26, 792
Echitamine chloride, 26, 792; 27, 1012
Echitamine derivatives, 26, 792
Echitamine iodide, 26, 792; 27, 1011
Eckermannite, 28, 322
n-Eicosane, 35B, 907
Einsteinium oxychloride, 34A, 179
Einsteinium trichloride, 34A, 179
Ekanite, 26, 538; 30A, 456; 31A, 225, 240
Elaeocarpine hydrobromide, 33B, 546
(±)-Elaeocarpine hydrobromide, 34B, 325
Elephantol *p*-bromobenzoate, 35B, 912
Ellagic acid, 33B, 232
Elpidite, 28, 276; 29, 404
Emerald, 33A, 480
Enanthic fibres, 26, 641
Enargite, 35A, 19
4-Endo,6'-endo-bis(π-azulene) iron, 34B, 473
3-Endo-phenyl-2-endo-norbornanol, 32B, 490
3-Endo-phenyl-2-endo-norbornanol-5-exo-7-anti-D(2), 32B, 497
Enniatin B complex with K^+, 34B, 288
Enstatite, 26, 510; 29, 429
Eosin, 26, 828
l-Ephedrine copper(II) chelate, 1/3 cyclohexane, 30B, 424
l-Ephedrine copper(II) chelate, 1/6 hexane, 30B, 424
l-Ephedrine copper(II) chelate, 2/3 mesitylene, 30B, 424
Ephesite, 32A, 477
Epibromohydrin, 26, 594
Epididymite, 30A, 456; 31A, 252; 35A, 478
3,5-Ephidithio-2,5-diphenylpenta 2,4-dienylidene-3-aminoquinoline, 35B, 914
2,4-Epidithio-1-nitro-1-nitroso-4-phenylbutadiene, 35B, 913
3,5-Epidithio-2-nitroso-1,5-diphenylpenta-2,4-dien-1-one, 34B, 193
Epidote, 30A, 455; 31A, 247
Epilimonol iodoacetate, 26, 801
6-Epimesembranol methiodide, 35B, 383
5,10-Episeleno-5,10-dihydroarsanthren, 35B, 921
Epistilbite, 26, 519; 32A, 488
Epistolite, 31A, 240
5,10-Epitelluro-5,10-dihydroarsanthren, 35B, 921
5,10-Epithio-5,10-dihydroarsanthren, 35B, 467
1,6-Epoxy(10)annulene, 32B, 488
9-10-Epoxy derivative of testosterone, 26, 799
1,4-Epoxy-1,4-dihydronaphthalene photodimer, 35B, 215
Epoxypropane derivatives, 26, 594
Epsomite, 29, 348
Erbium, 26, 141, 234; 27, 333
Erbium borate, 26, 491; 28, 170
Erbium digermanate, 35A, 317
Erbium dihydride, 27, 237
Erbium fluorooxalate tetrahydrate, 35B, 908
Erbium-gallium, 26, 146
Erbium-iridium, 26, 236
Erbium-iron, 26, 235
Erbium lithium oxide, 27, 516
Erbium-magnesium, 27, 195
Erbium-magnesium-sulphur, 27, 327
Erbium-manganese, 26, 192; 27, 196
Erbium-manganese-sulphur, 27, 327
Erbium orthochromite, 32A, 283
Erbium oxalate trihydrate, 35B, 636
Erbium niobate, 27, 521
Erbium-nitrogen, 27, 335
Erbium oxide, 27, 497; 31A, 246
Erbium phosphate, 33A, 309
Erbium-phosphorus, 27, 335
Erbium pyrosilicate, 33A, 465
Erbium-rhodium, 26, 236
Erbium-ruthenium, 26, 237
Erbium sesquioxide, 31A, 246
Erbium-silicon, 27, 336
Erbium sodium sulphide, 29, 348
Erbium-strontium-sulphur, 27, 327
Erbium tantalate, 27, 521
Erbium-tin, 31A, 42
Erbium titanate, 30A, 317
Erbium trihydride, 27, 237
Erbium vanadate, 33A, 309
Ergoflavin, 28, 686
Erionite, 32A, 511; 34A, 375
Erythristemine 2-bromo-4,6-dinitrophenolate, 35B, 913
Erythrocruorin, 33B, 527
D-Erythro-1,3-diacetoxy-2-acetamido-4-*trans*-octadecene, 34B, 346
Erythrosin, 26, 828
Eschenmoser's 'pseudo-corrin', 32B, 233
Esperite, 30A, 455
Estradiol monohydrate, 34B, 290
1,3,5(10),16-Estratetraen-3,17-diol diacetate, 28, 677
1,3,5(10)-Estratrien-2-ol-17-one 3-acetate, 28, 677
1,3,5(10)-Estratrien-3-ol-17-one 3-benzoate, 28, 677
1,3,5(10)-Estratrien-3-ol-17-one 3-ethyl ether, 28, 677
1,3,5(10)-Estratrien-3-ol-17-one 3-methyl ether, 28, 677
Estriol, 34B, 291
Estrone-related compounds, 29, 764
Estro-*p*-quinol-methyl ether, 35B, 924
Ethane, 27, 732; 30B, 401; 33B, 529
Ethanol, 33B, 539

Ethanolamine derivative of 7,9-dihydroxy-6α-methoxy-2-phenyl-trans-m-dioxano[5,4-e][1,4]-dioxepan, 26, 627
Ethanolatotetra-μ-thiobenzoato-dinickel(II), 35B, 800
Ethanol deoxycholic acid adduct, 27, 1039
Ethionamide hydrobromide, 33B, 114
Ethionamide hydrochloride, 33B, 114
Ethoxy-9-anthracene, 26, 748
Ethoxyantimony(V) tetrachloride, 33B, 287
1-Ethoxy-3-oxo-3a-phenyl-3,3a-dihydroazulene chromium tricarbonyl, 31B, 372
1-Ethoxy, 1,2,3,4-tetraphenylcyclobutenyl palladium(II) chloride dimer, 30B, 326
9-Ethyladenine:1-methyl-5-bromouracil, 30B, 396; 31B, 163
9-Ethyladenine-1-methyl-5-fluorouracil, 32B, 207
4-(α-Ethylamine)-acenaphthene, 27, 933
10-(2-Ethylaminoethyl)phenothiazine hydrochloride, 32B, 488
9-Ethyl-2-aminopurine, 1-methyl-5-fluorouracil complex, 31B, 164
Ethylammonium-ethylimmonium undecahydro-octaborane, 28, 581
Ethylammonium tetrachlorocuprate(II), 35B, 880
syn- and *anti*-Ethyl benzohydroximates, 35B, 913
9-Ethyl-8-bromoadenine-9-ethyl-8-bromohypoxanthine, 34B, 235
9-Ethyl-8-bromo-2,6-diaminopurine: 1-ethylthymine, 35B, 275
9-Ethyl-8-bromo-2,6-diaminopurine: 1-methyluracil, 35B, 275
1-Ethyl-5-bromouracil, 35B, 911
Ethyl carbamate, 32B, 5
Ethyl chloride, 27, 732
Ethyl 9-cyano 10-anthracene, 26, 748
1-Ethyldecarborane, 29, 524
N-Ethyl-2,2'-dimethylsulphonyl-vinylideneamine, 26, 604
Ethyl 3,5-dinitrobenzoate, 26, 711
cis-Ethylene amine dibromoplatinum, 29, 586
Ethylenebidiguanide copper(II) chloride monohydrate, 35B, 670
Ethylenebidiguanide silver(III) nitrate, 33B, 554
1,1'-Ethylene-2,2'-bipyridylium dibromide monohydrate, 34B, 158
N,N'-Ethylenebis-(acetylacetoneiminato)copper(II), 28, 640
NN-Ethylenebis(acetylacetoneiminato) copper(II) hemihydrate, 33B, 389
NN'-Ethylenebis(acetylacetoneiminato) copper(II) *mono*hydrate, 31B, 428
NN'-Ethylenebis(acetylacetone-iminato)copper(II)-methylammonium perchlorate, 35B, 615
N,N'-Ethylenebis(acetylacetone-iminato)oxovanadium(IV), 35B, 610
NN'-Ethylenebis(acetylacetone-iminato)-phenylaquocobalt(III), 35B, 912
Ethylenebis(biguanidine)nickel(II) dichloride monohydrate, 35B, 667
N,N'-Ethylenebis(*l*-ephedrine) copper(II) dihydrate, 33B, 447
N,N'-Ethylene bis(salicylaldehydaiminato)cobalt(II), 34B, 520
NN'-Ethylenebis(salicylaldehyde-iminato)-cobalt(III) derivatives, 35B, 913
N,N'-Ethylene bis(salicylaldiminato) copper(II) chloroform complex, 32B, 495
N,N'-Ethylene bis(salicylaldiminato) copper(II) *p*-nitrophenol complex, 32B, 495
NN'-Ethylenebis(salicylidene-iminato)cobalt(II)-oxygen-dimethylformamide adduct, 35B, 779
NN'-Ethylenebis(salicylideneiminato)-copper(II) chloroform adduct, 35B, 414
NN'-Ethylenebis(salicylidene-iminato)copper(II) *p*-nitrophenol adduct, 35B, 413
NN'-Ethylenebis(salicylideneiminato)-diaquochromium(III) chloride, 35B, 621
3,3'-Ethylenebis-sydnone, 34B, 196
Ethylene-*cis*-butene-2, 27, 801
1:1 Ethylene-carbon monoxide copolymer, 26, 632
Ethylene-cyclopentene, 27, 801
Ethylene-cyclopentene copolymer, 29, 759
Ethylenediamine, 35B, 911
Ethylenediamine-amylose complex, 35B, 906
1,10-Ethylene-7,8-dimethylisoalloxazinium iodide monohydrate, 35B, 196
N-Ethylenediamine-bis[trimethyl(acetylacetonato)platinum(IV)], 30B, 339
Ethylenediaminetetraacetato complexes, 29, 544
Ethylenediaminetetraacetato complexes of rare earths, 30B, 396
Ethylenediaminetetraacetato manganese (II) complex, 26, 692
Ethylenediamine tetracetic acid, 27, 775
Ethylenediaminetetra-acetic acid dirubidium salt dihydrate, 31B, 28
Ethylenediaminetetraacetic acid lanthanum (III) chelate, 30B, 397

Ethylenediamine tribromoplatinum, 26, 689
Ethylenediammonium chloride, 28, 391
Ethylenediammonium N-oxo-bis(*N*-hydroxyethylenediaminetriacetatoiron(III) hexahydrate, 32B, 441
Ethylenediammonium sulphate, 26, 589
Ethylenediammonium tetrachlorocuprate, 32B, 485
1,10-Ethylene-7,8-dimethylisoalloxazinium iodide monohydrate, 34B, 156
6,6-Ethylenedioxyheptachloropentacyclo-(5.2.0.0(2,5).0(3,9).0(4,8))-nonane-3-carboxylic acid, 34B, 114
6,6-(Ethylenedithio)fulvene, 35B, 221
Ethyleneiron tricarbonyl, 35B, 895
Ethylene oxide hydrate, 30B, 267
Ethylene sulphate, 33B, 552
Ethylene trithiocarbonate oxide, 33B, 553
Ethylenimine, 26, 590
Ethyleniminodimethylaluminum trimer, 35B, 481
Ethyl ester hydrobromide of 1-methyl-4-phenyl-piperidine-4-carboxylic acid, 26, 769
Ethyl ester hydriodide of 1-methyl phenyl-piperidine-4-carboxylic acid, 26, 769
Ethyl 2-S-ethyl-1,2-dithio-α-D-mannofuranoside, 35B, 911
1-Ethylgermatrane, 33B, 553; 35B, 484
Ethyl green, 26, 828
9-Ethylguanine-1-methylcytosine complex, 32B, 209
9-Ethylguanine-1-methyl-5-fluorocytosine complex, 32B, 210
N-Ethyl-2,2,4,4,6,6-hexaphenyl-3-azacyclotrisiloxane, 32B, 303
9-Ethylhypoxanthine and 5-fluorouracil, 1:1 complex, 32B, 206
Ethylidene-N,N-diacetamide, 34B, 8
Ethylidene tri(cobalt tricarbonyl) 32B, 341
5-Ethyl-5-isoamylbarbituric acid, 34B, 220
Ethyl lithium 28, 593
Ethylmagnesium bromide dietherate, 29, 533; 33B, 300
Ethylmagnesium bromide-triethylamine complex, 32B, 492
Ethyl mercaptan, 33B, 539
5-Ethyl-5-(1-methylbutenyl)-barbituric acid, 34B, 218
α-2-Ethyl-5-methyl-3,3-diphenyltetrahydrofuran, 34B, 176
5-Ethyl-1-methyl-5-phenyl-barbituric acid, 34B, 214
1-Ethyl-1-methyl-4-phenylpiperidinium perchlorate, 35B, 173
1-Ethyl-2-methylquinolinium iodide, 35B, 188
5-Ethyl-6-methyl-uracil, 31B, 166
N-Ethylnortropine methobromide, 29, 772
α-Ethyl-α-phenyl-α'-bromoglutarimide, 27, 969
α-Ethyl-α-phenyl-α'-iodoglutarimide, 26, 768
Ethylsodium, 35B, 473
Ethyl stearate, 27, 744; 30B, 397; 33B, 10
2-Ethyl-4-thiocarbamoylpyridine hydrobromide, 31B, 125; 33B, 114
2-Ethyl-4-thiocarbamoylpyridine hydrochloride, 31B, 126; 33B, 114
Ethyl-1-thio-α-D-glucofuranoside, 32B, 216
Ethylthioirontricarbonyl dimer, 28, 654
1-Ethylthymine: 9-ethyl-8-bromo-2,6-diaminopurine, 35B, 275
N-Ethyltropine methiodide, 29, 764
Ettringite, 31A, 247; 35A, 378
Euchroite, 28, 196; 31A, 196
Euclase, 27, 579; 29, 429
Eucryptite, 30A, 459
β-Eucryptite, 35A, 452
Eudidymite, 28, 322; 31A, 252
Eulytin, 28, 324
Eulytite, 31A, 227
Eunicellin dibromide, 33B, 555; 35B, 396
Eunicin iodoacetate, 33B, 547
Euparotin bromoacetate benzene solvate, 32B, 490
Euphenyl iodoacetate, 31B, 234
Eupteleogenin iodoacetate, 30B, 397
Europium, 26, 234; 27, 197
Europium borate, 26, 491
Europium(II) carbonate, 29, 383
Europium chalcogenides, 27, 197
Europium chloride, 27, 724
Europium chloride hexahydrate, 30A, 275
Europium dialuminate, 34A, 253
Europium-germanium, 29, 50
Europium hexaboride, 26, 69
Europium(II) hydroxide monohydrate, 31A, 256
Europium-indium, 29, 3
Europium(II) iodide, 26, 336; 34A, 182
Europium-lead, 35A, 58
Europium lithium oxide, 35A, 234
Europium-magnesium, 35A, 59
Europium molybdate, 30A, 458
Europium niobate, 27, 521
Europium oxides, 27, 475, 476, 546; 31A, 251
Europium oxychloride, 27, 724
Europium oxyiodide, 26, 336
Europium oxytelluride, 27, 424

Europium-selenium, 27, 197
Europium sodium sulphide, 29, 342
Europium(II) sulphate, 29, 353
Europium-sulphur, 27, 197
Europium tantalate, 27, 521
Europium-tellurium, 27, 197
Europium-thallium, 29, 3
Europium trichloride, 33A, 204
Europium wolframate, 28, 226
Euxenite, 26, 387
Euxenite-polycrase, 26, 387
Eveite, 33A, 421
Evonin adducts, 27, 1039

Fabianite, 35A, 286
Factor V_{1a} cobyrnic acid *abcfg* hexamide monaquo-cyanide, 27, 1040
Faheyite, 29, 425
Fairfieldite, 35A, 333
Faujasite, 29, 404; 34A, 371
Faujasites, 32A, 484; 33A, 497
Fayalite, 28, 265, 346; 29, 439; 31A, 255; 33A, 468
Feather keratin, 26, 824; 27, 1057
Febefsite, synthetic, 35A, 473
Felspars, 27, 676-684
Fenghuangite, 27, 599
Ferberite, 32A, 315; 33A, 346
Fergusonite, 26, 386; 28, 330; 30A, 456
β-Fergusonite, 26, 386
m'-Fergusonite, 32A, 311
Ferric chloride, 27, 440
Ferric chloride hexahydrate, 32A, 183
Ferric chromates, 31A, 247
Ferrichrome A, 29, 772
Ferrichrome-A tetrahydrate, 31B, 414
Ferric hydroxide tetraphenylporphine monohydrate, 29, 617
Ferricinium picrate, 32B, 493
Ferricinium tribromoferrate(III), 35B, 908
Ferricinium trichloroferrate(III), 35B, 908
Ferricinium triiodide, 33B, 345
Ferric oxides, 26, 395
Ferric vanadate(V), 29, 319
Ferricynium ferrichloride, 30B, 424
Ferrierite, 31A, 226, 259
Y Ferrites, 35A, 265
Ferrites, hexagonal, 30A, 349
Ferrithorite, 31A, 255
Ferritin, 29, 758
Ferritungstite, 26, 394
Ferrocene, 26, 650; 33B, 536
Ferrocene complex, 35B, 564
Ferrocene derivatives, 29, 763
Ferrocenedisulphonyl chloride, 29, 552
Ferrocene-tetracyanoethylene complex, 32B, 279; 33B, 552
[1,1,1]Ferrocenophane, 35B, 563
trans-β-Ferrocenylacrylonitrile, 35B, 544
Ferrocenyl ruthenocenyl ketone, 29, 551
Ferrocyanhydric acid, 31A, 114
Ferrolizardite, 29, 429
Ferrosilite, 30A, 456
Ferrosilite(III), 31A, 254
Ferrotremolite, 31A, 242
Ferrous fluosilicate hexahydrate, 27, 461
Ferrous *bis*nioxime *bis*imidazole, 29, 554
Ferrous sulphate, 26, 437
α-Ferrous sulphate, 35A, 372
Ferrous sulphate heptahydrate, 29, 351
Ferrucite, 32A, 158
Fersmanite, 28, 328
Fersmite, 26, 381
Fescue alkaloid hydrochloride, 29, 772
Fibroin, 33B, 527
Five-covalent copper(II) complex, 35B, 705
Flavaspidic acids, 35B, 906
Flinkite, 28, 348; 32A, 385
Florencite, 27, 600
Fluellite, 31A, 194
Fluoborite, 28, 350
mono Fluroacetamide, 27, 756
Fluoranthrene, 27, 940
Fluor-chlor-oxy-apatite, 31A, 242
Fluorescein, 26, 828
Fluorine, 33A, 157
α-Fluorine, 35A, 129
β-Fluorine, 29, 229
Fluorine strontium phosphate, 26, 469
Fluorite, 30A, 266
Fluor-magnesio-richterite, 27, 698
Fluormica, 27, 695
Fluoroacetonitrile, 26, 587
p-Fluoroaniline (microwave spectrum), 35B, 912
1-Fluoroanthraquinone, 28, 675
m-Fluorobenzamide, 30B, 61
p-Fluorobenzamide, 30B, 60
Fluorobenzene, 33B, 540
o-Fluorobenzoic acid, 26, 831
Fluoroberyllates, 26, 302
2'-Fluorobiphenyl-4-carboxylic acid, 30B, 427
Fluoroborates, 26, 304
[Fluoroborotris-(2-carboxalidimino-6-pyridyl)phosphine]nickel(II) tetrafluoroborate, 35B, 915
5-Fluoro-2'-deoxy-β-uridine, 29, 710
Fluoroethylenes, 26, 544
Fluoroformyl chloride, 26, 543
Fluoromanganates of alkali metals, 26, 308
Fluoropentachlorocyclotriphosphazene, 34A, 196
Fluoroperovskites, 34A, 203
p-Fluorophenol, 33B, 543
4'-Fluoropyrrolnitrin, 33B, 118
Fluorosulphate radical, 34A, 218

5-Fluorouracil and 9-ethylhypoxanthine, 1:1 complex, 32B, 206
5-Fluorouracil-1-methylcytosine complex, 34B, 224
Formaldehyde, 28, 664
Formaldehyde-aniline, 35B, 910
Formaldoxime, 28, 664
Formamide-potassium iodide, 27, 723
Formamide-potassium triiodide, 27, 723
Formamide solutions, 33B, 528
Formamidoxime, 30B, 52
Formate dihydrates, 28, 686
μ-Formato-monodiethylenetriamine copper(II) formate, 35B, 915
Formic acid dimer, 35B, 890
Formycin hydrobromide monohydrate, 31B, 194
Formyl fluoride, 26, 543
4-Formylpyridine thiosemicarbazone, 35B, 166
5-Formylvanillic acid, 29, 763; 30B, 422
Fornacite, 27, 602; 32A, 391
Forsterite, 33A, 468
Fortisan, 27, 799
Foshagite, 26, 519; 31A, 244
Francevillite, 31A, 256
Frankeite, 26, 432
Franklinite, 30A, 459
Fresnoite, 32A, 442, 511
Freudenbergite, 26, 378; 28, 322
Fuchsine, 26, 828
Fukinolidol bisbromoacetate, 35B, 386
Fumagillin derivative, 26, 800
Fumaric acid, 30B, 414
α-Fumaric acid, 31B, 4
β-Fumaric acid, 31B, 5
Fumaronitrilebis(triphenylphosphine)-platinum, 35B, 859
3,4-Furandicarboxylic acid, 29, 729
Furane-α,α'-dicarboxylic acid, 33B, 131
anti-Furfuraldoxime, 32B, 172
α-Furoic acid, 27, 947
trans-β-2-Furylacrylic acid, 32B, 174
Fusicoccin, 33B, 549
Fusidic acid intermediate, 35B, 338
Fusidic acid methyl ester 3-*p*-bromobenzoate, 33B, 204

Gadolinite, 30A, 455
Gadolinium, 26, 234
Gadolinium borate, 26, 491
Gadolinium chloride, 33A, 204
Gadolium dihydroxychloride, 34A, 221
Gadolinium-gallium, 26, 146, 235
Gadolinium-germanium, 26, 235; 29, 5
Gadolinium hexaboride, 26, 69
Gadolium-indium, 26, 237, 238
Gadolinium-iron, 26, 142; 27, 198; 35A, 60
Gadolinium iron garnet, 26, 413
Gadolinium-lanthanum, 27, 330
Gadolinium-magnesium-sulphur, 27, 327
Gadolinium-manganese, 26, 236; 29, 50
Gadolinium-manganese-sulphur, 27, 327
Gadolinium-nickel, 26, 143, 235, 236, 238, 239, 27, 198; 29, 51
Gadolinium niobate, 27, 521; 29, 329
Gadolinium orthovanadate, 27, 589
Gadolinium oxalate decahydrate, 30B, 408
Gadolinium oxalate hydrate, 35B, 906
Gadolinium oxide, 26, 366; 27, 723
Gadolinium oxide orthosilicate, 34A, 378
Gadolinium oxychloride, 27, 724
Gadolinium-oxygen-zirconium, 26, 365; 27, 542
Gadolinium oxytelluride, 27, 424
Gadolinium-platinum, 26, 235, 236
Gadolinium pyrosilicate, 32A, 436
Gadolinium sesquichloride, 35A, 150
Gadolinium-silver, 26, 237
Gadolinium sodium silicate, 35A, 299
Gadolinium sodium sulphide, 29, 342
Gadolinium-strontium-sulphur, 27, 327
Gadolinium-sulphur, 33A, 82
Gadolinium tantalate, 27, 521; 29, 329
Gadolinium-thorium, 26, 237, 238
Gadolinium trichloride, 30A, 457; 32A, 185
Gadolinium trichloride hexahydrate, 26, 322
Gadolinium trifluoride, 27, 724
Gadolinium-yttrium, 27, 330
Gadolinium-zinc, 32A, 62
Gadolinium zirconium oxide, 27, 535
Gagarinite, 26, 333, 27, 473
Gageite, 34A, 385
Gahnite, 29, 324
Galactitol, 33B, 162
D-Galactono-1,4-lactone, 32B, 214
α-D-Galactose, 26, 623
β-D-Galactose, 26, 623
Galanthamine methiodide, 29, 772
Galeite, 28, 326
Galena, 29, 439
Galenobismutite, 27, 81
Gallium, 26, 147, 279; 27, 199, 385; 30A, 48; 34A, 82
Gallium arsenide, 27, 381
Gallium fluoride, 26, 300
Gallium-gehlenite, 32A, 439
Gallium-germanium-lithium, 35A, 61
Gallium-germanium-nickel, 34A, 82
Gallium-hafnium, 27, 206, 378; 28, 27
Gallium-holmium, 26, 146
Gallium(III) hydrogen ethylenediamine-tetraacetic acid monohydrate, 28, 673
Gallium indium oxide, 33A, 281
Gallium-indium-selenium, 26, 42
Gallium-iridium, 32A, 68
Gallium-iron, 30A, 48
Gallium-iron-nitrogen, 27, 306
Gallium iron oxide, 30A, 322

Gallium-iron-titanium, 27, 161, 206
Gallium-iron-vanadium, 27, 161
Gallium-lanthanum, 26, 146
Gallium lithium oxide, 29, 316; 30A, 314
Gallium lithium silicate, 26, 523
Gallium-magnesium, 28, 23; 34A, 84; 35A, 62
Gallium-magnesium-nickel, 27, 261
Gallium magnesium oxide, 27, 488
Gallium-magnesium-sulphur, 32A, 65
Gallium-manganese, 27, 379
Gallium-manganese-oxygen, 26, 412
Gallium-manganese-sulphur, 27, 203
Gallium-neodymium, 26, 146
Gallium-nickel-oxygen 26, 412
Gallium-nickel-titanium, 27, 161, 206
Gallium-nickel-vanadium, 27, 161
Gallium niobate, 30A, 332
Gallium-niobium, 27, 379; 30A, 49
Gallium phosphate dihydrate, 31A, 183
Gallium-plutonium, 30A, 50
Gallium-oxygen-zinc, 26, 412, 413
Gallium-praseodymium, 26, 146
Gallium-rare earths, 26, 146
Gallium-rhodium, 32A, 68
Gallium-samarium, 26, 146
Gallium-scandium, 29, 53
Gallium selenide, 26, 147
Gallium-selenium, 26, 147; 27, 200
α-Gallium sesquioxide, 32A, 257
Gallium sesquisulphide, 26, 149
Gallium silicates, 26, 523
Gallium-silver, 26, 248; 29, 53
Gallium sodium oxide, 27, 546
Gallium sulphate, basic, 28, 219
Gallium-sulphur, 26, 147; 28, 24
Gallium-sulphur-zinc, 27, 201
Gallium tantalum oxide, 27, 527
Gallium-tellurium, 26, 149; 27, 200; 28, 26
Gallium-tellurium-thallium, 34A, 88
Gallium-terbium, 26, 146
Gallium thiohalides, 31A, 246
Gallium-thorium, 26, 257
Gallium-titanium, 27, 203, 377; 28, 27
Gallium transition-metal alloys, 27, 203
Gallium trichloride, 30A, 274
Gallium triiodide, 31A, 239
Gallium-vanadium, 27, 379; 29, 54; 30A, 52
Gallium-ytterbium, 29, 3
Gallium-yttrium, 26, 146; 29, 53
Gallium-zirconium, 27, 203, 378
Gallous bis-dioxan chloride, 26, 693
Galvinoxyl, 34B, 55
Garcinia acid calcium salt tetrahydrate, 34B, 18
Garnet, 29, 430, 439; 34A, 262
Garnet minerals, 27, 514
Garnets, 26, 413-417; 27, 510-515, 30A, 324, 456, 458; 33A, 282
Garnierite, 26, 522
Gases, 27, 716
Gaspeite, 31A, 242
Gaylussite, 33A, 435
Gedrite, 26, 528; 27, 689; 35A, 460
Gehlenite, 26, 521; 32A, 439
Geijerene-silver nitrate, 33B, 548
Gelsemicine derivatives, 26, 793
Gelsemicine hydrobromide hemihydrate, 26, 793
Genine D, 35B, 918
Genthelvine, 28, 324
Genthelvite, 26, 534
Gerhardite, 26, 478
Germacratriene-silver nitrate complex, 32B, 494
Germanate orthoclase, 31A, 250
Germanates, 26, 397-399; 29, 5
Germanides, 26, 84, 149, 156, 157; 29, 39, 90, 91
Germanium, 26, 279; 27, 208, 381; 29, 55
Germanium difluoride, 31A, 82
Germanium dioxide, 29, 303
Germanium garnets, 26, 415
Germanium-hafnium, 27, 379
Germanium-hafnium-nickel, 27, 162
Germanium-holmium, 32A, 77
Germanium-iridium, 32A, 80; 33A, 84; 34A, 89
Germanium-iron, 26, 149; 27, 208; 32A, 82; 34A, 90
Germanium-iron-manganese, 33A, 85
Germanium-iron-molybdenum, 26, 174
Germanium-iron-nitrogen, 27, 306
Germanium iron(III) oxides, 31A, 246
Germanium-lanthanum, 27, 341
Germanium lead aluminate, 35A, 263
Germanium-lithium, 30A, 53; 35A, 64
Germanium-lithium rhodium oxide, 27, 486
Germanium lithium zinc oxide, 27, 486
Germanium-magnesium-nickel, 27, 261
Germanium-magnesium-phosphorus, 26, 227
Germanium-magnesium-silicon, 27, 212
Germanium-manganese, 26, 149
Germanium-manganese-nickel, 26, 175; 27, 212, 261
Germanium-manganese-niobium, 33A, 86
Germanium-manganese-nitrogen, 27, 305
Germanium-manganese-oxygen-zinc, 26, 409
Germanium-manganese-sulphur, 34A, 91
Germanium-manganese-tellurium, 26, 151
Germanium-manganese-thorium, 29, 9
Germanium-molybdenum, 30A, 19; 32A, 70
Germanium-molybdenum-nickel, 26, 174
Germanium-molybdenum-niobium-tin, 27, 213
Germanium-neodymium, 26, 175
Germanium-nickel, 27, 292
Germanium-nickel-niobium, 27, 162
Germanium-nickel-oxygen, 26, 413
Germanium-nickel-scandium, 27, 162, 377
Germanium-nickel-tantalum, 27, 162
Germanium-nickel-thorium, 29, 91
Germanium-nickel-titanium, 27, 162, 377

Germanium-nickel-vanadium, 26, 174
Germanium-nickel-zirconium, 27, 162
Germanium-niobium-tin, 27, 213
Germanium-oxygen-thorium, 26, 403
Germanium-phosphorus, 35A, 20
Germanium-platinum, 33A, 87
Germanium-potassium, 26, 84, 156; 34A, 122
Germanium-rhodium, 32A, 83; 34A, 124
Germanium-rubidium, 26, 84, 157
Germanium-ruthenium, 27, 214
Germanium-samarium, 32A, 87
Germanium-scandium, 27, 341
Germanium selenide, 30A, 54
Germanium-selenium, 27, 215, 30A, 54
Germanium-selenium-tellurium, 26, 153
Germanium-selenium-zirconium, 29, 39
Germanium-silicon-wolfram, 26, 156
Germanium-sodium, 26, 156; 29, 57
Germanium sodium oxide, 28, 147
Germanium sodium tin oxide hydroxide, 32A, 248
Germanium-strontium-sulphur, 35A, 65
Germanium-sulphur, 30A, 55
Germanium-sulphur-zirconium, 29, 39
Germanium-tellurium, 31A, 42
Germanium-tellurium-tin, 26, 153; 27, 211
Germanium-tellurium-uranium, 34A, 92
Germanium-tellurium-zirconium, 29, 39
Germanium-thorium, 26, 257; 27, 216
Germanium-vanadium, 32A, 70
Germanium-ytterbium, 30A, 56
Germatranes, 35B, 908
Germyl fluoride, 26, 301
Gersdorffite, 27, 59, 724; 28, 50, 33A, 359
Gibberellic acid, 29, 772
Gibberellic acid methyl ester di-*p*-bromobenzoate, 28, 540
Gibbsite, 26, 347
Giessenite, 28, 334
Ginkgolide A mono-*p*-bromobenzoate ethanol solvate, 32B, 490
Gismondite, 28, 279
Glass, 29, 393
Glasses, 28, 292
Glauberite, 30A, 371; 31A, 201; 32A, 424
Glaucarubin *p*-bromobenzoate, 29, 753
Glauconic acid *m*-iodobenzoate, 27, 1028
Glaucophane, 33A, 495
Glitoxin, 32B, 239
D-Glucitol-pyridine, 35B, 911
4-O-(α-D-Glucopyranosyl)-β-D-glucopyranose monohydrate, 35B, 266
α-D-Glucosamine hydrobromide, 34B, 243
2-D-Glucosamine hydrobromide, 30B, 180
α-D-Glucosamine hydrochloride, 30B, 180; 32B, 489; 34B, 243
Glucosamine hydrohalides, 29, 766
α-D-Glucose, 30B, 179
β-D-Glucose, 28, 516; 33B, 161
D-β-Glucose-*p*-bromophenylhydrazone, 30B, 181
α-D-Glucose monohydrate, 27, 795; 35B, 922
α-D-Glucose-urea 35B, 911
β-D-Glucurono-1,4-lactone, 32B, 212
$(+)_{495}$-[L-Glutamatobis(ethylenediamine)cobalt(III)] perchlorate, 35B, 600
Glutaramide, 30B, 396; 31B, 7
N-(α-Glutarimido)-4-bromophthalimide, 30B, 395
Glutaronitrile copper(I) chloride, 32B, 496
L-1-*mono*Glyceride of 11-bromoundecanoic acid, 31B, 21
Glycerol, 33B, 157
Glycerolatocobalt(II), 35B, 911
Glycerolatozinc(II), 35B, 911
L-α-Glycerophosphorylcholine cadmium chloride trihydrate, 30B, 191
L-α-Glycerylphosphorylcholin, 31B, 180
Glycidol, 26, 594
γ-Glycine, 26, 588
Glycine addition compounds of Ba,Sr, and Mn chlorides, 35B, 919
Glycine silver nitrate, 29, 763; 33B, 552
Glycocyamine hydrobromide, 33B, 554
Glycolic acid, 32B, 497
α-Glycylglycine, 33B, 179
Glycylglycine cobalt(III) complexes, 35B, 913
Glycylglycine hydrochloride monohydrate, 32B, 489; 34B, 267
Glycylglycylglycine-calcium chloride complex trihydrate, 34B, 355
Glycylglycylglycinocopper(II) chloride sesquihydrate, 27, 1049; 29, 602; 33B, 543
Glycyl-L-histidinato copper(II) sesquihydrate, 32B, 420
Glycyl-phenylalanyl-glycine monohydrate, 26, 724
Glyoxal, 33B, 530
Glyoxime, 31B, 53
Gmelinite, 31A, 227
Gmorite, 28, 346
Goethite, 27, 423; 33A, 266
Gold, 26, 279; 27, 386
Gold fluorides, 35A, 162
Gold-hafnium, 27, 225, 234, 378
Gold-hydrogen-palladium, 33A, 120
Gold-iron, 27, 218
ζ-Gold-iron-tin, 27, 311
Gold-lithium, 26, 157
Gold-magnesium, 30A, 57; 33A, 89
Gold-manganese, 26, 158; 27, 183, 219, 226; 29, 60; 30A, 59; 31A, 42; 32A, 89; 35A, 66
ζ-Gold-manganese-tin, 27, 310

Goldmanite, 29, 430; 30A, 455
Gold-mercury, 35A, 66
Gold-nickel, 26, 159
ζ-Gold-nickel-tin, 27, 311
Gold oxides, 26, 366
Gold-platinum, 26, 159
Gold-potassium, 26, 160
Gold-rubidium, 26, 161
Gold-scandium, 27, 342
Gold-silver-sulphur, 31A, 43
Gold-sodium, 26, 161
Gold-tantalum, 26, 161
Gold-tellurium, 27, 221
Gold-terbium, 32A, 91; 33A, 91
Gold-thorium, 26, 256
Gold-tin, 26, 187
ζ-Gold-tin, 27, 310
Gold-titanium, 27, 222, 223, 378
Gold transition-metal alloys, 27, 223
Gold trifluoride, 32A, 155
Gold-vanadium, 27, 226
Gold-zinc, 27, 190, 228; 30A, 60
Gold-zirconium, 27, 224, 234, 378
Gorgeyite, 30A, 453
β-Gorgonene-silver nitrate, 33B, 372, 549
Gotzenite, 31A, 244
Goyazite, 27, 600
Graftonite, 33A, 401
Gramicidin S, 35B, 915
Gramicidin S hydrochloride, 28, 680, 681
Gramicidin S iodomercurate, 29, 764
Grandidierite, 30A, 456; 33A, 502
Grantsite, 26, 381
Graphite, 27, 395; 31A, 247
Graphite-alkali metal-dimethoxyethane complexes, 35B, 907
Graphite-bromine compounds, 30A, 251
Graphite compounds, 33B, 528
Graphite-lithium, 31A, 246
Graphite-molybdenum pentachloride, 35B, 905
Graphite nitrates, 31A, 240
Graphite oxide, 26, 297; 27, 396
Graphite-sulphur trioxide, 31A, 247
Gratonite, 28, 154; 34A, 23
Greigite, 30A, 458; 31A, 242
Grignard reagents, 29, 532
Grisorixin silver salt, 35B, 917
Grossuralite, 31A, 228
Groutite, 33A, 265
Gryolite, 30A, 456
Guaiazulenedimolybdenum hexacarbonyl, 33B, 379
Guanidinium aluminum sulphate hexahydrate, 32B, 57
Guanidinium chloride, 30B, 44
Guanidinium chloride: N,N-dimethylacetamide(3:1 complex), 29, 765; 30B, 258
Guanidinium chromium sulphate hexahydrate, 32B, 57
Guanidinium perchlorate, 30B, 407
Guanidoacetic acid hydrobromide glycocyamine hydrobromide, 32B, 497
Guanine, 33B, 545
Guanine-copper chloride, 35B, 920
Guanine hydrochloride dihydrate, 28, 686; 30B, 175
Guanosine, 33B, 545
Guanosine dihydrate, 35B, 283
Guanosine-5'-phosphate trihydrate, 34B, 260
Guanylic acid, 27, 1055
Guanylic acid 3'-polymer, 27, 1055
Guanylic acid 5'-polymer, 27, 1055
Guggenite, 28, 320
Gugiaite, 27, 703
Guilleminite, 30A, 456
Gumbelite, ($2M_2$ mica), 31A, 248
Gyrolite, 26, 513

Haemanthamine *p*-bromobenzoate, 35B, 921
Haematite, 27, 423; 30A, 450; 31A, 124
Haemoglobin, 26, 810, 812; 33B, 527
Hafnium carbide, 26, 91
Hafnium germanate, 26, 398
Hafnium halides, 27, 442
Hafnium iodide, 27, 442
Hafnium-iron, 27, 232
Hafnium-manganese, 26, 162, 192; 27, 196
Hafnium molybdate, 33A, 338
Hafnium-molybdenum, 26, 162
Hafnium-nickel, 26, 205, 238; 27, 232
Hafnium-nickel-silicon, 26, 175; 27, 376
Hafnium-nitrogen, 35A, 68
Hafnium-osmium, 26, 237
Hafnium oxides, 26, 363; 30A, 457
Hafnium-palladium, 27, 234
Hafnium phases, 26, 161
Hafnium phosphide, 27, 233
Hafnium-phosphorus, 27, 233; 33A, 92
Hafnium-rhenium, 26, 237
Hafnium-rhodium, 34A, 124
Hafnium-sulphur, 31A, 43; 35A, 69
Hafnium-technetium, 26, 255; 27, 360
Hafnium tetrachloride, 27, 718
Hafnium-tin, 26, 164; 27, 379; 29, 61
Hafnium titanium oxide, 33A, 306
Hafnium triiodide, 29, 349
Hafnium-vanadium, 26, 162
Hafnium-wolfram, 26, 162; 27, 234
Hafnium-yttrium, 27, 381
Hägg carbide, 31A, 30
Haidingerite, 27, 575; 31A, 194; 32A, 387
Hallimondite, 30A, 454
5-Halobenzfurazan-1-oxide, 27, 973
Halogen chalcogenide spinels, 33A, 235
Halogen derivatives of acenaphthene, 26, 746
1-Halogenoadamantanes, 33B, 541

mono-Halogeno benzenes, 26, 693
Halogenodiphenylarsines, 27, 867
Haloxine hydrobromide, 33B, 221
Halurgite, 29, 427
Hambergite, 28, 171; 30A, 454
Hannayite, 28, 352
(+)-Haplophytine dihydrobromide, 35B, 365
Haradaite, 32A, 441
Harmotome, 26, 517, 532
Harunganin, 27, 1007; 28, 684; 29, 684
Hastingsite, 31A, 253
Hatchite, 26, 427; 31A, 261; 32A, 230
Hauyne, 26, 537; 33A, 483
Hauyne-type structures, 30A, 431
Hedleyite, 27, 90
Heidornite, 32A, 504
Helium, 26, 279, 285; 27, 391
α-Helium, 27, 391
γ-Helium, 27, 391
Hellyerite, 28, 352
Helvite, 31A, 256
Heme protein from *Rhodospirillum Rubrum*, 26, 817
Hemihedrite, 35A, 461
Hemimorphite, 27, 699; 28, 324; 32A, 435
Hendersonite, 27, 603
Hendricksite, 31A, 242
Hen egg-white lysozyme, 27, 1058
Henritermierite, 34A, 387
Heptafulvalene, 32B, 486
Heptamethylbenzene tetrachloroaluminate, 33B, 41
Heptasodiumfluoridebisarsenate-19-hydrate, 35A, 199
Heptasulphurimide, 30A, 257
(D-*altro*-3-Heptulose), 35B, 259
Hetaerolite, 30A, 459
Heteratisine derivatives, 29, 764
Heteratisine hydrobromide monohydrate, 28, 685; 30B, 234
12-Heteropoly salts, 27, 672
Hetidine methiodide, 35B, 913
(+)-Hetisine hydrobromide, 28, 552
Heulandite, 32A, 511; 33A, 484
Heumilite, 31A, 242
Heusler alloys, 33A, 95
Hewettite, 26, 380
Hexa(acetylacetonato)aquotricobalt(II), 33B, 396
Hexaamminecobalt(III) iodide, 34A, 214
Hexaamminecopper(II) halides, 32A, 194
Hexaantipyrine calcium perchlorate, 30B, 425
Hexaantipyrine magnesium perchlorate, 30B, 425
Hexaantipyrine zinc perchlorate, 30B, 425
Hexaaquomagnesium nitrate, 34A, 340
1,12:2,3:4,5:6,7:8,9:10,11-Hexabenzocoronene, 26, 758
Hexabromobenzene, 28, 670
Hexabromobenzene-1,2,4,5-tetrabromobenzene, 29, 644
Hexabromocyclotriphosphazene, 35A, 145
Hexa(bromomethyl)benzene, 30B, 80
Hexacarbonyl-3,4,5,5,6,6-hexafluorocyclohexa-1-yne-3-enedicobalt, 33B, 325
Hexacene, 27, 938
Hexachlorobenzene, 26, 713
Hexachloroberkelate, 34A, 209
Hexachlorobis(triethylphosphine) dirhenium(III), 33B, 504
Hexachlorocyclopropane, 30B, 399
1,2,3,7,8,9-Hexachlorodibenzo-*p*-dioxin, 34B, 343
Hexachloroethane, 29, 449
Hexacrylonitrile, 28, 687
Hexacrylonitrile 1,1,4,4-tetra-(2-cyanoethyl)-1,4-dicyano-*trans*-2-butene, 29, 478
Hexacyanobenzene, 30B, 425
Hexacyanochromates, 26, 341
Hexacyanocobaltic acid, 33A, 242
Hexacyanoferrates(III) of silver and thallium, 31A, 254
Hexacyanoferrous acid, 34A, 242
Hexacyanoiridates, 26, 341
Hexacyanomanganates(III), 29, 292
Hexacyclo[10,3,1,$0^{2,10}$,$0^{3,7}$,$0^{6,15}$,$0^{9,14}$]hexadecane, 35B, 914
Hexadecaborane, 35A, 130
Hexadecamethoxycyclooctaphosphonitrile, 33B, 281
5,6,11,12,17,18-Hexadehydrotribenzo[*a*,*e*,*i*]cyclododecene, 35B, 133
1,5-Hexadiene-silver perchlorate, 33B, 376
Hexaethyl borazine, 30B, 270
Hexaethylcyclohexane, 33B, 83
Hexafluoroacetone, 35B, 891
Hexafluoro acetylacetonato-bis(carbonyl)rhodium(I), 32B, 495
Hexafluoroazomethane, 35B, 891
Hexafluorobenzene, 29, 642
Hexafluorobenzene, anthracene (1:1), 30B, 417
Hexafluorobenzene, perylene (1:1), 30B, 417
Hexafluorobenzene, pyrene (1:1), 30B, 417
Hexafluorochromates, 26, 309
Hexafluoroisobutene, 35B, 891
Hexafluoropropylimine, 35B, 891
Hexagonal ferrites, 29, 323; 30A, 349
Hexahydroebelin bromoacetate, 28, 680
1,2,3,4,4a,9a-Hexahydro-4a,9-propanocarbonazolium hydrobromide, 35B, 197
Hexaimidazole zinc(II) dichloride tetrahydrate, 32B, 436

Hexaiodobenzene, 30B, 79; 33B, 68; 35B, 77
Hexakis(diethoxyphosphonylacetyl-methano)trinickel, 35B, 907
Hexakis(imidazole)nickel(II) nitrate, 34B, 572
Hexakis(thiourea) nickel(II) bromide, 34B, 523
Hexakis(trimethylsilyl)-2,4-diamino-1,3,2,4-diazadiboretin, 34B, 374
Hexakis(urea)titanium(III) iodide, 35B, 630
Hexalithium diaurate(III), 35A, 214
Hexameric phosphonitrilic dimethyl-amide, 33B, 282
Hexa(methylamine) tetra-arsenic, 32B, 315
Hexamethylbenzene chromium tricarbonyl, 29, 767
Hexamethylcyclohexadienyl rhenium tricarbonyl, 32B, 494
Hexamethyl(Dewar benzene), 35B, 902
Hexamethyl Dewar benzene complexes of platinum and palladium, 33B, 548
N,N'-Hexamethylene bispropionamide, 27, 776
Hexamethylenebis[trimethylammonium bromide] dihydrate, 30B, 34
Hexamethylenediamine dihydroiodide, 28, 396
Hexamethylenediammonium adipate, 31B, 27
Hexamethylene di(trimethylammonium) dichloride dihydrate, 32B, 485
Hexamethylenetetramine, 26, 603; 28, 478
Hexamethylenetetramine-borine, 34B, 162
Hexamethylenetetramine hexahydrate, 30B, 269
Hexamethylenetetramine-iodoform, 35B, 202
Hexamethylenetetramine triphenol, 35B, 423
3,3,6,6,9,9,-Hexamethyl-1,2,4,5,7,8-hexa-oxacyclononane, 34B, 170
2,2,4,4,6,6-Hexamethyl-1,3,5-tris-(trimethylsilyl)cyclotrisi-lazane, 35B, 443
Hexammine chromium(III) penta-chlorocuprate(II), 33A, 207
Hexamminecobalt(II) chloride, 31A, 246
Hexamminecobalt chlorocuprate, 31A, 258
Hexamminecobalt(III) iodide, 31A, 246
Hexamminecobalt(III) tetrachloro-zincate(II) chloride, 35A, 410
Hexamminecopper(II) salts, 31A, 249
n-Hexane, 26, 559
Hexanitrobenzene, 31B, 66
Hexanitroethane, 35B, 907
Hexanitroferrates(II), -cobaltates(II), nickelates(II), and -cuprates(II), 31A, 248
Hexanitrosobenzene, 31B, 152
Hexaphenylbenzene, 33B, 76
2,2,4,4,6,6-Hexaphenylcyclo-triphosphazatriene, 34B, 393
Hexa-*m*-phenylene, 35B, 134
Hexapylidine iron(II) salt of tetra nuclear iron carbonyl anion, 31B, 433
Hexapyrazolenickel(II) nitrate, 35B, 710, 922
Hexaquomagnesium hexabromotellurate (IV), 29, 251
Hexaquomagnesium hexachlorotellurate (IV), 29, 251
Hexaquomagnesium hexahalogenotellu-rates (IV), 29, 251
Hexaquomagnesium hexaiodotellurate (IV), 29, 251
Hexaquozinc nitrate, 32A, 352
Hexa-rhodium carbonyl, 28, 71
Hexathiocyanato-cobalt-(II)-dimercury (II) benzene, 29, 629
1,3,5-*cis*-Hexatriene, 33B, 533
1,3,5-*trans*-Hexatriene, 33B, 533
3-Hexyne-chlorosulphonyl isocyanate adduct, 35B, 923
Hibiscus acid calcium salt tetra-hydrate, 34B, 18
Hibschite, 26, 521
High-bismuth oxide, 27, 476
High chalcocite, 28, 19
High-quartz lithium aluminum silicate, 35A, 453
High tridymite, 32A, 251
Himachalene monohydrochloride, 33B, 212
Himbacine hydrobromide, 27, 1003
Himbosine hydrobromide, 29, 772
Hinsdalite, 28, 326
Hirsutic acid *p*-bromophenacyl ester, 31B, 220
Hisingerite, 27, 695
Histamine, 26, 761; 29, 764
Histamine dihydrobromide, 29, 771
Histamine dihydrochloride, 27, 992
Histamine-iodine complex, 26, 615
DL-Histidine hydrochloride di-hydrate, 35B, 316
Histidine hydrochloride *mono*hydrate, 29, 701
Histidine hydrochloride-mercury(II) chloride, 35B, 647
D-Histidino-L-histidinocobalt(II) dihydrate, 35B, 664
Hodgkinsonite, 27, 706; 28, 230, 264
Hodgkinsine trimethiodide, 32B, 491
Hogbomite, 28, 334
Holmium, 26, 234; 27, 333
Holmium borate, 26, 491
Holmium compounds, 26, 164
Holmium dihydride, 27, 237
Holmium-iridium, 26, 236

Holmium iron heptasulphide, 33A, 362
Holmium-iron-sulphur, 33A, 98, 362
Holmium lithium oxide, 27, 536
Holmium-nickel, 26, 239
Holmium niobate, 27, 521
Holmium-nitrogen, 27, 335
Holmium oxide, 27, 497
Holmium-palladium, 26, 165
Holmium-phosphorus, 26, 164; 27, 335
Holmium-platinum, 26, 165
Holmium pyrosilicate, 33A, 465
Holmium-rhodium, 26, 236
Holmium-selenium, 26, 164
Holmium-silicon, 27, 336
Holmium sodium sulphide, 29, 342
Holmium stannate, 26, 399
Holmium-sulphur, 26, 164; 32A, 92
Holmium tantalate, 27, 521
Holmium-tellurium, 26, 164
Holmium trihydride, 27, 237
Holmium trisacetylacetonate trihydrate, 33B, 556
Holmium-zinc, 31A, 44; 33A, 100
Holmquistite, 26, 528; 34B, 390
Homochroman derivatives, 27, 988
Homocubane carboxylic acid *p*-bromoanilide, 32B, 487
D,L-Homocysteine thiolactone hydrochloride, 29, 766; 30B, 210
Homophthalic acid, 35B, 919
Hopeite, 26, 462; 27, 586; 28, 190
Hornblende, 30A, 433
Hornblendes, 29, 430
Hornesite, 31A, 253
Horse haemoglobin, 26, 812; 27, 1064
Horse liver alcohol dehydrogenase, 35B, 887
Hortonolite, 33A, 468
Hoshiite, 29, 426
Howlite, 30A, 459; 34A, 358; 35A, 481
Huanghoite, 26, 484, 28, 322
Hubnerite, 32A, 315
Hühnerkobelite, 30A, 454
Hulsite, 30A, 456
Human enamel matrix, 26, 826
Humite, 27, 725
Humulene bis (silver nitrate), 31B, 213
Humulene bromohydrin, 33B, 211
Humulene-silver nitrate, 29, 769
Hunteracine bromide, 35B, 914
Hunterburnine β-methiodide, 30B, 235
Hunterburnine methiodides, 27, 1008
Huntite, 27, 562; 29, 426
Hureaulite, 29, 425
Hutchinsonite, 30A, 355
Hyacinth, 26, 387
Hyalosiderite, 33A, 468
Hydrated nickel salts, 27, 642
Hydrated potassium manganese chlorides, 33A, 212
Hydrated potassium tetraperoxydiwolframate, 29, 332
Hydrated sodium peroxoborate, 26, 487
Hydrates, 26, 344; 33A, 159
Hydrazine-bis(ethanol) complex, 32B, 269
Hydrazine bismethanol, 33B, 37
Hydrazine complexes, 26, 343
Hydrazine fluoroborate, 26, 291
Hydrazine hydrate, 27, 422
Hydrazine *mono*hydrate, 29, 233
Hydrazine nitrogorm, 32A, 510; 32B, 485
Hydrazine-tetra-methanol, 32B, 269
Hydrazinium bis(dihydrogenphosphate), 31A, 182
Hydrazinium dihydrogen phosphate, 30A, 387
Hydrazinium fluoroborate, 29, 262; 35A, 392
Hydrazinium hydrazinedithiocarboxylate, 34B, 34
Hydrazinium hydrogen oxalate, 33B, 36
Hydrazinium pentachlorocadmate, 31A, 103
Hydrazinium perchlorate, 35A, 392
Hydrazinium perchlorate hemihydrate, 32A, 318
Hydrazinium sulphate, 35A, 366
Hydrazinium trinitromethide, 35B, 18
Hydrazinium tris (hydrazinecarboxylato-N,O) nickelate(II) monohydrate, 31B, 417; 32A, 151; 32B, 427
Hydrazinium(1+)zinc sulphate, 26, 441
3-Hydrazino-5-thiol-1,2,4-triazole, 32B, 487
Hydridobis(trichlorosilyl)carbonyl-π-cyclopentadienyliron, 35B, 535
Hydridobromocarbonyl*tris* (triphenylphosphine) osmium (II), 27, 873
Hydrido-carbonyl(fumaronitrile) bis(triphenylphosphine)iridium, 34B, 615
Hydridochlorobis(diphenylethylphosphine)-platinum, 30B, 388
Hydridochloro(trichlorosilyl)bis-(triphenylphosphine)rhodium, 35B, 873
Hydridochlorotris(triphenylphosphine)ruthenium(II) benzene solvate, 33B, 549
μ-Hydrido-μ-diphenylphosphido-bis (tetracarbonylmanganese), 32B, 466
Hydridotetrakis(trifluorophosphine) cobalt(I), 35A, 413
Hydroboracite, 27, 551, 28, 161
Hydrochlorothiazide, 35B, 910
Hydrogarnets, 31A, 254
Hydrogen, 26, 285; 27, 393; 31A, 240
Hydrogen bonding, study of, 26, 829
Hydrogen bromide hydrates, 35A, 194
Hydrogen bromide tetrahydrate, 33A, 229
Hydrogen bronze, 32A, 314
Hydrogen chloride, 32A, 180
Hydrogen chloride hydrates, 32A, 181

Hydrogen cyanide tetramer, 26, 600
Hydrogen fluoride, 34A, 171
Hydrogen-iron, 26, 166
Hydrogen-lanthanum-oxygen, 27, 332
Hydrogen-lanthanum-tellurium, 27, 332
Hydrogen-lithium, 27, 397
Hydrogen lithium fluoride, 27, 451
Hydrogen lithium paraperiodate, 27, 640
Hydrogen manganese sodium periodate hydrate, 27, 640
Hydrogen-nickel, 26, 166; 28, 27
Hydrogen-nickel-zirconium, 27, 237; 29, 62
Hydrogen-niobium, 33A, 81
Hydrogen-niobium-titanium, 27, 238
Hydrogen-oxygen-praseodymium, 27, 332
Hydrogen-palladium, 26, 166; 27, 239
Hydrogen peroxide, 30A, 303
Hydrogen peroxide dihydrate, 26, 298
Hydrogen-praseodymium-tellurium, 27, 332
Hydrogen sulphide, 26, 423; 27, 399; 29, 341
Hydrogen-thorium, 27, 239
Hydrogen-titanium, 26, 167
Hydrogen-vanadium, 26, 168; 27, 240
Hydrogen-ytterbium, 26, 169; 27, 241
Hydrogen-yttrium, 27, 241
Hydrogen-zirconium, 26, 167; 27, 241
Hydrograndite, 29, 430
Hydrogrossular, deuterated, 33A, 273
Hydrolysed cocarboxylase trihydrate, 31B, 182
Hydromagnesite, 30A, 458
Hydronium nitranilate, 32B, 107
Hydronium perchlorate, 27, 637
14*c*-Hydro-5a-phenylbenz[*a*]indeno[2,1-*c*]-fluorene-5,10-dione, 33B, 93
γ-Hydroquinone, 31B, 72
Hydrosodalite, 28, 312; 32A, 490
Hydrotungstite, 28, 332
Hydrous calcium borate, 26, 489
Hydrous strontium borate, 26, 489
Hydrous tin (II) oxide, 26, 360
Hydroxobis(pentafluorophenyl)-thallium(III), 35B, 510
Hydroxonium perchlorate, 26, 496
Hydroxotris(carbamido)uranyl polyiodide, 33B, 556
Hydroxyacetato-oxyacetato-aquo-erbium(III) monohydrate, 34B, 527
o-Hydroxyacetophenone-isobutyl-iminecopper(II), 35B, 730
Hydroxyapatite, 29, 369
Hydroxyascharite, 31A, 247
Hydroxybastnaesite, 31A, 255
m-Hydroxybenzamide, 31B, 63
11β-Hydroxy-12α-bromo-progesterone, 32B, 490
6-Hydroxybuphanidrine methiodide, 35B, 921
p-Hydroxy-*trans*-cinnamic acid, 35B, 44
4-Hydroxycoumarin monohydrate, 30B, 395; 31B, 141
6-Hydroxycrinamine, 32B, 256
Hydroxydammarenone, 27, 1036
α-(2-Hydroxy-3,5-dibromobenzylidene)-γ-butyrolactone, 34B, 13
2-(2'-Hydroxy-3',5'-dibromophenyl)-3,4,5-tribromopyrrole, 31B, 117
2-(2-Hydroxyethylamino)-2-thiazoline, 29, 771
Hydroxyethylated cellulose, 26, 628
Hydroxyhalides, 26, 327-329; 29, 250
7-Hydroxy-β-isosparteine perchlorate, 32B, 254
Hydroxylamine-O-sulphonic acid, 32A, 322
Hydroxylammonium chloride, 32A, 197
Hydroxylammonium perchlorate, 34A, 359
Hydroxylapatite, 30A, 455
Hydroxyl-benzene, 30B, 397
2'-Hydroxymethanesulphonanilide, 35B, 80
1-Hydroxy-6-methoxy-phenazine-5,10-dioxide, 33B, 187
5-Hydroxy-6-methoxy-1,3,8,10-tetrathia-2,4,7,9-tetra-azadecane-4,7-diato-S(1)-S(10)-nickel(II), 32B, 426
7α-(1-(R)-Hydroxy-1-methylbutyl)-6,14-*endo*-ethenotetrahydro-thebaine hydrobromide, 32B, 260
2-C-Hydroxymethyl-2,2'-O-cyclohexyli-dene-3-deoxy-5-O-(*p*-bromobenzene-sulphonyl)-D-*erythro*-pentono-1,4-lactone, 35B, 270
(1-(5-Hydroxymethyl-2-furyl)-9*H*-pyrido[3,4*b*]indole hydrobromide dihydrate, 35B, 233
γ-Hydroxy-γ-methylglutamic acids, 33B, 542
2-Hydroxymethylpyridine-*N*-oxide, 35B, 923
5α-Hydroxy-6,4'-(5'-methylpyrimidin-2'-one)-dihydrothymine monohydrate, 34B, 240
12-Hydroxy-6-methyltricyclo[5,3,1,$1^{2,6}$] dodec-3-yl *p*-iodobenzoate, 33B, 102
β-Hydroxy-α-naphthaldehyde-[β-hydroxy-α-naphthyl-methylimid]copper(II), 35B, 781
Hydroxypelenolide *p*-bromobenzoate, 35B, 389
2-(O-Hydroxyphenyl)benzothiazole, 35B, 241
2-Hydroxy-2-phenylbicyclo[1.1.1] pentane (urethan derivative), 33B, 99
2-*o*-Hydroxyphenylethanesulphonic acid, 32B, 488
3-Hydroxy-5-phenylisooxazole, 30B, 393
3-Hydroxy-5-phenylisooxazole (β form), 34B, 199

2-(*o*-Hydroxyphenyl)-1-phenylpropane sulphonic acid sultone, 32B, 201
4-Hydroxyphenyl-2,2,4-trimethylchroman, 35B, 213
(2-Hydroxy-1,3-propanediamine) copper(II) nitrate, 35B, 909
5(α-Hydroxy-α-2-pyridylbenzyl)-7(α-2-pyridylbenzylidene)-5-norbornene-2,3-dicarboximide, 31B, 210
8-Hydroxyquinolinatopalladium(II), 31B, 432
8-Hydroxyquinoline-chloranil complex, 32B, 283
8-Hydroxyquinoline-*N*-oxide, 35B, 923
8-Hydroxyquinoline-5-sulphonic acids, 35B, 185
8-Hydroxyquinolinium trichloroxyselenate, 32B, 158
23-Hydroxy-2,3-secours-12-ene-2,3,28-trioic acid (2→23)-lactone 3,28-dimethyl ester, 35B, 915
Hydroxyselenates, 27, 623
Hydroxystannates, 33A, 289
Hydroxysulphates, 27, 623
Hydroxy-N,N,N',N'-tetrakis(2-aminoethyl)ethylenediaminecobalt(III) iodide, 28, 673
Hydroxy-N,N,N',N'-tetrakis(2-aminoethyl) ethylenediamine cobalt(III) perchlorate, 28, 673
2-Hydroxy-1,3,4,7-tetramethoxyxanthone, 34B, 338
o-Hydroxy-α-toluenesulphonic acid, 32B, 488
7-Hydroxy-1,9,10-trimethoxy-4-azabicyclo[5.2.2]undeca-8,10-dien-3-one, 34B, 161, 35B, 204
α-4-Hydroxy-2:2:6-trimethylpiperidine, 26, 762
β-4-Hydroxy-2:2:6-trimethylpiperidine, 26, 762
4-Hydroxy-2,2,6-trimethylpiperidine hydrochloride, 28, 672
4-Hydroxy-2,2,6-trimethylpiperidine hydrochloride monohydrate, 28, 672
5-Hydroxy tryptamine creatinine sulphate monohydrate, 30B, 214
5-Hydroxy-DL-tryptophan, 35B, 923
Hydroxyurea, 31B, 44
N-Hydroxyurea, 32B, 52
Hydrozincite, 29, 381; 30A, 454
Hypochlorous acid, 34A, 220
Hypophosphorous acid, 31A, 259
Hypoxanthine hydrochloride monohydrate, 34B, 236

Ice, 26, 343; 28, 118; 29, 294; 30A, 450; 33A, 255
Ice Ic, 33A, 258
Ice II, 29, 294; 33A, 255
Ice III, 33A, 256
Ice V, 32A, 250
Ice VI, 29, 439
Ice VII, 29, 439
Ice IX, 33A, 256
Icosaborane-16, acetonitrile adduct, 31B, 292
Illite, 30A, 456
Imenine, 35B, 356
Imhofite, 30A, 459
Imidazole, 28, 687; 31B, 115; 34B, 123
Imidazole-iodine, 26, 615
2-(1-Imidazoline-2-yl)benzophenone hydrobromide, 33B, 111
γG-Immunoglobulin, 33B, 527
Indene adduct, 33B, 103
π-Indenyl-π-duroquinonerhodium, 35B, 922
Indenyl radical, 33B, 536
Inderborite, 29, 427
Inderite, 28, 160
Indirubine, 26, 783
Indite, 28, 320
Indium, 26, 169, 279; 27, 243
Indium acetylacetonate, 35B, 908
Indium borate, 26, 491
Indium chalcogen halides, 27, 465, 466
Indium(I) chloride, 31A, 96
Indium cyclopentadienyl, 29, 625
Indium fluoride, 26, 300
Indium hydroxysulphate, 27, 625
Indium iodide, 29, 288
Indium(III) iodide-bis(dimethyl sulphoxide), 35B, 916
Indium-iron-nitrogen, 27, 306
Indium lithium oxide, 35A, 225
Indium-magnesium, 28, 23
Indium-magnesium-nickel, 27, 11, 261
Indium-manganese, 27, 243
Indium-manganese-nickel, 26, 169
Indium-manganese-nitrogen, 27, 273
Indium-mercury, 26, 169
Indium-mercury-tellurium, 27, 244
Indium molybdate, 33A, 331
Indium-nickel, 30A, 61
Indium-nickel-nitrogen, 27, 306
Indium-nickel-titanium, 27, 11, 296
Indium niobate, 27, 522
Indium-niobium, 27, 244
Indium oxide, 32A, 260
Indium oxide hydroxide, 29, 310; 35A, 278
Indium phosphate dihydrate, 26, 464
Indium-phosphorus-selenium, 27, 247
Indium-platinum, 26, 225
Indium-scandium, 27, 245
Indium-selenium, 26, 170; 27, 200; 30A, 62; 31A, 44
Indium-selenium-sulphur, 26, 172
Indium-selenium-tellurium, 26, 42
Indium sesquioxide, 31A, 120
Indium-silver, 26, 172, 248
Indium-silver-tellurium, 27, 181
Indium sodium oxide, 35A, 213
Indium sodium silicate, 32A, 458
Indium-sodium-sulphur, 26, 172

Indium sodium wolframate, 35A, 253
Indium-strontium, 29, 3
Indium sulphide, 26, 173
Indium-sulphur, 26, 173; 27, 245; 30A, 62; 31A, 45; 32A, 94
Indium-sulphur-zinc, 27, 246; 32A, 95; 34A, 93
Indium tantalate, 27, 522
Indium-tellurium, 26, 173; 27, 200
Indium ternary alloys, 27, 247
Indium(I) tetrachloroindate(III), 31A, 250
Indium-titanium, 27, 248
Indium trifluoride, 31A, 260
Indium trihydroxide, 32A, 260
Indium-ytterbium, 29, 3
Indium-zirconium, 27, 379
Indole-trinitrobenzene, 29, 689
3-Indolylacetic acid, 29, 726
racem-α-(3-Indolyl)-propionic acid, 26, 776
Ingenol-triacetate, 35B, 923
Innelite, 26, 537; 28, 334
Inosine, 33B, 545; 35B, 288, 906
Inosine dihydrate, 35B, 283
Insulin, 26, 820; 27, 1058, 33B, 527; 35B, 887
Insulin citrate, 27, 1058
Insulin citrate I, 26, 820
Insulin citrate II, 26, 820
Insulin sulphate, 27, 1058
Intensity statistics, 27, 1065
Iodide of 4,4'bis(dimethylamino) diphenylamine radical, 31B, 59
Iodine, 32A, 149
Iodine *mono*bromide, 33A, 230
Iodine *mono*chloride, 27, 448
Iodine *mono*chloride complex of pentamethylenetetrazole, 32B, 270
Iodine heptafluoride, 27, 446; 30A, 288; 31A, 260
Iodine oxide trifluoride, 32A, 508
Iodine pentoxide, 35A, 208
Iodine-1,4-selenothiane complex, 27, 964
Iodine-tetrahydroselenophene, 29, 522
Iodinin, 34B, 155
5-Iodoacenaphthene, 28, 438
17β-Iodoacetoxy-4,4-dimethyl-5α-androstan-3-one, 35B, 915
17β-Iodoacetoxy-4,4-dimethyl-19-nor-5α-androstan-3-one, 35B, 915
1-Iodoanthraquinone, 28, 675
β-Iodobarene, 30B, 410
7-(*p*-Iodobenzenesulphonyl)-7-azabicyclo(4,1.0]heptane, 30B, 144
cis-11-(*p*-Iodobenzenesulphonyl)-11-azabicyclo[8.1.0]undecane, 35B, 190
1-(*p*-Iodobenzenesulphonyl)-1-azaspiro[2.5]octane, 35B, 203
6-(*p*-Iodobenzenesulphonyl)-3-oxa-6-azabicyclo [3.1.0] hexane, 31B, 154
p-Iodobenzonitrile, 30B, 38
p-Iodobenzoylamine derivative of *aci*-ergotamine (ethanol solvate), 31B, 250
2'-Iodobiphenyl-4-carboxylic acid, 30B, 427; 35B, 90
3'-Iodobiphenyl-4-carboxylic acid, 35B, 924
Iodobis-(2,2'-bipyridyl)copper(II) iodide, 28, 643
Iodo-[*NN*-bis-(2-diphenylphosphinoethyl)-2-methoxyethylamino]-nickel(II) iodide, 35B, 853
Iodocarbonyl-Π-cyclopentadienyl-pentafluoroethylrhodium, 30B, 313
Iodo-carbonylnitrosyl-bis(triphenylphosphine) iridium tetrafluoroborate benzene solvate, 34B, 614
Iodocyanoacetylene, 27, 758
π-Iodocyclopentadienyltetraphenylcyclobutadienylcobalt, 33B, 542
*mono*Iodocecaborane, 32A, 213
5-Iodo-2'-deoxyuridine, 29, 771; 30B, 204
Iodo-*M*-dicarbadodecaborane, 32B, 491
p-Iodo-N,N-dimethylaniline hydrochloride triiodide, 27, 958
Iododiphenylarsine, 28, 588
Iodoform-1,4-diselenane(2:1) addition compound, 28, 571
Iodoform - hexamethylenetetramine, 33B, 544
Iodoform-quinoline complex, 27, 986
17-Iodoheptadecanoic acid, 29, 762
7-Iodo-8-hydroxyquinoline-5-sulphonic acid, 35B, 186
L-5-Iodomethylpyrrolid-2-one, 35B, 920
(-)-1-Iodo-3-methyltricosane, 34B, 25
B-Iodoneobarene, 30B, 412
4-Iodo-2-nitrobenzaldehyde, 29, 652
p-Iodophenol, 33B, 543
Iodopentaborane-9, 30A, 254
9-(*p*-Iodophenyl)-9-azatetracyclo-$[5.3.1.0^{2,6}0^{8,10}]$undec-4-ene, 35B, 207
1-*p*-Iodophenyl-3-phenylpyrazoline adduct of isocolumbin (acetone solvate), 31B, 230
5-Iodo-2-phthalimidobenzoic acid monohydrate, 28, 474
Iodophyllanthin, 33B, 555
p-Iodopolystyrene, 29, 759
Iodotrinitromethane, 31B, 51
Iodo[tris-(2-diphenylphosphinoethyl)-amino-*NP*]-nickel(II) iodide, 35B, 862
5-Iodouridine, 35B, 276, 922
Iranite, 28, 328, 354
Irarsite, 31A, 257
Iridium(III) benzlacetophenone dichloride bis(dimethylsulphoxide), 32B, 493
Iridium boride, 27, 93
Iridium(III) chloride, 26, 324; 30A, 277
Iridium-magnesium, 27, 249
Iridium-niobium, 29, 70
Iridium-phosphorus, 26, 48, 270; 34A, 117
Iridium-rhenium, 27, 337

Iridium-scandium, 27, 342
Iridium-silicon, 27, 373; 35A, 70
Iridium-thorium, 26, 236, 239, 256; 27, 367
Iridium-titanium, 27, 249
Iridium trichloride, 26, 324; 30A, 277
Iridium-uranium, 26, 165
Iridium-vanadium, 30A, 63
Iridium-zirconium, 26, 237
Iridomyrmecin, 27, 998, 29, 742
isoIridomyrmecin, 27, 998
Iridomyrmecins, 27, 997
Iron, 27, 250
Iron arsenate, 33A, 418
Iron arsenide, 27, 725
Iron(III)benzhydroxamate trihydrate, 34B, 583
Iron-bis-salicylidene-ethylenediamine-oxo bis(pyridine), 33B, 554
Iron boracite, 27, 551
Iron borate, 30A, 417
Iron carbonate, 26, 481
Iron carbonyl carbide, 27, 647
Iron carbonyl complexes, 33B, 337; 35B, 584, 914
Iron carbonyl diphenylacetylene complex, 30B, 297
Iron carbonyl phenylacetylene complexes, 27, 877
Iron carbonyls, 30A, 259
Iron(II) chloride dihydrate, 30A, 274
Iron(III)chloride-O-methoxyphenyl-diazonium chloride, 28, 686
Iron π-complexes, 35B, 582
Iron(III) copper(I) oxide, 31A, 137
Iron cupferron, 30B, 358
Iron dimanganese carbonyl, 32A, 395
Iron fluoride, 26, 300
Iron(II) fluoride, 31A, 247, 249
Iron fluoroborate hexahydrate, 26, 304
Iron fluorocarbon complexes, 26, 680
Iron—gadolinium, 26, 236
Iron gallium oxide, 31A, 133, 163
Iron gallium spinels, 31A, 163
Iron(III) germanates, 31A, 257
Iron-germanium-thorium, 29, 91
Iron hexacyanochromate, 26, 343
Iron hexacyanochromate hexahydrate, 27, 653
Iron hexacyanoiridate, 26, 343
Iron hexathiodiphosphate, 35A, 138
Iron(II)hydrazinecarboxylate, 30B, 413
Iron(III) hydrogen ethylene diamine-tetraacetic acid monohydrate, 28, 673
Iron hydroxychloride, 26, 327
Iron hydroxysulphate, 27, 623
Iron lithium molybdates, 35A, 250
Iron lithium wolframate, 35A, 248
Iron—magnesium—oxygen, 26, 396
Iron malate hydrates, 33B, 542
Iron-manganese-molybdenum, 30A, 40
Iron—manganese—oxygen—zinc, 26, 409
Iron-manganese-silicon, 27, 375
Iron manganese spinels, 30A, 451
Iron—manganese—sulphur—zinc, 26, 251
Iron mica, 28, 249
Iron-microcline, 27, 684
Iron molybdate, 31A, 259; 33A, 331
Iron-molybdenum, 26, 177; 28, 27
Iron molybdenum oxide, 31A, 249
Iron-molybdenum-silicon, 27, 376
Iron neodymium titanium pentoxide, 35A, 237
Iron-nickel, 26, 411; 27, 251
Iron-nickel-nitrogen, 27, 252
Iron nickel oxides, 27, 495
Iron-nickel-phosphorus, 33A, 102; 35A, 71
Iron-nickel-selenium, 34A, 94
Iron-nickel-silicon, 26, 124
Iron niobate, 26, 389; 27, 522
Iron-niobium-oxygen-strontium, 26, 420
Iron-nitrogen, 27, 253
Iron-nitrogen-tin, 27, 306
Iron nitrosyl dithiolene complex, 32A, 509
Iron oxide, defect, 34A, 246
Iron(III) oxide hydroxide, 31A, 241
Iron oxides, 26, 540; 27, 497, 499; 30A, 450
Iron oxide-titanium oxide, 27, 500
Iron oxide-vanadium oxide, 27, 500
Iron oxychloride, 35A, 181
Iron-palladium, 27, 253
Iron pentacarbonyl, 27, 650; 29, 241, 30A, 259; 34A, 163
Iron-phosphorus, 26, 178; 27, 319
Iron-phosphorus-silicon, 27, 321
Iron-phosphorus-zirconium, 31A, 45; 33A, 103
Iron phthalocyanine, 30B, 435
Iron phyllosilicates, 30A, 456
Iron-platinum, 27, 254
Iron potassium fluoride, 26, 305
Iron potassium molybdate, 35A, 254
Iron potassium sulphide, 33A, 3
Iron-praseodymium, 31A, 46; 33A, 105
Iron-rhenium, 27, 255
Iron rhodite, 26, 399
Iron-rhodium, 27, 256
Iron rubidium sulphide, 33A, 3
Iron ruthenite, 26, 399
Iron-ruthenium, 26, 179
Iron-sanidine, 26, 525; 27, 684
Iron saponite, 30A, 456
Iron-scandium, 26, 237
Iron-selenium, 26, 179-182; 33A, 106; 34A, 96; 35A, 72
Iron-silicon, 26, 183; 28, 28
Iron-silicon-titanium, 35A, 73
Iron-silicon-vanadium, 26, 174, 270; 27, 376
Iron-silicon-wolfram, 26, 269; 27, 376
Iron-silver-tellurium, 27, 256; 29, 62
Iron sodium metagermanate, 32A, 493
Iron sodium oxide, 27, 546

β-Iron sodium oxide, 28, 133
Iron(sodium) trifluoride, 33A, 184
Iron strontium silicate, 26, 523
Iron strontium tantalate, 27, 509
Iron sulphate, 27, 526
Iron sulphate tetrahydrate, 27, 612
Iron-sulphur, 27, 257; 29, 63; 33A, 107; 34A, 97; 35A, 75
Iron—sulphur—yttrium, 26, 277
Iron-sulphur-zinc, 27, 257
Iron-technetium, 27, 361
Iron-tellurium, 35A, 75
Iron tellurium oxide, 27, 526
(-)-Iron tetracarbonyl-fumaric acid, 33B, 546
Iron(II) thiocyanate tetra(pyridine), 32B, 440
Iron(thio-*p*-toluoyldisulphide)bis(dithio-*p*-toluate), 33B, 552
Iron-thorium, 34A, 75
Iron thulium oxide, 33A, 323
Iron-tin, 27, 257
Iron titanium oxide, 27, 499
β-Iron(III) trifluoride trihydrate, 29, 257
Irontri(1,10-phenanthroline)-di(mercury tricobalt carbonyl), 35B, 907
Iron tris-acetylacetonate, 32B, 385
Iron(III) tris(O-ethylxanthate), 35B, 923
Iron(II)tris(pyridine-2-carboxaldehyde)-1,1,1-tris(aminomethyl)ethane diperchlorate, 35B, 921
Iron—uranium, 26, 30; 27, 36, 259
Iron uranium oxide, 31A, 255
Iron vanadate, 27, 522
Iron-vanadium, 27, 260
Iron-wolfram, 26, 177
Iron wolfram oxide, 27, 527; 31A, 249
Iron-yttrium, 26, 236, 276
Iron yttrium garnet, 27, 510
Iron yttrium garnets, 26, 414
Iron-zinc, 27, 173; 33A, 108
Iron-zirconium, 27, 260
Iscorite, 34A, 389
Isobutyl chloride, 29, 450
Isobutylene, 26, 553; 28, 666
Isobutyroyl hexachloroantimonate(V), 33B, 547
Isocitric acid, 27, 735
Isoclovene hydrobromide, 26, 796
Isoclovene hydrochloride, 26, 796
Isocollybolide, 35B, 917
Isocyanatocyclopentadienyldinitrosylchromium, 35B, 537
Isocytosine, 30B, 172
Isodibenzanthrone (isoviolanthrone), 29, 699
Iso-eremolactone, 33B, 214
Isoguanine sulphate tetrahydrate, 32B, 489
L-Isoleucine, 35B, 906
(-)-Isolunine hydrobromide dihydrate, 34B, 324
Isonicotinic acid hydrazide dihydrochloride, 29, 724
Isophorone photodimer reduction product, 35B, 128
Isophorone trimer, 35B, 923
Isoprene, 26, 558
β-Isoprene sulphone, 35B, 217
1,2-*dihapto*:7,8,11,12-*tetrahapto*-3-Isopropenyl-4-methyl-4-vinylcyclohexenedisilver nitrate, 35B, 520
Isopropyl alcohol, 29, 453
Isopropylamine octahydrate, 35B, 422
Isopropyl carboxaldehyde, 30B, 402
1-Isopropyl-4,8,12-trimethyl-2,4,7,11-cyclotetra decatetraene, 34B, 315
Isoproterenol sulphate dihydrate, 35B, 911
Isoquinocycline A, 29, 772
Isoquinoline hydrochloride, 27, 985; 30B, 137
Isoquinoline hydrochloride monohydrate, 30B, 137
Isorubratoxin B bis-*p*-bromophenylhydrazide methyl ester, 35B, 921
1,2-Isotactic poly-4-methyl penta-1,3-diene, 33B, 522
Isotactic *trans*-1,4-polypentadiene, 26, 637
Isotactic polyvinylethylsilane, 34B, 365
Isothiathiophthens, 35B, 912
(Isothiocyanato)dicarbonyl-2,2'-bipyridine-π-allyl molybdenum, 34B, 443
(Isothiocyanato)dicarbonyl-1,10-phenanthroline-(2-methylallyl)-molybdenum, 35B, 527
Isothiocyanato-dinitro-ethylenediamine-ammino-cobalt(III), 35B, 919
Isothiocyanatopentamminecobalt(III), chloride, 32A, 197
Isothiocyanatothiocyanato-[1,2-bis(diphenylphosphino)ethane]-palladium(II), 35B, 916
Isothiocyanatothiocyanato(1-phenylphosphine-3-dimethylaminopropane)palladium(II), 35B, 840
Isoviolanthrone, 29, 699
Isozygosporin A mono-*p*-bromobenzoate, 35B, 910
Itaconic acid, 26, 558
Ixiolite, 27, 548; 28, 144, 324

Jachimovite, 27, 673; 28, 277
Jacobine, 26, 831
Jacobine bromodilactone, 28, 560
Jacobine-bromohydrin, 28, 550
Jacobsite, 29, 417; 31A, 244
Jadeite, 28, 350; 31A, 228

Jadeite, green, 28, 350
Jadeite, white, 28, 350
Jakob's salt, 34A, 229
Jamine, 28, 684; 29, 747
Jarosite, 27, 634; 30A, 453
Jaroslavite, 31A, 255
Jatrophone dihydrobromide, 35B, 920
Jennite, 31A, 242
Jimboite, 28, 354; 30A, 415
Johannsenite, 29, 430, 439; 32A, 457
Jordanite, 31A, 164
Jouravskite, 30A, 456; 31A, 174; 34A, 313
Juddite, 28, 334

Kaersutite, 28, 279
Kainosite, 26, 534; 31A, 229
Kaliborite, 31A, 172
Kalistrontite, 27, 635
Kalsilite, 29, 430, 30A, 425
Kämmerite, 32A, 475
Kanamycin selenate monohydrate, 33B, 555
Kanamycin sulphate monohydrate, 33B, 555
Kaolinite, 26, 529; 27, 697
Karelianite, 28, 320
Kasoite, 30A, 459
Kasolite, 27, 673; 28, 278; 29, 405; 30A, 459
Kempite, 29, 251
Kennedyite, 26, 379
Kentrolite, 27, 714
Kentrolite-melanotekite, 31A, 243
Keratin, 35B, 908
α-Keratin, 26, 824; 27, 1056; 29, 758
Kermesite, 32A, 222
Kernite, 31A, 260; 32A, 501
2-Keto-3-ethoxybutyraldehyde-bis(thiosemicarbazone), 31B, 454; 34B, 24
2-Keto-3-ethoxybutyraldehyde bis (thiosemicarbazone) cupric complex, 31B, 454
β-Ketoglutaric acid, 32B, 10
6-Ketononanolide, 33B, 132
α-Keto-1,5-tetramethyleneferrocene, 32B, 358
α-Keto-1,1'-trimethyleneferrocene, 30B, 307
Kilchoanite, 26, 512
Kischtinite, 30A, 456
Knebelite, 31A, 244, 256
Kobsine methiodide, 35B, 91
Koenenite, 33A, 277
(-)-Kopsanone N(b)-methiodide, 33B, 548; 34B, 330
Kornerupine, 30A, 456; 33A, 492
Krausite, 30A, 372
Krauskopfite, 32A, 459
Kreittonite, 27, 488
Kreysiginine methiodide, 33B, 555
Kröhnkite, 26, 449
Kromycin, 35B, 921
Krypton, 26, 286
Kurchatovite, 31A, 255
Kurnakovite, 28, 320
Kurrol B, 28, 186
Kutnahorite, 31A, 243
Kyanite, 26, 505; 28, 239; 30A, 440; 31A, 225

Labradorite, 31A, 242, 254
Lactate dehydrogenase, 35B, 888
β-Lactoglobulin, 29, 758
α-Lactose monohydrate, 26, 625
Lamprophyllite, 30A, 459
Lanarkite, 35A, 374
Landauite, 31A, 248; 35A, 445
Landesite, 29, 425
Langite, 30A, 458
Lanthanide dihydroxide chloride, 34A, 221
Lanthanide *mono*nitrilotriacetate hydrates, 33B, 544
Lanthanide oxalate hydrates, 35B, 910
Lanthanide oxyiodides, 26, 338
Lanthanon chloride hexahydrate, 31A, 241
Lanthanon chloroacetates, 33B, 543
Lanthanon formates, 27, 815
Lanthanon molybdenum oxides, 31A, 250
Lanthanon sodium sulphides, 29, 341
Lanthanum, 26, 233
Lanthanum aluminate, 31A, 132
Lanthanum aluminate-lead titanate, 27, 505
Lanthanum borate, 26, 491; 27, 725; 35A, 282
Lanthanum borates, 35A, 282
Lanthanum carbonate octahydrate, 33A, 433
Lanthanum chloride, 27, 724; 33A, 204
Lanthanum digermanate, 34A, 261
Lanthanum ditelluride, 30A, 64
Lanthanum ferrate-lead titanate, 27, 505
Lanthanum(III) fluoride, 29, 258; 30A, 267
Lanthanum hexaboride, 26, 69
Lanthanum magnesium nitrate, 26, 475
Lanthanum-magnesium-sulphur, 27, 327
Lanthanum-manganese-silicon-sulphur, 35A, 76
Lanthanum-manganese-sulphur, 27, 327
Lanthanum niobate, 27, 521; 32A, 293
Lanthanum niobium oxides, 31A, 257
Lanthanum-nitrogen, 27, 304
Lanthanum oxalate decahydrate, 30B, 408
Lanthanum oxide, 26, 366; 30A, 309
Lanthanum oxide telluride, 32A, 308
Lanthanum oxide-uranium oxide, 27, 545
Lanthanum—oxygen—ytterbium, 26, 365
Lanthanum oxyselenide, 35A, 235
Lanthanum oxytelluride, 27, 424

Lanthanum—palladium, 26, 165
Lanthanum—platinum, 26, 236, 239; 27, 322
Lanthanum rhenium oxide, 33A, 353
Lanthanum rhodite, 26, 399
Lanthanum-rhodium, 35A, 77
Lanthanum ruthenite, 26, 399
Lanthanum sodium sulphide, 29, 341
Lanthanum strontium manganite, 26, 419
Lanthanum-strontium-sulphur, 27, 327
Lanthanum sulphide, 34A, 294
Lanthanum sulphide halides, 31A, 247
Lanthanum-sulphur, 34A, 98
Lanthanum tantalate, 32A, 293; 33A, 341
Lanthanum-tellurium, 27, 335, 336; 29, 64; 31A, 47
Lanthanum titanate, 31A, 138
Lanthanum trifluoride, 31A, 84
Lanthanum ytterbium oxide, 34A, 253
Lanthanum-yttrium, 27, 330
Lappaconine hydrobromide, 34B, 330; 35B, 361
Larderellite, 34A, 351
Larnite, 31A, 243
Larsenite, 32A, 428
Latrappite, 29, 439
Lattice distortions, 28, 292
Lattice parameters, 28, 294; 33B, 541 ff; 35B. 906 ff
Laueite, 30A, 395
Laumontite, 32A, 482, 483
Laurencin, 30B, 393, 34B, 340
Lauric acid, 26, 562
Lauric acid, form A_1, 28, 373
Laurinterol acetate, 32B, 490; 34B, 310
Lautite, 29, 23; 30A, 18; 31A, 240
Laveite, 34A, 331
Lâvenite, 30A, 456; 31A, 248
Laves phases, 26, 269; 27, 260
Lawsonite, 26, 520
Lazulite, 27, 690
Lead, 26, 280; 27, 385
Lead acetate trihydrate-isatin, 26, 828
Lead acetate trihydrate mixed crystals with various dyes, 26, 828
Lead arseno-sulphides, 27, 402
α-Lead azide, 28, 55; 34A, 170
Lead *n*-butylxanthate, 33B, 319
Lead chloride, 26, 324; 28, 80
Lead chromate, 27, 609
Lead chromate—lead molybdate, 26, 391
Lead cyanamide, 29, 239
Lead diethyldithiocarbamate, 32B, 492
Lead(II)O,O-di-iospropyl phosphorodithioate, 34B, 412
Lead dipotassium difluoroberyllate, 35A, 173
Lead ethyl xanthate, 31B, 331
Lead fluoride, 26, 301; 31A, 251; 32A, 156
Lead fluoroberyllate, 26, 302; 33A, 179
Lead germanate, 27, 541
Lead halides, 34A, 176
Lead hexa-antipyrine perchlorate, 31B, 334
Lead(II) hydroxide bromide, 31A, 107
Lead(II) hydroxide iodide, 26, 327; 31A, 108
Lead isopropylxanthate-pyridine complex, 34B, 413
Lead—lithium—magnesium, 26, 185; 28, 28
Lead lithium oxide, 29, 316
Lead magnesium carbonate, 31A, 253
Lead magnesium wolframate, 27, 510
Lead 8-mercaptoquinolinate, 32B, 492
Lead metaniobate, 31A, 149, 245, 250; 32A, 295
Lead metatantalate, 26, 388; 31A, 245
Lead molybdenum oxide, 30A, 328
Lead niobates, 31A, 256
Lead-niobium-oxygen, 27, 540
Lead niobium titanium oxides, 31A, 249
Lead orthoarsenate, 27, 587
Lead orthophosphate, 27, 587
Lead orthovanadate, 27, 587
Lead oxide, 27, 382, 477; 34A, 240
Lead *monoxide*, 26, 360
Lead—oxygen—tantalum, 26, 388
Lead oxyselenides, 26, 432
Lead periodates, 30A, 458
Lead phosphate, low, 35A, 330
Lead phosphite nitrate, 31A, 243
Lead polyphosphate, 29, 367
Lead, red, 30A, 313
Lead selenate, 27, 609
α-Lead selenate, 27, 609
β-Lead selenate, 27, 609
Lead—selenium—tin, 26, 153
Lead sesquioxide, 34A, 240
Lead sulphate, 26, 446; 28, 208
Lead tantalates, 26, 388; 31A, 256
Lead-tellurium-tin, 27, 211
Lead tetraborate, 31A, 169
Lead tetragermanate, 26, 398
Lead thioborate, 33A, 363
Lead thiocyanate, 31A, 258
Lead thiotitanate, 31A, 257
Lead—thorium, 26, 186
Lead-tin, 26, 187
Lead titanate, 26, 504; 30A, 457; 33A, 305
Lead titanate-potassium niobate, 27, 504
Lead uranate, 26, 405; 27, 543; 31A, 256; 32A, 317
Lead uranates, 31A, 256
Lead—uranium, 26, 186
Lead vanadium bronze, 34A, 266
Lead vanadium oxides, 31A, 260
Lead wolframate, 35A, 228
Lecithins, 33B, 527
Lecontite, 28, 354; 32A, 336
Legrandite, 28, 326, 354

Leightonite, 27, 634
Lengenbachite, 26, 427
Leonhardite, 26, 612; 28, 304; 32A, 482
Lepidolite, 26, 525
Lesserite, 27, 557
Lessingite, 30A, 456
L-Leucine hydrobromide, 32B, 229
Leucite, 29, 430; 33A, 458
Leucophane, 28, 350
Leucophanite, 32A, 444
Leucophoenicite, 35A, 466
D-Leucylglycine hydrobromide, 31B, 190; 34B, 267
Leurocristine, 30B, 397
Leurocristine methiodide, 30B, 397
Leurocristine methiodide dihydrate, 31B, 258
Liberite, 29, 430; 31A, 230
Libethenite, 27, 601; 28, 194
Lifschitz nickel II complexes, 26, 830
Lillianite, 29, 424
Limestone minerals, 30A, 456
Limonin, 26, 801
Linarite, 26, 451; 27, 623
Lipid-water systems, 27, 1063
Liquid crystals, 33B, 529
Lirioresinol-B,dibromodimethyl ether derivative, 27, 997
Liroconite, 27, 584; 32A, 388; 33A, 424
Lithian fluorphlogopite, 31A, 231
Lithiophyllite-tephroite, 31A, 252
Lithium, 26, 3, 279; 27, 264
Lithium acetate dihydrate, 35B, 11
Lithium acetylide, 32A, 233
Lithium aluminate, 26, 368, 394; 27 519; 30A, 313; 31A, 131
α-Lithium aluminate, 27, 519
γ-Lithium aluminate, 27, 519
δ-Lithium aluminate, 27, 519
Lithium aluminum fluoride, 33A, 181
Lithium aluminum hydride, 32A, 149
Lithium amidoaluminate, 31A, 247
Lithium ammonium hydrogen citrate *mono*hydrate, 32B, 26
Lithium antimonate, 34A, 273
Lithium azide, 33A, 168
Lithium bis(oxo-μ-trioxo-chromium(VI)) chromium(III), 32A, 277
Lithium bromate, 27, 643
Lithium cadmium hydroxy molybdate, 33A, 337
Lithium copper(II) chloride dihydrate, 28, 83
Lithium copper vanadate, 31A, 143, 251
Lithium cryolite, 31A, 258
Lithium cuprate, 35A, 216
Lithium deuteride, 34A, 151
Lithium diborate, 27, 548; 33A, 444
Lithium dichromate dihydrate, 35A, 387
Lithium digermanate, 33A, 507
Lithium dihydrogen citrate, 30B, 28
Lithium dihydrogen phosphite, 35A, 345
Lithium dimedonesulphonate, 30B, 421
Lithium disilicate, 34A, 363
Lithium dithionate dihydrate, 32A, 323
Lithium ferrite, 26, 394, 540; 29, 322; 30A, 451
Lithium ferrite-aluminate, 28, 285
Lithium ferrite-gallate, 28, 285
Lithium ferrite-lithium vanadate, 29, 326
Lithium ferrites, 30A, 451
Lithium ferrites-chromites, 30A, 457
Lithium fluoride, 26, 297, 299; 27, 451
Lithium fluoroantimonate, 27, 460
Lithium fluoroberyllate hydrate, 27, 606
Lithium fluorophlogopite, 32A, 471
Lithium fluorozirconate, 30A, 449
Lithium germanate, 30A, 325; 32A, 491; 34A, 258
Lithium glycolate, 31B, 25
Lithium glycolate monohydrate, 30B, 16
Lithium-(S)-glycolate-2-*d*, 31B, 25
Lithium halides, 34A, 171
Lithium heptagermanate, 35A, 318
Lithium hexafluorostannate(IV), 31A, 253
Lithium hydrazinium sulphate, 29, 343; 31A, 254; 32A, 330
Lithium hydride: carbon, 35A, 134
Lithium hydride-lithium deuteride, 31A, 257
Lithium hydrogen oxalate *mono*hydrate, 35B, 9
Lithium iodate, 31A, 214
Lithium iodide *mono*hydrate, 30A, 287
Lithium iron hexafluoride, 33A, 182
Lithium iron wolframate, 34A, 280
Lithium lutecium oxide, 27, 516
Lithium magnesium imide, 27, 420
Lithium manganate(III), 34A, 291
Lithium—manganese—phosphorus, 26, 187
Lithium—manganese—selenium, 26, 189
Lithium mesouranate, 27, 544
Lithium metaborate, 26, 486; 27, 718; 29, 386; 31A, 165
Lithium metaborate dihydrate, 31A, 166
Lithium metasilicate, 33A, 452
Lithium molybdate, 27, 724
Lithium—molybdenum—nitrogen, 26, 187
Lithium molybdenum oxides, 27, 724; 29, 316; 31A, 245; 32A, 296
Lithium molybdenyl arsenate, 35A, 356
Lithium nickel fluoride, 33A, 184
Lithium nickel vanadate, 26, 409
Lithium niobate, 29, 316; 30A, 331; 31A, 147

Lithium niobium oxide, 29, 316; 30A, 331; 31A, 147
Lithium niobium titinate, 35A, 241
Lithium nitrate, 26, 472
Lithium—nitrogen—wolfram, 26, 187
Lithium orthophosphate, 32A, 355
Lithium orthosilicate, 33A, 467
Lithium oxalate, 29, 460
Lithium oxalate monoperhydrate, 34B, 15
Lithium oxide, 27, 718, 724; 29, 316; 34A, 238
Lithium oxomercurate, 29, 338
Lithium—oxygen—vanadium—zinc, 26, 417
β-Lithium oxythallate, 26, 401
γ-Lithium oxythallate, 26, 401
Lithium pentafluorouranate(IV), 31A, 93
Lithium perchlorate trihydrate, 33A, 450
Lithium periodate trihydrate, 27, 640
Lithium-platinum, 27, 323
Lithium potassium sulphate, 27, 605
Lithium rare earth oxides, 27, 516; 28, 134
Lithium rhodite, 26, 399
Lithium-rhodium, 30A, 65
Lithium rhodium oxide, 26, 390
Lithium rhodium titanium oxide, 27, 486
Lithium rubidium fluoride, 30A, 269
Lithium ruthenite, 26, 399
Lithium samarium oxide, 35A, 214
Lithium scandate, 26, 370
Lithium selenate hydrate, 27, 606
Lithium silicate, 26, 506
Lithium silicon, 30A, 65, 31A, 48
Lithium sodium fluorohexathoriate(IV), 34A, 195
Lithium sodium sulphate, 32A, 328
Lithium sodium tetragermanate, 34A, 259
Lithium stannate, 31A, 256; 34A, 263
β-Lithium stannate, 35A, 220
Lithium stearate, 27, 1061
Lithium-strontium, 27, 265
Lithium sulphate, 27, 606
Lithium sulphate *mono*hydrate, 28, 207; 30A, 363; 33A, 365
Lithium tantalate, 29, 316; 31A, 249; 32A, 312
Lithium tetrachloro zincate, 34A, 205
Lithium tetrafluoroberyllate, 31A, 86
Lithium tetrafluoroberyllate(II) octafluorozirconate(IV), 29, 262
Lithium tetrafluoroytterbate, 35A, 166
Lithium tetrahydroxyborate, 31A, 166
Lithium tetramethylberyllate, 33B, 299
Lithium tetramethylzincate, 33B, 299
Lithium thulium oxide, 27, 516
Lithium-tin, 34A, 99
Lithium tin oxide, 33A, 288
Lithium titanate, 27, 527; 29, 318; 34A, 254
Lithium trichromate(III, VI), 31A, 150
Lithium uranate, 27, 543; 29, 316, 340; 30A, 347
α-Lithium uranate, 26, 404
Lithium uranium octafluoride, 32A, 179
Lithium uranium oxide, 29, 316
Lithium vanadium bronze, 30A, 343
Lithium vanadium oxide, 29, 316
Lithium vanadium wolfram oxide, 31A, 247
Lithium wolframate, 26, 391; 27, 724; 33A, 351; 34A, 280
Lithium wolframate hydrate, 31A, 156
Lithium wolfram bronze, 26, 357
Lithium ytterbium oxide, 27, 516
Lithium ytterbium wolframate, 35A, 249
Lithium yttrium oxide, 27, 516
Lithium yttrium fluoride, 26, 311
Lithium zirconium wolframate, 32A, 317
Lizardite, 30A, 434
Loeweite, 35A, 441
Lomonosovite, 31A, 240
Long chain alcohols, 26, 563
Longifolene hydrochloride, 30B, 225
Loparite, 26, 378; 28, 334
Lopezite, 32A, 278
Low cordierite, 31A, 223
Low cristobalite, 30A, 420
Lucidusculine hydroiodide, 31B, 254
Ludlamite, 31A, 187
Ludwigite, 29, 430; 31A, 253
Lumiphorbol 12,13,20-triacetate 4-(*p*-bromobenzoate), 33B, 545
Lunarine hydrobromide *mono*hydrate, 30B, 236
Lunarine hydroiodide *mono*hydrate, 30B, 236
Lüneburgite, 26, 494
Luranine, 35B, 390
Lutecium, 26, 234
Lutecium borate, 26, 491
Lutecium dihydride, 27, 237
Lutecium-manganese, 31A, 48
Lutecium niobate, 27, 521
Lutecium oxide, 27, 724
Lutecium-silicon, 27, 336
Lutecium stannate, 26, 399
Lutecium trihydride, 27, 237
2,6-Lutidine, urea (1:1 complex), 30B, 259
2,6-Lutidinium 2,4-dinitrophenyl phosphate, 32B, 487
Luzonite, 31A, 256, 261
L-Lysine hydrochloride dihydrate, 27, 767
Lysozyme, 26, 819; 33B, 527; 35B, 888
Lythridine methiodide, 33B, 552
β-Lyxose, 26, 621; 31B, 168

Macallisterite, 30A, 455

Macdonaldite, 33A, 489
Mackinawite, 29, 424
Macrosalhine bromide, 32B, 491
Macusine A, 26, 830
Macusine A-iodide, 28, 558
Maddrell salts, 28, 304; 32A, 383
Magnesium, 27, 382
Magnesium aluminate, 27, 519
Magnesium boracite, 27, 551
Magnesium(II) bromide, 34A, 180
Magnesium bromide diethyletherate, 32B, 322
Magnesium bromide-methanol system, 33B, 542
Magnesium carbonate, 26, 481
Magnesium chloride dodecahydrate, 31A, 93
Magnesium chloride sulphate, 31A, 251
Magnesium chromite, 27, 492
Magnesium chromium spinel, 31A, 251
Magnesium citrate decahydrate, 30B, 30
Magnesium diarsenate, 35A, 357
Magnesium diborate, 31A, 167
Magnesium dioxalatoplatinate, 33B, 406
Magnesium ferrite, 26, 411
Magnesium fluoride, 26, 299; 27, 430
Magnesium fluoroborate hexahydrate, 26, 304
Magnesium formate dihydrate, 29, 455
Magnesium gallate, 31A, 133
Magnesium germanate, 26, 397; 27, 501; 35A, 272
Magnesium hexa-antipyrine perchlorate, 30B, 399; 32B, 324
Magnesium hydroxychloride, 26, 327
Magnesium—manganese—oxygen—titanium—zinc, 26, 409
Magnesium manganite, 27, 493
Magnesium molybdate, 32A, 509
Magnesium—neodymium, 26, 189
Magnesium-neodymium-sulphur, 27, 327
Magnesium niobate, 26, 389
Magnesium niobium oxyfluoride, 35A, 179
Magnesium nitrate hexahydrate, 26, 474
Magnesium nitride fluorides, 34A, 189; 35A, 153
Magnesium nitride trifluoride, 34A, 189
Magnesium oxide-chromium oxide, 31A, 257
Magnesium—palladium, 26, 225; 27, 268; 33A, 110
Magnesium palladium hexafluoride, 33A, 188
Magnesium phosphate, 33A, 394
Magnesium—platinum, 26, 225; 27, 322
Magnesium potassium sulphate hexahydrate, 30A, 375
Magnesium potassium trimetaphosphate, 32A, 379; 33A, 413
Magnesium-praseodymium-sulphur, 27, 327
Magnesium pyroarsenate, 28, 195
Magnesium pyrophosphate, 26, 466; 30A, 385
α-Magnesium pyrophosphate, 32A, 372
Magnesium pyrophosphate-zinc pyrophosphate, 30A, 385
Magnesium rare-earth compounds, 27, 268
Magnesium rhodite, 26, 399
Magnesium ruthenite, 26, 399
Magnesium-samarium-sulphur, 27, 327
Magnesium selenate, 29, 357
Magnesium selenate hexahydrate, 27, 618
Magnesium selenite hexahydrate, 31A, 209
Magnesium silicate, 26, 510; 31A, 248
Magnesium silicate-nickel germanate, 27, 488
Magnesium-silver, 26, 190; 27, 269
Magnesium-silver-zinc, 32A, 97
Magnesium sodium fluoride, 26, 304
Magnesium stearate, 27, 827
Magnesium-strontium, 26, 175,176; 27, 94, 269; 30A, 23
β-Magnesium sulphate, 27, 606
Magnesium sulphate heptahydrate, 29, 348
Magnesium sulphate hexahydrate, 27, 618; 29, 347
Magnesium sulphate tetrahydrate, 27, 612; 29, 346
Magnesium-sulphur-ytterbium, 27, 327
Magnesium-sulphur-yttrium, 26, 277; 27, 327
Magnesium tantalate, 26, 389
Magnesium tellurate(VI), 35A, 242
Magnesium-thallium, 35A, 62
Magnesium thiosulphate hexahydrate, 27, 631; 34A, 314
Magnesium—throium, 26, 176
Magnesium thorium nitrate hydrate, 30A, 403
Magnesium tin borate, 33A, 443
Magnesium titanium spinel, 27, 528
Magnesium trihydrogen hexaoxoiodate, 35A, 396
Magnesium wolframate, 33A, 285; 34A, 347
Magnesium uranate, 27, 543
Magnesium-yttrium, 27, 195
Magnetite, 29, 419; 30A, 456; 31A, 250; 33A, 267
Magnocolumbite, 28, 330
Magnodravite, 31A, 251
Malachite, 31A, 260; 32A, 419
Maleic anhydride, 27, 945
Malonamide, 30B, 413; 35B, 23
Maltose monohydrate, 35B, 266
Mandelic acid:germanium dioxide complex dihydrate, 32B, 331
Manganarfuedsonite, 28, 334
Manganates, 33A, 318
α-Manganese, 32A, 98; 34A, 101; 35A, 78
Manganese ammonium sulphate hexahydrate, 31A, 204

Manganese boracite, 27, 551
Manganese(II) bromide dihydrate, 32A, 204
Manganese carbonate, 26, 481
Manganese(II) chloride dihydrate, 30A, 274
Manganese chromite, 27, 492
Manganese(II) croconate trihydrate, 31B, 440
Manganese cyclopentadienide, 29, 767
Manganese decacarbonyl, 28, 67
Manganese dichloride tetrahydrate, 29, 273
Manganese dioxide, 26, 357
Manganese dirhenium tetradecacarbonyl hydride, 32A, 396
Manganese ferrite, 26, 411; 27, 495; 30A, 451
Manganese fluoride, 26, 300; 28, 72
Manganese fluoroborate hexahydrate, 26, 304
Manganese(II) formate dihydrate, 33B, 402
Manganese germanate, 33A, 505
Manganese hexacyanochromate, 26, 343
Manganese hexacyanochromate hexahydrate, 27, 653
Manganese(II)hexacyanocobalt(II) hydrate, 35A, 405
Manganese hexacyanoiridate, 26, 343
Manganese hexahydroxostannate(IV), 34A, 236
Manganese hydraxine-carboxylate dihydrate, 30B, 409
Manganese hydrazine chloride, 27, 472
Manganese(II) hydroxide, 30A, 305
Manganese hydroxychloride, 26, 327
Manganese isoselenocyanato dimethylformamide complex, 29, 764
Manganese isothiocyanato dimethylformamide complex, 29, 764
Manganese-leonite, 26, 447; 33A, 369
Manganese-mercury, 26, 191; 27, 270
Manganese-mercury-nitrogen, 27, 272
Manganese molybdate, 30A, 327
Manganese-molybdenum-silicon, 26, 269, 270
Manganese-neodymium-sulphur, 27, 327
Manganese-nickel-silicon, 26, 175; 27, 165, 374, 376; 28, 28
Manganese niobate, 26, 389
Manganese-niobium-silicon, 26, 270; 33A, 86
Manganese-nitrogen, 27, 271, 273; 33A, 111
Manganese-nitrogen-tin, 27, 273
Manganese orthostannate, 33A, 285
Manganese orthotitanate, 29, 325; 33A, 320
Manganese oxide, 27, 497; 32A, 285
α-Manganese(III) oxide, 32A, 256
Manganese-oxygen, 26, 407
Manganese-oxygen-titanium, 26, 408
Manganese-palladium, 27, 274
Manganese pentacarbonyl hydride, 29, 240
Manganese(II) phosphate, 34A, 330
Manganese phosphide, 27, 275
Manganese-phosphorus, 27, 319
Manganese phthalo-cyanine, 30B, 435
Manganese-platinum, 27, 254
Manganese potassium chlorides hydrated, 33A, 212
Manganese potassium fluoride, 26, 305
Manganese potassium zinc silicate disilicate, 34A, 386
Manganese-praseodymium-sulphur, 27, 327
Manganese(II)-pyridoxylidenevaline chelate, 28, 639
Manganese (II)-pyridoxylidenevaline tetrahydrate, 29, 767
Manganese pyrochlore fluoroantimonate, 28, 200
Manganese pyrophosphate, 26, 466
Manganese rare-earth alloys, 26, 192
Manganese-rhenium, 26, 192
Manganese rhodite, 26, 399
Manganese rubidium chloride dihydrate, 32A, 188; 35A, 198
Manganese ruthenite, 26, 399
Manganese-samarium-sulphur, 27, 327
Manganese--scandium, 26, 237
Manganese selenate, 29, 357
Manganese selenite dihydrate, 31A, 243
Manganese-silicon, 27, 375; 29, 64; 32A, 99, 102; 34A, 101
Manganese silicon carbide, 35A, 134
Manganese-silicon-tantalum, 26, 270
Manganese-silicon-titanium, 26, 269, 270
Manganese-silicon-wolfram, 26, 269, 270
Manganese-silver, 27, 183
Manganese-silver-tin, 27, 277
ζ-Manganese-silver-tin, 27, 310
Manganese sodium diorthosilicate, 32A, 440
Manganese(II) sodium disilicate, 34A, 384
Manganese sodium fluoride, 34A, 201
Manganese(IV) spinels, 31A, 251
Manganese strontium silicate, 26, 523
Manganese sulphate, 26, 436; 30A, 369
Manganese-sulphur-ytterbium, 27, 327; 32A, 104
Manganese-sulphur-yttrium, 26, 277; 27, 327
Manganese tantalate, 26, 389
Manganese-technetium, 27, 361
Manganese telluride, 26, 151
Manganese tetracarbonyl bromide, 28, 69
Manganese thiogermanate, 30A, 441
Manganese thiosilicate, 30A, 441
Manganese-thulium, 26, 192; 27, 196

Manganese-tin, 26, 192, 193; 27, 259
Manganese-titanium, 26, 193; 27, 278
Manganese titanium spinel, 27, 528
Manganese(II)tris(pyridine-2-carboxaldehyde)-1,1,1-tris(aminomethyl)-ethane diperchlorate, 35B, 921
Manganese uranate, 26, 406; 27, 523; 31A, 159
Manganese-vanadium, 27, 279
Manganese vanadium oxide, 27, 490
Manganese wolframate, 31A, 260
Manganese-yttrium, 26, 236; 27, 279
Manganese-zinc, 27, 173, 270
Manganese zinc ferrite, 28, 143
Manganic acetate, anhydrous, 34B, 531
Manganite, 28, 131; 30A, 456
Manganites, 27, 493
Manganoan lipscombite, 27, 603
Manganoneptunite, 30A, 456
Manganostibite, 35A, 443
Manganotantalite, 31A, 244
Manganous fluoride, 28, 290
Manganous formate dihydrate, 29, 455
Manganous iodide, 27, 434
β-Manganous orthogermanate, 34A, 262
Manganous oxide, 27, 474
Mangapyrosmalite, 26, 536
D-Mannitol, 33B, 163, 543
α-D,L-Mannose, 32B, 489
Margarite, 30A, 459
Margarosanite, 28, 354; 34A, 383
Marialite, 27, 688
Marialite scapolite, 30A, 436
Marine brown algae, 27, 1056
Marokite, 29, 321; 31A, 136
Marrite, 28, 50; 32A, 229
Martensite, 27, 155, 251; 30A, 194
ε-Martensite, 27, 155
Maucherite, 27, 58, 60
Mawsonite, 30A, 453
Maximum microcline, 27, 678
Mayenite, 28, 418
Maytoline methiodide, 35B, 921
Mckelveyite, 30A, 456
Mckinstryite, 31A, 257
Meionite, 27, 688
Meisenheimer salts, 33B, 50
Melaleucic acid, 28, 684
Melanophlogite, 28, 322
Melanostibite, 33A, 325
Melanotekite, 27, 714
Melanterite, 29, 351
Melilite, 30A, 455
Melilitic silicates, 26, 523
Meliphanite, 32A, 444
Mellite, 30A, 454
Mellitic acid, 26, 713
Menthyl *p*-iodobenzenesulphinate, 29, 766
Menthyl trimethylammonium iodide, 27, 999
Mercallite, 29, 346
Mercaptocobalt carbonyl complexes, 33B, 488
2-Mercapto-6-methyl-purine monohydrate, 33B, 156
6-Mercaptopurine monohydrate, 34B, 234
8-Mercaptoquinolinatoplatinum, 35B, 908
cis-Mercuric bromide iron tetracarbonyl, 32A, 399
Mercuric chloride complexes, 29, 632
Mercuric iodide, 32A 211; 34A, 182
Mercuric oxides, 29, 310
Mercuric selenates, 27, 610
Mercuric sulphate, 26, 443; 27, 610; 28, 210, 212
Mercuric sulphate monohydrate, 26, 443, 444; 29, 356
Mercuric sulphates, 27, 610
Mercuric thiocyanate-thiourea, 28, 113
2-Mercuri-4-methylphenol-2-nitroso-4-methylphenol, 35B, 771
Mercuripapain, 26, 819
Mercury, 26, 194, 279; 27, 386
Mercury(I) acetate, 30B, 414
Mercury(II) acetate, 30B, 414
Mercury(II) acetamide, 34B, 548
Mercury aluminum oxide, 31A, 247
Mercury antimony bromide, 31A, 256
Mercury(I) bromate, 32A, 319
Mercury bromide iodide, 34A, 211
Mercury *tert*-butyl mercaptide, 29, 769
Mercury chalcogen halides, 27, 465, 467; 32A, 191
Mercury chloride thiourea complex, 26, 575
Mercury(II) chromate, 26, 444
Mercury complexes, 30A, 301
Mercury(II) cyanide, 34A, 228
Mercury dibenzyl, 30B, 428
Mercury dimethoxide-phenylisothiocyanate adduct, 32B, 456
γ-Mercurydisulphonium chloride, 33A, 513
Mercury ethylmercaptide, 30B, 374
Mercury(II) iodide, 32A, 211; 34A, 182
Mercury 8-mercaptoquinoleate, 33B, 543
Mercury methylmercaptide, 29, 630
Mercury(I) *o*-phthalate, 32B, 417
Mercury-potassium, 27, 280
Mercury orthotellurate, 28, 223
Mercury oxybromide, 33A, 235
Mercury oxychloride, 29, 283
Mercury(II) oxycyanide, 28, 116
Mercury-palladium, 26, 195, 30A, 66
Mercury(I) perchlorate tetrahydrate, 31A, 211
Mercury-platinum, 26, 225
Mercury(I) selenate, 34A, 300
Mercury(II) selenate, 26, 444
Mercury selenide, 27, 281
Mercury-selenium, 27, 281, 382
Mercury-selenium-tellurium, 26, 195
Mercury-silver, 26, 248; 34A, 103

Mercury-silver-tin, 26, 195
Mercury-sodium, 27, 280
Mercury(I) sulphate, 34A, 300
Mercury(II) sulphate, 26, 444
Mercury sulphide chloride, 33A, 228
Mercury sulphide halides, 31A, 253
Mercury-sulphur, 29, 67
Mercury tellurate, 27, 635
Mercury-tellurium, 27, 382
Mercury(II) tetrathioureatetrathiocyanatocobaltate(II), 30B, 342
Mercury thiocyanates, 33A, 253
Mercury(II) thiocyanate thiourea complex, 31B, 401
Mercury-tin, 26, 195
Mercury vanadium oxide, 33A, 312
Merocyanine, 29, 738
Merocyanine-iodine, 33B, 248
Merriheuite, 30A, 455
Merwinite, 32A, 431
Mesaconic acid, 35B, 919
Mesitaldehyde-perchloric acid, 29, 769
Mesitaldehyde, perchloric acid (2:1 adduct), 30B, 265
Mesitylene, 27, 1064
Mesitylmethylsulphone, 29, 770; 32B, 68
Meso-α,α'-dimethylglutaric acid, 28, 671; 29, 468
Mesomorphic phases, 27, 1061
Mesotartaric acid, 29, 765
Metaankoleite, 31A, 244
Meta-autunite I, 26, 468; 28, 354
Meta-autunite II, 26, 468
Metaborates, 27, 718
Metaboric acid, 26, 486; 28, 155; 29, 384
Metaboric acid, cubic, 28, 155
Metaboric acid, monoclinic, 28, 156
Metaborite, 29, 427
Metacinnabarite, 29, 67
Metahewettite, 26, 380
Metajennite, 31A, 242
Metal alkoxides, 33B, 301
Metal-carbamate complexes, 26, 655
Metal hydrazinium sulphates, 31A, 250
Meta-lomonosovite, 26, 536; 27, 689; 31A, 240
Metal oxide phosphorus oxychloride complexes, 26, 692
Metal tetraphenylporphines, 29, 617
Metal titanates, 33A, 294
Metal(I) vanadate-phases, 35A, 274
Metanilic acid, 30B, 65
Metaphosphates, 31A, 245
Metastrengite, 29, 425; 31A, 185
Metatorbernite, 26, 468; 29, 375
Metavariscite, 31A, 185
Metavauxite, 32A, 367
Metavoltine, 26, 448
Metazellerite, 31A, 243
π-Methallyl[bis-1,2-(diphenylphosphino)ethane]nickel bromide, 35B, 53
Methane, 26, 543; 35B, 922
Methanesulphinic acid, 34B, 42
Methanesulphonamide, 31B, 32
Methanesulphonanilide, 33B, 55
1,6-Methano-cyclodecapentaene-2-carboxylic acid, 30B, 117
Methanol-calcium chloride, 35B, 918
Methanol solvate, 32B, 171
Methinyltricobalttriphenylphosphine octacarbonyl, 35B, 846
L-Methionine, 35B, 906
(+-)-Methionine-palladium chloride complex, 32B, 496
(2S,SR)-Methionine sulphoximine, 35B, 312
p-Methoxybenzenediazonium fluoroborate, 29, 770
p-Methoxybenzoic acid, 32B, 84
3-Methoxycarbonyl-*trans*-3,5-dimethyl-Δ'-pyrazoline hydrobromide, 30B, 123
L-5-Methoxycarbonyl-7-formyl-1,2,5,6-tetrahydro-3H-pyrrolo(1,2,a)azepin-3-one, 32B, 162
Methoxy carbonyl mercury (II) chloride, 27, 816
5-Methoxycarbonylmethylene-2-piperidino-Δ^2-1,3-thiazol-4-one, 35B, 915
p-Methoxychalcone, 35B, 56
3-Methoxy-5β,19-cyclo-5,10-secoandrosta-1(10),2,4-trien-17β-ol *p*-bromobenzoate, 32B, 490; 33B, 198
4-Methoxy-2,6-dimethylbenzonitrile *N*-oxide, 33B, 550
2-Methoxy-(1'-hydroxyethyl)benchrotrene, 35B, 918
p-Methoxyindophenol N-oxide, 31B, 54
4-Methoxy-3-iodobiphenyl-4'-carboxylic acid, 35B, 910
Methoxyiron(III) mesoporphyrinIX dimethyl ester, 29, 767; 30B, 218
(Methoxy(methyl)carbene) triphenylphosphinetetracarbonyl chromium, 34B, 433
3-Methoxy-7α, 8α-methylene-1,3,5(10)-oestratrien-17β-yl bromoacetate, 35B, 335
3-Methoxy-8β-methyloestradiol-17-monobromoacetate, 34B, 299
2-Methoxynaphthalene, 33B, 542
6-Methoxy-8-nitro-5-(1H)-quinoline, 34B, 151
(+)-*m*-Methoxyphenoxypropionic acid, 31B, 16
α-Methoxyphenylacetic acid complexes, 35B, 909
o-Methoxyphenyl diazonium chloride-ferric chloride double salt, 28, 686; 30B, 51
2-*p*-Methoxyphenyl-3,4-dibenzyl-1,3,4-thiadiazolidine-5-thione, 30B, 154
N-(*p*-Methoxyphenyl-3-propyl)-*p*-bromo-benzamide, 34B, 58
Methoxytetraphenylantimony, 33B, 290

N-Methylacetamide, potassium iodide, potassium triiodide complex, 30B, 257
N-Methylacetanilide, 30B, 398; 32B, 96
2-(Methyl-acetyl-methylene)-4-acetyl-4-methyl-1,3-dithiacyclobutane, 33B, 142
N-Methylacridinium iodide, 35B, 915
9-Methyladenine, 29, 708
9-Methyladenine dihydrobromide, 27, 977
Methylallyl[bis-1,2-(diphenylphosphino)ethane]nickel bromide, 33B, 546
π-2-Methylallylpalladium chloride dimer, 33B, 333
Methylamine, 30B, 400
2-Methylaminobenzothiazole, 35B, 236
Methyl 2'-(2-aminoethyl)-2,4'-bithiazole-4-carboxylate monohydrobromide (a bleomycin), 33B, 555
1-Methylamino-7-methylimino-1,3,5-cycloheptatriene, 29, 766; 32B, 118
3-*N*-Methylaminomethylpinane hydrobromide, 35B, 139
2-Methyl-4-amino-1-naphthol hydrochloride, 32B, 486
2-Methyl-3—amino-1,4-naphthoquinone, 31B, 100
2-Methyl-3-amino-1,4-naphthoquinone hydrate, 31B, 101
3-Methyl-4-amino-1,2-naphthoquinone hydrate, 35B, 918
p-Methylaminophenol sulphate, 35B, 67
N-Methylaminotrimethylsilane, 27, 770
Methyl ammine nickel trichloride, 31B, 29
Methylammonium alum, 26, 585; 29, 537
Methylammonium bromide, 26, 585
Methylammonium chromium alum, 33B, 519
5'-Methylammonium-5'-deoxyadenosine iodide monohydrate, 35B, 919
Methylammonium N-methyl boroacetamidate, 27, 763
Methylammonium perchlorate, 33B, 21
9-Methylanthracene, 34B, 106
Methyl α-arabofuranoside, 26, 624
(-)-17-*O*-Methyl-aspidodispermine N*b*-hydrobromide, 35B, 923
N-Methylated diethylenetriamine cobalt (II) chloride complex, 30B, 394
Methylated purine salts, 26, 831
Methylated pyrimidine salts, 26, 831
Methyl-4,6-benzal-α-D-iodside, 30B, 429
Methyl-4,6-benzal-α-D-idoside hemihydrate, 30B, 429
Methyl-4,6-benzal-3-methyl-β-idoside, 30B, 429
m-Methylbenzamide, 28, 412
2'-Methyl-1,2-benzanthraquinone, 28, 447
5-Methyl-1,2-benzanthraquinone, 28, 449
2-*p*-Methylbenzilidene-1,3-indandione, 35B, 124
N-Methyl-N-benzilthioformamide, 29, 770
Methyl benzobicyclo[2.1.1]hexene-exo-8-bromo-endo-2-carboxylate, 35B, 910
2-Methyl-1,2-benzoisothiazolin-3-thione, 33B, 554
6-Methylbenzo[*g*]naphtho[1,2-*c*]-cinnoline-7,12,14-trione, 35B, 201
3-Methylbenzothiazoline-2-thione, 26, 719
p-Methyl-π-benzyl-π-cyclopentadienyl-dicarbonylmolybdenum, 33B, 350
Methyl-4,6-O-benzylidene-2-O-*p*-bromobenzene-sulphonyl-3-cyano-3-deoxy-α-*D*-altropyranoside, 35B, 273
p-Methylbenzylidene-*p*-nitroaniline, 33B, 545; 35B, 61
N-Methyl-bisdeoxo-cyclo-showdomycin acetonide hydrobromide, 32B, 490; 34B, 283
N-Methyl-*N*,*N*-bis[methyleneamino-*N*'-(dithionitrite)]amino-nickel(II), 35B, 786
16-Methyl-15-bromoacetato-enmein derivative, 34B, 314
3-Methyl-4-bromo-5-carboxamide-pyrazole, 35B, 157
2-Methyl-2-bromo-6-carboxypyridine oxide, 29, 763
Methyl *m*-bromocinnamate, 30B, 8
Methyl *p*-bromocinnamate, 30B, 8
Methylbromodiazirine, 35B, 903
Methyl bromogibberellate, 27, 1017; 31B, 223
Methyl 6α-bromo-12-methoxy-7-oxo-podocarpate, 35B, 129
Methyl 6-bromo-12-methoxy-7-oxo-podocarpa-8,11,13-trien-16-oate, 33B, 546
2-Methyl-3-bromo phenol, 31B, 73
4-Methyl-3-(*p*-bromophenyl)-1,2,5-oxadiazole 2-oxide, 34B, 197
4-Methyl-3-(*p*-bromophenyl)-1,2,5-oxadiazole 5-oxide, 34B, 198
1-Methyl-5-bromouracil, 30B, 416
1-Methyl-5-bromouracil-9-ethyl-2-aminopurine complex, 34B, 228
1-Methyl-5-bromouracil-9-ethyl-8-bromoadenine complex, 34B, 227
1-Methyl-5-bromouracil, 9-methyl-adenine complex, 30B, 416; 31B, 161
2-Methyl-1-butene, 33B, 531
N-Methyl-N-*t*-butyl-3-hydroxy-azetidinium methane-sulphonate, 34B, 125
4-Methylcarbamoylcholine bromide, 35B, 918

O-Methyl-(Co-C)-carboxymethyl-bis (dimethylglyoximato) pyridinato-cobalt, 32B, 497
Methyl β-cellobioside-methanol, 35B, 268
Methylcephalonate bromoacetate, 34B, 318
N-Methyl-*p*-chlorobenzaldoxime, 28, 686; 29, 657
Methyl *p*-chlorocinnamate, 30B, 423
Methyl 2-chloro-2-deoxy-D-galactopyranoside, 34B, 249
Methyl-2-chloromercury-2-deoxy-α-D-talopyranoside, 30B, 420
Methylchloromethanes, 29, 765
2-Methyl-4-chloromethyl-5-*p*-nitrophenyl oxazole, 28, 495
Methyl cinnamate, 30B, 423
1-Methyl-2-cyclopentadienylidene-1,2-dihydropyridine, 27, 957
(+)S-Methyl-L-cysteine sulphoxide, 27, 765
N-Methylcytosine, 29, 771
1-Methylcytosine hydrobromide, 27, 971
p-Methyldiazoaminobenzene, 33B, 63
N-Methyl-3,5-dibromopyridone, 31B, 120
4-Methyl-2,6-di-t-butylphenol, 29, 769
Methyl 4,6-dichloro-4,6-dideoxy-α-D-galactopyranoside, 34B, 249
Methyl 4,6-dichloro-4,6-dideoxy-α-D-glucopyranoside, 33B, 165
Methyldicyanoarsine, 31B, 312
10-(2'-Methyl-2'-diethylaminoethyl)-phenothiazine hydrochloride, 34B, 210
1-Methyl-1,4-dihydronicotinamide, dimeric acid product of, 32B, 168
1-Methyl-5,6-dihydrouracil, 33B, 546
2-Methyl-1,4-dihydroxynaphthalene, 34B, 97
(RS)-*trans*-2-Methyl-3-(2,6-dimethyl-4-chlorophenyl)oxaziridine, 32B, 192
N-Methyl-5,-dimethyloxazolidine-2,4-dione, 35B, 228
Methyldi(neopentyl)sulphonium tri-iododi(neopentyl) sulphonium methyl zincate, 34B, 47
Methyl diphenyl-selenophosphinite, 34B, 388
Methyl diphenyl-thiophosphinite, 34B, 388
Methylene blue chloride, 26, 828
Methylenecyclopropane, 35B, 904
Methylene dibromide adduct of sporides-min, 30B, 240
6,6'-Methylenediphenanthridine, 33B, 127
Methylene diphosphonic acid, 29, 766
cis, D,L-8,9-Methyleneheptadecanoic acid, 28, 375
6β,7β-Methylene-17β-hydroxyandrost-4-en-3-one 17-acetate, 35B, 336
1,5-*endo*-Methylene-quinolizidinium *p*-toluene sulphonate, 34B, 163
Methylenetriphenylphosphorane, 34B, 389
Methyl ethylene phosphate, 30B, 194
1-Methyl-2-fluorenylidene-1,2-dihydropyridine, 34B, 111
Methylfluorocyclotetraphosphonitriles, 35B, 914
1-Methyl-5-fluorouracil-9-ethyl-2-aminopurine complex, 34B, 230
Methyl α-fucofuranoside, 26, 624
α-Methyl D-galactoside 6-bromohydrin, 28, 687; 30B, 177
N-Methyl-gelsemicine hydrobromide tetrahydrate, 26, 793
(-)-N-Methyl-gelsemicine hydroiodide, 26, 830; 27, 1021
Methyl α-D-glucopyranoside, 33B, 168
Methyl 4-O-α-D-glucopyranosyl-β-D-glucopyranoside monohydrate, 33B, 218
Methylglyoxal, 26, 550
Methylglyoxal bis(guanylhydrazone) di-hydrochloride monohydrate, 29, 765; 33B, 31
9-Methylguanine, 30B, 416
9-Methylguanine hydrobromide, 29, 706
3-Methylhepta-1,6-dienyl cobalt buta-diene, 32B, 493
o-Methylhydroxylamine hydrochloride, 29, 474
N-Methylhydroxylammonium chloride, 32B, 28
Methyl 2-hydroxy-5-methoxy-3,4-dimethoxycarbonylphenylacetate, 35B, 916
2-Methyl-8-hydroxyquinoline-5-sulphonic acid monohydrate, 35B, 185
1-Methyl-2-hydroxy-tetracyclo (4.4.2(2,5)) dodeca-6,11-diene-8-one mono-bromoacetate, 31B, 111
3-Methyl-3*H*-imidazo[2,1-*i*]purine-8(7*H*)-one, 35B, 254
10-Methylisoalloxazine hydrobromide dihydrate, 33B, 550; 34B, 153; 35B, 191
10-Methylisoalloxazinium bromide dihydrate, 35B, 191
10-Methylisoalloxazinium bromide-sesqui-naphthalene-2,7-diol monohydrate, 35B, 911
Methyl-3-isopropyl-2-phenol, 30B, 394
O-Methylisourea hydrochloride, 35B, 34
N-Methyl-DL-leucylglycine hydro-bromide, 34B, 272
Methyllithium, 35B, 471
o-Methyllythrine hydrobromide, 30B, 237
Methyl α-lyxofuranoside, 26, 624
Methyl α-D-lyxofuranoside, 33B, 160
Methylmalaleucate iodoacetate, 30B, 228
Methylmalonic acid, 35B, 4
Methylmanganese pentacarbonyl, 35B, 897
Methyl α-D-mannopyranoside, 35B, 258
1-Methyl-5-mercaptouracil disulphide, 35B, 280

Methylmercury(II) cyanide, 33B, 320
Methyl metadithiophosphonate, 27, 941; 29, 539
2-Methyl-3-N-methylanilinomethyl-1,4-naphthoquinone, 35B, 120
Methyl(methyl-4,5 di-O-benzoyl-3,7-dideoxy-7-iodo-D-arabino heptulopyranosid)onate, 32B, 489
1-Methyl-6[5-(1-methyl-1H-1-pyrindinyl)]-1-azoniaindan iodide, 31B, 133
Methylmicromerol bromoacetate, 28, 543
*mono*Methyl-*trans*, *trans*-muconate, 32B, 15
Methyl-2-α-naphthohydroquinone, 30B, 395
(+)-(R)-N-Methyl-1-((1-naphthyl)ethyl-(R)-*o*-methylmandelamide, 34B, 99
Methyl (2,2',3-nitrilodiethoxypropyl)silane, 34B, 366
2'-Methyl-4'-nitroazobenzene-2-sulphenylcyanide, 35B, 907
4-Methyl-2-nitrobiphenyl-4'-carboxylic acid, 35B, 910
p-Methyl-*m*-nitro-*N*-methylaniline, 33B, 551
1(10→5)*abeo*-3β-Methyl-24β-nor-18-α-oleanane, 33B, 554
1(10→5)-*abeo*-3β-Methyl-24β-nor-25α-18-α-oleanane, 35B, 350
Methyloxocarbonium hexafluoroantimonate, 33B, 38
DL-2-Methyl-7-oxododecanoic acid, 33B, 6
3-Methyl-4-oxo-1,3-thiazine-2-thione, 35B, 916
Methylparathion, 35B, 446
1-Methylpentaborane(9), 33B, 538
Methyl pentafuranosides, 26, 624
4-Methylpentaleno[6,6*a*,1,2-*def*]heptalene, 33B, 545
N-Methylphenazinium tetracyanoquinodimethanide, 31B, 136
Methylphenylbarbital, 33B, 551
2-Methyl-3-phenyl-4-bromo-isoxazolin-5-one, 32B, 191
N-Methyl-3-phenyl-4-bromoisoxazolin-5-one, 34B, 201
N-Methyl phenylcarbamate, 26, 701
3-Methyl-5-phenyl-1,2-dithiolium perchlorate, 28, 670
N-Methyl-4-phenylisoxazolin-5-one, 34B, 203
Methyl phenyl phospholanium iodide, 32B, 308
Methylphenylsulphone, 30B, 83
1-Methyl-2-picryliminoindolin, 35B, 190
Methyl pinacol phosphate, 31B, 179
Methyl potassium, 33B, 545; 35B, 472
D-Methyl-*n*-propyl-phenyl-benzylphosphoniumbromide, 30B, 398
2-Methyl pyridine, 26, 763
3-Methyl pyridine, 26, 763
4-Methyl pyridine, 26, 763
N-Methyl-pyridine-2-aldoxime bromide, 31B, 124
N-Methyl-pyridine-2-aldoxime chloride, 31B, 123
N-Methyl-pyridine-2-aldoxime iodide, 31B, 123, 124
4-Methylpyridine bis-(*o*-hydroxyacetophenonato) copper(II), 34B, 512
4-Methylpyridine-copper(II)chloride complex, 32B, 496
1-Methyl-2-pyridone photodimer, 29, 771
N-Methylpyrrole, 27, 759
2-Methyl-8-quinolol-aluminum(III) complexes, 33B, 542
1-Methyl-2-quinolonium dihydrogen arsenate, 34B, 148
N-Methylrhoeagenine iodide, 35B, 359
Methyl β-ribofuranoside, 26, 624
Methyl silyl ether, 35B, 888
Methyl stearate, 35B, 7
4-Methylsulphonoxyadamantanone, 35B, 914
8β-Methyltestosterone 17β-monobromoacetate, 35B, 337
N-Methyl-N,2,4,6-tetranitroaniline, 32B, 94
α-Methyltetronic acid, 28, 489
5-Methyl-1-thia-5-azacyclo-octane-1-oxide perchlorate, 34B, 208
2-Methylthiobenzothiazole, 27, 980
2-Methyl-4-thiocarbamoylpyridine, 34B, 136
2-Methyl-thio-isonicotinamide, 34B, 130
4-Methylthiomorpholine-1,1-dioxide, 34B, 203
Methylthiostannic acid, 33B, 553
1-Methyl-4-thiouracil-9-methyladenine, 35B, 922
S-Methyl*iso*thiourea sulphate, 27, 770
cis-9-Methylthioxenthene-10-oxide, 34B, 191
Methyl-1-thio-β-D-xylopyranoside, 31B, 169
5-Methylthiuronium benzoate, 26, 733
5-Methylthiuronium *p*-bromobenzoate, 26, 733
5-Methylthiuronium *p*-chlorobenzoate, 26, 734; 28, 573
5-Methylthiuronium *p*-iodobenzoate, 26, 733
1-Methylthymine, 28, 505
1-Methylthymine *trans-anti* dimer, 32B, 489
1-Methylthymine(+)9-methyladenine, 28, 507

(+)-Methyl *p*-tolyl sulphoxide, 35B, 83
Methyl-3,4,6-tri-O-acetyl-2-chloromercuri-2-deoxy-β-D-glucopyranoside, 27, 796
Methyltribromosilane, 33B, 538
N-Methyltrichlorophosphinimine dimer, 31B, 299
Methyl trichloro-trideoxy-hexoside, 29, 762
7-Methyl-tridecane complexes, 26, 575
*mono*Methyltriethyl titanate, 30B, 399
Methyltrimethylsiloxo-mercury tetramer, 34B, 368
cis-syn-6-Methyluracil dimer, 35B, 922
1-Methyluracil: 9-ethyl-8-bromo-2,6-diaminopurine, 35B, 275
1-Methyluracil hydrobromide, 29, 705
5-Methyluridine hemihydrate, 34B, 263
Methyl ursolate bromoacetate, 28, 543
Methyl vanadate, 31B, 440
Methyl α-xylofuranoside, 26, 624
Methyl β-xyloside, 31B, 170
Metol, 35B, 67
Mexicanin-E, 27, 1028
Miargyrite, 29, 21
Mica, 26, 517, 525; 27, 695; 30A, 455; 31A, 250
Mica, iron, 28, 249
Micas, 30A, 456
Micrococcinic acid bis-4-bromoanilide (anisole solvate), 31B, 198
Microline, 29, 407
Microcline-sanidine, 27, 678
Microlite, 27, 535
Micromerol (ursolic acid), 29, 772
Microwave, 35B, 903
Microwave spectra, 33B, 538; 35B, 903, 905
Milarite, 26, 534
Millerite, 27, 294, 724
Mimetite, 31A, 243
Mineral olivines, 33A, 468
Mirabilite, 27, 620
Mircrolite, 31A, 244
Mitomycin A, N-bromobenzenesulphonyl derivative, 27, 1015
Mitragynine hydroiodide, 29, 772; 30B, 237
Mixed ammonium chloride thionate compounds, 34A, 224
Mixed bismuth oxides, 27, 538
Mixed rare-earth oxides, 27, 546
Mixed sulphides, 26, 251-252
Miyaconitine hydrobromide dihydrate, 35B, 923
Mizzonite, 31A, 231
Moctezumite, 30A, 454
Mohrite, 29, 424
Molecular addition compounds of aluminum chloride and some nitro derivatives, 26, 703-704
Molecular orientation, 27, 1063
Molybdates, double, 35A, 254
Molybdenite, 29, 68
Molybdenum, 27, 281
Molybdenum(II) acetate, 30B, 397
Molybdenum(V)-L-cysteine chelate, 33B, 549
Molybdenum dioxide, 32A, 259
Molybdenum ferrite, 26, 412
Molybdenum fluoro-bronzes, 35A, 179
Molybdenum-nickel, 26, 197; 28, 29
Molybdenum-nickel-silicon, 26, 269; 27, 376
Molybdenum niobium oxyfluoride, 33A, 189
Molybdenum-niobium-silicon-tin, 27, 213
Molybdenum nitride trichloride, 35A, 154
Molybdenum-nitrogen, 26, 197, 198; 28, 32
Molybdenum octahydrotriborotetracarbonyl, 32A, 510
Molybdenum-osmium, 27, 281
Molybdenum(VI) oxalate binuclearcomplex, 29, 621
Molybdenum oxide hydroxides, 26, 354; 34A, 236
Molybdenum oxides, 28, 125; 29, 308
Molybdenum oxide tetrafluoride, 32A, 508; 33A, 172
Molybdenum(V) oxide trichloride, 35A, 184
Molybdenum-oxygen-zirconium, 27, 542
Molybdenum(V) oxyphosphate, 29, 372
Molybdenum oxtribromide, 35A, 184
Molybdenum-palladium, 27, 319
Molybdenum pentachloride-graphite, 32A, 508
Molybdenum pentafluoride, 27, 445
Molybdenum-phosphorus, 28, 33; 30A, 67
Molybdenum-plutonium-uranium, 26, 198
Molybdenum potassium sodium oxyfluoride, 34A, 216
Molybdenum-rhenium, 28, 27
Molybdenum-selenium, 27, 343
Molybdenum-silicon-vanadium, 27, 284
Molybdenum-silicon-wolfram, 27, 284
Molybdenum-sulphur, 26, 199-201; 29, 68; 35A, 79

Molybdenum-technetium, 27, 285, 361
Molybdenum-tellurium, 26, 201; 27, 343
Molybdenum-titanium, 26, 202
Molybdenum-uranium, 26, 203-204; 31A, 49
Molybdenum vanadium oxides, 31A, 145, 252; 32A, 273
Molybdenum-wolfram, 27, 285
Molybdenyl metaphosphate, 27, 597
Molybdenyl phosphate, 35A, 337
Molybdenyl pyrophosphate, 27, 591
Molybdite, 28, 320
Molybdomenite, 30A, 458
Molybdooxalates, 33B, 542
Monalbite, 30A, 456
Monazite, 26, 454; 29, 425; 32A, 358; 33A, 402
Monensin silver complex, 35B, 754
Monochloroacetonitrile, 26, 588
Monohalogenobenzenes, 26, 693
2-Monolaurin, 30B, 11
Monomethyltriethyl titanate, 29, 766
Monopyridinecopper(II) acetate, 26, 776; 29, 606
Monorden, 30B, 431
Monorden, hemichloroform, 30B, 431
Monosodium phosphoramidate, 29, 360
1-Monostearin, 30B, 12
Montebrasite, 31A, 244
Montgomeryite, 29, 425
Monticellite, 30A, 424
Montmorillonite, 26, 518, 529
Montmorillonite-alcohol complexes, 33B, 543
Moorhouseite, 31B, 458
Moraesite, 28, 324
Mordenite, 26, 515, 519
Morensite, 26, 449
Morphine salts, 29, 749
Morpholine β-iodophenylacetylene, 29, 744
Motiol iodoacetate, 33B, 218
cis,cis-Muconic acid, 27, 734
trans,trans-Muconic acid, 27, 734
trans, trans-Muconodinitrile, 32B, 50
Mullite, 27, 708; 28, 233; 30A, 459; 31A, 252; 34A, 366
Mullite-sillimanite, 28, 236
Murmanite, 31A, 240
Murmanite-lomonosovite minerals, 31A, 240
Muscovite, 28, 240, 334; 29, 430; 30A, 459; 31A, 243, 250; 32A, 469; 33A, 477
Mycosamine hydrochloride, 27, 796
Myoglobin, 26, 810; 33B, 527
Myoglobin (binding of anaesthetics), 33B, 527
Myo-inositol, 29, 487
Myo-inositol dihydrate, 28, 425
Myxin, 33B, 187
Nacrite, 28, 352; 34A, 368
Naphthacene (tetracene) derivatives, 27, 939
Naphthalene, 26, 738, 739; 27, 925
Naphthalene-antimony trichloride complex, 34B, 357
Naphthalene chromium tricarbonyl, 32B, 370
β-Naphthalene derivatives, 35B, 908
Naphthalene-1,2,4,5-tetracyanobenzene 1:1 complex, 32B, 281
Naphthalene-tetracyanoethylene complex, 32B, 278
Naphthalic anhydride derivatives, 35B, 910
Naphthazarin, 27, 927-930
Naphthazarin C, 26, 744
Naphthazarin (form II), 26, 743
asym-αβ-Naphthazine, 28, 687; 30B, 143
1:5-Naphthiridine dihydrate, 26, 779
Naphtho[b]cyclobutene, 29, 763
α-Naphthohydroquinone, 30B, 395
1,4-Naphthohydroquinone, 32B, 121
2-Naphthoic acid, 26, 739
α-Naphthol, 29, 677
β-Naphthol, 30B, 423
α-Naphtholpicrate, 35B, 906
Naphthol yellow, 26, 828
1,4-Naphthoquinone, 26, 741; 29, 770
1,4-Naphthoquinone derivatives, 30B, 100
1,4-Naphthoquinone, 2.3-derivatives of, 32B, 123
α-Naphthyl-4-chlorophthalamic acid, 31B, 96
1-(α-Naphthyl)-germatrane, 35B, 922
α-Naphthyl mercuric iodide, 28, 662, 672
(±)-α-Naphthylphenylmethoxy-(-)-menthoxysilane, 35B, 907
α-Naphthylphenylmethylfluosilane, 29, 770
(+)-α-Naphthylphenyl-methyl-germane, 32B, 492
α-Naphthylphenylmethylsilanes, 29, 763
racem-α-(1-Naphthyl)-propionic acid, 26, 716
N-α-Naphthyl-1,2,3,6-tetrahydrophthalamic acid, 32B, 122
1,8-Naphthyridine complexes, 35B, 919
Narcissidine, 35B, 920
Narsarsukite, 27, 699
Natrolite, 28, 244; 29, 409; 31A, 241
Natrolite, dehydrated, 28, 354
Natrophilite, 30A, 454
Natural fibers, 26, 641
Natural glass, 26, 431
Na-type A zeolite, 33A, 488
Nazarov's compound, 31B, 111
Neighborite, 26, 304
Nemalite, 26, 348
Neobarene, 30B, 412

Neodymium, 26, 233; 27, 334
Neodymium borate, 26, 491; 27, 725
Neodymium chloride, 27, 724; 33A, 204
Neodymium (II) chloride, 26, 321
Neodymium diaquo-tris(acetylacetonate) monohydrate, 32B, 495
Neodymium dihydride, 27, 237
Neodymium fluoride, 31A, 257
Neodymium formate, 27, 815
Neodymium gallate, 32A, 270
Neodymium hexaboride, 26, 69
Neodymium iminodiacetate, 33B, 544
Neodymium (II) iodide, 26, 321
Neodymium molybdate, 34A, 278
Neodymium niobate, 26, 386; 27, 521
Neodymium nitrilotriacetate trihydrate, 31B, 409
Neodymium orthochromite, 32A, 283
Neodymium orthovanadate, 27, 589
Neodymium oxalate, 33B, 544
Neodymium oxalate decahydrate, 30B, 408
Neodymium(III) oxalate 10.5-hydrate, 35B, 639
Neodymium oxytelluride, 27, 424
Neodymium-platinum, 26, 239
Neodymium potassium wolframate, 34A, 282
Neodymium praseodymium tantalate, 27, 521
Neodymium rhenium(V) oxide, 35A, 256
Neodymium rhodium oxide, 28, 133
Neodymium samarium oxides, 27, 546
Neodymium-selenium, 31A, 50
Neodymium sesquioxide, 31A, 125
Neodymium sesquiwolframate, 35A, 261
Neodymium sodium fluoride, 30A, 272
Neodymium sodium sulphide, 29, 341
Neodymium strontium ferrate, 35A, 228
Neodymium-strontium-sulphur, 27, 327
Neodymium tantalate, 27, 521
Neodymium-tellurium, 31A, 47, 50; 32A, 104
Neodymium *mono*telluroxide, 32A, 309
Neodymium wolframate, 35A, 239
Neodymium wolframate hydroxide, 34A, 287
Neon, 26, 285; 27, 391, 721: 31A, 256
Neophorbol-13,20-diacetate-3-*p*-bromobenzoate, 32B, 491
Neostigimine bromide, 35B, 922
Neothiobinupharidine, 30B, 393
Neothiobinupharidine dihydrobromide tetrahydrate, 32B, 261
Nepheline, 35A, 454
Nepheline hydrate I, 29, 430
Neptunite, 28, 326; 30A, 456; 31A, 232, 240
Neptunium-plutonium, 26, 228
Newberyite, 26, 457; 31A, 243; 32A, 360
Nickel, 26, 205; 27, 285
Nickel β-alanine dihydrate, 29, 481
Nickel aluminate, 27, 519
Nickel biguanide chloride dihydrate, 30B, 415
Nickel-1,8-bis(diphenylphosphine)-3,6-dioxaoctane diiodide, 35B, 920
Nickel biuret, 30B, 409
Nickel boracite, 27, 551
Nickel bromate dihydrate, 27, 643
Nickel bromate hexahydrate, 27, 642
Nickel carbonate, 26, 481; 27, 561, 724
Nickel(II) chelates, 35B, 808
Nickel chlorate dihydrate, 27, 643
Nickel chlorate hexahydrate, 27, 642
Nickel chlorate tetrahydrate, 27, 643
Nickel(II) chloride dihydrate, 32A, 184
Nickel chloride hexahydrate, 26, 318; 34A, 178
Nickel(II) chloride-pyrazole complex, 32B, 432
Nickel 5-chlorosalicylaldoximate, 27, 871
Nickel chromite, 26, 409
Nickel cyanide ammoniate, 31A, 115
Nickel(II)-dehydroacetic acid monoimide complex, 32B, 434
Nickel(II) 2,4-diacetyldeuteroporphyrin-IX-dimethyl ester, 29, 768; 30B, 219
Nickel dicarbollide complexes, 35B, 429
Nickel(II)-*N*-β-diethyl aminoethylsalicylaldimine catecholate, 32B, 495
Nickel(II) diethyldiselenocarbamate, 32B, 495
Nickel diethyldithiocarbamate, 30B, 378
Nickel(II) diethylphosphinodithioic acid chelates, 33B, 548
Nickel diketone bisthiosemicarbazones, 35B, 912
Nickel(II)-2,12-dimethyl-3,7,11,17-tetra-azabicyclo(11.3.1)-heptadeca-1(17),13,15-triene perchlorate, 34B, 560
Nickel(II) N,N,-di-*n*-propyldithiocarbamate, 32B, 414
Nickel(II) dithiocarbamate, 32B, 411
Nickel dithionate, 27, 643
Nickel dithionate hexahydrate, 35A, 380
Nickel dithiosemicarbazide sulphate trihydrate, 27, 663
Nickel dithizonate, 35B, 923
Nickel(II) dithizonate bipyridyl, 35B, 912
Nickel ethylenediaminetetraacetate decahydrate, 35B, 924
Nickel etioporphyrin-I, 28, 528
Nickel ferrite, 26, 411
Nickel fluoride, 31A, 249
Nickel fluoroborate hexahydrate, 26, 304
Nickel(II) formate, 28, 628
Nickel germanate-nickel silicate, 27, 489

Nickel glycine peptides, 33B, 547
Nickel group antimonides, 27, 286
Nickel group bismuthides, 27, 286
Nickel hexacyanochromate, 26, 343
Nickel hexacyanochromate hexahydrate, 27, 653
Nickel(II)hydrazinecarboxylate, 30B, 414
Nickel hydroxychloride, 26, 327
Nickel iodate dihydrate, 27, 643
Nickel iodide hexahydrate, 32A, 509
Nickel isoselenocyanato dimethylformamide complex, 29, 764
Nickel isothiocyanato dimethylformamide complex, 29, 764
Nickel manganite, 27, 495
Nickel(II) mercury(II) thiocyanate dihydrate, 28, 111
Nickel molybdate, 27, 522
Nickel niobate, 26, 389
Nickel-niobium, 27, 287
Nickel-niobium-silicon, 26, 175, 269; 27, 376; 28, 28
Nickel nitrate hexahydrate, 27, 571
Nickel nitrate tetrahydrate, 32A, 348
Nickel nitrates, 27, 571
Nickel nitrate thiourea complex, 26, 576
Nickel-nitrogen, 27, 288
Nickel-nitrogen-zinc, 26, 219
Nickel-olivine, 28, 350
Nickel orthouranate, 31A, 254
Nickel oxide, 26, 358
Nickel (III) oxide, 26, 358
Nickel-oxygen-titanium, 26, 413; 28, 20
Nickel-palladium-sulphur, 27, 290
Nickel(II)-1,8,8,13,13-pentamethyl-5-cyano-*trans*-corrin chloride, 30B, 394
Nickel perchlorate dihydrate, 27, 642
Nickel perchlorate hexahydrate, 27, 642
Nickel perchlorate tetrahydrate, 27, 642
Nickel perchlorate thiourea complex, 26, 576
Nickel(II)-S-phenylisocyanato aminodithionitride S-aminodithionitride, 31B, 451
Nickel-phosphorus, 26, 270; 27, 290; 30A, 70; 34A, 117
Nickel phthalocyanine, 30B, 435
Nickel(II) picolinate tetrahydrate, 32B, 496
Nickel-platinum-titanium, 34A, 105
Nickel potassium fluoride, 26, 305; 27, 455
Nickel(IV) potassium hexaoxidoiodate (VII), 30A, 447
Nickel potassium sulphate hexahydrate, 26, 449
Nickel pyrophosphate, 33A, 409
α-Nickel pyrophosphate, 32A, 374
Nickel-pyroxene, 31A, 254
Nickel-rare-earth, 32A, 106; 34A, 105
Nickel rhodium oxide, 26, 390
Nickel-rhodium-selenium, 34A, 109
Nickel rubidium bromide, 34A, 211
Nickel rubidium chloride, 34A, 211
Nickel rubidium fluoride, 27, 454; 35A, 158
Nickel-ruthenium, 26, 207
Nickel salicylalbutyliminate, 26, 678
Nickel salicylalethylenediiminate, 28, 676
Nickel-scandium, 26, 236; 27, 342
Nickel-scandium-silicon, 27, 162, 376
Nickel selenate hexahydrate, 35A, 383
Nickel selenide minerals, 29, 440
Nickel-selenium, 26, 208-210; 34A, 73
Nickel-selenium-sulphur, 27, 168
Nickel-selenium-vanadium, 31A, 51
Nickel-silicon, 26, 211; 27, 292
Nickel-silicon-tantalum, 26, 175, 269; 27, 376
Nickel-silicon-titanium, 26, 175, 269; 27, 376; 30A, 75
Nickel-silicon-vanadium, 26, 174, 175, 269, 270; 27, 376; 28, 28
Nickel-silicon-wolfram, 26, 269; 27, 376
Nickel-silicon-zinc, 33A, 112
Nickel-silicon-zirconium, 26, 175; 27, 376
Nickel sodium fluoride, 27, 454
Nickel sodium stannate, 27, 518
Nickel sodium titanate, 27, 518
Nickel solid solutions, 26, 213
Nickel strontium molybdate, 27, 509
Nickel strontium tellurate, 35A, 242
Nickel strontium wolframate, 27, 509
Nickel squarate, 33B, 542
Nickel sulphate, 26, 436; 27, 526
Nickel sulphate hexadeuterate, 31A, 200
Nickel sulphate hexahydrate, 27, 643
Nickel sulphide, 26, 214
Nickel sulphite hexahydrate, 27, 633; 29, 353; 34A, 316
Nickel-sulphur, 26, 208-210; 27, 293; 34A, 73
Nickel-talc, 28, 350
Nickel-tantalum, 27, 287, 294; 29, 69
Nickel tellurate, 32A, 306
Nickel-tellurium, 27, 295; 28, 34; 30A, 75
Nickel-terbium, 35A, 80
Nickel tetrakis-piperidine N-difluorophosphine, 28, 677
Nickel thallium fluoride, 27, 454
Nickel thiocyanate derivatives, 28, 109
Nickel thiocyanato-di(ethylenediamine) iodide, 34B, 481
Nickel-thorium, 26, 256

Nickel-tin, 27, 259, 292; 32A, 107
Nickel-tin-titanium, 27, 296
Nickel titanate, 27, 507
Nickel-titanium, 27, 296; 28, 20
Nickel-titanium-oxide, 28, 140
Nickel-titanium-zinc, 27, 296
Nickel(II) tris(pyridine-2-carboxaldehyde)-1,1,1-tris(aminomethyl)-ethane diperchlorate, 35B, 92
Nickel uranium oxide, 33A, 356
Nickel uranyl vanadate tetrahydrate, 35A, 364
Nickel-vanadium, 30A, 77
Nickel(II) xanthate, 28, 627
Nickel-yttrium, 26, 237, 239, 276; 30A, 77; 32A, 106
Nickel-zinc, 33A, 108
Nickel-zirconium, 26, 205-207; 27, 297
Nidulin, 28, 772
Nigericin, 33B, 545
Nigerite, 30A, 456
Ninhydrin, 34B, 108
Niobates, 33A, 327; 35A, 257
Niobian aeschynite, 27, 537
Niobium, 31A, 54
Niobium bromide, 31A, 105, 239
Niobium chloride, 26, 323; 27, 444; 30A, 263
Niobium chromate, 27, 521
Niobium dioxide, 27, 479
Niobium diselenides, 30A, 79
Niobium double oxides, 31A, 255
Niobium fluoride, 30A, 261
Niobium iodide, 31A, 92, 259
Niobium molybdate, 27, 521
Niobium monoxide, 27, 475; 31A, 122
Niobium-nitrogen, 26, 215, 216
Niobium(IV,V) oxide, 34A, 241
Niobium oxide ethoxide, 33B, 546
Niobium oxide fluorides, 29, 269, 271; 31A, 123
Niobium oxides, 27, 475; 28, 122; 29, 305; 30A, 334; 31A, 122, 243, 251, 258; 34A, 241; 35A, 209
Niobium(IV) oxyfluorides, 29, 269, 271; 31A, 123
Niobium-oxygen-zirconium, 27, 303
Niobium oxynitrides, 26, 293; 30A, 457
Niobium-palladium, 27, 318; 29, 70
Niobium pentachloride-phosphorus oxychloride, 28, 95
Niobium pentafluoride, 29, 358
Niobium pentaiodide, 28, 101
Niobium pentoxide, 29, 305; 31A, 243, 251; 35A, 209
Niobium pentoxide-wolfram trioxide system, 31A, 251
Niobium-phosphorus, 31A, 51
Niobium phosphorus oxides, 30A, 336; 31A, 147
Niobium-platinum, 26, 216; 29, 70
Niobium potassium sodium oxyfluoride, 34A, 216
Niobium potassium trioxyfluoride, 27, 467
Niobium-rhenium, 26, 217; 29, 72
χ-Niobium-rhenium, 29, 72
Niobium-rhodium, 29, 73
Niobium-selenium, 27, 343; 29, 73; 33A, 113; 34A, 110
α-Niobium-selenium, 27, 343
β-Niobium-selenium, 27, 343
Niobium-selenium-thallium, 29, 91
Niobium-silicon, 26, 247
Niobium-silicon-tin, 35A, 84
Niobium sodium oxide, 30A, 448
Niobium subcarbide, 32A, 507
Niobium subiodide, 32A, 210
Niobium suboxides, 27, 483
Niobium sulphide chloride, 31A, 99
Niobium-sulphur, 33A, 115
Niobium-sulphur-thallium, 29, 91
Niobium tantalum oxide, 27, 538
Niobium-technetium, 26, 255; 27, 361
Niobium-tellurium, 27, 343; 29, 77
Niobium tetrachlorofluoride, 33A, 196
α-Niobium tetraiodide, 27, 443
Niobium-tin, 27, 302; 28, 34; 30A, 80
Niobium-titanium, 26, 218
Niobium titanium oxides, 26, 382-386; 30A, 338
Niobium triiodide, 27, 440
Niobium-uranium, 26, 218
Niobium-uranium-zirconium, 34A, 112
Niobium vanadate, 27, 521
Niobium vanadium oxide, 27, 475
Niobium wolfram oxides, 30A, 340; 31A, 148; 33A, 348; 34A, 288
Niobium-zinc, 26, 218
Niobium-zirconium, 27, 303
Niobophyllite, 29, 412
Niocalite, 31A, 232
Nitranilic acid hexahydrate, 32B, 107
Nitrato-bis(2,2'-bipyridine)-copper(II) nitrate monohydrate, 35B, 911
Nitric anhydride, 27, 718
Nitric oxide dimer, 26, 345
Nitrides in ternary systems, 27, 305
Nitrides, ternary, 35A, 136
Nitridodichlorobis(triphenylphosphine)rhenium(V), 32B, 477
Nitridodichloro tris(diethylphenylphosphine) rhenium(V), 32B, 477
Nitrilotriacetic acid, 32B, 13
Nitrilotrimethylenetriphosphoric acid, 32B, 304
Nitrito-bis(2,2'-bipyridyl)copper(II)-nitrate, 34B, 569
Nitritobis(ethylenediamine) nickel perchlorate, 27, 831
Nitritobis(ethylenediamine)nickel tetrafluoroborate, 30B, 331
4-Nitroaniline, 26, 696

β-*o*-Nitroaniline, 30B, 419
γ-*o*-Nitroaniline, 30B, 419
p-Nitroaniline, 27, 1063
p-Nitroaniline cadmium chloride, 33B, 552; 35B, 408
4-Nitroaniline (deuterated), 34B, 69
2-Nitrobenzaldehyde, 29, 651
o-Nitrobenzaldehyde, 27, 1063; 29, 651
Nitrobenzene, 33B, 528
1-*p*-Nitrobenzeneazo-2-naphthol, 34B, 39
o-Nitrobenzenesulphenic acid (methyl ester), 29, 769
2-Nitrobenzenesulphonyl chloride, 35B, 52
p-Nitrobenzoic acid, 30B, 399; 31B, 61
o-Nitrobenzoic acid, 32B, 76
Nitrobisethylenediamine nickel chloride, 31B, 381
Nitrobis(ethylenediamine)nickel iodide, 35B, 924
Nitrocellulose, 26, 629
α-(2-Nitro-4-chlorophenylazo) acetoacetanilide, 33B, 545
Nitrodimethylamine, 33B, 545
Nitrogen, 27, 721
α-Nitrogen, 26, 287; 29, 229; 35A, 129
β-Nitrogen, 27, 393; 35A, 129
γ-Nitrogen, 35A, 130
Nitrogen-oxygen-phosphorus-thorium, 34A, 135
Nitrogenpentammineruthenium(II) dichloride, 31A, 258; 33A, 211
Nitrogen-phosphorus-uranium, 34A, 135
Nitrogen-phosphorus-vanadium, 33A, 130
Nitrogen-praseodymium, 27, 304
Nitrogen-selenium, 26, 220
Nitrogen-selenium-thorium, 34A, 135
Nitrogen-selenium-uranium, 34A, 135
Nitrogen sulphide fluoride, 26, 314
Nitrogen-sulphur-thorium, 34A, 135
Nitrogen-sulphur-uranium, 34A, 135
Nitrogen-tantalum, 26, 253; 27, 307
Nitrogen-terbium, 27, 335
Nitrogen-thulium, 27, 335
Nitrogen-titanium, 26, 264; 27, 307; 34A, 115
Nitrogen triiodide-ammonia, 31A, 256; 33A, 229
Nitrogen-wolfram, 26, 220-223; 27, 309; 28, 32; 31A, 54
3-Nitro-4-hydroxy phenylarsonic acid, 28, 671
α-(2-Nitro-4-methylphenylazo)-acetoacetanilide, 33B, 545
trans-1-(nitromethyl)-3-phenylphthalan, 35B, 923
2-Nitronitrosoethane *trans*-dimer, 34B, 34
Nitron tetafluoroarsenate acetonitrile, 29, 766
Nitropentamminecobalt(III) halides, 33A, 209
3-Nitroperchlorylbenzene, 33B, 48
o-Nitroperoxybenzoic acid 30B, 58
o-Nitroperoxybenzoic acid, 29, 769
p-Nitroperoxybenzoic acid, 33B, 552; 35B, 41
p-Nitrophenol, 29, 769; 30B, 71
α-*p*-Nitrophenol, 27, 1063
β-*p*-Nitrophenol, 27, 1063
p-Nitrophenyl azide, 29, 770; 30B, 50
tris-*p*-Nitrophenylmethyl, 27, 915
m-Nitrophenyl(2,2',2"-nitrilotriethoxy) silane, 34B, 367
Nitrosobenzene, 27, 888
Nitroso compounds, 27, 888
Nitrosodimethylamine, 33B, 545
Nitroso(dimethyldithiocarbamato) cobalt, 27, 846
Nitrosomethane, 33B, 538
Nitrosopentammineruthenium(IV) trichloride monohydrate, 30A, 282
Nitrosyl chloride, 26, 327
Nitrosyl chlorosulphate, 32A, 198
Nitrosyldicarbonyl bis(triphenylphosphine)manganese, 32B, 480
Nitrosylfluoride, 34A, 215
Nitrosyl iron bis(*N*,*N*'diethyldithiocarbamate), 32B, 495
Nitrosyliron bis(*NN*-dimethyldithiocarbamate), 33B, 546
*mono*Nitrosylmolybdenum tris-(*NN*-di-*n*-butyldithiocarbamate), 35B, 912
Nitrosylpentamminecobalt dichloride, 30A, 280
Nitrosyl pentammineruthenium trichloride monohydrate, 28, 93
Nitrosyl tetracarbonyl manganese, 34A, 160
Nitrosyltricarbonyltriphenylphosphinemanganese, 33B, 429
3-Nitro-4-toluidine, 30B, 420
5-Nitrouracil monohydrate, 32B, 205
Nitrous oxide, 26, 344
Nitryl trisulphate(VI), 29, 344
Nobleite, 26, 489
Noble-metal zeta phases, 27, 310
Nonactin potassium thiocyanate complex, 32B, 242
Nonacyclodocosane, 33B, 553
Nonadecan-10-one, 35B, 908
Nonane dicarboxylic acid complex, 26, 575
n-Nonanoic acid hydrazide, 26, 562
Norbergite, 30A, 417; 34A, 366
Norbormide (*p*-bromobenzoyl derivative), 33B, 208

Norbornadiene-bis-*cis*-(1,2-perfluoromethylethene-1,2-dithiolato)nickel, 35B, 807
Norbornadiene copper(I) chloride, 29, 587
Norbornadiene palladium chloride, 30B, 324
Norbornadiene palladous chloride, 29, 768
Norbornadiene-silver nitrate, 31B, 367
2π,5-Norbornadien-7-yl-2,3-dimethylcarboxylate(π-cyclopentadienyl) nickel(II), 28, 606
Norbornane, 33B, 534
endo-Norbornene-*cis*-5,6-dicarboxylic acid, 35B, 924
endo-Norbornene-*cis*-5,6-dicarboxylic anhydride, 35B, 910
5-Norbornene-2,3-*endo*-dicarboxylic anhydride, 33B, 555; 34B, 304
anti-7-Norborneyl *p*-bromobenzoate, 30B, 224
Nor-dihydro-guaiaretic acid, 34B, 343
Nordite, 26, 536; 35A, 477
Nordstrandite, 31A, 253; 33A, 260
Norodrenaline hydrochloride, 32B, 263
Norsethite, 26, 483; 30A, 454; 32A, 417; 33A, 436
Noseane, 30A, 432
Novakite, 26, 48
Nowackiite, 31A, 261; 32A, 225
Nsutite, 30A, 456
Nucleic acid constituents, 35B, 294
Nucleic acids, 27, 1055
Nylon, 35B, 883, 907
α Nylon 4, 31B, 31
Nylon 5.6, 27, 811
γ-Nylon 5.7, 27, 811
γ-Nylon 5.9, 27, 811
Nylon 6, 26, 640; 29, 759
α-Nylon 6, 27, 811
δ-Nylon 6.5, 27, 811
Nylon 6.6, 26, 640
α-Nylon 6.6, 27, 811
Nylon 6.10, 27, 811
Nylon 6.6 salt, 31B, 27

Obtusifolon, 35B, 923
Ochotensine methiodide, 31B, 249
Octaborane (12), 29, 230
Octabutyrate of the condensation product of resorcinol and *p*-bromobenzaldehyde, 33B, 95
Octacalcium phosphate, 27, 575
Octacarbonyl (dicarbonyl) diethylacetylene tetracobalt, 27, 826
Octacarbonyldiphenylvinylidenediiron, 33B, 339
Octachlorocarborane, 29, 529
Octachlorocyclobutane, 30B, 84
2,2,2,4,6,8,8,8-Octachloro-1,3,5,7,9,10-hexamethyl,1,3,5,7,9,10-hexaza-2,4,6,8-tetraphospha(2,4,6,8-p(V)dispiro(3.1.3.1)decane, 32B, 307
Octachloronaphthalene, 28, 429
Octacyanowolframic acid, 35A, 423
n-Octadecane, 27, 743
Octa-*cis*-2-*trans*-6-diene-1,8-dioic acid, 34B, 9
Octa-*trans*-2-*trans*-6-diene-1,8-dioic acid, 34B, 9
Octa*kis*dimethylaminocyclotetraphosphaza-tetraene, 27, 989
Octafluorocyclobutane, 33B, 531
Octafluorocycloheza-1,3-dieneiron tricarbonyl complex, 26, 680; 32B, 371
Octafluorotetramethyleneiron tetracarbonyl, 26, 680
1,2,3,4,4a,5,6,8aOctahydro-3,5-bis-bromomethyl-3,5-dihydroxy-2,6-dioxo-1,4-ethenonaphthalene, 35B, 141
Octahydrodipyrido[1,2-*a*:1',2'-*c*] imidazol-10-ium bromide, 35B, 192
Octahydro-1,3,4,7-tetranitro-1,3,5,7-te-trazocine, 28, 467
Octahydro-1,3,5,7-tetranitro-1,3,5,7-tetrazocine, 32B, 487
Octahydroxycyclobutane, 33B, 77
Octamethylcyclotetraphosphonitrile, 26, 607
Octamethylcyclotetrasilazane, 28, 584
Octamethyl dialuminum monomagnesium, 34B, 398
Octamethyl-octa(siloxane), 31B, 298
Octamethyltetra-amino-2,3-diphosphinyl buta-1,3-diene, 34B, 386
2,2,4,4,6,6,8,8-Octamethyl-2,4,6,8-tetrasilabicyclo(3.3.0) oct-1(5)-ene, 34B, 373
Octammine-μ-dihydroxodicobalt(III) tetrachloride tetrahydrate, 28, 91
n-Octane, 26, 560; 32B, 40
2,2,4,4,6,6,8,8-Octaphenyl-cyclotetraarsenazatetraene, 32B, 492
Octaphenylcyclo-octatetraene, 29, 766, 770; 30B, 93
2,2,4,4,6,6,8,8-Octaphenyl-cyclotetraphosphazatetraene, 32B, 492
cis-Octa-2,4,6-triene-1:4,5:8 diolide, 29, 765; 31B, 140
trans-Octa-2,4,6-triene-1:4,5:8-diolide, 31B, 140
Oestradiol 3-*p*-bromobenzoate, 33B, 195
Oestrone, 35B, 918
18αH-Oleanane, 35B, 922
Olefinic distereoisomers of Pt(II), 30A, 394
Oleic acid, 27, 740

Oligomeric amides, 29, 759
Olivenite, 27, 601
Olivines, 29, 245
Olivines, mineral, 33A, 468
Olivine structures, 30A, 422
Omphacite, 29, 430; 31A, 260; 33A, 459
Ophiobolin methoxybromide, 30B, 244; 33B, 217
Organogermanium compounds, 35B, 889
Organosilicon compounds, 26, 615-616
Organosilicon polymers, 30B, 527
Ornithine hydrochride, 33B, 555
DL-Ornithine hydrobromide, 32B, 489; 35B, 314
L-Ornithine hydrochloride, 32B, 227; 34B, 266
Orthanilic acid, 32B, 91
Orthoboric acid, 31A, 165
Orthoclase, 26, 524
Orthoenstatite, 31A, 252; 34A, 365
Orthofluoroberyllates hexahydrates, 34A, 186
Ortho-lomonosovite, 27, 689
Orthoperiodic acid, 31A, 212
Orthophosphoric acid hemihydrate, 34A, 319
Orthotobernite, 26, 468
Osarizawaite, 26, 453; 27, 634; 28, 326
Osmium boride, 27, 93
Osmium carbonyl, 27, 651
Osmium carbonyl 2,3-dimethylbuta-1,3-diene reaction product, 28, 674
Osmium dioxide, 34A, 246; 35A, 211
Osmium oxide pentafluoride, 27, 665; 33A, 174
Osmium potassium oxide-nitride, 33A, 271
Osmium-rhenium, 27, 313
Osmium-ruthenium, 27, 314
Osmium-selenium, 33A, 34
Osmium-silicon, 27, 372; 35A, 84
Osmium-sulphur, 32A, 126
Osmium-technetium, 27, 361
Osmium tetroxide, 30A, 311; 31A, 239
Osmium-thorium, 26, 236, 256
Osmium-wolfram, 26, 223
Osumilite, 34A, 388
Osumilite and petalite type minerals, 27, 685
Oxacyclodeca-2,7-dione, 33B, 132
Oxalatobis(ethylenediamine)chromium(III) bis(oxalato)ethylenediamine-chromate(III) dihydrate, 35B, 604
Oxalatobis(1,10-phenanthroline)-iron(II) pentahydrate, 33B, 543
Oxalatodiamminecopper(II), 33B, 548
cis-Oxalato-*C*-*rac*(5,7,7,12,14,14-hexamethyl-1,4,8,11-tetra-aza-cyclotetradecane)cobalt(III) perchlorate, 35B, 773
Oxalic acid, 35B, 889
Oxalic acid dihydrate, 27, 1066; 32B, 3
α-Oxalic acid dihydrate, 34B, 3,4
Oxalic acid dihydrate (deuterated), 30B, 395
α-Oxalic acid dihydrate, deuterated, 34B, 4,6
β-Oxalic acid dihydrate, deuterated, 34B, 6
Oxalyl bromide, 26, 546; 27, 729
Oxalyl chloride, 27, 729
12-Oxa[4.4.32propella-2,4,7,9-tetraene bis-tricarbonyliron, 35B, 923
1-Oxa-4-selenacyclohexane-iodine, 31B, 272
1-Oxa-4-selenacyclohexane-iodine monochloride, 33B, 149
Oxidodiperoxido-2,2'-dipyridyl-chromium(VI), 33B, 439
8,16-Oxido-*cis*-[2.2]metacyclophane, 33B, 135
8,16-Oxido-*cis*[2,2]metacyclophane-1,9-diene, 34B, 117
Oxindigo, 27, 983
μ-Oxobis[bis(diethyldithiophosphate)-oxomolybdenum(V)] 1,2-dichloro-benzene, 34B, 598
μ-Oxo-bis[chlorobis(2,4-pentane-dionato)titanium(IV)-chloroform], 32B, 388
μ-Oxo-bis[cyclopentadienyltitanium(IV) dichloride], 28, 610
μ-Oxobis(phthalocyanatopyridine) manganese(II) dipyridinate, 32B, 238
μ-Oxo-bis(tetraphenyl porphine iron (III), *p*-xylene solvate, 32B, 283
μ_3-Oxo-μ_3-carbonyltricyclopentadienyl tricobalt, 34B, 466
Oxochlorobis-[1,2-bis(diphenylphos-phino)ethane]molybdenum(IV) trichloro(acetone)zincate, 35B, 917
μ-Oxo-μ-chloro-di-μ-propionato-bis-(chlorotriphenylphosphine-rhenium), 34B, 532
1-Oxo-1-chlorophosphacyclopent-3-ene, 35B, 894
1-Oxo-1-chloro-2-phosphacyclopentene, 35B, 919
Oxodiperoxo-1,10-phenanthroline-chromium(VI), 30B, 350
Oxodiperoxopyridinechromium(VI), 29, 541
Oxofluorometallates, 34A, 216
μ_4-Oxo-hexa-μ-chloro-tetrakis[(2-methylpyridine)copper(II)] hydrate, 35B, 775
μ_4-Oxo-hexa-μ-chloro-tetrakis(triphen-ylphosphine oxide)copper(II)], 31B, 446; 32B, 449
μ,μ',μ",μ'"-Oxo-hexakis(1,3-diphenyl-triazenido)tetrazinc(II), 35B, 919
2-Oxohexamethylenimine, 27, 987

3-α-Oxo-3-β-hydrido-7-β-hydroxy-2,4-dioxa-3-phosphabicyclo[3.3.1] nonane, 33B, 279
Oxomercurates, 29, 338
1-Oxo-4-methyl-2,6,7-trioxa-1-phosphabicyclo[2.2.2]octane, 33B, 278
Oxomolybdenum(V) xanthate complex, 29, 622
Oxopentachloropropionatobis-(triphenylphosphine)dirhenium(IV), 33B, 416
2-Oxo-2-phenoxy-1,3,2-dioxaphosphorinane, 32B, 222
Oxotetrachloro-*o*-phenylenebis-(dimethylarsine)tungsten(VI), 35B, 913
trans-Oxotrichloro-bis-diethyl-phenylphosphine rhenium(V), 28, 656
trans-μ_3-Oxo-tris{bisacetatoaquochromium(III)} chloride hexahydrate, 35B, 738
Oxotuberstemonine, 33B, 555
α-Oxo-γ-valerolactone-γ-carboxylic acid oxime, 33B, 542
Oxygen, 31A, 240
α-Oxygen, 32A, 147
β-Oxygen, 27, 394
γ-Oxygen, 29, 229; 31A, 259
Oxygen-bridged copper(II) complexes, 35B, 723
Oxygen-bridged nickel(II) complexes, 35B, 722
Oxygen-rare-earth-zirconium, 27, 542
Oxygen-silicon-throium, 26, 403
Oxygen-sulphur-zirconium, 27, 348
Oxygen-tantalum, 26, 252
Oxygen-thorium-titanium, 26, 403
Oxygen-titanium, 26, 261
Oxygen-titanium-zinc, 26, 413
Oxygen-ytterbium-zirconium, 26, 364; 27, 542
Oxygen-zircomium, 33A, 118
Oxyhaemoglobin, 29, 758
Oxyhemoglobin, H, 28, 681
Oxyhornblende, 29, 439
Oxyindigo, 26, 830
N-Oxyphenazine, 26, 781; 35B, 121
Oxythallates of alkali metals, 26, 400

Pabstite, 30A, 455
Pachypasa otus, 26, 827
Palermoite, 30A, 454
Palladium acetylacetonate, 35B, 605
Palladium amino-halides, 27, 667
Palladium benzoylacetonate, 32B, 387
Palladium(II) bromide, 31A, 106
Palladium(II) chloride bis(cyclohexanone oxime), 32B, 432
Palladium 5-chlorosalicylaldoximate, 27, 871
Palladium dibenzoylmethanate, 33B, 462
Palladium (II) 2-(*o*-hydroxyphenyl) benzoxazole, 27, 883; 29, 768
Palladium iodide, 33A, 240
Palladium-phosphorus, 26, 225, 270; 28, 35; 31A, 54
Palladium-plutonium, 33A, 121
Palladium *n*-propyl mercaptide, 33B, 497
Palladium-rhenium, 27, 337
Palladium-ruthenium, 27, 314
Palladium salicylalethylenediiminate, 28, 676
Palladium-scandium, 26, 165; 27, 342
Palladium-selenium, 27, 315
Palladium-silicon, 31A, 56; 35A, 86
Palladium-sulphur, 27, 315; 33A, 121
Palladium-tantalum, 26, 318; 29, 70
Palladium-tellurium, 35A, 87
Palladium tetraphenylporphine, 29, 617
Palladium-thallium, 26, 225; 33A, 124
Palladium-thorium, 26, 226; 27, 317; 28, 36; 29, 78
Palladium-titanium, 27, 234; 33A, 126
Palladium transition-metal alloys, 27, 318
Palladium-vanadium, 27, 318
Palladium-yttrium, 26, 165
Palladium-zinc, 26, 225; 34A, 115
Palladium zinc hexafluoride, 33A, 188
Palladium-zirconium, 27, 227, 234, 319
Palmarin derivative *p*-bromophenacyl ester, 34B, 346
Palygorskite, 28, 312
Panamine diperchlorate, 31B, 244
Pandaite, 30A, 456
Papain, 33B, 527
Parabutlerite, 35A, 375
Parachloral, 27, 967
α-Parachloral, 27, 967
β-Parachloral, 27, 967
[3.3] Paracyclophane, 30B, 121
Paradamite, 31A, 242
n-Paraffin complexes, 26, 575
Paraffins, lower, 32B, 40
Paragonite, 29, 430
Paragonite, $2M_1$, 28, 350
Paragonite, 3T, 28, 352
Parahopeite, 33A, 395
Paraperiodic acid, 26, 499
Paratacamite, 29, 250
Paratellurite, 30A, 458
Paraterphenyl, 34B, 83
Paravauxite, 34A, 332
Parawollastonite, 32A, 511, 33A, 453
Pascoite, 31A, 142
Patchouli alcohol diester of chromic acid, 28, 680
Paulingite, 31A, 260
Pearceite, 28, 288
Pectolite, 32A, 461
Pederin di-*p*-bromobenzoate, 33B, 556
Pegmatite minerals, 30A, 459

Pentaaminochloro ruthenium (III) dichloride, 27, 669
Pentaantipyrine copper perchlorate, 30B, 425
Pentabarium dihydroxy tetrametasilicate, 35A, 312
Pentabromochlorobenzene, 31B, 77
Pentabromofluorobenzene, 29, 769
Pentacarbonyl-[cyclohexylamino-(1-methoxyvinyl)-carbene] chromium (o), 35B, 501
Pentacarbonyl(2,3-diphenylcyclopropenylidene) chromium (O), 34B, 426
β-Pentacarbonyl manganese hydride, 34A, 158
Pentacarbonyl methyl(methylamino) carbene chromium, 32B, 493
Pentacarbonyl-tetraplatinum-tetra-(dimethylphenylphosphine), 34B, 614
Pentacene, 26, 756; 27, 938
Pentachloroaniline, 34B, 69
Pentachloro antimony(V) μ-oxo-dichloro selenium(IV), 32A, 200
Pentachloro-bis(1,5-dithiahexane) dirhenium, 33B, 483
Pentachlorobromobenzene, 31B, 77
1,3,3,3,3-Pentachloro-2,4-dimethyl-3-phospha-2,4-diazacyclobutene, 34B, 378
Pentachloroethoxycodide, 33B, 548; 34B, 166
Pentachlorofluorobenzene, 29, 769
Pentachlorophenol, 27, 903
4,5,6,7-*tetrahapto*-Pentacyclo[8,3,3-$0^{2,9}$,$0^{3,8}$,$0^{11,16}$]hexadeca-4,6,12,14-tetraenetricarbonyliron, 35B, 915
Pentaerythritol tetranitrate, 28, 405
Pentaethylcyclopentaphosphine-molybdenum tetracarbonyl, 33B, 506
Pentafluorobenzene, 33B, 540
Pentafluorobenzonitrile, 33B, 541
Pentafluoroxenon(IV) hexafluoroplatinate (V), 32A, 160
Pentafluoroxenonyl hexafluoroplatinate-(V), 31A, 258
Pentakis-(cyclopentadienyl)-dicobalt, 26, 653
Pentakis(methylisonitrile) cobalt(I) perchlorate, 30B, 295
Pentakis(2-picoline *N*-oxide)cobalt(II) perchlorate, 35B, 749
Pentakis-(trifluoromethyl)-cyclopentaphosphine, 26, 620
Pentakis-(2,8,9-trioxa-1-phospha-adamantane) nickel(II) perchlorate, 32B, 491; 35B, 852
Pentalithium aurate(III), 35A, 215
Pentamethonium iodide, 28, 398
Pentamethylene di(trimethylammonium) iodide, 28, 398
Pentamethylethylalcohol hemihydrate, 28, 671
1,3,7,8,10-Pentamethylisoalloxazinium iodide monohydrate, 34B, 156
2,2,3,4,4-Pentamethyl-1-phenylphosphetan 1-oxide, 33B, 549; 35B, 456
1,2,3,5,5-Pentamethyl-2-pyrazolinium perchlorate, 33B, 551
Pentane, 29, 449
n-Pentane, 32B, 40
DL-Pentane-2,4-diol diacetate, 31B, 14
Pentaoxyphosphorane, 30B, 395
Pentaphenylantimony, 27, 885; 29, 641; 33B, 292
Pentaphenylarsenic, 27, 885
Penta-*m*-phenylene, 35B, 134
Pentaphenylphosphorus, 27, 885; 29, 639
Pentlandite, 26, 121
Pepsin, 33B, 527
Perchlorates, 31A, 250
Perchloric acid hydrate, 33A, 449
Perchloro-1,4-dimethyl-5,8-dimethylenecycloocta-1,3,6-triene, 33B, 86, 555
Perchloro-(4,8-dimethylenetricyclo [3,3,3,$0^{1,5}$]deca-2,6-diene, 32B, 148
Perchloro-dimethyl-trisulphide, 26, 592
Perchlorofulvalene, 26, 643
Perchloro[4]radialene, 35B, 99
Perchloro-(3,4,7,8-tetramethylenetricyclo[4,2,0,$0^{2,5}$]octane), 32B, 135
Perchloryl fluoride, 26, 316
Perdeuterated violuric acid monohydrate, 29, 721
Perdeutero-α-glycylglycine, 35B, 304
Perdeuteronaphthalene, 34B, 97
Perfluorobutadiene iron tetracarbonyl, 32B, 493
Perfluoro-tricyclo[3,3,0.$0^{2,6}$]-octane, 29, 766
Perhydroanthracene, 32B, 136
cis-Perhydro-cyclopenta[1,2-*c*:3,4-*c*']-dithiophene-S,S,S',S'-tetroxide, 33B, 143
Perhydrodibenzo[1,2,3,6,7,8]hexathiecin, 35B, 911
trans-Perhydrotriphenylene, 32B, 142
trans, anti, trans, anti, trans-Perhydrotriphenylene - chloroform, 32B, 491
trans-Perhydrotriphenylene cyclohexane solvate, 32B, 292
trans, anti, trans, anti, trans-Perhydrotriphenylene-*N*-heptane, 32B, 491

Perkinamine hydrochloride, 27, 956; 29, 735
Perloline mercurichloride hydrate, 31B, 246
Perlolyrine, 35B, 233
Perlon, 26, 640
Perovskites, 26, 418-423; 27, 503-510; 29, 326, 439; 30A, 452, 458; 31A, 254
Perovskite structures with A_2BCO_6 formula, 27, 508
Peroxychlorotantalates, 31A, 244
Peroxypelargonic acid, 30B, 7
Perrhenates, 31A, 250
Perrierite, 31A, 242, 253
Pertechnates, 31A, 250
Pertechnetates, 27, 671
Perylene, 29, 694; 30B, 404
Perylene bis-*cis*-(1,2-perfluoromethylethene-1,2-dithiolato) nickel, 34B, 354
1:1 Perylene-fluoranil complex, 28, 576
Perylene, β form, 28, 451
Perylene, pyromellitic dianhydride (1:1) complex, 30B, 254
Perylene-tetracyanoethylene, 35B, 417
Perylene-tetracyanoethylene complex, 32B, 284
Perxenates, 29, 335
Petalite, 26, 506, 515
Pethidine hydrobromide, 35B, 175
Pharmacolite, 31A, 243 ; 34A, 335
Pharmacosiderite, 32A, 384
P Phases, 30A, 83
R Phases, 30A, 83
χ Phase, 35A, 50
Phases with boron carbide structure, 26, 67
Phenacites, 27, 606
Phenaksite, 29, 430
Phenanthrene, 26, 752; 28, 439
Phenanthrene chromium tricarbonyl, 29, 542; 33B, 366
9,10-Phenanthrene quinone, 26, 751
Phenanthridone, 35B, 193
1,10-Phenanthrolinegallium bromide, 35B, 906
o-Phenanthroline hydrate, 30B, 425
1,10-Phenanthrolinemercury(I) nitrate, 32B, 430
Phenarsazine bromide, 31B, 313
Phenarsazine chloride, 31B, 312
Phenaxite, 35A, 480
Phenazine, 35B, 121
Phenazine-5,10-dioxide, 26, 781; 28, 477
Phenazine-iodine complex, 32B, 273
Phengite, 29, 429
Phenobarital-8-bromo-9-ethyladenine complex, 33B, 254
Phenobarbitone, 33B, 543
Phenol, 28, 423; 32B, 102
Phenol red, 26, 828
Phenoquinone, 33B, 65
Phenothiazine, 30B, 426; 33B, 551
Phenothiazine-3,5-dinitrobenzoic acid (1:1 complex), 33B, 548
Phenothiazine-trinitrobenzene, 34B, 351
1-(10'-Phenothiazinyl)-2-propyltrimethylammonium methylsulphate, 35B, 918
Phenoxthionine, 31B, 159
α-Phenoxy*cyclo*butane, 27, 888
Phenoxymethyl-Δ^2-desacetoxyl cephalosporin, 35B, 384
Phenoxymethyl penicillin, 28, 530
Phenoxymethyl penicillin sulphoxide methanol solvate, 34B, 285
Phenylalanine, 26, 826
L-β-Phenylalanine, 35B, 906
L-Phenylalanine hydrochloride, 28, 672; 29, 700
(±)Phenylalanine-(pyridoxylidine 5-phosphate)copper(II), 33B, 548
9-Phenyl-9-arsafluorene, 28, 589
Phenylarsine bis(diethyldithiocarbamate), 32B, 316
Phenylazoazobenzene, 32B, 485
Phenylazotribenzoylmethane hemietherate, 35B, 908
2-Phenylazulene, 26, 722
N-Phenyl-N'-benzoylselenourea, 30B, 43
Phenyl-β-benzoylvinylsulphoneiron tetracarbonyl(*cis* and *trans*), 33B, 556
1-Phenyl-6-benzyl-7,7-dichloro-3,5-diaza-bicyclo[1.2.2]hept-5-en-2-one, 33B, 542
Phenyl bis(thiourea) tellurium chloride, 31B, 340
(1-Phenylbutane-1,3-dionato)(ethylene glycol)sodium, 35B, 912
4-Phenylcarbamoylcholine bromide, 35B, 918
Phenyl*mono*chlorosilane, 30B, 404
Phenylcyclobutane, 33B, 536
Phenylcyclobutenedione, 29, 665
1-(1-Phenylcyclohexyl)piperidine hydrochloride, 35B, 176
Phenylcyclopropane, 30B, 404
Phenylcyclosilanes, 33B, 542
2-Phenyl-1,3-dithiane, 31B, 149
2-(5'-Phenyl-1',2'-dithiole-3'-ylidene)-6-5"-t-butyl-1",2"-dithiole-3"-ylidene)-cyclohexane-thione, 35B, 910
4-Phenyl-1,2-dithiolium bromide, 28, 670; 34B, 184
3-Phenyl-1,2-dithiolium iodide, 28, 670; 30B, 147
4-Phenyl-1,2-dithiolium iodide, 28, 670; 31B, 147
3-Phenyl-1,2-dithiolium perchlorate, 28, 670

4-Phenyl-1,2-dithiolium perchlorate, 28, 670
4-Phenyl-1,2-dithiolium thiocyanate, 34B, 185
o-Phenylenebisdimethylarsine titanium tetrachloride complex, 26, 687
mono-*o*-Phenylenebisdimethylarsine tricarbonyl iron(O), 32B, 464
o-Phenylenebis-(*o*-dimethylarsinophenylmethylarsine) palladium chloride perchlorate benzene complex, 32B, 472
[*o*-Phenylene-bis(methylsulphide)]-wolfram tetracarbonyl, 35B, 919
2,2'-*p*-Phenylenebis-(5-phenyloxazole), 30B, 152
p-Phenylenediamine dihydrochloride, 34B, 72
o-Phenylenediarsine oxychloride, 27, 984
β-Phenylethylamine hydrobromide, 26, 702
β-Phenylethylamine hydrochloride, 26, 702
tris-(Phenylethynyl) antimony, 27, 916
tris-(Phenylethynyl) arsenic, 27, 916
Phenylethynylcopper(I), 29, 768
Phenylethynyl(isopropylamine)gold(I), 32B, 343
tris-(Phenylethynyl) phosphorus, 27, 916
Phenylethynyl-(trimethylphosphine)-copper(I), 31B, 345
Phenylethynyl (trimethylphosphine) silver(I), 31B, 344
β-Phenylglutaric acid, 35B, 916
1-Phenyl-1,2,2,3,4,4-hexamethyl phosphetanium bromide, 34B, 387
Phenylhydrazine, 33B, 54
Phenyl hydrazine derivative of 7,9-dihydroxy-6α-methoxy-2-phenyl-*trans*-*m*-dioxano [5,4-e][1,4]-dioxepan, 26, 627
Phenylhydrazine hydrochloride, 30B, 47
2-Phenylimino-3-phenyl-4-thioxothiazolo[3,2-a]tetrahydro-s-triazine, 32B, 197
2-Phenyl-1,3-indanedione, 35B, 924
3-Phenylisoxazolin-5-one, 34B, 201
Phenylmagnesium bromide dietherate, 28, 685; 29, 533
Phenyl mercuric bromide, 33B, 532
Phenylmercuric halides, 29, 628
Phenyl mercury 2,6-diiodo-4-methylphenolate, 33B, 544
Phenylmethoxycarbenepentacarbonylchromium, 33B, 336
3-Phenyl-1,4-naphthoquinone, 34B, 102
Phenyl-(2,2',2"-nitrilotriethoxy) silane, 33B, 270
Phenyl(2,2',2"-nitrilotriphenoxy) silane, 33B, 270
5-Phenyl-3-(5-Phenyl-1,2-dithiol-3-ylidenmethyl)-1,2-dithiolium iodide, 30B, 149
Phenyl(phenylimino) aluminum tetramer, 28, 599
1-Phenyl-2-(2-pyridyl)-ethanedione-1,2, 31B, 128
Phenyl rhodium tetraphenylporphine chloride, monohydrate, 32B, 490
4-Phenyl-1,2,3,4-tetrahydroisoquinoline derivative, 35B, 923
1-Phenyl-3-(2-thiazolin-2-yl)-2-thiourea, 35B, 237
1-Phenyl-2-thiomethylethenyl-2,4,6-trinitrobenzenesulphonate(E), 34B, 50
Phenyltrichlorosilane, 29, 538
Phenyltrimethylenemethane-iron tricarbonyl, 34B, 445
(Phenyltrimethylenemethane)tricarbonyliron, 33B, 547
N-Phenyl-2,4,6-trimethylpyridinium perchlorate, 34B, 141
Phillipsite, 27, 692
Phlogopite, 27, 696; 28, 288
Phloroglucinol, 30B, 77
Phomin-silver fluoroborate, 35B, 916
Phorbol bromofuroate-chloroform solvate, 33B, 215
Phosphobenzene, 29, 637
Phosphobenzene B, 30B, 279
Phosphobenzene hexamer, 31B, 309
Phosphobenzene, sulphur derivative, 30B, 418
Phosphocristobalite, 31A, 245
Phosphoferrite, 28, 191
Phosphoferrite-reddingite, 26, 464
Phospholanic acid, 34B, 379
Phospholipid extracts, 26, 826
Phosphomolybdate, 27, 672
Phosphonitrilic chloride, 27, 411
Phosphonitrilic chloride, pentameric, 33A, 198
Phosphonitrilic chloride, tetrameric, 33A, 197
Phosphonitrilic fluroide, 28, 59
Phosphonitrilic fluoride tetramer, 26, 314
Phosphonitrilic isothiocyanate, trimeric, 35A, 137
Phosphonium iodide, 32A, 216
Phosphophyllite, 26, 463
Phosphoric acid, 31A, 259
Phosphoric anhydride, 27, 718
Phosphoric oxide, 29, 360
Phosphoric triamide, 34A, 155
Phosphorus, 31A, 57; 34A, 149
Phosphorus, black, 30A, 251
Phosphorus heptabromide, 32A, 203
Phosphorus-nitrogen compounds, 33B, 273

Phosphorus oxides, 29, 300; 31A, 120; 34A, 240
Phosphorus oxybromide, 34A, 227
Phosphorus pentabromide, 35A, 151
Phosphorus pentafluoride, 30A, 266
Phosphorus-platinum, 32A, 108
Phosphorus-potassium, 26, 56
Phosphorus-rhenium, 26, 227; 31A, 57
Phosphorus-rhodium, 26, 48, 270; 34A, 117
Phosphorus-ruthenium, 27, 319
Phosphorus-silicon, 31A, 58; 32A, 110; 33A, 36
Phosphorus sulphides, 30A, 353
Phosphorus-sulphur, 30A, 353; 34A, 118
Phosphorus-sulphur solid solutions, 27, 401
Phosphorus-tantalum, 31A, 58
Phosphorus tantalum oxychloride, 27, 468
Phosphorus-terbium, 26, 255; 27, 335
Phosphorus-tin, 33A, 146; 35A, 87
Phosphorus-tin-zinc, 33A, 126
Phosphorus-titanium, 30A, 83; 32A, 111; 33A, 128
Phosphorus titanium oxychloride, 27, 468
Phosphorus trichloride, 27, 717
Phosphorus tricyanide, 26, 339; 29, 290
Phosphorus trifluoride, 34A, 172
Phosphorus trifluroide tris(difluoroboryl) borane, 34A, 191
Phosphorus trioxide disulphide, 32A, 220
Phosphorus triphenyl, 28, 684
Phosphorus-vanadium, 35A, 89
Phosphorus-wolfram, 27, 319
Phosphorus-zinc, 28, 37; 30A, 83
Phosphorus-zirconium, 33A, 128
Phosphosulphates, 31A, 246
Photocaryophyllene A bromoketone, 34B, 308
Photocaryophyllene D bromoketone, 34B, 308
cis-syn Photodimer of uracil, 35B, 256
Photoisopyrocalciferyl *m*-bromobenzoate, 26, 808; 29, 772; 33B, 202
Photolysis products from *N*-chloroacetylmescaline, 35B, 204
Phthalimide, 28, 671, 686
Phthalimidocyclohexane, 34B, 143
Phyllochlorine ester, 34B, 281
Phyllochrysine iodomethylate, 30B, 397
Phyllochrysine methiodide, 31B, 240
Phytin acid, sodium salt hydrate, 33B, 543
Phytolaccagenin, 29, 772
Picolic acid chlorhydrate, 28, 685
Picolinamide, 26, 762
α-Picolinamide, 31B, 121
α-Picoline, 26, 763
β-Picoline, 26, 763
γ-Picoline, 26, 763
4-Picoline and iodine, 26, 763
mono-α-Picolinecopper(II) chloroacetate, 35B, 774
4-Picoline-N-oxide, 26, 763
β-Picoline-N-oxide fumaric acid adduct, 29, 771
Picolinic acid, 26, 762
Picolinic acid hydrochloride, 30B, 130
α-Picolinium nonabromoantimonate(V), 33B, 289
2-Picolinium tetrabromobismuthate, 32B, 157
2-Picolinium tetraiodobismuthate, 32B, 157
3-Picolylamine dihydrochloride, 30B, 129
Picropharmacolite, 26, 470
Picrylazide, bis-8-hydroxyquinolinatocopper(II), (2:1 complex), 30B, 260
Piemontite, 31A, 245
Pigeonite, 29, 431; 35A, 475
Pillaronone monobromoacetate, 35B, 388
Pilocarpine trichlorogermanate(II) hemihydrate, 33B, 219
Pimelamide, 30B, 396
Pimelic acid, 26, 561; 31B, 10
Pimelic acid-urea, 26, 572
Pinacol hexahydrate, 35B, 29
α-Pinene, 29, 766
Pinnoite, 32A, 496
Pinoresinol(dibromo dimethylether), 33B, 551
Piperazine, 28, 666
1,4-Piperazine-γ,γ-dibutyric acid, 31B, 127
1,4-Piperazine-γ,γ'-dibutyric acid, 27, 960
Piperazine hexahydrate, 33B, 113
Piperidinium europium(III)-tetra-(benzoylacetonate), 34B, 508
Piperidinium gadolinium tetrakisbenzoylacetonate, 33B, 556
Pirssonite, 32A, 416
Pistomesite, 29, 439
Plancheite, 26, 522; 31A, 260
Platicodigenin, 33B, 555
Platinum alloys, 27, 321
Platinum amino-halides, 27, 665
Platinum biguanide perchlorate, 30B, 416
Platinum carbonyl, 26, 539
Platinum(II) chloride, 30A, 276
Platinum(II)diethyldithiocarbamate, 32B, 495
Platinum hexafluoride, 26, 301
Platinum(IV)-iodide, 34A, 183
Platinum oxides, 33A, 270
Platinum phthalocyanine, 33B, 432
Platinum-praseodymium, 26, 239
Platinum-scandium, 26, 165; 27, 322, 342
Platinum-silicon, 29, 79
Platinum-strontium, 27, 323
Platinum-tantalum, 29, 70; 30A, 85

Platinum-tellurium, 34A, 120
Platinum-thallium, 33A, 124
Platinum thiooxinate, 35B, 908
Platinum-thorium, 26, 256, 27, 325
Platinum-titanium, 33A, 126
Platinum-yttrium, 26, 165, 237; 27, 322
Platinum-zinc, 34A, 120; 35A, 90
Plumbic fluoride, 27, 442
Plumbojarosite, 31A, 242
Plumbopyrochlore, 31A, 248
Plutonium, 26, 228, 231; 27, 325; 28, 38
Plutonium alloys, 26, 228-233
Plutonium(III) chloride, 31A, 251
Plutonium metaphosphate, 27, 598
Plutonium oxaltes decahydrate, 30B, 408
Plutonium oxalate dihydrate, 30B, 409
Plutonium oxalte dihydrate (acetone solvate), 30B, 409
Plutonium oxalate hexahydrate, 30B, 408
α'-Plutonium oxide, 27, 477
β-Plutonium oxide, 27, 477
Plutonium oxides, 27, 477; 29, 315
Plutonium-silicon, 29, 81
Plutonium-titanium, 30A, 86
Plutonium-uranium, 27, 326
Plutonium-zinc, 26, 231; 32A, 114; 33A, 131
Plutonium-zirconium, 26, 231
Plutonyl oxalate trihydrate, 30B, 408
Poitevinite, 29, 440
Pollucite, 30A, 455; 32A, 480, 511; 34A, 370
Pollucite, iron analogue of, 28, 281
Polonium, 31A, 61
Polyacetaldehyde, 26, 636
Poly-*N*-acetyl-D-glucosamine, 33B, 522
Polyacrylates, 35B, 886
Polyacrylonitrile, 26, 634; 27, 804; 33B, 527
Polyadenylic acid, 26, 820
α-Poly-L-alanine, 31B, 193
Polyaldehydes, 26, 635; 29, 759
trans-Polyalkenamers, 32B, 45; 33B, 523
Polyamides, 27, 811
Poly-ε-amino-caproic acid, 26, 640
Polyantimonic acid, 33A, 425
Polybasite, 28, 288; 30A, 459
Poly-γ-benzyl-L-glutamate, 26, 809
Poly-S-benzylthio-L-cysteine, 27, 1053
Poly-3,3-bis(chloromethyl)oxacyclobutane, 29, 759
Polybis(μ-(2-picoline *N*-oxide)-chlorocopper(II) di-μ-chloro-)diaquocopper(II), 33B, 458
Polyborate, 34A, 350
trans-1,4-Polybutadiene, 33B, 526; 35B, 884
Polybutene-1, 29, 759
cis-1,4-Poly(2-*t*-butyl-butadiene), 29, 759
Poly-*n*-butyraldehyde, 26, 636
Polycaproamide, 33B, 527
Poly-ε-caprolactone, 35B, 885
Polycarbonate from 4,4'-dioxydiphenyl-2,2-propane, 27, 809
Poly-β-(*p*-chlorobenzyl)-L-aspartate, 35B, 886
Poly(chloromethyl ethylene oxide), 35B, 883
Poly(1,4-cyclohexylenedimethylene terephthalate), 26, 639
cis-Poly(1,4-cyclohexylenedimethylene terephthalate, 26, 629
trans-Poly(1,4-cyclohexylenedimethylene terephthalate), 26, 639
trans-Polydecenamer, 32B, 45, 46
Poly(difluorophosphazene), 35B, 915
Polydiketene, 26, 638
Polydimethyl silicone rubber, 27, 813
Poly[(2,6-dioxo-1,4-piperidenediyl)-trimethylene], 33B, 527
Poly(2,6-diphenyl *p*-phenylene oxide), 35B, 886
Polydymite, 27, 294
Polyepichlorohydrin, 26, 631; 35B, 883
Polyether-alkali metal complexes, 35B, 917
Polyetherin A, 33B, 550
Polyethers, 33B, 527
Polyethylene, 26, 629, 631; 27, 799; 33B, 526; 35B, 882, 907, 908
Polyethylene adipate, 27,806
Polyethylene chloride, 29, 759
Polyethylene glycol, 27, 810
Polyethylene oxide, 26, 631; 29, 759
Poly(ethylene oxide)-mercuric chloride, 33B, 524
Polyethylenes, 33B, 527
Polyethylene suberate, 27, 806
Polyethylene sulphide, 27, 811; 33B, 553
Polyethylene terephthalate, 27, 808; 29, 759
Polyformaldehyde, 27, 809
Poly(gly-ala-pro), 35B, 922
Polyglycolide, 33B, 521
Polyglycolides, 29, 759
Polygorskite minerals, 30A, 456
Poly-*l*-guluronic acid, 27, 1056
Polyhalite, 26, 449; 29, 431
Poly-*para*-halogeno-styrene, 27, 803
Polyhedral clathrate hydrates, 27, 778; 28, 678, 679
trans-Polyheptenamer, 32B, 44
Polyinosinic acid-polydeoxycytidylic acid, 35B, 887
Polyketone, 26, 632
Polyketones, 27, 805
Poly(S-lactic acid), 33B, 527
Poly-*d*-mannuronic acid, 27, 1056

Poly(3-methyl-1-butene), 35B, 884
Poly(methylenebutene-1), 29, 759
Poly-γ-methyl-L-glutamate, 26, 810; 27, 812
Polymethylmethacrylate, 33B, 526
Poly-4-methylpentene-1, 29, 759
Polymorphism, 28, 289
Polynuclear ferrocene complex, 35B, 564
trans-Polyoctenamer, 32B, 45
Polyolefins, 29, 759
Polyoxacyclobutene hydrate, 35B, 884
Polyoxymethylene, 26, 638; 27, 810; 29, 759
Polypentadiene, 26, 636
trans-Polypentenamer, 32B, 44
Polypeptides, 26, 809
Poly-4-phenyl-1-butene, 33B, 524
Poly(*p*-phenylene), 33B, 526
Poly(*p*-phenylene oxide), 34B, 177
Polypivalolactone, 29, 759
Poly-L-proline, 27, 1054
Poly-L-proline II, 33B, 185
Poly-β-propiolactone, 26, 638
Poly(1-propyl-*trans*-buta-1,4-diene), 29, 759
Polyporpylene, 26, 631; 27, 801; 29, 759; 33B, 526, 527
Poly(propylene oxide), 31B, 37
Poly(L-propyl-glycyl-glycine), 31B, 193
Polystyrene, 27, 802
Polystyrene-polyvinylpyridine, 35B, 907
Polytetrafluorethylene, 27, 804
Polytetrahydrofuran, 27, 812
Polythiazide, 35B, 910
Poly-ε-thiocaprolactone, 33B, 526
Polythiomethylene, 29, 759
Polytrifluorochloroethylene, 29, 759
Polytrifluoro-3,3,3-epoxy-1,2-propane, 35B, 909
Poly(trithiane), 33B, 527
Polyvinyl alcohol, 27, 809
Polyvinyl chloride, 26, 633; 29, 759
Polyvinyl formate, 27, 806
Poly(vinylidene fluoride), 31B, 37; 33B, 526, 527; 35B, 882
Poly(vinylmethyl ether), 33B, 521
Poly(vinyl trifluoroacetate), 26, 634
Poly-2,6-xylenol, 27, 810
Poly(*p*-xylylene), 33B, 527
Poly(*m*-xylylene adipamide), 29, 759
Porcine pancreatic elastase, 33B, 527
Porcine transferrin, 35B, 906
Porphine, 30B, 216
Porphyrin diacids, 33B, 185
Porphyrins, 28, 686
Portentol *p*-bromobenzoate, 35B, 914
Portlandite, 26, 349
Potassium, 26, 3, 271
Potassium acid chloromaleate, 26, 554
Potassium acid phthalate, 30B, 62
Potassium allantoinate, 33B, 175
Potassium aluminate, 31A, 133
Potassium alumino silicate, 32A, 478
Potassium aluminum germanate, 31A, 249
Potassium amide, 27, 420; 30A, 258
Potassium amidotriphosphate tetrahydrate, 30A, 392
Potassium *p*-aminobenzoate, 33B, 551
Potassium antimonate, 31A, 135
Potassium antimony tartrate, 29, 762; 30B, 395
*mono*Potassium aurate(III), 35A, 216
Potassium auricyanide monohydrate, 35A, 201
Potassium azide, 31A, 241
Potassium barium hexanitrocobaltate, (II), 31A, 181
Potassium barium hexathionate, 30A, 380
Potassium bicarbonate, 33A, 433
Potassium bichromate, 32A, 278
Potassium bifluoride, 29, 254
Potassium bis(acetylacetonato)-chloroplatinate(II), 34B, 495
Potassium bis(acetylacetone) chloroplatinate (II), 27, 822
Potassium bis-biureto cuprate (II) tetrahydrate, 26, 658
Potassium bis(iminodiacetato)-chromium(III) trihydrate, 35B, 906
Potassium bis(nitrilotriacetate)-zirconate(IV) monohydrate, 33B, 440
Potassium bisoxalato cuprate dihydrate, 27, 817
Potassium bisulphate, 29, 346
Potassium borates, 26, 488
Potassium boratohexafluoroberyllate, 35A, 188
Potassium bromide, 26, 297; 27, 426
Potassium bromide-chloride, 27, 425
Potassium caprate, 34B, 21
Potassium chloride, 26, 297; 27, 426
Potassium chloride-potassium bromide, 31A, 241
Potassium chlorotitanate, 26, 325
Potassium chromate permanganate, 33A, 317
Potassium chromato iodate, 32A, 320
Potassium chrome alum, 30A, 453
Potassium cobalticyanide, 26, 339
Potassium cobalt(II) thiocyanate trihydrate, 28, 113
Potassium copper(II) chloride, 28, 85
Potassium copper(II) ethylenediamine tetra-acetate trihydrate, 32B, 495
Potassium cuprous cyanide hydrate, 27, 416
Potassium cyanide, 26, 339; 27, 415
Potassium diammine dinitrocarbonato-cobaltate, 32A, 510
Potassium disaquotetrasulphatothorate-(IV), 31A, 205
Potassium diborate, 26, 488
Potassium dichromate, 29, 319; 33A, 316

Potassium 5,5'-diethylbarbiturate, 27, 973; 30A, 163
Potassium difluorophosphate, 31A, 192
Potassium dihydrogen isocitrate, 27, 735; 33B, 21
Potassium dihydrogentrifluoride, 28, 71
Potassium dihydroxyhexacarbonatedizirconate hexahydrate, 33A, 438
Potassium dihydroxytetrafluoroarsenate, 32A, 173
Potassium O,O-dimethyl phosphordithioate, 27, 755
Potassium dinitrobromocarbonate, 34A, 349
Potassium 4,4-dinitro-2-butenamide, 35B, 39
Potassium dinitroosmiate, 32A, 510
Potassium dioxalatomanganate(II) dihydrate, 33B, 554
Potassium dioxalatoplatinate(II) dihydrate, 29, 586
Potassium dioxatetrafluoromolybdate monohydrate, 31A, 91; 32A, 176
Potassium diperoxotricyanochromate(IV) 30A, 292
Potassium di-μ-tartrato-diantimonate-(III) trihydrate, 35B, 469
Potassium dithorium triphosphate, 31A, 259
Potassium dititanate, 26, 373
Potassium dizirconate, 34A, 257; 35A, 238
Potassium dodecawolframate hydrate, 31A, 258
Potassium enneafluorodiuranate(IV), 34A, 199
Potassium ethylenediaminetetraacetatoaluminum(III) dihydrate, 35B, 602
Potassium (ethylenetrichloroplatinum(II)) monohydrate, 34B, 437
Potassium feldspars, 33A, 478
Potassium ferricyanide, 34A, 230
Potassium ferrocyanide trihydrate, 29, 292
Potassium fluoride, 26, 297
Potassium fluoride hexafluosilicate, 27, 464
Potassium fluoride tetrahydrate, 29, 254
Potassium fluoroberyllate, 26, 302; 30A, 270
Potassium *p*-fluoro-bis(triethylaluminum, 28, 600
Potassium fluoromanganate, 26, 309
Potassium fluorosulphate, 32A, 334
Potassium *mono*fluoroxenate(VI), 34A, 219
Potassium germanide, 33A, 162
Potassium glucuronate dihydrate, 28, 513
Potassium heptafluorodizirconate(II), 31A, 254
Potassium heptafluoroniobate(V), 31A, 90
Potassium heptafluoroprotoactinate(V), 31A, 260
Potassium hexabromoplatinate, 35A, 168
Potassium hexabromorhenate, 35A, 168
Potassium hexabromostannate, 27, 462
Potassium hexabromotellurate, 27, 463; 29, 286
Potassium hexachloromanganate(IV), 31A, 101
Potassium hexachloromolybdate, 27, 463; 34A, 207
Potassium hexachlororhenate, 35A, 168
Potassium hexacyanomanganate, 27, 653; 35A, 401
Potassium hexacyanoplatinate, 32A, 236
Potassium hexacyanorhenate, 27, 653
Potassium hexacyanotechnetate, 27, 653
Potassium hexaethyldialuminum fluoride, 28, 600
Potassium hexafluoropalladate (IV), 26, 313
Potassium hexafluoroplatinate, 27, 666
Potassium hexafluorouranate (IV), 34A, 199
Potassium hexaiodotellurate, 27, 463
Potassium hexaisothiocyanatomolybdate(III) hydrate-acetic acid, 33B, 422
Potassium hexanitrodiphenylamide, 27, 866
Potassium hexatitanate, 27, 529
Potassium hydride, 27, 398
Potassium hydrogen bisacetylsalicylate, 33B, 47
Potassium hydrogen bisglycollate, 33B, 18
Potassium hydrogen bisphenylacetate, 33B, 45
Potassium hydrogen chloromaleate, 29, 765; 30B, 25
Potassium hydrogen dianisate, 33B, 46
Potassium hydrogen diaspirinate, 32B, 87
Potassium hydrogen di-*p*-chlorobenzoate, 28, 416
Potassium hydrogen diformate, 33B, 13
Potassium hydrogen di-*p*-hydroxybenzoate hydrate, 33B, 44
Potassium hydrogen di-*p*-nitrobenzoate, 26, 707
Potassium hydrogen disilicate, 32A, 511
Potassium hydrogen di-trifluoroacetate, 30B, 13
Potassium hydrogen fumarate, 26, 563
Potassium hydrogen fumarate, acid, 35B, 16

Potassium hydrogen maleate, 26, 555
Potassium hydrogen malonate, 33B, 546, 554; 35B, 12
Potassium hydrogen oxalate, 33B, 11
Potassium hydrogen silicate, 34A, 364
Potassium α-hydroxybenzylsulphonate, 32B, 67
Potassium hydroxylamine-*N*-sulphonate, 31A, 207
Potassium *mono*hydroxypentafluoroarsenate, 32A, 172
Potassium imidodisulphate, 28, 215
Potassium iodate, 26, 497; 27, 643; 33A, 451
Potassium iodide, 26, 297; 27, 425
1:2 Potassium iodide-diacetamide complex, 34B, 352
Potassium iodide-mercuric cyanide, 28, 104
Potassium iodide, potassium triiodide, N-methylacetamide complex, 30B, 257
Potassium lead hexanitrocuprate(II), 34A, 346
Potassium lead pyrogermanate, 30A, 326
Potassium malate hydrate, 33B, 542
Potassium manganate, 32A, 284
Potassium mercuric tribromide, 34A, 212
Potassium mesotartrate dihydrate, 30B, 26
Potassium metaborate, 35A, 285
Potassium metaphosphate, 27, 592; 28, 187
Potassium metasilicate, 29, 395
Potassium metasilicate, acid, 28, 233
Potassium metastannate, 34A, 258; 35A, 222, 223
Potassium metathorate, 31A, 245
Potassium metazirconate, 34A, 258; 35A, 222
Potassium methoxide, 28, 595
Potassium methylate, 35B, 472
Potassium *syn*-methyldiazotate, 30B, 48
Potassium methylenedisulphonate, 27, 752
Potassium molybdate, 34A, 277
Potassium molybdenum bronze, 30A, 341; 31A, 154
Potassium molybdenum dioxyoxalate, 28, 669
Potassium myristate, 30B, 429
Potassium myronate, 35B, 263
Potassium myronate hydrate, 28, 518
Potassium nickel(IV) hexaoxidoiodate-(VII), 31A, 215
Potassium niobate, 32A, 271
Potassium nitrate, 27, 565; 34A, 338
Potassium-μ-nitridooctachlorodiaquodiruthenate, 34A, 210
Potassium nitridopentachloroosmate(VI), 33A, 227; 34A, 225
Potassium nitridotetraisocyanidorhenate (V) monohydrate, 34A, 229
Potassium nitrilotrisulphonate dihydrate, 35A, 379
Potassium nitrite, 31A, 181; 34A, 346
Potassium *p*-nitrophenyldicyanomethanide, 32B, 57
Potassium nitrosodisulphonate, 33A, 386
Potassium nitrosohydroxotetranitroruthenate, 29, 380
Potassium nitrosopentachlororuthenate, 26, 330
Potassium nitrosopentacyanamolybdate, 33A, 246
Potassium octachlorodimolybdate, 34A, 207
Potassium octacyanomolybdate (IV) dihydrate, 33A, 247
Potassium octafluororhenate, 29, 269
Potassium orthoarsenate, 27, 724
Potassium osmate, 26, 399
Potassium osmyl chloride, 26, 331
Potassium oxalate *mono*hydrate, 29, 463; 34B, 16; 35B, 8
Potassium oxalate *mono*perhydrate, 32B, 21
Potassium oxalato palladate (II) tetrahydrate, 31B, 406
Potassium μ-oxido-bis(oxidodiperoxidoaquomolybdenum(VI)) dihydrate, 33A, 330
Potassium oxodiperoxooxalatomolybdate(VI), 35B, 635
Potassium oxomercurate, 27, 661; 29, 338
Potassium oxopentafluoromolybdate *mono*hydrate, 31A, 92; 33A, 177
Potassium oxoperoxotetrafluoromolybdate monohydrate, 32A, 174
Potassium oxycyanorhenate, 26, 342
Potassium oxypentafluoroniobate(V), 31A, 91
Potassium oxythallate, 26, 400
Potassium ozonide, 28, 118
Potassium palmitate, 30B, 19
Potassium pentaacetatocerate(III) monohydrate, 33B, 556
Potassium pentaborate, 30A, 411
γ-Potassium pentaborate, 26, 488
Potassium pentaborate tetrahydrate, 28, 168
Potassium pentabromocarbonyliridate (III), 29, 286
Potassium pentachlororhenium (IV)-μ-oxy-pentachlororhenate (IV) monohydrate, 27, 670
Potassium pentacyanonitrosyl, 32A, 237
Potassium pentacyanonitrosylchromate, 31A, 116
Potassium pentacyanonitrosylmanganate dihydrate, 31A, 258
Potassium pentacyanonitrosylvanadate (I), 35A, 398
Potassium pentafluorotellurate, 34A, 200; 35A, 167

Potassium pentasulphate, 34A, 298
Potassium perbromate, 34A, 361
Potassium perchlorylamide, 33A, 450
Potassium permanganate, 32A, 283
Potassium peroxochromate, 28, 224
Potassium pertechnetate, 27, 671
Potassium perxenate nonahydrate, 29, 335
Potassium phenyldinitromethane, 35B, 30
Potassium phosphide, 27, 420
Potassium picrate, 34B, 67
Potassium polyphosphates, 34A, 322
Potassium polyuranates, 26, 403
Potassium praseodymium fluoride, 26, 312
Potassium rhenium hydride, 27, 671; 29, 235
Potassium salt of isocitric lactone, 28, 385
Potassium salt of *o*-nitrophenol hemihydrate, 26, 704
Potassium salts of fumaric acid, 26, 563
Potassium scandium molybdate, 35A, 254
Potassium selenate, 27, 605; 35A, 382
Potassium selenocyanate, 30A, 296
*mono*Potassium sesqui[cyclohexane-1,4-dicarboxylate], 35B, 112
Potassium silicide, 33A, 162
Potassium-silicon, 26, 84, 156; 32A, 117; 33A, 162; 34A, 122
Potassium silver carbonate, 28, 175
Potassium silver thiocyanate, 27, 419
Potassium silyl, 26, 291
Potassium sodium sulphate, 27, 605
Potassium sodium tetracyanoplatinate trihydrate, 33A, 249
Potassium sodium thorium hexafluroide, 35A, 168
Potassium sodium titanium oxyfluoride, 34A, 216
Potassium sodium vanadium oxyfluoride, 34A, 216
Potassium squarate monohydrate, 29, 674
Potassium stannous chloride hydrate, 27, 457
Potassium stearate, 27, 1061
Potassium strontium arsenate, 26, 470
Potassium strontium phosphate, 26, 469
Potassium strontium tetrametaphosphate, 32A, 382
Potassium strontium vanadate, 26, 470
Potassium sulphamate, 32A, 325
Potassium sulphate, 26, 436; 27, 605; 33A, 367
α-Potassium sulphate, 31A, 244
β-Potassium sulphate, 31A, 244
Potassium-sulphur, 27, 350
Potassium-sulphur-vanadium, 29, 82
Potassium-superoxide, 27, 477
Potassium tantalate-sodium niobate, 27, 504
Potassium technetium hydride, 27, 671
Potassium tellurate, 29, 359
Potassium tellurate(VI) hydrate, 31A, 211
Potassium tetraborate tetrahydrate, 28, 166
Potassium tetrachloroaquonitride osmium hydrate, 35A, 431
Potassium tetrachloroiodide monohydrate, 28, 82
Potassium tetrachromate, 35A, 391
Potassium tetracyanocuprate, 33A, 245
Potassium tetracyanonickelate(II), 29, 293
Potassium tetracyanoplatinate hydrate, 33A, 250
Potassium tetracyanozincate, 31A, 117
Potassium tetradecacarbonylhexacobaltate, 34A, 159
Potassium tetraethinylcadmate(II), 33B, 324
Potassium tetraethinylzincate(II), 33B, 324
Potassium tetrafluoroargentate(III), 35A, 162
Potassium tetrafluoroaurate(III), 34A, 202; 35A, 162
Potassium tetrafluoroborate, 34A, 189
Potassium tetrafluorobromate(III), 34A, 202
β-Potassium tetrafluorocerate (III), 34A, 192
Potassium tetrahydrogen pentafluoride, 35A, 152
Potassium tetrahydroxyosmite, 27, 665
Potassium-2,2,5,5-tetramethyl-3-carboxypyrroline-1-oxyl, 35B, 159
Potassium tetranitratoaurate(III), 35A, 326
Potassium tetranitrohydroxynitrosorutheniate, 27, 669
Potassium tetraoxalate, 29, 465
Potassium tetroxalate dihydrate, 32B, 25
Potassium thiocyanate, 28, 108; 33A, 252
Potassium thorium sulphate dihydrate, 28, 214
Potassium-tin, 29, 83; 34A, 122
Potassium titanoniobate, 29, 327
Potassium 2-*p*-toluidinyl-6-naphthalenesulphonate, 35B, 116
Potassium triborate, 26, 488
Potassium tribromoethylene platinoate, 30B, 291
Potassium tribromozincate dihydrate, 31A, 260
Potassium trichloroammineplatinate(II) monohydrate, 35A, 430
Potassium trichloroethylene platinate, 30B, 291
Potassium *cis*-trichlorotrinitroplatinate, 27, 668

Potassium trichlorozincate(II) dihydrate, 29, 278
Potassium trichloro zincate monohydrate, 34A, 206
Potassium tricyanomethanide, 28, 107
Potassium trifluorochromate(II), 31A, 247
Potassium trifluoro-stannate-hemihydrate, 33A, 192
Potassium trinitromethane, 32A, 347; 32B, 64
Potassium triperoxo-O-phenanthroline)-niobate trihydrate, and its hydrogen peroxide adduct, 35B, 912
Potassium triphosphatodithorate, 33A, 398
Potassium tris(acetylacetonato) cobaltate (II), 27, 822
Potassium triselenocyanate hemihydrate, 28, 116
Potassium tris-oxalato vanadate(II) trihydrate, 32B, 496
Potassium trisulphatochromate(III), 26, 446
Potassium trithiocarbonate monohydrate, 35A, 483
Potassium trivanadate, 31A, 141
Potassium 12 tungstocobaltiate, 26, 393
Potassium uranate, 26, 403; 27, 543
Potassium violurate dihydrate, 28, 454; 30B, 164
Potassium wolframate, 34A, 282
Potassium wolfram bronze, 26, 356
Potassium xanthate, 28, 407
Potassium yttrium molybdate, 32A, 303
Potassium yttrium triorthosilicate, 33A, 473
Potassium yttrium wolframate, 33A, 345
Potassium zinc bromide dihydrate, 33A, 233
Potassium zinc decavanadate, 31A, 142
Potassium zinc fluoride, 26, 307
Potassium zinc fluoroberyllate, 32A, 163
Potassium zinc oxide, 33A, 325
β-Potassium zinc tribromide dihydrate, 32A, 207
Potassium zinc trichloride dihydrate, 28, 86
Potassium zinc trifluoroberyllate, 31A, 87
Potassium zinc trihalides, 35A, 161
Potassium zirconium disilicate, 35A, 310
Powellite, 27, 522; 33A, 334
Praseodymium, 26, 233
Praseodymium chloride, 27, 724
Praseodymium dihydride, 27, 237
Praseodymium dihydroxychloride, 34A, 221
Praseodymium disilicate, 35A, 307
Praseodymium fluorides, 26, 312
Praseodymium formate, 27, 815
Praseodymium hexaboride, 26, 69
Praseodymium niobate, 26, 386; 27, 521
Praseodymium nitrate, 26, 476
Praseodymium nitrate hexahydrate, 29, 379
Praseodymium oxides, 27, 481
Praseodymium oxydisulphide, 32A, 223
Praseodymium oxytelluride, 27, 424
Praesodymium oxywolframite, 35A, 255
Praseodymium rubidium fluoride, 26, 312
Praseodymium sodium flurodie, 26, 312
Praseodymium sodium sulphide, 29, 341
Praseodymium-strontium-sulphur, 27, 327
Praseodymium trichlorohexaoxowolframate, 34A, 286
Pregeijerene-silver nitrate, 35B, 912
Pregnadienes, 29, 764
5α-Pregnan-3β,20α-diol, 28, 679
5α-Pregnan-3β,20β-diol, 28, 679
5β-Pregnan-3α,20α-diol, 28, 679
5β-Pregnan-3α,20β-diol, 28, 679
5β-Pregnan-3β,20β-diol, 28, 679
5α-Pregnan-3β,20α-diol diacetate, 30B, 441
5α-Pregnan-3β,20β-diol diacetate, 30B, 441
5α-Pregnan-3β,17α-diol-30-one, 30B, 440
5α-Pregnan-17α,21-diol-3,11,20-trione 21-acetate, 30B, 440
5α-Pregnan-3,20-dione, 30B, 440
5α-Pregnan-21-01-3,20-dione acetate (form "a"), 30B, 440
5α-Pregnan-21-01-3,20-dione acetate (form "b"), 30B, 440
Pregnane derivatives, 35B, 344
Pregnanediol, 28, 679
5α-Pregnane-"ol-ones", 27, 1038
5α-Pregnanes, 29, 764
5α-Pregnan-3β-01-20-one 3-acetate, 30B, 440
5α-Pregnan-3β,11α,17β,20β, 21-pentol 3,20,21 triacetate, 30B, 441
5α-Pregnan-3β,11β,20β, 21-tetrol, 30B, 440
5α-Pregnan-3β,17α,20β-triol, 30B, 440
5-Pregnan-17α-bromo-3β-01-20-one 3-acetate, heptane, 30B, 439
4-Pregnen-17α-01-3, 20-dione, 30B, 440
Pregnenes, 29, 764
Prehnite, 29, 431; 30A, 434, 456; 32A, 468
Premontite, 28, 322
Pristimerol di-p-bromobenzoate, 35B, 918
Probertite, 28, 172; 30A, 419
Procaine-bis-p-nitrophenyl phosphate, 35B, 416
Prochlorite, 26, 534
Proline, 26, 826

L-Proline, 30B, 209
DL-Proline hydrochloride, 34B, 268
Promethium phosphate, 30A, 458
Propane-1,3-diol cyclic phosphate, 35B, 446
1,6:8,13-Propanediylidene[14]-annulene, 35B, 911
Propargyl alcohol, 29, 453; 33B, 554
Propargyl amine, 33B, 554
Propargyl bromide, 26, 549
Propargyl 2-bromo-3-nitrobenzoate, 31B, 64
Propargyl chloride, 26, 549
Propionic acid, 27, 736
β-5-n-Propoxy-*o*-quinone-2-oxime, 29, 647
n-Propyl alcohol, 26, 550; 29, 453
n-Propylammonium tetrachlorocuprate(II), 35B, 877
N-(*n*-Propyl)dihydronicotinamide, 28, 461
Propylene, 26, 549
N,N'-Propylenebis(salicylaldiminato)-oxovanadium(IV), 35B, 920
Propylenethiourea tellurium(II) salts, 26, 584
2-Propyloxynaphthalene, 33B, 542
Propyl stearate, 33B, 10
19-Propylthevinol hydrobromide, 32B, 260
2-Propyl-4-thiocarbamoylpyridine, 35B, 168
2-N-Propyl-4-thiocarbamoyl pyridine, 34B, 137
2-Propyl-thioisonicotinamide, 34B, 145
Prosopite, 34A, 215; 35A, 451
Prostaglandin F_{2-1} derivatives, 28, 566
Prostaglandin F_{2-1} methyl ester tri-*p* bromobenzoate, 28, 567
Prostaglandin F_{2-1} methyl ester tri-*p* iodobenzoate, 28, 566
Protactinium oxides, 29, 417
Protactinium oxydichloride, 33A, 223
β-Protactinium pentabromide, 34A, 180
Protactinium pentachloride, 32A, 186
Protein matrix of bovine dental enamel, 26, 826
Proteins, 35B, 887
Protoaesciginin 21-tiglate 22-acetate, 33B, 545
Protopine, 33B, 224
Proustite, 31A, 19
Pseudo-akuammigine iodomethylate, 30B, 432
Pseudo-autinite, 29, 426, 439
Pseudobrookites, 30A, 459
Pseudoclovene A-diol mono-*p*-bromobenzensulphonate, 34B, 312
Pseudocorrin, 30B, 396
Pseudolaveite, 34A, 331
Pseudomalachite, 28, 192
Pseudorutile, 31A, 252
Pseudotropine, 32B, 253
α-Pseudouridine monohydrate, 35B, 294
Psilomelane, 30A, 451; 31A, 252
Pumpellyite, 30A, 456; 34A, 393
Purine, 29, 771; 30B, 173
Purpurogenone bromoacetate, 35B, 917
Pycnite, 28, 350
Pyracene, 26, 753
Pyrargyrite, 31A, 19
βPyrazinamide, 26, 762
γ-Pyrazinamide, 26, 762
Pyrazinamidecobalt(II) sulphate monohydrate, 35B, 909
Pyrazine-silver nitrate complex, 27, 857; 31B, 417
Pyrazinic acid, 26, 762
Pyrazole, 35B, 152, 153
Pyrazoline hydrochloride, 27, 951
Pyrene, 30B, 109; 35B, 131
Pyrene-tetracyanoethylene, 33B, 250
Pyrene-tetracyanoethylene complex, 31B, 278
Pyrene 1:1-1,3,7,9-tetramethyluric acid complex, 26, 754
Pyridazino(4,5-D)pyridazine, 34B, 147
2H-Pyridaz-3-thione, 31B, 117
2,2'-Pyridil, 26, 788; 35B, 169
β-Pyridinecarbinol hydrochloride, 26, 764
mono-Pyridinecopper(II) acetate, 26, 776
Pyridine-cyanogen iodide complex, 32B, 491
*mono*Pyridine-*NN'*-ethylenebis(salicylideneiminato)cobalt(II), 35B, 625
Pyridine hydrochloride, 27, 959
Pyridine hydrogen nitrate, 28, 684; 30B, 129
Pyridine-lithium chloride, 31B, 274
Pyridine-N-oxide copper(II) chloride, 29, 768; 30B, 358
Pyridineoxide hydrochloride, 26, 703
Pyridine peroxochromate, 27, 857
Pyridine picrate, 29, 770
*mono*Pyridine-zinc(II) tetra(4-pyridyl)-porphine, 32B, 490
Pyridinium dichlorobromate(I), 32B, 487
Pyridinium dicyanomethylide, 30B, 131
Pyridinium dimetatetrafluoroarsenate, 29, 770
Pyridinium hexafluoroantimonate V, 31B, 119
Pyridinium hexafluoroarsenate, 31B, 119
Pyridinium hexafluorophosphate, 31B, 120
Pyridinium nonabromodiantimonate(III) dibromide, 35B, 876
Pyridinium rhenium(II) tetrachloride, 28, 459
Pyridinium tetrabromorhenate(II), 30B, 293

Pyridinium tetrachloroantimonate(III), 32B, 487; 35B, 875
Pyridinium tetraiodothallate(III) pyridine solvate, 35B, 919
Pyridinium tris(*o*-phenylenedioxy)-siliconate, 34B, 369
α-Pyridoin, 29, 771; 30B, 132
Pyridomycin dihydrobromide monohydrate ethyl acetate solvate, 32B, 490
2-Pyridone-MCl_2 complexes, 33B, 543
Pyridoxal phosphate oxime dihydrate, 34B, 254
Pyridoxonium chloride, 33B, 232
Pyridyl-3-benzthiazol, 26, 783
β-Pyridylcarbinol hydrochloride, 29, 725
5-[Pyridyl-(2)]-3-[5-chloropyridyl-(2)]-3*H*-1,2,3,4,-oxathiadiazole-S-oxide, 35B, 232
2-(2-Pyridylmethyldithio)benzoic acid, 34B, 64
2-(Pyridyl-4)-oxazole, 26, 776
Pyrimidine-2-one, 35B, 248
Pyroaurite, 28, 322; 32A, 422; 33A, 439
Pyrocatechol, 29, 769; 30B, 394
Pyrochlore, 26, 387, 400; 27, 534; 30A, 458
Pyrochlore-microlite, 26, 388
Pyrochlores, 33A, 190, 286; 35A, 244
Pyrogermanates, 29, 399
Pyromellitic dianhydride, 30B, 422
Pyromellitic dianhydride, benzene(1:1), 30B, 423
Pyromellitic dianhydride, naphthalene (1:1), 30B, 423
Pyromellitic dianhydride, pyrene (1:1), 30B, 423
Pyromorphite, 31A, 243
Pyrope, 26, 416; 28, 246; 30A, 426; 31A, 253
Pyrophanite, 29, 419
Pyrophosphates, 26, 466
Pyrosilicates, 29, 399
Pyrosmalite, 29, 431
Pyrostilpnite, 33A, 22
Pyroxene, 29, 431; 30A, 425, 456; 31A, 243
Pyroxenes, 26, 540; 32A, 455
Pyrrhotite, 34A, 137
L-Pyrrolid-2-one-5-carboxamide, 35B, 920
Pyrrolnitrin, 33B, 555
Pyruvidene-β-alaninatoaquocopper(II) dihydrate, 33B, 418
Pyrvinium pamoate, 35B, 908

Quartz, 27, 725; 28, 289; 29, 440; 31A, 244, 246, 255
α-Quartz, 27, 674; 28, 119; 30A, 420
β-Quartz, 27, 675
Quasi-racemate of (-)-1,2-dithiane-3,6-dicarboxylic acid and (+)-1,2-diselenane-3,6-carboxylic acid, 26, 774
Quasi-racemate of α-(1-naphthyl)-propionic acid and α-(3-thianaphthanyl)-propionic acid, 26, 775
Quaternary ammonium compounds, 26, 702-703
Quaterrylene, 27,925
Quinhydrone, 30B, 72; 33B, 65
Quinidine-(-)-1,1'-dimethylferrocene-3-carboxylate monohydrate, 32B, 363
Quinone-resorcinol, 35B, 409
Quinuclidinyl benzilate hydrobromide, 34B, 168
Quinuclidinyl di-α,α-thienylglycollate, 35B, 208

Racemates, 26, 774-776
Racemic α-(amminechlorotriethylenetetramine)cobalt(III) nitrate, 35B, 679
Racemic α,α'-dimethylglutaric acid, 33B, 4
Radial distribution curves, 27, 1064
Radicicol, 30B, 431
Radium, 33A, 134
Raffinose pentahydrate, 35B, 270
Ralstonite, 30A, 451
Ramie, 26, 628; 27, 799
Rammelsbergite, 27, 58, 59
Ramsagite, 34A, 381
Ransomite, 35A, 436
Rare-earth aluminates, 30A, 452
Rare-earth arsenides, 35A, 93
Rare-earth borates, 26, 491-494; 31A, 172
Rare-earth carbides, 32A, 118
Rare-earth carboxylates, 33B, 412; 35B, 642
Rare-earth chalcogenide fluorides, 35A, 155
Rare-earth chlorides, 33A, 204
Rare-earth copper compounds, 35A, 94
Rare-earth disilicates, 35A, 300
Rare-earth fluorosilicate, 31A, 256
Rare-earth formates, 29, 762; 35B, 908
Rare-earth hydride-chalcogenides, 27, 332
Rare-earth hydrides, 27, 235
Rare-earth hydroxides, 31A, 241
Rare-earth iron garnets, 27, 511
Rare-earth Laves phases, 27, 261
Rare-earth manganites, 28, 138
Rare-earth metals, 26, 233
Rare-earth metals and compounds, 27, 335
Rare-earth nickel compounds, 35A, 95
Rare-earth nickel silicides, 35A, 96
Rare-earth niobates, 26, 386; 29, 329
Rare-earth nitrides, 27, 304
Rare-earth orthoferrites, 35A, 218
Rare-earth orthovanadates, 27, 589
Rare-earth oxide hydroxides, 30A, 310; 31A, 241

Rare-earth oxides, 26, 364, 540; 27, 497; 30A, 450, 457; 31A, 245, 246
Rare-earth oxybromides, 30A, 289
Rare-earth oxyselenides, 31A, 241
Rare-earth oxytellurides, 27, 424
Rare-earth phosphates, 29, 439; 30A, 454
Rare-earth pyrosilicates, 33A, 464
Rare-earth sesquioxides, 29, 439; 31A, 245, 248
Rare-earth silicates, 30A, 455
Rare-earth silicides, 27, 336
Rare-earth stannates, 26, 399
Rare-earth-strontium apatite, 28, 188
Rare-earth sulphides, 29, 341
Rare-earth tantalates, 29, 329; 31A, 248
Rare-earth ternary oxides, 31A, 253
Rare-earth transition metal compounds, 26, 235-240
Rare-earth wolframates, 30A, 452
Rare-earth wolfram bronzes, 31A, 248
Rare-earth zinc phases, 35A, 97
Rare-earth zirconates, 30A, 452
Rare-earth zirconium oxides, 27, 535
Rathite I, 27, 402, 405; 30A, 357
Rathite II, 26, 428; 27, 403, 405; 35A, 143
Rathite III, 27, 403, 405
Rauvoxinine methiodide, 33B, 228
Rayon, 26, 628
Rectorite, 31A, 242
Red lead, 30A, 313
Redledgeite, 26, 537
Red tin (II) oxide, 26, 360
Reedmergnerite, 30A, 435
Reserpine, 33B, 230
Retamine, 30B, 430
Retamine hydrobromide, 30B, 430
trans-Retinal$_1$ (Vitamin A$_1$ aldehyde), 33B 553
Retusamine α'-bromo-D-camphor-*trans*-π-sulphonate, 27, 1002
Retusamine α'-bromo-D-camphor-*trans*-π-sulphonate monohydrate, 32B, 256
Retzian, 32A, 385, 386
Reyerite, 26, 513; 29, 431; 30A, 456
Rezhikite, 26, 528
Rhabdite, 35A, 71
Rhabdophane, 30A, 454
Rhenium, 26, 240
Rhenium(IV) bis(acetylacetonate) dichloride rhenium(IV) bis(pentane-2,4-dionate) dichloride, 32B, 495
Rhenium(III) cluster halides, 34A, 213
Rhenium (III) hydride bis(triphenylphosphine)-1,2-bis(diphenyl phosphino) ethane, 31B, 464
Rhenium (IV) oxide, 26, 365
Rhenium(VII) oxide, 34A, 244
Rhenium oxide dihydrate, 33A, 269
Rhenium oxide tetrafluoride, 32A, 508; 33A, 173
Rhenium(V) oxyhalide complex, 30B, 394
Rhenium pentachloride, 33A, 205
Rhenium perovskites, 26, 418
Rhenium-rhodium, 27, 337
Rhenium-ruthenium, 27, 338
Rhenium-selenium, 30A, 86
Rhenium-silicon-titanium, 29, 84
Rhenium-tantalum, 26, 241
Rhenium-technetium, 27, 361
Rhenium-thorium, 26, 237, 241
Rhenium-uranium, 26, 241
Rhenium-vanadium, 26, 242
Rhenium-yttrium, 26, 237
Rhodamine, 26, 828
Rhodan hydrate, 31B, 155
Rhodanine, 27, 952
Rhodium acetate, 27, 816; 35B, 920
Rhodium(III) chloride, 29, 275
Rhodium (I) chloride-1,5-*cyclo*-tadiene, 27, 853
Rhodium oxides, 26, 390; 28, 133; 33A, 270; 35A, 211
Rhodium dioxide, 33A, 270
Rhodium-scandium, 26, 165; 27, 342
Rhodium-selenium, 33A, 135
Rhodium sesquioxide, high-pressure form, 35A, 211
Rhodium-silicon, 27, 373; 28, 40; 30A, 88; 33A, 136
Rhodium-sulphur, 27, 338; 32A, 120
Rhodium-tantalum, 29, 70
Rhodium telluride, 31A, 61
Rhodium-thorium, 26, 165,255,256; 27, 340, 367
Rhodium-titanium, 27, 340; 34A, 124
Rhodium-uranium, 26, 165
Rhodium-yttrium, 26, 237
Rhodizite, 30A, 455; 31A, 260; 32A, 495
Rhodochrosite, 32A, 415
γ-Rhodomycinone, 35B, 132
Rhodonite, 28, 258
Rhodonite-wollastonite, 26, 512
Ribitol, 34B, 245
Riboflavin dihydrobromide hydroquinone, 35B, 914
Riboflavin hydrobromide monohydrate, 32B, 488
3-(β-D-Ribofuranosyl)-3-pyrroline-2,5-dione, 35B, 260
Ribonuclease, 26, 817; 27, 1059
Ribonuclease-parachloromercuribenzoate, 27, 1059
Ribonucleic acid, 26, 817; 27, 1055
Ribose *p*-bromophenylhydrazone, 29, 489
Richterite-asbestos, 30A, 459
Riebeckite, 28, 258
Rifamycin, 28, 687
Rifamycine *p*-iodoanilide, 29, 772
Rimuene, 35B, 349

Rinkite, 28, 247; 31A, 243; 32A, 447; 33A, 481
RNA, 29, 758; 33B, 527
tRNA, 35B, 922
Robustic acid methyl ether, 30B, 432
Rockbridgeite, 26, 462; 35A, 337, 339
Rodalquilarite, 33A, 393; 34A, 318
Roeblingite, 31A, 242
Roedderite, 31A, 242
Roemerite, 34A, 439
Rongalite, 27, 754
Rosein III, 35B, 914
Rosenbuschite, 28, 254, 322
Rosenite, 27, 613
Rossite, 28, 204
Rozenite, 28, 326
Rubeanic acid, 30B, 36
Rubidium, 26, 3
Rubidium acetylenediolate, 29, 467
Rubidium acid chloromaleate, 26, 555
Rubidium aluminum germanate, 31A, 249
Rubidium americyl hexafluoride, 27, 463
Rubidium amide, 27, 420; 30A, 258
Rubidium antimonyl tartrate, 29, 762
*mono*Rubidium aurate (III), 35A, 216
Rubidium azide, 30A, 256
Rubidium borane, 33A, 157
Rubidium bromide, 26, 297, 298
Rubidium chloride, 26, 297, 298; 27, 425
Rubidium chromium alum, 34A, 312
Rubidium 'cryptate', 35B, 912
Rubidium dichromate, 35A, 385
Rubidium dihydrogen phosphate, 31A, 252
Rubidium dimercaptomaleonitrile, 35B, 922
Rubidium di(meta)fluoroberyllate, 26, 303
Rubidium α, α-dinitroacetamide, 35B, 19
Rubidium dinitroacetonitrile, 32B, 48; 33B, 30
Rubidium ethylenediaminetetraacetato-aquaferrate, 26, 680
Rubidium ferric oxide, 26, 395
Rubidium fluoride, 27, 425; 29, 255
Rubidium fluoroberyllate, 26, 302; 27, 452; 30A, 270
Rubidium fluoromanganate, 26, 309
Rubidium 5-fluoroorotate monohydrate, 29, 721
*mono*Rubidium furantetracarboxylate, 30B, 398; 32B, 174
Rubidium glucuronate dihydrate, 28, 513
Rubidium heptachlorodiantimonite monohydrate, 31A, 260
Rubidium heptachlorodibisumthite monohydrate, 31A, 260
Rubidium hexabromomolybdate, 27, 464
Rubidium hexachloromolybdate, 27, 464
Rubidium hexafluoroactinates, 27, 463
Rubidium hexafluoroprotactinate(V), 33A, 195
Rubidium hexafluorouranate, 27, 463
Rubidium hexatitanate, 27, 529
Rubidium homophthalate, 35B, 919
Rubidium hydrogen bis(acetylsalicylate), 32B, 89
Rubidium hydrogen bisglycollate, 30B, 17
Rubidium hydrogen croconate, 31B, 38
Rubidium hydrogen diaspirinate, 32B, 89
Rubidium hydrogen di-*p*-chlorobenzoate, 28, 416
Rubidium hydrogen di-*o*-nitrobenzoate, 26, 705
Rubidium hydrogen di-trifluoroacetate, 30B, 13
Rubidium hydrogen fumarate, 34B, 17
Rubidium iodide, 26, 297; 27, 425
Rubidium(+)-3-isopropyl-3-methoxycarbonyl-3-methylpropionate hydrate, 32B, 485
Rubidium mesotartrate dihydrate, 30B, 26
Rubidium metaphosphate, 29, 360
Rubidium methylate, 35B, 472
Rubidium nitrate, 27, 568
Rubidium octadecahydroeicosaborate, 33A, 158
Rubidium oxalate monohydrate, 30B, 24
Rubidium oxalate monoperhydrate, 32B, 21
Rubidium oxomercurate, 29, 338
Rubidium oxythallate, 26, 400
Rubidium pallado-cyanide, 35A, 200
Rubidium pentaiodo(tetra)argentate(I), 32A, 216
Rubidium phosphide, 27, 420
Rubidium plumbate(IV), 35A, 223
Rubidium selenate, 27, 605
Rubidium-silicon, 26, 84, 156
Rubidium sodium erbium hexafluoride, 33A, 186
Rubidium sulphate, 27, 604
α-Rubidium sulphate, 31A, 245
β-Rubidium sulphate, 31A, 245
Rubidium tetrachloromanganate (II) dihydrate, 29, 275
Rubidium tetracyanopalladate monohydrate, 32A, 236
Rubidium tetrafluoroaurate(III), 35A, 162
Rubidium tetrafluoroborate, 34A, 189
Rubidium 1,1,2,2-tetranitroethanide, 32B, 486
Rubidium-1,1,3,3-tetranitro-heptane-6-one, 34B, 54
Rubidium tetraphenylboronate, 26, 735; 27, 884
Rubidium *mono*thioperrhenate, 34A, 292
Rubidium titanate, 29, 318
Rubidium trifluorochromate(II), 31A, 247
Rubidium trifluoroferrate(II), 31A, 249

Rubidium trinitromethane, 31B, 52; 33B, 30
Rubidium uranate, 26, 404; 27, 543
Rubidium uranyl nitrate, 30A, 406
Rubidium violurate, 30B, 164
Rubidium wolfram bronze, 26, 356
Rubrene, 27, 940
Rubrofusarin, 26, 830; 27, 1005
Ruby, 29, 299
Ruthenium, 27, 382
Ruthenium arsine complexes, 35B, 845
Ruthenium biguanide sulphate, 28, 674
Ruthenium biguanide sulphate heptahydrate, 30B, 416
Ruthenium boride, 27, 93
Ruthenium bromide, 33A, 236
Ruthenium bromocarbonyl, 33A, 512
Ruthenium carbonyl, 33A, 167
Ruthenium(III) chloride, 31A, 259
Ruthenium dioxide, 31A, 124; 35A, 212
Ruthenium(III) halides, 31A, 251
Ruthenium pentafluoride, 29, 260
Ruthenium-samarium, 26, 236
Ruthenium-scandium, 26, 237; 27, 342
Ruthenium-selenium, 33A, 34
Ruthenium-silicon, 26, 243; 27, 372; 34A, 125
Ruthenium-sulphur, 32A, 126
Ruthenium-tantalum, 26, 254
Ruthenium-technetium, 27, 361
Ruthenium tetracarbonyldiiode, 27, 646
Ruthenium-thorium, 26, 236, 256; 27, 367
Ruthenium-tin, 29, 85
β-Ruthenium trichloride, 32A, 185
Ruthenium-uranium, 26, 243
Ruthenium-yttrium, 26, 237
Ruthenocene, 33B, 536
Ryanodol-*p*-bromobenzyl ether, 32B, 546; 35B, 392

Saccharin, 33B, 147
Sainfeldite, 29, 426
Sakhaite, 31A, 255
Salesite, 27, 644
Salicylaldehyde azine, 34B, 35
Salicylaldimine metal chelates, 31B, 400
Salicylatotetra-amminecobalt(III) chloride monohydrate, 35B, 913
Salicylic acid, 30B, 56
N-Salicylidene-*p*-aminoacetophenone, 35B, 907
Salicylideneanilines, 29, 763
N-Salicylideneglycinatoaquocopper(II) hemihydrate, 32B, 391
N-Salicylideneglycinatoaquocopper(II) tetrahydrate, 34B, 513
Salutaridine methiodide, 32B, 491
Samandaridine hydrobromide methanol solvate, 28, 555
Samandarine derivatives, 26, 798
Samandarine hydrobromide, 26, 798; 27, 1034
Samandarine hydrochloride, 26, 798
Samandarine hydrohalides, 26, 798; 27, 1034
Samandarine hydroiodide, 26, 798
Samarium basic chloride, 35A, 195
Samarium borate, 26, 491
Samarium chloride, 27, 724
Samarium dihydride, 27, 237
Samarium dihydroxychloride, 34A, 221
Samarium disilicate, 35A, 305
Samarium fluoride, 27, 724
Samarium formate, 27, 815
Samarium hexaboride, 26, 69
Samarium niobate, 27, 521; 29, 329
Samarium oxychloride, 27, 724
Samarium oxyiodide, 26, 338
Samarium oxytelluride, 27, 424
Samarium-selenium, 33A, 140
Samarium sodium germanate, 35A, 315
Samarium sodium orthosilicate, 32A, 430
Samarium sodium sulphide, 29, 342
Samarium-strontium-sulphur, 27, 327
Samarium sulphides, 26, 244; 33A, 140
Samarium-sulphur, 26, 244; 33A, 140
Samarium tantalate, 27, 521; 29, 329
Samarium tribromate enneahydrate, 34A, 362
Samarium trihydride, 27, 237
Samarium-zinc, 35A, 98
Samarium zirconium oxide, 27, 535
Samarskite, 26, 387; 30A, 455
Samarskites, 30A, 456
Samsonite, 34A, 296
Sanidine, 28, 242; 29, 431; 30A, 456; 31A, 243
Sanidine, high pressure, 32A, 478
Sanidinized orthoclase, 28, 243
Sapphire, 27, 725
Sapphirine, 26, 517; 34A, 372
Sarcopside, 29, 425; 30A, 454
(-)(589)-Sarcosinato bis(ethylenediamine) cobalt(III) di-iodide dihydrate, 32B, 496
(+)$_{436}$-Sarcosinatotetra-amminecobalt(III) nitrate, 33B, 548
Sarcosine hydrochloride, 33B, 554
Sarcosine hydrochloride *N*-methyl glycine hydrochloride, 32B, 489
Sarcosinuric acids, 33B, 546
Sartorite, 27, 403, 405
Saryarkite, 29, 431
Sborgite, 27, 558
Scandenin, 28, 679
Scandium, 26, 234, 244
Scandium acetylacetonate, 35B, 908
Scandium borate, 31A, 242
Scandium bromide, 26, 298
Scandium chloride, 26, 298
Scandium fluoride, 26, 298

Scandium formate, 33B, 402
Scandium germanides, 27, 341
Scandium halides, 26, 298
Scandium iodide, 26, 298
Scandium niobate, 27, 522
Scandium orthovanadate, 27, 589
Scandium oxide, 33A, 263
Scandium oxide fluoride, 31A, 86
γ-Scandium oxide *mono*hydrate, 27, 423
α-Scandium oxide hydroxide, 32A, 244
Scandium sesquioxide, 26, 350; 32A, 254
Scandium sesquisulphide, 26, 245
Scandium silicides, 27, 341
Scandium-silicon, 27, 341
Scandium-silver, 27, 342
Scandium-sulphur-yttrium, 35A, 101
Scandium tantalate, 27, 522
Scandium-technetium, 26, 255; 27, 360
Scandium telluride, 26, 245
Scandium-tellurium, 26, 245; 30A, 91
Scandium-titanium, 27, 342
Scandium transition-metal compounds, 27, 342
Scandium wolframate, 31A, 158
Scandium zirconium oxides, 33A, 291
Scapolite, 26, 531; 27, 688; 30A, 457
Scapolites, 27, 688
Scarbroite, 30A, 459
Schairerite, 28, 326
Scheelite, 29, 331
Schelhammerine hydrobromide, 33B, 549
Schertelite, 28, 352
Schiff bases, 33B, 542
Schirmerite, 30A, 453
Schoderite, 27, 601
Scholzite, 35A, 333
Scleroclase, 26, 426
Scorodite, 31A, 252
Searlesite, 27, 698; 29, 409; 31A, 244
Sebacamide, 31B, 18
Sebacic acid, 26, 575; 31B, 17
Securinine hydrobromide, 30B, 427
Securinine hydrobromide dihydrate, 29, 772; 30B, 239
Securinine hydrobromide monohydrate, 30B, 427
Sedovite, 30A, 453
Seidozerite, 30A, 437
Selenium, 34A, 127
Selenium bis(diethyldiselenophosphinate), 34B, 381
Selenium bis(diethyldithiocarbamate), 35B, 500
Selenium dicyanide, 31A, 114
Selenium(II) diethyldithiocarbamate, 35B, 910
Selenium di(morpholydithiocarbamate), 35B, 499
Selenium nitride, 26, 293
Selenium-rare-earth, 32A, 18
Selenium-silicon-zirconium, 27, 348; 29, 39
Selenium-silver, 27, 344; 30A, 92
Selenium-silver-thallium, 30A, 93
Selenium-sodium, 27, 351
Selenium-sulphur-tin, 26, 245; 27, 345
Selenium-tantalum, 27, 343; 29, 73; 30A, 94; 31A, 62
α-Selenium-tantalum, 27, 343
β-Selenium-tantalum, 27, 343
Selenium-tantalum-thallium, 29, 91
Selenium-tellurium-tin, 26, 153
Selenium-terbium, 26, 255
Selenium-thallium-vanadium, 29, 91
Selenium-tin, 26, 245
Selenium-titanium, 27, 345
Selenium trioxide, 27, 482; 30A, 306
Selenium-vanadium, 27, 345
Selenium-wolfram, 27, 343
Selenium-zinc, 26, 246
Selenium-zirconium, 33A, 141
Selenocarbohydrazide, 27, 751
Selenodithiazyl, 27, 409
Selenoferrates, 31A, 261
Selenophene, 33B, 540; 35B, 904
α-Selnophenic acid, 27, 948
trans-Selenophthene, 34B, 186
2-Selenouracil, 32B, 489
Selenourea, 27, 750; 28, 669; 29, 762; 34B, 29
Seligmannite, 35A, 11
Sellaite, 29, 414
Semicarbazide-copper(II) chloride, 30B, 407
Semicarbazide hydrochloride, 30B, 46
Semicarbazide-zinc(II) chloride, 30B, 407
Semiresite, 30A, 457
Sepiolite, 28, 312; 30A, 457
Serine, 26, 826
O-Serine phosphate, 35B, 299
DL-O-Serine phosphate monohydrates, 35B, 299
DL-Serine phosphate monohydrate, 33B, 554
Serotonin-creatinine sulphate monohydrate (1:1 complex), 30B, 214
Serpentine minerals, 30A, 457; 31A, 244; 33A, 494
Serpierite, 33A, 382
Seselin, 35B, 369
Shanorellin chloride, 35B, 73
Shattuckite, 29, 431; 31A, 233, 260; 32A, 460
Shellolic bromolactone, 27, 1019
Shellolic chlorolactone, 27, 1019
Shirmodiol acetate *p*-bromobenzoate, 35B, 912
Showdomycin, 35B, 260
Siderophyllite, 29, 429
Siderotile, 27, 633; 28, 326

Sigloite, 27, 603
Sigma phases, 26, 19; 29, 86
2-Silaphenalane, 30B, 426
Silazanes, 33B, 272
Silica, 26, 502; 27, 674
Silica glass, 27, 676
Silica, vitreous, 34A, 363
Silicic acids, 29, 393
Silicon, 26, 247, 287; 27, 208, 347; 29, 55, 114
Silicon carbide, 27, 407; 28, 64; 29, 238; 30, 254; 31A, 77; 32A, 506; 33A, 160; 34A, 157
Silicon carbides, 28, 64
Silicon chloride isocyanate, 34A, 185
Silicon difluoride, 31A, 239
Silicon diphosphate, 34A, 327
Silicon nitride, 26, 292; 34A, 166
Silicon oxynitride, 29, 244
Silicon pyrophosphate, 35A, 341
Silicon-sodium, 26, 156; 29, 57; 30A, 96
Silicon-strontium, 27, 348; 30A, 97; 31A, 63; 32A, 122
Silicon-sulphur, 30A, 55
Silicon-sulphur-zirconium, 27, 348; 29, 39
Silicon-tantalum, 26, 246
Silicon-tellurium-zirconium, 29, 39
Silicon-terbium, 27, 336
Silicon tetrachloride, 34A, 173
Silicon-thorium, 26, 256
Silicon-thulium, 27, 336
Silicon-titanium, 26, 18
Silicon-uranium, 26, 271
Silicon-vanadium, 31A, 64
Silicon-ytterbium, 27, 336
Silicon-yttrium, 27, 336
Silicon zinc phosphide, 35A, 102
Silicon-zirconium, 30A, 97; 31A, 65; 33A, 143
Silicowolframate, 27, 672
Silk fibroins, 26, 827
Sillimanite, 26, 505; 27, 707; 28, 237; 31A, 252
Silver, 26, 279; 27, 386
Silver acetylide-silver nitrate, 29, 238
Silver alloys, 26, 248
Silver azide, 27, 414
Silver bifluoride, 31A, 258
Silver(II)-bis(pyridine-2,3-dicarboxylate) dihydrate, 35B, 919
Silver(II) bis(pyridine-2,6-dicarboxylate) monohydrate, 35B, 770
Silver bromide, 27, 427
Silver carbonate, 27, 725
Silver chloride, 27, 427; 28, 290
Silver chlorite, 26, 495
Silver cyanamide, 29, 239
Silver cyanate, 30A, 294
Silver cyanide nitrate, 30A, 294
Silverdecamolybdate, 35A, 270
Silver di(carboxymethyl)glycinate, 29, 483
Silver difluoride, 31A, 84
Silver 5,5-dimethylcyclohexane-1,3-dione-2-sulphonate, 33B, 553
Silver 5,5-dimethyl-1,3-cyclohexane-dione-2-sulphonate monohydrate, 28, 672
Silver(I) di-*n*-propyldithiocarbamate, 34B, 525
Silver fluorides, 35A, 162
Silver fulminate, 27, 747; 31A, 289
Silver halides, 28, 290; 31A, 250
Silver hexacyanocobaltate, 33A, 242
Silver-holmium, 34A, 129
Silver iodide, 27, 425, 427; 28, 100, 290; 31A, 244; 33A, 239
Silver ketenide, 35B, 912
Silver mesoperiodate, 26, 502
Silver molybdenum oxide phosphate, 30A, 394
Silver nitrate, 26, 475; 27, 573; 31A, 178
Silver nitrate, basic, 28, 178
Silver nitrate-*trans*-cyclodecene adduct, 32B, 378
Silver nitrate-pyrazine complex, 31B, 417
Silver oxide, 26, 359; 27, 724; 30A, 308
Silver (II) oxide, 26, 540
Silver permanganate, 33A, 319
Silver pertechnetate, 27, 671
Silver polyetherin A, 35B, 755, 910
Silver polyphosphate, 26, 459
Silver silicate, 26, 505-509
Silver-sodium, 26, 250
Silver subfluoride, 31A, 83
Silver sulphide halide, 30A, 289
Silver sulphide iodide, 26, 423
Silver-sulphur, 26, 250; 32A, 123
Silver-sulphur-ytterbium, 31A, 65
Silver-tellurium, 26, 250; 27, 221; 31A, 67
Silver tellurium nitrates, 31A, 256
Silver-tellurium-thallium, 29, 87
Silver-terbium, 33A, 91
Silver tetraborate, 30A, 412
Silver thiocyanate dinitrate, 35A, 327
Silver thiocyanate tri-*n*-propyl phosphine complex, 28, 655
Silver-thorium, 26, 256
Silver-tin, 26, 249; 27, 310
ζ-Silver-tin, 27, 310
Silver tricyanomethanide, 31B, 41
Silver tris(acetylacetonato) nickelate (II) di(silver nitrate) monohydrate, 31B, 381
Silver vanadium oxide bronze, 30A, 344
Silver-ytterbium, 35A, 103
Silver-zinc, 27, 349

Silver-zirconium, 27, 234
Silydianin, 35B, 923
Silyl acetylene, 28, 664
N-Silyldimethylamine, 35B, 891
Silyl isothiocyanate, 27, 719
Simarolide *m*-iodobenzoate, 29, 772
Simarolide 4-iodo-3-nitrobenzoate, 29, 772
Simpsonite, 27, 540
Sinhalite, 30A, 418
Sinigrin, 28, 518; 35B, 263
Sinnerite, 29, 23
Sjogrenite, 31A, 172; 32A, 422
Skatole-trinitrobenzene, 29, 689
Sklodowskite, 27, 673, 710; 29, 410
Skutterudite, 27, 48
Smithite, 29, 25
Soaps, 26, 564-568
Soap-water systems, 27, 1063
Sodalite, 26, 518; 29, 440; 32A, 490
Soda melilite, 35A, 455
Sodium, 26, 3, 279
Sodium acetate trihydrate, 30B, 409
Sodium aluminate, 26, 368; 27, 519
Sodium aluminotitanate, 32A, 266
Sodium aluminum hydrides, 31A, 246
Sodium alunite, 27, 634
Sodium *p*-aminobenzoate, 33B, 551
Sodium ammonium oxalatotrioxomolybdate(VI) dihydrate, 28, 629
Sodium-amphibole, 32A, 464
Sodium ascorbate, 34B, 247
Sodium azide, 31A, 241; 33A, 168
Sodium bicarbonate, 27, 559; 30A, 559
Sodium biselenite, 28, 221
Sodium bromide, 26, 297
1:2 Sodium bromide-diacetamide complex, 34B, 351
Sodium bromide dihydrate, 27, 427; 29, 284
Sodium carbonate, 29, 380; 34A, 347
Sodium chloride, 26, 297; 27, 425; 28, 290
Sodium chloride-urea hydrate, 28, 570
Sodium cyanamide, 29, 239
Sodium, 3,5-diacetylimino-1,2-dithiole hydrate, 31B, 145
Sodium dibenzenesulphonamide, 35B, 84
Sodium dichromate, 34A, 275
Sodium dichromate dihydrate, 35A, 388
Sodium 5,5'-diethylbarbiturate, 27, 973
Sodium *NN*-diethyldithiocarbamate trihydrate, 33B, 547
Sodium dihydrogen citrate, 30B, 28
Sodium dioxocadmate, 34A, 250
Sodium dioxodifluoroniobate(V), 34A, 217
Sodium disilicate, 33A, 461
Sodium dithionate dihydrate, 28, 220; 32A, 323
Sodium ferrate, 32A, 288
Sodium ferrite, 26, 368
β-Sodium ferrite, 28, 133
Sodium fluoride, 26, 297
Sodium fluoride-scandium trifluoride, 31A, 249
Sodium fluoroberyllate, 33A, 176
Sodium fluoromanganate, 26, 309
Sodium fluorosilicate, 26, 305; 29, 264
Sodium fluorotitanate-sodium bifluoride, 31A, 89
Sodium germanate hydrate, 32A, 449
Sodium glycylglycylglycino-cuprate monohydrate, 27, 1049; 30B, 348
Sodium hexafluorochromate(III), 34A, 198
Sodium hexafluoromolybdate (V), 26, 311
Sodium hexafluoroplumbate(IV), 31A, 253
Sodium hexafluorostannate(IV), 31A, 253, 260
Sodium hexafluorozirconate(IV), 34A, 193
Sodium hexametaphosphate hexahydrate, 30A, 393
Sodium hexamolybdochromate(III) octahydrate, 35A, 419
Sodium hexaoxo-μ-ethylenediaminetetraacetato-dimolybdate octahydrate, 34A, 279
Sodium hexatitanate, 27, 528
Sodium hydride, 27, 398
Sodium hydridodiethylberyllate etherate, 30B, 392
Sodium hydrogen diacetate, 26, 556
Sodium hydrogen fumarate, 33B, 552; 35B, 13
Sodium hydrogen oxalate monohydrate, 35B, 9
Sodium hydrogen phosphate dihydrate, 27, 581
Sodium hydrogen sulphate monohydrate, 30A, 365
Sodium hydrogen trifluoroborate, 35A, 189
α-Sodium hydroxide, 32A, 241
β-Sodium hydroxide, 32A, 241
Sodium hydroxide tetrahydrate, 29, 397
Sodium hydroxymethanesulphinate, 27, 754
Sodium hypochlorite pentahydrate, 27, 645
Sodium inosine 5'-*mono*phosphate, 33B, 548
*mono*Sodium inosine-5'-phosphate octahydrate, 34B, 261
Sodium iodate, 35A, 396
Sodium iodide acetone solvate, 28, 575
Sodium iodide dihydrate, 35A, 195
Sodium iodide, methanol (1:3 complex), 30B, 264
Sodium α-ketobutyrate, 28, 384
Sodium lanthanide oxides, 31A, 250
Sodium laurate, 26, 564
Sodium linolenate—water, 26, 564
Sodium mesouranate, 27, 544
Sodium metaaluminate, 33A, 280

Sodium metaborate, 26, 486; 27, 718; 28, 157
Sodium metaphosphate, 28, 186; 29, 363
Sodium metaphosphate glass, 28, 293
Sodium metasilicate, 32A, 448
Sodium methanethiosulphonate monohydrate, 29, 484
Sodium methoxide, 29, 452
Sodium molybdate, 27, 724; 31A, 250
Sodium molybdate dihydrate, 34A, 276, 281
Sodium molybdenum bronze, 31A, 153
Sodium molybdenyl phosphate, 27, 582
Sodium myristate, 26, 564
Sodium myristate—water—ethylbenzene, 26, 564
Sodium naphthionate, tetrahydrate, 31B, 95
Sodium niobate, 26, 381; 27, 503; 31A, 148; 34A, 268; 35A, 257
Sodium niobomanganate, 34A, 271
Sodium nitrate, 27, 564; 34A, 338
Sodium nitrite, 26, 473; 27, 572
Sodium nitroprusside, 28, 106
Sodium octacyanowolframate, 35A, 424
Sodium octafluoroprotactinate(V), 34A, 197
Sodium octafluorouranate, 31A, 246
Sodium oxalate perhydrate, 29, 462
Sodium 2-oxocaproate, 33B, 21
Sodium 2-oxocaprylate, 29, 471
Sodium oxomercurate, 27, 661; 29, 338
Sodium 2-oxovalerate, 34B, 19
Sodium oxythallate, 26, 400-402
Sodium ozonide, 27, 482
Sodium palmitate, 26, 564
Sodium pentaborate pentahydrate, 31A, 242
Sodium pentafluorodistannate, 29, 266
Sodium perchlorate-bis-[*NN*'-ethylene-bis(salicylideneiminato)copper(II)], 33B, 549
Sodium peroxide, 26, 348
Sodium pertechnetates, 27, 671
Sodium perxenate hexahydrate, 28, 228; 29, 335
Sodium perxenate octahydrate, 28, 229; 29, 335
Sodium phenylthiazolidinedione, 29, 733
Sodium plumbate(IV), 34A, 264
Sodium polyarsenatophosphate, 32A, 383
Sodium polyphosphates, 26, 459; 461; 33A, 411
Sodium potassium dioxotetracyanomolybdate hexahydrate, 33A, 248
Sodium pyrosphosphate decahydrate, 29, 360; 32A, 370
Sodium pyruvate, 26, 551
Sodium salt of cobalt(III) ethylene diammine tetra-acetate tetrahydrate, 26, 675
Sodium salt of hafnium tetrakis-oxalato complex, 26, 691
Sodium salt of zinc tetrakis-oxalato complex, 26, 691
Sodium scandium disilicate, 34A, 376
Sodium selenate, 28, 222; 35A, 381
Sodium selenides, 30A, 358
Sodium sesquicarbonate, 27, 560
Sodium silicate, 26, 505
α-Sodium silicate, 26, 506
Sodium silicate hydrate, 31A, 241; 32A, 449
Sodium stearate, 26, 564; 27, 1061
Sodium stearate-cyclohexane, 26, 564
Sodium stearate-tetradecane, 26, 564
Sodium strontium arsenate, 26, 469
Sodium strontium phosphate, 26, 469
Sodium strontium vanadate, 26, 470
Sodium sulphanilamide monohydrate, 33B, 551
Sodium sulphate, 27, 604
Sodium sulphate dechydrate, 26, 433
Sodium sulphides, 30A, 358
Sodium-sulphur, 27, 350; 30A, 358
Sodium tantalate, 27, 503
Sodium D-tartrate dihydrate, 33B, 15
Sodium tartrate tetrahydrate, 33B, 542
Sodium tellurides, 30A, 358
Sodium terbate, 27, 520
Sodium terbium(III) ethylenediamine-tetra-acetate octahydrate, 32B, 495
Sodium terbium sulphide, 29, 342
Sodium tetraborate, 32A, 497
Sodium tetrachloroaurate(III) dihydrate, 30A, 444
Sodium tetrachloroferrate(III), 30A, 277
Sodium tetracyanopalladate trihydrate, 32A, 235
Sodium tetraethylammonium bis(benzilato)oxovanadium(IV) 2-propanol solvate, 34B, 532
Sodium tetrafluoroaluminate, 34A, 192
Sodium tetrafluoroargentate(III), 35A, 163
Sodium tetrafluoroaurate(III), 35A, 162
Sodium tetrafluoroberyllate, 31A, 241
Sodium tetrafluoroborate, 31A, 88; 33A, 181
Sodium tetrafluorocuprate(II), 30A, 271
Sodium tetrahydroxoborate dihydrate, 28, 164
Sodium tetrametaphosphate tetrahydrate, 26, 455; 29, 365
Sodium tetrathionate dihydrate, 29, 345
Sodium tetrazolate monohydrate, 28, 56, 453

Sodium thallate, 26, 402; 29, 339
Sodium-thallium, 32A, 124
Sodium thiocyanate, 33A, 253
Sodium thiosulphate, 26, 435
Sodium thiosulphate pentahydrate, 27, 630
Sodium thymidylyl-(5'--3')-thymidylate-(5') dodecahydrate, 30B, 432
Sodium tin germanate, basic, 35A, 320
Sodium titanate, 26, 375; 27, 518
Sodium titanium oxide bronze, 27, 532
Sodium titanogermanate, 35A, 316
Sodium *p*-toluenesulphonate complexes, 35B, 908
Sodium tricyanomethanide, 32A, 234; 32B, 49
Sodium tridecafluorozirconate(IV), 30A, 272
Sodium trihydrogen selenite, 33A, 389
Sodium trihydroxyzincate (II), 26, 396
Sodium trimetaphosphate, 30A, 387
Sodium trimetaphosphate monohydrate, 30A, 387
Sodium trimetaphosphimate tetrahydrate, 30A, 389
Sodium triperoxouranyl hydrate, 31A, 258
Sodium triphosphate, 29, 360; 34A, 321
Sodium triphosphate hexahydrate, 31A, 259
Sodium uranate, 27, 543; 29, 340; 35A, 219, 220
α-Sodium uranate, 26, 405
β-Sodium uranate, 26, 404
Sodium uranyl triperoxide nonahydrate, 33A, 358
Sodium vanadium molybdate, 33A, 336
Sodium vanadium oxide, 31A, 140
Sodium warfarin, 2-propanol, water (clathrate), 30B, 432
Sodium wolframate, 27, 724; 30A, 329; 31A, 250
Sodium wolframate dihydrate, 26, 392; 34A, 281
Sodium wolframyl phosphate, 27, 582
Sodium xanthate dihydrate, 34B, 41
Sodium yttrium germanate, 35A, 229
Sodium yttrium sulphide, 29, 342
Sodium zeolite ZK-4, 26, 516
Sodium-A zeolite-bromine, 31A, 218
Sodium zincate, 30A, 325
Sodium zinc diorthosilicate, 32A, 437
Sodium zinc germanate, 33A, 506
Sodium zinc silicate, 31A, 218
Sodium zirconium disilicate, 35A, 309
Sodium zirconium fluoride, 33A, 185
Sodium zirconium oxyorthosilicate, 35A, 464
Sodium zirconium phosphate, 33A, 397
Söhngeite, 30A, 450
Solanite, 30A, 459
Solaphyllidine, 35B, 362
Solstitialin, 35B, 345
Sonolite, 28, 354
trans, trans-Sorbamide, 32B, 12
Sorbic acid-iron tricarbonyl, 35B, 518
α-L-Sorbose, 32B, 215
Sorensenite, 30A, 459
Spandite, 28, 334
Sparteine-N(16)-oxide sesquiperchlorate, 33B, 555; 34B, 322
Spencerite, 31A, 190
Spencite, 27, 558
Spermidine phosphate trihydrate, 34B, 336
Spermidine trihydrochloride, 31B, 260
Spermine phosphate hexahydrate, 29, 772; 30B, 187
Spermine tetrahydrochloride, 31B, 261
Spermwhale myoglobin, 26, 811
Spessartine, 28, 334
Spessartine-almandine garnet, 26, 416
Spessartite, 30A, 457
Sphalerite, 31A, 254
Spinaceamine dihydrochloride monohydrate, 29, 772; 33B, 554
Spinacine dihydrate, 29, 772
Spinel solid solutions, 31A, 252; 33A, 324
Spiradine A methiodide, 33B, 555
3,6-Spirodicycloheptylidene-1,2,4,5-tetraoxacyclohexane, 29, 776
3,6-Spirodicyclohexylidene-1,2,4,5-tetraoxacyclohexane, 29, 766; 32B, 177
3,6-Spirodicyclooctylidine-1,2,4,5-tetra-oxa-cyclo-hexane, 30B, 395
Spiroffite, 27, 716; 28, 350; 30A, 458
Spodiosites, synthetic, 35A, 192
Spodumene, 26, 515; 31A, 244; 33A, 456
Sporidesmin, 30B, 240
Stannic fluoride, 27, 442
Stannic iodide-sulphur, 27, 472
Stannous chloride, 26, 320
Stannous chloride dihydrate, 26, 321
Stannous fluoride, 27, 430
Staurolite, 31A, 234; 33A, 486
'Stelazine', 30B, 432
Stemofoline hydrobromide monohydrate, 35B, 910, 915
Stemonine hydrobromide hemihydrate, 35B, 382
Stentonite, 27, 564
Stephanite, 34A, 12; 35A, 15
Stepanovite, 28, 322
Stephavanine hydrobromide, 35B, 921
Sterigmatocystin *p*-bromobenzoate, 35B, 911
Stewartite, 28, 324
Stibiotantalite, 29, 440
Stilbellite, 29, 411
μμ'(*cis*-Stilbene-α,β-dithiolato)-bis (tricarbonyliron), 32B, 458
Stilbite, 31A, 234, 246; 35A, 456

Stillwellite, 30A, 459; 31A, 253; 32A, 433
Stishovite, 27, 675
Stokesite, 28, 262; 31A, 252
Straight-chain carboxylic acid, 26, 564
Stranskiite, 32A, 389
Strengite, 26, 462
Streptomycin oxime selenate, 33B, 555
Strontian chabazite, 30A, 459
Strontianite, 30A, 457
Strontioborite, 29, 427
Strontiogehlenite, 26, 521
Strontioginorite, 35A, 446
Strontiohilgardite, 29, 427
Strontium apatite, 29, 370
Strontium D-arabonate, 27, 788
Strontium D-arabonate pentahydrate, 27, 788
Strontium azide, 33A, 168
Strontium borate, 29, 387; 30A, 415
Strontium bromide, 28, 98
Strontium bromide monohydrate, 29, 284
Strontium caprylate hydrate, 29, 765
Strontium carbonate, 35A, 295
Strontium chromate, 27, 609; 31A, 150; 32A, 279
Strontium cyanamide, 31A, 78; 31B, 39
Strontium 3-deoxy-2-C-hydroxymethyl-D-erythro-pentoate, 34B, 250
Strontium diborate, 29, 389
Strontium diborate tetrahydrate, 31A, 249
Strontium dichromate(IV), 31A, 150; 32A, 280
Strontium di-lead hexaiodide, 26, 334
Strontium fluoride, 31A, 243
Strontium fluorides, ternary, 35A, 163
Strontium fluoroborate, 27, 459
Strontium formate dihydrate, 26, 548
Strontium germanate, 26, 397; 27, 501, 541
Strontium haidingerite, 27, 576; 31A, 194
Strontium hexaboride, 26, 69
Strontium hydroxide, 34A, 235
Strontium hydroxide monohydrate, 31A, 256; 32A, 245
Strontium hydroxyphosphate, 26, 469
Strontium iodide, 34A, 181
Strontium iron uranium oxide, 31A, 161
Strontium metaphosphates, 31A, 245
Strontium metavanadate tetrahydrate, 30A, 318
Strontium molybdate, 32A, 302
Strontium oxalate polhyhdrate, 30B, 22
Strontium palladate, 35A, 224
Strontium permanganate, trihydrate, 31A, 159, 248
Strontium phosphate, 34A, 326
Strontium pyroborate, 31A, 168
Strontium pyrophosphate, 33A, 407
α-Strontium pyrophosphate, 32A, 377
Strontium rare-earth oxides, 33A, 340
Strontium ruthenite, 26, 399
Strontium selenate, 27, 609
Strontium silicate, 30A, 440
Strontium-sulphur-ytterbium, 27, 327
Strontium-sulphur-yttrium, 27, 327
Strontium tartrate trihydrate, 33B, 16; 34B, 19
Strontium tetraborate, 31A, 169
Strontium tetragermanate, 26, 398
Strontium-thallium, 29, 3
Strontium tin borate, 33A, 443
Strontium uranate, 31A, 160
Strontium violurate tetrahydrate, 32B, 488
Strontium wolframate, 31A, 157
Strontium ytterbate, 35A, 230
Strontium-zinc, 29, 87
Strontium zincate, 26, 396
Strontium zinc molybdate, 27, 509
Strontium zinc wolframate, 27, 509
Strontium zirconium sulphide, 28, 151
Structural generalizations, 27, 1063
Styrene-palladium chloride, 26, 689
Suberamide, 29, 765; 30B, 413; 31B, 12
Suberic acid, 26, 575; 29, 470; 30B, 6
Suberic aicd-urea, 26, 572
Substituted alunite, 27, 633
Substituted benzoic acids, 26, 831
Substituted chromium yttrium garnets, 27, 513
N-Substituted 2-halogenoalkylammonium halides, 33B, 542
Substituted iron yttrium garnets, 27, 511
Substituted jarosite, 27, 633
Succinic acid-urea, 26, 571
Succinic anhydride, 29, 765; 30B, 4
Succinimide, 26, 597
Succinonitrile-silver nitrate complex, 31B, 418
Succinylcholine iodide, 33B, 542; 35B, 329
Succinylcholine iodide, perchlorate, and picrate, 33B, 542
Sucrose, 28, 519; 29, 766
Sucrose calcium chloride tetrahydrate, 28, 674
tris-(3-Sulphamylamido-6-methoxy-pyridazine) bismuth (III) chloride, 27, 881
Sulphanilamide, 30B, 66
β-Sulphanilamide, 29, 770
γ-Sulphanilamide, 29, 770
Sulphanilamide monohydrate, 30B, 392; 33B, 55

Sulphanilic acid monohydrate, 27, 894
α-Sulphanuric chloride, 27, 409; 31A, 80
Sulphate minerals, 27, 633
Sulphide minerals, 31A, 257
trans-Sulphiteisothiocyanate-bis-(ethylene diamine)cobalt(III)-dihydrate, 34B, 483
Sulphitotriamminepalladium(II), 32A, 326
p-Sulphobenzenediazonium inner salt, 35B, 78
O-Sulphobenzoic imide, 33B, 147
o-Sulphobenzoimide, 34B, 207
Sulphoborite, 26, 494; 28, 326
6-Sulpho-6-deoxy-α-D-glucopyranosyl-(1,1')-D-glycerol rubidium salt, 29, 740
Sulphohalite, 28, 326; 33A, 377
Sulphosalts, 26, 426
Sulphur, 26, 288; 27, 397; 28, 52; 29, 230; 30A, 252; 32A, 508; 34A, 149
γ-Sulphur, 27, 397
Sulphur-bridged molybdenum complexes, 35B, 907
Sulphur derivative of phosphobenzene, 30B, 418
Sulphur dicyanide, 31A, 113
Sulphur difluoride, 34A, 173
Sulphur dioxide, 34A, 242
Sulphur, fibrous, 31A, 77
Sulphur (high-pressure forms), 31A, 254
Sulphuric acid, 30A, 360
Sulphuric acid dihydrate, 35A, 365
Sulphuric acid monohydrate, 33A, 364
Sulphuric anhydride, 30A, 362
Sulphur imides, 34A, 156
Sulphur nitride, 28, 54
Sulphur nitride-boron trifluoride, 32A, 161
Sulphur nitride derivatives, 33A, 361
Sulphur-tantalum, 29, 77; 34A, 129; 35A, 105
Sulphur-tantalum-thallium, 29, 91
Sulphur-tellurium, 32A, 126
Sulphur-tellurium-zirconium, 27, 348
Sulphur—terbium, 26, 255
Sulphur tetrafluoride, 27, 717
Sulphur-thallium, 35A, 106
Sulphur-thallium-vanadium, 29, 91
Sulphur-tin, 27, 352; 32A, 128
Sulphur-titanium, 27, 353; 31A, 67; 32A, 130; 35A, 107
γ-Sulphur trioxide, 32A, 253
Sulphur-uranium, 29, 88
Sulphur-vanadium, 28, 41; 29, 89; 34A, 131
Sulphur-ytterbium, 32A, 132
Sulphur-yttrium, 26, 277
Sulphur-zinc, 27, 382; 32A, 135; 33A, 146, 34A, 133
Sulphur-zirconium, 35A, 69
Sulvanite, 31A, 41
Suolunite, 31A, 236
Suprasterol II, 26, 830
Suprasteryl II 4-iodo-5-nitro-benzoate, 30B, 223
Sursassite, 29, 431
Svabite, 31A, 247
Synadelphite, 35A, 360
5,6,5,6-Syn-bis(1,3-dimethyl-thymine), 34B, 241
Synchisite, 26, 483
Syndiotactic polyvinyl chloride, 26, 634
Syngenite, 32A, 335; 33A, 518
Synthetic fibres, 26, 641
Synthetic iron mica, 29, 406
Synthetic polypeptides, 27, 1053
Szaibelyite-sussexite, 30A, 455
Szajbelyite, 29, 440

Taaffeite, 28, 149
α-D-Tagatose, 34B, 242
Talc, 31A, 260
α-D-Talose, 26, 623
Tamarugite, 31A, 243; 34A, 310
Tantalates, 35A, 257
Tantalum, 27, 356
Tantalum bromide, 31A, 239; 33A, 203
Tantalum carbide, 26, 91
Tantalum chloride, 27, 444; 33A, 203
Tantalum deuteride, 35A, 133
Tantalum dioxide, 32A, 262
Tantalum diselenides, 30A, 79
Tantalum *mono*ethoxide tetra-*t*-butoxide, 29, 769
Tantalum halides, 33A, 203
Tantalum iodide, 30A, 264
Tantalum nitrides, 34A, 168
Tantalum oxides, 26, 354; 27, 482; 29, 307; 30A, 312; 31A, 251; 32A, 262
Tantalum oxynitrides, 26, 294; 34A, 168
Tantalum pentafluoride, 29, 258
Tantalum pentoxide, 27, 482
Tantalum-platinum metals, 29, 70
Tantalum-selenium, 30A, 79; 33A, 113
Tantalum subchloride heptahydrate, 31A, 97
Tantalum suboxide, 27, 483; 30A, 312; 31A, 251
Tantalum sulphides, 27, 357
Tantalum-technetium, 26, 255; 27, 361
Tantalum-tellurium, 27, 343; 29, 77
Tantalum(V) tetrachloride fluoride, 31A, 85
Tantalum-vanadium, 31A, 68
Tantalum-zirconium, 27, 359
Tapiolite, 29, 440; 31A, 252; 33A, 343
Taramellite, 30A, 437
Tarbuttite, 31A, 190, 242
Tartar emetic, 35B, 469
D-Tartaric acid, 29, 765; 31B, 5
*meso*Tartaric acid, 32B, 7

*meso*Tartaric acid monohydrate, 32B, 8
Tartaronic acid, 30B, 3
Taurine, 28, 524; 29, 483; 31B, 187
Taxa-4(16),11-diene-5α, 9α, 10β, 13α-tetraol *p*-bromobenzoate dihydro-anhydroacetonide, 32B, 252
Technetium, 26, 99, 254
Technetium carbonyl, 27, 651
Technetium(IV) chloride, 31A, 98
Technetium oxides, 29, 322; 34A, 244
Technetium oxide tetrafluoride, 32A, 508; 35A, 185
Technetium—titanium, 26, 255; 27, 360
Technetium transition-metal alloys, 27, 360
Technetium-vanadium, 27, 361
Technetium-wolfram, 27, 361
Technetium-zirconium, 26, 255; 27, 360
Teineite, 27, 635
Tellurates, 27, 635
Telluric acid, 35A, 281
Tellurite, 32A, 261; 33A, 391
Tellurite mineral, 33A, 391
Tellurium, 32A, 137
Tellurium bis(diethylthioselenophosphinate), 34B, 380
Tellurium(II) bromide-tetrathiourea, 26, 577
Tellurium(IV) catecholate, 32B, 339
Tellurium(II) chloride-tetrathiourea, 26, 576
Tellurium dibromide, ethylenethiourea (1:1 complex), 30B, 287
Tellurium dibromide-tetramethylthiourea complex, 26, 580
Tellurium dichloride-tetramethylthiourea complex, 26, 580
Tellurium(IV) diethyldithiocarbamate, 35B, 906
Tellurium di(ethylxanthate), 32B, 338
Tellurium(II) difluoride-tetrathiourea dihydrate, 26, 580
Tellurium(II) dihydrogen fluoride-tetrathiourea, 26, 579
Tellurium(II) dihydrogen fluoride-trithiourea, 26, 580
Tellurium diiodide, ethylenethiourea (1:1 complex), 30B, 287
Tellurium dimethanethiosulphonate, thiourea (1:1 complex), 30B, 284
Tellurium(II) dimethyldithiophosphate, 31B, 336
Tellurium di(morpholydithiocarbamate, 35B, 498
Tellurium dioxide, 26, 362; 27, 382
α-Tellurium dioxide, 33A, 263
Tellurium ethylenethiourea complexes, 26, 577
Tellurium(II) fluorides-thiourea, 26, 579
Tellurium hexafluoride, 31A, 239
Tellurium nitrate, basic, 31A, 175
Tellurium(II) nitrate-thiourea, 26, 577
Tellurium oxide, 31A, 257
Tellurium(II) perchlorate-thiourea, 26, 577
Tellurium-rare-earth, 32A, 18
Tellurium-terbium, 26, 255
Tellurium tetrabromide-tetramethylthiourea complex, 26, 580
Tellurium tetrachloride-tetramethylthiourea complex, 26, 580
Tellurium tetrafluoride, 33A, 170
Tellurium tetra-propylenethiourea dichloride dihydrate, 26, 584
Tellurium tetra-propylenethiourea diperchlorate, 26, 584
Tellurium-thallium, 35A, 108
Tellurium(II) thiocyanate-tetrathiourea, 26, 577
Tellurium thorium nitride, 35A, 113
Tellurium tin oxide, 27, 527
Tellurium-titanium, 26, 267; 27, 345, 364
β-Tellurium trioxide, 33A, 262
Tellurium tris-propylenethiourea diperchlorate, 26, 585
Tellurium-uranium, 35A, 110
Tellurium uranium nitride, 35A, 113
Tellurium-vanadium, 27, 345
Tellurium-wolfram, 27, 343
Tellurium-zinc, 27, 382
Tellurium zirconium oxide, 27, 527
Tephroite, 26, 515
Terbium, 26, 234; 27, 334
Terbium chloride, 29, 274
Terbium compounds, 26, 255
Terbium dihydride, 27, 237
Terbium hexaboride, 26, 69
Terbium niobate, 27, 521
Terbium orthoferrite, 32A, 268
Terbium oxides, 26, 364; 33A, 269
Terbium sesquioxide, 33A, 269
Terbium stannate, 26, 399
Terbium tantalate, 27, 521
Terbium trihydride, 27, 237
Terephthalic acid, 32B, 82
Terferrocenyl, 27, 860; 30B, 396
Terlinguaite, 33A, 221
Ternary borides, 26, 81; 27, 92
Ternary borides with tantalum, 26, 81
Ternary carbides with uranium, 26, 93
Ternary copper fluorides, 27, 453
Ternary hydrazine chlorides, 27, 471
Ternary nickel fluorides, 27, 453
Ternary oxides of titanium, 27, 527
Ternary perovskites, 31A, 247
Ternary silver halides, 27, 427
Ternary uranates, 27, 543

p-Terphenyl, 35B, 95
2,2'-,2"-Terpyridyl(dimethyl)chlorotin(IV) dimethyl(trichloro)stannate (IV), 33B, 314
Terpyridylzinc chloride, 31B, 431
Terramycin hydrochloride, 30B, 221
Tert-butyllithium, 30B, 415
Tert-butyllithium, tert-lithium butoxide (addition compound), 30B, 415
N-Tertiary butyl proplylamine, 28, 397
Tertiary phosphine metal halides, 35B, 911
Tert-lithium butoxide, 30B, 415
Testosterone, 35B, 906
Testosterone acetate, 30B, 439
Testosterone 17β-*p*-bromobenzoate, 35B, 337
1:1 Testosterone-*p*-bromophenol complex, 34B, 358
Testosterone butyrate, 30B, 439
Testosterone formate, 30B, 439
Testosterone *mono*hydrate, 35B, 906
Testerone-mercuric chloride, 33B, 203
Testerone propionate, 30B, 439
Testerone valerate, 30B, 439
Tetra-acetylethane, 31B, 33
Tetra-O-acetyl-D-gluco-pyranose-2,4-dinitro-phenylhydrazone, 30B, 432
Tetra-allyl dichromium, 34B, 441
Tetraamminecopper(II) dihalocuprates (I), 31A, 259
Tetra *iso*-amyl ammonium fluoride hydrate, 26, 613
Tetraamylphosphonium iodide, 35B, 465
Tetra-*p*-anisylethylene bis(dichloroiodate(I)), 32B, 75
Tetraaquo-dimonoethanolamine-cuprosulphate, 30B, 392
β-Tetraargentotellurium nitrate, 34A, 344
Tetra-arsenic trisulphide, 35A, 140
Tetraazapentacyclooctacosadodecaene, 28, 679
1,3α,4,6α-Tetraazapentalene derivatives, 28, 456
1,8,15,22-Tetra-aza-2,7,16,21-tetraoxocyclo-octacosane, 35B, 181
1,3,6,8-Tetra-azatricyclo[4,4,1,1^{3,8}]-dodecane, 35B, 915
Tetrabenz[*a*,*c*,*h*,*j*]anthracene, 33B, 554
γ-1,2,4,5-Tetrabromobenzene, 29, 642
Tetrabromo-μ-bis(diethyl sulphide)-diplatinum(II), 33B, 481
trans-Tetrabromobis(teramethylthiourea)tellurium(IV), 34B, 416
trans-*syn*-*trans*-1,2,5,6-Tetrabromocyclo-octane, 33B, 546
1,2,3,4-Tetrabromo-1,2,3,4-diphthaloylcyclobutane, 33B, 93
Tetrabromoethylene, 33B, 237
Tetrabromoethylene-pyrazine, 33B, 237
3,3,6,6-Tetra(bromomethyl)-1,2,4,5-tetroxane, 32B, 173
1,3,5,7-Tetrabromonaphthalene, 27, 924
1,4,5,8-Tetrabromonaphthalene, 30B, 394; 33B, 87
1,1,4,4-Tetrabromo-2,3,5,6-tetraphenyl-1,4-digermacyclohexa-2,5-diene, 30B, 434
Tetrabromo-*m*-xylene, 26, 712
Tetrabromo-o-xylene, 26, 712
Tetra-*n*-butyl ammonium benzoate hydrate, 27, 778
Tetra-*n*-butylammonium bis(maleonitriledithiolato)iron(III), 32B, 459
Tetra-*n*-butylammonium bis(1,2,3,4-tetrachlorobenzene-5,6-dithiolato)cobaltate, 33B, 499
Tetra-*n*-butylammonium bis(toluene-3,4-dithiolato)-1,10-phenanthrolinecobaltate, 35B, 826
Tetra-*n*-butylammoniumcopper(III) bis-(maleonitrile dithiolate), 29, 560
Tetra-*n*-butyl ammonium fluoride hydrate, 28, 578
Tetra-*n*-butylammonium hexadecacarbonyltetrarhenium, 33B, 329
Tetra-*n*-butylammonium hexaisothiocyanatoerbate(III), 33B, 553
Tetra-*n*-butylammonium tribromo-(quinoline)nickelate(II), 33B, 430
Tetrabutylammonium tris(acetylacetonato cobalt(II), 32B, 495
1,2,4,5-Tetra-*t*-butylbenzene, 33B, 555
Tetra-*n*-butyl phosphonium wolframate, 27, 780
Tetra-μ-*n*-butyratodirhenium(III) diperrhenate, 35B, 643
Tetra-*n*-butyratodiruthenium chloride, 34B, 536
Tetracalcium phosphate, 30A, 458
Tetracalcium trialuminate, 35A, 290
Tetracalcium trihydro trialuminate, 35A, 291
1,1,2,2-Tetracarbomethoxyethane, 33B, 5
Tetracarbonylacrylonitrileiron, 26, 684; 27, 824
Tetracarbonylbis(1,1,2,2-tetrafluoroethyl)iron(II), 32B, 341
Tetracarbonylcobalt-diphenyltinmanganesepentacarbonyl, 33B, 316
Tetracarbonyl[diphenyl-2-(prop-*cis*-1-enyl)phenylphosphine]-molybdenum(O), 34B, 451
DL-Tetracarbonyl fumaric acid iron, 32B, 343
(-)-Tetracarbonyl(fumaric acid)-iron, 33B, 407
Tetracarbonyl(hexamethylbicyclo-[2,2,0]hexa-2,5-diene)chromium, 33B, 546

Tetracarbonylmethylamine N-methylcarboxamidomanganese, 34A, 164
Tetracarbonyl[o-phenylene-bis(dimethylarsine)] chromium, 30B, 428
Tetracarbonyl[o-phenylene-bis(dimethylarsine)] molybdenym, 30B, 428
Tetracarbonyl[o-phenylene-bis(dimethylarsine)] wolfram, 30B, 428
Tetracene, 26, 754; 27, 938
9,9,10,10-Tetrachloroanthracene, 29, 770; 30B, 399
1,2,4,5-Tetrachloro-benzene, 30B, 418
Tetrachloro-*p*-benzoquinone, 27, 899
1,2,3,4-Tetrachlorobenzo(g)sesquifulvalene, 31B, 106
trans-Tetrachlorobis-(tetrahydrothiophen)-tin(IV), 30B, 422
trans-Tetrachlorobis(tetramethylthiourea)tellurium(IV), 34B, 416
Tetrachloro-1,4-bistriethyl stannyloxybenzene, 26, 721
2,4,4,6-Tetrachlorocyclohexa-2,5-diene-1-one, 33B, 79
1,1,4,4-Tetrachloro-1,4-digermaniacyclohexa-2,5-diene, 30B, 414; 33B, 304
9,9,10,10-Tetrachloro-9,10-dihydroanthracene, 31B, 103
11,11,12,12-Tetrachloro-9,10-dihydro-9,10-ethanoanthracene, 28, 676
trans-syn-trans-2,3,5,6-Tetrachloro-1,4-dioxane, 33B, 130
1,2,3,4-Tetrachloro-5,6-diphenylcalicene, 32B, 486; 35B, 103
1,2,3,4-Tetrachloro-5,6-di-*n*-propylcalicene, 34B, 88
1,1,3,3-Tetrachloro-1,3-disilacyclobutane, 33B, 529
2,2,4,4-Tetrachloro-1,3-dithiacyclobutane, 34B, 178
Tetrachloro-hydroquinone, 27, 901; 32B, 101
Tetrachloro-1-keto-naphthalene dihydride, 26, 745
1,4,1',4'-Tetrachloronaphthalene, 26, 745
1,4,5,8-Tetrachloronaphthalene, 26, 831; 27, 921
1,4,6,9-Tetrachlorophenazine, 29, 771; 30B, 141
2,3,7,8-Tetrachlorophenazine, 29, 771; 30B, 139
Tetrachlorophosphonium tetra chloroferrate(III), 33A, 215
Tetrachlorophthalic anhydride, 35B, 913
Tetra-μ_3-chlorotetrakis[triethylplatinum(IV)], 35B, 922
2,6,11,15-Tetrachloro-2,6,11,15-tetramethylhexadecane, 33B, 28
1,2,5,6-Tetrachloro-3,4,7,8-tetramethyl-tricyclo[4,2,0,0^{2,5}]octa-3,7-diene, 32B, 134
2,4,6,8-Tetrachloro-2,4,6,8-tetraphenylcyclotetraphosphazatetraene, 35B, 916
2,3,7,8-Tetrachloro-5,10,11,12-tetrathia-1,4,6,9-tetraazatricyclo[5,3,1,1^{2,6}]dodeca-3,8-diene, 32B, 193
Tetracobalt dodecacarbonyl, 31A, 259
Tetracyanobenzene-hexamethylbenzene, 33B, 250
1,2,4,5-Tetracyanobenzene-naphthalene 1:1 complex, 32B, 281
cis,trans,cis-1,2,3,4-Tetracyanocyclobutane, 33B, 77
Tetracyano-1,4-dithiin, 30B, 151
Tetracyanoethylene, 29, 765; 33B, 544
Tetracyanoethylene bis(diphenylphosphine)platinum, 33B, 552
Tetracyanoethylenebis(triphenylphosphine)platinum(O), 35B, 860
Tetracyanoethylene-ferrocene complex, 32B, 279
Tetracyanoethylene-naphthalene complex, 30B, 417; 32B, 278
Tetracyanoethylene-perylene complex, 30B, 417; 32B, 284
Tetracyanoethylene, pyrene (1:1), 30B, 417
Tetracyanonickelate(II), 34A, 231
7,7,8,8-Tetracyanoquinodimethane, 30B, 40
7,7,8,8-Tetracyanoquinodimethane-anthracene, 33B, 252
7,7,8,8-Tetracyanoquinodimethane-copper(II) 8-hydroxyquinolate complex, 32B, 286
7,7,8,8-Tetracyanoquinodimethane, N,N,N',N'-tetramethyl-*p*-phenylene diamine (1:1 complex), 30B, 249; 33B, 254
Tetracyanothiophene, 33B, 138
Tetracyanotriamminecopper(I), copper(II) complex, 32A, 509
Tetracycline complexes, 29, 764
π-Tetracyclo[8,6,0,0^{2,9},0^{3,8}]hexadecapenta-4,6,11,13,15-enetricarbonyliron, 33B, 377
Tetracyclohexylcyclotetraphosphine (form I), 34B, 392
Tetracyclohexyl diphosphine disulphide, 29, 762
Tetra(cyclopentadienyl-cobalt) hexasulphide chloroform solvate, 34B, 457
Tetracyclopentadienyl(diphenylbutadiyne)tetranickel, 30B, 436
Tetracyclopentadienylhydridotrirhodium, 33B, 351
Tetracyclopentadienyltitanium, 35B, 920
Tetra(dimethylamino)tin, 35B, 901
Tetra(dimethylsulphoxide)iron(III) dichloride tetrachloroferrate (III), 32B, 446

Tetradotoxin, 29, 764
Tetradotoxin derivatives, 29, 772
Tetra(1,4-epoxybutane)disodium(I) tetramethylbis-1,4-dihydro-1,4-naphthylenedialuminate, 35B, 478
Tetra(ethylamine)platinum(II) dibromo-tetra(ethylammine) platinum(IV) tetrabromide, 31B, 423
Tetraethylammonium bis-(3)-1,2-dicarbollylcuprate(II), 32B, 297
Tetraethylammonium diacetyldi-hydrobis(2-mercaptoanil)nickelate, 35B, 817
Tetraethylammonium hydrogen decacarbonyl dichromate, 31B, 31
Tetraethylammonium neptunium(III) tetrakis (*NN*-diethyldithio-carbamate, 35B, 818
Tetraethylammonium oxotetrabromoaquo-rhenate, 30B, 34
Tetraethylammonium pentachloro-indate(III), 34B, 23
Tetraethylammonium rhodium pentaiodo-carbonyl, 35B, 919
Tetraethylammonium tetrabromo-μμ'-dibromoplatinum(II), 29, 580
Tetraethylammonium tetrachloro-indate, 34B, 619
Tetraethylammonium tetrachloro-nickelate(II), 32B, 34
Tetraethylammonium tetrahydro-aluminate(III), 35B, 910
1,1,7,7-Tetraethyldiethylenetriamine cobalt(II) chloride, 32B, 381
1,1,4,4-Tetraethyl-2,5-dimethyl-1,4-diphosphoniacyclohexadiene-1,4-dibromide, 35B, 455
Tetraethyldiphosphine copper(I) bromide, 35B, 835
Tetraethyl diphosphine disulphide, 26, 609
1,1,4,4-Tetraethylpiperazinium dichloride-4-(*p*-nitroaniline), 35B, 419
Tetraethylthiuram disulphide, 32B, 39
Tetrafluorobenzobicyclo[2.2.2] octatrieneiron tricarbonyl, 35B, 922
Tetrafluorobis(pyridine)silicon, 34B, 371
Tetrafluorocyclobutenebis(dimethyl-arsine)triiron decarbonyl, 35B, 866
2,2,4,4-Tetrafluoro-1,3-dimethyl-2,4-diphenyl-1,3,-diaza-2,4-diphosphacyclo-butane, 32B, 491
Tetrafluoromethylphosphorane, 30B, 400
Tetragonal bronzes, 30A, 346
Tetrahedrite, 28, 50; 29, 15
cis-5,6,13,13α-Tetrahydro-3,9-dihydroxy-1,2,10-trimethoxy-8H-dibenzo[*a*,*g*] quinolizine hydro-bromide monohydrate, 35B, 199
1,2,5A,7B-Tetrahydro-5A,5B-dimeth-oxy-5BH-cyclobuta(1,4)cyclobuta-(1,2,3GH)-pyrrolizin-4(5H)-one, 34B, 165
1,2,5a,7bβ-Tetrahydro-5aβ,5bα-dimethoxy-5bH-cyclobuta[1,4]-cyclobuta[1,2,3-*gh*]pyrrolizin-4(5H)-one, 33B, 553
Tetrahydrofurane-magnesium bromide, 33B, 301
Tetrahydrofuran, hydrogen sulphide double hydrate, 30B, 268
Tetrahydrofuran-mercuric bromide, 35B, 918
Tetrahydrofuran-mercuric cyanide, 35B, 918
3,3,4,4-Tetrahydrofurantetrol, 29, 518
1,2,3,4-Tetrahydro-1,2,2,3,4,4-hexa-methylphosphinoline 1-oxide, 35B, 457
6,6A,6B,11-Tetrahydro-11H-5,6,11-methenobenzo(A) fluorene, 32B, 486
1-*p*-(1,2,3,4-Tetrahydro-1-naphthyl)-phenol *p*-iodobenzoate, 35B, 117
Tetrahydropentalenolactone solvate, 35B, 923
Tetrahydroselenophene, 35B, 894
Tetrahydrothiapyrone-1,1-dioxide, 32B, 488
Tetrahydroxy-*p*-benzoquinone, 29, 769
Tetrahydroxy-*p*-benzoquinone dihydrate, 30B, 77
Tetraiododibenzoatodirhenium(III), 34B, 530
1,1,4,4-Tetraiodo-1,4-digermana-cyclohexa-2,5-diene, 30B, 414; 32B, 330
Tetraiodoethylene-pyrazine, 33B, 237, 544
Tetra-*B*-isothiocyanatotetra-*N*-*t*-butylborazocine, 31B, 296
Tetrakin (μ-aceto) diaquo dicopper(II), 32B, 496
β-Tetrakis(acetylacetonato)-cerium-(IV), 34B, 503
β-Tetrakis(acetylacetonato)neptunium-(IV), 35B, 906
β-Tetrakis(acetylacetonato)uranium(IV), 35B, 608
Tetrakis(2-aminoethanethiol)-trinickel (II) chloride, 35B, 789
(+)D-(*N*,*N*,*N'*,*N'*-Tetrakis(2'-aminoethyl)-1,2-diaminoethane)cobalt(III) hexa-cyanocobaltate(III) dihydrate, 35B, 707
N,N,N',N'-Tetrakis(2-aminoethyl)-ethylenediamine pentahydrochloride, 32B, 37
Tetrakis antimony cobalt tricarbonyl, 35A, 403
Tetrakiscyclopentadienylhafnium, 35B, 913, 924

Tetrakis(cyclopentadienyl)zirconium, 35B, 556
Tetrakis-(*NN*-diethyldithiocarbamato)-titanium(IV), 35B, 914
(Tetrakis-dimethylformamido)nickel di-isoselenocyanate, 32B, 433
Tetrakis(3,4-dimethylpyridine)nickel (II) perchlorate, 33B, 449
Tetrakis-[1,3-dimethyltriazenocopper-(I)], 33B, 547
Tetrakis(diphenylbromoplumbyl)methane, 35B, 492
Tetrakis-(1,3-diphenyltriazenido)-dinickel(II), 33B, 550
Tetrakis(ethylenethiourea) tellurium (II) dibromide, 26, 578
Tetrakis(ethylenethiourea) tellurium (II) dibromide dihydrate, 26, 578
Tetrakis(ethylenethiourea) tellurium (II) dichloride, 26, 578
Tetrakis(ethylenethiourea) tellurium (II) dichloride dihydrate, 26, 578
Tetrakis(ethylenethiourea) tellurium (II) hexachlorotellurite, 26, 579
Tetrakis(ethylenethiourea) tellurium (II) perchlorate, 26, 579
Tetrakis(ethylenethiourea) tellurium (II) salts, 26, 578
Tetrakis-(*N*-ethylsalicylaldiminato)-zirconium(IV), 35B, 913
Tetrakis-(1,8-naphthyridine)iron(II) perchlorate, 35B, 913
Tetrakis(pyridine oxide)copper(II) tetrafluoroborate and perchlorate, 33B, 548
Tetrakis-(8-quinolinolato)bisdimethyl-sulphoxidethorium(IV), 35B, 913
Tetrakis(tetracarbonyliron)tin, 32A, 411
Tetrakis thioacetamidecopper(I) chloride, 27, 841
Tetrakis(thiourea)palladium(II) chloride, 35B, 628
P,P',P",P"'-Tetrakis(tricarbonyl-nickel)tetraphosphorus hexaoxide, 31A, 77; 31B, 467
Tetrakis(trifluoromethyl)cyclopenta-dienone-π-cyclopentadienyl cobalt, 29, 556
Tetrakis(trifluoromethyl)hydrazine, 30B, 402
Tetrakis(trifluorophosphine)nickel and platinum (gas phase), 35B, 914
Tetrakis(trimethylsilyl)methane, 29, 762
Tetrakis(trimethylsilyl)silane, 35B, 902
Tetrakis(tropolonato)-*N,N'*-dimethyl-formamidethorium(IV), 35B, 751
Tetralithium dihydrogen disilicate, 35A, 311
Tetrameric cyclopentadienyliron sulphide, 31B, 356
Tetrametaphosphimic acid dihydrate, 30A, 391
Tetramethanol dichlorocobalt(II), 31B, 438; 32B, 445
1,2,3,4-Tetramethoxydibenzo[*de,g*] quinazoline-7-one, 35B, 356
1,3,5,7-Tetramethoxy-2,6-dioxa-adamantane, 33B, 135
Tetra-μ3-methoxy-tetrakis[salicyl-aldehydato(ethanol) nickel(II)], 34B, 517
3,5,8,10-Tetramethylaceheptylene, 33B, 545
2,2,4,5-Tetramethyladipic acid, 35B, 6
π-1,1,3,3-Tetramethylallylpalladium chloride dimer, 33B, 334
Tetramethylammonium bichloride, 35B, 919
Tetramethylammonium bis(*o*-phenylene-dioxy)phenylsiliconate, 33B, 271
Tetramethylammonium bis-π-(5,9,10-tribromo-(1)-2,3-dicarbollyl)cob-altate(III), 33B, 265
Tetramethylammonium chloride, hydrogen chloride, 30B, 415
Tetramethyl ammonium chloride penta-selenyl-dichloride complex, 32B, 29
Tetramethylammonium dichloride, 29, 477
Tetramethylammonium fluoride tetra-hydrate, 32B, 291
Tetramethylammonium hexahydrohexa-borate, 30B, 271
Tetramethylammonium hydroxide pentahydrate, 31B, 282
Tetramethylammonium manganese(II) chloride, 32B, 30
Tetramethylammonium mercury tri-bromide, 28, 394
Tetramethylammonium mercury tri-chloride, 28, 393
Tetramethylammonium mercury tri-halides, 28, 393
Tetramethylammonium octahydrotriboro tetracarbonyl chromate, 32B, 485
Tetramethylammonium octahydrotri-borotetracarbonylchromium, 35B, 437
Tetra methylammonium μ4-oxo-hexa-μ-chloro-tetra(chlorocuprate(II)), 34B, 22
Tetramethylammonium perchlorate, 29, 476
Tetramethylammonium salt of tri-acontacarbonyl dodecarhodate dianion, 34B, 436
Tetramethylammonium silver iodide, 28, 392. 35B, 881
Tetramethylammonium sulphate tetra-hydrate, 33B, 27
Tetramethylammonium tetrachloro-cobaltate(II), 32B, 31
Tetramethylammonium tetrachloro-metal complexes, 32B, 31

Tetramethylammonium tetrachloronickelate(II), 32B, 31
Tetramethylammonium tetrachlorozincate(II), 32B, 32
Tetramethylammonium tetrahydroaluminate(III), 35B, 910
Tetramethylammonium tetraisocyanatocobaltate(II), 35B, 906
Tetramethylammonium tetraisocyanatozincate(II), 35B, 906
Tetramethylammonium tetraisoselenocyanatocobaltate(II), 35B, 906
Tetramethylammonium tetraisoselenocyanatozincate(II), 35B, 906
Tetramethylammonium tetraisothiocyanatocobaltate(II), 35B, 906
Tetramethylammonium tetraisothiocyanatozincate(II), 35B, 906
Tetramethylammonium tribromonickelate, 31B, 342
Tetramethylammonium trichloronickelate(II), 33B, 520
Tetramethylammonium tri-iodide, 32B, 33
Tetramethylammonium tri-iodomercurate, 32B, 485
Tetramethylammonium uranyl trisdiethyldithiocarbamate, 33B, 547
Tetramethylarsonium bromide, 28, 587
2,3,5,6-Tetramethyl-1,4-benzoquinone, 32B, 111
2,2,4,4-Tetramethyl-3-benzoyl-6-phenyl-2,4-disila-1,3,5-oxadiazine, 35B, 443
Tetramethylbiphosphine-bis(monoborane), 33B, 275
Tetramethyl-O,O'-bis-trimethylsilylcyclodialumoxane, 28, 673
Tetramethyl-N-N'-bistrimethylsilylcyclodisilazane, 27, 942
2,2,5,5-Tetramethyl-3-carboxypyrroline-1-oxyl potassium salt, 33B, 554
4,4,5,5-Tetramethyl-2-chloro-1,3,2-dioxaphospholane, 35B, 898
1,2,3,4-Tetramethylcyclobutadiene nickel chloride-benzene adduct, 27, 847
1,1,5,5-Tetramethylcyclodecane-8-carboxylic acid, 32B, 484, 486
1,3,5,7-Tetramethyl-cyclo-octa-*cis*,*cis*,*cis*,*cis*-1,3,5,7-tetraene, 34B, 95
(1,3,5,7-Tetramethylcyclooctatetraene) chromium tricarbonyl, 33B, 371
(1,3,5,7-Tetramethylcyclooctatetraene) diiron pentacarbonyl, 33B, 374
Tetramethylcyclopentadienone-*cyclo*pentadienylcobalt, 28, 862
N,*N*,*N'*,*N'*-Tetramethyl-*p*-diaminobenzene-chloranil, 33B, 247
N,*N*,*N'*,*N'*-Tetramethyl-*p*-diaminobenzene iodide, 33B, 61
Tetramethyldiborane, 33B, 531
1,1,4,4-Tetramethyl-1,4-digermanacyclohexa-2,5-diene, 30B, 422; 32B, 330
Tetramethyldiphosphine, 35B, 895
μ-Tetramethyldiphosphine-bis(tetracarbonyliron), 33B, 503
Tetramethyl diphosphine disulphide, 27, 770; 32B, 305
1,1'-Tetramethylene-2,2'-bipyridylium dibromide.H_2O.1/3HBr, 35B, 916
Tetramethylene diammonium adipate, 26, 644
Tetramethylenediammonium chloride, 28, 390
Tetramethylenedinitramine(diaquo)nickel(II), 31B, 419
Tetramethylenesilane, 27, 950
N,*N'*-Tetramethylenethiourea, 32B, 487
Tetra-*O*-methylergoflavin di-*p*-iodobenzoate, 31B, 267
(*N*,*N*,*N'*,*N'*-Tetramethylethylenediamine)-copper(II) bromide and chloride, 35B, 909
1,1'-(Tetramethylethylene)-ferrocene, 27, 86; 30B, 306
2,2,6,6-Tetramethyl-3,5-heptanedionatopraseodymium, 35B, 752
α,α,α',α'-Tetramethyl-β-ketoglutaric acid, 34B, 11
2,2'-7,7'-Tetramethyloctane, 27, 1064
2,2,4,4-Tetramethyl-3-oxo-glutaric acid, 33B, 554
N,*N*,*N'*,*N'*Tetramethyl-*p*-phenylenediamine-1,2,4,5-tetracyanobenzene complex, 32B, 282
2,2,6,6-Tetramethyl-4-piperidinol-1-oxyl, 33B, 117
Tetra-(4-methyl pyridine)cobalt(II) dithiocyanate, 28, 679
Tetra-(4-methyl pyridine)nickel(II) dithiocyanate, 28, 679
Tetra(4-methylpyridine)nickel(II) thiocyanate *p*-terphenyl clathrate, 35B, 919
Tetra(methylsilicon) hexasulphide, 33B, 549
Tetra-ethyl-*p*-silphenylenesiloxane dimer, 29, 766
Tetramethylstibonium tetrakistrimethylsiloxyaluminate, 28, 601
Tetramethylstibonium tetrakistrimethylsiloxyferrate, 28, 601
Tetramethylstibonium tetrakistrimethylsiloxygallate, 28, 601
Tetramethyl sulphonamide, 27, 769
Tetramethyltetraphenylcyclotetrasiloxane, 30B, 434
1,1,4,4-Tetramethyl-2,3,5,6-tetraphenyl-1,4-digermacyclohexa-2,5-diene, 30B, 435
1,1,4,4-Tetramethyl-2,3,5,6-tetraphenyl-1,4-disilicyclohexadiene-

2,5, 30B, 273, 435; 32B, 491
Tetramethylthiuram disulphide, 30B, 397
Tetra(methyltin)hexasulphide, 34B, 408
1,5,9,9-Tetramethyltricyclo-[6,2,1,$0^{4,11}$]-undec-5-ene (dibromo derivative), 35B, 912
Tetramethyltritintetra(irontetra-carbonyl), 32B, 332
1,3,7,9-Tetramethyluric acid, 28, 510
1,3,7,9-Tetramethyluric acid 3,4-benzpyrene complex, 31B, 111; 32B, 287
1,3,7,9-Tetramethyluric acid, pyrene (1:1 complex), 30B, 253
Tetramminecarbonatocobalt(III) bromide, 28, 100
Tetramminecopper(II) dihalocuprates(I), 35A, 415
Tetrammine copper(II) thiocyanate, 28, 110
Tetrammine nickel(II) thiocyanate, 28, 109
Tetramminezinc chloride, 30A, 255
Tetranickel hexacarbonyl tetra(tri-propionitrile-phosphine), 32B, 482
Tetranitrato complex ions, 35B, 756
2,3,4,6-Tetranitroaniline, 29, 770; 31B, 68
Tetranitrogentetraselenide, 31A, 82
n-Tetranonacontane, 27, 746
Tetraphenylantimony hydroxide, 34B, 396
Tetraphenylarsonium bis(*N*-cyanodithio-carbimato)nickelate(II), 33B, 498
Tetraphenylarsonium bis(trithio-carbonato) nickelate(II), 32B, 492
Tetraphenylarsonium *cis*-diaquotetra-chlororuthenate monohydrate, 31B, 315
Tetraphenylarsonium dodecarbonyl-dihydrogen-triruthenium, 33B, 553
Tetraphenylarsonium 3-fluoro-1,1,4,5-5-pentacyano-2-azapentadienide, 31B, 317
Tetraphenylarsonium hydrogen dinitrate, 32B, 492
Tetraphenylarsonium oxopentathio-cyanatoniobate(V), 35B, 782
Tetraphenylarsonium oxotetrabromoaceto-nitrilerhenate (V), 31B, 415
Tetraphenylarsonium oxotetrabromoaquo-molybdate, 32B, 317
Tetraphenylarsonium salts, 35B, 907
Tetraphenylarsonium tetrachloro-metallates, 35B, 906
Tetraphenylarsonium tetrakis (trifluoro-acetato) cobaltate (II), 31B, 410
Tetraphenylarsonium tetranitrato cobaltate(II), 29, 767; 31B, 322
Tetraphenylarsonium tetranitrato-cuprate(II), 35B, 758
Tetraphenylarsonium tetranitrato-manganate(II), 35B, 756
Tetraphenylarsonium tetranitrato-nickelate, 35B, 758
Tetraphenylarsonium tetranitrato-zincate, 35B, 758
Tetraphenylarsonium tetra(trifluoro-aceto) cobaltate(II), 29, 767
Tetraphenylarsonium tri-μ-chloro-octachlorotrirhenate (III), 31B, 319
Tetraphenylarsonium triiodo(triphenyl-phosphine)nickelate(II), 33B, 517
Tetraphenylarsonium tris(maleoni-trile-dithiolato) iron(IV), 32B, 492
cis,cis-1,2,3,4-Tetraphenylbutadiene, 30B, 37
Tetraphenylbutadienes, 35B, 920
Tetraphenyl compounds of group IV elements, 27, 875
Tetraphenylcyclobutadiene iron tri-carbonyl, 30B, 320
Tetraphenylcyclobutadiene molybdenum dicarbonyl bromide dimer, 34B, 479
1,2,3,4-Tetraphenylcyclobutane, 30B, 85
1,1,4,4-Tetraphenyl-1,4-digermano-cyclohexa-2,5-diene, 30B, 434; 31B, 333; 32B, 336
Tetraphenyl germanium, 27, 875
Tetraphenyl lead, 27, 875
5,6,11,12-Tetraphenylnaphthacene, 27, 940
Tetraphenylphosphonium bis(tetra-cyanoquinodimethanide, 33B, 254
Tetraphenylphosphonium π-cyclopenta-dienylbis-(1,2-dicyanoethylene-1,2-dithiolato)molybdenum, 35B, 566
Tetraphenylphosphonium tetrachloro-metallates(III), 35B, 906
Tetraphenylporphine, 29, 728
α,β,γ,δ-Tetraphenyl-porphine($2H^+$,Cl^-, $FeCl_4^-$), 33B, 185
α,β,γ,δ-Tetraphenylporphine dihydro-chloride, 32B, 490
Tetraphenylporphyrin, 32B, 235
Tetraphenyl silicon, 27, 875
2,3,3,5-Tetraphenyl-1,2,5-thiadia-zolidine-4-one, 35B, 911
Tetraphenyl tin, 27, 875; 35B, 490
Tetraphosphorus enneaoxide, 32A, 252
Tetraphosphorus heptasulphide, 28, 354
Tetraphosphorus triselenodiiodide, 35A, 157
Tetrapotassium dihydrogendecaoxo-diiodate(VII), 28, 350
Tetra-*n*-propylammonium bis(maleonit-riledithiolato)-1,10-phenanthrol-inecobaltate, 35B, 821
Tetra(*n*-propyl)ammonium bromide, 33B, 543
Tetrapyridinediisothiocyanatoiron-(II), 33B, 544
Tetra(pyridine oxide) copper(II) fluoroborate, 34B, 578

Tetra(pyridine oxide) copper(II) perchlorate, 34B, 580
α,β,γ,δ-Tetra(4-pyridyl)porphinatomonopyridinezinc(II), 35B, 717
α,β,γ,δ-Tetra-4-pyridylporphine dihydrochloride, 32B, 489
α,β,γ,δ-Tetra(4-pyridyl)porphine-6-(hydrogen chloride), 33B, 185
Tetraquodichlorochromium(III) chloride dihydrate, 31A, 94
Tetrarhodium dodecacarbonyl, 32A, 510
Tetrasodium cuprammonium bis copper dithiosulphate hydrate, 35A, 414
Tetrasodium decafluorotristannate, 35A, 173
Tetrasodium divanadyl(IV) *d*-tartrate *l*-tartrate dodecahydrate, 33B, 409
Tetrasodium tetrakisoxalatozirconate-(IV) trihydrate, 26, 486; 28, 632
Tetrasodium tetramminecopper(II) di-catena-di-μthiosulphato-cuprate(I), 31A, 208
Tetrasulphur tetraimide, 32A, 154
1,2,3,4-Tetrathiadecalin, 35B, 911
Tetrathiazylfluoride, 26, 315; 28, 58
Tetrathiotetracene, 32B, 188
Tetrathioureamercury(II) tetra-isothiocyanatocobaltate(II), 28, 299
Tetrathioureatellurium(II) dichloride, 30B, 285
Tetrathioureatellurium(II) dichloride dihydrate, 30B, 285
Tetrathioureatellurium(II) salts, 26, 576
Tetratitanium pentasulphide, 35A, 142
1,2,8,9-Tetraza-1-8-cyclotetradecadien-3-10-ylene diacetate, 33B, 545
2,3,6,7-Tetrazanaphthalene, 34B, 147
3-(1-Tetrazolyl)pyrazole hydrate, 35B, 154
Tetrodon acid, 28, 687
Tetrodonic acid hydrobromide, 31B, 239
Tetrodotoxin, 35B, 367
Tetrodotoxin hydrobromide, 35B, 368
Thalidomide, 35B, 914
Thallia, 33A, 262
Thallium, 26, 279; 27, 366
β-Thallium, 27, 367
Thallium acetate, 35B, 907
Thallium I benzoate-tetra(thiourea) complex, 34B, 356
Thallium bromide, 26, 297; 27, 426
Thallium(I) carbonate, 31A, 256
Thallium chalcogenides, 29, 91
Thallium chloride, 26, 297; 27, 426
Thallium chloride-thallium bromide, 31A, 257
Thallous cyanamide, 29, 239
Thallium dibromide, 28, 99
Thallium 5,5-dimethyl-1,3-cyclohexane-dione-2-sulphonate monohydrate, 28, 672
Thallium(I) di-*n*-propyldithio-carbamate, 34B, 404
Thallium formate hydrate, 35B, 907
Thallium germanates, 27, 541
Thallium hexachloromolybdate, 27, 464
Thallium iodide, 26, 297
Thallium methoxide, 27, 814
Thallium phosphite, 26, 469
Thallium sulphate, 27, 604
α-Thallium sulphate, 31A, 244
β-Thallium sulphate, 31A, 244
Thallium(I) *mono*thioperrhenate, 31A, 256
Thallium trifluoride, 31A, 260
Thallium trioxalatorhodiate, 28, 458
Thallium wolframate, 27, 724
Thallous dihydrogen phosphate-thiourea, 33B, 241
Thallous fluoride I, 33A, 171
Thallous nitrate, 26, 476; 27, 569
Thallous sulphate, 31A, 198
Thelepoeine methiodide, 28, 553
7-Thiabicyclo-(4,2,0)-1(8)-octene-7,7-dioxide, 32B, 488
1,2,5-Thiadiazole, 26, 760; 28, 664; 29, 733
1,3,4-Thiadiazole, 35B, 889
1-Thia-2,5-diazole-3,5-dicarboxamide, 27, 954
3-Thiadodecanoic acid, 28, 371
3-Thiadodecanoic acid 1,3-diglyceride, 28, 379
6-Thiaestrogen, 35B, 923
Thiamine hydrochloride monohydrate, 27, 994
Thiamine *n*-propyl disulphide, 34B, 238
Thiamine pyrophosphate hydrochloride hemihydrate, 31B, 181
Thiamine pyrophosphate tetrahydrate, 34B, 255
racem-α-(3-Thianaphthenyl)-propionic acid, 26, 775
Thiapyrylium iodide, 27, 963
1-Thia-4-selenacyclohexane-4,4-dibromide, 32B, 200
1-Thia-4-telluracyclohexane-4,4-diiodide, 35B, 495
Thiazoline dyes, 33B, 554
Thiazyl compounds, 27, 409
Thiene(3,4-D)thiepin, 34B, 182
Thieno(3,4-D)thiepin-6,6-dioxide, 34B, 182
trans-β-2-Thienylacrylic acid, 32B, 185
Thiepin 1,1-dioxide, 33B, 553; 35B, 219
Thiethylperazine, 35B, 242
Thioacetamide-S-oxide, 33B, 543
Thioacetanilide-S-oxide, 34B, 73
2-Thio-3-allyl-5-(2'-(3'-methylthiazolidinylidene))-thiazolidine-2,4-dione,

27, 954
Thiobarbituric acid, 32B, 488
Thiobis(dimethylamine), 35B, 895
Thiobis-*NN*-(thiocarbonylpiperidine), 35B, 170
Thioboric acid, 29, 424
1-Thiocarbamoylimidazolidine-2-thione, 35B, 155
4-Thiocarbamoyl pyridine, 31B, 121; 32B, 156
Thiocarbohydrazide, 34A, 349
DL-6-Thioctic acid, 32B, 186
Thiocyanato-bis(ethylenediamine) nickel(II) iodide, 29, 768
catena-μ-Thiocyanato-bis(thiosemicarbazide)silver(I), 33B, 548
Thiocyanopentamminocobalt nitrate, 32A, 354
Thiocytosine, 35B, 244
*mono*Thiodibenzoylmethanato-π-methallyl-palladium(II), 34B, 444
Thiodiglycollic acid, 30B, 398; 32B, 6
Thiodilactic acid, 35B, 922
Thiodipropionic acid, 27, 773
2-Thio-3-ethyl-5[2'-(3'-methylthiazolidinylidene)]-thiazolidine-2,4-dione, 26, 785
6-Thioguanine, 35B, 251
2-Thiohydantoin, 34B, 124
Thioisomaltol, 34B, 180
DL-Thiomalic acid, 29, 762
Thionyl chloride and sulphuryl chloride, 34A, 224
Thionylimide, 34A, 242
Thionyltetrafluoride, 34A, 218
Thiophene, 26, 760; 29, 771; 35B, 893
Thiophene-2-carboxylic acid(β-thiophenic acid), 29, 731
Thiophene chromium tricarbonyl, 29, 767; 30B, 318
α-Thiophenic acid, 27, 948
Thiophthalide, 33B, 138
6-Thiopurine riboside, 33B, 170
α-Thiopyran-1,1-dioxide, 32B, 182
Thiosemicarbazide, 34B, 29, 32; 35B, 28
*mono*Thiosemicarbazidesilver(I) chloride, 33B, 466
Thiostrepton, 35B, 922
Thiosulphate pentamminecobalt(II) chloride monohydrate, 35A, 408
Thiothiazyl nitrate, 27, 410
Thiothiophthene, 35B, 910
Thiothixene, 32B, 488
Thio-*p*-toluyl disulphidobis(dithio-*p*-toluato)iron(III), 34B, 599
Thiotrithiazyl nitrate, 29, 245; 30A, 256
Thiotrithiazyl salts, 26, 294
2-Thiouracil, 32B, 489
Thiourea, 26, 572, 573; 32B, 53; 33B, 30
*mono*Thiourea-cadmium sulphate dihydrate, 32A, 346
Thiourea complexes, 26, 575; 35B, 421
Thiourea dioxide, 27, 749
Thiourea inclusion compounds, 35B, 906
Thiourea, ionic complexes of, 32B, 54
Thiourea nickel(II) thiosulphate, 34A, 315
Thiourea nitrate, 33B, 31
Thiourea parabanic acid, 35B, 245
Thiourea-thallous chlorate, 35B, 791
Thiourea zinc nitrate complex, 26, 575
4-Thiouridine hydrate, 35B, 278
Thioxanthen-9-01-10-oxide, (β-form), 32B, 187
Thiuret hydrobromide, 30B, 157
Thiuret hydrochloride hemihydrate, 30B, 396; 31B, 156
Thomsenolite, 32A, 164
Thomsonite, 26, 519; 30A, 457
Thorbastnasite, 30A, 454
Thoreaulite, 35A, 444
Thorianite, 27, 673
Thoria rare-earth sesquioxide systems, 26, 493
Thorium borate, 26, 492
Thorium chelates, 27, 887
Thorium compounds, 27, 367
Thorium diaquo tetraformate monohydrate, 31B, 408
Thorium diiodide, 33A, 241
Thorium hexaboride, 26, 69
Thorium hydroxynitrate, dinuclear, 33A, 431
Thorium(8-hydroxyquinolinate).(8-hydroxyquinoline), 27, 887
Thorium(IV) metaarsenate, 31A, 245
Throium(2-methyl-8-hydroxyquinolinate), 27, 888
Thorium molybdate, 31A, 254; 33A, 338
Thorium niobium oxide, 31A, 150
Thorium nitrate pentahydrate, 31A, 179
Thorium nitrates, basic, 31A, 260
Thorium nitride, 31A, 68
Thorium nitride fluoride, 31A, 253
Thorium oxalate dihydrate, 30B, 409
Thorium oxides, 26, 403
Thorium perchlorate, dimethyl sulphoxide (1:12), 30B, 410
Thorium(IV) pyroarsenate, 31A, 245
Thorium ternary germanides, 29, 90
Thorium tetrabromide, 31A, 248
Thorium tetrachloride, 31A, 251; 34A, 174
Thorium tetrafluoride, 34A, 173
Thorium tetraformate hydrates, 33B, 543
Thorium tetraiodide, 29, 289
Thorium(IV) tetrakis-(*NN*-diethyldithiocarbamate), 35B, 813
Thorium titanium oxide, 33A, 307
Thorium—uranium, 26, 258
Thorium-vanadium, 27, 368
Thorium-zinc, 26, 256, 258

Thorium—zirconium, 26, 260
Thorolite, 32A, 509
Thortveitite, 27, 702
Three-pyraphene, 30B, 435
L-Threonyl-L-phenylalanine-*p*-nitrobenzyl ester hydrobromide, 34B, 274
Thujic acid *p*-bromophenacyl ester, 31B, 262
Thujic acid (7,7-dimethylcycloheptatriene-3-carboxylic acid), 27, 1025
Thulium, 26, 234; 27, 334, 368
Thulium borate, 26, 491; 28, 170
Thulium dihydride, 27, 237
Thulium hexaboride, 26, 82
Thulium niobate, 27, 521
Thulium oxyiodide, 26, 338
Thulium stannate, 26, 399
Thulium tetraboride, 26, 82
Thulium trihydride, 27, 237
Thymidine, 34B, 262
Thymine, 34B, 237
Thymine dimer E, 34B, 239
trans-(5,6'.5',6)-Thymine dimer, 34B, 239
Thymine *mono*hydrate, 26, 765
Thymine photodimer, 33B, 554
Thymine phototrimer, 35B, 922
Thymol blue, 26, 828
Thymol vicinal, 30B, 394
DL-Thyroxine pentahydrate, 32B, 489
Tikhonenkovite, 29, 415; 31A, 252; 32A, 166
Tilleyite, 35A, 472
Tin, 26, 260, 280; 27, 369
β-Tin, 27, 382
Tinaksite, 30A, 455
Tin(IV) chloride glutaronitrile, 33B, 307
Tin(IV)chloride-phosphorus oxychloride, 28, 94
Tin dioxide, 27, 382
Tin(II) fluoride, 32A, 508
Tin halogenoacetates, 33B, 543
Tin, high pressure SnII, 31A, 69
Tin(IV) nitrate, 32A, 350
Tin(II) oxide, 34A, 239
Tin oxide, 32A, 509
Tin(II) phthalocyanine, 35B, 913
Tin tetrabromide, 28, 99
Tin tetrachloride seleninyldichloride, 34A, 222
Titanaugite, 32A, 455
Titanium, 26, 261; 27, 370
Titanium-aluminium spinels, 31A, 250
Titanium(IV) bromide, 31A, 105
Titanium carbide, 30A, 457
Titanium(III) chloride, 26, 317
Titanium(IV) chloride-ethyl acetate complex, 31B, 442
Titanium dibromide, 26, 333
Titanium dichloride, 27, 434
Titanium dioxide, 26, 351; 31A, 248
Titanium dioxide(II), 32A, 255
Titanium(IV) ethoxide, 28, 651
Titanium(III) iodide, 31A, 260
Titanium molybdate, 27, 521
Titanium monoxide, 27, 484; 32A, 254
Titanium niobate, 27, 520
Titanium(IV) nitrate, 31A, 176
Titanium nitride bromide, 29, 248
Titanium nitride chloride, 29, 248
Titanium nitride iodide, 29, 248
Titanium oxide bronzes, 26, 352
Titanium oxides, 26, 350; 27, 484, 499; 28, 121; 29, 301; 31A, 259; 32A, 254; 33A, 265
Titanium selenides, 26, 264-266
Titanium sesquioxide, 28, 121
Titanium suboxide, 33A, 265
Titanium sulphides, 35A, 141
Titanium tantalate, 27 , 520
Titanium tetramethoxide, 33B, 455
Titanium transition-metal alloys, 27, 371
Titanium trichloride, 27, 441
Titanium—uranium, 26, 268
Titanium uranium diniobium decaoxide, 34A, 270
Titanium uranium oxide, 31A, 255
Titanium-vanadium, 26, 268
Titanium-yttrium, 27, 381
Titanium yttrium oxide, 33A, 306
Titanium-zinc, 27, 372
Titanium zirconium oxide, 33A, 326
Titanolivine, 30A, 459
Titanomaghemite, 26, 379
Titanomagnetites, 30A, 458; 31A, 247, 252
Titanous chloride-cyclopentadiene complex, 26, 652
Tobacco mosaic virus, 29, 772
Todorokite, 26, 370
Tohdite, 29, 299; 30A, 450
Tolane, 33B, 72
Tolane-diphenylmercury, 33B, 72
1-*o*-(-*p*-Toluenesulphonyl)-2,6-anhydro-3-deoxy-D-*gluco*-heptitol, 29, 762
N-*p*-Toluenesulphonyl-N'-cyclohexylurea, 26, 719
m-Toluic acid, 28, 671
o-Toluic acid, 32B, 83
p-Toluidine, 28, 422
ω-(*p*-Toluoyl)-acetophenone enol, 34B, 26
Toluquinone, 35B, 918
(+)-*trans*-2-*o*-Tolylcyclohexyl-3-nitro-4-bromobenzoate, 35B, 922
N-*p*-Tolyl-2,2-diphenylvinylideneamine, 35B, 64
trans-3-*p*-Tolylthiocinnamic acid, 35B, 51
p-Tolyl triphenylphosphoranylidenemethyl sulphone, 30B, 277

Tolypomycinone tri-*m*-bromobenzoate, 34B, 287
Tomatidine hydrobromide, 32B, 258
Tomatidine hydrohalides, 32B, 258
Tomatidine hydroiodide, 32B, 260
Topaz, 27, 711, 725
Torbernite, 26, 468
'Torecan', 30B, 432
N'-Tosyloxydiimide-*N*-oxides, 35B, 923
Tosyl-L-prolyl-L-hydroxyproline monohydrate, 27, 1051
Toundrite, 34A, 379
Tourmaline, 26, 523; 27, 712; 29, 431; 31A, 243
Transition-metal borides, 27, 92
Transition-metal carbides, 27, 121; 35A, 110
Transition-metal carbonyls, 27, 646
Transition-metal dichalcogenides, 31A, 261
Transition-metal ditellurides, 31A, 69
Transition-metal hexacyanocobaltates, 33A, 244
Transition-metal hexafluorides, 31A, 249
Transition-metal orthovanadates, 35A, 362
Transition-metal phases, 27, 377
Transition-metal phases with silicon, 26, 269
Transition-metal phosphides, 26, 270; 27, 319; 30A, 98
Transition-metal salicylalbutyliminates, 26, 678
Transition-metal selenates, 33A, 388
Transition-metal silicides, 27, 372
Transition metals-niobium sulphide, 35A, 111
Transition-metal stannides, 35A, 112
Transition-metal sulphate monohydrates, 31A, 199
Transition-metal ternary germanides, 29, 90
Trechmannite, 26, 427; 34A, 28
3,4,6-Tri-O-acetyl-2-bromo-2-deoxy-α-D-mannopyranosyl fluoride, 34B, 252
3,4,6-Tri-*O*-acetyl-2-bromo-2-deoxy-β-D-mannopyranosyl fluoride, 33B, 546
Tri-*O*-acetyl-*O*-iodoacetyl-granaticin, 33B, 190
Triacetylsphingosine, 34B, 346
Tri-O-acetyl-D-xylopyranose-2,4-phenylhydrazone, 30B, 431
β,β',β"Triaminotriethylamine-isothiocyanatocopper(II) thiocyanate, 32B, 381
β,β',β"-Triaminotriethylamineisothiocyanatozinc(II) thiocyanate, 34B, 486
2,2',2"-Triaminotriethylaminenickel(II) dithiocyanate, 35B, 682
2,2',2"-Triamino-triethylamine-trihydrochloride, 28, 687; 33B, 27
1,3,5,-Triamino-2,4,6-trinitrobenzene, 28, 686; 30B, 69
Triammonium orthophosphate trihydrate, 35A, 331
Triaquo[2,6-di(2'-pyridyl-4-(pyridinium-2'-yl-bromide)-1,3,5-triazine] nickel(II) bromide monohydrate, 34B, 565
Triaquo-hexa-acetate-μ(3)-oxo-tri-chromium(III) chloride hexahydrate, 32B, 496
Triarylarsines, 27, 868
3,13,17-Triaza-15,21-ethanopentacyclo[12,6,$1^{1,17}$,$0^{1,14}$,$0^{2,7}$,$0^{8,13}$]eicosane 17-methiodide, 32B, 171
1,8,10,9-Triazaboradecalin, 34B, 362
Triazidocarbonium hexachloroantimonate, 35B, 468
sym-Triazine, 32B, 155; 33B,543
1,2,4-Triazole, 30B, 127; 34B, 122
1,2,4-Triazole-5-methylamino-3-phenyl-1-bis(dimethylamido) phosphate, 32B, 487
1,2,4-Triazole-5-phenylamino-3-ethyl-1-bis(dimethylamido)phosphate, 32B, 487
1,2,4-Triazole-5-phenylamino-3-ethyl-2-bis(dimethylamido) phosphate, 32B, 487
1,2,4-Triazole-5-phenylamino-3-isopropyl-1-bis(dimethylamido) phosphate hemihydrate, 32B, 487
1,2,4-Triazole-5-phenylamino-3-isopropyl-2-bis(dimethylamido) phosphate, 32B, 487
1,2,4-Triazole-5-phenylamino-3-methyl-1-bis(dimethylamido) phosphate, 32B, 487
1,2,4-Triazole-5-phenylamino-3-methyl-2-bis(dimethylamido) phosphate, 32B, 487
Tri-(1,3,2-benzodioxaborol-2-yl)-amine, 35B, 439
Tribenzotalarene, 35B, 923
Tribenzo[B,F,J][1.5.9]triazacyclo-duodecine nickel nitrate dihydrate, 30B, 363
Tribenzyloxyaluminum, 35B, 907
Tribenzyltin acetate, 33B, 317
Triboron silicide, 27, 406
2,4,6-Tribromoaniline, 34B, 70
Tribromo-bis(phenyl(dimethyl)phosphine) nickel(III)-dibromo-bis-(phenyldimethyl-phosphine) nickel(II) complex benzene solvate, 34B, 601

1,2,3-Tribromo-6-(*o*-methoxyphenyl)-fulvene, 30B, 88
Tribromonitromethane, 27, 746
Tribromostannic acid methyl ester, 28, 602
Tribromotriindiumtetracobaltpentadecacarbonyl, 35A, 402
Tri-11-bromo-undecanoin, 29, 765
α-Tri-11-bromoundecanoin, 30B, 436
Tri-*n*-butyl sulphonium fluoride hydrate, 27, 781
Tricaesium pentachlorocobalt(II), 29, 277
Tricaesium pentaisothiocyanatodioxouranate(VI), 31A, 119
Tricalcium aluminate hexahydrate, 26, 519
Tricalcium aluminate hydrate, 29, 317
Tricalcium dichlorosilicate, 35A, 465
β-Tricaprin, 28, 686; 29, 765; 31B, 24
Tricarbonylbis(triphenylphosphine) osmium(O), 34B, 617
Tricarbonylchromiumanisole-1,3,5-trinitrobenzene complex, 31B, 275
Tricarbonylchromium-methylbenzoate, 32B, 369
Tricarbonylchromium-*o*-toluidine, 32B, 368
Tricarbonyl-*N*-cinnamylideneaniline-iron, 33B, 547
Tricarbonylcycloheptatriene chromium derivatives, 30B, 393
Tricarbonyl-cycloocta-1,3,5-trienyl chromium, 27, 851
Tricarbonyl-π-cyclopentadienylethylmolybdenum, 28, 612
Tricarbonyl-π-cyclopentadienylheptafluoropropylmolybdenum, 32B, 351
Tricarbonyl-π-cyclopentadienyl wolframotriphenylphosphine gold, 34B, 463
Tricarbonyl-[2,4-diethyl-1,5-bis-(diphenyl)-3-phenyl-1,3,5-triphospha-2,4-diazapentane-*PPP*]-molybdenum, 35B, 857
Tricarbonyliron complex of *N*-methoxycarbonylazepine, 35B, 569
Tricarbonyl iron N-*p*-tolylsulphonyl-1,4-dimethoxy-4-imino-1-but-2-enyl iron tricarbonyl, 34B, 451
Tricarbonyl-1-isopropoxycarbonyl-1,2-diazepineiron, 35B, 911
Tricarbonyl-1,6-methano-cyclodecapentaene chromium, 34B, 472
Tricarbonyl-*exo*-7-phenylcyclohepta-1,3,5-trienechromium, 33B, 373
Tricarbonyltetrakis-(trifluoromethyl)cyclopentadienone-iron, 31B, 352
Tricarbonyltrimethylene-1,2-cyclopentadienylrhenium, 33B, 347
Tricarbonyl(triphenylphosphine-σ-tetrafluoroethylcobalt(I), 32B, 470
Tricarbonyl vanadium complex, 26, 684
Trichloroacetonitrile, 26, 588
Trichloro(acetonitrile)bis(triphenylphosphine)rhenium(III), 35B, 914
Trichloro(*p*-acetylphenylimino)bis(diethylphenylphosphine)-rhenium(V), 33B, 511
Trichloroadeniniumzinc, 35B, 917
Trichloro-ammine-bis(triphenylphosphine)osmium(III), 34B, 611
mer-Trichlorobis-(*o*-dimethylaminophenyldimethylarsine)rhodium(III), 35B, 868
Trichlorobis-(*o*-methoxyphenyl-dimethyl arsine)rhodium(III), 34B, 604
Trichlorobis(tri-*n*-butylphosphine)trimethylphosphiterhodium(III), 35B, 916
Trichloro-bis(trimethylamine)vanadium(III), 34B, 549
Trichloro(*cis*-but-2-enylammonium) platinum(II), 35B, 513
Trichloro(*trans*-but-2-enylammonium) platinum(II), 35B, 513
Tri-π-chloro-chloropentakis(diethylphenylphosphine)diruthenium(II), 33B, 518
Trichloro-1,5-diamino-3-azapentane-chromium(III), 35B, 665
Tri-π-chloro-dichlorotetrakis(tri-*n*-butylphosphine)diruthenium, 33B, 517
1,1,1-Trichloroethane, 33B, 539; 35B, 908
Trichloroquaniniumzinc, 35B, 917
Tri-μ-chloro-hexakis(diethylphenylphosphine)diruthenium(II) trichlorotris(diethylphenylphosphine) ruthenate(II), 34B, 619
Trichloro(*p*-methoxyphenylimino)-bis(diethylphenylphosphine)-rhenium(V), 33B, 511
B-Trichloro-C-methylbarene, 30B, 411
α-Trichloromethyl-N-ethyleniminocarbinol, 32B, 153
Trichloro(methylimino)bis(diphenylethylphosphine)rhenium(V), 34B, 609
Trichloromethylmercury(II) bromide, 33B, 320
β-Trichloromethyl-N-methylolethylenimine, 30B, 122
Trichloro(octamethylcyclotetraphosphonitrilium)copper(II), 35B, 459
Tri-μ-chloro-pentacarbonyl (trichlorostannyl) diruthenium, 35A, 427
2,4,6-Tri(*p*-chlorophenyl)-*s*-tiazine, 30B, 135
Trichlorophone, 34B, 378
Trichloroquinaldine, 30B, 423
Trichlorosilyltetracarbonylcobalt, 32A, 402

Trichlorostannic acid methylester, 28, 602
Trichloro-(2,2',2"-terpyridyl)-gallium(III), 35B, 912
3,4,5-Trichlorotetracyclo[4.4.0.$0^{3,9}$.$0^{4,8}$]decan-2-one, 33B, 95
1,1,1-Trichloro-2,2,2-trifluoroethane, 33B, 539
Trichlorotrimethylaminealuminum(III), 34B, 2, 6
1,2,4-Trichloro-3,5,6-trimethylbenzene, 32B, 112
1,3,5-Trichloro-2,4,6-trimethyl-tri(arsenonitrile), 32B, 314
Trichodermol *p*-bromobenzoate, 29, 772; 31B, 218
Tricobalt enneacarbonyl sulphide, 32A, 406
Tricoordinated copper(II) complex, 26, 666
Tricopper arsenide, 30A, 268
Tricopper phosphide, 30A, 268
cis-1,2,3-Tricyanocyclopropane, 31B, 85
Tricyclic methyldisilanylenesiloxane, 32B, 302
Tricyclo[4.3.1.$0^{1,6}$]deca-2,4-diene-chromium(O) tricarbonyl, 35B, 911
Tricyclo[4.3.1.$0^{1,6}$]deca-2,4-diene-iron(O) tricarbonyl, 35B, 911
Tricyclo[5,2,1,$0^{2,6}$]-deca-4,8-dienyl *p*-bromobenzoate, 34B, 117
Tricyclo(5.3.0.0(2,6))decan-3,8-dione, 30B, 108
Tricyclo 3,3,1,$1^{3,7}$ decane, 30B, 115
Tricyclo[3.3.2.$0^{4,6}$]deca-2,7,9-triene, 33B, 96
Tricyclohexyltin acetate, 33B, 318
2,4,6-Tricyclohexyltrioxane, 28, 490
Tricyclo[3.3.0.$0^{2,6}$] octane, 33B, 535
Tricyclo(4.2.0.$0^{2,5}$)octane, 35B, 899
Tricyclo[4,2,0,$0^{2,5}$]octane compounds, 32B, 134
anti-8-Tricyclo[3,2,1,$0^{2,4}$]octyl *p*-bromobenzene-sulphonate, 30B, 113
Tricyclopentadienyl berkelium compounds, 35B, 907
Tricyclopentadienyl californium compound, 35B, 907
Tricyclopentadienylcurium, 35B, 907
Tricyclopentadienyl lanthanon compounds, 35B, 907
Tri(cyclopentadienyl manganese)tetranitrosyl, 32B, 493
π-Trichclopentadienyl uranium chloride, 30B, 315
Tricycloquinazoline, 34B, 159
Tricyclo[4,3,1,$1^{3,8}$]undecane-4,5-dione, 35B, 138
Tridecaboron diphosphide, 27, 406
Tridecan-1,13-dioic acid, 33B, 7
Trideuterium cobalt hexacyanide, 35A, 404
2,4,6-Tri(dimethylamino)-1,3,5-triazine-s-trinitrobenzene, 31B, 274
Tri-μ-dimethyl-germanium-bis(tricarbonyl-iron), 34B, 408
Tridymite, 27, 675; 28, 120; 29, 431, 440
Tridymite, high, 32A, 251
Triethylamine hydrochloride, 30B, 33
Triethylammonium bis-7,7,8,8-tetracyanoquinodimethanide, 35B, 76
Triethylammonium cyclic-uridine-3',5'-phosphate, 34B, 256
Triethylammonium uridine-2',3'-O,O-cyclophosphorothioate, 35B, 292
Triethylammonium uridine-2',3'-O,O-cyclothiophosphate, 34B, 258
Triethylenediamine, 29, 515
Triethylenediammoniumbis-dichloroiodide, 33B, 544
cis-α-Triethylenetetraminediisothiocyanatonickel(II), 35B, 683
p-(Triethylenetetramine-*O*-ethylglycylglycine)cobalt(III) perchlorate hydrate, 33B, 547
L(-)$_{589}$β$_2$-(*RRS*)-(Triethylenetetramine-(*S*)-prolinato)cobalt(III) diiodide dihydrate, 35B, 658
D-β$_2$-(*SSS*)-Triethylenetetramine-(*S*)-prolinato)cobalt(III) tetrachlorozincate, 35B, 660
Triethylphosphine-carbon disulphide, 26, 606
Triethyl phosphite, 30B, 401
Triethylprop-2-ynylammonium *p*-bromobenzenesulphonate, 35B, 50
Triethylscarphane, 27, 774
Trieuropium tetroxide, 31A, 125
1,1,1-trifluoroazomethane, 35B, 891
Trifluorodimethylphosphorane, 30B, 400
N-(β,β,β-Trifluoroethyl)-N,2,4,6-tetranitroaniline, 34B, 71
Trifluoromethylphosphine, 33B, 538
Trifluoromethyl phosphine tetramer, 27, 943
Trifluoronitromethane, 27, 746
Trifluoronitrosomethane, 30B, 400
Trifluoroselenium(IV) μ-fluoro bis (pentafluoroniobate(V)), 35A, 175
Trifluoroselenium(IV) hexafluoroniobate(V), 35A, 174
Trifluorosilyltetracarbonyl cobalt, 35A, 412
Trigermylphosphine, 34A, 155
Triglycerides, 28, 687; 35B, 909
Triglycine sulphate, 27, 1065
Trigonomagneborite, 30A, 455
Trihalogenamminoplatinates(II), 30A, 265

1,3,5-Trihydroxybenzene, 30B, 77
3α,11α,17β-Trihydroxy-11β-carboxy-13β-C-nor-5β-androstane-11β-carboxylic acid 11α,17-lactone, 33B, 196
3,4,8-Trihydroxy-2-methoxy-1-(3-methyl-2-butenyl)-xanthone, 28, 562
β-Triiodobarene, 30B, 441
Triiron dodecacarbonyl, 31A, 259; 34A, 160
2,2,2-Triisopropoxy-4,5-(2',2"-biphenyleno)-1,3,2-dioxaphospholene, 32B, 311
Triisopropylidenecyclopropane, 33B, 545; 35B, 100
Triketoindane, 30B, 105, 422
2,4,6-Triketo-1,3,5-trimethyltriazine-mercuric chloride complex, 27, 858
β-Trilaurin, 30B, 10
Trimagnesium nitride trifluoride, 34A, 189; 35A, 153
Trimeric acetone peroxide, 34B, 170
Trimeric bis(acetylacetonato)zinc(II), 33B, 395
Trimeric palladium(II) acetate, 35B, 914
Trimeric phosphonitrilic bromide, 32A, 207
Trimeric phosphonitrilic isothiocyanate, 35A, 137
Trimesic acid, 34B, 62
Trimetaphosphates, 31A, 245
1,8,8-Trimethoxy-2,6-dioxo-4-(2'-aminoethyl)bicyclo[3.2.1]-oct-3-ene perchlorate, 35B, 206
Tri-p-methoxyphenylmethyl hydrogen dibromide tetrahydrate, 30B, 55
Tri-p-methoxyphenylmethyl hydrogen dichloride tetrahydrate, 30B, 55
2,4,6-Trimethoxy-1,3,5-trimethyl-2,4,6-trioxocyclotriphosphazane, 30B, 393; 33B, 280
Trimethyl(acetylacetonyl)-2,2'-bipyridyl platinum (IV), 27, 886
Trimethyl aluminum, 29, 766
Trimethylaluminum dimer, 32B, 327; 35B, 911
Trimethylamine, 33B, 530
Trimethylamine boron trichloride, 34B, 360
Trimethylamine clathrate hydrate, 33B, 260
Trimethylamine gallane, 28, 596
Trimethylamine oxide, 29, 473
Trimethylamine oxide hydrochloride, 27, 762
Trimethylamine triborane, 26, 616
Trimethylammonium chloride, 33B, 25
Trimethylammonium iodide, 35B, 25
4,6,8-Trimethylazulenetetra-ruthenium enneacarbonyl, 33B, 552; 34B, 475
2,3,4-Trimethylbenzoic acid, 28, 672
2,4,6-Trimethylbenzoic acid, 35B, 46
3,4,5-Trimethylbenzoic acid, 35B, 45
β-Trimethylborazole, 31B, 288
Trimethylborine, 30B, 400
Trimethylcyanogermane, 31B, 329
2,4,6-Trimethyl-1,3,5-dioxaselenane monoselenoparaldehyde, 32B, 488
Trimethylenecyclopropane, 33B, 532
Trimethylenediammonium dihydrochloride, 33B, 26
Trimethylenemethaneiron tricarbonyl, 33B, 549
Trimethylene sulphite, 31B, 159; 35B, 893
Trimethylenethiourea, 29, 516
1,3,10-Trimethylisoalloxazinium iodide, 32B, 488; 35B, 195
(1,3,5-Trimethyl-7-methylene-1,3,5-cyclooctatriene)diiron pentacarbonyl, 33B, 375
Trimethyl(nonane-4,6-dionato)-platinum(IV), 33B, 547; 34B, 432
Trimethyloxosulphonium fluoborate, 28, 409
Trimethyloxosulphonium perchlorate, 28, 410
2,3,4-Trimethylpentane-2,4-phosphinic acid, 33B, 549
2,4,6-Trimethylphenyl-methyl-sulphone, 32B, 68
Trimethylphosphine oxide, 30B, 401
Trimethylplatinum(IV) hydroxide, 33B, 323
Trimethylplatinum(IV) iodide, 33B, 322
O,O',N-Trimethylpyoluteorin, 35B, 163
Trimethylpyridine mercury(II) chloride complex, 32B, 429
Trimethyl-(8-quinolinolato) platinum (IV), 29, 768; 30B, 364
Trimethyl(salicylaldehydato) platinum (IV), 29, 768; 32B, 399
Trimethylselenonium iodide, 31B, 58
Trimethylselenonium perchlorate, 29, 766
Trimethylsilyl isocyanato, 31B, 298
Trimethylsilyl isothiocyanate, 31B, 298
Trimethylsilyl perrhenate, 34B, 362
Trimethylsulphonium mercuritriiodide, 31B, 450
Trimethylthallium, 35B, 508
2,3,5-Trimethyl-1-thiaindenesulphone, 35B, 907
Trimethyltin cyanide, 31B, 330
1(Trimethyl tin)-5,5'-dimethyl hexane, 27, 1064
Trimethyltin fluoride, 28, 685; 29, 625
Trimethyltin hydroxide, 30B, 413
Trimethyltin isothiocyanate, 35B, 487

Trimethyltin-pentacarbonylmanganese, 33B, 308
B-Trimethyl N-triphenyl borazole, 29, 766
2,4,6-Trimethyl-1,3,5-triselenane triselenoparaldehyde, 32B, 488
2,4,6-Trimethyl-1,3,5-trithiane (β form), 34B, 188
Trinickel tellurate, 32A, 306
Trinitrato-1,2-di(pyridine-2-aldimino)ethanegadolinium(III), 35B, 916
meridional-Trinitro(ammine)ethylenediaminecobalt(III), 35B, 586
2-Trinitrobenzene-2-triaminobenzene, 35B, 410
Trinitro(diethylenetriamine) cobalt(III), 32B, 494
2,4,6-Trinitrophenetole, 33B, 49
2,4,6-Trinitrophenetole-caesium ethoxide, 30B, 424; 33B, 424
2,4,6-Trinitrophenetole-potassium ethoxide, 30B, 424; 33B, 52
Tri-*p*-nitrophenylmethyl, 32B, 74
Tri-*p*-nitrophenyl phosphate, 32B, 489; 35B, 274
1,3,5-Trinitro-2,4,6-trihydrotriazine-tetramethylene sulphone, 33B, 551
1,3,5-Trinitro-2,4,6-tri(triphenyl-phosphoranylideneamino) benzene toluene solvate, 32B, 492
Trioxane, 27, 966; 28, 480; 34B, 169
Trioxo(diethylenetriamine)molybdenum-(VI), 29, 620
2,6,8-Trioxypurine, 31B, 162
Triphenyl aluminum, 32B, 492
Triphenyl arsine, 27, 868
Triphenylarsine "hydroxyhalides", 33B, 549
Triphenylarsine oxide monohydrate, 34B, 39
Triphenylbismuth, 33B, 295
Triphenylbismuth dichloride, 33B, 296
Triphenylbromomethane, 26, 735
sym-Triphenylcyclopropenium perchlorate, 28, 684; 29, 770; 31B, 60
Triphenyldichlorostibine, 31B, 326
2,4,5-Triphenyl-1,3-dithiolium perchlorate, 28, 670
Triphenylene, 28, 440
Triphenylethylphosphonium pentanitratocerate(III), 35B, 912
Triphenylgallium, 35B, 482
Triphenylgermanium-manganese pentacarbonyl, 30B, 396
(Triphenylgermyl)tricarbonyl(triphenylphosphine)cobalt(I), 35B, 875
Triphenylindium, 35B, 482
Triphenyl methane, 28, 677; 30B, 403
Triphenylmethylarsonium bis(toluene-3,4-dithiolato)cobaltate hemiethanolate, 33B, 514
Triphenylmethyl bromide, 26, 735; 31B, 82
Triphenylmethyl free radical, 30B, 403
Triphenylmethyl perchlorate, 30B, 54
Triphenylmethylphosphonium bis((3)-1,2-dicarbollyl)cuprate(III), 33B, 267
Triphenylmethylphosphonium bis(1,2-dicyanoethylene-1,2-dithiolato) nickelate (III), 31B, 452
Triphenyloxonium bromide dihydrate, 28, 676
Triphenyloxonium chloride dihydrate, 28, 676
Triphenyloxonium fluoborate, 28, 677
Triphenyloxonium iodide, 28, 676
Triphenyloxonium tetraphenyl borate, 28, 677
Triphenyl phosphate, 27, 916; 30B, 193
Triphenylphosphinegold-cobalt tetracarbonyl, 30B, 396
Triphenylphosphine gold-tetracarbonyltriphenoxyphosphine manganese, 32B, 481
Triphenylphosphine oxide, 35B, 464
Triphenylphosphine oxide-seleninyl dichloride complex, 34B, 354
Triphenylphosphine (pentahaptocyclopentadienyl)copper(I), 35B, 561
Triphenylphosphine sulphide-iodine complex, 33B, 259
Triphenylphosphoranylideneketen, 31B, 303
Triphenylphosphoranlidenethioketen, 32B, 310
Triphenylphosphorus, 29, 638
Triphenylstannoxytetraphenylcyclopentadienyl tricarbonylmanganese, 32B, 367
Triphenyltin chloride, 30B, 431; 35B, 489
Triphenyltin pentacarbonyl manganese, 32B, 334; 33B, 542
Triphenyltin tetracarbonyltriphenylphosphinemanganese, 32B, 481
s-Triphenyltriazine, 30B, 134
1,2,3-Triphenyl-1,2,3-triphosphaindane, 31B, 306
2,4,6-Triphenyltropone-iron tricarbonyl, 27, 876
Triphosphonitrilic bromide, 27, 413; 32A, 207
Triploidite, 35A, 432
Triptycene, 35B, 149
Trirhenium nonaiodide, 33A, 238
Trirubidiumoctachlorodimolybdate, 34A, 209

Trirutiles, 27, 526; 33A, 312
Triirutiles, inverted, 33A, 344
Tris(μ-acetato-μ-acetoximato-palladium(II)]-0.5-benzene, 35B, 913
Tris(acetylacetonato)aquoytterbium-(III), 34B, 508
Tris(acetylacetonato)aquoytterbium-(III)benzene solvate, 34B, 510
Trisacetylacetonatochromium(III), 30B, 332
Trisacetylacetonatomanganese(III), 29, 543
Tris-acetylacetonato-oxo-molybdenum triethanolate, 32B, 390
Tris(acetylacetonato)vanadium(III), 35B, 606
Tris(acetylacetonato)yttrium(III) trihydrate, 32B, 386
Λ(+)$_{589}$-*trans*-Tris[(+)-3-acetyl-camphorato]chromium(III), 35B, 921
Tris(2,2'-bipyridine) chromium(O), 28, 680
Tris(2,2'-bipyridine) titanium(O), 28, 680
Tris(2,2'-bipyridine) vanadium(O), 28, 680
Tris[bis(diphenylphosphinyl)methane]-sodium bromide hydrate, 35B, 915
Tris(bis(trimethylsilyl)amino) aluminum, 34B, 370
Tris-*p*-bromophenylamine, 28, 676
Tris-*p*-chlorophenylamine, 28, 676
Tris-cyclooctatetraene-dititanium, 33B, 380
Tris(π-cyclopentadienylcarbonyl-rhodium, 32B, 360; 34B, 461
Tris(cyclopentadienyl) nitrosyl-molybdenum, 34B, 453
Tris(cyclopentadienyl)samarium(III), 34B, 459
5,11,17-Trisdehydro[18]annulene, 29, 766
(-)$_{589}$Tris(+*trans*-1,2-diaminocyclo-hexane)cobalt(III) chloride pentahydrate, 35B, 713
Tris(*trans*-1,2-diaminocyclopentane)-cobalt(III) chloride tetrahydrate, 35B, 919
Tris(1,2-diaminopropane)nickel perchlorate, 29, 763
Tris-(*NN*-di-*n*-butyldithiocarbamato)-iron(III), 33B, 550
Tris(*N*,*N*-di-*n*-butyldithiocarbamato)-nickel(IV) bromide, 35B, 921
Tris(diethyldithiocarbamato)cobalt-(III), 34B, 527
Tris(diethyldithiophosphato)indium(III), 35B, 912
Tris(β-diketonato)europium(III) chelates, 33B, 544
Tris(2-dimethylaminoethyl)aminecobalt (II) bromide, 32B, 383
Tris-(2-dimethylaminoethyl)amine-copper(II) bromide, 33B, 427
Tris-(2-dimethylaminoethyl)amine-iron(II) bromide, 33B, 427
Tris(2-dimethylaminoethyl)amine-manganese(II) bromide, 33B, 427
Tris-(2-dimethylaminoethyl)amine-nickel(II) bromide, 33B, 427
Tris-(2-dimethylaminoethyl)amine-zinc(II) bromide, 33B, 427
Tris-1,3,5-(dimethylamino)-1,3,5-triboracyclohexane, 34B, 363
Tris(dimethylammonium) hexabromobis-muthate(III), 33B, 293
2,4,6-Tris-(2,2'-dioxybiphenyl)cyclo-triphosphazene, 35B, 915
Tris-*O*-diphenylarsinophenyl)arsinemer-cury(II) bromide(dichloromethane adduct), 31B, 463
Tris-(*o*-diphenylarsinophenyl)arsine-platinum(II)-iodide tetraphenylboro-hydride, 26, 690
Tris-(*o*-diphenylarsinophenyl)-arsineruthenium dibromide, 30B, 391
Tris(*cis*-1,2-diphenylethene-1,2-dithiolato)rhenium, 30B, 394; 31B, 455
Tris (*cis*-1,2-diphenylethene-1,2-dithiolato)vanadium, 31B, 456; 32B, 462
Tris (diphenylphosphine)dibromo-cobalt(II), 31B, 458
Tris(diphenylphosphine)di-iodonickel (II), 31B, 458
Tris(diphenylpropanedionato)-aquo-holmium, 32B, 495
Tris(1,3-diphenyltriazenido)cobalt(III) toluene solvate, 32B, 497
Tris(dithioacetylacetonato)iron(III), 35B, 919
Tris(dithiobenzoato)chromium(III), 33B, 554
Tris(dithiophenylacetato)indium(III), 35B, 816
1,3,5-Triselenane, 33B, 136
Triselenourea dibromide hydrate, 35B, 910
Triselenourea dichloride hydrate, 35B, 910
Tris(ethylenediamine)chromium(III) hexacyanocobaltate(III) hexahydrate, 33B, 387
Tris(ethylenediamine)chromium(III) pentacyanonickelate(II) sesqui-hydrate, 33B, 386
Tris(ethylenediamine)cobalt(III) bromide monohydrate, 27, 830
(+)D-Tris(ethylenediamine)cobalt(III) chloride monohydrate, 34B, 485
Tris(ethylenediamine)cobalt pentacyano-nitrosylchromate dihydrate, 35B, 602
Tris(ethylenediamine)copper(II) sulphate, 35B, 594

Trisethylenediaminenickel silicate hydrate, 35B, 441
Tris(ethylenediamine)nickel(II) sulphate, 35B, 595
Tris(ethylenediamine)ruthenium chloride hydrate, 35B, 917
Tris(ethylenethiourea) tellurium(II) perchlorate, 26, 579
Trisethylsulphonyl methane, 34B, 46
Tris-(O-ethylxanthato)iron(III), 35B, 912
Tris-*p*-fluorophenylamine, 28, 676
Trisgallium(I) hexaiodogallate(III), 30A, 250
Tris(hexacarbonyldicobalt-π-ethynyl)-arsine, 35B, 914
Tris(hexafluoroacetylacetonato)-iridium(III), 35B, 908
Tris(hexafluoroacetylacetonato)-rhodium(III), 35B, 908
Tris(hydroxyacetato)gadolinium(III) hemihydrate, 34B, 528
Trisilver(I) iodide dinitrate, 35A, 190
Trisilylstibine, 34A, 155
Tris-*p*-iodophenylamine, 28, 676
Tris(*N*-methylenediamine)nickel(II) nitrate, 33B, 544
Tris(methylsulphonyl)methane, 30B, 53
Tris-(*p*-nitrophenylethynyl)-arsine, 33B, 285
Tris-octamethylpyrophosphoramide complex of cobalt(II) perchlorate, 35B, 742
Tris-octamethylpyrophosphoramide complex of copper(II) perchlorate, 35B, 742
Tris-octamethylpyrophosphoramide complex of magnesium(II) perchlorate, 35B, 742
Tris(octamethylpyrophosphoramide)-copper(II) perchlorate, 33B, 553
Trisodium nitriloacetate monohydrate, 32B, 25
Trisodium-tris(oxydiacetato)neodymiumate(III) di(sodium perchlorate) hexahydrate, 35B, 642
Trisodium-tris(oxydiacetato)ytterbiumate(III) di(sodium perchlorate) hexahydrate, 35B, 642
Trisodium tris(pyridine-2,6-dicarboxylato)-ytterbiumate(III)-14-hydrate, 35B, 644
α-Tris(2,4-pentanedionato) vanadium-(III), 34B, 507
β-Tris(2,4-pentanedionato) vanadium (III), 34B, 507
Tris(pentane-2,4-dione)iridium(III), 35B, 908
Tris(pentasulphido)platinum(IV) anion, 32A, 509
Tris(1,10-phenanthroline)-cobalt dibromide heptahydrate, 30B, 426
Tris(phenanthroline)cobalt(III) dihydrate, 35B, 906
Tris(phenanthroline)cobalt(III) *mono* hydrate, 35B, 906
Tris(1,10-phenanthroline)-iron dibromide heptahydrate, 30B, 426
Tris(1,10-phenanthroline)-nickel dibromide heptahydrate, 30B, 426
Tris(1-phenyl-1,3-butanedionato) aquoyttrium(II), 33B, 463
Tris(phenyldiethylphosphine)nonachloro-trirhenium(III), 29, 626
Tris(*o*-phenylenedioxy) phosphonitrile trimer inclusion compounds, 32B, 308
Tris(phenylethynyl)phosphine, 32B, 314
Tris(1,3-propanediamine)chromium(III) pentacyanonickelate(II) dihydrate, 35B, 919
Tris-l-propylenediamine cobalt(III) ion, 28, 684
Tris-l-propylenediaminecopper(II) bromide hydrate, 35B, 911
NNN'-Tris[2-(2'-pyridyl)ethyl]ethylene-diaminenickel(II) perchlorate (nitromethane solvate), 35B, 912
Tris-(8-quinolinolato)chromium(III) methanol adduct, 33B, 549
Tris(tetra-*n*-butylammonium)octa-cyanomolybdate(V), 35B, 879
Tris(thiosemicarbazide)disilver dibromide, 33B, 554
Tris(thiourea)cadmium sulphate, 35B, 627
Tris(thiourea)copper(I) chloride, 29, 593
Tris-thiourea copper(I) thiocyanate, 28, 670
Tristhioureamercury(II) chloride, 33B, 477
Tris(thiourea)zinc(II) sulphate, 33B, 479
Tris(thiourea)zinc thiosulphate monohydrate, 33B, 554
Tris(1,1,1-trifluoropentane-2,4-dione)-iridium(III), 35B, 908
Tris(1,1,1-trifluoropentane-2,4-dione)-rhodium(III), 35B, 908
Tristrimethylenebenzene(trindan), 29, 693
(-)D-Tris(trimethylenediamine cobalt-(III) bromine monohydrate, 33B, 545, 34B, 491
Tristriphenylphosphine rhodium carbonyl hydride, 30B, 389
Tristriphenylphosphinetriphenyl-arsine-rhodium(I) hydride, 35B, 915
Tris(tropolonato)*mono*chlorotin(IV), 35B, 751
Tris(tropolono)-*mono*chlorotin(IV) chloroform solvate, 32B, 493

Tris(tropolonato)*mono*hydroxotin(IV), 35B, 751
Trisulphurdinitrogen dioxide, 26, 345
1,3,5-Trithian copper(I) chloride, 33B, 550; 35B, 916
1,3,5-Trithiane, 32B, 181; 33B, 152; 34B, 181
1,3,5-Trithian silver(I) salts 1:1 complexes, 33B, 473
1,3,5-Trithian silver(I) salts 2:1 complex, 33B, 476
Trithiazylchloride, 31A, 79
Trithioacetaldehyde, 34B, 188
Trithiocarbonic acid, 30A, 410; 30B, 53
Tri-μ-thiocyanato-triamminecopper(I) copper(II), 34A, 233
Trithioureatellurium(II) hydrogen difluoride, 30B, 286
Trititanium tetrasulphide, 35A, 141
Tri-*p*-tolyarsine, 27, 869; 28, 588
Triuret, 30B, 41; 31B, 45
Trivalent cobalt acetylacetonate, 26, 676
Trivinyl phosphite, 30B, 401
Tri-*p*-xylylarsine, 27, 869; 28, 591
Trizinc diorthoborate, 35A, 284
Trizirconium oxide, 33A, 268
Troemelite, 29, 426
Troilite, 35A, 140
*mono*Tropein rubidium salt dihydrate, 32B, 490
Tropine ethiodide, 29, 764
Tropine ethobromide, 29, 764
Tropolonato complexes, 35B, 751
Tropolone, 35B, 923
Tropone iron carbonyl, 29, 552
Trypsin, 35B, 908
Tryptamine derivatives, 33B, 542
Tryptamine hydrochloride, 35B, 923
L-Tryptophan hydrobromide, 31B, 191; 32B, 231
L-Tryptophan hydrochloride, 31B, 191
Tschermakitic augite, 31A, 253
Tuberostemonine methobromide dihydrate, 32B, 491
Tucanite, 30A, 459
Tugtupite, 31A, 236
Tunellite, 26, 489; 28, 163; 29, 390
Tunorite, 28, 324
Turkey calamus, 26, 824
Turnip yellow mosaic virus, 29, 772
Turquois, 30A, 395
Tutton's salts, 29, 354; 30A, 374; 31A, 203
Tyretskite, 29, 427
Tyrosine, 26, 826
DL-Tyrosine, 35B, 906
Tyrosine ethyl ester, 35B, 325
L-Tyrosine hydriodide, 32B, 489
Tysonite, 26, 301
Uklonskovite, 29, 424; 30A, 377
Ulvospinel, 30A, 316
Umangite, 31A, 40
Umohoite, 26, 391; 28, 225
Undecacarbonyl(triphenylphosphine) triiron, 33B, 512
Undecachloropentacyclo-[5,3,0,$0^{2,6}$, $0,^{3,9}$,$0^{4,8}$]decan-5-chlorosulphonate, 32B, 146
α-Undecanedeioique acid, 30B, 399
Undecane dicarboxylic acid complexes, 26, 575
Undecanedioic acid, 31B, 19
Uracil, 32B, 203
cis-anti-Uracil dimer, 35B, 922
cis-syn-Uracil photodimer, 33B, 548
Uralolite, 29, 425
Uranates, 26, 403-406
Uraninite, 28, 334; 29, 431
Uranium, 26, 271; 27, 379; 28, 41; 31A, 72
α-Uranium, 26, 271; 27, 381
Uranium(IV) acetate, 29, 458
Uranium carbide, 26, 91
Uranium chloride π-tricyclopentadienyl, 30B, 315
Uranium(V) chromium(III) oxide, 30A, 347
Uranium dimolybdenum octaoxide, 32A, 305
Uranium dioxide, 31A, 247, 251
Uranium divanadium octaoxide, 32A, 276
Uranium fluorides, 32A, 157
Uranium hexafluoride, 30A, 269
Uranium imide chloride, 31A, 100
Uranium iron oxide, 32A, 292
Uranium metaphosphate, 27, 598
Uranium minerals, 26, 539; 27, 672
Uranium molybdate, 31A, 254; 32A, 305
Uranium nitride, 32A, 137
Uranium nitride bromide, 31A, 253
Uranium nitride iodide, 31A, 253
Uranium oxalate dihydrate, 30B, 409
Uranium oxalate hexa-hydrate, 30B, 408
Uranium oxides, 26, 366-368; 27, 481; 28, 129; 29, 312; 31A, 126, 247, 250, 251, 255; 34A, 242; 35A, 213
Uranium oxytelluride, 26, 432; 29, 342
Uranium pentachloride, 32A, 187
Uranium pentoxide, 31A, 255
Uranium rare-earth oxides, 31A, 162
Uranium tetrafluoride, 29, 261; 31A, 248; 34A, 173
Uranium trioxide, 31A, 126, 250, 251
Uranium vanadium oxide, 34A, 268
Uranium vanadium pentoxide, 35A, 236
Uranium wolframates, 31A, 249
Uranium yttrium oxide, 27, 544
Uranium-zinc, 26, 272-275
Uranocene, 34B, 470
β-Uranophane, 29, 432
Uranothorianite, 27, 673

Uranyl acetate, 35B, 908
Uranyl acetate dihydrate, 35B, 908
Uranyl acetylacetonate *mono*hydrate, 29, 769
Uranyl bromide dimethyl formamide adduct, 28, 672
Uranyl chloride, 33A, 221
Uranyl chloride hydrates, 33A, 221
Uranyl chloride-triethylammonium chloride, 35B, 911
Uranyl dihydroxide, 29, 313
Uranyl hydroxide chloride, 34A, 225
Uranyl hydroxyformate, 35B, 907
Uranyl nitrate dihydrate, 26, 477; 27, 570
Uranyl nitrate hexahydrate, 30A, 404; 32A, 355
Uranyl nitrate trihydrate, 26, 477; 28, 180
Uranyl oxalates, 29, 762
Uranyl oxalate tri-hydrate, 30B, 408
Uranyl 8-quinolinolate 8-quinolinol complex, 32B, 497
Uranyl(VI) solutions, 35B, 905
Urea, 26, 568, 573; 27, 748, 1066; 29, 473
Urea ammonium bromide, 28, 569
Urea cadmium bromide, 35B, 909
Urea cadmium nitrate, 26, 571
Urea calcium bromide, 35B, 909
Urea calcium nitrate, 26, 571
Urea inclusion compounds, 26, 575; 35B, 906, 907
Urea lithium iodide, 35B, 474
Urea nitrate, 34B, 30
Urea parabanic acid, 35B, 245
Urea uranate, 28, 669
Urea zinc bromide, 35B, 909
Ureyite, 30A, 455
Uric acid, 30B, 416; 31B, 162
Uric acid dihydrate, 30B, 416
Utilin, 35B, 916
Uvarovite, 28, 322, 336

Vaesite, 27, 59, 294
Valeric acid, 27, 739
Valine, 26, 826
DL-Valine, 34B, 269
L-Valine, 35B, 310
L-Valine hydrobromide, 31B, 193
L-Valine hydrochloride, 31B, 188; 32B, 226
L-Valine hydrochloride *mono*hydrate, 30B, 398; 34B, 265
Valleriite, 33A, 278
Vanadates, 35A, 274
Vanadinite, 27, 604; 31A, 243
Vanadium, 26, 275; 27, 380; 31A, 54
Vanadium bronzes, 29, 302; 31A, 252
Vanadium chlorimide trichloride, 31A, 258; 33A, 226
Vanadium chromate, 27, 521
Vanadium dioxide, 26, 352; 35A, 204
Vanadium fluoride, 26, 302, 31A, 239
Vanadium *mono*xide, 27, 475, 484
Vanadium oxides, 26, 352; 27, 475, 484, 499; 29, 301; 35A, 202, 204
Vanadium(IV) oxide sulphate, 30A, 368
Vanadium oxybromide, 34A, 228
Vanadium pentafluoride, 34A, 172
Vanadium pentoxide, 26, 353
Vanadium(III) perovskites, 31A, 247
Vanadium peroxy compounds, 27, 659
Vanadium sesquioxide, 35A, 205
Vanadium(III) spinels, 34A, 264
Vanadium spodiosite, 35A, 192
Vanadium tantalate, 27, 521
Vanadium tellurates, 34A, 317
Vanadium tourmaline, 31A, 243
Vanadium(III) tris(*OO*-diethyl phosphorodithioate), 35B, 799
Vanadium wolfram oxides, 33A, 314; 34A, 267; 35A, 264
Vanadium-zinc, 29, 92
Vanadyl acetylacetonate:1,4-dioxan (2:1 complex), 35B, 916
Vanadyl bisacetylacetonate, 26, 686
Vanadyl deoxophylloerythroetioporphyrin, 33B, 552
Vanadyl deoxophylloerythroetioporphyrin-1,2-dichloroethane solvate, 34B, 279
Vanadyl(IV) pyridine-2,6-dicarboxylate tetrahydrate, 33B, 437
α-Vanadyl sulphate, 35A, 373
Vanthoffite, 29, 350
Vanuralite, 28, 350
Vanuranilite, 30A, 457
Varulite, 31A, 244
Vaterite, 28, 174; 34A, 348
Vauquelinite, 33A, 405
Väyrynenite, 27, 577
p-Veatchite, 29, 427; 31A, 249
Veralkamine hydroiodide, 33B, 228
Veratrilphthalide, 35B, 918
Veratrobasine, 35B, 552
Vermiculite, 31A, 237; 35A, 461
Vermiculite; 6-amino hexanoic acid complex, 35A, 461
Vermiculite-hexamethylenediamine, 28, 575
Vernolide iodobenzoate, 35B, 923
Veronal salts, 27, 972
Verrucarin A, 30B, 397
Verrucarin A *p*-iodobenzensulphonate (acetone solvate), 31B, 196
Verticinone methobromide, 33B, 555
Verticinone methyl bromide, 35B, 394
Vesuvian, 26, 537
Vesuvianite, 31A, 242
Viacristine, 30B, 397
Villalstonine, 30B, 241
Vimsite, 33A, 446
Vinbarbital I, 34B, 218

Vincaleukoblastine, 30B, 397
Vinogradovite, 30A, 457; 32A, 511
C-Vinylbarenylmethylmercury, 30B, 412
Vinyl bromide, 28, 664
Vinylene sulphate, 33B, 552
Vinyl fluoride, 26, 544
Vinylidene chloride-vinyl chloride copolymer, 29, 759
Vinylidene fluoride, 26, 545; 28, 664
1-Vinylnaphthalene-iron tricarbonyl, 35B, 920
2-Vinylnaphthalene-iron tricarbonyl, 35B, 920
Vinyl polymers, 33B, 527
Vinyl silane, 26, 587
Vinyltrichlorosilane, 29, 450
Viocidic acid dihydrobromide, 35B, 366
Viocidic acid dihydrobromide trihydrate, 33B, 555
Violanthrone, 29, 698
Viomycidine hydrobromide, 33B, 548; 35B, 923
Vishnevite, 30A, 457
Visnagane, 34B, 338
Vitamin A, 26, 808
Vitamin-A acid, 28, 377
(Vitamin-A aldehyde)tricarbonyl iron, 35B, 529
Vitamin B_6 hydrochloride, 26, 808
Vitamin B_{12}, 27, 1042; 29, 754
Vitamin B_{12} coenzyme, 33B, 186
Vitamin B_{12}-5'-phosphate, 35B, 329
Vitamin C, 26, 808; 29, 771; 33B, 158
Vitamin K_3, 33B, 551
Vitreous silica, 27, 1064; 34A, 363
Vlasovite, 26, 531; 32A, 465
Voachalotine, 28, 679
Voaphylline hydroxyindolenine hydroiodide, 35B, 353
Vobasine methiodide hydrate, 29, 772
Volkovskite, 31A, 255
Vonsenite, 31A, 256
Vorobyevite, 30A, 457
Vousenite, 26, 489
Vulcanite, 26, 138
Vysotskite, 27, 290

Wadeite, 29, 440
Wagnerite, 32A, 362
Wallisite, 30A, 453, 458; 33A, 148
Walstromite, 33A, 460
Wardite, 26, 462
Water, 27, 722; 30A, 303
Wavellite, 28, 324; 31A, 189, 256; 33A, 404
Waylandite, 27, 600; 28, 324
Weddellite, 30A, 454; 30B, 22
Wegscheiderite, 26, 484; 28, 352
Weilerite, 31A, 255
Weilite, 28, 324; 35A, 354
Welinite, 34A, 387
Wenkite, 27, 688; 31A, 260
Westgrenite, 27, 599; 28, 330
Whewellite, 26, 484; 27, 819
Wightmanite, 27, 558
Willemite, 27, 699; 35A, 314
Winchite, 29, 432
Withaferin A acetate-*p*-bromobenzoate (ethyl acetate solvate), 33B, 202
Withamite, 29, 432
Wodginite, 28, 354; 30A, 457
Wohlerite, 27, 714
Wolffram's red salt, 26, 611
Wolfram, 27, 381
Wolfram bromide, 33A, 231
Wolfram bronzes, 26, 356
Wolfram carbonyl acetophenone complex, 30B, 294
Wolfram octahydrotriborotetracarbonyl, 32A, 510
Wolframooxalates, 33B, 542
Wolfram oxides, 26, 355; 27, 483; 30A, 312; 31A, 123, 248, 252
Wolfram oxide tetrabromide, 31A, 99
Wolfram oxide tetrachloride, 31A, 99
Wolgram oxide tetrafluoride, 32A, 508; 33A, 172
Wolfram oxychloride, 31A, 99, 246; 33A, 219
Wolfram oxychlorobromide, 33A, 220
Wolfram oxyphosphates, 29, 373
Wolfram(V) oxytrichloride, 31A, 246
Wolfram sulphide tetrabromide, 35A, 156
Wolfram sulphide tetrachloride, 35A, 155
Wolfram(VI) sulphide tetrahalides, 35A, 155
Wolfram trioxide, 26, 355; 27, 483
Wolfram trioxide whiskers, 26, 355
Wollastonite, 26, 519; 28, 246; 34A, 377
Wolsendorfite, 26, 407
Woodruffite, 28, 336
Wulfenite, 27, 522; 30A, 328
Würster's blue iodide, 33B, 61
Würster's red bromide, 33B, 60
Wurtzite, 27, 257
Wüstite, 29, 417; 30A, 457; 31A, 246, 251, 255

Xanthane hydride, 28, 497
Xanthan hydride, 26, 830
Xanthazole *mono*hydrate, 29, 771
Xanthiosite, 30A, 457
Xanthoconite, 33A, 37
Xanthophyllite, 31A, 238
Xenon, 26, 286; 27, 392
Xenon difluoride, 28, 73
Xenon difluoride: iodine pentafluoride, 35A, 177

Xenon difluoride-xenon tetrafluoride, 28, 74; 30A, 266
Xenon(II) fluoride fluorosulphate, 34A, 219
Xenon tetrafluoride, 28, 75
Xenon trioxide, 28, 130
Xenotime, 30A, 386
Xylan, 29, 759
Xylerythrin, 30B, 392
Xylerythrin dibromoacetate, 31B, 265
Xylitol, 34B, 246

Yeatmanite, 31A, 223
Y Ferrites, 35A, 265
Yoderite, 27, 686
Yohimbane hydrobromide, 35B, 354
Yoshimuraite, 26, 537
Ytterbium, 26, 233
Ytterbium borate, 26, 491
Ytterbium hexaboride, 26, 69
Ytterbium niobate, 27, 521
Ytterbium orthoferrite, 31A, 241
Ytterbium oxide, 26, 366; 27, 497
Ytterbium oxyiodide, 26, 338
Ytterbium trisacetylacetonate, 33B, 556
Ytterbium-zinc, 30A, 98
Yttria, 30A, 309
Yttrium spessartine, 31A, 244
Yttrium, 26, 234
Yttrium borate, 26, 491; 28, 170
Yttrium dihydride, 27, 237
Yttrium ferrous hydroxychloride, 33A, 216
Yttrium hexaboride, 26, 69
Yttrium hydroxide, 32A, 246
Yttrium hydroxides, 29, 305; 32A, 246
Yttrium hydroxychloride, 31A, 98, 257; 32A, 199
Yttrium-iron-gallium garnets, 31A, 163
Yttrium iron garnet, 31A, 249
Yttrium niobate, 27, 521
Yttrium orthoferrite, 30A, 323
Yttrium orthosilicate, 32A, 434
Yttrium orthovanadate, 27, 589
Yttrium oxide, 27, 497; 30A, 309
Yttrium oxide deuteroxide, 31A, 129
Yttrium oxide-titanium dioxide, 31A, 241
Yttrium oxychloride, 27, 724
Yttrium oxyfluoride, 35A, 182
Yttrium oxyselenide, 31A, 241
Yttrium oxysilicate, 33A, 471
Yttrium pyrosilicate, 33A, 463, 465
Yttrium silicate, 35A, 297
Yttrium tantalate, 27, 521; 31A, 257; 33A, 341
Yttrium tantalum oxide, 31A, 257
Yttrium titanate, 34A, 257
Yttrium trihydride, 27, 237
Yttrium vanadate, 33A, 310
Yttrium-zirconium, 27, 381
Yugawaralite, 32A, 511; 34A, 374
Yuzurimine hydrobromide, 31B, 255

Zeise salts, 30B, 291
Zellerite, 31A, 243
Zeolite L, 34A, 371
Zeolite Zk-5, 30A, 438
Zeolite hydrogen faujasite, 34A, 370
Zeolites, 26, 519; 27, 690; 29, 412; 30A, 438; 31A, 246; 32A, 479; 34A, 371
Zeolites (types A and X), 31A, 246
Zhemchuzhnikovite, 28, 322
Zinc, 26, 279
Zinc boracite, 27, 551
Zinc borates, 27, 559
Zinc-*n*-butylphenylphosphinate, 34B, 582
Zinc carbonate, 26, 481
Zinc-chkalovite, 31A, 240
Zinc chloride, 26, 319; 27, 722; 35A, 148
Zinc(II) chloride 2,9-dimethyl-1,10-phenanthroline complex, 32B, 496
Zinc croconate, 29, 669
Zinc cyanide, 27, 416
Zinc(II) di-*n*-butyl phosphinothionate, 35B, 810
Zinc diethyldithiocarbamate, 30B, 381; 32B, 496
Zinc diethyldithiophosphate, 34B, 592
Zinc(II) diethyldithiophosphinate, 33B, 548; 35B, 796
Zinc dimethyldithiocarbamate, 31B, 405
Zinc ethylxanthate, 31B, 404
Zinc ferrite, 26, 411
Zinc fluoride, 28, 72, 290
Zinc fluoroborate hexahydrate, 26, 304
Zinc formate dihydrate, 28, 669
Zinc glutamate, 30B, 395
Zinc glutamate dihydrate, 31B, 412
Zinc guanidinium sulphate, 35B, 733
Zinc hexacyanochromate, 26, 343
Zinc hexacyanochromate hexahydrate, 27, 653
Zinc hexahydroxystannate, 32A, 247
Zinc(II)hydrazinecarboxylate, 30B, 414
Zinc hydrazinecarboxylate hydrazine complex, 30B, 367
Zinc hydrazine chloride, 27, 472
Zinc hydroxide, 29, 303; 34A, 235
Zinc hydroxide carbonates, 31A, 248
β-Zinc hydroxide-2,4-dinitrophenolate, 26, 738
Zinc hydroxychloride, 26, 327
Zinc hydroxychloride II, 26, 330
Zinc hydroxychloride *mono*hydrate, 33A, 217
Zinc hydroxychlorosulphate, 31A, 243

Zinc 8-hydroxyquinolinate dihydrate, 29, 614
Zinc hydroxysulphate, 27, 629
Zincian chromite, 29, 419
Zinc iodo-arsenite, 27, 467
Zinc iodophosphite, 27, 467
Zinc iron tellurite, 32A, 307
Zinc manganite, 27, 493
Zinc metaborate, 26, 490
Zinc methacrylate, 33B, 527
Zinc mica, 31A, 252
α-Zinc molybdate, 32A, 299
Zinc molybdenum(IV) oxide, 31A, 155
Zinc nitrate, basic, 35A, 325
Zinc nitrate dihydrate, 34A, 341
Zinc nitrate tetrahydrate, 35A, 324
Zincobotryogen, 29, 424
Zincocopiapite, 29, 424
Zinc orthoborate, 33A, 441
Zinc orthogermanate, 35A, 314
Zinc orthophosphate, 30A, 383
β-Zinc orthophosphate, 32A, 357
γ-Zinc orthophosphate, 28, 189
Zinc orthosilicate, 35A, 314
Zinc orthotitanate, 26, 378; 31A, 244
Zinc oxide, 27, 475; 34A, 238
Zinc phosphates, 30A, 385, 454; 35A, 343
Zinc phosphate tetrahydrate, 28, 190
Zinc phthalocyanine, 30B, 436
Zinc pyrophosphate, 30A, 385
α-Zinc pyrophosphate, 35A, 343
Zinc rhodite, 26, 399
Zinc ruthenite, 26, 399
Zinc salicylal-*o*-anisidinate, 33B, 556; 35B, 627
Zinc salicylalbutyliminate, 26, 678
Zinc selenate, 31A, 254
Zinc selenate *mono*hydrate, 31A, 256
Zinc selenite dihydrate, 29, 358
Zinc stannate, 27, 490
Zinc sulphate, 26, 437, 438
Zinc sulphate *mono*hydrate, 26, 439
Zinc sulphate tetrahydrate, 31A, 244
Zinc tellurate, 31A, 260
Zinc tellurite, 32A, 307
Zinc tetraisoselenocyanatomercurate, 31A, 255
Zinc tetraphenylporphine dihydrate, 29, 617
Zinc titanates, 26, 376-377
Zinc(II) tris(pyridine-2-carboxaldehyde)-1,1,1-tris(aminomethyl)ethane diperchlorate, 35B, 921
Zinc vanadates, 30A, 454; 31A, 247, 33A, 313
Zinc vanadium oxides, 31A, 247; 33A, 313
Zinc wolframate, 33A, 347
Zinc-zirconium, 26, 277
Zircon, 29, 411
Zirconian schorlomite, 27, 514
Zirconium bis(monohydrogen orthoarsenate) mono-hydrate, 34A, 336
Zirconium boronhydride, 32A, 508; 34A, 151
Zirconium bromide, 27, 442
Zirconium chloride, 27, 442; 34A, 174
Zirconium cupferate, 35B, 748
Zirconium dioxide, 27, 480
Zirconium dioxide, amorphous, 31A, 240
Zirconium halides, 27, 442
Zirconium hydrogen phosphate, 34A, 328
Zirconium iodate, 26, 498
Zirconium iodide, 27, 442
Zirconium molybdenum oxide, 31A, 256
Zirconium nitride bromide, 29, 248
Zirconium nitride chloride, 29, 248
Zirconium nitride fluoride, 31A, 253
Zirconium nitride hydride, 35A, 115
Zirconium nitride iodide, 29, 247
Zirconium oxygen fluoride, 35A, 181
Zirconium oxide nitrides, 27, 415
Zirconium oxide sulphide, 27, 423
Zirconium periodate, 27, 724
Zirconium phosphate *mono*hydrate, 31A, 259
Zirconium pyroarsenate, 30A, 458
Zirconium pyrophosphate, 26, 467
Zirconium silicate, 26, 531
Zirconium sulphate, 34A, 301
Zirconium sulphate hydrates, 34A, 301
Zirconium sulphate hydroxide hydrate, 31A, 205
Zirconium sulphates, 35A, 367
Zirconium tetrabromide, 27, 444
Zirconium tetrachloride, 27, 718; 35A, 149
Zirconium(IV) tetrakisacetylacetonate, 28, 619
Zirconium titanate, 31A, 256; 32A, 267
Zirconium tribromide, 29, 249
Zirconium trichloride, 29, 249; 31A, 96
Zirconium trihalides, 29, 249
Zirconium triiodide, 29, 249
Zirconium ytterbium oxide, 35A, 259
Zirconolite, 28, 336
Zirconyl chloride octahydrate, 33A, 218
Zirconyl perchlorate, dimethyl sulphoxide (1:8), 30B, 410
Zirconyl phosphates, 26, 467
Zirkelite, 28, 336
Zoisite, 26, 521; 31A, 247; 33A, 475
Zunyite, 26, 535
Zussmanite, 30A, 459; 34A, 392

CUMULATIVE FORMULA INDEX OF METALS AND INORGANIC COMPOUNDS

The *symbols* in each formula are arranged alphabetically, and the formulae are then arranged alphabetically. Within groups of formulae containing the same elements the order is governed by the subscripts of the first, second, ... symbols in turn.

Polymers *etc.* with indefinite subscripts *p, n, x* are placed in approximately the positions they would occupy if the subscript letter had the value 1.

Solvents of crystallization are not included in the alphabetical formula, but placed seperately at the end. Inorganic compounds containing water possibly or certainly as OH groups may be entered twice, once with the water alphabetized with the rest of the formula, and again with the water separated at the end.

When solid solutions are indexed, it is sometimes necessary to ignore second or third elements in parenthesis. Thus for example $(Al,Si)_2Mo$ would follow AlMo in the index and not AlSi.

$ABCa_2O_6$, 30A, 452
ABO_6Pb_2, 30A, 452
Ac, 26, 3
AcF_5M, 31A, 249
$AcF_{31}M_6$, 31A, 249
Ag, 26, 279; 27, 386; 29, 142; 30A, 184
Ag—Al, 26, 187; 27, 33
Ag_2AlHf_3, 29, 96
$AgAl_6Mg_4$, 30A, 100
$Ag_{57}Al_{57}O_{384}Si_{135}$, 32A, 484
Ag_2AlZr_3, 29, 96
Ag—As, 26, 249
$Ag_4As_3Cu_4$, 26, 48
$Ag_{0\cdot6}As_{15\cdot7}Pb_{11\cdot6}S_{36}$, 34A, 17
$AgAsPbS_3$, 28, 50; 32A, 229
$AgAs_2(Pb,Tl)_2S_5$, 31A, 261
$AgAs_2PbS_5Tl$, 32A, 230
$(Ag,As)As_4(Pb,Tl)_3S_{10}$, 30A, 357
$AgAsS_2$, 29, 25, 96; 34A, 28
Ag_3AsS_3, 31A, 19; 33A, 37
Ag-Au-Cd, 27, 107
$AgAuCd_2$, 27, 107
β-Ag_3AuS_2, 31A, 43
Ag_3AuTe_2, 30A, 100
$Ag_{3\cdot52}Au_{0\cdot472}Te_2$, 30A, 100
$Ag_{3\cdot72}Au_{0\cdot28}Te$, 30A, 100
$Ag_{3\cdot9}Au_{0\cdot1}Te_2$, 30A, 100
$AgAuZn_2$, 28, 42
AgB_2, 26, 73
$Ag_{1\cdot2}B_8Na_{0\cdot8}O_{13}$, 30A, 412
$Ag_2B_8O_{13}$, 30A, 412
AgBS, 29, 96
AgBSe, 29, 96
Ag_2Ba, 30A, 100
Ag_2BaGe_2, 35A, 24, 116
$AgBa_2IO_6$, 29, 421
$(Ag,Bi,Pb)Bi_5Cu_4S_{11}$, 33A, 41
$(Ag,Cu)_2Bi_6Pb_7S_{17}$, 27, 405
$(Ag,Cu)_3Bi_5Pb_2S_{11}$, 30A, 453
$Ag_4Bi_4PbS_9$, 30A, 100, 453
$Ag_4Bi_4PbSe_9$, 30A, 100
$AgBiS_{1\cdot5}Se_{0\cdot5}$, 33A, 44
$AgBiTe_2$, 29, 31, 96
AgBr, 27, 427
Ag_3BrS, 30A, 289
$Ag_3C_2NO_3$, 29, 238
$Ag_3C_6CoN_6$, 30A, 450; 33A, 242
$AgC_6CoN_6Tl_2$, 30A, 450
$Ag_2C_6CoN_6Tl$, 30A, 450
$AgC_3CuH_6N_5S_3$, 31A, 118
$Ag_3C_6FeN_6$, 31A, 254
$AgC_6FeN_6Tl_2$, 31A, 254
$Ag_2C_6FeN_6Tl$, 31A, 254
$AgC_2KN_2S_2$, 27, 419
$AgCKO_3$, 28, 175
Ag_2CN_2, 26, 296; 29, 239
AgCNO, 27, 747; 30A, 294
$Ag_3CN_3O_6$, 30A, 294
$Ag_3C_2NO_3$, 29, 238
$Ag_3CN_3O_6S$, 35A, 327
Ag_2CO_3, 27, 725
Ag_3Ca_5, 27, 119
Ag_8Ca_3, 29, 33, 96
Ag_2CaGe_2, 35A, 24, 116
Ag—Cd, 26, 83, 248; 27, 111, 191
AgCd, 26, 83; 30A, 42
Ag_5Cd_8, 26, 83
Ag-Cd-Sn, 29, 142
Ag—Cd—Zn, 26, 83
Ag(Cd,Zn), 26, 83
$Ag_{51}Ce_{14}$, 35A, 116
$AgCe_2Se_{3\cdot5}$, 30A, 100
AgCl, 27, 427; 28, 290; 29, 415
Ag-Cl-O, 28, 320
$AgClO_2$, 26, 495
$AgCoF_3$, 28, 296
$AgCrS_2$, 33A, 61
$AgCrSe_2$, 33A, 61
$AgCs_2F_6K$, 31A, 260
AgCsO, 31A, 253; 33A, 280
$Ag_{0\cdot25}Cu_{0\cdot75}InTe_2$, 27, 181
$Ag_{0\cdot50}Cu_{0\cdot50}InTe_2$, 27, 181
$Ag_{0\cdot75}Cu_{0\cdot25}InTe_2$, 27, 181
$Ag_{1-x}Cu_xInTe_2$, 27, 181
$Ag_{1\cdot18}Cu_{0\cdot82}S$, 31A, 257
Ag-D-Pd, 33A, 120
AgDy, 26, 237
Ag_2Dy, 34A, 81
$Ag_{51}Dy_{14}$, 35A, 116
Ag_2Er, 30A, 100
$Ag_{51}Er_{14}$, 35A, 116
Ag_2Eu, 30A, 100
Ag_5Eu, 30A, 100; 35A, 116
AgF_2, 31A, 84, 258
Ag_2F, 31A, 83

AgF_6Ir, 28, 298
AgF_4K, 26, 309; 35A, 162
AgF_4Na, 35A, 162
AgF_6Nb, 28, 298
Ag-F-O, 28, 320
AgF_6Os, 28, 298
AgF_6Ru, 28, 298
AgF_6Sb, 28, 298
AgF_6Ta, 28, 298
AgF_6U, 30A, 449
AgF_6V, 28, 296
$AgFeO_2$, 29, 419
$AgFeTe_2$, 27, 256; 29, 62, 96
Ag—Ga, 26, 248; 29, 53, 96
Ag-Ga-Zn, 29, 142
AgGd, 26, 237; 29, 96; 30A, 100
$Ag_{51}Gd_{14}$, 35A, 116
$AgGe_2$, 29, 96
Ag_8GeS_6, 30A, 100
$AgGeSe_2$, 34A, 138
Ag_8GeSe_6, 30A, 100
Ag_2Ge_2Sr, 35A, 24, 116
Ag_2HIO_5, 26, 502
AgHf, 29, 96
$AgHf_2$, 29, 96
Ag--Hg, 26, 248
Ag_2Hg_3, 34A, 103
Ag_5Hg_8, 26, 196
Ag—Hg—Sn, 26, 196
Ag_2Ho, 30A, 100; 34A, 129
$Ag_{51}Ho_{14}$, 35A, 116
$AgHoSe_2$, 34A, 138
AgI, 27, 425, 427, 28, 100 290; 29, 416; 30A, 449; 31A, 244; 33A, 239
$Ag_3IN_2O_6$, 35A, 190
Ag_4I_5Rb, 32A, 216
Ag_3IS, 26, 423; 30A, 289
Ag—In, 26, 248; 30A, 184
Ag_3In, 26, 172
$AgInS_2$, 34A, 138
$AgInSe_2$, 34A, 138
$AgInTe_2$, 27, 181; 30A, 151
Ag-In-Zn, 29, 143
AgKO, 31A, 253; 33A, 280
Ag_5La, 35A, 116
$Ag_{51}La_{14}$, 35A, 116
$AgLa_2Se_{3.5}$, 30A, 101
Ag*Ln*, 28, 42
Ag_4Lu, 35A, 116
AgMg, 26, 190; 27, 269
$Ag_{0.15}MgZn_{1.85}$, 32A, 97
Ag-Mn, 27, 183
$AgMn_3N$, 29, 96
$AgMnO_4$, 33A, 319
$Ag_2MnS_5Sb_2$, 34A, 296
Ag-Mn-Sn, 27, 277, 310
Ag-Mn-Zn, 29, 144
$Ag_6Mo_{10}O_{33}$, 35A, 270
$AgMoO_6P$, 30A, 394
AgN_3, 27, 414
$AgNO_2$, 27, 573
$AgNO_3$, 26, 475; 31A, 178
Ag_7NO_{11}, 28, 178, 322; 30A, 308
$Ag_4N_2O_6Te$, 31A, 256; 34A, 344
$Ag_7N_5O_{15}Te$, 31A, 256
Ag_2Na, 26, 250
Ag_3Nd, 34A, 138
$Ag_{51}Nd_{14}$, 35A, 116
$AgNd_2Se_{3.5}$, 30A, 101
AgO, 26, 359
Ag_2O, 27, 724; 28, 344
$(AgO_3P)_x$, 26, 459
$Ag_2O_5Si_2$, 26, 506
Ag_4O_4Si, 27, 706
AgO_4Tc, 27, 671
AgO_5TeV, 34A, 318
$Ag_{1-x}O_5V_2$, 30A, 344
AgP_2, 30A, 45, 101
$Ag_{16}P_{12}Pd_{12}$, 35A, 116
$Ag_{14}P_{14}Pt_{72}$, 35A, 116
Ag-Pd, 29, 145, 146
Ag_3Pr, 34A, 138
$Ag_{51}Pr_{14}$, 35A, 116
$AgPr_2Se_{3.5}$, 30A, 101
$Ag_{14}Pt_{72}Si_{14}$, 35A, 116
Ag_2S, 26, 250; 32A, 123
AgS_2Sb, 29, 21, 96
Ag_3S_3Sb, 31A, 19; 33A, 22
Ag_5S_4Sb, 34A, 12; 35A, 15
Ag_8S_6Si, 30A, 101
Ag_8S_6Sn, 30A, 101
AgS_2Yb, 31A, 65
Ag—Sb, 26, 249; 30A, 184
$AgSbTe_2$, 26, 46; 28, 42; 29, 96
$\sim Ag_{19}Sb_{29}Te_{52}$, 26, 46
AgSc, 27, 342
Ag_4Sc, 35A, 116
Ag_2Se, 27, 344; 30A, 92, 101
Ag_8Se_6Si, 30A, 101
$AgSe_{3.5}Sm_2$, 30A, 101
Ag_8Se_6Sn, 30A, 101
$AgSe_2Tl$, 30A, 93, 101
Ag-Si, 30A, 101
Ag_3Sm, 30A, 101
$Ag_{51}Sm_{14}$, 35A, 116
Ag—Sn, 26, 249; 27, 310; 30A, 184
Ag_3Sn, 26, 196
AgSnTe, 29, 96
Ag_2Sr, 30A, 101
AgTb, 29, 96
Ag_2Tb, 33A, 91
Ag_3Tb, 34A, 138
$Ag_{51}Tb_{14}$, 35A, 116
Ag-Te, 27, 221
AgTe, 29, 96
Ag_2Te, 26, 250; 27, 256; 30A, 101
Ag_5Te_3, 29, 96
Ag_7Te_4, 31A, 67
$AgTe_2Tl$, 29, 87, 96
Ag_2Th, 26, 256
AgTi, 29, 96; 30A, 101
$AgTi_2$, 29, 96; 30A, 101
Ag_3Tm, 34A, 138; 35A, 116
AgY, 28, 42
Ag_2Y, 29, 96; 30A, 101
$Ag_{51}Y_{14}$, 35A, 116
AgYb, 35A, 104
Ag_2Yb, 35A, 104
Ag_2Yb_3, 35A, 104
Ag_3Yb_5, 35A, 103
Ag_7Yb_2, 35A, 104, 116
Ag—Zn, 26, 83; 27, 191; 29, 146
AgZn, 26, 83; 27, 349
Ag_5Zn_8, 26, 83
Ag_2ZnZr_3, 30A, 101
$AgZr_2$, 27, 234
Al, 26, 3, 279; 27, 381, 385; 30A, 185
$Al_2(Al,Fe)Ca_2HO_{13}Si_3$, 26, 521
$Al_4(Al,,Si)_{12}H_{12}Mn_9O_{45}$, 29, 431
$Al_4(Al,Si)_{12}Mn_9O_{39}\cdot 6H_2O$, 29, 431
$Al_4Ar_{2.5}Mg_2O_{18}Si_5$, 27, 694
AlAs, 27, 3; 29, 96
$AlAsCu_2H_4O_8\cdot 4H_2O$, 32A, 388; 33A, 424
$AlAsO_4$, 28, 350
$Al_{14}As_{14}Pt_{72}$, 35A, 116
Al-Au, 29, 147
Al_2Au, 29, 96, 146, 147
$AlAu_2Hf_3$, 30A, 102
$AlAu_2Mn$, 28, 42
$AlAu_2Zr_3$, 30A, 102
Al—B, 26, 4
$Al_{0.9}B_2$, 29, 4, 96
AlB_2, 26, 3
AlB_{10}, 28, 50; 31A, 3; 32A, 3; 34A, 3
α-AlB_{12}, 26, 3, 29, 4
γ-AlB_{12}, 26, 5; 29, 4
AlBBe, 27, 3
$Al_4B_{11}Be_4CsH_4O_{29}$, 30A, 455
$Al_4B_{12}Be_4CsO_{28}$, 31A, 260; 32A, 495
$AlB_{24}C_4$, 29, 4, 97; 34A, 3
$AlB_{40}C_4$, 35A, 6
$Al_{1.3}B_{24}C_4$, 34A, 3
$Al_{2.1}B_{51}C_8$, 34A, 3
$Al_3B_{48}C_2$, 29, 4, 97; 30A, 102
$Al_{3-x}B_{48-y}C_{2-z}$, 29, 97

$Al_{3+x}B_{48+y}C_{2+z}$, 29, 97
$Al_4B_{1-3}C_4$, 29, 97
$AlBCaO_4$, 33A, 444
Al-B-Co, 28, 42
$Al_3B_6Co_{20}$, 27, 4, 92
$Al_{2.7}B_{104}Cu_2$, 35A, 3
$Al_3B_4DyO_{12}$, 27, 562
$Al_3B_4ErO_{12}$, 27, 562
$Al_3B_4EuO_{12}$, 27, 562
$Al_6B_3FFe_3NaO_{30}Si_6$, 31A, 242
AlB_2Fe_2, 34A, 6
$Al_3BFe_{0.1}Mg_{0.9}O_9Si$, 33A, 502
$Al_3B_4GdO_{12}$, 27, 562
$Al_3B_4HoO_{12}$, 27, 562
$AlB_{14}Mg$, 35A, 4
$AlBMgO_4$, 30A, 418
AlB_2Mn_2, 31A, 3
AlBMo, 30A, 102; 31A, 4
$Al_3B_4NdO_{12}$, 27, 562
Al-B-Ni, 28, 42
$Al_3B_{6-12}Ni_{20}$, 27, 4, 92
$Al_2B_2Ni_4O_{10}$, 35A, 288
$Al_3B_4O_{12}Sm$, 27, 562
$Al_3B_4O_{12}Tb$, 27, 562
$Al_3B_4O_{12}Y$, 27, 562
$Al_3B_4O_{12}Yb$, 27, 562
$Al_3B_{48}Si$, 34A, 137
$Al_3B_xTa_5$, 29, 97
(Al,Ge)Ba, 34A, 138
Al_2Ba, 29, 97
$Al_2BaH_4O_6$, 35A, 294
$Al_2Ba_3H_{12}O_{12}$, 34A, 252
Al-Ba-K-O-Si, 30A, 455
Al_2BaO_4, 30A, 316
$Al_2Ba_2O_4.5H_2O$, 35A, 292
$Al_4Ba_2O_{32}Si_{12}.12H_2O$, 26 532
$(Al,Fe)Be_5$, 27, 4
$Al_{2-x}Be_{2+x}Ca_4H_{2+x}O_{28}Si_9$, 31A, 221
$Al_{2.6}Be_{1.4}Ce$, 29, 97
$Al_{2.8}Be_{1.2}Ce$, 29, 97
$Al_2Be_2(Cl,S)_2Na_8O_{24}Si_8$, 31A, 236
$Al_{2-x}BeCr_xO_4$, 29, 317
$Al_{1.9}Be_3Cr_{0.1}O_{18}Si_6$, 33A, 480
Al-Be-Fe, 27, 4
$Al_{2-x}BeFe_xO_4$, 29, 317
$AlBeHO_5Si$, 27, 579; 29, 429
Al_3BeLa, 34A, 138
Al_4BeMgO_8, 28, 149
Al_2BeO_4, 28, 141
$(Al,Cr,Fe)_2BeO_4$, 29, 317
$Al_2Be_3O_{18}Si_6$, 31A, 221; 33A, 480
$AlBiCl_4$, 26, 324
$Al_{5.2}Bi_xF_xO_{12-x}Y_3$, 28, 352
Al-Bi-Fe-La-O, 29, 326
$Al_4Bi_2O_9$, 33A, 321
$AlBrC_6Cl_3H_4NO_2$, 26, 704
$Al_2Br_6C_6H_6$, 26, 653
Al-Br-Ce, 30A, 102
$Al_2Br_2Na_2O_4Si_2.\ 4.4H_2O$, 31A, 218
AlBrO, 28, 300
$AlBr_6Sb$, 34A, 213
Al_4C_3, 28, 4, 296; 30A, 6
$(Al,Si)_{24}(CO_3,Cl,SO_4)(Ca,K,Na)_8O_{48}$, 31A, 231
$Al_6CCaNa_6O_{27}Si_6.2H_2O$, 30A, 428
$Al_6CCa_4O_{21}Si_6$, 27, 688
Al_2CCoTa_3, 30A, 185
$AlCCr_2$, 28, 3
Al_2CCrTa_3, 30A, 185
Al_2CCuTa_3, 30A, 185
$AlC_{0.7}Dy_3$, 30A, 102
$AlCF_5O_3Sr$, 27, 564
ε-$AlCFe_3$, 29, 147
Al_2CFeTa_3, 30A, 185
$AlC_{0.9}Gd_3$, 30A, 102
$Al_2CH_{16}Mg_6O_{19}.4H_2O$, 32A, 421
$AlCH_6N_3O_8S_2.6H_2O$, 33A, 159
$AlCH_2NaO_5$, 32A, 419
$AlC_{0.7}Ho_3$, 30A, 102
$(Al,Fe)C_6MgNaO_{12}$ 8.3 H_2O, 28, 322
$AlCMn_3$, 27, 272
Al_2CMnTa_3, 30A, 185
Al_2CMo_3, 28, 42
Al_5C_3N, 28, 4; 31A, 6
$Al_6C_3N_2$, 28, 4; 31A, 6
$Al_7C_3N_3$, 28, 3; 31A, 6
$Al_8C_3N_4$, 28, 4; 31A, 6
$Al_9C_3N_5$, 28, 4
$Al_{10}C_3N_6$, 28, 4
$AlCNb_2$, 28, 42
Al_2CNb_3, 28, 42; 29, 97
$AlC_{0.9}Nd_3$, 30A, 102
$AlC_{0.25}Ni_3$, 30A, 102
Al_2CNiTa_3, 30A, 186
Al_2CO, 26, 295
$Al_2C_{12}O_{12}.18H_2O$, 30A, 454
Al_4CO_4, 28, 66
$Al_{14}C_{12}O_{33}$, 34A, 251
AlC_xPt_3, 26, 225
$AlCSc_3$, 30A, 102
α-Al_4C_4Si, 26, 5
β-Al_4C_4Si, 26, 5
$AlCTa_2$, 28, 42
Al_2CTa_3, 28, 42
Al_2CTa_3V, 30A, 186
Al_2CTa_3Zn, 30A, 186
$AlCTi_2$, 28, 42
$AlCTi_3$, 29, 97
$AlCV_2$, 28, 42
Al-C-Y, 30A, 102
$AlCY_3$, 30A, 102
$Al_{59}Ca_{7.6}Ce_{12}Na_{7.8}O_{384}Si_{133}.270H_2O$, 32A, 484
$Al_4Ca_{1.32}Cl_{12}H_{22}Mg_{7.68}Na_4O_{22}$, 33A, 277
$Al_2Ca_4Cl_2O_6.\ 10H_2O$, 31A, 253
$Al_4Ca_8Cl_2O_{16}S.\ 24H_2O$, 31A, 253
$Al_{20}CaCr_2$, 33A, 151
$Al_2Ca_3D_{12}O_{12}$, 33A, 273
$Al_2Ca(F,HO)_8$, 34A, 215
$Al_2CaF_4H_4O_4$, 35A, 451
$Al_2Ca_3F_{10}H_2O_2.H_2O$, 31A, 255
$AlCaF_6Na.H_2O$, 32A, 164
$Al_6Ca_4F_2O_{12}$, 27, 519
$Al_{1.94}Ca_2Fe_{0.06}H_2O_{12}Si_3$, 32A, 468
$Al_{0.67}Co_2Fe_{1.33}O_5$, 27, 547
$Al_{1.38}Ca_2Fe_{0.62}O_5$, 27, 547
$Al_{2x}Ca_2Fe_{2-2x}O_5$, 27, 547
$(Al,Fe,Si)_{15}(Ca,K,Na)(Fe,Mg)_2O_{30}.H_2O$, 28, 322
$Al_2Ca_3Ge_3O_{12}$, 26, 415
$Al_4(Ca,Sr)H_4(Li,Na)_2O_{20}P_4$, 30A, 454
$Al_{3.67}Ca_{1.10}H_2Mg_{2.18}O_{12}Si_{1.05}$, 31A, 238
$Al_{0.5}Ca_3H_4Mn_{1.5}O_{12}Si_2$, 34A, 387
$AlCa_2HMn_2O_{13}Si_3$, 28, 322
$Al_2Ca_3H_{12}O_{12}$, 29, 317; 32A, 426
$Al_3Ca_2HO_7.H_2O$, 35A, 291
$Al_7(Ca, Pb, Sr)_2H_{16}O_{28}P_3.3H_2O$, 28, 354
$Al_2Ca_6H_{12}O_{24}S_3.26H_2O$, 35A, 378
$Al_2Ca_2H_2O_{12}Si_3$, 30A, 434
$Al_2Ca_3H_2O_{10}Si_2$, 26, 521
$Al_2Ca_3H_{3.36}O_{12}Si_{2.16}$, 32A, 426
$Al_3Ca_2HO_{13}Si_3$, 33A, 475
$Al_{5.2}Ca_{1.6}K_{0.7}Mg_{0.1}Na_{1.1}O_{48}Si_{18.8}.12.7H_2O$, 28, 252
$Al_9(Ca,K_2,Mg,Na_2)_{4.5}O_{72}Si_{27}.27H_2O$, 34A, 375
$Al_6Ca_2KNa_5O_{30}S_{1.5}Si_6$, 33A, 483
$Al_{6.29}Ca_{2.59}K_{0.10}Na_{1.06}O_{48}Si_{17.71}.15.74H_2O$, 32A, 488
$Al_8(Ca,K,Na)_8O_{32}Si_8$, 35A, 454
$Al_{0.5}CaMo_6Si_{1.5}$, 32A, 455
$Al_{9.9}Ca_{3.4}Mg_{0.2}Na_{2.7}O_{72.0}Si_{26.4}.24.5H_2O$, 35A, 456
$Al_4Ca_3MgO_{10}$, 29, 422

$Al_{3.4}CaMg_{1.5}O_{9.6}Si.H_2O$, 26, 521
$Al_6Ca_2Na_6O_{32}S_2Si_6$, 30A, 431
$AlCaNaO_7Si_2$, 30A, 455; 35A, 455
$Al_{1.8}Ca_{0.8}Na_{0.2}O_8Si_{2.2}$, 31A, 222
$Al_2(Ca,Na_2)O_{12}Si_4.6H_2O$, 31A, 227
$Al_2CaNa_2O_{18}Si_7.6H_2O$, 32A, 511; 33A, 484
$Al_{10}Ca_4Na_2O_{72}Si_{26}.32H_2O$, 31A, 235
$Al_{12}Ca_4Na_4O_{48}Si_{12}$, 32A, 479
$Al_{12}Ca_4Na_4O_{48}Si_{12}.5.65I_2$, 32A, 479
$(Al,Fe)_2Ca_2O_4$, 29, 418
$Al_2Ca_3O_6.6H_2O$, 26, 519; 28, 151; 29, 317; 34A, 251
$Al_2Ca_4O_7.13H_2O$, 28, 320
$Al_2Ca_4O_7.19H_2O$, 26, 369
α-$Al_2Ca_4O_7.19H_2O$, 26, 369
α_1-$Al_2Ca_4O_7.19H_2O$, 26, 369
$(Al,Si)_3Ca_2O_7$, 32A, 439
Al_4CaO_7, 35A, 290
$Al_6Ca_4O_{13}$, 35A, 290
$Al_6Ca_4O_{13}.3H_2O$, 26, 369, 518
$Al_{12}CaO_{19}$, 33A, 274
$Al_{14}Ca_{12}O_{33}$, 29, 418
$Al_{14}Ca_{12}O_{33}.H_2O$, 34A, 251
$Al_2Ca_4O_{10}S.12H_2O$, 30A, 458
$Al_2Ca_6O_{18}S_3.3H_2O$, 31A, 247
$Al_{12}Ca_8O_{32}S_2$, 30A, 431
Al_2CaO_6Si, 31A, 243
$Al_2CaO_8Si_2$, 26, 518; 27, 679
$Al_2CaO_8Si_2.4H_2O$, 28, 279
$Al_2CaO_8Si_2.2H_2O$, 26, 519, 520
$Al_2CaO_{10}Si_3.5H_2O$, 26, 519
$Al_2CaO_{12}Si_4.2H_2O$, 26, 518
$Al_2CaO_{12}Si_4.<4H_2O$, 32A, 483
$Al_2CaO_{12}Si_4.nH_2O$, 32A, 482
$Al_2CaO_{16}Si_6.5H_2O$, 26, 519
$Al_2CaO_{24}Si_{10}.7H_2O$, 26, 519
$Al_2Ca_2O_7Si$, 26, 521
$Al_2Ca_3O_8Si_2$, 31A, 228
$Al_2Ca_3O_{12}Si_3$, 26, 519
$Al_{3.9}Ca_{1.95}O_{24}Si_{8.1}.13(?).H_2O$, 28, 250
$Al_{3.9}Ca_2O_{24}Si_{8.1}$, 27, 691
$Al_4Ca_2O_{24}Si_8.12H_2O$, 31A, 242
$Al_4Ca_2O_{24}Si_8.?13H_2O$, 29, 401
$Al_4Ca_2O_{32}Si_{12}.8H_2O$, 34A, 374
$Al_{0.5}CaO_{2.75}Ti_{0.5}$, 29, 419
$Al_{12}Ca_8O_{32}W_2$, 30A, 431
Al_2CaSi_2, 32A, 5
$Al_2Cd_3Ge_3O_{12}$, 26, 415
$AlCdInO_4$, 30A, 451
Al_2CdO_4, 31A, 247
Al_4CdO_7, 31A, 247
$Al_{12}CdO_{19}$, 31A, 247
Al_2CdS_4, 33A, 151
Al_2CdSe_4, 33A, 151
AlCe, 26, 174; 29, 97; 30A, 102; 32A, 9
Al_2Ce, 30A, 186
$(Al,Si)_2Ce$, 28, 42
Al_3Ce, 30A, 3, 102
$(Al,Zn)_4Ce$, 29, 101; 30A, 102
$Al_{11}Ce_3$, 32A, 7
$Al_2Ce_2Co_{15}$, 27, 7
$Al_{20}CeCr_2$, 33A, 151
Al-Ce-Cu, 30A, 102
Al_3CeCu, 29, 97
$Al_{3.25}CeCu_{0.75}$, 29, 97
Al_4CeCu, 29, 97
Al_8CeCu_4, 26, 175; 27, 5
$Al_{9.7}Ce_2Cu_{7.3}$, 27, 6
$Al_{10}Ce_3Cu_7$, 27, 6
$Al_{10.5}Ce_2Cu_{6.5}$, 27, 6
$Al_{24}Ce_3Cu_8Mn$, 32A, 139
Al-Ce-Fe, 27, 6
Al_8CeFe_4, 26, 175; 27, 5
$Al_{60}Ce_{10}Fe_{30}$, 34A, 138
$Al_{1.5}CeGa_{2.5}$, 29, 97
$Al_{2.0}CeGa_{2.0}$, 29, 97
Al-Ce-Mn, 28, 42
Al_8CeMn_4, 26, 175; 27, 5
$Al_{10}Ce_2Mn_7$, 27, 6
Al-Ce-Ni, 27, 8
AlCeNi, 35A, 74
$Al_{3.33}Ce_6S_{14}$, 33A, 4
Al—Ce—Th, 26, 5
$Al_8Ce_2Th_{14}$, 26, 5
$Al_8Ce_3Th_{13}$, 26, 5
$Al_8Ce_4Th_{12}$, 26, 5
$AlCl_3.6H_2O$, 33A, 202
Al_2Cl_8Co, 27, 459
$Al_6Cl_{0.4}H_{2.7}Na_{7.1}O_{24.7}Si_{5.5}$, 32A, 490
$Al_6Cl_{0.67}(HO,O)_{25}Na_{10}O_{1.33}Si_6$, 30A, 312
$(Al,Si)_{12}ClNa_4O_{24}$, 30A, 436
$Al_3ClNa_4O_{24}Si_9$, 27, 688
$Al_6Cl_2Na_8O_{24}Si_6$, 32A, 490
AlClO, 30A, 300
$AlCl_6Ti$, 29, 415
Al_5Co_2, 26, 6
$Al_{13}Co_4$, 27, 8
Al-Co-Cu-Fe-Ni, 30A, 186
Al-Co-Cu-Fe-Ni-Ti, 30A, 187
AlCoDy, 27, 264
Al-Co-Fe-Ni, 30A, 187
$Al_8Co_2H_4O_{24}Si_4$, 30A, 459
$AlCo_2Hf$, 29, 97
$AlCoInO_4$, 30A, 451
Al-Co-Mn, 27, 10
$AlCo_2Mn$, 26, 174
$AlCo_2Nb$, 29, 97
Al-Co-Ni, 30A, 103
$Al_2Co_3O_{12}Si_3$, 27, 515
$Al_{5-x}Co_xO_{12-x}Y_3$, 28, 352
AlCoSc, 30A, 103
$AlCo_2Si_2$, 27, 11
$Al_3Co_3Si_4$, 27, 11
$AlCo_2Ta$, 29, 97
$AlCo_2Ti$, 27, 12
$AlCo_2U_2$, 33A, 151
$AlCo_3U_2$, 29, 97
AlCoV, 30A, 103
AlCoW, 30A, 103
$AlCo_2Zr$, 29, 97
Al_2CoZr_6, 35A, 7
$Al_{16}Co_7Zr_6$, 34A, 138
$AlCr_2$, 30A, 103
$(Al_xSi_{1-x})_2Cr$, 26, 22
$(Al_xSi_{1-x})_3Cr_5$, 26, 22
AlCrCu, 30A, 103
$AlCrGa_3O_{12}Y_3$, 27, 513
$Al_2CrGa_2O_{12}Y_3$, 27, 513
$Al_{2.5-0.5}Cr_2Ga_{0.5-2.5}O_{12}Y_3$, 27, 513
$Al_3CrGaO_{12}Y_3$, 27, 513
AlCrHf, 29, 97
$(Al,Cr)_2Mg_3O_{12}Si_3$, 35A, 258
Al—Cr—Mo—Ni—Ti—W, 26, 213
Al—Cr—Mo—Si, 26, 22
$AlCr_xMo_{1-x}Si$, 26, 23
AlCrNb, 29, 97
Al—Cr—Ni, 26, 213
AlCrNi, 29, 97; 30A, 103
Al—Cr—Ni—Ti, 26, 213
Al—Cr—Ni—Ti—W, 26, 213
Al-Cr-O, 27, 500
$Al_{1.92}Cr_{0.08}O_3$, 29, 299
$(Al,Cr)_2O_3$, 30A, 320
$Al_3Cr_2O_{12}Y_3$, 27, 513
$Al_4CrO_{12}Y_3$, 27, 513
Al_2CrS_4, 26, 117
Al—Cr—Si, 26, 21
$Al_xCr_3Si_{1-x}$, 26, 22
$AlCrTi_2$, 29, 97
$AlCsGeO_4$, 31A, 249

$Al_{x+y}Cs_xNa_yO_{96}Si_{48-x-y}.(16-x)H_2O$, 34A, 370
$AlCs_{0.7}Na_{0.3}O_6Si_2.0.3H_2O$, 32A, 480
$AlCsO_2$, 29, 318
$AlCsO_8S_2.12H_2O$, 31A, 202
$Al_2Cs_2O_{12}Si_4.xH_2O$, 30A, 455
$Al_xCs_xO_4Ti_{2-x}$, 33A, 304
$AlCs_5O_{40}W_{12}.nH_2O$, 27, 672
Al—Cu, 26, 6; 27, 185; 28, 6; 30A, 187
AlCu, 27, 12
β_1'-$AlCu_3$, 29, 97
$AlCu_4$, 28, 42
μ-Al_4Cu, 29, 97
Al_4Cu_9, 30A, 3, 103; 33A, 72
α-(Al—Cu—Fe), 26, 7
$Al_6CuH_8O_{24}P_4.4H_2O$, 30A, 395
$(Al,Cu)_3H_6O_{14}PbS_2$, 28, 326
$AlCu_2Hf$, 29, 97
$AlCu_2Hf_3$, 29, 97
$Al_{10}Cu_2La$, 34A, 138
Al_5CuLi_3, 28, 42
Al-Cu-Mg, 27, 12
Al-Cu-Mn, 28, 42
K-AlCuMn, 26, 8
$AlCu_2Mn$, 27, 15
$AlCu_3Mn_2$, 29, 97
$Al_{40}Cu_{20}Mn_{40}$, 27, 15
$Al_{50}(Cu,Mn)_{50}$, 27, 15
Al_2CuMo, 30A, 103
Al_5CuMo_2, 29, 97
AlCuNb, 29, 98
$AlCu_2Nb$, 30A, 103
$Al_{40}Cu_{20}Nb_{40}$, 29, 98
β_2-Al-Cu-Ni, 29, 98
γ'-Al-Cu-Ni, 29, 98
$Al(Cu,Ni)_3$, 31A, 39
Al_2CuO_4, 26, 396
$Al_2Cu_2O_4$, 26, 396
$AsCu_3S_4$, 35A, 19
AlCuSc, 30A, 103
$AlCu_2Sc$, 30A, 103
$(Al_{0.5}Cu_{0.5})Ta$, 29, 98
AlCuTa, 29, 98
$AlCu_2Ta$, 30A, 103
$(Al,Cu)_2Ta$, 29, 98
AlCuTi, 29, 98; 30A, 103
$AlCu_2Ti$, 27, 296; 30A, 103
$(Al, Cu)_2Ti$, 29, 98
~$Al_3Cu_9Ti_4$, 29, 98
Al_5CuTi_2, 29, 98; 30A, 103
$Al_{11}CuTi_4$, 30A, 103
$Al_{23}CuTi_8$, 30A, 103
$Al_{71}Cu_4Ti_{25}$, 29, 98
$Al_{73}Cu_2Ti_{25}$, 29, 98
$Al_3(Cu,V)$, 30A, 104
$Al_4(Cu,V)_9$, 30A, 104
$Al_8(Cu,V)_5$, 30A, 104
AlCuZr, 29, 98
$AlCu_2Zr$, 27, 296; 29, 98; 30A, 104
$AlCu_2Zr_3$, 29, 98
Al_2CuZr, 30A, 104
Al_5CuZr_2, 29, 98
AlD_3, 34A, 153, 155
$AlD_4NO_8S_2.12D_2O$, 32A, 339
AlD_2Th_2, 26, 27
AlD_3Th_2, 26, 27
AlD_4Th_2, 26, 27
AlDy, 26, 237; 30A, 104; 32A, 8
Al_2Dy, 26, 236; 30A, 104
$(Al,Ni)_2Dy$, 27, 264
Al_2Dy_3, 29, 5, 97; 30A, 104
Al_3Dy, 29, 6, 97
AlDyFe, 27, 264
$AlDyO_3$, 28, 136; 29, 422
$Al_5Dy_3O_{12}$, 29, 423
$Al_{10}Dy_6O_{24}$, 29, 421
AlEr, 30A, 104
$AlEr_2$, 30A, 104
Al_2Er, 26, 174; 28, 42; 30A, 104
Al_2Er_3, 30A, 104
Al_3Er, 28, 42; 29, 98; 30A, 104
$AlErO_3$, 28, 136
$Al_5Er_3O_{12}$, 29, 423
Al_2Eu, 30A, 105
Al_4Eu, 29, 98; 30A, 104
$AlEuO_3$, 28, 136
$Al_2Eu_4O_9$, 34A, 253
AlF_3, 27, 438
$(Al,Fe)(F,HO)_2(Fe, Mg)_4Na_3O_{30}Si_8$, 28, 322
$Al_{0.2}F_3Ga_{0.8}.3H_2O$, 26, 300
$Al_{0.5}F_3Ga_{0.5}.3H_2O$, 26 300
$Al_{0.6}F_3Ga_{0.4}.3H_2O$, 26, 300
$(Al_{1-x}, Ga_x)F_6H_{12}N_3$, 29, 415
$AlF_3.H_2O$, 29, 253
β-$AlF_3.H_2O$, 26, 300
α-$AlF_3.3H_2O$, 26, 300
AlF_3H_2O, 29, 253
$Al_2F_2HO_5P.7H_2O$, 31A, 194
$Al_2(F, HO)_2O_4Si$, 27, 725
$AlF_4HOSr.H_2O$, 31A, 252
$Al_2F_8H_2O_2Sr_2.2H_2O$, 32A, 166
$Al_4(F, HO)O_{13}Ta_3$, 27, 540
$(Al,Si)_4F_2K(Li,Mg)_3O_{10}$, 31A, 231
AlF_4Li, 30A, 449
AlF_6Li_2, 30A, 449
AlF_6Li_3, 31A, 258
$Al_{1.12}(F,HO)_6Mg_{0.88}Na_{0.88}.H_2O$, 30A, 451
AlF_4Na, 34A, 192
Al—Fe, 26, 11; 30A, 187
AlFe, 29, 98
α'-AlFe, 26, 11
$AlFe_3$, 26, 15; 27, 19
α"-$AlFe_3$, 26, 11
Al_6Fe, 27, 18; 30A, 4, 105
$Al_3Fe_2Gd_3O_{12}$, 26, 414
$AlFe_{13}H_{14}KO_{56}Si_{17}$, 34A, 392
$Al_2(Fe,Mg)_3H_2Li_2O_{24}Si_8$, 34A, 390
$Al_{2.6}Fe_{3.4}H_8Mg_{1.3}O_{18}Si_{2.7}$, 30A, 428
$Al_2Fe(HO, O)O_8P_2.8H_2O$, 27, 603
$Al_2FeH_{14}O_{16}P_2.2H_2O$, 34A, 332
$Al_2FeH_{18}O_{18}P_2$, 32A, 367
$Al_{1.84}Fe^{3+}_{0.15}H_4O_9Si_2$, 26, 529
$Al_9Fe_2HO_{24}Si_4$, 31A, 234
$Al_{18}Fe_4H_4O_{48}Si_8$, 33A, 486
Al—Fe—Li—O, 26, 395
$Al_xFe_{2.5-x}Li_{0.5}O_4$, 28, 285
$(Al,Fe)_4(Fe,Mg,Mn)_{14}O_{29}Sb_2Si_2$, 31A, 223
$Al_4(Fe,Mg,Mn)_2O_{18}Si_5.nH_2O$, 31A, 223
$Al_{15.94}Fe_{1.98}Mg_{0.07}O_{32}Si_{0.13}Zn_{5.76}$, 28, 354
$Al_{15.95}Fe_{2.56}Mg_{0.14}O_{32}Si_{0.11}Zn_{5.14}$, 28, 352
$Al_xFe_{2-x}MnO_4$, 30A, 451
$AlFe_2Ni$, 30A, 188
$AlFeO_3$, 30A, 451; 31A, 244
$Al_{0.5}Fe_{1.5}O_{12}S_3.9H_2O$, 35A, 437
$Al_4Fe_2O_8Si_5$, 27, 694
$(Al,Fe)_2O_9W_2.3H_2O$, 26, 394
$Al_3Fe_2O_{12}Y_3$, 26, 414
Al_3FeSi_2, 30A, 105
AlFeTa, 30A, 105
$(Al,Fe)_2U$, 27, 20; 29, 149
Al_7Fe_5Zr, 34A, 138
Al-Ga-Li, 29, 147
Al—Ga—Li—O—Si, 26, 523
$Al_{0.5}Ga_{0.5}LiO_6Si_2$, 26, 523
$AlGaO_3$, 31A, 244
$AlGa_4O_{12}Y_3$, 33A, 282
$Al_2Ga_3O_{12}Y_3$, 33A, 282
$Al_3Ga_2O_{12}Y_3$, 33A, 282
$Al_4GaO_{12}Y_3$, 33A, 282

AlGd, 26, 237; 29, 98; 30A, 105
Al_2Gd, 26, 236; 30A, 105
Al_2Gd_3, 29, 5, 97; 30A, 105
Al_3Gd, 29, 98
$AlGdO_3$, 28, 136
$AlGeKO_4$, 31A, 249
$AlGeLiO_4$, 28, 346; 30A, 459
$Al_5Ge_9Li_5O_{36}Zn_8$, 28, 348
AlGeMn, 26, 9; 34A, 9
$Al_2Ge_3Mn_3O_{12}$, 26, 415
$AlGeNaO_4$, 32A, 492
Al_3GeNb, 35A, 116
$Al_{10}GeO_{20}Pb_3$, 35A, 263
$AlGeO_4Rb$, 31A, 249
AlGeSr, 34A, 138
AlH_3, 34A, 153
$Al_5H_{20}(K, Na)_5O_{45}Si_{11}$, 27, 692
$Al_2H_6K_2O_7$, 31A, 133
$Al_3H_6KO_{14}S_2$, 27, 634; 30A, 376
$AlH_2KO_9Si_3$, 29, 428
$Al_3H_2KO_{12}Si_3$, 28, 240; 32A, 469; 33A, 477
AlH_4Li, 32A, 149
AlH_8LiN_4, 31A, 247
$Al_4H_2LiNaH_{12}Si_2$, 32A, 477
$Al_2H_4Mg_2O_9Si$, 27, 690
$(Al,Si)_4H_2Mg_{3 \cdot 4}O_{12} \cdot 3.72H_2O$, 31A, 237
$(Al,Si)_{11}HMg_3O_{22}$, 33A, 492
$AlH_4NO_8S_2.12H_2O$, 32A, 339
$AlH_4NO_8Si_3.\frac{1}{2}H_2O$, 29, 429
$AlH_5NO_{8 \cdot 5}Si_3$, 29, 429
AlH_4Na, 31A, 246
AlH_6Na_3, 31A, 246
$Al_3H_6NaO_{14}S_2$, 27, 634
$AlHNaO_{4 \cdot 5}Si$, 29, 430
$Al_2H_4Na_2O_{12}Si_3$, 29, 409
$(Al,Fe)HO_2$, 28, 283
AlH_3O_3, 31A, 253; 32A, 243; 33A, 260
$Al_{10}H_2O_{16}$, 29, 299
$Al_2H_3O_7P$, 33A, 403
$Al_3H_3O_{11}P_2.5H_2O$, 28, 324; 33A, 404
$Al_3H_3O_{11}P_2.6\ H_2O$, 31A, 189
$Al_3H_6O_{14}PPbS$, 28, 326
$Al_4H_{32}O_{32}P_2V_2$, 27, 601
$Al_2H_{18}O_{18}S_2.2H_2O$, 27, 625
$Al_2H_{22}O_{20}S_2$, 27, 625
$Al_2H_{18}O_{18}Se_2.2H_2O$, 27, 625
$Al_2H_{22}O_{20}Se_2$, 27, 625
$Al_2H_4O_9Si_2$, 26, 504; 28, 352, 34A, 368
$Al_4H_8O_{18}Si_4$, 26, 505
$Al_{59}H_{59}O_{384}Si_{133}$, 34A, 370
$AlHO_{13}U_2V_2.8H_2O$, 28, 350
$Al_4H_8Th_8$, 28, 6
$Al_4H_xTh_8$, 28, 6
Al-Hf, 27, 16
$(Al_{0.35}Si_{0.65})Hf$, 27, 18
$(Al_{\sim 0.5}Si_{\sim 0.5})Hf$, 26, 11
AlHf, 26, 9; 27, 16, 378
$AlHf_2$, 26, 10
$(Al,Si)Hf_2$, 26, 10; 27, 18
Al_2Hf, 26, 237; 27, 16
Al_2Hf_3, 27, 16
Al_3Hf, 26, 165; 27, 16, 18
Al_3Hf_2, 27, 16
Al_3Hf_4, 27, 16
Al_3Hf_5, 27, 18
$(Al_{0.5}Si_{0.5})_3Hf_5$, 27, 18
$(Al_{0.85}Si_{0.15})_3Hf_5$, 27, 18
$(Al,Si)_3Hf_5$, 27, 18
AlHfMo, 30A, 105
$Al_{0.1}HfNi_{1.9}$, 34A, 138
$AlHfNi_2$, 29, 99
$Al_{1.4}HfNi_{0.6}$, 34A, 138
Al—Hf—Si, 26, 10, 11
Al_2HfZn, 30A, 105
$Al_{15-50}Hf_{25}Zn_{60-25}$, 29, 99
$Al_{12}HfO_{19}$, 31A, 247
Al_2HgS_4, 33A, 151
Al_2HgSe_4, 33A, 151
AlHo, 30A, 105
Al_2Ho, 29, 99; 30A, 105
Al_2Ho_3, 30A, 105
Al_3Ho, 29, 99; 30A, 105; 31A, 12
AlHoNi, 33A, 13
$AlHoO_3$, 28, 136
$Al_5Ho_3O_{12}$, 29, 423
AlIO, 28, 302
$AlI_3O_{12}.12H_2O$, 26, 500
$AlInMgO_4$, 30A, 451
$AlInMnO_4$, 30A, 451
$AlInNiO_4$, 30A, 451
$AlInO_4Zn$, 30A, 451
$AlKMo_2O_8$, 35A, 254
$Al_5(K, Na)_5O_{35}Si_{11}.10H_2O$, 27, 692
$Al_9K_6Na_3O_{72}Si_{27}.21H_2O$, 34A, 371
$Al(K,Na)O_6Si_2$, 33A, 458
$Al_{10.67}K_2O_{17}$, 33A, 277
$Al_{10.83}K_{1.50}O_{17}$, 33A, 276
$AlKO_7P_2$, 29, 425
$AlKO_8S_2.12H_2O$, 32A, 339
$AlKO_4Si$, 30A, 425
$AlKO_8Si_3$, 26, 524; 27, 678; 28, 242, 243; 29, 407; 31A, 243; 32A, 478; 33A, 478
$AlKO_8Si_3.H_2O$, 29, 428
$Al_{57}K_{57}O_{384}Si_{135}$, 32A, 484
AlLa, 30A, 105, 106
$AlLa_3$, 30A, 106
Al_2La, 30A, 105
Al_3La, 30A, 105
Al_4La, 30A, 105; 32A, 8
$Al_{11}La_3$, 32A, 6
$Al_{55}La_{16.3}Na_{13.4}O_{388}Si_{137}.270H_2O$, 32A, 484
$AlLaO_3$, 29, 326, 30A, 452; 31A, 132
Al-La-O-Pb-Ti, 27, 505
AlLi, 29, 6
β-AlLi, 27, 21
$(Al,Zn)_2Li$, 28, 42
Al_4Li_9, 33A, 6
Al(Li, Zn)Mg, 29, 99
Al-Li-Mg-Zn, 27, 21; 29, 6
$Al_{2.5-x}Li_{0.5}Mn_xO$, 30A, 451
$AlLiMo_2O_8$, 35A, 247
$AlLiO_2$, 26, 368, 394; 30A, 313; 31A, 131
α-$AlLiO_2$, 27, 519
γ-$AlLiO_2$, 27, 519
$(Al,Ti)LiO_4$, 28, 344
Al_5LiO_8, 26, 394, 413; 27, 519; 28, 320
δ-Al_5LiO_8, 27, 519
$AlLiO_4Si$, 30A, 459; 35A, 452
$AlLiO_6Si_2$, 26, 523; 33A, 456, 457
$AlLiO_{10}Si_4$, 26, 506
$Al_2Li_2O_8Si_2$, 30A, 455
$Al_2Li_2O_{10}Si_3$, 35A, 453
$Al_2Li_2O_{12}Si_4$, 26, 515
$Al_2Li_2O_{20}Si_8$, 26, 515
$Al_5Ln_3O_{12}$, 30A, 452
Al-Mg, 27, 33; 29, 147, 148; 30A, 188
Al_3Mg_2, 30A, 5, 106
β-Al_3Mg_2, 27, 115
$Al_{30}Mg_{23}$, 33A, 8
$Al_{10}MgNa_2O_{17}$, 34A, 250
$Al_{5.5}Mg_2Na_{1.5}O_{72}Si_{30.5} \cdot 18H_2O$, 31A, 226
Al-Mg-O, 27, 545; 30A, 451
Al_2MgO_4, 26, 396, 413; 27, 487; 30A, 451; 33A, 272

$Al_{26}MgO_{30}$, 27, 519
$Al_2MgO_{12}Si_4$, 27, 686
$Al_2Mg_3O_4Si_3$, 26, 416
$Al_2Mg_3O_{12}Si_3$, 28, 246; 30A, 426
$Al_4Mg_2O_8Si_5$, 27, 694
$Al_4Mg_2O_{10}Si$, 26, 517
$Al_4Mg_2O_{18}Si_5$, 27, 724, 725
$Al_9Mg_{3.5}O_{20}Si_{1.50}$, 34A, 372
(Al, Mg)Zn, 27, 22
Al-Mn, 30A, 106
$Al_{0.89}Mn_{1.11}$, 28, 7
β-AlMn, 27, 23
γ-AlMn, 27, 23
Al_3Mn, 26, 16
Al_8Mn_5, 30A, 106
$Al_{10}Mn_3$, 29, 99
AlMnNi, 26, 8
$Al_4Mn_2O_8Si_5$, 27, 694
$AlMnPd_2$, 33A, 97
$Al_{0.75}MnSi_{1.25}$, 29, 99
$Al_{1.3}MnSi_{0.7}$, 29, 99
$Al_4Mn_3Si_2$, 29, 99
$Al_5Mn_2Si_5$, 30A, 106
$Al_9Mn_2Si_{1.8}$, 31A, 8
$Al_{27}Mn_{33}Si_{40}$, 29, 99; 30A, 106
$Al_{40}Mn_{33}Si_{27}$, 30A, 106
$Al_{20}Mn_{3.5}Si_{1.1}Zn_{1.4}$, 26, 16
$(Al, Mn)_2Th$, 29, 149
Al-Mn-U, 29, 148
$(Al, Mn)_2U$, 29, 148
$Al_{102}Mn_{26}Zn_{24}$, 26, 17
$(Al, P)_2Mo$, 28, 48
$(Al, Si)_2Mo$, 26, 19; 27, 26
Al_4Mo, 27, 25; 29, 7, 99
Al_8Mo_3, 27, 23, 377
$Al_{12}Mo$, 29, 9, 99
Al_5Mo_2Ni, 29, 99; 30A, 106
Al—Mo—Si, 26, 19, 21
$Al_{0.77}Mo_3Si_{0.23}$, 27, 26
AlMoSi, 27, 26
$Al_{1.3}MoSi_{1.7}$, 26, 19
AlN, 31A, 6
Al_2NNb_3, 29, 99
Al—N—Ni, 26, 219
$AlN_3O_9.9H_2O$, 30A, 402, 454
$AlNTi_2$, 28, 42
$AlNaO_2$, 26, 368; 33A, 280
β-$AlNaO_2$, 26, 368
γ-$AlNaO_2$, 26, 368
Al_5NaO_8, 27, 519
$Al_{10.83}Na_{1.50}O_{17}$, 33A, 276
$Al_{22}Na_2O_{34}$, 32A, 265; 33A, 275
$AlNaO_8S_2.6H_2O$, 31A, 243; 34A, 310
$AlNaO_8S_2.12H_2O$, 32A, 339
$Al_6Na_8O_{28}SSi_6$, 30A, 432
$Al_{26}Na_2O_{64}Se_8.73H_2O$, 28, 217
$AlNaO_4Si.0.5H_2O$, 29, 430
$AlNaO_6Si_2$, 28, 350; 31A, 228
$AlNaO_6Si_2.H_2O$, 28, 241; 30A, 424; 31A, 219
$AlNaO_8Si_3$, 27, 676; 29, 398; 34A, 368
$Al_2Na_2O_{10}Si_3$, 27, 690; 28, 354
$Al_2Na_2O_{10}Si_3.2H_2O$, 28, 244; 29, 409; 33A, 159
$Al_6Na_6O_{24}Si_6$, 28, 322
$Al_6Na_{7.14}O_{26.73}Si_{7.08}.4.05\ H_2O$, 35A, 458
$Al_8Na_8O_{96}Si_{40}.24H_2O$, 26, 515
$Al_9Na_9O_{48}Si_{15}.27H_2O$, 26, 516
$Al_{12}Na_{12}O_{48}Si_{12}$, 33A, 488
$Al_{30}Na_{30}O_{192}Si_{66}.98H_2O$, 30A, 438
$Al_{57}Na_{57}O_{384}Si_{135}$, 32A, 484
$Al_5NaO_{12}Ti_2$, 32A, 266
Al—Nb, 26, 19
$AlNb_2$, 26, 19
σ-$AlNb_2$, 26, 19
Al_3Nb, 26, 19
AlNbNi, 29, 99
$AlNbNi_2$, 29, 99; 30A, 106
$AlNbNi_3$, 27, 26
Al_2NbNi, 29, 99
Al_5Nb_2Ni, 29, 99
$Al_{37}Nb_{25}Ni_{38}$, 30A, 106
Al-Nb-O, 30A, 452
$AlNbO_4$, 27, 523; 28, 346
$AlNb_{11}O_{29}$, 31A, 255
$AlNbO_6Pb_2$, 28, 332
Al—Nb—Si, 26, 21
Al-Nb-Zr, 27, 3
AlNd, 30A, 106
Al_2Nd, 26, 236; 30A, 106
Al_4Nd, 26, 238
$Al_{11}Nd_3$, 32A, 7
Al—Ni, 26, 213
$(Al_{0.5}, Si_{0.5})Ni$, 27, 27
$AlNi_3$. 27, 26; 30A, 106
Al_3Ni, 27, 28
Al_3Ni_2, 27, 27
$(Al, Si)_7Ni_3$, 27, 27
$Al(Al,Ni)O_4$, 31A, 251
Al_2NiO_4, 26, 413; 28, 289
$Al_4Ni_2O_8$, 27, 519
Al_8NiO_{13}, 27, 519
$Al_{24}NiO_{37}$, 27, 519
AlNiSc, 30A, 106
Al-Ni-Si, 27, 27
AlNiTa, 29, 99
$AlNi_2Ta$, 29, 99; 30A, 107
Al_2NiTa, 29, 99
$Al_{37}Ni_{38}Ta_{25}$, 30A, 107
$AlNi_4Th$, 32A, 139
Al-Ni-Ti, 30A, 107
AlNiTi, 29, 99; 30A, 107
$AlNi_2Ti$, 27, 296
Al_2NiTi, 29, 99; 30A, 107
Al_3NiTi_2, 29, 99
$Al_{67}Ni_8Ti_{25}$, 29, 99
$Al_{71}Ni_4Ti_{25}$, 29, 99
AlNiZr, 29, 100; 30A, 107; 32A, 139; 35A, 74
$AlNi_2Zr$, 27, 296; 29, 100; 30A, 107
Al_5NiZr_2, 29, 100; 30A, 107
Al_2O_3, 26, 346, 518; 27, 545; 30A, 450
α-Al_2O_3, 26, 347; 27, 499
γ-Al_2O_3, 26, 347; 29, 417
δ-Al_2O_3, 26, 346
ε-Al_2O_3, 29, 417
κ-Al_2O_3, 26, 347
λ-Al_2O_3, 26, 347
θ-Al_2O_3, 26, 347
χ-Al_2O_3, 26, 347
$Al_2O_3.H_2O$, 26, 518
$Al_{10}O_{15}.H_2O$, 29, 299; 30A, 450
AlO_4P, 29, 366; 31A, 183, 245
$AlO_4P.2H_2O$, 31A, 185
$Al_4O_{16}P_2V_2.16H_2O$, 27, 601
$Al_{12}O_{19}Pb$, 30A, 452
$Al_{10}O_{20}Pb_3Si$, 35A, 264
AlO_6Pb_2Ta, 28, 332
AlO_3Pr, 29, 418
AlO_2Rb, 29, 318
$AlO_8RbS_2.12H_2O$, 32A, 339
AlO_8RbSi_3, 28, 350
$Al_{26}O_{57}S_6.79H_2O$, 28, 218
$Al_{12}O_{32}S_2Sr_8$, 30A, 431
Al-O-Si, 28, 236
Al_2O_5Si, 26, 502; 27, 707; 28, 237, 239, 352; 30A, 440; 31A, 225
$Al_{3.42}O_{7.13}Si$, 27, 708
$Al_{3.66}O_{7.49}Si$, 34A, 366
$Al_{3.84}O_{7.76}Si$, 28, 233
Al_4O_8Si, 27, 708
AlO_3Sm, 29, 418
Al_2O_4Sr, 28, 322
$Al_{12}O_{32}Sr_8W_2$, 30A, 431
AlO_3Tb, 28, 136
$Al_5O_{12}Tb_3$, 29, 423
Al_2O_6Te, 27, 526

$Al_5O_{12}Y_3$, 27, 513
Al-O-Zn, 30A, 451
Al_2O_4Zn, 27, 488; 29, 324
Al_2Os, 30A, 10, 171
Al_3Os_2, 30A, 10, 107
$Al_{13}Os_4$, 29, 10, 100
Al_3OsU_2, 29, 100
AlP, 28, 42, 296
$Al_{14}P_{14}Pt_{72}$, 35A, 116
Al-Pd, 29, 12, 100
AlPd, 29, 148
$AlPd_2$, 28, 7
AlPr, 29, 100; 30A, 107
$AlPr_3$, 29, 100
Al_2Pr, 30A, 107
$Al_{11}Pr_3$, 32A, 7
Al-Pt, 27, 321
AlPt, 28, 42
$AlPt_3$, 26, 225; 27, 321; 29, 100
Al_2Pt, 28, 42
Al_3Pt_2, 28, 42; 29, 12, 100
Al_3Pt_5, 29, 13, 100
Al_3Pt_{13}, 27, 321
$Al_{21}Pt_5$, 29, 100
$Al_{21}Pt_8$, 31A, 10
AlPu, 27, 28
δ-Al—Pu, 26, 231; 29, 149
$AlPu_3$, 27, 28
Al_2Pu, 27, 28
Al_3Pu, 30A, 11, 107
Al_4Pu, 30A, 107
Al_5Pu, 30A, 107
χ-Al—Re, 26, 23
AlRe, 27, 29
Al_xRe_y, 26, 23
$Al_5O_{12}Tm_3$, 29, 423
$Al_2O_{12}W_3$, 33A, 352
AlO_3Y, 28, 136
$Al_2O_9Y_4$, 27, 520
Al_2Re, 26, 23; 27, 29
Al_2Re_3, 26, 24
$Al_{3.02}Re$, 26, 24
Al_4Re, 28, 42
Al_5Re_{24}, 27, 29
Al_6Re, 27, 30; 32A, 11
$Al_{12}Re$, 26, 23; 27, 29, 30; 29, 9, 100
$Al_{36.5}Re_{63.5}$, 29, 100
Al_xRe_y, 27, 29
Al-Rh, 29, 149
Al_5Rh_2, 29, 100, 32A, 12
Al_9Rh_2, 33A, 15
Al_2Ru, 28, 42; 31A, 16
Al_6Ru, 33A, 16
$Al_{13}Ru_4$, 30A, 12, 108
Al_2S_3, 28, 8
Al_2S_4Zn, 32A, 13
AlSb, 28, 42
AlSc, 30A, 108
$AlSc_2$, 34A, 138
Al_2Sc, 29, 100
Al_3Sc, 29, 100
Al_2Se_3, 31A, 17
Al_2Se_4Zn, 33A, 151
Al—Si—Ta, 26, 25
$(Al, Si)_2Ta$, 26, 25
Al-Si-Ti, 30A, 108
$(Al, Si)_2Ti$, 26, 19
$(Al,Si)_3Ti$, 28, 108
$AlSi_7Ti_4$, 29, 100
$Al_5Si_{12}Ti_7$, 30A, 108
Al—Si—V, 26, 21
$Al_xSi_{1-x}V$, 30A, 188
Al—Si—W, 26, 25
$(Al_{0.3}Si_{0.7})_2W$, 26, 26
$(Al_{0.7}Si_{0.3})_2W$, 26, 26
$(Al, Si)_2W$, 26, 26
Al-Si-Zr, 26, 10, 11; 27, 3; 30A, 108
$(Al_{0.2}Si_{0.8})Zr$, 27, 32
AlSiZr, 30A, 108
$(Al, Si_3)Zr$, 26, 11
$AlSi_3Zr$, 30A, 108
$(Al, Si)Zr_2$, 26, 10; 27, 32
$(Al_{0.9}Si_{0.1})_3Zr$, 27, 32
$(Al,Si)_3Zr_5$, 27, 31
$(Al_{0.15-0.8}Si_{0.85-0.2})_3Zr_5$, 27, 32
$Al_4Si_5Zr_3$, 30A, 14, 108
AlSm, 30A, 109
$AlSm_2$, 30A, 109
Al_2Sm, 30A, 109
Al_3Sm, 29, 100; 30A, 109
Al_4Sm, 29, 100; 30A, 109
Al—Ta, 26, 25
σ-(Al—Ta), 26, 25
$AlTa_2$, 26, 25; 30A, 109
$AlTa_3$, 26, 25
$AlTa_4$, 26, 26
Al_3Ta, 26, 25; 30A, 109
$Al_{12}Ta_{17}$, 29, 100; 30A, 109
AlTb, 30A, 109
Al_2Tb, 30A,109
Al_2Tb_3, 30A, 109
Al_3Tb, 29, 100; 31A, 15
$AlTc_2$, 30A, 109
Al_3Tc_2, 28, 43
Al_4Tc, 28, 42
Al_6Tc, 27, 30; 32A, 11
$Al_{12}Tc$, 28, 42; 29, 9, 100,; 30A, 109
$AlTh_2$, 26, 27
Al_2Th, 26, 256; 29, 149
Al_7Th_2, 33A, 17
Al_8Th_{16}, 26, 5
AlTi, 30A, 109
$AlTi_2$, 27, 30
$AlTi_3$, 27, 30; 30A, 110
Al_2Ti, 27, 35, 378
δ-Al_2Ti, 26, 28
Al_3Ti, 30A, 109
γ-Al_3Ti, 26, 28
$Al_{11}Ti_5$, 29, 101; 30A, 109
$Al_{23}Ti_9$, 29, 100; 30A, 109
$Al_{0.05}Ti_{0.80}V_{0.15}$, 27, 35
Al-Ti-Zn, 30A, 110
$Al_{11}Ti_4Zn$, 29, 101; 30A, 110
Al-Ti-Zr, 30A, 188
Al_2Tm, 28, 43; 30A, 110
Al_3Tm, 28, 43
Al_2U, 26, 30; 27, 36; 29, 149
Al_3U, 30A, 188
Al_4U, 30A, 110
Al—U—Zr, 26, 30
AlV_3, 28, 43
AlVZr, 30A, 110
AlY, 28, 43; 32A, 16
AlY_2, 30A, 110; 32A, 16
AlY_3, 32A, 16
Al_2Y, 28, 43; 32A, 16
Al_2Y_3, 28, 43; 30A, 110; 32A, 16
Al_3Y, 29, 101; 31A, 14; 32A, 15
Al_4Y, 26, 240
Al_2Yb, 29, 101
Al_3Yb, 29, 101
Al-Zn, 27, 33; 29, 149
Al-Zn-Zr, 30A, 110
$Al_{25-55}Zn_{50}Zr_{25}$, 29, 101
Al-Zr, 27, 37
AlZr, 27, 36, 378
$AlZr_2$, 26, 10, 31; 27, 32, 37
$AlZr_3$, 27, 37
Al_2Zr, 26, 30; 27, 37
Al_2Zr_3, 27, 37
Al_3Zr, 27, 37
Al_3Zr_2, 26, 31; 27, 36
Al_3Zr_4, 27, 37
Al_3Zr_5, 27, 32, 37
Al_4Zr_5, 27, 37, 378
Am, 26, 32; 27, 39
Am_2C_3, 35A, 116
$AmCCsO_5$, 30A, 450
$AmCKO_5$, 29, 427
$AmCl_3$, 35A, 151
AmF_2O_2Rb, 30A, 450
AmF_6Rb_2, 27, 463
AmX_3 (X=halogen), 30A, 449
Ar, 26, 279; 27, 386; 30A, 449
As, 30A, 189; 31A, 239; 34A, 13
As_4, 31A, 239
AsB, 29, 150
AsB_6, 29, 150
As_2B_{13}, 26, 67
$As_4BC_{78}H_{62}IPt$, 26, 690
$AsBCa_2H_4O_8$, 26, 471
$AsBaKO_4$, 26, 470
$AsBaNaO_4$, 26, 469
$As_2BaNi_2O_8$, 34A, 338

$As_6Be_2Ca_3O_{20}Si_2(Sn,Ti)$, 34A, 381
$As_3Bi_8H_5O_{22}$, 28, 348
$AsBr_3$, 30A, 285; 32A, 206
$AsBrC_4H_{12}$, 28, 587
$AsBrC_{12}H_{10}$, 27, 867
$As_2Br_3Cd_4$, 28, 300
$AsBr_2Hg_2$, 30A, 449
$As_2Br_3Hg_4$, 30A, 449
AsC_3N_3, 26, 339; 28, 105
$AsCNb_2$, 33A, 152
$AsCV_3$, 32A, 24
$As_3Ca_5(Cl,F,OH)O_{12}$, 31A, 247
$AsCa_2ClO_4$, 35A, 193
$AsCaCuHO_5$, 28, 198
$AsCaHO_4$, 28, 324; 35A, 354
$AsCaHO_4.H_2O$, 27, 576; 31A, 194; 32A, 387
$AsCaHO_4.2H_2O$, 31A, 243; 34A, 355
$AsCaH_3O_5$, 27, 576
$AsCaKO_4$, 26, 470
$As_2(Ca, Mg)_3O_8.6H_2O$, 26, 470
$AsCaNaO_4$, 26, 469
$As_5CaO_{26}Th_5$, 29, 426
As_2Cd, 33A, 27; 35A, 18
As_2Cd_3, 29, 150; 33A, 25, 151; 34A, 15
α-As_2Cd_3, 29, 101
As_2CdGe, 28, 43; 29, 101; 30A, 110
$AsCd_4I_3$, 28, 302
As-Cd-In-Sn, 27, 47
$As_8CdIn_2Sn_3Zn_2$, 30A, 189
$As_8CdIn_4Sn_2Zn$, 30A, 189
$As_8Cd_2In_2Sn_3Zn$, 30A, 189
As-Cd-In-Te, 29, 152
$As_2Cd_3O_8$, 35A, 352
$As_5CdO_{26}Th_5$, 29, 426
$As_{13}Cd_{15}Pd_{72}$, 35A, 116
$As_{14}Cd_{14}Pt_{72}$, 35A, 116
As_2CdSi, 35A, 19
As_2CdSn, 27, 47
$As_2Cd_{1-x}SnZn$, 30A, 189
As-Cd-Zn, 29, 150
$As_2(Cd, Zn)_3$, 29, 150; 30A, 189
As_2Ce, 35A, 117
As_3Ce_4, 34A, 138; 35A, 94, 117
$AsClHg$, 30A, 449
$AsCl_2Hg_2$, 30A, 449
$As_2Cl_3Hg_4$, 30A, 449
$As_3ClO_{12}Pb_5$, 31A, 243
As_2Co, 27, 61; 31A, 21
$As_{2.950-2.960}Co$, 27, 49
$As_{3-x}Co$, 27, 48
As_3Co, 26, 47; 27, 48
$As_{3-y}(Co, Fe)$, 27, 48
$As(Co, Ni)S$, 28, 344
$As_2Co_3O_8$, 35A, 350
$As_3Co_8O_{16}$, 35A, 359
$AsCoS$, 27, 56; 28, 50; 30A, 17, 110
$AsCr$, 30A, 110
$AsCr_2$, 30A, 17, 110, 111; 33A, 151
$AsCr_3$, 30A, 110
$(As,P)Cr(Cu,Fe)HO_9Pb_2$, 32A, 391
$AsCrFe$, 30A, 111
$As(Cr_xFe_{1-x})_2$, 30A, 189
As-Cr-Fe-Ni, 29, 151
$AsCr_3N$, 33A, 28
$AsCs_3$, 26, 54
As-Cu, 27, 185
$AsCu_3$, 30A, 268
$As(Cu,Fe)_2$, 30A, 111
$As_3Cu_2Fe_4H_7O_{19}.6H_2O$, 29, 426
$As_3Cu_2Fe_4H_{19}O_{25}$, 29, 426
As_3CuGe_2, 26, 48
As—Cu—Ge—Se, 26, 48
$As_3Cu_5Ge_4Se_6$, 26, 48
$As_3Cu_9Ge_6Se_{12}$, 26, 48
$AsCu_2HO_5$, 27, 601
$AsCu_2HO_5.3H_2O$, 28, 196; 31A, 196
$AsCu_3H_3O_7$, 30A, 397
As-Cu-In-Te, 29, 151
$AlCu_2Mn$, 33A, 95
$AlCu_3Mn_2$, 33A, 96
$As_2Cu_3O_8$, 33A, 419
$As_2CuO_8Zn_2$, 32A, 389
$AsCuPbS_3$, 35A, 11
As_2CuPbS_5Tl, 33A, 148
$As_{0.9}Cu_{1.4}S_{2.1}$, 29, 24, 101
$AsCuS$, 29, 23, 101; 30A, 18, 111
$(As,Sb)Cu_3S_4$, 31A, 261
$As_4Cu_{12}S_{13}$, 31A, 164
$As_{0.644}Cu_3S_4Sb_{0.356}$, 32A, 25
$As_4Cu_6S_{12}Zn_3$, 31A, 261; 32A, 225
$AsCuSe_2$, 33A, 30
$AsCu_3Se_4$, 30A, 111
$AsDyO_4$, 28, 306
$AsErO_4$, 28, 306
$AsEuO_4$, 28, 304
$AsF_4H_2KO_2$, 32A, 173
AsF_5HKO, 32A, 172
$As_2FH_{36}Na_7O_{26}.H_2O$, 35A, 199
$As_2F_8H_8N_2O_2$, 28, 296
$As_2F_8K_2O_2$, 28, 296
AsF_6Li, 28, 298
AlF_6Li_3, 33A, 181
AsF_6Na, 28, 296
$As_2FNA_7O_8.19H_2O$, 35A, 199
$As_2F_8O_2Rb_2$, 28, 296
AsF_6Tl, 28, 298
$AsFe$, 27, 725; 34A, 16
$AsFe_2$, 30A, 111
As_2Fe, 33A, 63; 35A, 10
$As_2Fe_2H_2O_{10}Pb$, 27, 602; 28, 197
$As_3Fe_{1-x}Ni_x$, 27, 48
$AsFeO_4$, 33A, 418
$AsFeO_4.2H_2O$, 31A, 252
$As_3Fe_4O_{16}.5H_2O$, 32A, 384
As—Fe—S, 26, 50
$AsFeS$, 26, 52; 27, 56
$As_{1.1}FeS_{0.9}$, 26, 51
$AsFeSe$, 30A, 111
$AsFeTe$, 30A, 111
As-Ga, 29, 151
$AsGa$, 27, 50, 51, 381; 28, 43; 30A, 190
As_xGaP_{1-x}, 30A, 190
$As_{15}Ga_{10}Pd_{75}$, 35A, 117
As-Ga-S, 27, 51
As_xGaSb_{1-x}, 30A, 190
$AsGdO_4$, 28, 304
As-Ge, 29, 152
$AsGe$, 32A, 139; 35A, 20
As_2Ge, 27, 52
As-Ge-In-Sn, 29, 152
$AsGe_{x/2}In_{1-x}Zn_{x/2}$, 26, 49
As—GeSe, 26, 39
As—GeTe, 26, 39
AsH_2KO_4, 27, 724; 30A, 454
AsH_6KO_9U, 29, 377
$As_2H_3KO_{13}U_2.6H_2O$, 29, 377
$As_2H_{15}KO_{19}U_2$, 29, 377
$AsHMn_2O_5$, 33A, 421
$AsH_4Mn_2O_8$, 28, 348
$AsH_4Mn_3O_8$, 32A, 385
$As_2H_8Mn_7O_{16}$, 33A, 422
$As_{13}HgMn_9O_{60}.2H_2O$, 35A, 360
$AsH_4Mn_2O_8Y$, 32A, 385
AsH_6NO_4, 30A, 454
$AsH_9N_2O_4$, 35A, 353
$AsH_4NO_6U.3H_2O$, 29, 377
$AsH_{10}NO_9U$, 29, 377
$AsH_{12}N_3S_4$, 28, 153
$AsHNa_2O_4.7H_2O$, 35A, 335
$As_2H_6O_8.H_2O$, 33A, 418
$As_3H_5O_{10}$, 31A, 121
$AsHO_4Sr.H_2O$, 27, 576; 31A, 194
AsH_3O_5Sr, 27, 576
$AsHO_5Zn_2$, 31A, 242
$AsHO_5Zn_2.H_2O$, 28, 326, 354
$As_2H_2O_8Zr.H_2O$, 34A, 336
$AsHf$, 27, 55
As_2Hf, 27, 54
$As_2Hg_3I_4$, 30A, 449
$As_2Hg_4I_3$, 30A, 449
$As_{10}Hg_{20}Pd_{70}$, 35A, 117
$AsHo$, 26, 164
$AsHoO_4$, 28, 306
AsI_3, 30A, 287; 31A, 239

$As_2I_{1.65}S_3$, 27, 402
AsI_3Zn_3, 27, 467
AsIn, 26, 42; 27, 50, 246; 28, 43; 29, 101, 152
AsIn II, 29, 101
$As_{12}In_{16}Pd_{72}$, 35A, 117
$As_{14}In_{14}Pt_{72}$, 35A, 117
As-In-Se, 26, 42; 27, 247
As-In-Sn-Zn, 29, 152
$AsIn_{1-x}Sn_{x/2}Zn_{x/2}$, 26, 49
As-In-Te, 27, 246
As_2Ir, 26, 58; 27, 61
As_3Ir, 26, 48, 55, 176
$(As,S)_2(Ir,Pd,Pt,Rh)$, 30A, 111
As(Ir,Pt,Rh,Ru)S, 31A, 257
$As_5KO_{24}Pb_9Si$, 28, 324
$AsKO_4Sr$, 26, 470
$AsKO_6U.3H_2O$, 29, 377
As_2La, 35A, 117
$As_4Li_{4.6}Mn_{4.0}$, 33A, 31
$As_4Li_{6.45}Mn_{3.0}$, 33A, 32
$As_4Li_{6.4-5.0}Mn_{1.6-3.0}$, 26, 188
$As_4Li_{7-6.5}Mn_{1.0-1.5}$, 26, 189
$As_4Li_{7.33-7.20}Mn_{0.67-0.80}$, 26, 188
$As_4Li_{7.93}Mn_{2.2}$, 33A, 32
$As_4Li_{8.3}Mn_{2.0}$, 33A, 32
$As_4Li_{9.41}Mn_{1.4}$, 33A, 32
$AsLiMoO_6$, 35A, 356
$AsLiO_4Zn$, 26, 417
AsLu, 29, 101
$AsLuO_4$, 28, 306
As_2Mg_3, 29, 101
β-As_2Mg_3, 29, 152
As-Mg-Mn, 29, 152
$As_2Mg_{1.72-2.43}Mn_{1.37-0.61}$, 29, 152
$As_2Mg_2O_7$, 28, 195; 35A, 357
$As_2Mg_3O_8.8H_2O$, 31A, 253
$As_{14}Mg_{14}Pt_{72}$, 35A, 117
AsMn, 26, 53; 29, 25, 101
$AsMn_2$, 27, 57
$AsMn_7O_{12}Sb$, 35A, 443
$As_{0.9}MnP_{0.1}$, 33A, 33
$As_{0.92}MnP_{0.08}$, 35A, 21
$As_{0.3}Mn_2Sb_{0.7}$, 27, 58
AsMo, 29, 101; 31A, 24
$As_2Mo_{0.4}$, 30A, 21, 111
As_2Mo, 29, 19, 101; 30A, 21, 111
As_3Mo_2, 30A, 19, 111; 31A, 22
As_4Mo_5, 29, 101; 30A, 111; 31A, 24
$As(N,O)_2Th_2$, 34A, 135
AsN_2U_2, 34A, 135
$(As,P)NaO_3$, 32A, 383
$[(As,P)NaO_3]_x$, 28, 304
$(As,P)_3Na_3O_9$, 28, 304
$AsNaO_4Sr$, 26, 469
AsNb, 28, 9; 29, 19, 101
$AsNb_3$, 34A, 138
As_2Nb, 29, 19, 101; 30A, 14, 111
As_3Nb_5, 34A, 138
As_4Nb_7, 34A, 138
$As_2Nb_{18}O_{50}$, 29, 422
As_2Nd, 35A, 117
$AsNi_{1-x}$, 27, 58
AsNi, 27, 59
As_2Ni, 27, 58; 33A, 34, 63
$(As,S)_2Ni$, 27, 59; 33A, 359
As_2Ni_{5-x}, 27, 60
As_2Ni_5, 29, 101; 35A, 14
As_8Ni_{11}, 27, 58
$As_{0.77-1.77}NiS_{1.23-0.23}$, 27, 59
$As_{0.82}NiS_{1.18}$, 32A, 26
AsNiS, 27, 59, 724; 28, 50; 32A, 26
As_2O_3, 28, 320
$As_2O_8Pb_3$, 27, 587
$As_2O_{10}Pb_2U$, 30A, 454
AsO_4Sc, 28, 304
AsO_4Sn, 28, 304
$As_2O_{50}Ta_{18}$, 29, 422
AsO_4Tb, 28, 306
As_2O_7Th, 31A, 245
$As_4O_{12}Th$, 31A, 245
$AsO_4Tl.2H_2O$, 26, 464
AsO_4Tm, 28, 306
AsO_4Y, 28, 304
AsO_4Yb, 28, 306
$As_2O_9Zn_4.3H_2O$, 28, 354
As_2O_7Zr, 30A, 458
As_2Os, 26, 55; 28, 43; 33A, 63
AsOsP, 28, 43
AsOsS, 29, 101
AsOsSe, 29, 101
AsOsTe, 29, 101
AsPPt, 28, 43
AsPRu, 28, 43
As_2PbS_4, 26, 426
$As_2Pb_2S_5$, 31A, 261; 32A, 227
$As_4Pb_9S_{15}$, 28, 154; 34A, 23
$As_5(Pb,Tl)_2S_9$, 30A, 111, 355
$As_5Pb_3S_{10}$, 27, 403
$As_9Pb_5S_{18}$, 26, 428; 27, 402
$As_9Pb_7S_{20}$, 27, 402
$As_{10}PbS_{16}$, 26, 431
$As_{10}Pb_2S_{17}$, 26, 431
$As_{13}Pb_9S_{28}$, 26, 428; 27, 403
$As_{14}Pb_{26-28}S_{46-48}$, 31A, 164
$As_{50}Pb_{37}S_{112}$, 35A, 143
$AsPd_2$, 26, 55; 28, 43; 34A, 25
α-$AsPd_2$, 29, 102
β-$AsPd_2$, 29, 102
$AsPd_3$, 26, 55; 28, 43; 29, 102
$AsPd_5$, 29, 102
As_2Pd, 26, 55; 28, 43; 29, 101; 30A, 111
As_2Pd_5, 29, 101
AsPdSb, 28, 43
$AsPd_5Tl$, 35A, 22
As_2Pr, 35A, 117
As_3Pr_4, 34A, 138; 35A, 94, 117
As_2Pt, 26, 55; 30A, 111; 33A, 34
$As_{14}Pt_{72}Tl_{14}$, 35A, 117
$As_{14}Pt_{72}Zn_{14}$, 35A, 117
$AsRb_3$, 26, 54
As_7Re_3, 30A, 112; 34A, 27
AsRh, 26, 55; 28, 43
$AsRh_{1.6-1.3}$, 27, 61
AsRh1.7, 27, 61
$AsRh_2$, 26, 55
α-$AsRh_2$, 27, 61
β-$AsRh_2$, 27, 61
As_2Rh, 26, 55; 27, 61
As_3Rh, 26, 48, 55
AsRu, 26, 55
As_2Ru, 26, 55; 28, 43; 33A, 63
AsRuS, 29, 102
AsRuSb, 28, 43
AsRuSe, 29, 102
AsRuTe, 29, 102
As-S, 27, 61
As_2S_3, 26, 431; 27, 62, 402; 28, 293; 30A, 453
As_4S_3, 35A, 140
AsS_3Sb, 30A, 112
As_2S_4Sr, 28, 43
As-Sb, 30A, 191
$(As,Sb)_4Th_3$, 30A, 191
As_2Se_3, 28, 293, 302; 29, 102; 30A, 22, 112
AsSi, 30A, 22, 112
As_2Si, 32A, 140; 33A, 36
As_2SiZn, 29, 102
As_2Sn_3, 30A, 112
$As_3Sn_{3.8}$, 33A, 146
As_3Sn_4, 33A, 146
As_2SnZn, 27, 47; 28, 43; 29, 102; 30A, 112
AsTa, 29, 19, 102; 30A, 112
$AsTa_2$, 34A, 138
As_2Ta, 29, 19, 102; 30A, 112

As_4Ta_5, 34A, 138
AsTb, 26, 255; 27, 335
As_2Te_3, 28, 9, 43, 292
As_4Th_3, 30A, 112
AsTi, 29, 102
As_2Ti, 29, 27, 102
As_3Ti_5, 30A, 112
$As_4(Ti,W)_5$, 29, 102
AsTm, 29, 102
AsU, 29, 102
As_2U, 29, 102; 30A, 112
As_4U_3, 29, 102
As_2V, 29, 102; 30A, 112
As_2W, 29, 102; 30A, 112
As_3W_2, 30A, 19, 112; 31A, 22
AsY, 29, 102
AsYb, 29, 102
As_3Yb_4, 35A, 94, 117
As_3Yb_5, 35A, 93
As_2Zn_3, 29, 150
α-As_2Zn_3, 29, 103
Au, 26, 279; 27, 226, 386; 30A, 191
AuB_2, 26, 73
$AuC_4KN_4.H_2O$, 35A, 201
$(Au,Ga)_2C_xNb_3$, 30A, 112
Au_5Ca, 29, 103
$(Au, Ca, Mn)_2O_4Te_3$, 28, 350
Au-Cd, 27, 107
α_2-AuCd, 26, 83
Au_3Cd, 30A, 32, 112
α-Au_3Cd, 26, 83
Au-Cd-Sn, 27, 108
Au-Ce, 30A, 191
Au_3Ce, 34A, 138
$AuCl_4Na.2H_2O$, 30A, 444
$AuCl_4P$, 27, 471
Au-Co, 27, 160
Au-Cr, 27, 226
Au_4Cr, 27, 226
AuCs, 26, 56
AuCsO, 33A, 280
Au-Cu, 26, 134; 27, 185
AuCu, 27, 180; 33A, 75
AuCu I, 27, 179
AuCu II, 26, 135; 27, 179; 31A, 37
$AuCu_3$ I, 27, 177
$AuCu_3$ II, 27, 176
Au_3Cu, 26, 136
$AuCuPd_2$, 30A, 113
$AuCu_2Pd$, 31A, 38
Au_2CuPd, 30A, 112
Au-Cu-Zn, 29, 152
$AuCuZn_2$, 31A, 39
β'-$AuCuZn_2$, 29, 103
θ-$AuCuZn_2$, 29, 103
Au-D-Pd, 33A, 120
Au_2Dy, 34A, 81
Au_2Er, 30A, 113
AuF_3, 29, 414; 32A, 155
AuF_4K, 34A, 202; 35A, 162
AuF_4Na, 35A, 162
AuF_4Rb, 35A, 162
ζ-Au-Fe-Sn, 27, 311
$(Au,Ga)_2N_xNb_3$, 30A, 113
$(Au,Ga)_2N_xTa_3$, 30A, 113
$(Au,Ga)_2N_xV_3$, 30A, 113
$(Au,Ga)_2Nb_3$, 30A, 113
AuGd, 29, 103; 30A, 113
Au-Ge, 30A, 113, 192
Au-H-Pd, 30A, 192; 33A, 120
Au-Hg, 27, 225
AuHf, 27, 225, 378
$AuHf_2$, 27, 225, 234, 378
Au_2Hf, 27, 225
Au_3Hf, 27, 225
Au_4Hf, 27, 225
$Au_{10}Hf_7$, 27, 226
Au_2Hg, 35A, 66
Au_6Hg_5, 35A, 67
α-Au-In, 29, 154
α_1-Au-In, 29, 154
ε-Au-In, 29, 154
ζ-Au-In, 29, 154
ζ_1-Au-In, 29, 154
ψ-Au-In, 29, 154
$AuIn_2$, 29, 103
$AuInTe_2$, 27, 181
AuK, 26, 160
AuK_2, 26, 160
Au_2K, 26, 160
Au_4K, 26, 160
Au_5K, 29, 103
$AuKN_4O_{12}$, 35A, 326
$AuKO_2$, 35A, 216
Au-La, 30A, 192
Au_3La, 34A, 138
β'-AuLi, 26, 158
β_1'-AuLi, 26, 158
β_2'-AuLi, 26, 158
$AuLi_3$, 26, 158
δ_1-Au_3Li_5, 26, 158
Au_4Li_{15}, 26, 157
Au_5Li_4, 26, 158
α-Au_5Li_4, 26, 158
α_1-Au_5Li_4, 26, 158
$AuLi_5O_4$, 35A, 215
$Au_2Li_6O_6$, 35A, 214
Au*Ln*, 28, 43
Au-Mg, 30A, 113
$(Au_{0.9}Si_{0.1})Mg_2$, 33A, 151
$AuMg_3$, 33A, 151
$Au_{41}Mg_{13}$, 33A, 89
$Au_{74}Mg_{24}$, 30A, 58
$Au_{74}Mg_{26}$, 30A, 57
$Au_{77}Mg_{23}$, 33A, 89
Au-Mn, 27, 183, 226; 32A, 89
AuMn, 27, 220
$AuMn_2$, 27, 226; 35A, 66
$AuMn_3$, 30A, 59
Au_2Mn, 26, 159; 27, 227
Au_3Mn, 27, 219, 227; 29, 103; 31A, 42; 32A, 89
Au_4Mn, 27, 219; 32A, 89
Au_5Mn_2, 29, 60, 103
ζ-Au-Mn-Sn, 27, 310
$Au_2N_xNb_3$, 30A, 113
AuNa, 26, 161
$AuNa_2$, 26, 161
Au_2Na, 26, 161
$Au_xNb_3Rh_{1-x}$, 30A, 192
Au-Nd, 30A, 192
Au_3Nd, 34A, 138
Au—Ni, 26, 159
AuNi, 26, 159
ζ-Au-Ni-Sn, 27, 311
AuO_2Rb, 35A, 216
Au-Pd, 29, 154; 30A, 193
AuPd, 29, 103
Au_3Pr, 34A, 138
Au-Pt, 29, 155
AuPt, 26, 160
$AuPt_3$, 26, 160
AuRb, 26, 161
Au_2Rb, 26, 161
Au_4Rb, 26, 161
Au_5Rb, 29, 103
Au—Sb, 26, 42; 30A, 113
$AuSb_2$, 29, 103; 30A, 113
$AuSb_3$, 26, 42
AuSc, 27, 342
Au-Si, 30A, 114
Au-Sm, 30A, 193
Au—Sn, 26, 187
ζ-Au-Sn, 27, 310
AuSn, 28, 43
AuTa, 26, 161
$AuTa_2$, 26, 161
$AuTa_3$, 26, 161
Au_2Tb, 32A, 91; 33A, 91
Au_3Tb, 34A, 138
Au-Te, 27, 221
Au_2Th, 26, 256
Au-Ti, 27, 223
AuTi, 27, 224, 378
$AuTi_3$, 27, 224
α-$AuTi_3$, 29, 103
β-$AuTi_3$, 29, 103
Au_2Ti, 27, 223
Au_4Ti, 27, 223
Au_3Tm, 34A, 138
Au-V, 27, 226
α-AuV_3, 29, 103
β-AuV_3, 29, 103
Au_2V, 27, 226
Au_4V, 27, 226
Au-Y, 30A, 193
Au_2Y, 30A, 114
AuYb, 34A, 138
$AuYb_2$, 34A, 138
Au_2Yb, 34A, 138
Au_3Yb, 34A, 138
Au_3Yb_5, 34A, 138

Au_3Yb_7, 34A, 138
Au_4Yb, 34A, 138
Au_4Yb_5, 34A, 138
Au-Zn, 27, 191; 29, 155
Au_3Zn, 27, 229
Au_5Zn_3, 30A, 60, 114
Au-Zr, 27, 224
$AuZr_2$, 27, 224, 234, 378
$AuZr_3$, 27, 224
Au_2Zr, 27, 224
Au_3Zr, 27, 224
Au_4Zr, 27, 224
Au_4Zr_5, 27, 224
$Au_{10}Zr_7$, 27, 224

B, 27, 395; 28, 296; 30A, 251, 449; 35A, 127
α-B, 26, 67
B_6Ba, 26, 66
$BBaF_5$, 29, 414
B_2BaF_8, 29, 414
$B_2BaH_8O_8$, 28, 302; 30A, 415
$B_2BaH_8O_8.H_2O$, 33A, 445; 34A, 358
B_2BaO_4, 29, 393; 31A, 171
$B_2BaO_4.4H_2O$, 31A, 249
$B_2BaO_4.5H_2O$, 32A, 504
B_4BaO_7, 29, 392; 30A, 413
B_8BaO_{13}, 34A, 356
BBe_2, 26, 57; 27, 67
BBe_4, 27, 65
BBe_6, 26, 57
B_2Be, 26, 57
B_6Be, 26, 57
$BBe_2F_2KO_3$, 35A, 188
BBe_2HO_4, 28, 171; 30A, 454
$BBe_2HO_4.H_2O$, 32A, 50; 33A, 447
B—Be—Hf, 26, 64
B—Be—Zr, 26, 64
$B_{10}Br_2C_4H_{14}$, 29, 531; 31A, 77
$BBr_2C_{12}H_{10}P$, 26, 729
B_4C, 26, 68
B_8C_3, 33A, 151
$B_{12}C_3$, 26, 67
$B_4C_2Ca_4H_6MgO_{18}$, 33A, 440
$B_{10}C_2Cl_8H_4$, 29, 529; 31A, 77
BC_4Cr_7, 31A, 26
$(B,C)_2Cr_3$, 31A, 26
$B_2C_{10}H_7Mn_3O_{10}$, 31A, 259
B—C—Hf, 26, 68
B-C-Mo, 27, 127
α-$B_{0.1}C_{1.9}Mo_3$, 27, 127
$BCMo_2$, 28, 12; 34A, 36
B_3CNO_2, 27, 746
B_2C_2Sc, 30A, 28, 114
B—C—Th, 26, 68
BCTh, 26, 68
BC_2Th, 26, 68
BC_2Th_2, 26, 68
B_2CTh, 26, 68
B—C—Ti, 26, 70
B—C—U, 26, 70
BCU, 26, 70
B—C—Zr, 26, 68
B—Ca, 26, 75
B_4Ca, 26, 75
B_6Ca, 26, 66
$B_5Ca_3ClH_2Na_2O_{18}S_2$, 32A, 504
B_2CaF_8, 27, 459
$B_6CaH_6MgO_{14}.3H_2O$, 27, 551; 28, 161
$B_6CaH_{12}MgO_{17}$, 27, 551
$B_5CaH_4NaO_{11}.3H_2O$, 28, 172; 30A, 419
$B_2CaH_4O_6$, 33A, 446
$B_2CaH_6O_7.3H_2O$, 28, 302
$B_2CaH_8O_8$, 32A, 502
$B_2CaH_8O_8.2H_2O$, 28, 166; 29, 387
$B_2CaH_{12}O_{10}$, 29, 387
B_3CaHO_6, 27, 554; 35A, 286
$B_3CaH_3O_7.H_2O$, 31A, 172
$B_6CaH_2O_{11}.2H_2O$, 31A, 255
$B_6Ca_2H_2O_{12}$, 29, 391
$B_6Ca_2H_{10}O_{16}$, 29, 391
$B_6Ca_2H_{14}O_{18}$, 29, 391
$B_6Ca_2H_{18}O_{20}$, 29, 391
$B_6Ca_2H_{26}O_{24}$, 29, 391
$B_{14}(Ca,Sr)_2H_2O_{26}.5H_2O$, 35A, 446
$BCaHO_5Si$, 32A, 425
$B_5Ca_2H_5O_{14}Si$, 34A, 358; 35A, 481
$BCaO_3$, 26, 491
B_2CaO_4, 34A, 354, 355
$(B_2CaO_4)_x$, 28, 162
$B_2Ca_3O_6$, 34A, 356
$B_6CaO_{10}.4H_2O$, 26, 489
$B_6Ca_2O_{11}.H_2O$, 29, 391
$B_6Ca_2O_{11}.5H_2O$, 29, 391
$B_6Ca_2O_{11}.7H_2O$, 29, 391
$B_6Ca_2O_{11}.9H_2O$, 29, 391
$B_6Ca_2O_{11}.13H_2O$, 29, 391
$B_{14}Ca_2O_{23}.8H_2O$, 28, 346
$B_{10}Ca_8O_{35}Si_6.6H_2O$, 27, 706
$B_2CdF_8.6H_2O$, 26, 304
B_2CdO_5, 27, 549
B_4CdO_7, 27, 548; 31A, 166
$B_7Cd_3O_{12.65}S_{0.85}$, 35A, 288
B_4Ce, 30A, 114
B_6Ce, 26, 66; 29, 103; 30A, 114
B_2CeCo_3, 34A, 39
$BCeO_5Si$, 29, 411; 30A, 459; 32A, 433
BCl_3, 31A, 239
B_2Cl_4, 34A, 173
B_8Cl_8, 31A, 77
$B_7ClCo_3O_{13}$, 27, 551
$B_7ClFe_3O_{13}$, 27, 551
$B_7ClMg_3O_{13}$, 27, 551
$B_7ClMn_3O_{13}$, 27, 551
BCl_6NS_2, 34A, 206
$B_7ClNi_3O_{13}$, 27, 551
$B_7ClO_{13}Zn_3$, 27, 551
B_2Co_3Dy, 34A, 39
B_2Co_3Er, 34A, 39
$B_2CoF_8.6H_2O$, 26, 304
$B_{20}Co_{70}Ga_{10}$, 27, 92
B_2Co_3Gd, 34A, 39
B_2Co_3Hf, 34A, 38
$B_6Co_{21}Hf_2$, 30A, 114
B_2Co_3Ho, 34A, 39
$B_6Co_{21}In_2$, 30A, 114
B_2Co_3Lu, 34A, 39
B-Co-Mo, 30A, 114
BCoMo, 33A, 47
B_3CoMo_3, 34A, 41
$B_6Co_{21}Mo_2$, 30A, 114
$B_6Co_{21}Nb_2$, 30A, 114
BCo_3O_5, 31A, 253
B-Co-P, 27, 102
$B_{0.1}CoP_{0.9}$, 27, 321
ε-$B_{0.5}Co_3P_{0.5}$, 27, 102
$B_{0.7}Co_3P_{0.3}$, 27, 102
B_2Co_5P, 27, 102
$B_6Co_{21}Sb_2$, 30A, 114
$B_6Co_{22}Sb$, 30A, 114
$B_{21.5}Co_{24.5}Sb$, 30A, 114
B_2Co_3Sc, 34A, 39
B_2Co_3Sm, 34A, 39
$B_6Co_{21}Sn_2$, 27, 92, 95
B—Co—Ta, 26, 81
B_2CoTa, 26, 81
B_2Co_5Ta, 26, 81
$B_6Co_{21}Ta_2$, 29, 103
B_2Co_3Tb, 34A, 39
$B_6Co_{20}Ti_3$, 30A, 114
B_2Co_3Tm, 34A, 39
$B_6Co_{21}U_2$, 30A, 115
$B(Co_xV_{1-x})_2$, 30A, 193
$B_6Co_{21}V_2$, 30A, 115
$B_6(Co_xV_{1-x})_{23}$, 30A, 193
BCoW, 34A, 138
B_2CoW_2, 31A, 27
B_3CoW_3, 34A, 39
$B_6Co_{21}W_2$, 30A, 115
B_2Co_3Y, 34A, 39
B_2Co_3Zr, 34A, 38
$B_6Co_{21}Zr_2$, 30A, 115
B_4Cr, 33A, 46
B_4Cr_3, 26, 71
$B_{41}Cr$, 35A, 34
B_4CrDy, 35A, 117
B_4CrEr, 35A, 117
$B_{0.9}Cr_{0.9}Fe_{1.1}$, 29, 31, 103

B_4CrGd, 35A, 117
$B_4Cr_3GdO_{12}$, 27, 562
B_4CrHo, 35A, 117
B_4CrLu, 35A, 117
$BCr_{1.76}Mo_{0.24}$, 34A, 139
BCr_3O_3, 29, 393
B–Cr–P, 27, 101
ε-$B_{0.2}Cr_3P_{0.8}$, 27, 101
B_4CrTb, 35A, 117
$BCr_{\sim 1.85}V_{\sim 0.15}$, 34A, 138
B_4CrY, 35A, 36
$BCsF_4$, 34A, 189
$B_{18}Cs_2O_{28}$, 32A, 499
$B_{10}Cu_2H_{10}$, 27, 399
BCuS, 29, 103
BCuSe, 29, 103
$BCuSe_2$, 28, 43; 29, 103
$B_{10}D_{14}$, 34A, 151
BD_3O_3, 31A, 165
$BDyO_3$, 26, 491
$BErO_3$, 26, 491; 28, 170
$BEuO_3$, 26, 491
BF_3, 31A, 239
$BF_3.2H_2O$, 29, 252
$B_2F_8Fe.6H_2O$, 26, 304
BF_3H_2KNO, 30A, 449
BF_4H_4N, 34A, 189
$BF_4H_5N_2$, 26, 291; 29, 262; 35A, 392
BF_3HNaO, 35A, 189
$BF_3H_4O_2$, 29, 252
BF_4K, 34A, 189, 190
$B_2F_8Mg.6H_2O$, 26, 304
$B_2F_2Mg_4O_6$, 30A, 455
$B_2F_6Mg_6O_6$, 30A, 455
$B_2F_8Mn.6H_2O$, 26, 304
$BF_3N_4S_4$, 32A, 161
BF_4Na, 31A, 88; 32A, 158; 33A, 181
$B_2F_8Ni.6H_2O$, 26, 304
B_4F_9P, 34A, 191
BF_4Rb, 34A, 189
BF_7S, 29, 414
B_2F_8Sr, 27, 459
$B_2F_8Zn.6H_2O$, 26, 304
BFe, 27, 96
BFe_2, 27, 96
$B(Fe, Ga)O_3$, 28, 302
B_2FeMo_2, 29, 103
BFeNb, 32A, 139; 35A, 74
BFe_2O_5, 26, 489
BFe_3O_6, 30A, 417
$B_7Fe_3O_{12.65}S_{0.85}$, 35A, 288
B–Fe–P, 27, 96
ε_1-$B_xFe_3P_{1-x}$, 29, 156
ε-$B_{1-x}Fe_3P_x$, 29, 155
B_2Fe_5P, 27, 96, 99
B–Fe–Ta, 26, 81
BFeTa, 32A, 139
B_2FeTa, 26, 81
B_2Fe_5Ta, 26, 81
B_2FeW_2, 33A, 151
$B_2Fe_{1.25}W_{1.75}$, 34A, 139
B–Fe–Zr, 27, 102
$B_{12}Ga$, 26, 72
$B_xGa_3Nb_5$, 30A, 115
$B_{20}Ga_{10}Ni_{70}$, 27, 92
$B_xGa_3Ta_5$, 30A, 115
$BGdO_3$, 26, 491
BGe_3Nb_5, 30A, 115
B_4GeNi_{15}, 29, 103
$B_{20}Ge_{10}Ni_{70}$, 27, 92
$B_xGe_3Ta_5$, 30A, 115
BGe_3V_5, 30A, 115
B_2H_6, 30A, 252; 31A, 259
B_4H_{10}, 31A, 77
B_5H_{11}, 31A, 77
B_6H_{10}, 31A, 77
B_8H_{12}, 29, 230; 31A, 77
B_9H_{15}, 26, 290; 31A, 77
$B_{10}H_{14}$, 28, 360
$B_{16}H_{20}$, 35A, 130
$B_{18}H_{22}$, 28, 60, 62; 31A, 77
$B_{20}H_{16}$, 28, 296; 29, 231
B_5H_8I, 30A, 254
$B_{10}H_{13}I$, 32A, 213
$B_{12}H_{11}KMg_2O_{26}.4H_2O$, 31A, 172
$BH_2KMg_3O_{12}Si_3$, 27, 695
$B_5HK_2O_9.2H_2O$, 34A, 350
$B_5H_4KO_{10}.2H_2O$, 33A, 159
BH_4LiO_4, 31A, 166
$BHMgO_3$, 28, 158; 30A, 458
$B_{1.87}H_{0.39}Mg_2O_5.H_2O$, 31A, 247
$B_2H_{16}Mg_9O_{20}$, 27, 558
$B_3H_5MgO_8.5H_2O$, 27, 557; 28, 160
$B_3H_{15}MgO_{13}$, 27, 557
$B_{14}H_{20}Mg_3O_{34}$, 27, 558
$B_2H_6Mg_3O_{14}P.6H_2O$, 26, 492
$B_2H_2Mg_3O_{10}S.4H_2O$, 26, 492
BH_6N, 28, 296
BH_7N_2, 32A, 150
$B_3H_6N_3$, 34A, 152
$B_3H_7N_4$, 34A, 153
$B_5H_6NO_9.H_2O$, 34A, 351
$B_5H_8NO_{10}.2H_2O$, 26, 353
$BH_4NaO_4.2H_2O$, 28, 164
$B_2H_4Na_2O_8.6H_2O$, 26, 487
$B_4H_2Na_2O_8.3H_2O$, 31A, 260; 32A, 501
$B_5H_{10}NaO_{13}$, 27, 558
$BH_2NaO_7Si_2$, 27, 698; 29, 409
BHO_2, 26, 486; 29, 384, 427
$(BHO_2)_n$, 28, 155, 156
$B_3H_3O_3$, 34A, 153
$B_2H_8O_8Sr$, 29, 388, 427; 30A, 415
$B_6H_8O_{14}Sr$, 29, 390
$B_9H_9Rb_2$, 33A, 157
$B_{20}H_{18}Rb_2$, 33A, 158
$B_3H_3S_6$, 29, 424
$B_4H_{16}Zr$, 34A, 151
BHf, 26, 68, 73
B_2Hf, 26, 64, 68, 73; 29, 156; 30A, 449
B–Hf–N, 26, 73
$B_6Hf_3Ni_{20}$, 30A, 115
B–Hf–Si, 26, 64
$BHoO_3$, 26, 491
BI_3, 27, 438
$B_7INi_3O_{13}$, 35A, 288
π-$B_{0.5}InNi_3$, 29, 104
τ-$B_6In_2Ni_{21}$, 29, 104
$B_{20}In_{10}Ni_{70}$, 27, 92
$BInO_3$, 26, 491
$B_{\sim 1.1-1.5}Ir$, 27, 93
$B_{1.35}Ir$, 28, 13
$B_3K_3O_6$, 35A, 285
$B_4K_2O_7$, 26, 488
$B_4K_2O_7.4H_2O$, 28, 166
$B_5KO_8.4H_2O$, 28, 168
$B_6K_2O_{10}$, 26, 488
$B_{10}K_2O_{16}$, 30A, 411
γ-$B_{10}K_2O_{16}$, 26, 488
B–La, 26, 74
B_4La, 26, 74
B_6La, 26, 66, 75
$BLaO_3$, 26, 491; 27, 725; 35A, 282
B_3LaO_6, 35A, 282
B_yLi_x, 27, 126
$BLiO_2$, 26, 486; 27, 718; 28, 302; 29, 386; 31A, 165
$B_4Li_2O_7$, 27, 548; 33A, 444
$BLiS_2$, 30A, 453
$BLnO_3$, 31A, 172
B_2Lu, 28, 43
$BLuO_3$, 26, 491
B_2MO_6Sn, 31A, 245
$B_7MO_{13}X$ (M=M^{2+}, x= halogen), 30A, 450
B-Mg-Ni, 28, 43
τ-$B_6Mg_{3.5}Ni_{19.5}$, 29, 104
$B_{20}Mg_{10}Ni_{70}$, 27, 92
$B_2(Mg,Mn)_2O_5.xH_2O$, 30A, 455
$B_2MgO_4.3H_2O$, 32A, 496
$B_2Mg_9O_{12}.8H_2O$, 27, 558
B_4MgO_7, 29, 427; 31A, 167
$B_6MgO_{10}.7\frac{1}{2}H_2O$, 30A, 455
$B_6Mg_2O_{11}.15H_2O$, 28, 320
$B_{12}Mg_2O_{20}.15H_2O$, 30A, 455
$B_{14}Mg_3O_{24}.10H_2O$, 27, 558
$B_7Mg_3O_{12.65}S_{0.85}$, 35A, 288

B_2MgO_6Sn, 33A, 443
B_xMn, 35A, 121
B_4Mn, 35A, 38
$B_6Mn_2Ni_{21}$, 30A, 115; 34A, 139
$B_6Mn_8Ni_{15}$, 34A, 139
$B_2Mn_3O_6$, 28, 354; 30A, 415
$B_7Mn_3O_{12.65}S_{0.85}$, 35A, 288
B-Mn-P, 27, 102
ϵ-$B_{0.5}Mn_3P_{0.5}$, 27, 102
B_2Mn_5P, 27, 102
B_2MnS_4, 29, 104
B_4Mo, 26, 75
B_5Mo_2, 26, 76
B_2Mo_2Ni, 31A, 27
B_3Mo_3Ni, 34A, 41
BN, 26, 75, 291; 27, 408; 28, 43, 296; 29, 156; 31A, 250
B—N—Ti, 26, 70
B—N—Zr, 26, 73
$B_{15}Na$, 35A, 39
$BNaO_2$, 27, 718
$B_3Na_3O_6$, 28, 157
$B_5NaO_8.5H_2O$, 27, 558; 31A, 242
$B_8Na_2O_{13}$, 32A, 497
$BNaO_6Si_2.H_2O$, 31A, 244
$BNaO_8Si_3$, 30A, 435
B_2Nb, 29, 156
$B_{0.32}Nb_4Ni_2$, 32A, 139
BNbNi, 33A, 152
B_6Nd, 29, 104
$BNdO_3$, 26, 491; 27, 725
BNi_3, 32A, 32
B_3Ni_4, 32A, 32, 33
$B_2Ni_3O_6$, 28, 302
B-Ni-P, 27, 102
$B_6Ni_{21}Sb_2$, 30A, 115
$BNi_{4.6}Si_2$, 35A, 38
B-Ni-Sn, 28, 43
$B_{5-9}Ni_{21}Sn_2$, 27, 92
$B_6Ni_{21}Sn_2$, 27, 103; 29, 104
B—Ni—Ta, 26, 81
BNiTa, 33A, 152
B_2NiTa, 26, 81
B_2Ni_5Ta, 26, 81
$B_6Ni_{20}Ti_3$, 30A, 115
$B_6Ni_{21}U_2$, 30A, 115
$B_6Ni_{21}V_2$, 30A, 115
B_3NiW_3, 34A, 41
B-Ni-Zn, 27, 103; 28, 43
$B_{5-8}Ni_{19-20}Zn_{3-4}$, 27, 92
$B_6Ni_{20}Zr_3$, 30A, 115
B_2O_3, 27, 718; 33A, 258; 35A, 202
B_7O, 26, 67
B_4O_7Pb, 29, 392; 31A, 169
BO_3Sc, 31A, 242
BO_3Sm, 26, 491
B_2O_6SnSr, 33A, 443
B_2O_4Sr, 29, 393
$B_2O_4Sr.4H_2O$, 29, 388; 31A, 249
$B_2O_5Sr_2$, 31A, 168
B_4O_7Sr, 29, 257, 392; 31A, 169
$B_6O_{10}Sr.4H_2O$, 26, 489; 28, 163; 29, 390
BO_3Tb, 26, 492
B_2O_5Th, 26, 494
BO_3Ti, 29, 393
BO_3Tm, 26, 491; 28, 170
BO_3V, 29, 393
BO_3Y, 26, 491; 28, 170
BO_3Yb, 26, 491
$B_2O_6Zn_3$, 33A, 441; 35A, 284
$B_6O_{13}Zn_4$, 26, 490; 27, 559
B—Os, 26, 76
$B_{\sim 1.2-2.2}Os$, 27, 93
B_2Os, 26, 76; 27, 93; 28, 13
B_3Os, 26, 75
B_5Os_2, 26, 76
BP, 27, 96; 28, 43, 296
$B_{13}P_2$, 26, 67, 69; 27, 406; 28, 296
BPS_4, 28, 153, 302
$B_2Pd_2S_5$, 29, 104; 33A, 363
BPd_3, 26, 76
B_2Pd_5, 26, 76
$B_5(Pd,W)_2$, 32A, 139
B_6Pr, 26, 66; 29, 104
BPt_3, 30A, 116
B_2Re, 27, 103
B—Ru, 26, 76
$B_{1.1-2.1}Ru$, 27, 93
$B_{1.5}Ru$, 32A, 34
B_2Ru, 26, 76; 27, 93; 28, 13
B_3Ru, 26, 76
B_3Ru_2, 34A, 41
B_3Ru_7, 27, 93; 30A, 36
B_5Ru_2, 26, 76
B_8Ru_{11}, 27, 93
$B_{12}S$, 26, 79
B_2Sc, 29, 156
$B_{12}Sc$, 28, 43; 30A, 116
$B_{10.40}Si_{3.60}$, 27, 406
$B_{12}Si_3$, 26, 67
$B_{14}Si$, 30A, 116
$(B_xSi_{1-x})V_3$, 30A, 193
B—Si—Zr, 26, 64
B_6Sr, 26, 66
BTa, 26, 80
BTa_2, 26, 80
B_2Ta, 27, 26; 29, 157
B_2Ta_3, 26, 80
$B_{12}Tb$, 28, 43
BTc_3, 29, 104
B_2Tc, 29, 104
B_3Tc_7, 29, 104
B_4Th, 26, 68, 81; 30A, 116
B_6Th, 26, 68, 81; 30A, 116
B—Th—U, 26, 81
$B_4(Th, U)$, 26, 81
B_2Ti, 29, 157; 30A, 449
B_4Tm, 26, 82
B_6Tm, 26, 82
B_2U, 26, 70, 81; 27, 105; 30A, 116
B_4U, 26, 70, 81; 27, 105; 30A, 116
$B_{12}U$, 26, 70, 81; 30A, 116
B-U-Zr, 27, 105
$B_{12}(U,Zr)$, 27, 105
BW, 29, 104
$B_{2.0}W$, 34A, 41
B_4W, 26, 75
$B_{4-5}W$, 31A, 29
B_5W_2, 26, 76; 32A, 139
B_4Y, 30A, 31, 116
B_6Y, 26, 66
$B_{12}Y$, 28, 43; 30A, 29, 116

$B_{66}Y$, 34A, 43
$B_{25}Zn$, 29, 414
B_2Zr, 26, 74, 27, 105; 29, 157; 30A, 449
Ba, 28, 43; 29, 104; 34A, 139
Ba II, 29, 104
$BaBeF_4$, 26, 302
$(Ba,Pb)BeF_4$, 26, 302
$BaBeFO_4P$, 32A, 368
$Ba_4Be_5F_4O_{17}P_4.0.35H_2O$, 31A, 248
$Ba_2Be_3O_5$, 29, 418
$BaBe_2O_7Si_2$, 27, 704; 29, 398
$BaBi_3$, 26, 62
Ba_2BiLaO_6, 28, 346
$BaBi_2Nb_2O_9$, 27, 538
$BaBi_3NbO_{12}Ti_2$, 26, 402
$BaBi_2O_9Ta_2$, 27, 538
$BaBi_4O_{15}Ti_4$, 26, 402; 27, 539
$Ba_2Bi_4O_{18}Ti_5$, 27, 539
$Ba_2Bi_2Se_5$, 30A, 116
$BaBr_2$, 28, 77
$BaBr_2.2H_2O$, 26, 335
$BaBrH_2O_4$, 27, 643
$BaBrO_3.H_2O$, 27, 643
$Ba_{10}Br_2O_{24}P_6$, 26, 469
$(Ba,Ca,Sr,TR)_3C_5(Ca,Na)_3O_{15}$, 33A, 437
$Ba_6C_{13}Ca_7O_{39}$, 26, 483; 27, 563
BaC_2CeFO_6, 26, 484; 28, 322
BaC_2FLaO_6, 26, 484
BaC_2MgO_6, 26, 483; 30A, 454; 32A, 417; 33A, 436
$BaC_4N_4Ni.4H_2O$, 34A, 231
$BaCS_3$, 30A, 453
$(Ba,Ca,Sr)_2FeMoO_6$, 31A, 250
$BaCa_4H_2O_{38}Si_{16}10.4H_2O$, 33A, 489
$Ba(Ca,Nb)O_3$, 28, 330
$Ba(Ca, Ta)O_3$, 28, 332
$BaCa_{0.5}O_3Re_{0.5}$, 26, 418
$BaCa_2O_9Si_3$, 26, 524; 33A, 460
$Ba_4Ca_4O_{30}Sr_2Ti_{10}$, 28, 326
$BaCaO_6SrW$, 28, 310
$Ba_3CaO_9Ta_2$, 26, 420
Ba_2CaO_6Te, 29, 420
$Ba_{1-x}Ca_xO_3Ti$, 26, 372
$BaCa_2O_6W$, 28, 310
$Ba_{1.5}Ca_{1.5}O_6W$, 28, 310
Ba_2CaO_6W, 28, 310
$BaCaO_6WZn$, 28, 310
$(Ba_{0.95}Ca_{0.05})Pb_3$, 31A, 16
$Ba(Cd, Nb)O_3$, 28, 330
$BaCdO_2$, 27, 517
$Ba(Ca, Ta)O_3$, 28, 332
$BaCd_{0.5}O_3Re_{0.5}$, 26, 418
Ba_2CdO_6W, 28, 310
$Ba_yCe_{2+x}S_{3+x+y}$, 27, 63
$BaCe_2S_4$, 27, 327
$BaCe_2Se_4$, 29, 104
$BaCl_2$, 28, 77, 78, 79
$BaCl_2.2H_2O$, 28, 79
$Ba_5ClCr_3O_{12}$, 30A, 451
$Ba_4ClFe_{0.64}Nb_{1.44}O_{28}Si_4Ti_{5.92}$, 34A, 382
$Ba_4Cl(Fe, Nb)_2O_{28}Si_4Ti_6$, 28, 274
$Ba(Cl, H, O)Fe_2O_8Si_2Ti$, 28, 270
$Ba_4ClNbO_{28}Si_4Ti_7$, 26, 536
$BaCl_2O_6.H_2O$, 33A, 159, 448
Ba_4Cl_6O, 35A, 186
$Ba_{10}Cl_2O_{24}P_6$, 26, 469
$BaCoF_4$, 35A, 165
Ba_2CoF_6, 32A, 168
$BaCo_{0.3}F_{0.3}Fe_{11.7}O_{18.7}$, 28, 352
$BaCo_{0.4}F_{0.4}Fe_{11.6}O_{18.6}$, 28, 352
$BaCo_{1.3}F_{1.3}Fe_{16.7}O_{23.7}$, 28, 352
$BaCo_{2.1}F_{2.1}Fe_{15.9}O_{24.9}$, 28, 352
$BaCo_{0.5}F_{0.5}Ga_{11.5}O_{18.5}$, 28, 352
Ba-Co-Fe-O, 30A, 451
$Ba_3Co_2Fe_{24}O_{41}$, 30A, 451
$BaCoK_2N_6O_{12}$, 31A, 181
$Ba(Co, Ta)O_3$, 28, 332
$BaCo_{0.5}O_3Re_{0.5}$, 26, 418
Ba_2CoO_6W, 32A, 272
$Ba_5Cr_3FO_{12}$, 28, 308
$Ba_5Cr_3HO_{13}$, 28, 308
$BaCrO_4$, 27, 609
$BaCr_2O_4$, 28, 328, 344
$Ba_3Cr_2O_8$, 28, 308; 30A, 451
Ba_2CuF_6, 32A, 168
BaD_3Li, 33A, 160
Ba_2DyNbO_6, 27, 509
$BaDy_2S_4$, 27, 327
Ba_2ErNbO_6, 27, 509
$BaEr_2S_4$, 27, 327
Ba_2EuNbO_6, 27, 509
BaF_2, 29, 439; 31A, 243
BaF_4Fe, 31A, 245
Ba_2F_6Fe, 31A, 245
$BaF_{0.5}Ga_{11.5}Ni_{0.5}O_{18.5}$, 28, 352
BaF_4Mn, 35A, 164
BaF_4Ni, 26, 309
Ba_2F_6Ni, 32A, 168
$Ba_{10}F_2O_{24}P_6$, 26, 469
BaF_8Tm_2, 35A, 172
Ba-F-Y, 30A, 449
Ba_2F_6Zn, 32A, 168
$Ba_2(Fe,Mg,Ti)_2H_2O_{14}Si_4$, 30A, 437
$BaFe_3H_2O_{10}Si_2Ti$, 28, 268
$Ba(Fe, Nb)O_3$, 28, 330
Ba_2FeNbO_6, 26, 424; 27, 509
$Ba_6FeNb_9O_{30}$, 28, 330
$BaFeO_3$, 26, 420
$BaFeO_{3-x}$, 31A, 249
$BaFe_2O_4$, 26, 395; 28, 328
$BaFe_{12}O_{19}$, 32A, 290
$Ba_2(Fe,U)O_6$, 33A, 356
$BaFe_{0.5}O_3Re_{0.5}$, 26, 418
$Ba_2FeO_7Si_2$, 26, 523
Ba_2FeO_6Ta, 26, 422
$Ba_3Fe_2O_9U$, 31A, 161
$Ba_3Fe_2O_9W$, 34A, 286
$Ba_2Fe_{12}O_{22}Zn_2$, 35A, 265
$Ba_8Fe_{60}O_{104}Zn_6$, 29, 323
$Ba_{10}Fe_{56}O_{102}Zn_8$, 28, 328
$Ba_{10}Fe_{72}O_{126}Zn_8$, 29, 323; 30A, 349
$Ba_{12}Fe_{68}O_{124}Zn_{10}$, 28, 328
$Ba_{12}Fe_{84}O_{148}Zn_{10}$, 29, 323
$Ba_{14}Fe_{80}O_{146}Zn_{12}$, 28, 328
$Ba_{14}Fe_{96}O_{170}Zn_{14}$, 29, 323
$Ba_{16}Fe_{92}O_{168}Zn_{14}$, 28, 328
$BaGa_4$, 30A, 116
$Ba_{0.8}Ga_{11.8}La_{0.2}Ni_{0.2}O_{19}$, 28, 352
$Ba_{0.8}Ga_{11.8}La_{0.2}O_{19}$, 28, 352
$BaGa_2O_4$, 28, 322
$BaGaSe_3$, 30A, 116
Ba_2GdNbO_6, 27, 509
$BaGd_2S_4$, 27, 327
$BaGe_2$, 33A, 152
BaGeMg, 35A, 24
$BaGe_2Mg_2$, 35A, 24, 117
Ba-Ge-O, 30A, 451
$BaGeO_3$, 26, 397; 27, 502
$BaGe_4O_9$, 26, 398; 27, 541; 31A, 134; 33A, 509
$Ba_2Ge_7O_{16}.7H_2O$, 26, 398
$BaGe_3O_9Ti$, 26, 397
BaH_3Li, 29, 104; 33A, 160
$BaH_2O_2.H_2O$, 31A, 256
$BaH_2O_2.8H_2O$, 28, 119; 29, 297
$BaH_{18}O_{10}$, 29, 297
$BaH_2O_4S_2$, 27, 630
$Ba_5H_2O_{14}Si_4$, 35A, 312
$Ba_{0.3}HfO_3Pb_{0.7}$, 28, 332

Ba_2HoNbO_6, 27, 509
BaI_2, 28, 77
$BaIn_4$, 30A, 116
Ba_2InNbO_6, 27, 509
$BaIn_2O_4$, 29, 419
Ba_2InO_6Sb, 29, 420
$BaIr_xO_3Ru_{1-x}$, 31A, 137
$Ba_{1-x}IrO_3Sr_x$, 31A, 137
$BaIr_{0.33}O_3Ti_{0.67}$, 26, 371
$(Ba, K)Mn_{0.5}O_3Re_{0.5}$, 26, 418
$BaKO_4P$, 26, 468; 27, 582
$Ba_2KO_{12}PS_2$, 31A, 246
$BaK_2O_{12}S_{12}$, 30A, 380
$BaKO_4V$, 26, 470
Ba_2LaNbO_6, 26, 423; 27, 509
Ba_2LaO_6Ta, 26, 423
$Ba_{1-x}La_xO_3Ti$, 26, 372
$BaLa_2S_4$, 27, 327
$BaLa_2Se_4$, 29, 104
$BaLi_4$, 30A, 25, 116
$BaLi_{0.5-0}Na_{0-0.5}O_3Re_{0.5}$, 26, 418
$Ba_3Li_4Sb_4$, 35A, 9, 117
$BaLiSi$, 33A, 39
$BaLi_2Si$, 32A, 27
$BaLn_2O_4$, 29, 421
$BaLn_6O_{12}$, 29, 421
$Ba_3Ln_4O_9$, 33A, 341
Ba_2LuNbO_6, 27, 509
Ba_3MO_5, 30A, 458
Ba_2Mg_{17}, 26, 175; 27, 64; 28, 43
Ba_6Mg_{23}, 26, 176; 30A, 117
$Ba(Mg, Nb)O_3$, 28, 330
$Ba_9MgNb_{14}O_{45}$, 28, 330
$Ba(Mg, Ta)O_3$, 28, 332
$BaMg_{0.5}O_3Re_{0.5}$, 26, 418
$BaMg_xO_3Ru_{1-x}$, 31A, 137
$BaMgO_{10}Si_4$, 27, 699
$BaMgO_6SrW$, 28, 308
Ba_2MgO_6Te, 29, 420
Ba_2MgO_6W, 28, 308
$BaMgSi$, 35A, 24
$BaMg_2Si_2$, 35A, 24
Ba_2MnNbO_6, 26, 423
$BaMnO_3$, 26, 394
α-$BaMnO_3$, 27, 662
$Ba(Mn, Ta)O_3$, 28, 332
$BaMn_2O_8$, 27, 662
$BaMn_{0.5}O_3Re_{0.5}$, 26, 418
$BaMn_xO_3Ru_{1-x}$, 31A, 137
$Ba_2MnO_7Si_2$, 26, 523
Ba_2MnO_6Ta, 26, 423
$BaMo_{0.5}Ni_{0.5}O_3$, 26, 420
Ba—Mo—Ni—O—Sr, 26, 420
Ba—Mo—O, 26, 391
$BaMoO_4$, 35A, 226
BaN_6, 30A, 449; 34A, 169; 35A, 26
BaN_2O_6, 32A, 351
$BaNa$, 29, 104
$BaNa_4$, 29, 104; 30A, 117
$Ba_{(4+x)}Na_{(2-2x)}Nb_{10}O_{30}$, 34A, 269
$BaNaO_2$, 26, 486
$BaNaO_4P$, 26, 469
$BaNaO_{14}Si_4Ti_2$, 27, 705
$BaNaO_4V$, 26, 470
Ba_2NbNdO_6, 26, 423; 27, 509
$Ba(Nb, Ni)O_3$, 28, 330
$Ba(Nb, Pb)O_3$, 28, 330
$Ba(Nb, Zn)O_3$, 28, 330
$(Ba, Pb)(Nb, Ti)O_3$, 27, 506
$BaNb_2O_6$, 33A, 328
$(Ba,Sr)Nb_2O_6$, 33A, 327
$Ba_{0.6}Nb_{0.5}O_3Pb_{0.15}Ti_{0.5}$, 27, 506
Ba_2NbO_6Pr, 26, 423
$Ba_3Nb_6O_{26}Si_4$, 35A, 269
Ba_2NbO_6Sm, 26, 423; 27, 509
$Ba_{0.27}Nb_2O_{5.78}Sr_{0.75}$, 33A, 327
$Ba_6Nb_8O_{30}Ti_2$, 30A, 346; 33A, 328
Ba_2NbO_6Tm, 27, 509
Ba_2NbO_6V, 31A, 247
Ba_2NbO_6Y, 26, 423
Ba_2NbO_6Yb, 27, 509
Ba_2NdO_6Ta, 26, 428
$BaNd_2S_4$, 27, 327
$BaNd_2Se_4$, 29, 104
$Ba(Ni, Ta)O_3$, 28, 332
$BaNi_{0.5}O_3Re_{0.5}$, 26, 418
$BaNi_xO_3Ru_{1-x}$, 31A, 137
$Ba_xNiO_3Sr_{2-x}W$, 27, 509
Ba_2NiO_6W, 27, 509; 32A, 272
$BaNiS_2$, 35A, 25
BaO, 26, 391; 29, 422
$Ba_2O_9P_2Ti$, 26, 467
α-$Ba_2O_9P_2Ti$, 26, 467
β-$Ba_2O_9P_2Ti$, 26, 467
$(Ba,K)_2O_{12}P_2U_2.6H_2O$, 31A, 244
$BaO_3(Pb, Ta)$, 28, 332
$Ba_{0.1}O_3Pb_{0.9}Ti$, 27, 506
Ba_2O_6PrTa, 26, 423
BaO_3Pu, 28, 334
Ba_3O_6Pu, 29, 315
$BaO_3Re_{0.5}Sr_{0.5}$, 26, 418
$BaO_3Re_{0.5}Zn_{0.5}$, 26, 418
BaO_3Ru, 30A, 329
$Ba_{1-x}O_3RuSr_x$, 31A, 137
$BaO_3S_2.H_2O$, 27, 630
BaO_4S, 28, 78, 208, 209; 32A, 327
$BaO_6S_2.2H_2O$, 31A, 208
Ba_2O_6SbSc, 29, 420
BaO_4Se, 27, 609
$BaO_3Si.6H_2O$, 26, 513
$BaO_5Si_2.3H_2O$, 32A, 459
Ba_2O_4Si, 30A, 421
$Ba_2O_{10}Si_4$, 27, 699
$(Ba,Sr)_3O_5Si$, 31A, 254
$BaO_9Si_3Sn_{0.77}Ti_{0.23}$, 30A, 455
$Ba_3O_{23}Si_4Ta_6$, 35A, 267
BaO_9Si_3Ti, 34A, 380
$Ba_2O_8Si_2Ti$, 32A, 442
BaO_4Sm_2, 28, 346
Ba_2O_6SmTa, 26, 423
BaO_3Sn, 28, 324
$Ba_3O_9SrTa_2$, 26, 420
BaO_6Sr_2W, 28, 310
BaO_6SrWZn, 28, 310
Ba—O—Ta, 26, 391
BaO_6Ta_2, 33A, 342
$Ba_5O_{13}Ta_4$, 26, 422
$Ba_5O_{15}Ta_4$, 26, 421, 35A, 261
$Ba_{5.45}O_{60}Ta_{21.8}$, 32A, 313
Ba_2O_6TaTl, 29, 421
Ba_2O_6TaV, 31A, 247
Ba_2O_6TaY, 26, 423
$BaO_3(Ta, Zn)$, 28, 332
BaO_3Tb, 31A, 253
BaO_3Tc, 29, 322
$BaO_{2.76}Ti$, 26, 371
BaO_3Ti, 26, 370, 373; 27, 506, 507; 28, 326; 29, 418; 30A, 451
$(Ba,Pb)O_3Ti$, 26, 372
$BaO_{11}Ti_5$, 34A, 255
Ba_2O_4Ti, 26, 371
$Ba_6O_{40}Ti_{17}$, 35A, 271
Ba-O-Ti-U, 30A, 457
BaO_4U, 34A, 289
BaO_7U_2, 30A, 447
$BaO_{10}U_3.4H_2O$, 28, 310
$BaO_{19}U_6.10\text{-}11H_2O$, 27, 673
Ba_3O_6U, 31A, 160
$(Ba,Pb)O_{12}U_2V_2.5H_2O$, 31A, 256
$Ba_3O_{4-\alpha}V$, 30A, 319
$Ba_3O_8V_2$, 35A, 363
Ba—O—W, 26, 391
BaO_4W, 26, 391; 35A, 226
Ba_2O_5W, 26, 290
Ba_3O_6W, 26, 391
Ba_2O_6WZn, 28, 310
BaO_3Zr, 29, 419
Ba_3P_2, 33A, 40
$Ba_4P_{2.67}$, 35A, 28

$BaPb$, 29, 29, 104
$BaPb_3$, 29, 29,104
Ba_5Pb_3, 29, 29, 104
$BaPb_2.4Tl_{0.6}$, 35A, 23
$BaPd_5$, 35A, 27
$BaPr_2S_4$, 27, 327
$BaPr_2Se_4$, 29, 104
$BaPt_5$, 35A, 27
BaS, 27, 63
$BaS_4.H_2O$, 34A, 294
Ba-S-Sm, 27, 330
BaS_4Sm_2 27, 327
BaS_3Ta, 29 104
BaS_3Ti, 28, 151; 29, 104
BaS_3U, 35A, 28
BaS_3V, 34A, 295
BaS_4Y_2, 27, 327
BaS_4Yb_2, 27, 327
Ba_2S_3Zn, 26, 424
BaS_3Zr, 28, 151; 29, 104
$BaSe_3Sn$, 30A, 117
$BaSe_3Ti$, 30A, 117
$BaSi_2$, 28, 11; 33A, 39
Ba_5Si_3, 31A, 25
$BaTl_2$, 29, 3, 105
Be, 26, 56; 27, 65; 29, 105
$BeBr_2$, 28, 300
$Be_{13}Ca$, 27, 68
$BeCaFNaO_6Si_2$, 32A, 444
$Be(Ca,Na)FNaO_6Si_2$, 32A, 444
$Be_2Ca_3H_2O_{12}Si_3$, 32A, 466
$Be_3Ca_2O_5$, 27, 546
$Be_{17}Ca_{12}O_{29}$, 31A, 130
$BeCa_2O_7Si_2$, 27, 703
$Be_{13}Ce$, 27, 71; 29, 105
$BeCl_2.4.5H_2O$, 30A, 449
$BeCl_2.4NH_3$, 30A, 449
$BeCl_4K_2$, 30A, 449
Be-Cl-Na, 30A, 449
Be_xCo, 35A, 117
$BeCoF_4.6H_2O$, 34A, 186
Be—Co—Mn, 26, 58
$Be_{1.2}Co_{0.8}Mn$, 26, 58, 174
$BeCr_{2-x}Fe_xO_4$, 29, 317
$BeCsF_3$, 27, 552; 33A, 176
$BeCs_2F_4$, 33A, 177
$Be_2CuLnO_{10}Si_2$, 30A, 455
$Be_{15}Cu_8Nb_6$,, 30A, 117
$Be_{15}Cu_8Ta_6$, 30A, 117, 193
$Be_{15}Cu_8Ti_6$, 30A, 117
$Be_{15}Cu_8Zr_6$, 30A, 117
$Be_{13}Dy$, 29, 105
$Be_{13}Er$, 29, 105
$Be_{13}Eu$, 28, 43; 29, 105
BeF_2, 27, 430
$Be(F,HO)_2Fe_3O_9Si_3$, 35A, 473
$BeF_4H_2Li_2O$, 27, 606
BeF_3H_4N, 27, 552
$BeF_4H_8N_2$, 26, 303; 30A, 457
$Be_2F_5H_4N$, 26, 302
$Be_2F_8H_8N_2Pb$, 35A, 174
$Be_3F_{12}H_8N_2Zn_2$, 33A, 180
$Be_2F_{10}HNa_5$, 29, 414
BeF_4K_2, 30A, 270
Be_2F_5K, 26, 302
$Be_2F_8K_2Pb$, 35A, 173
Be_3F_9KZn, 31A, 87
$Be_3F_{12}K_2Zn_2$, 32A, 163
BeF_4Li_2, 31A, 86
$BeF_4Li_2.H_2O$, 27, 606
$BeF_{12}Li_6Zr$, 29, 262
$Be_2F_8M^{+}{}_2M^{2+}.6H_2O$, 34A, 186
BeF_4Na_2, 30A, 422; 31A, 241; 33A, 176
$Be_2F_{14}Na_6Si$, 28, 296
$BeF_4Ni.6H_2O$, 34A, 186
BeF_4Pb, 26, 302; 33A, 179
$Be_2F_8PbRb_2$, 35A, 174
$Be_2F_8PbTl_2$, 35A, 174
BeF_3Rb, 27, 452
BeF_4Rb_2, 30A, 270
Be_2F_5Rb, 26, 303
γ-Be_2F_5Rb, 26, 302
$BeF_4Zn.6H_2O$, 34A, 186
Be_xFe, 35A, 117
ζ-$Be_{11}Fe$, 27, 4
$Be_2FeLnO_{10}Si_2$, 30A, 455
$Be_3(Fe, Mn, Zn)_4O_{12}SSi_3$, 26, 534; 28, 324
$BeGe_6O_{21}Pb_8$, 29, 399
$BeHNaO_8Si_3$, 28, 322; 35A, 478
$Be_2H_4Na_4O_{20}Si_6Sn$, 30A, 459
$Be_2HO_5P.4H_2O$, 28, 324
$Be_4H_2O_9Si_2$, 26, 509
Be_2Hf, 26, 60
Be_5Hf, 26, 59
$Be_{13}Hf$, 26, 58
α-$Be_{17}Hf_2$, 26, 59
β-$Be_{17}Hf_2$, 26, 59
Be—Hf—Si, 26, 60
$BeHfSi$, 26, 60
$Be_{13}Ho$, 29, 105
BeI_2, 28, 300
Be_xIr, 35A, 117
$Be_{13}La$, 27, 71; 29, 105
$Be_2La_2O_5$, 29, 421; 33A, 271
$Be_2La_6O_{11}$, 29, 421
$BeLi_2O_4Si$, 31A, 230
$Be_{13}Ln$, 28, 43
$Be_2LnMnO_{10}Si_2$, 30A, 455
$Be_{13}Lu$, 29, 105
$Be_{13}Mg$, 27, 68
$(Be, Si)_2Mo$, 28, 43
$Be_{22}Mo$, 27, 70; 28, 43
Be_3N_2, 34A, 29
BeN_2O_6, 26, 474
BeN_6O_{14}, 26, 474
$Be_4N_6O_{19}$, 26, 474
BeN_2Si, 26, 61; 32A, 28
$BeNaO_4P$, 26, 456; 27, 580
$Be_{13}Nd$, 29, 105
BeO, 27, 473; 29, 296; 30A, 304, 450; 34A, 238
$BeO_{21}Pb_8Si_6$, 29, 399
$BeO_4S.4H_2O$, 34A, 298

$Be_2O_7SiY_2$, 34A, 377
Be_3O_4Sr, 34A, 249
$Be_3O_5Sr_2$, 27, 547
BeO_4Y_2, 32A, 264
$Be_2O_5Y_2$, 27, 547
$Be_{17}Os_3$, 27, 70
$Be_{13}Pr$, 29, 105
Be-Pt, 27, 322
$BePt_{15}$, 27, 322
$Be_{22}Re$, 27, 71
$Be_{6.6}Rh$, 35A, 29
$Be_{17}Ru_3$, 27, 74
Be_5Sc, 29, 105
$Be_{13}Sc$, 27, 70
Be-Si-Zr, 26, 60
BeSiZr, 26, 60
$Be_{13}Sm$, 28, 43; 29, 105
$Be_{13}Sr$, 29, 105
$BeTa_2$, 30A, 117
Be_2Ta, 26, 60
Be_2Ta_3, 26, 60
Be_3Ta, 26, 59
$Be_{17}Ta_2$, 26, 59
$Be_{13}Tb$, 29, 105
Be_3Ti, 26, 59
$Be_{12}Ti$, 26, 59; 29, 105
$Be_{17}Ti_2$, 26, 175
α-$Be_{17}Ti_2$, 26, 59
β-$Be_{17}Ti_2$, 26, 59; 29, 105
$Be_{13}Tm$, 29, 105
$Be_{22}W$, 27, 70; 28, 43
$Be_{13}Y$, 27, 71; 28, 43; 29, 105
$Be_{13}Yb$, 28, 44; 29, 105
Be_2Zr, 26, 60
$Be_{13}Zr$, 26, 60
Bi, 26, 62, 282; 27, 76, 385; 32A, 29; 35A, 9
Bi II, 29, 105
(Bi, Ca)H(Nb, $Ta)_2O_7$, 28, 330
$Bi_2CaNb_2O_9$, 27, 538
$Bi_4CaO_{15}Ti_4$, 27, 538
BiCe, 27, 335
$Bi_{12}Cl_{14}$, 28, 81
$Bi_{24}Cl_{28}$, 27, 450
$Bi_{0.5}Cl_6Cs_2Sb_{0.5}$, 28, 300
$Bi_2Cl_7Rb.H_2O$, 31A, 260
$BiCrHO_5$, 27, 657; 29, 320
$BiCrO_3$, 30A, 451
$BiCs_3$, 26, 54, 63
Bi_2Cs, 26, 54
$Bi_2Cs_3I_9$, 33A, 241
Bi_2Cu_5, 26, 63
Bi_3CuGd_5, 34A, 34
Bi_4CuO_7, 31A, 243
$BiCuPbS_3$, 35A, 30
BiF_3, 27, 724
BiFO, 29, 271
(Bi,Fe,La,Mn)O_3, 30A, 458
(Bi,Fe,Mn,Sn,Sr)O_3, 30A, 451
(Bi,Fe,Nb,Pb)O_3, 30A, 458
$BiFeO_3$, 27, 506; 28, 135; 29, 323; 326, 418; 31A, 248, 249; 32A, 269
$Bi_2Fe_4O_9$, 29, 419; 33B, 321
Bi-Fe-O-Pb-Ti, 27, 506
Bi-Fe-O-Sr-Ti, 30A, 451
$Bi_5GaO_{15}Ti_3$, 27, 539
Bi_3Gd_4, 29, 105
Bi_3Gd_5, 34A, 34
Bi-Gd-Sb, 29, 157
$Bi_4Ge_3O_{12}$, 28, 324
$Bi_{12}GeO_{20}$, 28, 324; 32A, 292
$Bi_{14}GeO_{23}$, 29, 419
$Bi_{24}Ge_2O_{40}$, 28, 324
Bi_2GeTe_4, 30A, 28, 117
$Bi_2Ge_3Te_6$, 35A, 31
Bi_4GeTe_7, 33A, 42
$Bi_6H_5N_5O_{24}.\frac{1}{2}H_2O$, 28, 302
$BiHO_4.5P$, 27, 588
Bi_2Hf, 29, 105
BiHo, 26, 164
Bi-In, 26, 42
$BiIn_2$, 27, 80
Bi_2In_3, 26, 42
Bi_3In_5, 34A, 31
Bi-In-Sb, 26, 42
Bi-In-Te, 26, 43
Bi_2Ir, 30A, 117
Bi_3Ir, 30A, 117
BiK_3, 28, 44
$BiK_2Nb_5O_{15}$, 28, 330
$Bi_{0.5}K_{0.5}O_3Ti$, 27, 508
$Bi_{4.5}K_{0.5}O_{15}Ti_4$, 27, 539
$BiLi_3O_4$, 29, 316
$BiLi_7O_6$, 29, 316
Bi*Ln*, 28, 44
BiLu, 29, 105
$Bi_{0.8}Mg_{0.5}O_3Ti_{0.65}$, 33A, 305
$BiMnO_3$, 30A, 451
$Bi_2Mn_4O_{10}$, 33A, 321
Bi_2MoO_6, 31A, 250
$Bi_2Mo_2O_9$, 31A, 261
$Bi_2Mo_3O_{12}$, 29, 420; 31A, 242
$BiN_3O_9.5H_2O$, 30A, 401
BiN_2Th_2, 35A, 113
$Bi(N,O)_2Th_2$, 35A, 113
BiN_2U_2, 35A, 113
$BiNa_2Nb_5O_{15}$, 28, 330
$Bi_{0.5}Na_{0.5}O_3Ti$, 27, 508
$Bi_{4.5}Na_{0.5}O_{15}Ti_4$, 27, 539; 32A, 268
$BiNb_3$, 29, 105
β-$BiNb_3$, 29, 105
$Bi_{0.5}NbO_3$, 27, 506
$BiNbO_4$, 28, 354; 32A, 509
$Bi_2Nb_2O_9Pb$, 26, 402; 27, 539
$Bi_3NbO_{12}PbTi_2$, 26, 402
$Bi_2Nb_2O_9Sr$, 27, 538
Bi_3NbO_3Ti, 27, 538
Bi_3NbO_9Ti, 27, 539
Bi-Ni-Rh, 27, 287
Bi-Ni-Sb, 27, 287
BiO, 30A, 446
$Bi_2O_{2.7-2.8}$, 29, 311
Bi_2O_3, 28, 320; 35A, 210
δ-Bi_2O_3, 27, 476
BiO_4P, 27, 588
$BiO_4P.0.5H_2O$, 27, 588
$BiO_{12}P_2Pb_3SiTl$, 28, 324
$Bi_2O_{12}Pb_3Si_3$, 28, 324
$Bi_2O_9PbTa_2$, 26, 402
$Bi_4O_{15}PbTi_4$, 26, 402; 27, 539
$Bi_4O_{18}Pb_2Ti_5$, 27, 539
$Bi_6O_{18}Pb_2Ti_5$, 27, 539
δ-$Bi_2O_{15}Si$, 27, 477
$Bi_4O_{12}Si_3$, 28, 324; 31A, 227
$Bi_{12}O_{20}Si$, 28, 324
$Bi_{24}O_{40}Si_2$, 28, 324
$Bi_{12}O_{20}Sn$, 28, 324
$Bi_2O_9SrTa_2$, 27, 538
$Bi_2O_{18}Sr_2Ti_5$, 27, 539
$Bi_4O_{15}SrTi_4$, 27, 538
$Bi_4O_{18}Sr_2Ti_5$, 27, 539
BiO_4Ta, 28, 354; 32A, 509
Bi_3O_3TaTi, 27, 538
$Bi_4O_{12}Ti_3$, 26, 402
$Bi_4O_{12}Ti_4$, 27, 538
BiO_4V, 28, 354
δ- Bi_2O_9W, 27, 477
Bi-Pb, 27, 76, 81; 28, 44
Bi-Pb-Pt, 27, 287
$BiPb_3Pt$, 34A, 137
Bi_3PbPt_2, 34A, 139
Bi_7PbPt_4, 34A, 137
Bi_2PbS_4, 27, 81
$Bi_2Pb_2S_5$, 26, 432
$Bi_2Pb_3S_6$, 29, 105, 424
Bi-Pb-Se, 26, 63
$BiPbSe_2$, 26, 63
Bi_2PbSe_4, 33A, 43
$Bi_2Pb_2Se_5$, 31A, 25
Bi-Pb-Te, 34A, 33
$BiPbTe_4$, 26, 63
$Bi_2Pb_2Te_5$, 34A, 32; 35A, 33
Bi_4PbTe_7, 27, 83; 29, 105; 34A, 32; 35A, 32, 33
Bi-Pb-Te-Tl, 27, 363
Bi-Pd-Pt, 27, 287
Bi-Pd-Sb, 27, 287
BiPt, 27, 85
Bi_2Pt, 27, 85; 30A, 117; 34A, 137

α-Bi_2Pt, 27, 85
β-Bi_2Pt, 27, 85
γ-Bi_2Pt, 27, 85
Bi_3Pt_2, 34A, 139
Bi-Pt-Sb, 27, 287
BiPtSb, 28, 44
$BiRb_3$, 26, 54
Bi_2Rb, 26, 54
BiRh, 26, 65
Bi_2Rh, 26, 35, 65; 27, 87
α-Bi_2Rh, 26, 65
Bi_4Rh, 26, 65; 27, 87
α-Bi_4Rh, 26, 65
β-Bi_4Rh, 26, 65
Bi_2STe_2, 27, 90
$Bi_3S_5Te_7$, 27, 90
Bi-Sb, 27, 76
$Bi_{2-x}Sb_xSe_yTe_{3-y}$, 26, 33
$(Bi,Sb)_2(Se, Te)_3$, 28, 44; 30A, 193
$(Bi, Sb)_2Te_3$, 27, 40
Bi-Sb-Te-Tl, 27, 362
BiSc, 27, 88
BiSe, 27, 89; 29, 105; 30A, 117
Bi_2Se, 27, 88
Bi_2Se_2, 29, 105; 32A, 30
Bi_2Se_3, 26, 63; 27, 88; 28, 12, 44; 29, 105; 30A, 117
Bi_2Se_3 II, 29, 105
Bi_4Se_3, 30A, 117; 33A, 44
$Bi_{4.26}Se_3$, 27, 89
$Bi_{11}Se_9$, 28, 44
Bi_2SeTe_2, 26, 65; 28, 12; 29, 105; 33A, 152
Be_2Se_2Te, 28, 12
$BeSe_2Tl$, 27, 89
Bi-Sn, 27, 76
BiSn, 33A, 152
Bi-Sn-Te-Tl, 27, 363
Bi_3Sr, 26, 62
BiTb, 26, 255
Bi-Te, 27, 76
BiTe, 29, 105
Bi_2Te_3 26, 43; 27, 91, 211; 28, 12, 29, 105; 30A, 117; 33A, 45
$BiTe_2Tl$, 26, 46; 27, 211
BiTm, 29, 105
Bi_2U, 32A, 21
BiY, 27, 92; 29, 105
Bi_2Zr, 33A, 24
$BkCl_6Cs_2$, 34A, 209
$BkCl_6Cs_2Na$, 34A, 209
$Br_2.8.6H_2O$, 28, 296
Br_2nC, 28, 65
$BrCCoH_{12}N_4O_3$, 28, 100
$BrCCoH_{12}N_4O_{12}$, 27, 654
$BrCCoH_{15}N_5O_3.H_2O$, 30A, 286
$Br_3C_{15}Co_4In_3O_{15}$, 35A, 402
$Br_2C_4FeHg_2O_4$, 32A, 399
Br_5CIrK_2O, 29, 286
$BrCKN_2O_4$, 34A, 349
$Br_{0.3}C_4K_2N_4Pt.2.5H_2O$, 33A, 250
$Br_2C_8Mn_2O_8$, 28, 69, 296
$Br_2C_3O_3Ru$, 33A, 512
Br_2Ca, 28, 97
BrCaFe, 28, 300
$Br_2CaO_5Ta_2$, 27, 534
Br_2Cd, 27, 431
$Br_2CdH_8N_4$, 30A, 457
$Br_3Cd_4P_2$, 28, 300
BrCeS, 31A, 247
$Br_7Cl_3CsRe.2H_2O$, 34A, 213
$BrClH_4IN$, 32A, 219
$BrCl_3OW$, 33A, 220
BrClPb, 34A, 176
$Br_2Co.2H_2O$, 30A, 449; 32A, 204
$Br_2Co.6H_2O$, 26, 334
$Br_3CoH_{18}N_6$, 27, 657
$Br_2CoH_{15}N_6O_2$, 33A, 209
$Br_3Co_2H_{21}N_6O_3$, 32A, 208
BrCoHO, 26, 327; 27, 470
α-$BrCo_2H_3O_3$, 29, 250
β-$BrCo_2H_3O_3$, 29, 250
$BrCoN_2O_2$, 28, 300
$Br_2CoO_6.6H_2O$, 28, 306
Br_2Cr, 27, 433
Br_3Cr, 27, 439
$BrCr_2CuSe_3$, 33A, 235
$Br_3CrH_{18}N_6$, 27, 657
$Br_6Cr_4H_{42}N_{12}O_6.2H_2O$, 35A, 399
BrCrO, 31A, 107
BrCrS, 31A, 253
BrCs, 26, 297
BrCs.1/3 Br_2H_4O, 33A, 201
Br_3Cs, 34A, 180
$BrCsI_4$, 31A, 112
Br_6Cs_2Mo, 27, 464
Br_3CsPb, 31A, 110
Br_6Cs_2Re, 29, 415
$Br_8Cs_2Re_2$, 35A, 171
$Br_{11}Cs_2Re_3$, 31A, 110
$Br_{12}Cs_3Re_3$, 34A, 213
Br_6Cs_2Te, 31A, 109
$Br_2CuH_{18}N_6$, 32A, 194
$Br_4Cu_3H_{12}N_4$, 35A, 416
$BrCu_2H_3O_3$, 26, 327; 29, 250
$BrCuO_2$, 34A, 252
$Br_2CuO_6.6H_2O$, 28, 306
BrEuO, 30A, 289
$BrEu_3O_4$30A, 450
BrF_4K, 34A, 202
BrF_8Sb, 34A, 198
Br_2Fe, 28, 300
$BrFe_2H_3O_3$, 29, 250
$Br_{11}Ga_9S_8$, 31A, 246
$BrH.H_2O$, 35A, 194
$BrH.2H_2O$, 35A, 194
$BrH.3H_2O$, 35A, 194
$BrH.4H_2O$, 33A, 229
$Br_6H_{16}HgN_4$, 30A, 449
$Br_3H_3KNPt.H_2O$, 30A, 265
$Br_6H_{12}MgO_6Te$, 29, 251
$BrH_3Mn_2O_3$, 29, 251
$Br_5H_8MoN_2O$, 35A, 423
BrH_4N, 26, 298; 29, 284; 30A, 284; 35A, 145
BrH_4N I, 29, 416
BrH_4N II, 29, 415, 416
BrH_4N III, 29, 416
$(Br,Cl)H_4N$, 31A, 241
BrH_7N_2, 29, 416
$BrH_{13}N_4$, 26, 298
$Br_6H_{12}N_4Pt_2$, 27, 667
$Br_4HN_3S_2$, 26, 294
$Br_{12}H_{16}N_4Sb_2$, 31A, 108
$Br_6H_8N_2Te$, 31A, 109
$Br_2H_8N_4Zn$, 30A, 457
BrH_4NaO , 29, 284
BrH_4NaO_2,27, 427
$BrH_3Ni_2O_3$, 29, 251
$Br_2H_4NiO_8$, 27, 643
BrHOPb, 31A, 107
Br_2H_2OSr, 29, 284
Br_4Hf, 28, 300
BrHgI, 34A, 211
$Br_3HgK.H_2O$, 28, 300; 34A, 212
$Br_2Hg_2O_6$, 32A, 319
$Br_2Hg_5O_4$, 30A, 301; 33A, 235
$Br_2Hg_3S_2$, 28, 344; 31A, 253
Br_2Hg_2Sb, 31A, 256
$Br_2Hg_3Te_2$ 27, 465
Br_3HgTl, 30A, 449
BrI, 33A, 230
BrIPb, 34A, 176
BrInS, 27, 466
BrInSe, 27, 466
BrInTe, 27, 466
Br_3Ir, 33A, 237
BrK, 26, 297; 27, 426; 29, 416
$BrKO_4$, 34A, 361
$Br_{\sim 0.9}K_{\sim 0.9}O_3U$, 29, 282
Br_6K_2Pt, 35A, 168
Br_6K_2Re, 28, 300; 35A, 168
Br_6K_2Sn, 27, 462
Br_6K_2Te, 27, 463; 29, 286
$Br_3KZn.2H_2O$, 31A, 260;

32A, 207; 33A, 233; 35A, 161
BrLaS, 31A, 247
$BrLiO_3$, 27, 643
$BrLnO$, 30A, 289
BrMg, 34A, 180
$Br_2MgO_6.6H_2O$, 28, 306
$Br_2Mn.2H_2O$, 30A, 449; 32A, 204
Br_3MoO, 35A, 184
Br_6MoRb_2, 27, 464
$Br_6N_3P_3$, 27, 413; 32A, 207; 35A, 145
BrNTi, 29, 248
BrNU, 31A, 253
α-BrNZr, 29, 248
BrNa, 26, 297
BrNa. 5 1/7 H_3N, 30A, 260
$BrNa.2H_2O$, 27, 427; 29, 284
$Br_{3.04}Nb$, 30A, 449
Br_4Nb, 29, 416
Br_5Nb, 28, 300; 31A, 239
Br_8Nb_3, 30A, 449; 31A, 105
$Br_2Ni.2H_2O$, 28, 300
$Br_2Ni.6H_2O$, 28, 300
$Br_2NiO_6 2H_2O$, 27, 643
$Br_2NiO_6.6H_2O$, 28, 306
Br_3NiRb, 34A, 211
$Br_2O_{24}P_6Sr_{10}$, 26, 469
BrOSm, 30A, 289
BrO_4Sm_3, 30A, 450
$Br_3O_9Sm.9H_2O$, 34A, 362
BrOV, 34A, 228
Br_2O_2W, 30A, 450
Br_4OW, 31A, 99
Br_4O_2W, 30A, 450
BrOYb, 30A, 289
BrO_4Yb_3, 30A, 450
$Br_2O_6Zn.6H_2O$, 28, 306
Br_5P, 35A, 151
Br_7P, 32A, 203
$Br_2P_2S_6$, 30A, 355
Br_5Pa, 34A, 180
Br_2Pd, 31A, 106
BrRb, 26, 297, 298
$Br_6Rb_2Re_2$, 29, 415
Br_6Rb_2Sb, 33A, 515
Br_3Rh, 33A, 237
Br_3Ru, 31A, 251; 33A, 236
Br_4SW, 35A, 156
α-Br_3Sb, 29, 285
β-Br_3Sb, 27, 439
Br_3Sc, 26, 298; 29, 416
Br_4Sn, 28, 99
Br_2Sr, 28, 98
Br_4Ta, 29, 416
Br_5Ta, 28, 300; 31A, 239
$Br_{15}Ta_6$, 31A, 259; 33A, 203
Br_4Te, 29, 416
Br_4Th, 31A, 248
Br_2Ti, 26, 333
Br_4Ti, 31A, 105
BrTl, 26, 297; 27, 426
Br_2Tl, 28, 99
Br_5W, 30A, 450
Br_6W, 30A, 450
$Br_{16}W_6$, 31A, 259; 33A, 231
Br_3Zr, 27, 442; 29, 249
Br_4Zr, 27, 444; 28, 300

C, 26, 286; 27, 395, 32A, 147; 34A, 149
C_2Ca, 26, 91; 27, 117
$C_2Ca(III)$, 26, 91
$C_2Ca(IV)$, 26, 91
$C_2(Ca,Ce)F_2O_6Th.3H_2O$, 30A, 454
$CCa_2(Er, Y)_2O_{15}Si_4.H_2O$, 26, 534
C_2CaFO_6Y, 27, 563
$CCa_3H_6MnO_{13}S.12H_2O$, 31A, 174; 34A, 313
$CCaH_{12}O_9$, 35A, 296
$C(Ca,Ln,Na,Sr)O_3$, 31A, 261
$CCa_2(Ln,Y)_2O_{15}Si_4.H_2O$, 31A, 229
C_2CaMgO_6, 26, 481
$C_4CaMg_3O_{12}$, 27, 561
C_2CaMnO_6, 26, 481
$CCaN_2$, 27, 419
$C_4CaN_4Ni.5H_2O$, 33A, 517; 34A, 231
$C_4CaN_4Pd.5H_2O$, 33A, 517
$C_2(Ca,Na)MO_6$, 32A, 417
$C_2CaNa_2O_6.2H_2O$, 32A, 416
$C_2CaNa_2O_6.5H_2O$, 33A, 435
$C_3CaNa_2O_{11}U.6H_2O$, 28, 177
$CCaO_3$, 26, 481; 27, 561; 28, 174; 29, 426; 30A, 408; 34A, 348
$CCaO_3.6H_2O$, 35A, 296
$C_2CaO_4.H_2O$, 26, 484
$C_2CaO_4.2H_2O$, 30A, 454
$CCa_2O_{15}(RE, Y)_2Si_4.H_2O$, 28, 275
$C_2Ca_5O_{13}Si_3$, 35A, 472
$C_2Ca_{0.1}O_3Sr_{0.9}$, 26, 443
$C_2CaO_8U.3H_2O$, 31A, 243
$C_2CaO_8U.5H_2O$, 31A, 243
$C_{12}Cd_3Co_2N_{12}.12H_2O$, 33A, 244
$C_{12}Cd_3Cr_2H_{12}N_{12}O_6$, 27, 653
$C_{12}Cd_3Cr_2N_{12}.6H_2O$, 26, 341; 27, 653
$C_{12}Cd_3H_{12}Mn_2N_{12}O_6$, 29, 292
$C_8Cd_4H_2MoN_{12}.4H_2O$, 34A, 229
$CCdH_2N_4O_2S_2.2H_2O$, 32A, 346
$C_2CdH_8N_6S_2$, 26, 343
$C_4CdH_8N_6S_4$, 34A, 234
$C_4CdH_8N_6S_2Se_2$, 34A, 234
$C_8CdH_{16}N_{12}S_8Zn$, 32A, 240
C_4CdHgN_4, 33A, 254
$C_4CdHgN_4S_4$, 33A, 253
$C_4CdHgN_4Se_4$, 31A, 255
C_2CdMgO_6, 26, 481
$C_{12}Cd_3Mn_2N_{12}.6H_2O$, 28, 344; 29, 292
$CCdO_3$, 26, 481
$CCdTi_2$, 29, 106
CCe, 29, 106
C_2Ce, 26, 91; 29, 106; 31A, 32; 32A, 118
C_3Ce_2, 26, 92; 29, 106; 32A, 120
C-Ce-Th, 29, 157
C-Ce-U, 29, 157
$C_2Cl_3Co_3H_{26}N_{12}O_{18}$· $ClNaO_4$. $2H_2O$, 35A, 394
$CCl_2CoH_{15}N_6S$, 32A, 197
$C_4Cl_3CoO_4Si$, 32A, 402
$CCl_4Cs_2H_2O_2Ru$, 35A, 429
$C_2Cl_4CuH_{12}N_2$, 29, 278
$C_4Cl_4CuH_{16}N_2$, 29, 278
$CCl_2H_{14}Mg_2O_{10}$, 29, 426
$C_6Cl_2H_{24}N_{12}NiO_8S_6$ 26, 475
$CCl_5H_5N_2OTe$, 28, 300
$C_{10}Cl_6H_{12}N_2Te$, 28, 300
$C_3Cl_5H_9POSb$, 28, 95
$C_4Cl_{0.3}K_2N_4Pt.2.5H_2O$, 33A, 250
$CCl_2Mg_2O_3.7H_2O$, 29, 426
$C_{15}ClMn_3O_{15}Sn$, 32A, 510
CCl_3NOSi, 34A, 185
$C_2Cl_2N_2O_2Si$, 34A, 185
$C_3ClN_3O_3Si$, 34A, 185
C_2ClO_2Rh, 26, 332
$C_6Cl_4O_6Ru_2$, 34A, 165
$C_5Cl_6O_5Ru_2Sn$, 35A, 427
CCo_2, 26, 95
CCo_3, 26, 97
$C_{12}Co_3Cr_2H_{12}N_{12}O_6$, 27, 653
$C_{12}Co_3Cr_2N_{12}.6H_2O$, 26, 341; 27, 653
$C_6CoCs_2LiN_6$, 34A, 231
$C_{15}Co_6Cs_2O_{15}.3H_2O$, 33A, 165
$C_6CoD_3N_6$, 35A, 404
$C_4CoF_3O_4Si$, 35A, 412
$C_{12}Co_2Fe_3N_{12}.12H_2O$, 33A, 244
$C_8CoH_{16}HgN_{12}S_8$, 28, 302; 30A, 299
$CCoH_{15}IN_5O_3.H_2O$, 32A, 218
$C_{12}Co_3H_{12}Mn_2Ni_{12}O_6$, 29, 292
$C_6CoH_3N_6$, 33A, 242
$CCoH_{15}N_8O_6S$, 32A, 354
$C_4Co_2H_{30}N_{14}O_2S_4$, 30A, 297
$C_2CoH_8N_6S_2$, 26, 343
$C_4CoH_8N_6S_4$, 34A, 234
$C_4CoH_8N_6S_2Se_2$, 34A, 234
$C_4CoHgN_4S_4$, 33A, 253
$C_4CoHgN_4Se_4$, 26, 343
$C_8Co_2HgO_8$, 33A, 164
$C_{12}Co_3Ir_2N_{12}.12H_2O$, 26, 340

$C_6CoK_3N_6$, 26, 339
$C_4CoK_2N_4S_4.3H_2O$, 28, 113
$C_{14}Co_6K_4O_{14}.6H_2O$, 34A, 159
$C_{14}Co_6K_4O_{14}.8H_2O$, 33A, 166
$C_{12}Co_2Mn_3N_6.13H_2O$, 35A, 405
$C_{12}Co_2Mn_3N_{12}.12H_2O$, 33A, 244
$C_{12}Co_3Mn_2N_{12}.6H_2O$, 28, 344; 29, 292
CCo_6Mo_6, 34A, 139
C_2CoN_2, 28, 302
$C_{12}Co_5N_{12}.12H_2O$, 33A, 244
$C_{12}Co_2N_{12}Ni_3.12H_2O$, 33A, 244
$C_6CoN_6Tl_3$, 30A, 450
$C_{12}Co_2N_{12}Zn_3.12H_2O$, 33A, 244
C_xCoNb_2, 29, 106
CCo_3Nb_3, 30A, 117
C-Co-Nb-Ta-W, 30A, 193
$CCoO_3$, 26, 481
$C_8Co_2O_8$, 29, 241
$C_{12}Co_4O_{12}$, 31A, 259
$C_{16}Co_6O_{16}$, 33A, 166
$C_9Co_3O_9S$, 32A, 406
$C_{14}Co_6O_{14}S_4$, 32A, 404
$C_{12}Co_4O_{12}Sb_4$, 35A, 403
$C_8Co_2O_8Zn$, 32A, 510
CCo_3Ta_3, 30A, 118
C-Co-Ti-W, 26, 97
C-Co-W, 26, 97
C-Co-Zn, 26, 99
$C_{17}Co_{51}Zn_{32}$, 26, 99
C_2Cr_3, 27, 122; 34A, 56
$(C,N)_2Cr_3$, 31A, 26
C_3Cr_7, 26, 96; 30A, 36; 34A, 139
C_6Cr_{23}, 26, 96; 27, 122
$C_{12}Cr_2Cu_3N_{12}.6H_2O$, 26, 341; 27, 653
$C_{12}Cr_2Cu_3H_{12}N_{12}O_6$, 27, 653
$C(Cr,Fe)_3$, 28, 44
$C_{12}Cr_2Fe_3H_{12}N_{12}O_6$, 27, 653
$C_{12}Cr_2Fe_3N_{12}.6H_2O$, 26, 341; 27, 653
CCr_2Ga, 28, 44
CCr_2Ge, 28, 44
$C_{12}Cr_2H_{12}Mn_3N_{12}O_6$, 27, 653
$C_{12}Cr_2H_{12}N_{12}Ni_3O_6$, 27, 653
$CCrH_6NO_8S_2.12H_2O$, 33A, 372
$C_{12}Cr_2H_{12}N_{12}O_6Zn_3$, 27, 653
$C_{12}Cr_3Ir_2N_{12}.12H_2O$, 26, 340
$C_3CrK_3N_3O_4$, 30A, 292
$C_5CrK_3N_6O$, 31A, 116
$C_{12}Cr_2Mn_3N_{12}.6H_2O$, 26, 341; 27, 653
$C_{12}Cr_2N_{12}Ni_3.6H_2O$, 26, 341; 27, 653
$C_{12}Cr_2N_{12}Zn_3.6H_2O$, 26, 341; 27, 653
C-Cr-Nb, 30A, 194
C_6CrO_6, 32A, 393
$CCr_{2.4}Pt_{1.7}$, 30A, 118
CCr_4Si_3, 26, 103
C-Cr-Si-W, 26, 103
C-Cr-Ta, 30A, 194
C_2Cr_2V, 31A, 26
$C_5Cs_3N_5O_2S_5U$, 28, 114; 31A, 119
$C_8Cs_4N_8S_8Th.2H_2O$, 28, 320
$C_8Cs_4N_8S_8U.2H_2O$, 28, 320
$C_{12}Cu_3H_{12}Mn_2N_{12}O_6$, 29, 292
$CCuH_3N_2$, 30A, 290
$CCuH_4N_3$, 31A, 116
$C_4Cu_3HgN_7$, 32A, 509
$CCuH_6N_2O_3$, 26, 660; 28, 176
$C_2CuH_6N_4S_2$, 28, 108; 31A, 118; 32A, 239
$C_2CuH_{12}N_6S_2$, 28, 110
$C_3Cu_2H_9N_6S_3$, 34A, 233
$CCu_2H_2O_5$, 31A, 260; 32A, 419
$C_2(Cu,Zn)_5H_6O_{12}$, 30A, 454
$CCu_2H_6O_{21}Pb_5S_3$, 35A, 376
$C_4CuH_{10}O_8S_2.4H_2O$, 34A, 306
$C_4CuHgN_4S_4$, 27, 660; 28, 112
$C_4CuK_3N_4$, 33A, 245
$C_3Cu_2KN_3.H_2O$, 27, 416
$C_{12}Cu_3Mn_2N_{12}.6H_2O$, 28, 344; 29, 292
CDNO, 34A, 233
C_2Dy, 33A, 52
C_2Eu, 29, 106
$CEuO_3$, 29, 383
$C(F,OH)LnO_3$, 31A, 255
C-Fe, 29, 157; 30A, 194
ε-CFe_x, 27, 124
$CFe_{1.37}$, 30A, 118
CFe_3, 26, 99; 27, 122; 28, 15; 29, 35, 106
C_2Fe_5, 28, 15; 31A, 30; 32A, 136
C_3Fe_7, 29, 106; 30A, 36, 118; 32A, 139; 34A, 139
$C_6FeH_6K_4N_6O_3$, 29, 292
$CFe_2H_{16}Mg_6O_{19}.4H_2O$, 31A, 174; 32A, 422; 33A, 439
$C_3Fe_6H_{48}Mg_{18}O_{57}.12H_2O$, 28, 322
$C_{12}Fe_3H_{12}Mn_2N_{12}O_6$, 29, 292
$C_6FeH_4N_6$, 29, 416; 31A, 114; 33A, 242
$C_6Fe_2H_4N_2O_6$, 33A, 163
$C_2FeH_8N_6S_2$, 26, 343
C_2FeHf_2S, 30A, 35, 118
$C_{12}Fe_3Ir_2N_{12}.12H_2O$, 26, 342
$C_6FeK_3N_6$, 34A, 230
$C_6FeK_4N_6.3H_2O$, 29, 292
$C_2FeMgNaO_4.8.5H_2O$, 28, 322
$C(Fe,Mg,Ni)O_3$, 31A, 242
C-Fe-Mn, 30A, 194
$CFe_{2.7}Mn_{0.3}$, 30A, 37, 118
$C(Fe,Mn)_3$, 31A, 31
$C_2(Fe,Mn)_5$, 31A, 31
C_2FeMn_5, 26, 100
$C_3Fe_7Mn_7$, 26, 100
$C_{12}Fe_3Mn_2N_{12}.6H_2O$, 28, 344; 29, 292
$C_{14}FeMn_2O_{14}$, 32A, 395
CFe_2Mo, 29, 106
$C_5FeN_6Na_2O.2H_2O$, 28, 106
$C_6FeN_6Tl_3$, 31A, 254
$CFeO_3$, 26, 481
$CFe_2(O?)$, 29, 106
C_5FeO_5, 27, 650; 29, 241; 30A, 259; 34A, 163
$C_{12}Fe_3O_{12}$, 31A, 259; 34A, 160
$C_{16}Fe_5O_{15}$, 27, 647
$C_6Fe_2O_6S_2$, 30A, 259
$C_8Fe_2O_{10}S$, 32A, 400
$C_9Fe_3O_9Se_2$, 28, 70
$C_{16}Fe_4O_{16}Sn$, 32A, 411
C-Fe-Si, 26, 100
C-Fe-Sn, 26, 100
$C_{15}Fe_{70}Sn_{15}$, 26, 100
C_2FeU, 27, 125
CFe_3V_3, 30A, 118
CFe_3W_3, 30A, 38; 32A, 45
CFe_6W_6, 29, 35; 32A, 45
$CGaMn_3$, 30A, 118
$CGaMo_2$, 28, 44
$CGaNb_2$, 29, 106
$C_xGa_3Nb_5$, 28, 44
C_xGaPt_3, 26, 225
$CGaTa_2$, 29, 106
$CGaTi_2$, 29, 106
$CGaV_2$, 28, 44
$C_2(Gd,U)$, 29, 158
$C_3(Gd,U)_2$, 29, 158
$CGeMn_3$, 33A, 29
$C_xGe_3Mo_5$, 29, 106
$C_xGe_3Nb_5$, 29, 106
$C_x(Ge\ Si_{1-x})_3V_5$, 30A, 207
$C_xGe_3Ta_5$, 28, 44
$CGeTi_2$, 28, 44
$CGeV_2$, 28, 44
$CHKO_3$, 33A, 433
$C_6H_2K_6O_{20}Zr_2.6H_2O$, 33A, 438
$(C,H)Li$, 35A, 134
$CH_2Mg_2O_5.3H_2O$, 30A, 408
$C_8H_6Mg_{11}O_{30}.7H_2O$, 30A, 458
$C_7H_9MnN_2O_5$, 34A, 164
C_5HMnO_5, 34A, 158
$C_9H_7MnO_3$, 34A, 162
$C_{14}HMnO_{14}Re_2$, 32A, 396, 510

$C_4H_4N_4$, 30A, 291
$C_2H_3N_3Ni$, 31A, 115
$C_3H_{14}N_8NiO_6 \cdot H_2O$, 32A, 151
$C_4H_{16}N_8NiO_3S_6 \cdot H_2O$, 34A, 315
$C_4H_8N_6NiS_4$, 34A, 234
$C_4H_8N_6NiS_2Se_2$, 34A, 234
$CH_5N_5O_6$, 32A, 510
$C_2H_{14}N_8O_4Zn$, 30A, 409
$C_6H_{26}N_6O_{20}Zr_2 \cdot 3H_2O$, 35A, 417
CH_6N_4S, 34A, 349
$C_2H_8N_6S_2Zn$, 30A, 298
$C_8H_4N_8W \cdot 6H_2O$, 35A, 423
$CHNaO_3$, 30A, 408
$CH_2O_5Zn_2 \cdot H_2O$, 31A, 248
$C_2H_6O_{12}Zn_5$, 30A, 454
$C_3H_2O_{11}Zn_4 \cdot 4H_2O$, 31A, 248
$C_3H_6O_{15}Zn_5$, 31A, 248
CH_2S_3, 30A, 410
$C_{0 \cdot 3}HZr$, 34A, 137
$C_{0 \cdot 6}H_{0 \cdot 4}Zr$, 30A, 35
$C_{1-x}H_yZr_2$, 32A, 44
$CH_{1 \cdot 44}Zr_2$, 32A, 44
$C_{1-x}Hf$, 26, 68; 27, 121
CHf, 26, 68, 91, 93; 29, 158; 30A, 194
CHf_2In, 28, 44
CHf_2Pb, 29, 106
CHf_2Sn, 28, 44
CHf_2Tl, 29, 106
C-Hf-U, 26, 93
C_xHf_2Zn, 29, 106
C(Hf,Zr), 30A, 194
C_2HgIKN_2, 28, 104
C_2HgN_2, 34A, 228
$C_4HgN_4NiS_4 \cdot 2H_2O$, 28, 111
$C_2Hg_2N_2O$, 28, 116; 30A, 301
$C_2HgN_2S_2$, 30A, 450
$C_4HgN_4Se_4Zn$, 31A, 255
CHo_2, 29, 106
C_2Ho, 32A, 118
$C_4I_2O_4Ru$, 27, 646
$CInNb_2$, 29, 106
C_xInPd_3, 26, 225
$CInTi_2$, 28, 44; 29, 106
$CInTi_3$, 29, 106
$CInZr_2$, 28, 44
$C_{12}Ir_2Mn_3N_{12} \cdot 12H_2O$, 26, 340
$C_{12}Ir_4O_{12}$, 31A, 259
$C_{32}K$, 30A, 449
$C_6K_3MnN_6$, 35A, 401
$C_6K_5MnN_6$, 27, 653
$C_5K_3MnN_6O \cdot 2H_2O$, 31A, 258; 32A, 237
$C_8K_4MoN_8 \cdot 2H_2O$, 33A, 247
$C_4K_3MoN_4NaO_2 \cdot 6H_2O$, 33A, 248
$C_5K_4MoN_6O$, 33A, 246
CKN, 26, 339; 27, 415
C_4KN_3, 28, 107
$C_4KN_4NaPt \cdot 3H_2O$, 33A, 249
$C_4K_2N_4Ni$, 29, 293
CKN_3O_6, 32A, 347
$C_2K_2N_4O_8$, 33A, 432
$C_4K_3N_4O_2Re$, 26, 342
$C_4K_2N_4OS_4V \cdot 5H_2O$, 28, 302
$C_5K_3N_6OV \cdot 2H_2O$, 35A, 398
$C_6K_2N_6Pt$, 32A, 236
$C_4K_2N_5Re \cdot H_2O$, 34A, 229
$C_6K_5N_6Re$, 27, 653
$CKNS$, 28, 108; 33A, 252
$CKNSe$, 30A, 296
$C_3KN_3Se_3 \cdot \frac{1}{2}H_2O$, 28, 116
$C_4K_2N_4Zn$, 31A, 117
$CKNpO_5$, 29, 426
$CK_2O_3 \cdot 3H_2O_2$, 27, 560
$C_2K_4O_6 \cdot 3H_2O$, 32A, 414
CKO_5Pu, 29, 426
$CK_2S_3 \cdot H_2O$, 35A, 483
C_2La, 26, 91; 33A, 53
C_3La_2, 26, 92
$C_3La_2O_9 \cdot 8H_2O$, 33A, 433
C_nLi, 30A, 449
C_2Li_2, 27, 126; 30A, 39; 32A, 233
$C_{16}Li$, 30A, 118
$C_{40}Li$, 30A, 118
C_2Lu, 26, 91
C_2MgNiO_6, 29, 426
$CMgO_3$, 26, 481
$CMgO_3 \cdot Cl_2Mg \cdot 7H_2O$, 29, 426
$CMgO_3 \cdot 2H_2O$, 30A, 454
$CMgO_3 \cdot 3H_2O$, 30A, 454
C_2MgO_6Pb, 31A, 253; 33A, 437
$C_4Mg_{24}Pt_{72}$, 26, 225
C_2Mn_5, 26, 76
C_3Mn_7, 26, 100; 29, 106; 30A, 36, 118
CMn_4MoSi, 35A, 135
$C_{0 \cdot 25}Mn_4N_{0 \cdot 75}$, 27, 272
$C_{12}Mn_5N_{12} \cdot 6H_2O$, 28, 344; 29, 292
$C_{12}Mn_2N_{12}Ni_3 \cdot 6H_2O$, 28, 344; 29, 292
C_4MnNO_5, 34A, 160
$C_{12}Mn_2N_{12}Zn_3 \cdot 6H_2O$, 28, 344; 29, 292
$CMnO_3$, 26, 481; 32A, 415
$C_{10}Mn_2O_{10}$, 28, 67; 34A, 162
CMn_5Si, 35A, 134
C-Mo, 27, 127, 131
$C_{0 \cdot 25-0 \cdot 50}Mo_{0 \cdot 75-0 \cdot 50}$, 27, 130
CMo, 27, 126
α-CMo, 26, 101
CMo_2, 26, 101; 27, 127; 28, 15
β-CMo_2, 27, 131
α-C_2Mo_3, 27, 127
γ'-C_2Mo_3, 27, 131
η-C_2Mo_3, 27, 127
C-Mo-Nb, 26, 101
C_{1-x}(Mo-Nb), 26, 101
$C_{1-x}(Mo,Nb)_2$, 26, 101
$C(Mo,Nb)_2$, 26, 101
CMo_6Ni_6, 34A, 139
C(Mo,Re), 29, 107, 159
CMo_2-Re, 29, 159
CMo_3Re_2, 29, 106
$C_{0 \cdot 6}Mo_{4 \cdot 8}Si_3$, 30A, 39, 118
CMo_4Si_3, 26, 103
C-Mo-Si-W, 26, 103
$C_{1 \cdot 70}MoU$, 31A, 32
C_2MoU, 29, 36, 107
C-Mo-V, 27, 128
γ-C-Mo-Zr, 27, 131
$C_{0 \cdot 03-0 \cdot 51}Mo_{0 \cdot 03-0 \cdot 96}Zr_{0 \cdot 005-0 \cdot 54}$, 27, 130
C_2N_2, 28, 52, 401
CN_2Na_2, 29, 239
C_4N_3Na, 32A, 234
$C_4N_4Na_2Pd \cdot 3H_2O$, 32A, 235
$CNNaS$, 33A, 253
$C_8N_8Na_3W \cdot 4H_2O$, 35A, 424
C_3N_3P, 26, 339; 29, 290
$C_6N_9P_3S_6$, 35A, 137
CN_2Pb, 29, 239
$C_2N_2PbS_2$, 31A, 258
$C_4N_4PdRb_2 \cdot H_2O$, 32A, 236; 35A, 200
C_2N_2S, 28, 302; 31A, 113
$C_2N_2S_3$, 28, 302; 29, 291
$C_2N_2S_6$, 28, 302
C_2N_2Se, 28, 302; 31A, 114
CN_2Sr, 31A, 78
CN_2Tl_2, 29, 239
C-N-V, 29, 160, 161
(C,N)V, 30A, 195
C_2N_2Zn, 27, 416
$C_{32}Na$, 30A, 449
$C_{120}Na$, 30A, 449
CNa_2O_3, 29, 380
$CNa_2O_3 \cdot H_2O$, 33A, 159
$CNa_2O_3 \cdot 7H_2O$, 28, 302
$CNa_2O_3 \cdot 10H_2O$, 28, 302; 34A, 347
$C_8Na_4O_{16}Zr \cdot 3H_2O$, 26, 486
C-Nb, 26, 102
$C_{0 \cdot 75}Nb$, 34A, 58
$C_{0 \cdot 98}Nb$, 29, 160
$C_{1-x}Nb$, 27, 121
CNb, 26, 96, 101, 102; 29, 107; 30A, 118
CNb_2, 26, 101, 102; 29, 107; 32A, 507
C_xNb_2Ni, 29, 107
CNb_3Ni_3, 30A, 118
CNb_2P, 33A, 152
CNb_2S, 33A, 152
CNb_2Sn, 29, 107
C(Nb,Ta), 30A, 195
C-Nb-U, 26, 93
C_xNb_2Zn, 29, 107
C_2Nd, 31A, 32; 32A, 118
CNi_3, 26, 205
$CNiO_3$, 26, 481; 27, 561, 724; 28, 322

$C_{12}Ni_4O_{18}P_4$, 31A, 77
CNp, 34A, 139
C_3O_2, 29, 296
$C_{12}O_{12}Os_3$, 27, 651; 28, 344
$C_{12}O_{12}Ru_3$, 33A, 167
CO_3Sr, 35A, 295
CPV_2, 33A, 130
CPa, 34A, 139
$CO_4Pb_2.2H_2O$, 29, 383
$C_2O_7Pb_3$, 29, 426; 30A, 454
(C,O)Pu, 30A, 449
$C_{12}O_{12}Rh_4$, 31A, 259
$C_{16}O_{16}Rh_6$, 28, 71, 344
$C_6O_{12}RhTl_3.2H_2O$, 30A, 458
$C_{12}O_{12}Ru_3$, 28, 344
$C_{10}O_{10}Tc_2$, 27, 651; 30A, 443; 31A, 259
$C_x(O?)_yTh$, 29, 160
CO_3Tl_2, 31A, 256
CO_3Zn, 26, 481
C-Pb-Pd, 26, 225
$CPbTi_2$, 29, 107
$CPbZr_2$, 29, 107
C_xPd_3Sn, 26, 225
C_2Pr, 31A, 32; 32A, 118
C_3Pr_2, 26, 92
C-Pu, 26, 233; 29, 160, 161
$C_{0.84}Pu$, 29, 161
CPu, 26, 233; 27, 132
C_2Pu, 26, 233; 30A, 118
ζ-C_2Pu_3, 26, 233
C_3Pu_2, 26, 233
CPuU, 27, 132
C(Pu,U), 29, 161, 162, 163
$C_{58}PuU_{41}$, 29, 107
CRe_2, 34A, 139
C-Re-W, 26, 102
C(Re,W), 28, 44
CRe_2W_3, 28, 44
C_xRuZr_2, 29, 107
CS_2, 26, 289; 27, 719; 33A, 161
CSc, 26, 103
C_3Sc_4, 34A, 59
$CSc_{0.25}Ti_{0.75}$, 26, 103
CSi, 27, 407; 28, 64, 296; 29, 237; 30A, 254, 449; 31A, 77; 32A, 506; 33A, 160; 34A, 157
C_2SiTi_3, 32A, 46
$C_3Si_2U_3$, 30A, 40, 118
$C_3Si_{16}U_{20}$, 31A, 33; 33A, 54
$(C_xSi_{1-x})V_3$, 30A, 195
CSi_3W_4, 26, 103
$CSnTi_2$, 28, 44
$CSnZr_2$, 28, 44
C-Ta, 26, 253; 27, 133; 29, 163
ζ-$C_{0.64}Ta$, 27, 133
$C_{0.76}Ta$, 34A, 58
$C_{1-x}Ta$, 27, 121; 29, 107
CTa, 26, 91, 104, 294; 27, 133; 30A, 118
β-C_xTa, 29, 163
γ-C_xTa, 29, 163
CTa_2, 27, 133; 30A, 40, 119
β-CTa_2, 29, 107
CTa_{64}, 30A, 41
C-Ta-U, 26, 93
C_2Ta_2V, 35A, 43
C-Ta-W, 26, 133
$C_{1-x}(Ta,W)$, 27, 133
CTb_2, 34A, 60
C_2Tb, 26, 91; 32A, 118
C_3Tb_2, 26, 92
C-Tc, 26, 104
$C_{1-x}Th$, 27, 136
CTh, 26, 68, 105; 27, 134; 28, 44; 29, 107
$C_{1.97}Th$, 33A, 56
C_2Th, 26, 68, 105; 27, 134, 136; 29, 107, 163; 30A, 195
C-Th-U, 26, 104
C-Th-W, 27, 135
$C_{0.25}Ti$, 26, 264
$C_{1-x}Ti$, 27, 121
$C_{0.95}Ti$, 29, 164
CTi, 26, 97, 264; 30A, 195
C_xTi, 30A, 457
C_2Ti, 27, 136
CTi_2Sn, 29, 107
CTi_2Tl, 29, 107
CTi_3Tl, 29, 107
C_xTi_2Zn, 29, 107
$CTlZr_2$, 29, 107
CU, 26, 91, 93, 105; 27, 137; 28, 44; 29, 107
$C_{1.86}U$, 32A, 119
$C_{1.96}U$, 29, 107
C_2U, 26, 91, 105; 27, 137; 29, 107
C_3U_2, 29, 107; 31A, 34
C-U-W, 27, 136
C_2UW, 27, 136
C-U-Zr, 26, 93
$C_{0.97}U_{0.30}Zr_{0.70}$, 30A, 196
C-V, 27, 127, 138; 29, 164
$C_{0.48}V$, 29, 107
δ-$C_{0.63-0.79}V$, 26, 105
$C_{0.73}V$, 30A, 196
ε-$C_{0.79-0.92}V$, 26, 105
$C_{0.80}V$, 26, 105
$C_{0.84}V$, 33A, 57
$C_{0.88}V$, 27, 137
$C_{0.89}V$, 30A, 196
$C_{0.91}V$, 26, 105
$C_{1-x}V$, 27, 121
CV, 26, 105; 27, 127, 137
CV_2, 27, 127, 137, 138; 30A, 40, 119, 196
C_3V_4, 26, 96; 35A, 110
C_5V_6, 33A, 57
C_7V_8, 35A, 114
C-V-W, 27, 139
$C(V,W)_2$, 27, 139
CW, 26, 97, 102, 103, 105; 27, 141
CW_2, 27, 133, 141; 33A, 152
C_2W, 28, 44
C_xY, 34A, 60
CY_2, 34A, 61
C_2Y, 26, 91; 33A, 152
C_2Yb, 26, 91
C_xZnZr_2, 29, 107
C-Zr, 27, 132; 29, 165
$C_{0.45}Zr_{0.55}$, 27, 131
$C_{0.49}Zr_{0.51}$, 27, 131
$C_{0.95}Zr$, 29, 165
$C_{1-x}Zr$, 27, 121, 136
CZr, 30A, 197
CaB_2O_4, 35A, 283
(Ca,Cd)O_4W, 30A, 458
$(Ca,Ce)_5(F,O)_4Na_2(Nb,Ti)O_7Si_2$, 33A, 481
$(Ca,Ce)_5(F,O)_4Na_2(Nb,Ti)O_{14}Si_4$, 32A, 447
Ca-Ce-O-Zr, 30A, 452
$CaCe_2S_4$, 27, 327
$CaCl_2$, 28, 300
$CaCl_2.6H_2O$, 30A, 449
Ca_2ClCrO_4, 32A, 202
CaClFe, 28, 300
$(Ca,Na)_9(Cl,O)_{37}(Fe,Mn)_2(Nb,Zr)_2Si_{12}$, 28, 322
$CaClFeO_4$, 28, 306
CaClHO, 26, 327
$Ca_3Cl_2H_4O_6$, 33A, 217
$CaClLiO_6Ta_2$, 27, 534
$CaClNaO_6Ta_2$, 27, 534
Ca_2ClO_4P, 32A, 202
$Ca_{10}Cl_2O_{24}P_6$, 26, 468
$Ca_3Cl_2O_4Si$, 35A, 465
$CaCl_2O_5Ta_2$, 27, 534
Ca_2ClO_4V, 35A, 192
Ca_2CoO_6W, 28, 310
$Ca_3Cr_2Ge_3O_{12}$, 26, 415
$Ca_5Cr_3HO_{13}$, 28, 308
$CaCrO_4$, 28, 308, 328
$CaCr_2O_4$, 28, 344
$CaCu_2Ge_2$, 35A, 24, 117
$Ca_2CuH_4K_2O_{18}S_4$, 27, 634
$CaCu_4(H,O,S)_2.3H_2O$, 30A, 453
$CaCu_{2.64}H_6O_{14}S_2Zn_{1.36}.3H_2O$, 33A, 382
$Ca_2CuK_2O_{16}S_4.2H_2O$, 27, 634
$CaCu_2O_3$, 34A, 248
Ca_2CuO_3, 35A, 246
$CaCu_2Si_2$, 35A, 24, 117
CaD_2, 27, 119
$CaDy_2S_4$, 27, 327
$CaEr_2S_4$, 27, 327
CaF_2, 29, 414, 439; 30A, 266

$Ca_xF_xFe_{0.5}O_{12-x}Y_{3-x}$, 30A, 452
$Ca_4F_2H_6O_{10}Si_2$, 28, 256
$CaFKO_6Ta_2$, 27, 534
$CaFLiO_6Ta_2$, 27, 534
$Ca_{0.8}F_6Ln_{1.6}Na_{0.8}$, 27, 473
$CaFNaO_6Sb_2$, 27, 534
$Ca_7F_5Na_5O_{31}Si_8Zr_4$, 28, 322
$CaFNaO_6Ta_2$, 27, 534
$(Ca,Na)_{16}F_8Nb_{16}O_{48}$, 33A, 190
$Ca_{3.5}(F,O)Nb_{0.5}O_7Si_2$, 31A, 232
$Ca_{10}F_2O_{24}P_6$, 26, 468; 31A, 242
$Ca_{1.56}F_{0.44}O_{6.37}Sb_2$, 28, 200
$Ca_4F_2O_7Si_2$, 26, 513
$(Ca,Na)_3(F,OH)_2O_7Si_2Ti$, 31A, 244
$CaFO_6SmTi_2$, 27, 534
Ca_2FO_6TaTi, 27, 534
CaF_6Pd, 33A, 188
CaF_6U, 33A, 193
Ca-F-Y, 30A, 449
$Ca_3Fe_2Ge_3O_{12}$, 26, 415
$CaFeHO$, 28, 302
$CaFe_{12}H_{12}O_{44}P_8.4H_2O$, 35A, 337
$(Ca,Na)_2(Fe,Mg)_5H_2(K,Na)_{0.5}O_{24}Si_8$, 30A, 433
$Ca_{0.09}Fe_{0.52}Mg_{0.39}O_3Si$, 35A, 475
$(Ca,Fe,Mn,Na)O_4P$, 30A, 454
$CaFe_{0.5}Mn_{0.5}O_{2.75}$, 29, 419
$Ca_2Fe_{1.5}Mn_{0.5}O_5$, 29, 419
$(Ca,Na)_2(Fe,Mn)_3O_{12}P_3$, 31A, 244
$(Ca,Fe,Mn)_3O_8P_2$, 29, 425
$CaFeNO_3$, 28, 302
$CaFe_2O_4$, 28, 344; 31A, 241, 246
$CaFe_4O_7$, 27, 548
$Ca_2Fe_2O_5$, 27, 547, 29, 419
$Ca_2Fe_2O_6$, 35A, 243
$CaFe_{0.5}O_3Re_{0.5}$, 26, 418
$Ca_3Fe_2O_{12}Si_3$, 30A, 441; 31A, 219
$Ca_{0.40}Fe_{4.60}O_{12}Si_{0.40}Y_{2.60}$, 27, 511
$CaFe_{0.5}O_{2.75}Ti_{0.5}$, 29, 419
$Ca_3Fe_{3.5}O_{12}V_{1.5}$, 35A, 259
$CaGa_4$, 30A, 119
$Ca_3Ga_2Ge_3O_{12}$, 26, 415
$CaGa_2O_4$, 28, 144, 344
$Ca_2(Ga,Si)_3O_7$, 32A, 439
$CaGd_2S_4$, 27, 327
Ca_7Ge, 28, 14
$Ca_{33}Ge$, 28, 14
$Ca_{0.9}Ge_{1.97}Li_{0.13}$, 35A, 43
$CaGeMg$, 33A, 53
$CaGe_3Na_2O_{12}Sn_2$, 26, 416
$CaGe_3Na_2O_{12}Ti_2$, 26, 415
$CaGeO_3$, 27, 501
$CaGe_4O_9$, 27, 541
Ca_2GeO_4, 28, 350; 35A, 313
Ca_3GeO_5, 28, 304
$CaGe_6O_{21}Pb_8$, 29, 399
$Ca_2Ge_2O_7Pb_2$, 28, 310
$Ca_3Ge_3O_{12}Sc_2$, 26, 415
$CaGe_{0.5}O_3Si_{0.5}$, 32A, 463
$Ca_2Ge_{0.75}O_4Si_{0.25}$, 28, 350
Ca_2GeS_4, 29, 107
$CaGe_2Zn_2$, 35A, 24, 117
CaH_2, 27, 118
$Ca_2HK_2NaO_{20}Si_7Ti$, 30A, 455
$CaHK_3O_8P_2$, 27, 577; 34A, 329
$Ca_3H_6MnO_{14}S_2.3H_2O$, 33A, 381
CaH_4N_2, 28, 296
$CaH_4NO_9P_3$, 32A, 379
$CaH_6N_3O_9Tl$, 29, 426
$CaH_6NaO_6P_3$, 34A, 334
$Ca_2HNaO_9Si_3$, 32A, 461
$Ca_8H_{14}Na_2O_{26}Si_5$, 31A, 242
$Ca_8H_{22}Na_2O_{30}Si_5$, 31A, 242
$CaHNaO_7Ta_2$, 27, 534
$Ca_{0.5}H_4NaO_{18}V_6$, 27, 589
$(Ca,RE)(HO,O)_6(Nb,Ti)_2$, 28, 328
CaH_2O_2, 26, 349; 27, 422
$CaHO_4P$, 26, 458
$CaHO_4P.2H_2O$, 27, 574; 33A, 159
$CaH_4O_8P_2.H_2O$, 26, 458
CaH_5O_6P, 27, 574
$Ca_5HO_{13}P_3$, 29, 369
$Ca_8H_2O_{24}P_6.5H_2O$, 27, 575
$Ca_8H_{12}O_{29}P_6$, 27, 575
$Ca_{10}H_2O_{26}P_6$, 29, 369
$CaH_{12}O_9S_2$, 27, 643
$Ca_2H_2O_{11}Si_4$, 26, 513
$Ca_2H_2O_{11}Si_4.3H_2O$, 26, 513
$Ca_6H_2O_{13}Si_3$, 35A, 470
CaH_6O_6Sn, 32A, 247; 33A, 289
$Ca_2H_{16}O_{32}V_9$, 27, 603
$CaHf_4O_9$, 30A, 452
$CaHg_2$, 29, 107
$CaHg_3$, 29, 107
$CaHo_2Se_4$, 30A, 119
$CaIn$, 29, 107
$CaIn_2$, 29, 3, 107
Ca_3In, 29, 108
$CaIn_2O_4$, 29, 419
$CaIrO_3$, 30A, 333
Ca_2IrO_4, 31A, 138
$Ca_2K_2MgO_{16}S_4.2H_2O$, 26, 449
$CaKO_4P$, 26, 469
$CaKO_9P_3$, 31A, 246
$CaK_2O_8S_2.H_2O$, 33A, 518
$CaK_2O_8S_2.2H_2O$, 32A, 335
$Ca_5K_2O_{24}S_6.H_2O$, 30A, 453
$(Ca,Na)_2KO_{20}Si_8Th$, 31A, 225
$CaKO_4V$, 26, 470
$CaLa_2S_4$, 27, 327
$CaLa_2Te_4$, 34A, 139
Ca-*Ln*-O, 31A, 452
$CaLu_2O_4$, 35A, 231
$CaLu_2Se_4$, 30A, 119
$(Ca,Mg,Mn,Zn)O_5Te_2$, 26, 455
Ca-Mg-O-Si, 30A, 455
$CaMgO_4Si$, 30A, 424
$CaMgO_6Si_2$, 27, 700
$Ca_3MgO_8Si_2$, 32A, 431
$CaMgO_6SrW$, 28, 308
Ca_2MgO_6W, 28, 308
$CaMgSi$, 34A, 53
$CaMgSn$, 34A, 53
$CaMgZn$, 26, 90
$CaMn_2O_4$, 28, 344; 29, 321; 30A, 322; 31A, 136
$Ca_2MnO_8P_2.2H_2O$, 35A, 333
$(Ca,Mn)_2O_3PbSi$, 28, 354
$(Ca,Mn)O_3Si$, 28, 258
$CaMnO_6Si_2$, 27, 701; 32A, 457
$CaMn_6O_{12}Si$, 31A, 217
$(Ca,Mn,Zn)O_5Te_2$, 30A, 382
$CaMoO_4$, 33A, 334
$CaMo_{0.9}O_4W_{0.1}$, 27, 522
Ca_2N, 31A, 29; 33A, 163
Ca_3N_2, 33A, 50
$Ca_{11}N_8$, 34A, 54
$CaN_2O_6.4H_2O$, 28, 302
$CaN_3O_6\ Tl.3H_2O$, 29, 426
$CaNaO_4P$, 26, 469
$CaNa_2O_8S_2$, 30A, 371
$CaNa_2O_8Si_2$, 32A, 424
$Ca_{0.5}NaO_{16}V_6$, 27, 589
$Ca_{0.5}NaO_{16}V_6.2H_2O$, 27, 589
$CaNaO_4V$, 26, 470; 33A, 310
$CaNb_2O_6$, 26, 381, 382; 35A, 240
$Ca_2Nb_2O_7$, 26, 382
$Ca_3Nb_2O_8$, 26, 382; 27, 537
$(Ca,Nb)O_3Sr$, 28, 330
Ca_2NbO_6V, 31A, 247
$CaNd_2S_4$, 27, 327
$CaNd_2Te_4$, 34A, 139
Ca_2NiO_6W, 28, 310
CaO, 29, 422
$CaO_4P.0.2H_2O$, 26, 454
$(Ca,Th)O_4P.H_2O$, 31A, 244
$Ca_2O_7P_2$, 27, 591; 31A, 191; 33A, 407
$Ca_3O_8P_2$, 33A, 401
$Ca_4O_{19}P_6$, 29, 426
CaO_8P_2Th, 30A, 454

CaO_8P_2U, 30A, 454
$CaO_{12}P_2U_2.0\text{-}2H_2O$, 26, 468
$CaO_{12}P_2U_2.2\text{-}6H_2O$, 26, 468
$CaO_{12}P_2U_2.10\text{-}12H_2O$, 26, 468
$CaO_{12}P_2U_2.nH_2O$, 28, 354
$CaO_8P_2Zn_2.2H_2O$, 35A, 333
CaO_8P_2Zr, 28, 304
$CaO_{21}Pb_8Si_6$, 29, 399
$Ca_2O_7Pb_2Si_2$, 28, 310
$Ca_2O_9PbSi_3$, 34A, 383
Ca-O-Pb-Sn-Ti, 29, 326
Ca-O-Pb-Ti-Zr, 29, 326
CaO_6Pb_2W, 28, 310
Ca_2O_6PbW, 28, 310
CaO_4Pd_3, 29, 334
$CaO_3S_2.6H_2O$, 27, 643
CaO_4S, 27, 607; 28, 207
$CaO_4S.0.2H_2O$, 26, 454
$CaO_4S.\frac{1}{2}H_2O$, 30A, 453
$CaO_4S.0\text{-}2H_2O$, 30A, 453
Ca_2O_4S, 27, 604
CaO_4Sc_2, 29, 418; 30A, 316; 31A, 136
CaO_4Se, 28, 306
$CaO_4Se.\frac{1}{2}H_2O$, 28, 306
$CaO_4Se.2H_2O$, 28, 306
CaO_3Si, 26, 397, 512, 519; 33A, 453; 34A, 377
$CaO_3Si.H_2O$, 30A, 459
$CaO_4Si.0.2H_2O$, 26, 454
Ca_2O_4Si, 26, 511; 28, 310, 312, 350; 30A, 423
$\beta\text{-}Ca_2O_4Si$, 29, 394
$Ca_2O_4Si.8H_2O$, 28, 322
Ca_3O_5Si, 28, 304; 29, 427
$Ca_3O_7Si_2$, 26, 512
$Ca_4O_{10}Si_3.H_2O$, 26, 519
$Ca_6O_{12}Si_3.H_2O$, 26, 511
$Ca_8O_{18}Si_5$, 31A, 252
$CaO_9Si_3Sn.2H_2O$, 28, 262
$Ca_3O_{12}Si_3V_2$, 30A, 455
$CaO_4SiZn.H_2O$, 28, 261
$(Ca,Pb)O_4SiZn$, 30A, 455
CaO_3Sn, 28, 324; 29, 326
Ca_2O_4Sn, 32A, 305
$(Ca,Ta)O_3Sr$, 28, 332
$Ca_xO_3Sr_{1-x}Ti$, 26, 372
CaO_6Sr_2W, 28, 310
CaO_6Sr_3W, 28, 310
Ca_2O_6SrW, 28, 308
CaO_6Ta_2, 28, 201
$CaO_{11}Ta_4$, 35A, 258
$Ca_2O_7Ta_2$, 27, 534
$Ca_2O_{22}Ta_8$, 28, 202
CaO_6TaV, 31A, 247
$Ca_{0.5}O_4ThV$, 30A, 452
CaO_7Ti_2Zr, 29, 319
CaO_9TiZr_3, 26, 379
CaO_4U, 34A, 289
$(Ca,Pb)O_7U_2.2H_2O$, 26, 411
CaO_9U_3, 28, 350
$CaO_{19}U_6.10\text{-}11H_2O$, 27, 673
Ca_2O_5U, 28, 334; 34A, 289
Ca_3O_6U, 31A, 160; 34A, 289
CaO_4V_2, 28, 344
$CaO_6V_2.4H_2O$, 28, 204
$CaO_{16}V_6.3H_2O$, 26, 380
$CaO_{16}V_6.5\text{-}9H_2O$, 26, 380
$Ca_3O_8V_2$, 29, 419
$Ca_2O_{24}V_9.8H_2O$, 27, 603
$Ca_3O_{28}V_{10}.17H_2O$, 31A, 142
CaO_4W, 27, 542; 29, 331
Ca_3O_6W, 28, 308
Ca_6O_9W, 27, 542
Ca_2O_6WZn, 28, 310
CaO_4Yb_2, 32A, 310; 35A, 231
CaO_3Zr, 28, 330; 29, 326, 420
Ca_2Pb, 26, 87
Ca_3Pb, 28, 14
Ca_5Pb_3, 28, 14
$CaPr_2S_4$, 27, 327
Ca-Pt, 27, 322
$CaPt_2$, 27, 322
$CaPt_5$, 27, 322
Ca_2Pt_7, 27, 324
CaS, 26, 88; 27, 202; 29, 424
Ca_2S_4Si, 29, 109
CaS_4Sm_2, 27, 327
Ca_2S_4Sn, 29, 109
Ca-S-Y, 26, 88
CaS_4Y_2, 27, 327
$\alpha\text{-}CaS_4Y_2$, 26, 89
CaS_7Y_4, 26, 88
CaS_4Yb_2, 27, 327
CaS_3Zr, 28, 151
Ca_2Se_4Si, 29, 108
$CaSe_4Tm_2$, 30A, 119
Ca-Se-Y, 30A, 197
$CaSe_4Y_2$, 30A, 119
$CaSe_4Yb_2$, 30A, 119
$CaSi_2$, 32A, 139; 33A, 51
$Ca_{0.5}Si_2Sr_{0.5}$, 32A, 43
$CaSi_4Sr$, 35A, 135
Ca_2Sn, 26, 87
$CaZn_2$, 26, 90
$CaZn_5$, 33A, 131
Cd, 26, 279; 30A, 197
$Cd_{4.5}Ce$, 30A, 119
Cd_6Ce, 29, 108
$CdCl_2$, 28, 300
$CdCl_3Cs$, 29, 280
$CdCl_4Cs_2$, 29, 280
$CdCl_2H_8N_4$, 27, 471
$CdCl_5H_{15}N_6$, 31A, 103
CdClHO, 26, 327
$CdCl_4Na_2$, 26, 316
$CdCl_8Na_6$, 26, 316
$Cd_4Cl_3P_2$, 28, 300
$Cd_{1-x}Co_xO_3Ti$, 30A, 451
$Cd_{1-x}Cr_2Fe_xS_4$, 29, 165
$Cd_3Cr_2Ge_3O_{12}$, 26, 415
$CdCrInO_4$, 30A, 451
$CdCr_2S_4$, 28, 302
$(Cd,Zn)Cr_2S_4$, 28, 302
$CdCu_2$, 27, 106
Cd_3Cu_4, 27, 115; 30A, 119; 32A, 36
Cd_8Cu_5, 33A, 72
$CdCu_{1.5}Ga_{0.5}$, 27, 261
$CdCu_{1.5}Ge_{0.5}$, 27, 261
$CdCu_2GeS_4$, 34A, 47
CdCuIn, 27, 261
Cd-Cu-In-Te, 29, 165
$CdCu_2S_4Sn$, 30A, 119
$CdCu_2Se_4Sn$, 30A, 119
Cd_2Dy, 29, 108
Cd_6Dy, 29, 108
Cd_6Er, 29, 108
CdEu, 29, 108
Cd_2Eu, 30A, 119
Cd_6Eu, 29, 108
$Cd_{11}Eu$, 29, 108; 30A, 119
$CdFNaNb_2O_6$, 27, 534
$Cd_{2.0}F_{0.1}O_{6.9}Sb$, 28, 200
$CdF_2O_5Ta_2$, 27, 534
CdF_6Pd, 33A, 188
$Cd_3Fe_2Ge_3O_{12}$, 26, 415
Cd-Fe-Mn-S-Zn, 26, 252
Cd-Fe-S-Zn, 26, 252
$Cd_3Ga_2Ge_3O_{12}$, 26, 415
$CdGaInO_4$, 30A, 451
$CdGa_2O_4$, 26, 413
Cd-Ga-S, 27, 202
$CdGa_2S_4$, 27, 202
$Cd_6Ga_2S_9$, 27, 202
Cd_6Gd, 29, 108
$CdGeO_3$, 34A, 260
Cd_2GeO_4, 30A, 451
$Cd_3Ge_3O_{12}Sc_2$, 26, 415
$CdGeP_2$, 33A, 49
Cd_4GeS_5, 29, 108
$CdH_8I_2N_4$, 30A, 457
$CdH_3IO_6.3H_2O$, 35A, 397
$CdHLiMoO_5$, 33A, 337
$CdH_4NO_9P_3$, 32A, 379
$CdH_8N_2O_8S_2.6H_2O$, 31A, 203
CdH_2O_2, 31A, 130, 256
CdH_6O_6Sn, 29, 421
CdHf, 29, 108; 30A, 119
$CdHf_2$, 29, 108; 30A, 119
Cd-Hg-Se, 29, 165
Cd_6Ho, 29, 108
CdI_2, 27, 431, 29, 287; 30A, 288; 33A, 517
$Cd_4I_3P_2$, 28, 302
Cd-In, 27, 109; 29, 185; 30A, 198
$CdInS_2$, 34A, 52
$CdIn_2Se_4$, 32A, 41
$CdIn_2Te_4$, 33A, 152
$CdKO_9P_3$, 31A, 246
$CdLi_2O_8Ti_3$, 28, 348
Cd_6Lu, 29, 108
$CdMO_9P_3$, 31A, 245
Cd-Mg, 27, 109; 29, 166

$Cd_2Mn_3O_8$, 32A, 285
$CdMoO_4$, 31A, 156
CdN_2O_6, 29, 426
$CdN_2O_6.4H_2O$, 28, 179, 302; 31A, 178; 34A, 342
Cd_2Na, 27, 113
$Cd_{11}Na_2$, 30A, 33, 120
$CdNa_2O_2$, 34A, 250
$CdNa_2O_6Si_2$, 32A, 454
$Cd_2Na_4O_{10}Si_3$, 33A, 469
$Cd_3Na_2O_{10}Si_3$, 33A, 470
$(Cd,Nb)O_3Sr$, 28, 330
Cd_6Nd, 29, 108
CdNi, 30A, 33, 120
Cd_5Ni, 35A, 40
$CdNiO_4Tc$, 29, 322
$Cd_{11}Np$, 34A, 139
CdO_6P_2, 33A, 417
$Cd_2O_7P_2$, 34A, 327
$CdO_8P_2Zn_2$, 33A, 399
$Cd_2O_8P_2Zn$, 33A, 399
CdO_6Pb_2W, 28, 310; 30A, 452
Cd_2O_6PbW, 28, 310
$Cd_2O_7Re_2$, 30A, 446
CdO_4Rh_2, 26, 399
CdO_4S, 26, 441, 443; 28, 210
α-CdO_4S, 26, 442
β-CdO_4S, 26, 442
$CdO_4S.H_2O$, 35A, 372
CdO_3Si, 29, 397
Cd_2O_4Si, 29, 397
Cd_3O_5Si, 29, 397
CdO_3Sn, 28, 350
$Cd_{2x}O_4SnZn_{2-2x}$, 33A, 286
$Cd_{0.5}O_4ThV$, 30A, 452
$Cd_{0.43}O_{2.86}U$, 27, 545
$CdO_{3.63}U$, 31A, 257
β-CdO_4U, 26, 405
CdO_6U_2, 27, 543
CdO_9U_3, 28, 350
Cd_2O_5U, 29, 420
$Cd_2O_7U_2$, 27, 543
CdO_4V_2, 34A, 264
CdO_6V_2, 34A, 265
$Cd_2O_7V_2$, 32A, 274
CdO_4W, 26, 392; 31A, 158
CdP_2, 35A, 41
α-CdP_2, 34A, 49
β-CdP_2, 34A, 50
$Cd_{10}P_{15}Pd_{75}$, 35A, 117
$Cd_{14}P_{14}Pt_{72}$, 35A, 117
CdP_2Si, 29, 108
(Cd,Pb)Se, 30A, 199
Cd-Pb-Te, 29, 166
(Cd,Pb)Te, 30A, 199
$Cd_{14}Pd_{72}Se_{14}$, 35A, 117
Cd_6Pr, 29, 108
$Cd_{0.6}Pt_3Zn_{4.4}$, 35A, 42
$CdPt_2Zn$, 35A, 41
$Cd_4Pt_3Zn_2$, 35A, 42
Cd_6Pu, 30A, 120
$Cd_{11}Pu$, 30A, 120
CdS, 26, 251; 27, 109, 115, 202, 382; 28, 44; 29, 108; 30A, 198
α-CdS, 27, 115
β-CdS, 27, 115
CdS-SeZn, 27, 110
CdS_2Te, 34A, 52
CdS_7Y_4, 26, 277
Cd-S-Zn, 26, 251
CdSb, 26, 34; 27, 41
"Cd_3Sb_4", 29, 108
Cd_4Sb_3, 30A, 120
Cd-Sb-Zn, 26, 34; 29, 166
$CdSb_2Zn$, 27, 41
CdSc, 28, 44
Cd_3Sc, 27, 109, 29, 32, 108
CdSe, 27, 109, 383; 28, 44; 29, 167
Cd-Se-Te, 29, 167
$CdSe_2Tl$, 34A, 52
Cd_6Sm, 29, 108
Cd-Sn, 29, 167
Cd_3Sr, 30A, 120
Cd_6Tb, 29, 108
CdTe, 27, 383; 28, 44; 29, 108, 152, 167
$CdTe_2Tl$, 34A, 52
Cd_2Th, 26, 256
CdTi, 27, 116
$CdTi_2$, 27, 117
Cd_6Tm, 29, 108
CdV_3, 28, 44
CdY, 28, 44
Cd_2Y, 28, 44; 33A, 50
Cd_6Y, 29, 109
Cd_6Yb, 29, 109
Cd-Zn, 29, 167; 30A, 199
CdZr, 30A, 120
$CdZr_2$, 30A, 120
Cd_3Zr, 34A, 139
Ce, 26, 106, 234; 27, 334; 34A, 62
β-Ce, 27, 141
γ-Ce, 27, 142
$CeCl_3$, 27, 724
$CeCl_6Cs_2$, 31A, 104
CeClS, 31A, 247
$CeCo_2$, 30A, 120
$CeCo_3$, 30A, 120
$CeCo_5$, 26, 239
Ce_2Co_7, 35A, 117
Ce_2Co_{17}, 27, 7; 29, 109
$(Ce,Pu)_5Co_3$, 29, 37
$Ce_{24}Co_{11}$, 27, 142
$CeCo_3Pu_4$, 29, 109
CeCoSi, 35A, 117
$CeCoSi_2$, 34A, 63
$CeCrO_3$, 28, 328
$CeCs_2F_6$, 26, 312
$CeCs_3F_7$, 26, 312
CeCu, 26, 106; 30A, 120
$CeCu_2$, 26, 109
$CeCu_4$, 29, 109
$CeCu_5$, 29, 109
Ce-Dy, 27, 148
Ce-Er, 27, 148
CeF_3, 26, 301
$CeF_6H_8N_2$, 34A, 193
CeF_4K, 34A, 192
CeF_6K_2, 26, 312
CeF_6Na_2, 26, 312
CeF_7Na_3, 26, 312
CeF_6Rb_2, 26, 312
CeF_7Rb_3, 26, 312
CeFS, 30A, 450; 35A, 155
$CeFe_7$, 31A, 46
Ce_2Fe_{17}, 27, 6
CeFeSi, 35A, 44
$CeGa_2$, 26, 146
Ce-Ga-Se, 29, 109
$CeGaSe_3$, 29, 109
Ce-Gd, 27, 148
CeGe, 29, 109; 30A, 120
$CeGe_2$, 29, 109
Ce_5Ge_3, 27, 341; 29, 109
$CeGe_2Ni$, 35A, 117
$Ce_6Ge_{2.5}S_{14}$, 34A, 139
$Ce_2H_{48}Mg_3N_{12}O_{60}$, 27, 571
$CeH_8N_8O_{18}$, 33A, 426
$CeHNaO_{13}P_3Sr_3$, 29, 374
$Ce_yH_2Na_{3-x}O_{14}Si_2Ti_2.8H_2O$, 34A, 379
$CeH_{0.90}O$, 27, 332
$CeH_{0.20}Sb$, 27, 335
$CeH_{1.41}Te$, 27, 332
Ce_5Hg_{21}, 32A, 139
Ce-Ho, 27, 148
CeIS, 30A, 450; 31A, 247; 34A, 214
$CeIn_3$, 30A, 120
Ce-La, 27, 145, 148
$(Ce,La,Nd,Th)O_4P$, 33A, 402
Ce-La-Th, 27, 146
Ce-Lu, 27, 148
$CeMg_9$, 26, 190
$CeMg_{10.3}$, 32A, 50
$CeMg_{12}$, 29, 109
Ce_2Mg_{17}, 28, 44
Ce_5Mg_{41}, 32A, 47
Ce_5Mg_{42}, 32A, 47
$Ce_2Mg_3N_{12}O_{36}.24H_2O$, 27, 571; 28, 181
Ce_2MgS_4, 27, 327
Ce-Mn-S, 27, 328
Ce_2MnS_4, 27, 327
CeN, 27, 304
$CeNaS_2$, 29, 341
$CeNaSe_2$, 30A, 120
$CeNbO_4$, 26, 386; 27, 521; 28, 346
$CeNbO_6Ti$, 27, 537
$(Ce,Y)NbO_6Ti$, 28, 328
Ce-Nd, 27, 148
CeNi, 26, 109; 29, 109; 30A, 120
$CeNi_5$, 26, 239

Ce_2Ni_7, 35A, 118
Ce_7Ni_3, 26, 108
$CeNiSi_2$, 34A, 63; 35A, 96
$CeNi_2Si_2$, 31A, 35
$CeNi_{8.6}Si_{2.4}$, 34A, 64
$Ce_3Ni_6Si_2$, 31A, 35
CeO_2, 27, 480
$(Ce,Zr)O_2$, 30A, 457
CeO_4P, 32A, 358
CeO_7P_2, 28, 304
$Ce_2O_{12}S_3.8H_2O$, 30A, 453
$Ce_2O_{12}S_3.9H_2O$, 30A, 453; 33A, 374
$Ce_2O_7Si_2$, 35A, 302
$Ce_2O_7SiTi.4H_2O$, 28, 324
CeO_3Sr, 31A, 255
CeO_4Sr_2, 31A, 255
Ce_2O_2Te, 27, 424
CeO_4V, 27, 589, 724; 28, 306
$CePb_3$, 29, 109
Ce_5Pb_3, 30A, 120
$CePd_3$, 30A, 120
Ce-Pr, 27, 148
CePt, 30A, 121
$CePt_5$, 26, 239
δ-Ce-Pu, 26, 231
CeRh, 30A, 121
$CeRu_2$, 27, 149; 30A, 199
Ce-S, 29, 168
Ce_2S_3, 33A, 60
$Ce_{2+x}S_{3+x}$, 27, 63
CeSSb, 27, 335
Ce_2S_4Sr, 27, 327
CeSb, 27, 335
$CeSb_2$, 32A, 20
Ce-Sc, 27, 148
CeSe, 29, 109
$CeSe_2$, 27, 335; 29, 109; 32A, 20; 33A, 58
Ce_2Se_3, 29, 109
$(Ce,Sr)_3Se_4$, 29, 109
Ce_2Se_4Sr, 29, 109
CeSi, 29, 109; 30A, 121
$CeSi_{1.77-1.98}$, 29, 168
$CeSi_2$, 29, 109, 168
Ce_3Si_2, 30A, 121; 32A, 139
Ce_5Si_3, 30A, 121
$CeSn_3$, 29, 109; 30A, 121
Ce-Tb, 27, 148
CeTe, 29, 110
$CeTe_2$, 32A, 20
$CeTe_3$, 27, 336; 30A, 121
Ce_2Te_3, 28, 44; 29, 110
Ce_3Te_4, 28, 44
Ce-Th, 27, 145, 147; 29, 168
Ce-Tm, 27, 148
Cex, 29, 110
Ce-Y, 27, 148; 29, 169
CeZn, 30A, 121
$CeZn_2$, 30A, 121
$CeZn_3$, 30A, 121; 35A, 97
$CeZn_{3.67}$, 30A, 121
$CeZn_{\sim 4.5}$, 30A, 121
$CeZn_5$, 26, 175; 30A, 121
$CeZn_{8.5}$, 30A, 121
$CeZn_{11}$, 30A, 121
Ce-Zr, 29, 170
CfFO, 33A, 175
Cl_2, 27, 716; 28, 360; 30A, 252
Cl_2Co, 28, 298; 34A, 178
$Cl_2Co.2H_2O$, 27, 435; 28, 76; 30A, 449; 31A, 95
$Cl_2Co.6H_2O$, 27, 435; 35A, 402
Cl_3CoCs, 26, 326; 33A, 214
Cl_5CoCs_3, 29, 277
$Cl_{17}Co_4Cu_5H_{72}N_{24}$, 31A, 258
$ClCoF_6H_{15}N_5Si$, 32A, 167
$Cl_2CoH_{15}N_8$, 32A, 193
$Cl_2CoH_{18}N_6$, 31A, 246
$Cl_3CoH_{15}N_5$, 28, 90; 33A, 208
$Cl_3CoH_{18}N_6$, 27, 657
$Cl_5Co_2H_{26}N_9.4H_2O$, 35A, 406
$Cl_2CoH_{15}N_6O$, 30A, 280
$Cl_2CoH_{15}N_6O_2$, 33A, 209
$Cl_4Co_2H_{26}N_8O_2.4H_2O$, 27, 655; 28, 91
$Cl_4Co_2H_{26}N_{10}O_2.4H_2O$, 35A, 409
$Cl_4Co_2H_{34}N_8O_6$, 27, 655
$ClCoH_{15}N_5O_3S_2.H_2O$, 35A, 408
ClCoHO, 26, 327
α-$ClCo_2H_3O_3$, 29, 250
β-$ClCo_2H_3O_3$, 29, 250
$Cl_2CoH_4O_2$, 27, 434
$Cl_2CoH_{12}O_6$, 27, 435
$Cl_5CoH_{18}N_6Zn$, 35A, 410
$Cl_4Co_5Mo_4Na_2O_{16}$, 30A, 349; 35A, 262
$ClCoN_2O_2$, 28, 300
$[ClCoN_2O_2]n$, 32A, 195
$Cl_2CoO_6.6H_2O$, 28, 306
Cl_3CoRb, 32A, 190
Cl_5CoRb_3, 32A, 190
Cl_2Cr, 26, 317
Cl_3Cr, 26, 317; 29, 272
Cl_3CrCs, 27, 457
Cl_4CrCs_2, 27, 457
Cl_5CrCs_2O, 29, 281
$Cl_{4.5}CrCuH_{18}N_6$, 26, 326
$Cl_5CrCuH_{18}N_6$, 26, 326; 33A, 207
$Cl_9CrCu_2H_{18}N_6$, 26, 326
$Cl_3CrH_{15}N_5.H_2O$, 27, 657
$Cl_3CrH_{18}N_6$, 27, 657
$Cl_3CrH_{17}N_5O$, 27, 657
$Cl_6Cr_4H_{42}N_{12}O_6.4H_2O$, 33A, 206
$Cl_3CrH_8O_4.2H_2O$, 30A, 280; 31A, 94
Cl_6CrK_2Na, 30A, 449
ClCrO, 26, 329; 27, 469
$ClCrO_4Sr_2$, 30A, 451
$ClCr_3O_{12}Sr_5$, 30A, 451
ClCs, 26, 297; 29, 415
$ClCs.1/3Cl_2H_4O$, 33A, 201
Cl_3CsCu, 31A, 103
$Cl_4Cs_2Cu.2H_2O$, 31A, 244
Cl_3CsGe, 30A, 278
$Cl_3CsH_4MnO_2$, 27, 456
$Cl_4Cs_2H_4MnO_2$, 29, 275
$Cl_6Cs_2In_{0.5}Sb_{0.5}$, 28, 300
Cl_3CsMg, 35A, 160
$Cl_3CsMn.2H_2O$, 27, 456
$Cl_4Cs_2Mn.2H_2O$, 29, 275
Cl_6Cs_2Mo, 27, 464
Cl_5Cs_2MoO, 29, 281
Cl_6Cs_2Nb, 30A, 449
Cl_5Cs_2NbO, 29, 281; 30A, 450
$ClCsO_4$, 27, 725; 31A, 250
$Cl_{\sim 0.9}Cs_{\sim 0.9}O_3U$, 29, 282
$Cl_4Cs_2O_2U$, 31A, 105
Cl_5CsOW, 29, 281
Cl_6Cs_2Re, 29, 415
$Cl_{12}Cs_3Re_3$, 28, 87
$Cl_5Cs_2Ru.H_2O$, 31A, 101
$Cl_6Cs_2Sb_{0.5}Tl_{0.5}$, 28, 300
Cl_3CsSn, 35A, 159
$Cl_2Cu.2H_2O$, 35A, 197
$Cl_2CuH_{18}N_6$, 32A, 194
$Cl_4CuH_8N_2$, 29, 278
$Cl_6Cu_2H_8N_2$, 28, 85
$Cl_4Cu_3H_{12}N_4.H_2O$, 35A, 416
$Cl_4CuH_8N_2.2H_2O$, 34A, 204
$Cl_2CuH_{19}N_6O_8$, 26, 497
ClCuHO, 26, 327
α-$ClCu_2H_3O_3$, 29, 250
γ-$ClCu_2H_3O_3$, 29, 250
δ-$ClCu_2H_3O_3$, 29, 250
$Cl_6Cu_2K_2$, 28, 85
$Cl_4CuK_2.2H_2O$, 33A, 159; 35A, 197
$Cl_3CuLi.2H_2O$, 28, 83
$Cl_2CuO_8.6H_2O$, 26, 496
ClD, 32A, 180
Cl_3Es, 34A, 179
ClEsO, 34A, 179
Cl_3Eu, 27, 724; 33A, 204
$Cl_3Eu.6H_2O$, 30A, 275
ClEuO, 27, 724
$Cl_5FN_3P_3$, 34A, 196
$(Cl,F)Na_3O_4S$, 28, 326
$ClFNa_6O_8S_2$, 33A, 377
Cl_4FNb, 33A, 196
$ClFO_3$, 26, 316
Cl_4FSb, 31A, 85
ClF_8Sb, 35A, 178
Cl_4FTa, 31A, 85
Cl_2Fe, 28, 298

Cl_3Fe, 27, 440
Cl_6Fe_2, 28, 361
$Cl_2Fe.2H_2O$, 28, 274
$Cl_2Fe.4H_2O$, 33A, 159
$Cl_3Fe.6H_2O$, 32A, 183
$ClFeHO$, 26, 327
α-$ClFe_2H_3O_3$, 29, 250
β-$ClFe_2H_3O_3$, 29, 250
$ClFe_2H_3O_{12}Te_4$, 33A, 393
$ClFe_2HO_6Te_2.\frac{1}{2}H_2O$, 34A, 318
$Cl_2FeH_3O_3Y$, 33A, 216
Cl_4FeNa, 30A, 277
$ClFeO$, 35A, 181
Cl_8FeP, 33A, 215
Cl_3Ga, 30A, 274
$ClGaO$, 28, 300
$Cl_{11}Ga_9S_8$, 31A, 246
Cl_3Gd, 32A, 185; 33A, 204
Cl_3Gd_2, 35A, 150
$Cl_3Gd.6H_2O$, 26, 322
$ClGdH_2O_2$, 34A, 221
$ClGdO$, 27, 724
ClH, 32A, 180
$ClH.2H_2O$, 32A, 181
$ClH.3H_2O$, 32A, 181
$Cl_4H_2KNOOs.H_2O$, 35A, 431
$Cl_3H_3KNPt.H_2O$, 30A, 265; 35A, 430
Cl-H-K-O, 30A, 449
$Cl_{10}H_2K_4O_2Re_2$, 27, 670
$Cl_4H_2K_2OSn$, 27, 457
Cl_5HK_2OTc, 32A, 192
$Cl_3H_4KO_2Zn$, 29, 278
$ClHMgO$, 26, 327
α-$ClH_3Mg_2O_3$, 29, 251
β-$ClH_3Mg_2O_3$, 29, 251
$Cl_6H_{12}MgO_6Te$, 29, 251
$Cl_2H_8MnN_4$, 28, 89
$ClHMnO$, 26, 327
α-$ClH_3Mn_2O_3$, 29, 250
β-$ClH_3Mn_2O_3$, 29, 251
α-$Cl_2H_8MnO_4$, 29, 273
$Cl_4H_4MnO_2Rb_2$, 29, 275
$Cl_9H_{20}Mo_2N_5.H_2O$, 35A, 418
ClH_4N, 27, 426
$ClH_{13}N_4$, 26, 298
$Cl_5H_8N_2NbO$, 30A, 450
ClH_4NO, 32A, 197
ClH_4NO_3, 27, 641
ClH_4NO_4, 27, 639
ClH_4NO_5, 34A, 359
$ClH_5N_2O_4$, 35A, 392
$ClH_5N_2O_4.\frac{1}{2}H_2O$, 32A, 318
$Cl_3H_{15}N_6ORu.H_2O$, 28, 93; 30A, 282
$Cl_2H_{12}N_4O_2RuS$, 30A, 444
$ClH_{12}N_3O_4S_2$, 34A, 224
$Cl_6H_8N_2Os$, 27, 723
$Cl_2H_6N_2Pd$, 31A, 250
$Cl_4H_8N_2Pd$, 31A, 102
$Cl_6H_8N_2Pd$, 31A, 102
$Cl_6H_{12}N_4Pd_2$, 27, 667
$Cl_6H_{12}N_4(Pd,Pt)_2$, 27, 667
$Cl_2H_{15}N_3Ru$, 33A, 211
$Cl_2H_{15}N_7Ru$, 31A, 258
$Cl_3H_{15}N_5Ru$, 27, 669
$Cl_8H_{12}N_3Tc_2.2H_2O$, 35A, 426
$Cl_6H_8N_2Te$, 31A, 100
$ClHNU$, 31A, 100
$Cl_2H_8N_4Zn$, 27, 472; 28, 92
$Cl_2H_{12}N_4Zn$, 31A, 255
$ClH_{10}NaO_6$, 27, 645
$ClHNiO$, 26, 327
α-$ClH_3Ni_2O_3$, 29, 251
β-$ClH_3Ni_2O_3$, 29, 251
$Cl_2H_4NiO_8$, 27, 643
$Cl_2H_8NiO_{10}$, 27, 643
$Cl_2H_{12}NiO_{12}$, 27, 642
$ClHO$, 34A, 220
ClH_3O_5, 27, 637
$Cl_2H_4O_2$, 27, 722
$ClHO_4.2H_2O$, 33A, 449
$Cl_4H_6O_{23}Pb_6.H_2O$, 34A, 360
ClH_2O_2Pr, 34A, 221
$Cl_3H_{15}O_{27}S_3Zn_{12}.5H_2O$, 31A, 243
ClH_2O_2Sm, 34A, 221; 35A, 195
$Cl_2H_6O_6Sn_4$, 28, 300
$Cl_2H_{12}O_{15}Ti$, 29, 427
$Cl_2H_2O_6U_2.4H_2O$, 34A, 225
ClH_2O_2Y, 30A, 450; 31A, 98, 257
$Cl_2H_5O_6Y_3$, 32A, 199
$ClHOZn$, 26, 327
$Cl_2H_{12}O_6Zn$, 27, 722
$Cl_2H_{22}O_{11}Zn$, 27, 723
$Cl_2H_8O_8Zn_5.H_2O$, 26, 330; 33A, 217
$Cl_2H_8O_8Zn_5.H_2O$, 26, 330
$Cl_5HZn_2.2H_2O$, 35A, 148
Cl_7HfOP, 31A, 255
$Cl_4HgK_2.H_2O$, 33A, 159
$ClHg_2NO_4$, 30A, 454
$Cl_2Hg_3O_2$, 30A, 301
$Cl_2Hg_4O_2$, 30A, 301; 33A, 221
Cl_4Hg_3O, 29, 283
$Cl_2Hg_2O_8.4H_2O$, 31A, 211
$Cl_2Hg_3S_2$, 27, 465; 31A, 253; 32A, 191; 33A, 228, 513
$Cl_2Hg_3Se_2$, 27, 465
$Cl_2Hg_3Te_2$, 27, 465
Cl_3HgTl, 30A, 449
$Cl_{16}Hg_3Tl_{10}$, 30A, 449
ClI, 27, 448
$Cl_4IK.H_2O$, 28, 82
$ClIPb$, 34A, 176
$ClIn$, 31A, 96
Cl_4In_2, 31A, 250
$ClInS$, 27, 466
$ClInSe$, 27, 466
$ClInTe$, 27, 466
Cl_3Ir, 26, 324; 30A, 277
ClK, 26, 297; 27, 426
Cl_3KMn, 30A, 449
Cl_6K_2Mn, 31A, 101
$Cl_3KMn.2H_2O$, 33A, 212
$Cl_4K_2Mn.2H_2O$, 33A, 213
Cl_6K_2Mo, 27, 464
Cl_6K_3Mo, 34A, 207
$Cl_8K_4Mo_2.2H_2O$, 34A, 207
ClK_2NO_3, 33A, 450
$Cl_3K_2N_3O_6Pt$, 27, 668
Cl_5K_2NORu, 26, 330
Cl_5K_2NOs, 33A, 227; 34A, 225
$Cl_8K_3NRu_2.2H_2O$, 34A, 210
Cl_6K_2NaV, 30A, 449
Cl_6K_2Nb, 30A, 449
$Cl_4K_2O_2Os$, 26, 331
$Cl_{10}K_4ORe_2.H_2O$, 27, 670
$Cl_{\sim 0.9}K_{\sim 0.9}O_3U$, 29, 282
Cl_6K_2Re, 28, 300; 35A, 168
$Cl_8K_2Re_2.2H_2O$, 30A, 283
$Cl_4K_2Sn.H_2O$, 27, 457
Cl_6K_2Tc, 32A, 192
Cl_6K_2Ti, 26, 325
$Cl_3KZn.H_2O$, 34A, 206
$Cl_3KZn.2H_2O$, 28, 86; 29, 278
Cl_3La, 26, 3; 27, 724; 33A, 204
$ClLaS$, 31A, 247
$ClLi$, 34A, 171
$ClLiO_4.3H_2O$, 33A, 450
$Cl_4Li_2Zn.2H_2O$, 34A, 205
$Cl_3\mathit{Ln}.6H_2O$, 31A, 241
$Cl_5\mathit{M}_2NbO_2$, 30A, 450
Cl_2Mg, 28, 298
$Cl_2Mg.12H_2O$, 31A, 93
$Cl_2MgO_6.6H_2O$, 28, 306
$Cl_{10}MgO_6P_4$, 35A, 186
$Cl_2Mg_2O_4S.8H_2O$, 31A, 251
Cl_2Mn, 28, 298
$Cl_2Mn.2H_2O$, 30A, 274
α-$Cl_2Mn.4H_2O$, 29, 273
$Cl_3MnRb.2H_2O$, 32A, 188; 35A, 198
$Cl_4MnRb_2.2H_2O$, 29, 275
Cl_3Mo, 28, 300
Cl_3MoN, 35A, 154
Cl_2MoO, 28, 300
Cl_3MoO, 35A, 184
Cl_6MoRb_2, 27, 464
$Cl_8Mo_2Rb_3$, 34A, 209
Cl_8MoTl_2, 27, 464
$ClNO$, 26, 327
$Cl_4N_3O_2PS_2$, 34A, 177
$ClNO_4S$, 32A, 198
$Cl_3N_3O_3S_3$, 27, 410; 31A, 80
$Cl_8N_4P_4$, 27, 411; 33A, 197
$Cl_{10}N_5P_5$, 31A, 259; 33A, 198
$Cl_2N_2S_3$, 31A, 79
$Cl_3N_3S_3$, 27, 409; 31A, 79
$Cl_5N_2S_2Se$, 27, 409
$ClNTi$, 29, 248
Cl_4NV, 31A, 258; 33A, 226
$ClNZr$, 29, 248

ClNa, 26, 297; 27, 425; 28, 290
ClNa. 5 1/7 H_3N, 30A, 260
$ClNaO.5H_2O$, 27, 645
$ClNa_3O_{12}Pb_2S_3$, 32A, 333; 34A, 311
$Cl_3Na_{21}O_{40}S_{10}Zn$, 26, 453
Cl_4Nb, 28, 300
Cl_8Nb_3, 26, 323; 27, 444; 28, 300
$Cl_{14}Nb_6$, 30A, 263
Cl_2NbO, 28, 300
Cl_8NbOP, 28, 95
Cl_5NbORb_2, 30A, 450
Cl_6NbRb_2, 30A, 449
Cl_2NbS_2, 31A, 99
Cl-Nd, 26, 322
$Cl_{2.00}Nd$, 26, 322
$Cl_{2.27}Nd$, 26, 322
$Cl_{2.37}Nd$, 26, 322
Cl_3Nd, 26, 322; 27, 724; 33A, 204
Cl_2Ni, 28, 300
$Cl_2Ni.2H_2O$, 32A, 184
$Cl_2Ni.6H_2O$, 26, 318; 28, 300; 34A, 178
$Cl_2NiO_6.2H_2O$, 27, 643
$Cl_2NiO_6.4H_2O$, 27, 643
$Cl_2NiO_6.6H_2O$, 27, 642; 28, 306
$Cl_2NiO_8.nH_2O$, 27, 642
Cl_3NiRb, 34A, 211
ClO, 34A, 244
ClO_2, 34A, 243
$^{35}ClO_2$, 26, 346
$^{37}ClO_2$, 26, 346
Cl_2O_7, 27, 719; 30A, 445
$ClO_{12}P_3Pb_5$, 31A, 243
Cl_7OPSn, 31A, 255
$Cl_{10}O_2P_2Sn$, 28, 94
$Cl_{16}O_6P_4Sn_2$, 34A, 222
$Cl_2O_{24}P_6Sr_{10}$, 26, 469
Cl_8OPTa, 27, 468
Cl_7OPTi, 31A, 255
$Cl_{10}O_2P_2Ti$, 27, 468
Cl_7OPZr, 31A, 255
Cl_2OPa, 33A, 223
$Cl_2O_9Pb_2.2H_2O$, 30A, 458
$ClO_{12}Pb_5V_3$, 27, 604; 31A, 243
$Cl_3O_6Pr_3W$, 34A, 286
$Cl_{\sim 0.9}O_3Rb_{\sim 0.9}U$, 29, 282
Cl_2OS, 34A, 224
Cl_2O_2S, 34A, 224
Cl_7OSbSe, 32A, 200
$Cl_8O_2Se_2Sn$, 34A, 222
ClOSm, 27, 724
$Cl_2O_{11}Te_6$, 30A, 450
$Cl_4O_{16}Th.4H_2O$, 28, 306
$Cl_2O_9Ti.6H_2O$, 29, 427
ClO_4Tl, 31A, 250
Cl_2O_2U, 33A, 221
$Cl_2O_2U.H_2O$, 33A, 221
$Cl_2O_2U.3H_2O$, 33A, 221
$Cl_2O_9V.nH_2O$, 30A, 450
Cl_2O_2W, 33A, 219
Cl_3OW, 31A, 246
Cl_4OW, 31A, 99
ClOY, 27, 724
$Cl_2O_6Zn.6H_2O$, 28, 306
$Cl_2OZr.8H_2O$, 33A, 218
$Cl_2O_9Zr.2H_2O$, 28, 306
Cl_3P, 27, 717
Cl_5Pa, 32A, 186
Cl_2Pb, 26, 324; 28, 78, 80; 34A, 176
Cl_3Pr, 27, 724
Cl_2Pt, 30A, 276
ClRb, 26, 297, 298; 27, 425
Cl_6Rb_2Re, 29, 415
$Cl_7RbSb_2.H_2O$, 31A, 260
Cl_5Re, 33A, 205
Cl_3Rh, 29, 275
Cl_3Ru, 31A, 259; 32A, 185; 33A, 237
Cl_2S_2, 34A, 178
Cl_4SW, 35A, 155
Cl_3Sc, 26, 298
Cl_4Se, 29, 415
Cl_4Si, 34A, 173
Cl_3Sm, 27, 724
Cl_2Sn, 26, 320
Cl_4Sn, 28, 300
$Cl_2Sn.2H_2O$, 26, 321
Cl_2Sr, 28, 284
Cl_4Ta, 27, 444
$Cl_{15}Ta_6$, 31A, 259; 33A, 203
$Cl_{14}Ta_6.7H_2O$, 31A, 97
Cl_3Tb, 29, 274
Cl_4Tc, 31A, 98
Cl_4Te, 29, 415
Cl_4Th, 31A, 251; 34A, 174
Cl_2Ti, 27, 434
α-Cl_3Ti, 26, 317; 27, 441
β-Cl_3Ti, 27, 441
γ-Cl_3Ti, 26, 317
δ-Cl_3Ti, 26, 317; 27, 441
Cl_4Ti, 28, 298
ClTl, 26, 297; 27, 426
Cl_5U, 32A, 187
Cl_2Zn, 26, 319
α-Cl_2Zn, 26, 319
β-Cl_2Zn, 26, 319
γ-Cl_2Zn, 26, 319
$Cl_2Zn.\frac{1}{2}HCl.H_2O$, 35A, 148
$Cl_2Zn.2H_2O$, 27, 722
$Cl_2Zn.4/3H_2O$, 35A, 149
$Cl_2Zn.6H_2O$, 27, 722
$Cl_2Zn.11H_2O$, 27, 723
Cl_3Zr, 27, 442; 29, 249; 31A, 96
Cl_4Zr, 34A, 174; 35A, 149
Cm, 34A, 80
Cm_2O_3, 29, 418
CmX_3 (X-halogen), 30A, 449
Co, 26, 97; 29, 170; 35A, 49
Co_3Cr, 34A, 139
$CoCr_{2-x}Fe_xO_4$, 29, 326
$CoCrInO_4$, 30A, 451
CoCrNb, 30A, 121
$Co(Co_xCr_{1-x})_2O_4$, 27, 493
$Co_{0.87}Cr_{0.13}P$, 27, 320
Co-Cr-Pt, 29, 170
$CoCr_2S_4$, 30A, 121; 32A, 53
$Co_{2.5}Cr_{2.5}Si_3$, 30A, 121
$Co_3Cr_2Si_2$, 30A, 121
$Co_3Cr_3Si_2$, 27, 376
$Co_5Cr_3Si_2$, 27, 376; 30A, 121
χ-$Co_5Cr_3Si_2$, 27, 153
$Co_{29.0}Cr_{17.4}Si_{11.6}$, 27, 154
Co_4Cr_3W, 30A, 122
$CoCsF_3$, 34A, 203
$CoCs_2H_{12}O_{14}S_2$, 29, 425
$CoCs_2O_8S_2.6H_2O$, 29, 425
Co-Cu, 27, 160
$(Co,Cu,Fe)O_4S.5H_2O$, 28, 326
Co-Cu-Ge, 26, 130
$CoDO_2$, 34A, 237
Co_2Dy, 26, 236; 30A, 122
Co_3Dy, 30A, 122
Co_5Dy, 26, 239
Co_7Dy_2, 30A, 122; 35A, 118
Co_2Er, 30A, 122
Co_3Er, 30A, 122
Co_5Er, 26, 239
$Co_{5.75}Er$, 34A, 139
Co_7Er_2, 32A, 59; 35A, 118
$Co_{17}Er_2$, 34A, 139
Co_3Eu, 29, 383
CoF_2, 28, 290
$CoF_3.3H_2O$, 26, 300
CoF_3H_4N, 28, 296
$Co_2F_2H_{35}N_{14}O_{16}$, 26, 479
$CoF_{12}HP_4$, 35A, 413
CoF_3K, 26, 306; 28, 296
CoF_4K_2, 28, 296
CoF_3Na, 28, 296
CoF_3Rb, 28, 296
CoF_4Rb_2, 28, 298
CoF_3Tl, 28, 296
CoF_4Tl_2, 28, 298
CoFe, 26, 120; 27, 163
Co-Fe-Ni, 27, 49
Co-Fe-Ni-S, 26, 121; 27, 164
$Co_yFe_xNi_zS_8$, 26, 123
$CoFe_2O_4$, 33A, 324
$Co_{0.6}Fe_{0.4}P$, 27, 320
$(Co,Fe)_2P$, 27, 164
$(Co,Fe)S_2$, 30A, 122
Co_2FeSe_4, 30A, 122
$CoFe_2Se_4$, 30A, 122
Co-Fe-Si, 26, 124
$Co_{0.1}Fe_{0.9}Si_2$, 26, 125
$Co_{0.9}Fe_{0.1}Si_2$, 26, 125
$(Co,Fe)U_6$, 30A, 200
$(Co,Fe)_2U$, 30A, 200

$(Co,Fe)V_3$, 28, 45
$CoGa_3$, 32A, 139
$CoGaInO_4$, 30A, 451
Co-Ga-N, 27, 306
$Co_{1.5}Ga_{0.5}Nb$, 29, 170
Co_2GaNb, 29, 110
$CoGa_2O_4$, 26, 412
$(Co,Mg)Ga_2O_4$, 26, 412
$Co_{1.5}Ga_{0.5}Ta$, 29, 170
Co_2GaTa, 29, 110
Co_2GaTi, 27, 161, 206
CoGd, 26, 119; 27, 161
$CoGd_3$, 26, 119; 27, 161; 34A, 139; 35A, 49
Co_2Gd, 26, 119, 236; 27, 161; 30A, 122; 34A, 139
Co_3Gd, 26, 119; 27, 161; 30A, 122; 34A, 139
Co_3Gd_2, 26, 119
Co_3Gd_4, 34A, 139
Co_4Gd, 26, 120; 27, 161
Co_5Gd, 26, 239; 27, 161; 34A, 139
Co_7Gd_2, 30A, 42, 122; 34A, 139; 35A, 118
$Co_{17}Gd_2$, 27, 161; 28, 44; 34A, 139
$CoGd_2MnO_6$, 34A, 292
$CoGdO_3$, 29, 421
$(Co,Ni)_{1.67}Ge$, 27, 162
Co_2Ge, 28, 44
Co_5Ge_7, 27, 161
$Co_{16}Ge_7Hf_6$, 27, 162, 377; 30A, 122
$CoGe_3Li_2O_8$, 27, 486
Co_3GeMn_2, 28, 44
$Co_{1-x}GeMn_xO_3$, 29, 419
Co_2Ge_2Mo, 26, 174
$Co_3Ge_2Mo_5$, 30A, 122
Co-Ge-N, 27, 306
$Co_{59.1}Ge_{20.5}N_{20.4}$, 27, 306
$Co_{1.5}Ge_{0.5}Nb$, 29, 170
$Co_{16}Ge_7Nb_6$, 27, 162, 377
$CoGeO_3$, 28, 312; 29, 419; 30A, 451; 33A, 504
Co_2GeO_4, 26, 413
$CoGe_6O_{21}Pb_8$, 29, 399
$Co_{1.5}Ge_{0.5}Ta$, 29, 170
$Co_{16}Ge_7Ta_6$, 27, 162, 377
Co_2Ge_2Th, 29, 91, 110
CoGeTi, 35A, 74, 118
Co_3GeTi_2, 30A, 122
$Co_4Ge_7Zr_4$, 34A, 66
$Co_{16}Ge_7Zr_6$, 27, 162, 377
$Co(H_3N)_6I_2.Co(H_3N)_6I_3$, 29, 416
$CoH_{18}I_3N_6$, 31A, 246; 34A, 214
$Co_2H_{36}I_5N_{12}$, 29, 416
$Co_2H_{21}I_3N_6O_3$, 32A, 208
$Co_2H_3IO_3$, 29, 250
$Co_2H_3KMo_2O_{10}$, 32A, 301
$CoH_{12}K_2O_{14}S_2$, 29, 424
$CoH_{15}N_{14}$, 29, 246
$Co_2H_{26}N_{13}O_{14}$, 34A, 345
$Co_2H_{30}N_{15}O_{17}$, 26, 481; 28, 184; 33A, 427
$Co_2H_{32}N_{16}O_{15}$, 28, 183; 33A, 428
$CoH_4NO_9P_3$, 32A, 379, 381
$CoH_4NO_4P.H_2O$, 33A, 395
$CoH_{10}N_4O_8S_2$, 31A, 250
$Co_2H_{33}N_{10}O_{18}S_4$, 31A, 206
$CoH_{20}N_2O_{14}S_2$, 29, 425
$Co_2H_8N_2O_{12}S_3$, 29, 439
$CoH_8N_2O_8S_2.6H_2O$, 29, 425; 32A, 343
$Co_2H_{30}N_{10}O_{10}S_2.4H_2O$, 33A, 375
$CoHO_2$, 34A, 237
$CoH_{12}O_{14}Rb_2S_2$, 29, 425
$CoH_{12}O_{10}S$, 27, 616
Co-H-Th, 27, 240
$CoH_{4.2}Th$, 27, 240
Co_2Hf, 26, 162
Co_4Hf, 34A, 139
$Co_4Hf_2P_3$, 33A, 67
$Co_{16}Hf_6Si_7$, 27, 162, 377; 30A, 122
Co_2HfSn, 35A, 118
$CoHo_3$, 34A, 68, 139
Co_2Ho, 30A, 123; 34A, 139
Co_3Ho, 30A, 43, 123; 34A, 139
Co_3Ho_4, 34A, 72, 139
Co_5Ho, 27, 163
$Co_{5.5}Ho$, 34A, 139
Co_7Ho_2, 30A, 122; 34A, 139; 35A, 118
$Co_{17}Ho_2$, 34A, 139
$CoI_2.6H_2O$, 28, 102
$CoIN_2O_2$, 34A, 183
Co-In-N, 27, 306
$Co_{61.5}In_{20.2}N_{18.3}$, 27, 306
$Co_{71}In_{18}N_{11}$, 27, 306
$(Co,Ir)V_3$, 28, 45
$CoKO_9P_3$, 31A, 246
$CoK_2O_8S_2.6H_2O$, 29, 424
$CoK_5O_{40}W_{12}.20H_2O$, 26, 39
$CoLa_3$, 26, 127
Co_5La, 26, 239
Co_7La_2, 35A, 118
$Co_{13}La$, 32A, 139
CoLaSi, 35A, 118
$(Co,Sb)LiO_4$, 28, 344
$CoLiO_4P$, 30A, 454
$Co_{0.5}LiO_4Ti_{1.5}$, 28, 346
$CoLi_2O_8Ti_3$, 28, 348
$CoLiO_4V$, 26, 409; 28, 346
Co_2Lu, 30A, 123
Co-Mg, 26, 219
Co_2Mg, 26, 219; 29, 110
CoMgNi, 26, 174; 27, 165
$Co_xMg_{1-x}O_3Ti$, 30A, 451
Co-Mn, 27, 165
$(Co,Si)_{23}Mn_{14}$, 35A, 50
Co-Mn-Ni-O, 30A, 451
$CoMn_2O_4$, 26, 408; 27, 493
Co_2MnO_4, 26, 408
$CoMnO_4Tc$, 29, 322
$Co_{0.8}Mn_{0.2}P$, 27, 320
Co-Mn-Pt, 29, 171
Co-Mn-Si, 27, 165, 374
$Co_{1.1}MnSi_{0.9}$, 27, 376
Co_2MnSi, 27, 376
$Co_2Mn_5Si_3$, 27, 376
$Co_{20}Mn_{53}Si_{27}$, 29, 110
$Co_{55}Mn_{10}Si_{35}$, 29, 110
CoMo, 28, 44
Co_2Mo_3, 28, 17
Co_3Mo, 28, 44; 30A, 43, 123
μ-Co_7Mo_6, 27, 166
$Co_{2.31}Mo_3NaO_{12}$, 29, 329
$CoMoO_4$, 27, 522; 28, 348; 30A, 326
$CoMoO_6Sr_2$, 27, 509
$CoMo_2S_4$, 34A, 70
CoMoSi, 27, 376
$CoMo_3Si$, 32A, 57
Co_3Mo_2Si, 26, 269
$Co_3Mo_5Si_2$, 30A, 123
CoN_xNb_2, 29, 110
CoN_2O_6, 28, 426
$CoN_2O_6.6H_2O$, 30A, 454
Co-N-Sn, 27, 306
$Co_{63}N_{16}Sn_{21}$, 27, 306
CoNb, 30A, 200
$(Co,Rh)Nb_3$, 30A, 200
Co-Nb-O, 30A, 452
$CoNb_2O_x$, 29, 110
$Co_4Nb_2O_9$, 26, 389
$CoNb_3S_6$, 35A, 111
$Co_{1/4}NbSe_2$, 35A, 115
$Co_{1/3}NbSe_2$, 35A, 118
Co-Nb-Si, 29, 171
$CoNb_4Si$, 30A, 44, 123
Co_2Nb_3Si, 29, 110
Co_3Nb_2Si, 29, 110
$Co_3Nb_4Si_7$, 33A, 70
$Co_{16}Nb_6Si_7$, 27, 162, 377; 29, 110
CoNbSn, 35A, 118
Co_2NbSn, 35A, 118
Co_2Nd, 30A, 123
Co_3Nd, 30A, 123
Co_7Nd_2, 35A, 118
CoNdSi, 35A, 118
Co_2NiO_4, 33A, 290
$CoNiO_4Tc$, 29, 322
$Co_{2-x}Ni_xO_4V$, 27, 491
Co_2NiS_4, 33A, 290
Co_8NiS_8, 26, 122
$CoNi_2Se_4$, 30A, 123
Co_2NiSe_4, 30A, 123
Co-Ni-Si, 26, 124
$(Co,Ni)V_3$, 28, 45
$Co_{42.5}Ni_{15}Zn_{42.5}$, 26, 219
CoO, 27, 474; 31A, 251; 33A, 268
Co_3O_4, 29, 325; 33A, 290
$CoO_{21}Pb_8Si_6$, 29, 399
$Co_{1/2}O_3PbW_{1/2}$, 29, 420
CoO_6Pb_2W, 28, 310; 29, 420

$CoO_8Rb_2S_2.6H_2O$, 29, 425
CoO_4Rh_2, 26, 390
CoO_4S, 26, 436, 437, 438; 27, 526; 30A, 370
$CoO_4S.H_2O$, 26, 437; 28, 306; 31A, 199
$CoO_4S.4H_2O$, 30A, 458
$CoO_4S.6H_2O$, 27, 616; 30A, 458
$Co_{2.33}O_4Sb_{0.67}$, 26, 412
$Co_7O_{12}Sb_2$, 26, 410; 27, 527
CoO_4Se, 29, 357; 33A, 388
$CoO_4Se.H_2O$, 30A, 453
Co_2O_4Si, 28, 350; 29, 396
$Co_3O_{11}Si_4.H_2O$, 28, 350
Co_2O_4Sn, 28, 324
$Co_4O_8Sn_2$, 27, 490
$CoO_8Sn_2Zn_3$, 27, 490
$Co_2O_8Sn_2Zn_2$, 27, 490
$Co_3O_8Sn_2Zn$, 27, 490
(Co, Ta)O_3Sr, 28, 332
CoO_6Sr_2W, 27, 309
CoO_6Ta_2, 26, 412
$Co_4O_9Ta_2$, 26, 389
Co_2O_4Tc, 29, 322
CoO_3Ti, 27, 507; 29, 314
Co_2O_4Ti, 26, 413
CoO_4U, 26, 406; 27, 523
CoO_6U_2, 33A, 356
$CoO_{10}U_3$, 28, 334
CoO_4V_2, 28, 286
Co_2O_4V, 28, 286
$Co_3O_8V_2$, 35A, 362
CoP, 27, 319; 30A, 98, 123
CoP_3, 26, 47; 34A, 117
$CoPS_3$, 30A, 123
$CoP_{0.6}Si_{0.4}$, 27, 321
Co—Pd, 26, 128
Co_2Pr, 30A, 123
Co_3Pr, 30A, 123
Co_5Pr, 26, 239
Co_7Pr_2, 35A, 118
CoPrSi, 35A, 118
CoPt, 29, 42, 110
Co—Pu, 26, 231
$CoPu_2$, 26, 232
$CoPu_3$, 26, 232; 28, 17
$CoPu_6$, 26, 232
Co_2Pu, 26, 232
Co_3Pu, 26, 233
$Co_{17}Pu_2$, 26, 233
Co_3Re_2Si, 27, 376
$Co_{5.7}Re_6Si_{1.3}$, 29, 110
(Co, Rh)V_3, 28, 45
Co—S, 26, 208
CoS, 26, 125
β-CoS, 26, 210
β′-CoS, 26, 210
CoS_{1+x}, 30A, 123
CoS_2, 26, 210; 29, 171; 30A, 123
Co_3S_4, 26, 210; 30A, 123; 33A, 290
β-Co_4S_{3+x}, 26, 210
Co_9S_8, 26, 121, 210; 27, 169; 30A, 123
Co(S, Se)$_2$, 27, 168
CoS_4V_2, 30A, 124
Co—S—Zn, 26, 126
$CoSb_2$, 26, 35
$CoSb_3$, 26, 47
CoSbTi, 28, 44
CoSbV, 28, 44
CoSc, 29, 110
$CoSc_2$, 27, 342
Co_2Sc, 26, 236; 29, 110
$Co_{16}Sc_6Si_7$, 28, 45
$CoSe_2$, 34A, 73
Co_3Se_4, 29, 110
$CoSe_4Ti_2$, 30A, 124
$CoSe_4V_2$, 30A, 124
Co_2Se_4V, 30A, 124
CoSi, 26, 124; 27, 375; 29, 171
ε-CoSi, 29, 171, 172
CoSiSm, 35A, 118
$Co_{16}Si_7Ta_6$, 27, 162, 377
Co_2Si_2Th, 30A, 124
Co_3SiTi_2, 30A, 124
$Co_{16}Si_7Ti_6$, 27, 162, 377
Co_3SiU_2, 29, 110
(Co, Si)V_3, 28, 45
$Co_{1.1}Si_{0.9}V$, 26, 174; 27, 376
Co_2SiV, 27, 376
$Co_3Si_2V_5$, 27, 376
$Co_5Si_2V_3$, 27, 376
$Co_5Si_3V_4$, 26, 269
$Co_{5.2}Si_2V_{2.8}$, 26, 174
$Co_6Si_{2.5}V_{1.5}$, 26, 174
$Co_{40}Si_{25}V_{45}$, 26, 270
CoSiW, 27, 376
Co_3SiW_2, 26, 269
$Co_{16}Si_7Zr_6$, 27, 162, 377
$CoSm_3$, 33A, 152
Co_2Sm, 30A, 124; 33A, 152
Co_3Sm, 30A, 124; 33A, 152
Co_5Sm, 33A, 152
Co_7Sm_2, 33A, 152; 35A, 118
$Co_{17}Sm_2$, 33A, 152
Co-Sn, 27, 258
$Co_{1.4}Sn$, 27, 258
Co_2SnTi, 28, 45; 35A, 118
Co_2SnV, 28, 45
Co_2SnZr, 35A, 118
Co-Ta, 27, 171
CoTa, 32A, 139
$CoTa_2$, 32A, 139
α-Co_2Ta, 27, 171
β-Co_2Ta, 27, 170
γ-Co_2Ta, 27, 171
Co_3Ta, 32A, 139
α-Co_3Ta, 27, 171
β-Co_3Ta, 27, 171
Co_2Tb, 30A, 124
Co_3Tb, 30A, 124
Co_7Tb_2, 30A, 124; 35A, 118
$CoTe_2$, 35A, 75
CoTh, 27, 240
Co_5Th, 26, 239
Co_7Th_2, 35A, 53
$Co_{17}Th_2$, 34A, 75
$Co_{16}Ti_6Si_7$, 27, 162
Co_2Tm, 30A, 124
Co_3Tm, 30A, 124
Co_3V, 30A, 124
(Co, V)$_2$Zr, 28, 19
Co_2Y, 26, 236
Co_3Y, 30A, 78, 125
Co_3Y_2, 30A, 125
Co_5Y, 26, 239; 27, 163; 30A, 124
Co_7Y_2, 30A, 124
$Co_{17}Y_2$, 29, 110; 30A, 124
CoZn, 26, 99
$CoZn_{13}$, 27, 173
CoZr, 29, 110; 30A, 125
$CoZr_2$, 26, 128; 29, 110
$CoZr_3$, 30A, 125; 35A, 54
Co_2Zr, 29, 172
Co_4Zr, 29, 110
$Co_{23}Zr_6$, 30A, 125
Cr, 26, 110; 27, 150; 29, 172, 32A, 51
$CrCsIO_6$, 26, 500
Cr_3CsO_8, 31A, 150; 32A, 281
$CrCsO_8S_2.12H_2O$, 33A, 372
Cr—Cu—Fe—O, 26, 411
$CrCuHO_9PPb_2$, 33A, 405
$Cr_xCuMn_{2-x}O_4$, 26, 394
Cr_2CuO_4, 26, 411
$Cr_2Cu_2O_4$, 26, 396
$CrCuS_2$, 33A, 61
$CrCu_3S_3$, 30A, 44
$CrCuSe_2$, 33A, 61
Cr_2CuSe_4, 30A, 125; 32A, 53
Cr_2CuTe_4, 30A, 125
$CrDO_2$, 28, 223
$CrDyO_3$, 28, 328; 33A, 316
$CrDyO_4$, 28, 308
$CrErO_3$, 28, 328; 32A, 283
$CrErO_4$, 28, 308
$CrEuO_3$, 28, 328
$CrEuO_4$, 28, 308
CrF_5, 28, 296
Cr_2F_5, 29, 255
$CrF_3.3H_2O$, 26, 300
CrF_3K, 26, 306; 31A, 247
CrF_6Na_3, 34A, 198
$CrFO_4$, 28, 296
$Cr_6F_2O_{32}Pb_{10}Si_2Zn$, 35A, 461
$Cr_3FO_{12}Sr_5$, 28, 308
CrF_3Rb, 31A, 247
CrF_4Sr, 35A, 163
σ-CrFe, 27, 155
$Cr_2Fe_3H_6KO_{14}$, 31A, 247
$CrFeHO_5$, 31A, 247
$Cr_{2-x}Fe_xNiO_4$, 29, 326
Cr—Fe—Ni—Ti, 26, 110
(Cr, Fe, Ni)$_{24}Ti_5$, 26, 110
Cr-Fe-O, 27, 500
Cr_2FeO_4, 29, 38, 110

$Cr_{2-x}Fe_xO_4$, 29, 326
$(Cr,Fe,Ta)_{1/2}O_3Sr$, 31A, 251
Cr_2FeS_4, 29, 40, 110; 30A, 125
$CrFe_2Se_4$, 30A, 125
Cr_2FeSe_4, 32A, 54
Cr-Fe-Si, 27, 156
Cr-Fe-Ti, 29, 172
CrGa, 29, 110; 30A, 125
$CrGa_4$, 30A, 125
$CrGa_4La_{0.5}O_{12}Y_{2.5}$, 27, 513
Cr_4GaLiO_8, 31A, 245
$CrGaN_x$, 30A, 125
Cr_2GaN, 34A, 139
Cr_3GaN, 29, 111
$CrGa_4O_{12}Y_3$, 27, 513
$Cr_2Ga_3O_{12}Y_3$, 27, 513
$CrGa_2S_4$, 26, 117
$CrGdO_3$, 28, 328
$CrGdO_4$, 28, 308
Cr-Ge, 30A, 201
Cr_3Ge, 30A, 125, 200
$Cr_{11}Ge_8$, 29, 111
$Cr_{11}Ge_{19}$, 29, 111; 32A, 70
$Cr_2Ge_3Mn_3O_{12}$, 26, 415
Cr_2Ge_2Th, 29, 91, 111
CrH, 28, 45
$CrH_{18}I_3N_6$, 27, 657
CrH_4INO_6, 26, 500
$Cr_3HK_2O_9$, 35A, 389
$CrH_6Mo_6Na_3O_{24}.8H_2O$, 35A, 419
$CrH_8N_2O_4$, 27, 421; 34A, 274; 35A, 384
$CrH_9N_3O_4$, 29, 302
$CrH_{12}N_3O_8$, 27, 659
$CrH_{10}N_4O_8S_2$, 31A, 250
$Cr_2H_{14}N_4O_8Zn$, 34A, 276
$Cr_3HNa_2O_9$, 35A, 390
$CrHO_2$, 28, 223; 31A, 129
$Cr_3H_7O_{14}S_2$, 31A, 246
$Cr_3HO_{13}Sr_5$, 28, 308
Cr_2Hf, 26, 162; 28, 45; 29, 111; 33A, 152
$CrHg_4N_2O_4$, 30A, 454
$CrHgO_4$, 26, 444
$CrHgO_4.\frac{1}{2}H_2O$, 26, 444
$CrHg_3O_6$, 26, 444
$CrHoO_3$, 28, 328
$CrHoO_4$, 28, 308
CrI_2, 27, 432
$CrIKO_6$, 26, 500; 32A, 320
$CrIKO_6.\frac{1}{2}H_2O$, 26, 500
$CrIO_6Rb$, 26, 500
$CrIn_3$, 29, 111
Cr_4InLiO_8, 31A, 245
$CrInMgO_4$, 30A, 451
$CrInMnO_4$, 30A, 451
$CrInNiO_4$, 30A, 451
$CrInO_4Zn$, 30A, 451
Cr-In-S, 26, 117
$CrIn_2S_4$, 26, 117
CrK_3MnO_8, 33A, 317
CrK_3O_8, 27, 659; 28, 224
$Cr_2K_2O_7$, 29, 319; 32A, 278; 33A, 316
$Cr_4K_2O_{13}$, 35A, 391
$CrK_3O_{12}S_3$, 26, 446
$CrLaO_3$, 28, 328
$CrLiMnO_4$, 28, 346
$CrLi_9N_5$, 26, 187
$CrLiO_2$, 26, 390
$(Cr,Ti)LiO_4$, 28, 346
$CrLi_3O_4$, 30A, 451
$Cr_2Li_2O_7.2H_2O$, 35A, 387
Cr_3LiO_8, 31A, 150; 32A, 277
$CrLi_3O_8V_2$, 28, 348
$CrLiS_2$, 35A, 44
$CrLuO_3$, 28, 328
$CrLuO_4$, 28, 308
$(Cr,S)MgO_4$, 30A, 454
Cr_2MgO_4, 26, 396; 27, 492
$Cr_{12}Mg_8O_{130}Si_4Ti_{48}$, 26, 537
$Cr_6Mn_6Ni_{16}Si_7$, 26, 175
Cr_2MnO_4, 27, 492
$CrMnO_4Zn$, 26, 409
$Cr_{0.1}Mn_{1.9}Sb$, 27, 58
(Cr,Mn)Sb, 28, 45
$(Cr,Mn)_2Sb$, 28, 45
Cr-Mn-Si, 27, 375
Cr-Mo-Nb, 26, 110
$Cr_{13.5}Mo_{22.0}Ni_{20.5}$, 30A, 83
$Cr_{18}Mo_{42}Ni_{40}$, 30A, 40
$Cr_2Mo_3O_{12}$, 31A, 249
$Cr_{0.9}Mo_{0.1}O_{4.0}Pb$, 26, 391
$(Cr_{1-x}Mo_x)_5Si_3$, 26, 23
Cr-Mo-W, 26, 111; 27, 157
$CrN_3O_9.9H_2O$, 30A, 454
$Cr_2Na_2O_7.2H_2O$, 34A, 275; 35A, 388
$CrNaO_8S_2.12H_2O$, 33A, 372
$CrNaO_6Si_2$, 30A, 455
$CrNaS_2$, 33A, 62
$CrNaSe_2$, 33A, 62
Cr_2Nb, 26, 112; 30A, 125
Cr-Nb-Ni, 26, 113
CrNbNi, 26, 113; 30A, 125
Cr-Nb-O, 27, 521; 30A, 452
$CrNb_{49}O_{124}$, 31A, 255
$CrNbO_6Pb_2$, 28, 332
$Cr_{1/4}NbSe_2$, 35A, 118
$Cr_{1/3}NbSe_2$, 35A, 118
ρ-Cr-Nb-Si, 26, 113
β-CrNbSi, 26, 114
$CrNb_2Si_2$, 26, 114
$Cr_2Nb_3Si_3$, 26, 114
$Cr_4Nb_2Si_5$, 33A, 64
$Cr_5Nb_2Si_6$, 26, 114
θ-$Cr_7Nb_3Si_{10}$, 26, 114
τ-$Cr_{37}Nb_{27}Si_{36}$, 26, 114
ν-$Cr_{45}Nb_{15}Si_{40}$, 26, 114
$CrNdO_3$, 28, 328; 32A, 283
$CrNdO_4$, 28, 308
$CrNdO_5Ti$, 35A, 237
Cr-Ni, 26, 112; 29, 173
$CrNi_2$, 27, 157
Cr_2NiO_4, 26, 409; 30A, 451
σ-Cr-Ni-P, 27, 158
$(Cr, Ni)_3P$, 27, 158
Cr_2NiS_4, 31A, 36
(Cr, Ni)Si, 27, 374
$Cr_3Ni_5Si_2$, 27, 153, 376
$Cr_6Ni_{16}Si_7$, 28, 45
$Cr_{6.5}Ni_{2.5}Si$, 27, 152, 376
$Cr_{9.5}Ni_{12.5}Si_7$, 27, 376
$Cr_6Ni_{16}Si_7Ta_6$, 26, 175
$Cr_6Ni_{16}Si_7V_6$, 26, 175
CrO, 28, 320
CrO_2, 26, 354; 27, 478
CrO_3, 35A, 206
Cr_2O_3, 27, 499
$(Cr,V)_2O_3$, 30A, 457
Cr_5O_{12}, 30A, 304
CrO_4Pb, 27, 609; 30A, 320
$CrO_4Pb.H_2O$, 28, 328, 354
CrO_5Pb_2, 28, 328
CrO_3Pr, 28, 328
CrO_4Pr, 28, 308
CrO_5Py, 28, 685
$Cr_2O_7Rb_2$, 35A, 385
$CrO_8RbS_2.12H_2O$, 34A, 312
CrO_4Rh_2, 26, 309
$Cr_2O_{12}S_3$, 30A, 453
CrO_6SbSr_2, 29, 420
CrO_3Sm, 28, 328
CrO_4Sm, 28, 308
CrO_4Sr, 27, 609
CrO_4Sr_2, 31A, 150; 32A, 279
Cr_2O_7Sr, 31A, 150; 32A, 280
$Cr_2O_8Sr_3$, 28, 308; 30A, 451
CrO_3Tb, 28, 328
CrO_4Tb, 28, 308
Cr_2O_6Te, 27, 526; 28, 328; 33A, 344
$Cr_2O_{2n-1}Ti_{n-2}$, 29, 291
CrO_3Tm, 28, 328
CrO_4Tm, 28, 308
CrO_4U, 30A, 347
CrO_6U, 30A, 454
$Cr_2O_9U_2$, 26, 407
$Cr_{1-x}O_2V_x$, 26, 354
CrO_4V, 27, 521, 524
Cr_2O_6W, 27, 526; 31A, 259; 33A, 344
CrO_3Y, 28, 328
CrO_4Y, 28, 308
CrO_3Yb, 28, 328
CrO_4Yb, 28, 308
CrOZr, 29, 111
Cr-Os, 29, 173
CrOs, 29, 86

σ-CrOs, 29, 86
CrP, 27, 319; 30A, 98, 125
Cr_3P, 27, 158
ε-Cr_3P, 27, 101
Cr-Pt, 27, 323
$Cr_{1.2}Pt_{2.8}$, 27, 258
Cr_3Pt, 27, 323
σ-CrRe, 29, 86, 111
Cr_2Re_3, 29, 86
$CrRh_3$, 34A, 139
Cr_3Rh, 30A, 126
CrRu, 29, 86
σ-CrRu, 29, 86
Cr_2Ru, 26, 114; 31A, 36
Cr_3Ru, 26, 114
Cr_5S_6, 32A, 55
Cr_5S_8, 34A, 139
Cr_2S_4Ti, 30A, 126; 33A, 66
CrS_7Y_4, 26, 277
Cr_2S_5Zn, 28, 302
$CrSb_2$, 33A, 63; 35A, 10
Cr_3Sb, 26, 35
$Cr_{0.68}Se$, 35A, 45
$Cr_{1-x}Se$, 29, 111
CrSe, 26, 117
Cr_2Se_3, 29, 111; 35A, 45, 46
Cr_3Se_4, 26, 115; 27, 159; 29, 40, 111
Cr_5Se_6, 29, 111
Cr_5Se_8, 34A, 139
Cr_7Se_8, 26, 116; 29, 41, 111
$Cr_3Se_2Te_2$, 29, 111
Cr_2Se_4Ti, 33A, 66
$CrSe_4Zn$, 30A, 126
$CrSi_2$, 27, 156, 374
$(Cr,V)Si_2$, 27, 375
Cr_3Si, 27, 159
$(Cr,W)_3Si$, 27, 159
Cr_5Si_3, 27, 159
Cr_2Si_2Th, 30A, 126
Cr-Si-U, 29, 173
CrSiU, 29, 173
Cr_3SiU_2, 29, 111, 173
$Cr_5Si_3U_4$, 29, 111, 174
$Cr_{0.5}Si_2V_{0.5}$, 27, 375
$Cr_{0.75}Si_2V_{0.25}$, 27, 375
$CrSn_2$, 31A, 37
Cr_2Sn_3, 31A, 37
Cr-Ta, 30A, 201
Cr_2Ta, 26, 118; 30A, 126, 201
Cr-Tc, 27, 361
$Cr_{1-x}Te$, 28, 45; 29, 111
$CrTe_x$, 28, 45
Cr_2Te_3, 28, 45; 29, 111; 35A, 47
Cr_3Te_4, 28, 45; 29, 111; 30A, 126; 35A, 47
Cr_5Te_6, 28, 45; 29, 111; 35A, 47
Cr_5Te_8, 29, 111
Cr_7Te_8, 28, 45
$CrTe_4Ti_2$, 30A, 126
Cr_2Te_4Ti, 31A, 37
Cr_3Te_4Ti, 30A, 126
$CrTe_4V_2$, 30A, 126
Cr_2Te_4V, 30A, 126
$Cr_{2.4}W_{0.6}Si$, 27, 159
Cr_2Zr, 29, 111
Cs, 26, 3
Cs II, 29, 173
Cs III, 29, 173
$CsCuO_2$, 34A, 247
Cs_2ErF_6Na, 33A, 186
Cs_2F_6InNa, 35A, 158
CsF_2Li, 27, 451; 30A, 269
$Cs_4F_{10}Mg_3$, 34A, 188
CsF_3Mn, 26, 309; 27, 455; 29, 414
CsF_5Mn, 26, 309
Cs_2F_4Mn, 29, 414
CsF_3Ni, 34A, 203
CsF_2O_2P, 32A, 369
Cs_2F_6Pr, 26, 312
Cs_3F_7Pr, 26, 312
Cs_3F_7Tb, 26, 313
CsF_6U, 32A, 178
$Cs_2Fe_2O_3$, 26, 397
$Cs_2Fe_{14}O_{22}$, 26, 395
$CsFeO_8S_2$, 31A, 246
$CsFeO_6Si_2$, 28, 281
$Cs_xFe_xO_4Ti_{2-x}$, 33A, 304
$Cs_5FeO_{40}W_{12}.nH_2O$, 27, 672
$CsFeS_2$, 33A, 3
$CsFeSe_2$, 31A, 261
CsGe, 26, 84, 157
$Cs_2Ge_5O_{11}$, 26, 398; 31A, 252
$Cs_2Ge_6O_{13}$, 31A, 252
$CsH_{24}O_{20}S_2V$, 27, 723
Cs_2HgO_2, 29, 338
CsI, 26, 297
Cs_2I_6Re, 29, 415
Cs_2I_2Te, 27, 463
$Cs_2K_2O_{12}U_2$, 30A, 348
$(Cs_xK_{1-x})_3Sb$, 30A, 201
$Cs_{0.70}Mg_{0.35}O_4Ti_{1.65}$, 33A, 303
$CsMnO_4$, 31A, 259
$Cs_xMn_xO_4Ti_{2-x}$, 33A, 304
$Cs_4MnO_{40}W_{12}.nH_2O$, 27, 672
Cs_xMoO_3, 35A, 277
Cs_2MoO_4, 35A, 233
$CsMoS_4$, 30A, 453
CsN_3, 27, 414; 30A, 256
$CsN_3O_{11}U$, 26, 477
Cs-Nb-O, 30A, 452
$Cs_{0.10}NiO_2$, 34A, 293
CsO_3, 31A, 257
CsO_4Re, 31A, 250
Cs_2O_4S, 27, 605
$Cs_2O_6S_2$, 28, 306
$CsO_8S_2V.12H_2O$, 27, 723
Cs_2O_4Se, 27, 605
$Cs_xO_4Se_xTi_{2-x}$, 33A, 304
Cs-O-Ta, 30A, 452
CsO_4Tc, 27, 671; 31A, 250
Cs_xO_2Ti, 29, 318
Cs_2O_4U, 26, 404
CsO_8V_3, 31A, 141
$Cs_{0.31}O_3W$, 26, 357
Cs_2O_4W, 35A, 233
CsPb, 29, 111
Cs_2S_6, 33A, 60
CsS_4W, 30A, 453
Cs_3Sb, 26, 54
CsSi, 26, 84, 156
CsSn, 29, 111
Cu, 26, 3, 129; 30A, 201
CuDy, 26, 237; 29, 111
Cu_2Dy, 28, 45
Cu_5Dy, 34A, 139
$Cu_8Dy_6Ge_8$, 35A, 55, 118
$Cu_3Dy_2S_{4.5}$, 30A, 126
$Cu_6Dy_2S_6$, 30A, 126
Cu_2Er, 28, 45
Cu_5Er, 34A, 139
$Cu_8Er_6Ge_8$, 35A, 55, 118
$Cu_3Er_2S_{4.5}$, 30A, 44, 126
$Cu_6Er_2S_6$, 30A, 126
Cu_2Eu, 28, 45
$CuEuO_2$, 34A, 252
$CuF_2.2H_2O$, 27, 436
$Cu_{1.4}F_{0.4}Fe_{1.6}O_{3.6}$, 28, 352
$Cu_{1.65}F_{0.65}Fe_{1.35}O_{3.35}$, 28, 352
Cu_2FFeO_3, 28, 352
$CuF_5H_4NO_2.4H_2O$, 31A, 245
$CuF_7H_4NSi.4H_2O$, 31A, 245
$CuF_7H_4NSn.4H_2O$, 31A, 245
$CuF_4H_4NTi.4H_2O$, 31A, 258
$CuF_7H_4NTi.4H_2O$, 32A, 171
$CuF_2H_4O_2$, 27, 436
CuF_3K, 26, 306; 28, 298
CuF_4K_2, 28, 298
CuF_3Na, 27, 454; 28, 298
CuF_4Na_2, 27, 454; 28, 298; 30A, 271
CuF_3Rb, 28, 298
CuF_4Rb_2, 27, 454; 28, 298
CuF_6Sr_2, 35A, 163
$CuF_6Ti.4H_2O$, 31A, 258; 32A, 170
CuF_3Tl, 27, 454; 28, 298
CuF_4Tl_2, 27, 454; 28, 298
$Cu_2F_8Zr.12H_2O$, 32A, 508
Cu-Fe, 30A, 201, 202
Cu-Fe-Ge, 26, 131
$Cu_2(Fe,Zn)GeS_4$, 30A, 126
$CuFe_6H_8O_{24}P_4.4H_2O$, 30A, 454
Cu-Fe-N, 27, 306

$CuFeO_2$, 31A, 137
$CuFe_2O_4$, 26, 396, 411; 30A, 451
$Cu_2Fe_2O_4$, 26, 396
$CuFe_2O_{16}S_4.6H_2O$, 35A, 436
$Cu_{0.47}Fe_{0.44}O_4SZr_{0.09}.H_2O$, 27, 634
$CuFeS_2$, 29, 112
$(Cu,Fe)_3S_2$, 27, 182
Cu_5FeS_4, 26, 137; 29, 43, 111, 173, 174
$Cu_7Fe_2S_{10}Sn$, 30A, 126, 453
Cu_2FeSe_4Sn, 30A, 126
Cu-Ga, 27, 185; 30A, 202
ζ-Cu-Ga, 27, 175
$Cu_{1.5}Ga_{0.5}Li$, 30A, 127
$CuGaMO_4$, 31A, 251
CuGaMg, 26, 174
$Cu_{1.25}Ga_{0.75}Mn$, 26, 174
$Cu_{1.5}Ga_{0.5}Mn$, 26, 174
CuGaNb, 29, 112
Cu_2GaNb_3, 29, 112
Cu-Ga-Te, 30A, 202
$CuGaTe_2$, 30A, 127
$Cu_{1.25}Ga_{0.75}Ti$, 29, 112
Cu-Ga-Zn, 30A, 127
CuGaZr, 32A, 139
CuGd, 26, 237; 29, 112; 30A, 127
Cu_2Gd, 28, 45
Cu_5Gd, 26, 239; 34A, 139
Cu_6Gd, 35A, 118
$Cu_8Gd_6Ge_8$, 35A, 54
Cu-Ge, 27, 185
$Cu_2GeHgSe_4$, 30A, 127
$Cu_{1.5}Ge_{0.5}Li$, 30A, 127
Cu_2GeLi, 30A, 127; 34A, 139
$Cu_8Ge_8Lu_6$, 35A, 55, 118
CuGeMg, 35A, 119
$Cu_{1.55-1.25}Ge_{0.45-0.75}Mg$, 26, 174
Cu_3GeMg_2, 26, 131; 35A, 118
Cu-Ge-Mn, 26, 130
$CuGe_2Mn_2$, 30A, 127
$Cu_{1.5}Ge_{0.5}Mn$, 26, 174
$CuGeO_3$, 32A, 494
$CuGe_6O_{21}Pb_8$, 29, 399
$CuGe_2P_3$, 26, 227
Cu_2GeS, 30A, 127
Cu_2GeS_3, 26, 134; 28, 45; 30A, 127
Cu_8GeS_6, 30A, 127
Cu_2GeSe_3, 26, 48, 134 28, 45; 30A, 127
$Cu_{2.67}Ge_{1.33}Se_4$, 29, 112
Cu_2GeSe_4Zn, 30A, 127
Cu-Ge-Si, 26, 132
$CuGe_{0.6}Si_{0.4}Ti$, 33A, 152
Cu_2GeTe_3, 26, 134; 28, 45; 30A, 127
Cu_2Ge_2Th, 29, 91, 112
$Cu_3H_{12}I_4N_4$, 35A, 415
$CuHIO_4$, 27, 644
$Cu_2H_3IO_3$, 26, 327; 29, 250
CuH_4MoNS_4, 35A, 144
CuH_6N_8, 31A, 77
$Cu_2H_{12}N_4Na_4O_6S_4$, 31A, 208
$Cu_3H_{12}N_4Na_4O_{12}S_8.H_2O$, 35A, 414
$Cu_3H_{15}N_5Na_4O_{12}S_8$, 34A, 310
$CuH_{12}N_6O_4$, 26, 478
$Cu_2H_3NO_6$, 26, 478
CuH_4NO_3S, 33A, 384
$CuH_8N_2O_8S_2.½H_2O$, 30A, 453
$CuH_8N_2O_8S_2.6H_2O$, 26, 449; 30A, 374; 31A, 203; 34A, 305
$CuH_{10}N_4O_8S_2$, 31A, 250
$CuH_{12}N_4O_4S.H_2O$, 34A, 307
$CuH_{12}N_4O_4Se$, 34A, 307
CuH_2O_2, 26, 358
$CuHO_3P.2H_2O$, 34A, 333
Cu_2HO_5P, 27, 601; 28, 194
$Cu_3H_3O_7P$, 29, 371
$Cu_5H_4O_{12}P_2$, 28, 192
$CuH_{16}O_{20}P_2U_2$, 29, 375
CuH_2O_6PbS, 26, 451; 27, 623
$Cu_5H_6O_{28}Pb_2Se_6U_2.2H_2O$, 30A, 453
$CuH_{10}O_9S$, 27, 615
$Cu_3H_4O_8S$, 26, 439, 440; 28, 211
$Cu_2H_6O_9Si_2$, 33A, 472
$Cu_5H_2O_{14}Si_4$, 31A, 233, 260; 32A, 460
$Cu_8H_4O_{26}Si_8$, 31A, 260
$CuH_2O_{12}Si_2U_2.5H_2O$, 28, 277
CuH_4O_5Te, 27, 635
$CuHf_2$, 27, 234
$(Cu,Sn)_4Hf_5$, 30A, 127
$CuHf_5Pb_3$, 30A, 127
Cu_2Hf_3Zn, 30A, 127
Cu_7Hg_6, 34A, 77
$CuHgN_2O_7.3H_2O$, 34A, 343
Cu_2HgS_4Sn, 30A, 127
Cu_2HgSe_4Sn, 30A, 127
$Cu_2HgSnTe_4$, 30A, 128
$CuHgTe_2$, 28, 320
CuHo, 29, 112
Cu_2Ho, 28, 45
Cu_5Ho, 26, 239
$Cu_3Ho_2S_{4.5}$, 30A, 128
$Cu_6Ho_2S_6$, 30A, 128
CuI, 26, 335
Cu-In, 27, 185
Cu_4InMg, 26, 136, 174
Cu_2In_2Mn, 26, 174
$Cu_6In_2Mn_2$, 29, 112
Cu_6In_3Mn, 29, 112
$Cu_2In_2O_5$, 33A, 284
$CuInTe_2$, 27, 181; 33A, 152
Cu_2InTi, 27, 12, 296
$CuK_2N_6O_{12}Pb$, 32A, 510; 34A, 346
CuKO, 33A, 280
$CuKO_2$, 34A, 247
$CuK_2O_{12}P_4$, 35A, 346
Cu_2La, 28, 45
Cu_5La, 26, 239
$CuLaO_2$, 34A, 252
$CuLi_2O_2$, 35A, 216
$Cu_{0.5}LiO_4Ti_{1.5}$, 28, 346
$CuLiO_4V$, 31A, 143, 251
Cu_2LiSi, 30A, 128; 34A, 139
$CuLi_2Sn$, 34A, 139
Cu_2LiSn, 35A, 55
$Cu_{1.5}LiZn_{0.5}$, 30A, 128
Cu_2Lu, 28, 45
$Cu_3Lu_2S_{4.5}$, 30A, 128
$Cu_6Lu_2S_6$, 30A, 128
Cu_2Mg, 28, 45
$CuMgO_2$, 26, 396; 28, 320; 30A, 451
Cu_3MgO_4, 29, 419
Cu-Mn, 27, 183, 186
$CuMn_3N$, 29, 112
$CuMn_2O_4$, 27, 494
CuMnSb, 33A, 97
Cu_2MnSb, 27, 42
Cu_4MnSn, 26, 174
CuMnZn, 27, 261
$CuMoO_4$, 33A, 335
CuN_6, 32A, 232; 33A, 510
CuN_2O_6, 28, 361; 30A, 400
$CuN_2O_6.2.5H_2O$, 34A, 340; 35A, 322
$CuN_2O_6.3H_2O$, 31A, 177
CuNaO, 34A, 248
$CuNaO_2$, 34A, 247
$CuNa_2O_8S_2.2H_2O$, 26, 449
$Cu_{0.65}NbS_2$, 28, 45
Cu_3NbS_4, 26, 138; 28, 45
Cu_3NbSe_4, 26, 138; 28, 45
$Cu_3Nb_{½}Se_4V_{½}$, 26, 138
$Cu_4Nb_5Si_4$, 33A, 77
$Cu_3Nb_{½}Ta_{½}Te_4$, 26, 138
Cu_3NbTe_4, 26, 138
CuNd, 30A, 128
Cu_2Nd, 28, 45
Cu_6Nd, 35A, 119
$CuNdO_2$, 34A, 252
Cu-Ni, 27, 186
$CuNi_{0.5}O_2Sn_{0.5}$, 27, 518
$Cu_{0.75}Ni_{0.25}O_2Sr$, 35A, 245
$CuNi_{0.5}O_2Ti_{0.5}$, 27, 518
Cu-Ni-Pd, 30A, 202
CuNiTi, 33A, 152
Cu_2NiZn, 30A, 128

CuO, 27, 474; 35A, 207
$Cu_2O_7P_2$, 26, 466; 31A, 192; 32A, 376; 33A, 410
$CuO_{12}P_2U_2.8H_2O$, 29, 375
$CuO_{12}P_2U_2.12H_2O$, 26, 468
$CuO_{21}Pb_8Si_6$, 29, 399
CuORb, 34A, 248
CuO_2Rb, 34A, 247
CuO_2Rh, 26, 390
CuO_4Rh_2, 26, 390, 399
CuO_4S, 26, 436, 438
$CuO_4S.H_2O$, 26, 439; 31A, 246
$CuO_4S.3H_2O$, 33A, 368
$CuO_4S.5H_2O$, 27, 614
Cu_2O_5S, 26, 440; 27, 621; 28, 216
$Cu_3O_6S_2.2H_2O$, 30A, 380
CuO_4Se, 29, 357
$CuO_4Se.5H_2O$, 30A, 453
$Cu_6O_{16}Si_5.1\frac{1}{2}-2H_2O$, 26, 522
$Cu_6O_{18}Si_6.6H_2O$, 33A, 160
CuO_2Sm, 34A, 252
CuO_2Sr, 35A, 246
Cu_2O_2Sr, 35A, 217
$CuO_3Te.2H_2O$, 27, 635; 31A, 253
Cu_2O_4Te, 31A, 253
Cu_3O_6Te, 33A, 392
Cu_2OTi_4, 28, 20, 140
CuO_4U, 33A, 357
$CuO_{10}U_3$, 28, 334
$Cu_{0.60}O_5V_2$, 35A, 275
CuO_4W, 35A, 226
Cu_3O_6W, 34A, 284
CuP_2, 30A, 45, 128
Cu_3P, 30A, 268
CuPS, 29, 112
$CuPS_2$, 28, 45; 29, 112
Cu_3PS_4, 29, 112
$CuPbS_3Sb$, 35A, 11
CuPSe, 29, 112
$CuPSe_2$, 28, 45; 29, 112
CuP_3Si_2, 26, 227
$CuPb_3Zr_5$, 30A, 128
Cu-Pd, 27, 186
CuPr, 30A, 128
Cu_2Pr, 28, 45
Cu_5Pr, 26, 239
Cu_6Pr, 35A, 119
Cu-Pt, 27, 186
$CuPt_{37}Zn_{62}$, 35A, 119
$Cu_3Pt_2Zn_5$, 35A, 56, 119
$Cu_3Pt_{14}Zn_{83}$, 35A, 119
Cu-Rh, 29, 174
$CuRh_2S_4$, 29, 112
CuS, 33A, 79; 34A, 139
$Cu_{1.96}S$, 29, 46, 112
$Cu_{2-x}S$, 35A, 57
Cu_2S, 28, 19
Cu_7S_4, 35A, 57
$Cu_{7.2-x}S_4$, 28, 20
Cu_3S_4Sb, 26, 37
$Cu_{12}S_{13}Sb_4$, 28, 50; 29, 15, 112
Cu_3SSe_2TaTe, 26, 138
$Cu_3S_2Se_2V$, 26, 138
Cu_2S_3Si, 28, 45; 30A, 128
Cu_8S_6Si, 30A, 128
Cu_2S_3Sn, 26, 134; 28, 45; 30A, 128
Cu_2S_4SnZn, 30A, 128
CuS_3Ta, 29, 46
Cu_3S_4Ta, 26, 138; 28, 45
$Cu_6S_6Tb_2$, 30A, 129
$Cu_3S_{4.5}Tm_2$, 30A, 129
$Cu_6S_6Tm_2$, 30A, 129
Cu_3S_4V, 26, 137; 31A, 41
$Cu_3S_{4.5}Y_2$, 30A, 129
$Cu_6S_6Y_2$, 30A, 129
$Cu_3S_{4.5}Yb_2$, 30A, 129
$Cu_6S_6Yb_2$, 30A, 129
CuS_4Zr_2, 33A, 152
Cu-Sb, 27, 185
Cu_2Sb, 29, 174
$CuSbSe_2$, 29, 14, 112
Cu_3SbSe, 30A, 129
CuSc, 27, 342; 29, 112
CuSe, 27, 183
$CuSe_2$, 30A, 46, 129
Cu_3Se_2, 31A, 40
Cu_8Se_6Si, 30A, 129
Cu_2Se_3Sn, 26, 134; 28, 45; 30A, 129
Cu_2Se_4Sn, 28, 45
Cu_2Se_4SnZn, 30A, 129
Cu_3Se_4Ta, 26, 138; 28, 45
Cu-Se-Te, 30A, 129
Cu_3Se_4V, 26, 138; 28, 45
Cu-Si, 26, 132
Cu_3Si, 30A, 129
Cu_2Si_2Sr, 35A, 24, 119
Cu_2SiTe_3, 30A, 129
Cu_2Si_2Th, 30A, 129
$Cu_{0.1}Si_{1.9}Ti$, 34A, 139
CuSiTi, 34A, 139
Cu_2Si_2Y, 34A, 139
Cu-Si-Zn, 29, 174
CuSm, 29, 112
Cu_2Sm, 28, 45
Cu_6Sm, 35A, 94
Cu-Sn, 27, 185
β-CuSn, 29, 112
β′-CuSn, 29, 113
γ-Cu-Sn, 29, 113, 175
δ-Cu-Sn, 29, 112
γ-Cu_3Sn, 31A, 41
ε'-Cu_3Sn, 31A, 41
Cu_2SnTe_3, 26, 134; 28, 45; 30A, 129
Cu_2SnTe_4Zn, 30A, 129
Cu_2SnTi, 27, 296
$CuSn_3Zr_5$, 30A, 130
$(Cu,Sn)_4Zr_5$, 30A, 129
Cu_3TaTe, 26, 138
CuTb, 29, 113
Cu_2Tb, 28, 45
Cu_5Tb, 34A, 140
Cu-Te, 30A, 130
CuTe, 26, 138; 29, 48, 113
$Cu_{4-x}Te$, 34A, 79
Cu_7Te_4, 32A, 61
$Cu_{10.5}Te$, 33A, 80
Cu_3Te_4V, 26, 138
Cu_2Th, 26, 256
Cu_6Th, 35A, 119
Cu-Ti, 27, 186
$CuTi_2$, 26, 138; 27, 187; 28, 20
Cu_2Ti, 29, 113; 30A, 130
Cu_3Ti_2, 29, 113; 30A, 47, 130
Cu_4Ti, 33A, 152
Cu_4Ti_3, 29, 113; 30A, 47, 130
Cu_2TiZn, 27, 187, 296
Cu-Ti-Zr, 26, 138
CuTm, 29, 113
Cu_2Tm, 28, 45
Cu_5Tm, 34A, 140
Cu_5U, 34A, 140
CuY, 26, 276; 28, 45
Cu_2Y, 28, 21
Cu_4Y, 26, 276
Cu_5Y, 26, 239
Cu_2Yb, 28, 45
Cu-Zn, 27, 185, 190; 29, 176; 30A, 130, 202
α′-CuZn, 26, 139
β′-Cu_3Zn_2, 27, 175; 29, 113
Cu_5Zn_8, 33A, 72
Cu_2ZnZr, 27, 296
Cu_2ZnZr_3, 30A, 130
$CuZr_2$, 27, 234

D, 26, 285
D_2, 27, 393; 30A, 449
D_4EuV, 28, 308
D_4FN, 30A, 449
D_2Ho, 28, 45
$D_{2.96}Ho$, 29, 48
D_3Ho, 28, 50; 29, 48, 113
DLi, 34A, 151
D_2Mg, 28, 22
D_3N, 26, 290, 298
$D_{0.95}Nb$, 33A, 81
$D_{2.93}NiZr$, 27, 237
D_2O, 30A, 303; 33A, 255
DO_2Y, 31A, 129
DTa_2, 29, 50; 30A, 133; 35A, 133
β_1-DTa_2, 26, 139

β_2-DTa_2, 26, 139
D_2Tb, 28, 46
α-D_3U, 27, 194
D_2Zr, 26, 168
Dy, 26, 141, 234; 27, 333; 28, 46; 29, 176; 30A, 130, 202
$DyFe_2$, 26, 236
$DyFe_3$, 35A, 119
Dy_6Fe_{23}, 30A, 130; 35A, 119
$DyFeO_3$, 35A, 218
DyFeSi, 35A, 119
DyGa, 26, 235
$DyGa_2$, 26, 146
Dy_5Ga_3, 34A, 140
DyGe, 26, 235; 30A, 130
Dy_2Ge_3, 29, 113
Dy_5Ge_3, 29, 6, 113
DyH_2, 27, 237
DyH_3, 27, 237
DyHg, 29, 113
$DyHg_2$, 29, 113
$DyHg_3$, 29, 113
DyIn, 26, 237
$DyIn_3$, 26, 238; 30A, 130
$DyLiO_2$, 27, 516
β-Dy-Mg, 29, 176
Dy_5Mg_{24}, 27, 195; 29, 113
Dy_2MgS_4, 27, 327
$DyMn_2$, 26, 236; 27, 264
$(Dy,Tm)Mn_2$, 27, 264
$DyMn_{12}$, 31A, 48
Dy_6Mn_{23}, 30A, 130
$DyMn_2O_5$, 32A, 287
$Dy_2Mn_4O_9$, 31A, 249
Dy_2MnS_4, 27, 327
Dy_4MnS_7, 29, 113
DyN, 27, 335
$DyNaS_2$, 29, 342
$DyNaSe_2$, 30A, 131
$DyNbO_4$, 27, 521; 28, 348
Dy_3NbO_7, 29, 329
$DyNbO_6Ti$, 28, 326
Dy-Nd, 30A, 202
DyNi, 29, 113; 30A, 131
$DyNi_2$, 26, 236
$DyNi_3$, 35A, 119
$DyNi_5$, 26, 239
Dy_2Ni_7, 35A, 119
$DyNiSi_2$, 35A, 119
$DyNi_2Tb$, 30A, 131
Dy_2O_3, 26, 364; 30A, 450
DyO_4P, 28, 304
$Dy_2O_7Si_2$, 35A, 301
$Dy_2O_7Sn_2$, 26, 399
Dy_2OSr, 28, 346
DyO_4Ta, 27, 521; 28, 348
Dy_3O_7Ta, 29, 329
$Dy_2O_7Tc_2$, 29, 322
Dy_2O_2Te, 27, 454
Dy_2O_5Ti, 29, 421
DyO_4V, 28, 308
Dy-O-Zr, 26, 364
$Dy_{14}O_{27}Zr_3$, 27, 535
$Dy_{22}O_{51}Zr_9$, 27, 535
$DyPb_3$, 29, 113
Dy_5Pb_4, 33A, 152
$DyPd_3$, 30A, 131
Dy_5Pd_2, 29, 113
DyPt, 26, 235; 30A, 131
$DyPt_2$, 26, 236
$DyPt_3$, 26, 238
DyRh, 30A, 131
$DyRh_2$, 26, 236
Dy_2S_3, 27, 330
Dy_5S_7, 29, 113
Dy-S-Sr, 27, 330
Dy_2S_4Sr, 27, 327
$DySb_2$, 32A, 20
Dy_4Sb_3, 29, 113
$DySe_{1.40}$, 29, 113
$DySe_2$, 32A, 20
Dy_2Se_3, 28, 46, 50; 29, 113; 30A, 131
DySi, 29, 113; 30A, 131
$DySi_2$, 29, 113
Dy_5Si_3, 30A, 131
$DyTc_2$, 29, 113
$DyTe_2$, 32A, 20
$DyTe_3$, 30A, 131
Dy_2Te_3, 30A, 131
DyTh, 26, 237
$DyTh_3$, 26, 238
DyZn, 29, 113
$DyZn_5$, 29, 113
$DyZn_{12}$, 30A, 131
Dy_2Zn_{17}, 30A, 131
Dy_3Zn_{11}, 35A, 119
$Dy_{13}Zn_{58}$, 35A, 119

Er, 26, 141, 234; 27, 333
ErF_6NaRb_2, 33A, 186
ErFO, 30A, 450
$ErFe_2$, 26, 236; 34A, 140
$ErFe_3$, 34A, 140
Er_2Fe_{17}, 34A, 140
Er_6Fe_{23}, 30A, 131; 34A, 140
$Er_{1.5}Fe_5Gd_{1.5}O_{12}$, 26, 414
$ErFeO_3$, 31A, 260; 35A, 218
$Er_{1.5}Fe_5O_{12}Y$, 26, 414
Er_4FeS_7, 29, 113
$ErGa_2$, 26, 146
$ErGa_3$, 29, 113
Er_5Ga_3, 29, 113
$Er_3Ga_5O_{12}$, 27, 723
Er_2Ge_3, 29, 114
Er_5Ge_3, 30A, 131
$Er_2Ge_2O_7$, 35A, 317
ErH_2, 27, 237
ErH_3, 27, 237
$ErHO_2$, 30A, 310
ErHg, 29, 114
$ErHg_2$, 29, 114
$ErHg_3$, 29, 114
ErIn, 30A, 131
$ErIn_3$, 30A, 131, 132
ErIr, 30A, 132
$ErIr_2$, 26, 236
$ErLiO_2$, 27, 516
β-Er-Mg, 29, 176
$ErMg_2$, 29, 114
Er_5Mg_{24}, 27, 195
Er_2MgS_4, 27, 327
$ErMn_2$, 26, 192; 27, 196; 29, 114
$ErMn_{12}$, 31A, 48
Er_6Mn_{23}, 30A, 132
$ErMnO_3$, 28, 138
Er_2MnS_4, 27, 327
Er_4MnS_7, 29, 114
ErN, 27, 335
$ErNaS_2$, 29, 342
$ErNaSe_2$, 30A, 132
$ErNbO_4$, 27, 521
$ErNbO_6Ti$, 28, 326
ErNi, 29, 114; 30A, 132
$ErNi_3$, 35A, 119
Er_2Ni_7, 35A, 119
$ErNiSi_2$, 35A, 119
$ErO_{1.5}$, 29, 422
Er_2O_3, 27, 497; 30A, 457; 31A, 246
ErO_4P, 28, 304; 33A, 309
Er_2O_5Si, 30A, 455
$Er_2O_7Si_2$, 33A, 465; 35A, 300
ErO_4Ta, 27, 521
$Er_2O_7Tc_2$, 29, 322
$Er_2O_7Ti_2$, 30A, 317
ErO_4V, 28, 308; 33A, 309
ErP, 27, 335
$ErPb_3$, 29, 114
$ErPd_3$, 30A, 132
ErPt, 30A, 132
$ErPt_3$, 30A, 132
$ErRe_2$, 30A, 132
ErRh, 30A, 132
$ErRh_2$, 26, 236
$ErRu_2$, 26, 237
Er_5S_7, 29, 114
Er_2S_4Sr, 27, 327
Er_2S_4Yb, 29, 114
ErSb, 27, 334
$ErSb_2$, 32A, 20
ErSe, 29, 114; 30A, 132
$ErSe_2$, 30A, 133; 32A, 18, 20
Er_2Se_3, 28, 50; 29, 114, 30A, 132
Er_3Se_4, 30A, 132
ErSi, 29, 114; 30A, 133
$ErSi_2$, 27, 336; 28, 46
Er_5Si_3, 30A, 133
$ErSn_2$, 31A, 42

$ErTc_2$, 29, 114
Er-Te, 30A, 203
$ErTe$, 29, 114
$ErTe_2$, 32A, 20
$ErTe_3$, 30A, 133
Er_2Te_3, 30A, 133
$ErTl$, 30A, 133
$ErTl_3$, 30A, 133
$ErZn$, 29, 114
$ErZn_{12}$, 30A, 133
Er_6Zn_{23}, 30A, 133
Eu, 26, 234; 27, 197
$EuFS$, 30A, 450; 35A, 155
$Eu_3Fe_2Ga_3O_{12}$, 33A, 282
$EuFeO_3$, 35A, 218
$EuGa_2$, 29, 114
$EuGa_4$, 30A, 133
$EuGd_2S_4$, 30A, 133
$Eu_{0.5}Gd_{1.6}Se_{2.9}$, 29, 114
$EuGe$, 30A, 133
$EuGe_2$, 29, 50, 114
EuH_2, 29, 114
EuH_3, 26, 169
EuH_3Li, 29, 114
$EuHO_2$, 29, 417; 30A, 451
$EuH_2O_2.H_2O$, 31A, 256
EuH_3O_3, 29, 417
$EuHg$, 29, 114
$EuHg_2$, 29, 114
$EuHg_3$, 29, 114
EuI_2, 26, 337; 34A, 182
$EuIO$, 26, 337
$EuIn_2$, 29, 3, 114
$EuK_3O_7Si_2$, 30A, 455
$EuLiO_2$, 28, 134, 308, 344; 30A, 452
Eu_3LiO_4, 35A, 234
$EuMg_2$, 29, 114
$EuMg_5$, 35A, 59
$EuNaS_2$, 29, 342
$EuNaSe_2$, 30A, 133
$EuNbO_4$, 27, 521
$EuNbO_6Ti$, 27, 537
Eu_2Ni_{17}, 30A, 133
Eu-Np-O, 30A, 452
EuO, 27, 475
Eu_2O_3, 26, 493; 27, 546
Eu_3O_4, 27, 476; 31A, 125
EuO_4S, 29, 353; 30A, 453
Eu_2O_4Si, 29, 428
$Eu_2O_7Si_2$, 35A, 302
Eu_2O_4Sr, 27, 476
EuO_4Ta, 27, 521; 28, 332
Eu_2O_2Te, 27, 424
$Eu_2O_7Th_2$, 26, 493
Eu-O-U, 28, 286
$Eu_2O_{12}W_3$, 28, 226
EuP, 29, 114
$EuPb_3$, 29, 115; 32A, 139
Eu_2Pb, 32A, 139
Eu_5Pb_3, 35A, 58
$EuPd_3$, 30A, 134
$EuSe$, 27, 197
$EuSi$, 29, 115
$EuSn_3$, 30A, 134
$EuTe$, 27, 197; 30A, 134, 453
$EuTe_3$, 27, 336
$EuTl_2$, 29, 3, 115
$EuZn$, 29, 115
$EuZn_2$, 29, 115
$EuZn_5$, 29, 115; 30A, 134
$EuZn_{11}$, 30A, 134
$EuZn_{13}$, 29, 115

F_2, 33A, 157; 35A, 129
β-F_2, 29, 229
F_2Fe, 31A, 247
$F_3Fe.3H_2O$, 26, 300
β-$F_3Fe.3H_2O$, 29, 257
β-$F_3FeH_6O_3$, 29, 257
$F_6FeH_{12}O_6Si$, 27, 461
F_3FeK, 26, 306
F_3FeK_x, 30A, 449
F_4FeK_2, 30A, 449
$F_7Fe_2K_3$, 30A, 449
$(F,HO,O)_{3.5}(Fe,Mn)_4KMgNaO_{12}Si_4Ti$, 28, 322
F_6FeK_2Na, 26, 309
F_6Fe_2Li, 33A, 182
$F_3FeNa_{0.11}$, 33A, 184
$FFeO$, 30A, 450
$FFeO_3Sr_2$, 28, 350
$F_6FeSi.6H_2O$, 27, 461
$F_3Ga.3D_2O$, 26, 300
$F_3Ga.3H_2O$, 26, 300
$F_6GaH_{12}N_3$, 29, 266
F_6GaK_2Na, 26, 309
$F_xGa_{5-x}Ni_xO_{12-x}Y_3$, 28, 352
F_3Gd, 27, 724
F_2Ge, 31A, 82
$FGeH_3$, 26, 301
$(FH)_n$, 34A, 171
F_2HK, 29, 254
F_3H_2K, 28, 71
F_5H_4K, 35A, 152
FH_8KO_4, 29, 254
F_2HLi, 27, 451
F_6H_4LiNSi, 30A, 449
$F_{0.08}H_2Mg_{0.98}O_2$, 26, 348
$F_{1.3}H_{0.7}Mg_5O_{8.7}Si_2$, 35A, 469
F_3H_4MnN, 26, 309
$F_5H_8MnN_2$, 34A, 201
$(F,HO)_2Mn_3O_4Si$, 28, 354
$F_5H_{12}MoN_3O_3$, 35A, 421
FH_4N, 30A, 449; 34A, 172; 35A, 146
F_2HN, 28, 360
F_3H_4NNi, 27, 454
$F_4H_8N_2Ni$, 27, 454
$F_2H_4NO_2P$, 34A, 324
$F_2H_4NO_2Pu$, 30A, 450
FH_4NO_3S, 27, 421; 35A, 191
$F_5H_{12}N_3O_2U$, 31A, 258; 34A, 218
$F_4H_8N_2OV.H_2O$, 30A, 450
F_6H_4NP, 27, 421
$F_6H_8N_2Si$, 31A, 88
$F_8H_{16}N_4Th$, 34A, 194
F_6H_4NU, 30A, 449
$F_8H_{16}N_4U$, 35A, 170
$F_7H_{16}N_4Zr$, 35A, 169
F_2HNa, 29, 414
F_8HNa_3Ti, 31A, 89
F_6H_2V, 30A, 450
$FHoO$, 30A, 450
F_7I, 27, 446; 30A, 288; 31A, 260
F_7IXe, 35A, 177
F_3In, 26, 300; 31A, 260
F_6IrRb, 28, 298
FK, 26, 297
$FK.4H_2O$, 29, 254
F_3KMn, 26, 306, 309; 29, 414
F_4KMn, 26, 309
F_4K_2Mn, 29, 414
F_5KMn, 26, 309
$F_7K_3Mn_2$, 29, 415
F_6KMo, 28, 298
$F_3K_2MoNaO_3$, 34A, 216
$F_4K_2MoO_2.H_2O$, 28, 296; 30A, 450; 31A, 91
$F_4K_2MoO_3.H_2O$, 28, 296; 32A, 174
$F_5K_2MoO.H_2O$, 31A, 92; 32A, 177
$F_4K_2NaNbO_2$, 34A, 216
F_5K_2NaOTi, 34A, 216
F_5K_2NaOV, 34A, 216
F_6KNaTh, 35A, 168
F_7K_2Nb, 31A, 90
$FKNbO_2$, 29, 271
FK_2NbO_3, 27, 467
F_5K_2NbO, 31A, 91
F_3KNi, 26, 306
F_4K_2Ni, 27, 455
F_2KO_2P, 31A, 192
FKO_3S, 32A, 334
FKO_3Xe, 34A, 219
F_7K_2Pa, 31A, 260; 32A, 159
F_6K_2Pd, 26, 313
F_6K_2Pr, 26, 312
F_7K_3Pr, 26, 312
F_6KPt, 27, 666
F_6KRe, 28, 298
F_8K_2Re, 29, 269
F_6KSb, 28, 298
F_6K_2Si, 28, 296
F_7K_3Si, 27, 464
$F_3KSn.\frac{1}{2}H_2O$, 33A, 192
F_6K_2Tc, 28, 298
F_5KTe, 34A, 200; 35A, 167
F_6KU, 30A, 449

F_6K_2U, 34A, 199
F_9KU_2, 34A, 199
F_6KW, 28, 298
F_3KZn, 26, 306
$F_7K_3Zn_2$, 31A, 254
F_3La, 26, 301; 29, 258; 30A, 267; 31A, 84
$FLaS$, 30A, 450; 35A, 155
FLi, 26, 297, 299; 30A, 449; 34A, 171
F_6Li_2Mn, 26, 309
F_6LiMo, 28, 298
$F_{31}(Li,Na)_7Th_6$, 34A, 195
$F_{62}Li_8Na_6Th_{12}$, 34A, 195
F_6LiNb, 28, 298
$FLiNbO_2$, 29, 271
$FLiNb_6O_{15}$, 30A, 334
F_4Li_2Ni, 33A, 184
F_6LiP, 28, 296
F_6Li_2Pd, 33A, 188
F_6Li_2Pt, 33A, 188
F_2LiRb, 27, 451; 30A, 269
F_6LiRe, 28, 298
F_6LiSb, 27, 460; 28, 298
F_6LiTa, 28, 298
F_5LiU, 31A, 93
F_8Li_4U, 32A, 179
F_6LiV, 28, 296
F_6LiW, 28, 298
F_4LiY, 26, 311
F_4LiYb, 35A, 166
F_4Li_2Zn, 30A, 449
$F_6M^{4+}Na_2$, 30A, 449
F_2Mg, 26, 299; 27, 430; 29, 414
α-F_2Mg, 26, 299
α'-F_2Mg, 26, 300
β-F_2Mg, 26, 300
FMg_2N, 35A, 153
F_3Mg_3N, 34A, 189; 35A, 153
F_3MgNa, 26, 304
$F_2Mg_6Na_2O_{22}Si_8$, 27, 698
$F_2MgNb_{14}O_{35}$, 35A, 179
F-Mg-O-P, 30A, 451
FMg_2O_4P, 32A, 362
$F_2Mg_3O_4Si$, 34A, 366
F_6MgPd, 33A, 188
F_2Mn, 28, 72, 290; 30A, 449
$F_3Mn.3H_2O$, 26, 300
F_3MnNa, 26, 309; 34A, 201
F_3MnRb, 26, 309; 29, 414
F_4MnRb, 26, 309
F_4MnRb_2, 29, 414
F_5MnRb, 26, 309
F_3MnTl, 30A, 449
F_5Mo, 27, 445
F_6MoNa, 26, 311
$FMoNb_{15}O_{40}$, 33A, 189
$F_{0.6}MoO_{2.4}$, 35A, 179
$F_{0.8}Mo_4O_{11.2}$, 35A, 179
F_4MoO, 33A, 172
F_6MoTl, 28, 298
F_2N_2, 28, 360
F_4N_2, 28, 360
FNO, 34A, 215
$F_6N_3P_3$, 28, 59
$F_8N_4P_4$, 26, 314
FNS, 26, 314
F_3NS, 27, 411
$F_4N_4S_4$, 26, 315; 27, 409; 28, 58
$FNTh$, 31A, 253
$FNZr$, 31A, 253
FNa, 26, 297
F_6NaNb, 28, 298
$FNaNbO_2$, 29, 271
$FNaNb_6O_{15}$, 30A, 335
F_2NaNbO_2, 34A, 217
F_4NaNd, 30A, 272
F_3NaNi, 27, 454
FNa_3O_4S, 28, 326
$FNaO_3U$, 26, 406
F_8Na_3Pa, 34A, 197
F_6Na_2Pr, 26, 312
F_7Na_3Pr, 26, 312
F_6Na_2Si, 26, 305; 29, 264
F_5NaSn_2, 29, 266
F_6Na_2Sn, 31A, 260
$F_{10}Na_4Sn_3$, 35A, 173
F_6NaTa, 28, 298
F_8Na_2U, 31A, 246
F_6NaV, 28, 296
$F_6Na_{1.5}Y_{1.5}$, 28, 298
F_6Na_2Zr, 34A, 193
$F_{13}Na_5Zr_2$, 28, 298; 30A, 272
$F_{31}Na_7Zr_6$, 33A, 185
F_5Nb, 29, 258
$F_{15}Nb_6$, 30A, 261
FNb_3O_7, 29, 270
$FNb_{17}O_{42}$, 31A, 123
$FNb_{31}O_{77}$, 31A, 123
$FNbO_2Rb$, 29, 271
F_9NbSe, 35A, 174
$F_{14}Nb_2Se$, 35A, 175
F_6NbTl, 28, 298
F_3Nd, 31A, 257
F_2Ni, 31A, 249
F_3NiRb, 27, 454; 34A, 203; 35A, 158
F_4NiRb_2, 27, 454
F_4NiTl_2, 27, 454
F_2NpO_2Rb, 30A, 450
F_5OOs, 27, 665; 33A, 174
$F_2O_{24}P_6Sr_{10}$, 26, 469
F_6O_2Pt, 27, 666; 31A, 92
F_2O_2PuRb, 30A, 450
F_4ORe, 33A, 173
FO_3S, 34A, 218
F_4OS, 34A, 218
F_2O_3SXe, 34A, 219
$FOSc$, 31A, 86
$F_2O_5SrTa_2$, 27, 534
F_4OTc, 35A, 185
F_4OW, 33A, 172
FOY, 35A, 182
$FOYb$, 30A, 450
$F_{10}O_9Zr_7$, 35A, 181
F_6OsRb, 28, 298
F_3P, 28, 360; 34A, 172
F_5P, 28, 360; 30A, 266
F_6PaRb, 33A, 195
F_2Pb, 26, 301; 31A, 251; 32A, 156
F_4Pb, 27, 442
F_6PdZn, 33A, 188
F_6PrRb_2, 26, 312
F_7PrRb_3, 26, 312
F_6Pt, 26, 301
$F_{11}PtXe$, 31A, 258; 32A, 160
FRb, 27, 425; 29, 255
F_6RbU, 30A, 449
F_6Rb_2U, 27, 463
F_5Ru, 29, 260
F_6RuTl, 28, 298
F_2S, 34A, 173
F_2S_2, 28, 360
F_4S, 27, 717; 28, 361
F_6S, 28, 361
F_9SSb, 26, 301
F_3Sb, 35A, 147
F_3Sc, 26, 298
F_4Se, 28, 361
F_6Se, 28, 361
F_2Si, 31A, 239
F_3Sm, 27, 724
F_2Sn, 27, 430; 32A, 508
F_4Sn, 27, 442
F_2Sr, 29, 439; 31A, 243
F-Sr-Y, 30A, 449
F_5Ta, 29, 258
F_6TaTl, 28, 298
F_4Te, 33A, 170
F_6Te, 31A, 239
F_4Th, 34A, 173
F_6TiV, 28, 296
FTl, 33A, 171
F_3Tl, 31A, 260
F_6TlU, 30A, 449
F_3U, 32A, 157
F_4U, 29, 261; 30A, 449; 31A, 248; 34A, 173
$F_4U.4/3H_2O$, 30A, 449
F_6U, 30A, 269
F_9U_2, 32A, 157
$F_3V.3H_2O$, 26, 300
F_5V, 31A, 239; 34A, 172
F_2Xe, 28, 73
F_4Xe, 28, 75
F_6Xe, 31A, 259
F_6Xe_2, 30A, 266
F_2Zn, 28, 72, 290
$F_2Zn.4H_2O$, 30A, 449
F_4Zr, 30A, 449
β-F_4Zr, 29, 439

Fe, 29, 176
α-Fe, 26, 166
γ-Fe, 27, 250
ε-Fe, 29, 115
Fe-Ga, 30A, 203
$FeGa_3$, 30A, 48, 134
Fe_3Ga, 29, 115
Fe_3Ga_2, 30A, 134
Fe_3Ga_4, 30A, 134
Fe_7Ga_6, 30A, 134
Fe_8Ga_{11}, 30A, 134
$Fe_{2.5-x}Ga_xLi_{0.5}O_4$, 28, 285
Fe-Ga-N, 27, 306
$FeGaO_3$, 31A, 254
$Fe_xGa_{2-x}O_3$, 30A, 322
$Fe_{1.15}Ga_{0.85}O_3$, 31A, 133
$Fe_{3-x}Ga_xO_4$, 31A, 163
$(Fe,Ga)_5O_{12}Y_3$, 31A, 163
$Fe_{5-x}Ga_xO_{12}Y_3$, 30A, 452
Fe_2GaTi, 27, 161, 206
Fe_3GaZr_2, 29, 115
Fe-Gd, 26, 142; 27, 198
Fe_2Gd, 26, 142, 236
Fe_3Gd, 26, 143; 27, 198; 30A, 78, 134
Fe_3Gd_2, 26, 142
Fe_4Gd, 26, 143
Fe_5Gd, 26, 143; 27, 199
Fe_7Gd_2, 26, 143
Fe_9Gd, 27, 198
$Fe_{17}Gd_2$, 26, 142; 28, 46; 35A, 60
$Fe_{23}Gd_6$, 30A, 134
$Fe_5Gd_{3-x}Nd_xO_{12}$, 26, 414
$FeGdO_3$, 35A, 218
$Fe_5Gd_3O_{12}$, 26, 413
FeGdSi, 35A, 119
Fe-Ge, 29, 176; 30A, 204
α-Fe-Ge, 27, 208
FeGe, 28, 46; 29, 115; 30A, 135; 32A, 82
$FeGe_2$, 27, 210; 30A, 135
$Fe_{1+x}Ge$, 30A, 204
$Fe_{1.67}Ge$, 30A, 135
Fe_2Ge, 26, 149; 27, 210; 28, 46; 30A, 135
Fe_3Ge, 28, 46; 30A, 134, 135; 34A, 90
$Fe_{3+x}Ge$, 27, 210
$Fe_{3.25}Ge$, 26, 149; 30A, 134
Fe_5Ge_3, 30A, 135
$Fe_{13}Ge_3$, 27, 209
$Fe_5Ge_9Li_5O_{36}Zn_8$, 28, 348
$Fe_{0.9}GeMn_{0.9}$, 33A, 85
$Fe_{1-x}GeMn_xO_3$, 29, 419
$Fe_2Ge_3Mn_3O_{12}$, 26, 415; 34A, 262
$Fe_{4.75-4.00}Ge_{0.25-1.00}Mn_{0.25-1.00}O_{12}Y_{2.75-2.00}$, 27, 511
Fe_2Ge_2Mo, 26, 174
$Fe_{33}Ge_{17}Mo_{50}$, 30A, 135
Fe-Ge-N, 27, 306
$Fe_3GeN_{0.51}$, 33A, 29
$FeGe_2NaO_6$, 32A, 493
$FeGeO_3$, 28, 312
Fe_2GeO_5, 31A, 246
Fe_4GeO_8, 31A, 246
Fe_2Ge_2Th, 29, 91, 115
FeGeTi, 35A, 74, 119
Fe_3GeZr_2, 30A, 135
Fe-H, 26, 166
$Fe_2H_3IO_3$, 29, 250
$(Fe,Si)_6(HO,O)_{18}(K,Na)_2(Nb,Ti)_6$, 28, 322
$Fe_3H_6KO_{14}S_2$, 27, 634; 30A, 453
$Fe_4H_2KO_{12}Si_3$, 28, 249; 29, 406
$FeH\mathit{M}O_9S_2.7H_2O$, 32A, 338
Fe-H-Mg-Ni-O, 30A, 451
$Fe_{0.5}HMn_{1.5}O_5P$, 35A, 432
$Fe_2H_2MnO_{10}P_2.8H_2O$, 28, 324; 30A, 395; 34A, 331
$FeH_8N_2O_8S_2.6H_2O$, 32A, 343
$FeH_{36}N_9O_{18}S_6$, 29, 424
$Fe_2H_8N_2O_{12}S_3$, 29, 439
$(Fe,Zn)_2H_{2-x}Na_xO_9Te_3.yH_2O$, 32A, 307; 33A, 391
$FeHO_2$, 30A, 450; 31A, 241; 33A, 266
$Fe_3H_{2.5}O_{10.5}P_2.3H_2O$, 32A, 365
$Fe_6H_5O_{17}P_3$, 35A, 339
$FeHO_5S$, 27, 623
$FeHO_5S.2H_2O$, 35A, 375
$(Fe,Mg,Mn,Zn)HO_9S_2.7H_2O$, 33A, 379
FeH_8O_8S, 27, 613, 633
$FeH_{14}O_{11}S$, 29, 351
$Fe_3H_{28}O_{30}S_4$, 35A, 439
$Fe_4H_5O_9Si$, 27, 697; 29, 403
$(Fe,Mg)_7H_2O_{24}Si_4$, 31A, 224
FeHf, 27, 232
Fe_2Hf, 26, 162
$Fe_{16}Hf_6Si_7$, 30A, 135
Fe_3Ho, 33A, 152
$Fe_{17}Ho_2$, 30A, 135; 33A, 152
$Fe_{23}Ho_6$, 30A, 135; 33A, 152
$FeHoO_3$, 35A, 218
$FeHo_4S_7$, 29, 115; 33A, 98, 362
FeHoSi, 35A, 119
$Fe_2I_2N_4O_4$, 34A, 183
Fe-In-N, 27, 306
$FeInO_3$, 31A, 254
$FeIn_2S_4$, 28, 46, 320
Fe-Ir, 29, 177
$(Fe,Mg,Mn)_2KLiNa_2O_{24}Si_8Ti_2$, 31A, 232
$FeKMo_2O_8$, 35A, 254
$FeKNaO_{10}Si_4$, 35A, 480
$(Fe,Mg)_5(K,Na)_2O_{30}Si_{12}$, 30A, 455
$Fe_{12}K_2O_{19}$, 28, 328
$Fe_{14}K_2O_{22}$, 26, 395
$Fe_{22}K_2O_{34}$, 26, 395
$FeKO_8S_2$, 31A, 246
$FeKO_8S_2.H_2O$, 30A, 372
$FeKO_8S_2.xH_2O,yC_2H_6O$, 30A, 453
$FeKO_8Si_3$, 26, 525; 27, 683
$FeKS_2$, 33A, 3
$FeKSe_2$, 31A, 261
$Fe_5La_3O_{12}$, 27, 513
Fe-La-O-Pb-Ti, 27, 513
$Fe_5La_{0.5}O_{12}Y_{2.5}$, 26, 415
FeLaSi, 35A, 119
$Fe_{2.5-x}Li_{0.5}Mn_xO$, 30A, 451
$FeLi_3Mo_3O_{12}$, 35A, 250
$Fe_2Li_2Mo_3O_{12}$, 35A, 251
$FeLiO_2$, 26, 394; 27, 515; 29, 322
Fe_5LiO_8, 26, 394; 29, 326
α-Fe_5LiO_8, 29, 423
β-Fe_5LiO_8, 29, 423
γ-Fe_5LiO_8, 29, 423
δ-Fe_5LiO_8, 29, 423
$FeLiO_4Ti$, 28, 346
$(Fe,Li)_2O_4W$, 33A, 351
$FeLi_2O_8W_2$, 34A, 280
$FeLiO_8W_2$, 35A, 248
$Fe\mathit{Ln}O_3$, 30A, 452
$Fe_{10}\mathit{Ln}_6O_{24}$, 30A, 452
Fe_2Lu, 30A, 135
$Fe_{23}Lu_6$, 30A, 135
$FeLuO_3$, 35A, 218
$FeLu_2S_4$, 29, 115
$(Fe,Mg,Mn)(Nb,Ta,Ti)_2O_6$, 28, 328
$(Fe,Mg,Mn)_3O_8P_2$, 30A, 454
$(Fe,Mg,Mn)O_4W$, 33A, 346
Fe_2MgO_4, 26, 396, 411; 31A, 248
$Fe_{0.2}Mg_{1.8}O_4Si$, 30A, 422
$FeMgO_6Si_2$, 29, 431
$Fe_{1.07}Mg_{0.93}O_6Si_2$, 30A, 425
$Fe_{1.1}Mg_{4.6}O_{10}Si_4$, 29, 429
$(Fe,Mg)_2O_4Si$, 28, 245
$Fe_2Mg_3O_8Sn$, 33A, 324
$Fe_2MgO_{10}Ti_3$, 26, 379
$Fe_2Mg_xO_4Zn_{1-x}$, 33A, 324
$Fe_{52}Mn_{16}Mo_{32}$, 30A, 40

$Fe_{82.5}Mn_{25.5}Mo_{51}$, 30A, 83
(Fe,Mn,Nb,Sn,Ta)$_4O_8$, 28, 145
(Fe,Mn)(Nb,Sn,Ta,Ti)$_2O_6$, 28, 354
Fe_2MnO_4, 26, 411; 27, 495
$Fe_{3-x}Mn_xO_4$, 31A, 163
$Fe_{3-x}Mn_xO_{4+y}$, 28, 285
(Fe,Mn)$_3O_8P_2.3H_2O$, 28, 191
$FeMnO_9Pb_2Si_2$, 27, 714
$Fe_{1/2}MnO_3Sb_{1/2}$, 33A, 325
$Fe_{0.95}Mn_{0.05}O_3Si$, 30A, 455
$Fe_{0-1.4}Mn_{0-2.00}O_4Zn$, 26, 409
$Fe_{2.067}Mn_{0.645}O_4Zn_{0.288}$, 28, 143
Fe-Mn-S-Zn, 26, 251
Fe-Mn-Si, 27, 375
Fe-Mo, 26, 177; 30A, 204
Fe_3Mo_2, 26, 178
$FeMoO_4$, 28, 348; 33A, 332
$Fe_2Mo_3O_9$, 31A, 249
$Fe_2Mo_3O_{12}$, 30A, 458; 31A, 259; 33A, 331
FeMoSi, 27, 376
$Fe_3Mo_5Si_2$, 30A, 135
FeN, 27, 253
(Fe,Ni)$_2$N, 27, 252
(Fe,Ni)$_3$N, 27, 252
(Fe,Ni)$_4$N, 27, 252
$Fe_{24}N_{9.36}$, 29, 115
$Fe_{10}N_3O_{24}Y_3$, 29, 418
Fe-N-Sn, 27, 306
~Fe_3NSn, 27, 306
$Fe_2Na_2Nb_{0.19}O_{18}Ti_{6.81}$, 26, 378
$FeNaO_2$, 28, 133
β-$FeNaO_2$, 26, 368
γ-$FeNaO_2$, 26, 368
$Fe_5Na_3O_9$, 27, 546; 32A, 288
$FeNaO_8S_2$, 31A, 246
$FeNa_2O_{12}S_4Ti$, 28, 326
$FeNaO_4Sn$, 33A, 294, 295
$FeNaO_4Ti$, 33A, 294, 295
(Fe,Sc)NaO_4Ti, 33A, 295
$Fe_xNa_xO_4Ti_{2-x}$, 33A, 296
FeNb, 32A, 139
Fe_2Nb, 29, 115
$FeNbO_4$, 27, 522; 29, 419
$FeNb_{11}O_{29}$, 31A, 255
$FeNb_{49}O_{124}$, 31A, 255
$Fe_4Nb_2O_9$, 26, 389
(Fe,Nb)O_3Sr, 28, 330
$FeNbO_6Sr_2$, 26, 420; 28, 328
$FeNb_9O_{30}Sr_6$, 28, 330
$FeNb_{0.2}O_6Ta_{1.8}$, 33A, 343
$FeNb_3S_6$, 35A, 111
$Fe_{1/4}NbSe_2$, 35A, 119
$Fe_{1/3}NbSe_2$, 35A, 119
$Fe_{1/2}NbSe_2$, 35A, 119
$FeNb_4Si$, 30A, 44, 135
$Fe_2Nb_{0.4}Zr_{0.6}$, 34A, 140
Fe_7Nd, 31A, 46
$Fe_{17}Nd_2$, 28, 46; 30A, 136
$FeNdO_3$, 35A, 218
$Fe_5Nd_3O_{12}$, 27, 513
$Fe_5Nd_xO_{12}Sm_{3-x}$, 26, 414
$FeNdO_4Sr$, 35A, 228
$FeNdO_5Ti$, 35A, 237
$Fe_5Nd_xO_{12}Y_{3-x}$, 26, 414,415
$Fe_{10}Nd_3O_{24}Y_3$, 28, 328; 29, 418
FeNdSi, 35A, 119
$Fe_5Nd_{0.5}Y_{2.5}$, 26, 415
Fe-Ni, 26, 411; 27, 251
$FeNi_3$, 27, 251
Fe_2NiO_4, 26, 411
$Fe_{3-x}Ni_xO_4$, 27, 495
Fe_2NiP, 33A, 102; 35A, 71
$Fe_{0.96}Ni_{0.04}S$, 29, 424
$Fe_{4.5}Ni_{4.5}S_8$, 26, 121
$Fe_{1/2}Ni_{1/2}Sb_3$, 35A, 119
$FeNi_2Se_4$, 30A, 136
Fe_2NiSe_4, 30A, 136
$(Fe_{0.2}Ni_{0.8})_{22}Se_{16}$, 34A, 94
Fe-Ni-Si, 26, 124
$Fe_{0.902}O$, 34A, 246
FeO, 26, 379
FeO_x, 29, 417; 31A, 246
Fe_2O_3, 26, 379, 414; 28, 320; 30A, 450; 31A, 124; 34A, 239
α-Fe_2O_3, 27, 499
γ-Fe_2O_3, 27, 497
(Fe,Rh)$_2O_3$, 30A, 458
Fe_3O_4, 26, 414; 31A, 250; 33A, 267
$FeO_4P.2H_2O$, 31A, 185
$Fe_3O_8P_2$, 33A, 401
$FeO_8P_2Zn_2.4H_2O$, 26, 463
FeO_3Pr, 35A, 218
$Fe_5O_{12}Pr_3$, 27, 513
$Fe_5O_{12}Pr_{0.5}Y_{2.5}$, 26, 415
$Fe_5O_{12}Pr_{1.0}Y_{2.0}$, 26, 415
$Fe_2O_4Rb_2$, 26, 395
$Fe_{14}O_{22}Rb_2$, 26, 395
FeO_8RbS_2, 31A, 246
$Fe_{0.5}O_3Re_{0.5}Sr$, 26, 418
FeO_4Rh_2, 26, 399
FeO_4S, 26, 437; 27, 526; 35A, 372
$FeO_4S.H_2O$, 31A, 199
$FeO_4S.4H_2O$, 27, 613, 633; 28, 326
$FeO_4S.7H_2O$, 29, 351
$Fe_2O_9S_2.7H_2O$, 33A, 377
$Fe_2O_{12}S_3$, 30A, 453
FeO_4Sb_2, 31A, 241
FeO_6SbSr_2, 29, 420
$Fe_{0.75}O_3Sc_{1.25}$, 26, 350
FeO_3Sc, 26, 350
FeO_3Sc_2, 26, 350
$Fe_{1.25}O_3Sc_{0.75}$, 26, 350
Fe_2O_4Si, 28, 265; 30A, 422
$Fe_7O_{10}Si$, 34A, 389
$FeO_7Si_2Sr_2$, 26, 523
FeO_3Sm, 35A, 218
(Fe,Ta)O_3Sr, 30A, 452
(Fe,Ti)O_3Sr, 28, 282
FeO_6Sr_2Ta, 27, 509; 28, 328
$Fe_2O_9Sr_3U$, 31A, 161
FeO_3Tb, 32A, 268; 35A, 218
$Fe_{10}O_{24}Tb_6$, 29, 418
Fe_2O_6Te, 27, 526; 33A, 344
Fe-O-Ti, 26, 379; 27, 500
$Fe_{3/2}O_3Ti_{1/2}$, 27, 499
Fe_2O_4Ti, 30A, 316
(Fe,Ti,Zn)$_{\sim 2}O_7Ti_2$, 35A, 445
$Fe_2O_9Ti_3$, 31A, 252
$Fe_5O_{24}Ti_8U$, 26, 378
FeO_3Tm, 33A, 323; 35A, 218
FeO_4U, 32A, 292
$FeO_{10}U_3$, 31A, 255
Fe-O-V, 27, 500
FeO_4V, 27, 522; 29, 319
FeO_4V_2, 27, 491; 28, 286; 34A, 264
Fe_2O_4V, 28, 286
FeO_4W, 32A, 315; 33A, 346
Fe_2O_6W, 27, 527; 31A, 249
Fe-O-Y, 26, 414
FeO_3Y, 26, 414; 30A, 323
$Fe_5O_{12}Y_3$, 26, 415; 27, 510; 29, 423
FeO_3Yb, 31A, 241; 35A, 218
Fe_2O_4Zn, 26, 411
FeP, 27, 96, 319; 30A, 98, 136
FeP_2, 27, 96; 33A, 63
Fe_2P, 27, 96; 30A, 90; 35A, 74
Fe_3P, 27, 96
$Fe_{3.092}P$, 27, 100
ε-Fe_3P, 27, 97
Fe_4P, 26, 178
FePS, 30A, 136
$FePS_3$, 30A, 136
$Fe_2P_2S_6$, 35A, 138

FePSe, 30A, 136
$FePSe_3$, 30A, 136
$FeP_{0.5}Si_{0.5}$, 27, 321
FePZr, 31A, 45
$Fe_{12}P_7Zr_2$, 33A, 103
Fe-Pd, 28, 46
$FePd_3$, 27, 254
Fe_7Pr, 31A, 46
$Fe_{17}Pr_2$, 30A, 136; 33A, 105
FePrSi, 35A, 119
γ′′-FePt, 27, 255
γ′-$FePt_3$, 27, 254
$FeRbS_2$, 33A, 3
$FeRbSe_2$, 31A, 261
Fe-Re, 27, 256
Fe_2Re_3, 29, 115; 30A, 136
Fe_6Re_6Si, 29, 115
Fe-Rh, 27, 256; 29, 177, 178
$Fe_{0.47}Rh_{0.53}$, 28, 46
FeRh, 26, 178; 29, 178
Fe-Ru, 26, 179
$Fe_{0.9}S$, 28, 46
FeS, 26, 125, 251; 27, 257; 29, 112, 115; 30A, 136; 33A, 107; 35A, 140
FeS_2, 29, 178; 34A, 97; 35A, 75
Fe_3S_4, 28, 46; 29, 63, 115; 30A, 458; 34A, 140
Fe_7S_8, 34A, 137
Fe_9S_8, 26, 121
Fe_9S_{10}, 30A, 136
FeS_4Sc_2, 29, 115
FeS_7Tm_4, 29, 115
FeS_4V_2, 30A, 136
FeS_7Y_4, 26, 277; 29, 115
FeS_4Yb_2, 29, 115
FeS_7Yb_4, 29, 115
Fe-S-Zn, 26, 126
$FeSb_2$, 33A, 63; 35A, 10
FeSbSe, 30A, 136
FeSbTe, 30A, 136
FeSbTi, 28, 46
FeSbV, 28, 46
Fe_2Sc, 26, 237
Fe_3Se_4, 29, 115; 33A, 106; 34A, 96; 35A, 72
Fe_7Se_8, 26, 179; 29, 115
$FeSe_4Ti_2$, 30A, 136
Fe_2Se_4Ti, 30A, 136
$FeSe_4V_2$, 30A, 137
Fe_2Se_4V, 30A, 137
Fe-Si, 26, 182
FeSi, 26, 124; 28, 28
$FeSi_2$, 26, 125; 27, 156; 34A, 137
Fe_3Si, 26, 182
Fe_5Si_3, 29, 115; 30A, 137
FeSiSm, 35A, 119
FeSiTb, 35A, 119
Fe_2Si_2Th, 30A, 137
FeSiTi, 35A, 73
Fe_3SiU_2, 29, 115
$Fe_5Si_3U_4$, 29, 115
Fe_2SiV_2, 27, 376; 30A, 137
$Fe_4Si_4V_5$, 30A, 137
$Fe_5Si_2V_3$, 26, 174; 27, 376; 30A, 137
$Fe_{41}Si_{22}V_{37}$, 26, 270
FeSiW, 27, 376
$Fe_{50}Si_{10}W_{40}$, 26, 269
FeSiYb, 35A, 119
Fe_3SiZr_2, 30A, 137
Fe_7Sm, 31A, 46
Fe-Sn, 30A, 205
γ-Fe-Sn, 27, 258
$FeSn_2$, 27, 257
~$Fe_{1.3}Sn$, 27, 258
Fe_2SnTi, 28, 46
FeTa, 32A, 139
Fe_2Ta, 29, 115
Fe_2Tb, 30A, 137; 32A, 139
$Fe_{17}Tb_2$, 30A, 137
$Fe_{23}Tb_6$, 30A, 137
Fe-Tc, 27, 361
$FeTe_2$, 35A, 75
Fe_2Te_3, 27, 256
Fe_3Th, 30A, 78, 137
$Fe_{17}Th_2$, 34A, 75
$Fe_{23}Tm_6$, 30A, 137
Fe_2U, 26, 30; 27, 259; 29, 149
Fe-V, 29, 179
α-Fe-V, 27, 260; 29, 179
σ-Fe-V, 29, 179
Fe-W, 26, 177; 30A, 205
Fe_2W, 26, 178
Fe_3W_2, 26, 177
Fe_6Xe_2, 28, 74
Fe_2Y, 26, 236, 276
$Fe_{17}Y_2$, 29, 116
$Fe_{23}Y_6$, 30A, 137
Fe_2Yb, 30A, 137
$Fe_{23}Yb_6$, 30A, 137
Fe-Zn, 29, 179, 180
$FeZn_{13}$, 27, 173
Fe_4Zn_9, 33A, 108
Fe_8Zn_{18}, 33A, 108
$FeZr_2$, 27, 260; 30A, 137
$FeZr_4$, 27, 260
Fe_3Zr, 28, 46
$Fe_{23}Zr_6$, 30A, 137

Ga, 26, 145, 279; 27, 199, 385; 30A, 48, 137; 34A, 82
γ-Ga, 29, 116
Ga II, 29, 116
GaGd, 26, 235
Ga_2Gd, 26, 146
$GaGdO_3$, 31A, 252
GaGeLi, 35A, 61
$GaGeLiO_4$, 28, 346; 30A, 459
$Ga_5Ge_9Li_5O_{36}Zn_8$, 28, 348
$Ga_2Ge_3Mn_3O_{12}$, 26, 415
$GaGe_3Ni_4$, 34A, 82
$Ga_{1.2}Ge_{2.8}Ni_4$, 34A, 82
~$Ga_{0.3}Ge_{1.7}Ti$, 29, 116
$GaGe_3Ti_2$, 30A, 138
$Ga_{50}Ge_{17}Ti_{33}$, 30A, 138
Ga-Ge-Zr, 30A, 138
$Ga_{0.1}Ge_{0.9}Zr$, 29, 116
$Ga_{11}GeZr_4$, 30A, 138
GaH_3O_3, 30A, 450
GaHf, 27, 207
$GaHf_2$, 27, 207, 378
Ga_2Hf, 27, 206, 207, 379
Ga_3Hf, 27, 207, 379
Ga_3Hf_2, 27, 206, 207, 378
Ga_3Hf_5, 27, 207, 378; 28, 27
Ga_4Hf_5, 27, 207
$GaHfNi_2$, 29, 116
$GaHfZn_2$, 29, 116; 30A, 138
~Ga_2HfZn, 29, 116
$Ga_{11}Hf_4Zn$, 29, 116; 30A, 138
Ga_2Ho, 26, 146
Ga_3Ho, 29, 116
Ga_3Ho_5, 34A, 140
GaI_3, 31A, 239
Ga_4I_6, 31A, 250
$GaInMgO_4$, 30A, 451
$GaInNiO_4$, 30A, 451
$GaInO_3$, 33A, 281
$GaInO_4Zn$, 30A, 451
Ga-In-Se, 26, 43
$GaInTe_2$, 34A, 89
GaIr, 28, 46
Ga_5Ir_3, 32A, 69
GaLa, 30A, 138
Ga_2La, 26, 146
$Ga_{2.5-x}Li_{0.5}Mn_xO$, 30A, 451
$GaLiO_2$, 30A, 314
$GaLi_5O_4$, 29, 316
Ga_5LiO_8, 28, 348
$GaLiO_8Rh_4$, 31A, 245
$GaLiO_6Si_2$, 26, 523
$GaLiO_4Ti$, 28, 346
Ga_3Lu, 29, 116
GaMg, 28, 23
$GaMg_2$, 35A, 62
Ga_2Mg, 34A, 84
Ga_2Mg_5, 28, 23
Ga_5Mg_2, 34A, 86
$Ga_{0.4}MgNi_{1.6}$, 27, 261
$Ga_{0.5}MgNi_{1.25}$, 27, 261

Ga_2MgO_4, 27, 488; 31A, 133
Ga_2MgS_4, 32A, 65
Ga-Mn, 28, 46; 30A, 138, 205
$GaMn_2$, 29, 116; 30A, 139
$GaMn_3$, 30A, 139
$GaMn_4$, 30A, 139
Ga_2Mn_3, 29, 116; 30A, 138
Ga_3Mn_7, 30A, 139
Ga_4Mn, 30A, 138
Ga_5Mn_2, 27, 379
Ga_5Mn_8, 30A, 139
Ga_6Mn, 30A, 138
Ga_6Mn_7, 30A, 138
$Ga_{44.5}Mn_{55.5}$, 30A, 138
$GaMn_3N$, 29, 116; 33A, 152
Ga_2MnO_4, 26, 412
Ga-Mn-S, 27, 203
$Ga_2Mn_6S_9$, 27, 203
$Ga_3N_xNb_5$, 29, 116
$Ga_3N_xTa_5$, 29, 116
$GaNTi_2$, 29, 116
Ga_2NV_3, 29, 116
Ga_4Na, 34A, 140
$Ga_5Na_3O_9$, 27, 546
Ga_2Nb_3, 29, 116
Ga_3Nb, 27, 379; 30A, 139
Ga_3Nb_5, 29, 116; 30A, 139
Ga_4Nb_5, 29, 116; 30A, 139
$Ga_{13}Nb_5$, 30A, 49, 139
$Ga_{0.5}NbNi_{1.5}$, 29, 170
$GaNbNi_2$, 29, 116
$GaNbO_4$, 30A, 332
Ga-Nb-Zr, 27, 3
Ga_2Nd, 26, 146
$GaNdO_3$, 32A, 270
$Ga_5Nd_3O_{12}$, 27, 723
$GaNdSe_3$, 29, 116
Ga_3Ni_5, 34A, 140
Ga_4Ni_3, 34A, 137
Ga_9Ni_{13}, 34A, 137
Ga_2NiO_4, 26, 412
$Ga_{0.5}Ni_{1.5}Ta$, 29, 170
$GaNi_2Ta$, 29, 116
$GaNi_2Ti$, 27, 161, 206
Ga_2O_3, 32A, 257; 34A, 239
$GaO_4P.2H_2O$, 31A, 183
Ga_2O_5Pb, 28, 322
$Ga_{12}O_{19}Pb$, 30A, 452
$Ga_2O_{12}S_3$, 30A, 453
$Ga_6O_{21}S_4.9H_2O$, 28, 219
GaO_6SbSr_2, 29, 420
$Ga_5O_{12}Sm_3$, 27, 723
Ga_2O_4Sr, 28, 322
GaO_4Ta, 27, 527
$Ga_3O_xTi_5$, 27, 204
GaO_3Y, 31A, 252
$Ga_5O_{12}Y_3$, 27, 723; 29, 423
GaO_3Yb, 31A, 252
$Ga_5O_{12}Yb_3$, 27, 723
Ga_2O_4Zn, 26, 412, 413
GaP, 27, 51
$Ga_{14}P_{14}Pd_{72}$, 35A, 119
$Ga_{14}P_{14}Pt_{72}$, 35A, 119
Ga-P-S, 29, 180
Ga-P-Se, 29, 180
Ga-P-Te, 29, 180
GdPd, 33A, 152
Ga_2Pr, 26, 146
$GaPt_2$, 34A, 140
GaPu, 30A, 139
$GaPu_3$, 30A, 139
Ga_2Pu, 30A, 139
Ga_3Pu, 30A, 50, 139
Ga_3Pu_5, 30A, 139
Ga_4Pu, 30A, 51, 139
Ga_6Pu, 30A, 51, 139
$Ga_{17}Rh_{10}$, 32A, 68
GaRu, 28, 46
Ga_2Ru, 28, 46
GaS, 26, 147
Ga_2S_3, 27, 51, 201
α-Ga_2S_3, 26, 149; 28, 24
Ga-S-Zn, 27, 201
Ga_2S_4Zn, 27, 202
GaSb, 28, 46; 30A, 140, 205
Ga-Sb-Te, 30A, 205
GaSc, 30A, 140
Ga_2Sc_3, 30A, 140
Ga_3Sc_5, 29, 53, 116
$Ga_3Sc_5(O?)$, 29, 53
GaSe, 26, 147
Ga_2Se_3, 26, 42; 27, 200; 29, 109, 116; 30A, 140
$GaSe_3Sm$, 29, 117
$Ga_2Se_{15}Sm_8$, 29, 116
$Ga_2Se_{18}Sm_2$, 29, 117
$Ga_{10}Se_{18}Sm_2$, 29, 116
$(Ga,Si)V_3$, 28, 46
Ga_2Sm, 26, 146
Ga_4Sr, 30A, 140
Ga_2Ta_3, 28, 46
Ga_3Ta, 30A, 140
Ga_3Ta_5, 28, 46; 29, 117; 30A, 140
GaTb, 29, 117
Ga_2Tb, 26, 146
GaTe, 26, 149; 28, 26
$GaTe_3$, 26, 149
Ga_2Te_3, 26, 149; 27, 200; 28, 46; 30A, 140
Ga_3Te_2, 26, 149
$GaTe_2Tl$, 34A, 88
Ga_2Th, 26, 257; 29, 117
GaTi, 27, 204, 378
$GaTi_2$, 27, 204
$GaTi_3$, 27, 204
Ga_2Ti, 27, 203, 378
Ga_3Ti, 27, 204, 378
$(Ga,Zn)_3Ti$, 30A, 140
Ga_3Ti_2, 27, 203, 378
Ga_3Ti_5, 27, 204, 377; 28, 27
Ga_4Ti_5, 27, 203, 378
Ga_3Tm, 29, 117
Ga-V, 29, 180; 30A, 140
GaV, 30A, 140
GaV_3, 29, 180; 30A, 206
Ga_4V, 30A, 140
Ga_5V_2, 27, 379; 28, 46; 29, 117; 30A, 52, 140
Ga_5V_6, 29, 117
Ga_7V_6, 29, 117
GaY, 30A, 141
Ga_2Y, 26, 146; 28, 46
Ga_3Y_5, 29, 53, 117
Ga_2Yb, 29, 3, 117
Ga_4Yb, 30A, 141
Ga-Zn, 29, 181
~$Ga_{1.5}Zn_{1.5}Zr$, 29, 117
$(Ga,Zn)_3Zr$, 30A, 141
$Ga_{10}Zn_{45}Zr_{45}$, 29, 117
$Ga_{15}Zn_{45}Zr_{40}$, 30A, 141
Ga-Zr, 30A, 206
GaZr, 27, 205, 378
$GaZr_2$, 27, 205, 378
Ga_2Zr, 27, 205, 378
Ga_2Zr_3, 27, 203, 378
Ga_3Zr, 27, 205, 378
Ga_3Zr_2, 27, 205
Ga_3Zr_5, 27, 205; 28, 46
Ga_4Zr_5, 27, 205, 378
Ga_5Zr_3, 27, 205
Gd, 26, 234; 30A, 141, 206
Gd II, 29, 117
GdGe, 26, 235; 30A, 141
$GdGe_{2-n}$, 30A, 141
$GdGe_2$, 29, 117
Gd_2Ge_3, 30A, 141
Gd_5Ge_3, 29, 6, 117; 30A, 141
GdIn, 26, 237
$GdIn_3$, 26, 238; 29, 117; 30A, 141
Gd-La, 27, 330
$GdLiO_2$, 28, 134, 308; 29, 439
β-Gd-Mg, 29, 181
GdMg, 28, 46
$GdMg_2$, 28, 46
$GdMg_3$, 28, 46
Gd_2MgS_4, 27, 327
$GdMn_2$, 26, 236
$GdMn_{12}$, 31A, 48
Gd_6Mn_{23}, 29, 50, 117; 30A, 141
Gd-Mn-S, 27, 328
Gd_2MnS_4, 27, 327
$GdNaO_2$, 29, 439
$GdNaO_4Si$, 35A, 299
$GdNaS_2$, 29, 341
$GdNaSe_2$, 30A, 141

$GdNbO_4$, 27, 521; 28, 348
Gd_3NbO_7, 29, 329
$GdNbO_6Ti$, 27, 537; 28, 326
Gd-Ni, 27, 198
GdNi, 26, 143, 235; 27, 198; 29, 51, 117; 30A, 141
$GdNi_2$, 26, 143, 236; 27, 198
$GdNi_3$, 26, 144; 27, 198; 35A, 120
$GdNi_4$, 26, 144
$GdNi_5$, 26, 144, 239; 27, 198
Gd_2Ni_7, 26, 144; 27, 198; 34A, 106, 107; 35A, 119, 120
Gd_2Ni_{15}, 27, 198
Gd_2Ni_{17}, 26, 144; 34A, 105
Gd_3Ni, 26, 143; 27, 198; 32A, 106
Gd_3Ni_2, 26, 143
$GdNiSi_2$, 35A, 120
$GdNi_2Tb$, 30A, 141
Gd_2O_3, 26, 364; 27, 723; 30A, 450
GdO_4P, 29, 425
Gd_2O_5Si, 34A, 378
$Gd_2O_7Si_2$, 32A, 436; 35A, 301
Gd-O-Sm-Zr, 27, 535
GdO_4Ta, 27, 521; 28, 348
Gd_3O_7Ta, 29, 329
Gd_2O_2Te, 27, 424
Gd_2O_5Ti, 29, 421
Gd-O-U, 30A, 453
GdO_4V, 27, 589; 28, 308
Gd-O-Zr, 26, 364
$Gd_2O_7Zr_2$, 27, 535
$Gd_6O_{15}Zr_2$, 27, 542
$Gd_8O_{14}Zr$, 27, 535
$Gd_{22}O_{51}Zr_9$, 27, 535
$GdPb_3$, 29, 117; 34A, 140
Gd_5Pb_3, 34A, 140
Gd_5Pb_4, 34A, 140
GdPd, 33A, 152
$GdPd_3$, 30A, 142
GdPt, 26, 235; 30A, 142
$GdPt_2$, 26, 236
$GdRe_2$, 30A, 142
GdRh, 30A, 142
$GdRu_2$, 28, 46
Gd_2S_3, 33A, 82
$Gd_6S_{14}Si_{2.5}$, 34A, 140
Gd_2S_4Sr, 27, 327
$GdSb_2$, 32A, 20
Gd_4Sb_3, 29, 117
Gd-Sc, 29, 181
GdSe, 29, 117
$GdSe_2$, 32A, 20
Gd_2Se_3, 28, 46; 29, 117; 30A, 142
Gd_3Se_4, 29, 117
$Gd_{1.6}Se_{2.9}Y_{0.5}$, 29, 117
GdSi, 29, 117; 30A, 142
Gd_5Si_3, 30A, 142
$GdSn_3$, 30A, 142
$GdTc_2$, 29, 118
GdTe, 29, 118
$GdTe_2$, 29, 118; 32A, 20
$GdTe_3$, 30A, 142
$GdTe_4$, 29, 118
Gd_2Te_3, 29, 118; 30A, 142
GdTh, 26, 237
$GdTh_3$, 26, 238
Gd-Y, 27, 330
GdZn, 29, 118
$GdZn_3$, 35A, 120
Gd_2Zn_{17}, 30A, 142
Gd_3Zn_{11}, 35A, 97, 120
Gd_3Zn_{22}, 35A, 120
$Gd_{13}Zn_{58}$, 32A, 62; 33A, 131
Ge, 26, 279; 27, 208, 381; 28, 46, 50; 29, 182; 30A, 142, 206, 207
Ge II, 29, 55, 118
$GeH_2Na_2O_4.6H_2O$, 32A, 511
$Ge_4H_4Na_4O_{16}Sn_2$, 35A, 320
$Ge_{10}H_4Na_8O_{34}Sn_4$, 32A, 248
$Ge_7HO_{16}Tl_3.4H_2O$, 30A, 451
Ge_3H_9P, 34A, 155
$GeHf_2$, 27, 379
$GeHf_3$, 29, 118; 30A, 142
Ge_2Hf_3, 29, 118; 30A, 142
$Ge_7Hf_6Ni_{16}$, 27, 162, 377; 30A, 143
$GeHfO_4$, 26, 398; 28, 324
$GeHf_3O_8$, 28, 324
GeHo, 30A, 143
Ge_3Ho_2, 29, 118
Ge_3Ho_5, 29, 118; 30A, 143
$Ge_{10}Ho_{11}$, 32A, 77
Ge-In-Sb-Te, 29, 182
$GeIn_2Te$, 29, 182
Ge_4Ir, 34A, 89
Ge_5Ir_4, 32A, 80; 33A, 84
GeK, 26, 84, 156
Ge_4K, 26, 157
$Ge_{45}K_{7.4}$, 33A, 162
$Ge_{46}K_8$, 34A, 122
$Ge_5K_2O_{11}$, 26, 398
$Ge_6K_2O_{13}$, 31A, 252
$Ge_8K_2O_{17}$, 31A, 252
$Ge_2K_2O_7Pb_2$, 28, 310; 30A, 326
GeLa, 30A, 143
Ge_2La, 29, 118
Ge_3La_5, 27, 341; 29, 118
$Ge_2La_2O_7$, 34A, 261
$Ge_3La_6S_{14}$, 34A, 140
Ge_4Li_9, 35A, 64
Ge_4Li_{15}, 30A, 53, 143
Ge_5Li_{22}, 29, 118; 30A, 143
Ge_4LiNaO_9, 34A, 259
$GeLiNi_2$, 34A, 140
Ge-Li-O, 30A, 451
$GeLiO_4$, 32A, 491
$GeLi_2O_3$, 33A, 503
$GeLi_4O_4$, 34A, 258
$Ge_2Li_2O_5$, 30A, 325; 33A, 507
$Ge_7Li_2O_{15}$, 35A, 318
$GeLiO_4Rh$, 27, 486
$Ge_{1.5}Li_2O_5Si_{0.5}$, 33A, 508
$Ge_3Li_2O_8Zn$, 27, 486; 28, 143
$GeLi_2Zn$, 34A, 140
Ge_3Lu_2, 29, 118
Ge_3Lu_5, 29, 118
$GeMg_2$ II, 29, 118
$(Ge,Si)Mg_2$, 27, 212
$GeMgN_2$, 35A, 136
$Ge_{0.4}MgNi_{1.6}$, 27, 261
$Ge_7Mg_6Ni_{16}$, 27, 261; 29, 118
Ge-Mg-Ni-O-Si, 27, 488
$GeMgO_3$, 26, 397; 27, 501
$Ge(Mg,Zn)_2O_4$, 29, 428
$Ge_{10}Mg_{28}O_{48}$, 35A, 272
$Ge_6MgO_{21}Pb_8$, 29, 399
$GeMgP_2$, 26, 227
$GeMg_2S_4$, 29, 118
Ge-Mn, 26, 149
ζ-$GeMn_{2.65}$, 26, 150
$GeMn_3$, 27, 305
$GeMn_{3.25}$, 27, 305
ε-$GeMn_{3.4}$, 26, 150
ε_1-$GeMn_{3.4}$, 26, 150
κ-Ge_2Mn_5, 26, 150
Ge_3Mn_5, 28, 46; 30A, 143
η-Ge_3Mn_5, 26, 150
Ge_8Mn_{11}, 29, 118
Ge-Mn-N, 27, 305
$GeMn_3N_{0.75}$, 33A, 29
$Ge_{24.3}Mn_{72.9}N_{2.8}$, 27, 305
GeMnNb, 33A, 86
$Ge_{0.45}MnNi_{1.55}$, 27, 213
$Ge_{0.5}MnNi_{1.5}$, 27, 261
$Ge_{0.7}MnNi_{1.3}$, 27, 212
$Ge_{0.75}MnNi_{1.25}$, 27, 261
$Ge_7Mn_6Ni_{16}$, 26, 175
$GeMnO_3$, 28, 312, 352
$GeMn_2O_4$, 33A, 505; 34A, 262
$Ge_3Mn_3O_{12}Sc_2$, 26, 415
$GeMnO_4Zn$, 26, 409
$GeMn_2S_4$, 30A, 441; 34A, 91

Ge-Mn-Te, 26, 151
$Ge_{0.1-0.9}Mn_{0.95-0.1}Te$, 26, 151
Ge_2Mn_2Th, 29, 91, 118
$GeMo_3$, 27, 213
$Ge_{1.7}Mo$, 30A, 143
Ge_2Mo, 30A, 21, 143
$Ge_{23}Mo_{13}$, 29, 118; 32A, 72
Ge_2MoNi_2, 26, 174
$Ge_{23}Mo_{45}Ni_{32}$, 30A, 143
$Ge_3N_xNb_5$, 29, 118
$Ge_3N_xTa_5$, 29, 118
GeNa, 26, 156; 29, 57, 118
$GeNa_2O_3.6H_2O$, 32A, 449
$Ge_9Na_4O_{20}$, 28, 147
$Ge_2NaO_{12}P_3$, 33A, 397
$GeNaO_4Sm$, 35A, 315
$GeNa_2O_5Ti$, 35A, 316
$GeNaO_4Y$, 35A, 229
$GeNa_2O_4Zn$, 33A, 506
$GeNb_3$, 27, 213
$(Ge,Sn)Nb_3$, 27, 213
Ge_3Nb_5, 27, 213
$Ge_{0.5}NbNi_{1.5}$, 29, 170
$Ge_7Nb_6Ni_{16}$, 27, 162, 377
$GeNb_{18}O_{47}$, 29, 422
$Ge_{2-x}NbSi_x$, 28, 46
GeNd, 29, 118; 30A, 143
Ge_2Nd, 26, 175
Ge_3Nd_5, 29, 118
Ge-Ni, 27, 293
$GeNi_2$, 28, 46; 29, 182
$GeNi_3$, 28, 46; 29, 118
$GeNi_2O_4$, 26, 413
$Ge_6NiO_{21}Pb_8$, 29, 399
$Ge_xNi_2O_4Si_{1-x}$, 27, 489
$Ge_7Ni_{16}Sc_6$, 27, 162, 377; 28, 46
$Ge_{0.5}Ni_{1.5}Ta$, 29, 170
$Ge_7Ni_{16}Ta_6$, 27, 162, 377; 30A, 143
Ge_2Ni_2Th, 29, 91, 118
$Ge_7Ni_{16}Ti_6$, 27, 162, 377
$Ge_2Ni_5V_3$, 26, 174
$Ge_7Ni_{16}Zr_6$, 27, 162, 377
GeO, 30A, 455
GeO_2, 28, 289, 320; 29, 303
GeO_7P_2, 28, 304
GeO_9P_2, 28, 304, 324
Ge_4O_9Pb, 26, 398; 27, 541
$Ge_7O_{16}Pb_2.7H_2O$, 26, 398
$Ge_2O_7Pb_2Rb_2$, 28, 310
$Ge_6O_{21}Pb_8Sr$, 29, 399
$Ge_6O_{21}Pb_8Zn$, 29, 399
$Ge_5O_{11}Rb_2$, 26, 398
$Ge_6O_{13}Rb_2$, 31A, 252
$Ge_7O_{15}Rb_2$, 31A, 252
$Ge_8O_{17}Rb_2$, 31A, 252
GeO_3Sr, 26, 397; 27, 501
Ge_4O_9Sr, 26, 398; 27, 541
Ge-O-Th, 26, 403
GeO_4Th, 26, 403
$Ge_4O_9Tl_2$, 27, 541
$Ge_5O_{11}Tl$, 26, 398
$Ge_6O_{13}Tl_2$, 27, 541; 31A, 252
$Ge_{17}O_{15}Tl_2$, 29, 423
GeO_4Zn_2, 35A, 314
$Ge_3O_8Zn_2$, 28, 143
GeO_2Zr, 27, 542
GeO_4Zr, 28, 324
GeO_8Zr_3, 28, 324
GeP, 32A, 139; 35A, 20
Ge-Pb-Se-Te, 30A, 207
(Ge,Pb)Te, 30A, 208
GePr, 30A, 143
Ge_2Pr, 29, 119
Ge_3Pr_5, 29, 118
Ge_3Pt_2, 33A, 87
GeRb, 26, 84, 157
Ge_4Rb, 26, 157
Ge_3Rh_5, 34A, 124
$Ge_{22}Rh_{17}$, 32A, 83
GeRu, 27, 214
Ge_3Ru_2, 27, 214; 29, 119
GeS, 30A, 143
GeS_2, 30A, 55, 143
GeS_4Sr_2, 35A, 65
GeSZr, 29, 119
Ge-Sb-Te, 26, 39
$GeSb_2Te_4$, 31A, 18
$GeSb_4Te_7$, 33A, 19
$Ge_2Sb_2Te_5$, 33A, 20
GeSc, 30A, 143
Ge_2Sc, 29, 119
Ge_3Sc_5, 27, 341
GeSe, 27, 215; 30A, 54, 143
$GeSe_2$, 27, 215
Ge-Se-Sn-Te, 26, 153; 30A, 207
GeSeZr, 29, 119
Ge-Si, 29, 182; 30A, 207
$Ge_{2-x}Si_xTa$, 28, 46
Ge-Si-W, 26, 158
$Ge_{0.25}Si_{0.35}W_{0.4}$, 26, 156
GeSm, 30A, 143
Ge_2Sm, 29, 119
Ge_3Sm_5, 29, 119
Ge_4Sm_5, 32A, 87
Ge-Sn-Te, 26, 153
$Ge_{0.17}Sn_{0.83}Te$, 27, 211
(Ge,Sn)Te, 27, 211
GeSr, 32A, 139
Ge_2Sr, 33A, 152
GeSrZn, 34A, 140
$GeTa_3$, 29, 119; 30A, 143, 144
GeTb, 30A, 144
Ge_3Tb_2, 29, 119
Ge_3Tb_5, 29, 116
GeTe, 26, 151; 27, 211; 31A, 42
Ge_2Te_3, 30A, 144
GeTeU, 34A, 92
GeTeZr, 29, 119
$GeTh_2$, 28, 47
$Ge_{1.5}Th$, 28, 47
$Ge_{1.6}Th$, 28, 47
Ge_2Th, 27, 216; 28, 47
Ge_2Th_3, 28, 47
$GeTi_3$, 30A, 144
Ge_3Tm_2, 29, 119
Ge_3Tm_5, 29, 119
Ge_2U, 29, 119
Ge_3U, 29, 119
$(Ge_xSi_{1-x})V_3$, 30A, 208
$(Ge,Si)V_3$, 30A, 207
$(Ge_xSi_{1-x})_3V_5$, 30A, 207
GeV_3, 28, 47
$Ge_{1.8}V$, 29, 119
Ge_2V_3, 28, 47
Ge_3V_5, 28, 47
Ge_8V_{11}, 29, 119
$Ge_{31}V_{17}$, 32A, 74
GeY, 30A, 144
Ge_2Y, 29, 119
Ge_3Y_2, 29, 119
Ge_3Yb_2, 29, 116
Ge_5Yb_3, 30A, 56, 144
GeZr, 29, 119; 30A, 144
$GeZr_3$, 29, 119; 30A, 144
Ge_4Zr_5, 30A, 144

H_2, 26, 285; 27, 393, 30A, 449
H_yHf_xZr, 30A, 208
$H_2HgN_2O_3$, 30A, 454
$HHg_4N_2O_4P$, 30A, 454
H_2HgO_5S, 29, 356
$H_4Hg_3O_{11}S_2$, 27, 611
$HHoO_2$, 30A, 310
HI, 30A, 449
$H_4I_{1+x}K_xN$, 27, 421
H_7IKNO_6, 29, 427
$H_2I_2K_4O_{10}.8H_2O$, 27, 641; 30A, 350
$H_{18}I_2K_4O_{18}$, 27, 641
$H_3ILi_2O_6$, 27, 640
H_6ILiO_7, 27, 640
$H_{15}IMgO_{12}$, 35A, 396
$H_{12}I_6MgO_6Te$, 29, 251
$H_4I_3MnNa_7O_{18}.17H_2O$, 27, 640
$H_{38}I_3MnNa_7O_{35}$, 27, 640
$H_3IMn_2O_3$, 29, 251
$H_3I_3N_2$, 31A, 256; 33A, 229
H_4IN, 27, 425; 30A, 449; 31A, 243; 35A, 146
$H_4IN.2H_3N$, 29, 416
H_4I_3N, 35A, 147
H_7IN_2, 28, 300
$H_{10}IN_3$, 29, 416
$H_{13}IN_4$, 26, 298
$H_{16}IN_5$, 26, 298
$H_8I_6N_2Te$, 27, 463

$H_8I_2N_4Zn$, 30A, 457
$H_3INi_2O_3$, 29, 251
$H_4I_2NiO_8$, 27, 643
HI_3O_8, 31A, 212
H_5IO_6, 26, 499; 31A, 212
$HIOPb$, 26, 327; 31A, 108
H_4IP, 32A, 216
$HInO_2$, 29, 310; 35A, 278
H_3InO_3, 28, 320; 32A, 260
$HInO_4S$, 27, 625
H_5InO_7S, 26, 454
HK, 27, 398
$H_2K_2LiO_{10}P_3$, 27, 594
$H_2K_3Mg_{2.5}O_{12}Si_4$, 30A, 455
H_2KN, 30A, 258
α-H_2KN, 27, 420
$H_3KN_3O_6$, 30A, 454
$H_2K_4NO_9P_3.4H_2O$, 30A, 392
$HK_2N_5O_{10}Ru$, 27, 669; 29, 380
$HK_2NO_6S_2$, 28, 215
H_2KNO_3S, 32A, 325
H_2KNO_4S, 31A, 207
$H_9KNbO_{7.5}$, 29, 420
$H_4K_2O_6Os$, 27, 665
HKO_4S, 29, 346
HKO_5S, 30A, 453
$H_4K_4O_{18}S_4Th$, 28, 217; 31A, 205
HKO_3Si, 28, 233
HKO_5Si_2, 34A, 364
$H_4K_4O_{12}Si_4$, 29, 395
$H_2K_3O_{10}Si_3Y$, 33A, 473
$H_5KO_6Te.H_2O$, 29, 359
H_7KO_7Te, 29, 359
$H_2K_8O_{40}W_{12}.9H_2O$, 31A, 258
$H_8K_2O_{15}W_2$, 29, 332
$H_{18}K_4O_{15}Xe$, 29, 335
H_2KP, 27, 420
H_8K_2Re, 27, 671
H_9K_2Re, 27, 671; 29, 235
H_3KSi, 26, 291
H_8K_2Tc, 27, 671
H_9K_2Tc, 27, 671; 29, 236
H-La, 30A, 208
$H_{48}La_2Mg_3N_{12}O_{60}$, 27, 725
$H_{0.78}LaO$, 27, 332
$H_{0.91}LaTe$, 27, 332
HLi, 27, 397; 30A, 449
$H_3Li_2Mg_2N_3$, 27, 420
$H_5LiN_2O_4S$, 29, 343; 31A, 254; 32A, 330
H_2LiO_3P, 28, 304; 35A, 345
$H_2Li_2O_5Se$, 27, 606
$H_3LiO_6Se_2$, 28, 306
$H_2Li_4O_7Si_2$, 35A, 311
H_3LiSr, 29, 119
$H\mathit{Ln}O_2$, 30A, 451
$H_3\mathit{Ln}O_3$, 30A, 451
H_2Lu, 27, 237
H_3Lu, 27, 237
$H_{16}\mathit{M}_2O_{20}S_3$, 29, 424
$H_6\mathit{M}O_6Sn$, 30A, 451
$H_2Mg_{0.99}Mn_{0.01}O_2$, 26, 348
$H_{12}MgN_2O_{12}$, 34A, 340
$H_4MgNO_4P.6H_2O$, 35A, 329
$H_4MgNO_9P_3$, 32A, 379; 33A, 416
$H_{10}MgN_2O_8P_2.4H_2O$, 28, 352
$H_{12}Mg_3N_2O_{16}P_4.8H_2O$, 28, 352
$H_8MgN_2O_8S_2.6H_2O$, 27, 619; 29, 354
$H_8Mg_2N_2O_{12}S_3$, 29, 439
$H_{20}MgN_2O_{14}S_2$, 27, 619; 29, 354
$HMgNaO_5S.2H_2O$, 30A, 377
$H_4(Mg,Ni)_3O_9Si$, 26, 522
H_2MgO_2, 30A, 445; 32A, 242
$HMgO_4P.3H_2O$, 26, 457; 31A, 243; 32A, 360
H_8MgO_8S, 27, 612; 29, 346
$H_{12}MgO_9S_2$, 27, 631
$H_{12}MgO_{10}S$, 27, 618; 29, 347
$H_{14}MgO_{11}S$, 29, 348
$H_{12}MgO_{10}Se$, 27, 618; 29, 425
$H_4Mg_3O_9Si_2$, 26, 511; 30A, 434
$H_2MgO_{12}Si_2U_2.4H_2O$, 27, 710
$H_{10}MgO_{16}Si_2U_2$, 27, 710
$H_{14}MgO_{18}Si_2U_2$, 29, 410
$H_8MnN_2O_{12}P_4$, 35A, 347
$H_8MnN_2O_8S_2.6H_2O$, 31A, 204
$H_8Mn_2N_2O_{12}S_3$, 29, 439
$H_8Mn_2N_2O_{12}Se_3$, 29, 425
$HMnO_2$, 28, 131; 33A, 265
H_2MnO_2, 30A, 305
$H_2Mn_5O_{10}Si_2$, 35A, 467
$H_2Mn_7O_{14}Si_3$, 35A, 466
$H_2MnO_6SiZn_2$, 28, 230, 264
H_6MnO_6Sn, 34A, 236
$H_{24}Mo_7N_6O_{24}.4H_2O$, 32A, 296
$H_7Mo_2NO_{10}Zn_2$, 32A, 301
H_4MoNS_4, 30A, 453
$H_{0.5}MoO_3$, 26, 355
$H_2Mo_4O_{12}$, 26, 355; 34A, 236
H_3N, 26, 298; 29, 233
$H_3N.H_2O$, 26, 298
$H_4N_2.H_2O$, 27, 422; 29, 233
H_3NNaO_3P, 29, 360
$H_3N_3Na_3O_6P_3.4H_2O$, 30A, 389
$H_6N_2Na_2O_6PdS_2.6H_2O$, 34A, 317
$HN_5Na_2O_{10}Ru.2H_2O$, 30A, 407
$H_4NNaO_4S.2H_2O$, 28, 354; 32A, 336
$H_{32}N_8Na_2O_{38}W_{12}.13H_2O$, 29, 423
$H_{12}N_3NbO_8$, 29, 420
$H_{12}N_2NiO_{12}$, 27, 571
$H_8N_2NiO_8S_2.6H_2O$, 28, 213; 29, 354
$H_8N_2Ni_2O_{12}S_3$, 29, 439
$H_{10}N_4NiO_8S_2$, 31A, 250
$H_{20}N_2NiO_{14}S_2$, 29, 354
$H_2N_4NiS_4$, 33A, 361
$H_4N_2O_3$, 27, 566
$H_4N_2O_3.2HNO_3$, 35A, 322
H_5NO_2, 31A, 128
H_6N_2O, 27, 422; 29, 233
H_6N_3OP, 34A, 155
$H_7N_2O_4P$, 30A, 387
$H_8N_4O_8P_4.2H_2O$, 30A, 391
$H_{10}N_2O_6P_2$, 29, 361
$H_{10}N_2O_8P_2$, 31A, 182
$H_{12}N_3O_4P.3H_2O$, 35A, 331
$H_{16}N_4O_{12}P_4$, 29, 360; 35A, 348
$HN_2O_9PPb_2$, 31A, 243
H_8N_3OPS, 33A, 406
$H_{10}N_3O_2PS$, 34A, 325
$H_4NO_{12}P_4Sr$, 31A, 245
$H_4NO_9P_3Zn$, 32A, 379
$H_8N_2O_8PbSe_2$, 30A, 453
$H_9N_3O_3PdS$, 32A, 326
$H_{12}N_3O_{15}Pr$, 29, 379
$HNO_6Rb_2S_2$, 30A, 367
$(H_4N,K)O_4Re.\mathit{x}H_2O$, 30A, 458
$HNOS$, 34A, 242
H_3NO_4S, 32A, 322
H_5NO_4S, 29, 424
$H_6N_2O_4S$, 35A, 366
$H_8N_2O_3S.H_2O$, 30A, 378
$H_8N_2O_4S$, 28, 206; 31A, 197
$H_8N_2O_6S_2$, 28, 306
$H_8N_2O_8S_2$, 27, 421
$H_9N_3O_6S_2$, 30A, 367
$H_{10}N_4O_4S$, 30A, 366
$H_{12}N_4O_7S$, 30A, 454
$H_{16}N_6O_{10}S$, 27, 570
$H_{20}N_8O_{13}S$, 27, 569
$H_4NO_8S_2Sc$, 31A, 255
$H_{12}N_3O_{12}S_3Sc$, 31A, 255
$H_8N_2O_8S_2V.6H_2O$, 32A, 343
$H_8N_2O_8S_2Zn.6H_2O$, 29, 354

$H_{10}N_4O_8S_2Zn$, 26, 441; 31A, 250
$H_{20}N_2O_{14}S_2Zn$, 29, 354
$H_8N_2O_4Se$, 27, 606
$H_4N_2O_{10}Sn_3$, 26, 475
$H_{12}N_3O_8Ta$, 29, 420
H_4NO_4Tc, 27, 671
HNO_7Te, 31A, 175
$H_2N_2O_8Th.xH_2O$, 31A, 260
$H_2N_6O_{20}Th_2.8H_2O$, 31A, 260
$H_{14}N_6O_{26}Th_2.2H_2O$, 33A, 431
$H_3NO_6U_2.3H_2O$, 28, 310
$H_3NO_9U_3.5H_2O$, 28, 310
$H_4N_2O_{10}U$, 27, 570
$H_4N_2O_{10}U.4H_2O$, 30A, 404
$H_6N_2O_9U_3.4H_2O$, 28, 310
$H_{12}N_3O_8V$, 27, 659
$H_{40}N_{10}O_{37}W_{12}.5H_2O$, 29, 423
$H_{40}N_{10}O_{37}W_{12}.7H_2O$, 29, 422
$H_{40}N_{10}O_{37}W_{12}.11H_2O$, 29, 422
$H_{40}N_{10}O_{41}W_{12}.7H_2O$, 26, 393
$H_{40}N_{10}O_{41}W_{12}.11H_2O$, 26, 393
$H_{42}N_{10}O_{42}W_{12}.10H_2O$, 34A, 283
$H_{50}N_{10}O_{42}W_{12}$, 29, 423
$H_{54}N_{10}O_{44}W_{12}$, 29, 422
$H_{58}N_8O_{51}W_{12}$, 29, 423
$H_{62}N_{10}O_{48}W_{12}$, 29, 422
$H_8N_2O_{14}Zn_5.2H_2O$, 35A, 325
$H_{12}N_2O_{12}Zn$, 32A, 352
$H_3N_3PbS_2$, 31A, 81
$H_2N_4PdS_4$, 33A, 361
$H_8N_2PtS_{15}.2H_2O$, 34A, 296
H_2NRb, 30A, 258
α-H_2NRb, 27, 420
HNS_7, 30A, 257
$H_2N_2S_6$, 34A, 156
$H_3N_3S_5$, 31A, 81
$H_4N_4S_4$, 32A, 154
H_4NS_4W, 30A, 453
$H_8N_2S_4W$, 28, 227, 302
$H_8N_2Se_4W$, 33A, 387
H_7NSi_2, 34A, 155
H_4N_2Sr, 28, 296
$H_{0.80}N_{0.36}Zr$, 35A, 115
$H_{2.4}N_{0.6}Zr_2$, 32A, 44
HNa, 27, 398
$HNaNb_6O_{16}$, 30A, 335
HNaO, 32A, 241
$HNaO.4H_2O$, 29, 297
$H_2Na_2O_2.8H_2O$, 28, 322
H_9NaO_5, 29, 297
$HNa_2O_4P.2H_2O$, 27, 581
$HNa_2O_4P.7H_2O$, 35A, 335
$HNa_2O_9P_3$, 27, 593
$(HNa_2O_9P_3)_x$, 26, 459
$(HNa_3O_{12}P_4)_x$, 33A, 412
$H_2Na_nO_{3n+1}P_n$, 26, 459
$H_2Na_2O_{12}P_4$, 29, 363
$H_5Na_2O_6P$, 27, 581
$H_8Na_4O_{16}P_4$, 29, 367
$H_{20}Na_4O_{17}P_2$, 29, 360
$HNaO_4S.H_2O$, 30A, 365
$H_4Na_2O_8S_4$, 29, 345
$H_{10}Na_2O_8S_2$, 27, 630
$H_{20}Na_2O_{14}S$, 27, 620
$HNaO_3Se$, 28, 221
$H_3NaO_6Se_2$, 33A, 389
$H_2Na_2O_4Si.4H_2O$, 31A, 216
$H_2Na_2O_4Si.8H_2O$, 31A, 215
$H_6Na_2O_{18}Si_6Zr$, 29, 404
$H_{12}Na_4O_{12}Xe$, 29, 335
$H_{16}Na_4O_{14}Xe$, 29, 335
H_3NaO_3Zn, 26, 396
$H_{0.95}Nb$, 33A, 81
HNb, 26, 215
H_xNb, 29, 183
H-Nb-Ti, 27, 239
H_2Nd, 27, 237
H_3Nd, 27, 237
$HNdO_5W$, 34A, 287
$H_{0.6}Ni$, 28, 27; 29, 119
$H_{0.8}Ni$, 26, 166
$HNiO_2$, 34A, 293
$H_9Ni_5O_{10}$, 29, 417
H_8NiO_8S, 29, 424
$H_{12}NiO_9S$, 27, 633, 643; 29, 353
$H_{12}NiO_{10}Se$, 29, 425
$H_{2.7}NiZr$, 27, 237
$H_{2.981}NiZr$, 29, 62, 119
H_3NiZr, 27, 237
H_xNp, 30A, 209
$H_{2.98}Np$, 30A, 144
H_2O, 28, 118; 29, 294; 30A, 303; 32A, 250
H_2O_2, 30A, 303
$H_2O_2.2H_2O$, 26, 298
H_3O_2P, 31A, 259
H_3O_4P, 31A, 259
$H_3O_4P.\frac{1}{2}H_2O$, 34A, 319
$H_6O_8P_2.H_2O$, 34A, 319
$H_4O_{18}P_2Pb_2U_3.3H_2O$, 27, 583
$H_{10}O_{21}P_2Pb_2U_3$, 27, 583
H_2O_4PRb, 31A, 252
$H_2O_{26}P_6Sr_{10}$, 26, 469
HO_3PTl_2, 26, 471
HO_5PZn_2, 31A, 190, 242
$HO_5PZn_2.1.5H_2O$, 31A, 190
$H_8O_{12}P_2Zn_3$, 27, 586
$H_2O_8P_2Zr.H_2O$, 31A, 259; 34A, 328
H_2O_7PbSiU, 29, 405
$H_{0.56}OPr$, 27, 332
$H_4O_9Re_2$, 35A, 279
H_2O_4S, 30A, 360
$H_2O_4S.H_2O$, 33A, 364
H_6O_6S, 35A, 365
$H_{16}O_{20}S_3Y_2$, 29, 424
$H_2O_6SZn_2$, 27, 629
$H_2O_{14}S_3Zr_2.4H_2O$, 31A, 205
$H_8O_9Sb_2$, 33A, 425
HO_2Sc, 32A, 244
γ-HO_2Sc, 27, 423
H_4O_5SeZn, 29, 358
$H_2O_5Si_2$, 30A, 455
$H_2O_5Si_2$ I, 29, 394
$H_2O_5Si_2$ II, 29, 394
H_6OSi_2, 28, 360
$H_2O_9Si_2Zn_4.H_2O$, 27, 699; 28, 324; 32A, 435
$H_4O_{10}Si_2Zn_4$, 27, 699
H_6O_6SnZn, 32A, 247; 33A, 289
H_2O_2Sr, 30A, 450; 34A, 235
$H_2O_2Sr.H_2O$, 31A, 256; 32A, 245
H_6O_6Te, 35A, 281
H_2O_4U, 29, 313
$H_{0.23}O_3W$, 32A, 314
$H_2O_4W.H_2O$, 28, 332
HO_2Y, 29, 305; 30A, 310
H_3O_3Y, 29, 305; 32A, 246
HO_2Yb, 30A, 310
H_2O_2Zn, 29, 303; 34A, 235
H_2PRb, 27, 420
H-Pd, 30A, 144, 209
HPd, 27, 239
α-HPd, 26, 166
β-HPd, 26, 166
H_2Pr, 27, 237
H_3Pr, 27, 237
$H_{1.39}PrTe$, 27, 332
H_2S, 26, 423; 27, 399; 29, 341
H_9SbSi_3, 34A, 155
H-Sc, 29, 183
H_2Sm, 27, 237
H_3Sm, 27, 237
H_2Tb, 27, 237
H_3Tb, 27, 237
H_2Th, 27, 239
$H_{15}Th_4$, 27, 240
H_2Ti, 26, 168
H_2(Ti-Zr), 26, 168
H-V, 27, 240; 29, 184
$H_{1.45-1.77\pm5}V$, 26, 168
H_2Y, 27, 237, 241
H_3Y, 27, 237, 241
H_2Yb, 27, 241
$H_{2.55\pm5}Yb$, 27, 241
H_3Yb, 26, 169
HZr, 27, 241
γ-HZr, 27, 242
H_2Zr, 26, 168
He, 26, 279; 27, 391
He^3, 26, 285
β-He^2, 26, 285
Hf, 28, 47; 30A, 144, 209

β-Hf, 26, 163
Hf_2Hg, 32A, 139
Hf_2Ho, 27, 237
Hf_3Ho, 27, 237
HfI_3, 27, 442; 29, 249
$HfIn_2$, 30A, 144
Hf_3In_4, 30A, 14, 144
$Hf_5(In_{0.5}Sb_{0.5})_3$, 28, 47
$HfIr_{1+x}$, 29, 119
$Hf_{47}Ir_{53}$, 29, 119
$Hf_2La_2O_7$, 29, 421
$HfMn_2$, 26, 162, 192; 27, 196
ε_2-HfMo, 26, 163
$HfMo_2$, 26, 162
ε_1-$HfMo_2$, 26, 163
η-$HfMo_2$, 26, 163
$HfMo_2O_8$, 33A, 338
(Hf,Mo)O_3Pb, 30A, 452
HfN_{1-x}, 26, 73
Hf_2N, 26, 73
Hf_3N_2, 35A, 68
Hf_4N_3, 35A, 69
Hf_2N_xZn, 29, 119
$HfNaO_4Sc$, 33A, 294, 295
Hf-Nb, 29, 184
Hf-Ni, 26, 205
HfNi, 26, 206; 27, 232
$HfNi_5$, 26, 207, 238
Hf_2Ni, 26, 205; 27, 232
Hf_2Ni_7, 26, 207
Hf_7Ni_{10}, 26, 206; 27, 232
Hf_9Ni_{11}, 26, 206
$Hf_6Ni_{16}Si_7$, 26, 175; 27, 376
HfNiSn, 35A, 120
$HfNi_2Sn$, 35A, 120
Hf_5NiSn_3, 30A, 144
Hf-O, 30A, 450
$HfO_{0.10-0.50}$, 26, 363
HfO_2, 30A, 450
HfO_7P_2, 28, 304
(Hf,W)O_3Pb, 30A, 452
HfO_4Si, 26, 399
HfO_4Ti, 33A, 306
$HfOs_2$, 26, 237
HfP, 27, 233
HfP_2, 29, 119
Hf_2P, 33A, 92
Hf_3P_2, 33A, 92, 93
Hf_2Pd, 27, 234
HfPt, 30A, 144
Hf_2Pt_3, 29, 119
ε-Hf-Pu, 29, 184
Hf-Re, 28, 47
$HfRe_2$, 26, 237; 27, 233; 28, 47
Hf_3Re_2, 27, 234
Hf_5Re_{24}, 27, 233
$HfRh_{1+x}$, 29, 119
Hf_2Rh_3, 29, 119
Hf_3Rh_5, 34A, 124
$Hf_{42}Rh_{58}$, 29, 119
$Hf_{46}Rh_{54}$, 29, 119
HfS_2, 30A, 144
Hf_2S, 31A, 43; 32A, 139
Hf_3S_4, 35A, 69
HfSb, 29, 120; 30A, 145
$HfSb_2$, 30A, 145; 33A, 24
α-$HfSb_2$, 29, 120
$HfSb_3$, 30A, 145
Hf_3Sb, 29, 120; 30A, 144
$HfSe_2$, 30A, 145
Hf_2Se, 32A, 139
HfSi, 34A, 140
$HfSi_2$, 26, 63, 64; 27, 18; 34A, 140
Hf_3Si_2, 26, 11; 27, 18; 34A, 140
Hf_5Si_3, 27, 18; 34A, 140
Hf_5Si_4, 30A, 145; 34A, 140
Hf-Sn, 29, 184
HfSn, 29, 61, 120
$HfSn_2$, 27, 379
Hf_5Sn_3, 26, 164
Hf_5Sn_4, 27, 379; 30A, 145
Hf-Ta, 29, 184
Hf-Tc, 26, 255; 27, 360
Hf_2Tm, 27, 237
Hf_3Tm, 27, 237
HfV_2, 26, 162
HfW_2, 26, 162; 27, 234; 30A, 145
Hf-Y, 27, 381
$HfZn_2$, 30A, 145
$HfZn_3$, 30A, 145
$HfZn_5$, 30A, 145
Hf_2Zn, 30A, 145
Hg, 26, 279; 27, 386
β-Hg, 26, 194
HgHo, 29, 120
Hg_2Ho, 29, 120
Hg_3Ho, 29, 120
HgI_2, 32A, 211; 34A, 182
Hg_2INaO_2, 29, 310
$Hg_3I_2S_2$, 28, 344
HgI_6Tl_4, 28, 302
Hg-In, 26, 170
Hg-In-Te, 27, 244
$HgIn_2Te_4$, 33A, 152
HgK, 27, 280
Hg_2K, 27, 280
Hg_5K, 27, 280
$HgK_3N_5O_{11}$, 28, 302
HgK_2O_2, 27, 661; 29, 338
$HgLi_2O_2$, 29, 338
HgMn, 27, 270; 29, 120; 30A, 145
Hg_5Mn_2, 26, 191
Hg-Mn-N, 27, 272
(Hg,Mn)Te, 28, 47
$Hg_2N_2O_3$, 30A, 454
HgNa, 27, 280
Hg_5Na, 27, 280
$HgNa_2O_2$, 27, 661; 29, 338
$Hg_2Nb_2O_7$, 33A, 287
HgO, 29, 310
HgO_2Rb_2, 29, 338
HgO_4S, 26, 443, 444; 27, 610; 28, 210, 212; 30A, 301
$HgO_4S.H_2O$, 26, 443, 444; 29, 356; 30A, 301
Hg_2O_4S, 34A, 300
$Hg_3O_4S_3$, 28, 350
Hg_3O_6S, 26, 444; 27, 610
$Hg_3O_9S_2.2H_2O$, 27, 611
HgO_8S_2Te, 28, 350
$Hg_2O_7Sb_2$, 33A, 287
HgO_4Se, 26, 444
$HgO_4Se.H_2O$, 26, 444
Hg_2O_4Se, 34A, 300
Hg_3O_6Se, 26, 444; 27, 611
$Hg_2O_7Ta_2$, 33A, 287
Hg_3O_6Te, 27, 635; 28, 223
HgO_6V_2, 33A, 312
$Hg_{14}P_{14}Pd_{72}$, 35A, 120
$Hg_{14}P_{14}Pt_{72}$, 35A, 120
Hg-Pd, 26, 195
Hg_5Pd_2, 30A, 66, 145
$Hg_{14}Pd_{72}Se_{14}$, 35A, 120
$HgPt_3$, 26, 225
Hg_2Rh, 32A, 139
HgS, 28, 47; 29, 67, 120; 30A, 145
HgSc, 28, 47
Hg_3Sc, 28, 47
HgSe, 26, 195; 27, 281, 383; 28, 47; 29, 184
Hg-Se-Te, 26, 195
Hg-Se-Zn, 29, 184
Hg-Sn, 26, 195
γ-$HgSn_9$, 26, 196
Hg_2Sr, 29, 120
Hg_3Sr, 29, 120
HgTb, 29, 120
HgTe, 27, 382; 28, 47; 30A, 145
HgY, 28, 47
Hg_2Y, 29, 120
Hg_3Y, 28, 47
Ho, 26, 234; 27, 333; 30A, 145, 209
HoIn, 30A, 145
$HoIn_3$, 30A, 146
HoIr, 30A, 146
$HoIr_2$, 26, 236; 28, 47
$HoLiO_2$, 27, 516
β-Ho-Mg, 29, 185
$HoMg_2$, 29, 120

Ho_5Mg_{24}, 29, 120
$HoMn_{12}$, 31A, 48
Ho_6Mn_{23}, 30A, 146
$HoMnO_3$, 28, 138
Ho_4MnS_7, 29, 120
HoN, 27, 335
$HoNaS_2$, 29, 342
$HoNaSe_2$, 30A, 146
$HoNbO_4$, 27, 521
HoNi, 29, 120; 30A, 146
$HoNi_3$, 35A, 120
$HoNi_5$, 26, 239
Ho_2Ni_7, 35A, 120
$HoNiSi_2$, 35A, 120
Ho_2O_3, 27, 497; 30A, 457
HoO_4P, 28, 304
$Ho_2O_7Si_2$, 33A, 465; 35A, 300, 301
$Ho_2O_7Sn_2$, 26, 399
Ho_2O_7SnTi, 33A, 287
HoO_7SnTiY, 33A, 287
HoO_4Ta, 27, 521; 29, 421
HoO_4V, 28, 308
HoP, 26, 164; 27, 335
$HoPb_3$, 29, 120
$HoPd_3$, 26, 165; 30A, 146
Ho_5Pd_2, 29, 120
HoPt, 30A, 146
$HoPt_3$, 26, 165; 30A, 146
HoRh, 30A, 146
$HoRh_2$, 26, 236
HoS, 26, 165
Ho_2S_3, 32A, 92
Ho_5S_7, 29, 120
Ho_2S_4Yb, 29, 120
HoSb, 26, 164; 27, 334
$HoSb_2$, 32A, 20
Ho_5Sb_3, 33A, 21
HoSe, 26, 165
$HoSe_2$, 32A, 20
Ho_2Se_3, 30A, 146
HoSi, 29, 120; 30A, 146, 147
$HoSi_2$, 27, 336; 28, 47; 29, 120
Ho_5Si_3, 30A, 146
$HoTc_2$, 29, 120
$HoTe_2$, 32A, 20
$HoTe_3$, 30A, 147
Ho_2Te_3, 30A, 147
HoTl, 30A, 147
$HoTl_3$, 30A, 147
HoZn, 29, 120
$HoZn_2$, 31A, 44
$HoZn_3$, 33A, 100
Ho_2Zn_{17}, 30A, 147
Ho_6Zn_{23}, 30A, 147
$Ho_{13}Zn_{58}$, 35A, 120

I_2, 32A, 149
I_6In_2, 29, 288
IInS, 27, 466
IInSe, 27, 466
IInTe, 27, 466
I_3Ir, 33A, 237
IK, 26, 297; 27, 425, 723
$IKNiO_6$, 30A, 447; 31A, 215
IKO_3, 26, 497; 27, 643
$I_2K_4O_9$, 33A, 451
I_6K_2Te, 27, 463
$I_3KZn.2H_2O$, 35A, 161
ILaS, 31A, 247
$ILi.H_2O$, 30A, 287
$ILiO_3$, 31A, 214
$ILiO_4.3H_2O$, 27, 640
I_3Ln, 29, 416
I_2Mn, 27, 434
INTi, 29, 248
INU, 31A, 253
INZr, 29, 247
$INa.2H_2O$, 35A, 195
$INaO_4$, 35A, 396
I_3Nb, 27, 440
α-I_4Nb, 27, 443
I_5Nb, 28, 101
I_8Nb_3, 31A, 105
$I_{11}Nb_6$, 31A, 259; 32A, 210
I_2NbO, 28, 302
I-Nd, 26, 322
$I_2Ni.6H_2O$, 28, 300
$I_2NiO_6.2H_2O$, 27, 643
I_2O_5, 35A, 208
IOSm, 26, 340
IOTm, 26, 340
IOYb, 26, 340
$I_3O_{12}Zr$, 26, 498
$I_4O_{12}Zr$, 27, 724
$I_2P_4Se_3$, 35A, 157
I_3PZn_3, 27, 467
$I_6Pb_2Sr.7H_2O$, 26, 338
I_2Pd, 33A, 240
I_4Pt, 34A, 183
IRb, 26, 297; 27, 425; 29, 416
I_6Rb_2Re, 29, 416
I_9Re_3, 33A, 238
I_3Rh, 33A, 237
I_3Ru, 31A, 251; 33A, 237
ISSb, 31A, 112; 32A, 223
$I_3S_{24}Sb$, 28, 103
$I_4S_{16}Sn$, 27, 473
I_3Sb, 28, 361
I_3Sc, 26, 298; 29, 416
I_2Sr, 34A, 181
$I_{14}Ta_6$, 30A, 264
I_2Th, 33A, 241
I_3Th, 28, 302
I_4Th, 29, 289
I_3Ti, 31A, 260
ITl, 26, 297
I_3Y, 28, 300
I_3Zr, 27, 442; 29, 249

In, 26, 169, 279; 27, 243; 29, 120, 185; 30A, 209
In_4K, 34A, 140
In_3La, 30A, 147
$InLi_3O_3$, 35A, 225
$InLiO_8Rh_4$, 31A, 245
In_3Lu, 29, 121; 30A, 147
In-Mg, 28, 47; 30A, 210
$InMg_2$, 28, 23
$InMg_3$, 28, 23
In_2Mg_5, 28, 23
$InMgNi_2$, 27, 12, 261
In-Mn, 27, 243
$InMn_3$, 27, 243
$In_{0.1-0.4}Mn_{3.9-3.6}N_{1.04-1.00}$, 27, 273
$InMnNi_2$, 26, 169
$InMnPd_2$, 33A, 97
$In_2Mo_3O_{12}$, 33A, 331
In-N-Ni, 27, 306
$In_{21.9}N_{12.4}Ni_{65.7}$, 27, 306
$InNTi_2$, 29, 121
$InNTi_3$, 29, 121
$InNZr_2$, 29, 121
$InNaO_2$, 35A, 213
$InNaO_6Si_2$, 32A, 458
$InNaO_8W_2$, 35A, 253
$InNaS_2$, 26, 172
$InNaSe_2$, 26, 172
$InNb_3$, 27, 244; 29, 121
$InNbO_4$, 27, 522
In_3Nd, 30A, 147
InNi, 30A, 147
$InNi_3$, 30A, 147
In_3Ni_2, 30A, 147
$In_{7-x}Ni_3$, 30A, 61, 147
$In_{9.2}Ni_{12.8}$, 34A, 137
$InNi_2Ti$, 27, 12, 296
In_2O_3, 31A, 120, 260; 32A, 260; 34A, 239
$InO_4P.2H_2O$, 26, 464
InO_4Sb, 28, 348
InO_3Sc, 31A, 254
InO_4Ta, 27, 522
InO_3Tl, 31A, 254
InP, 27, 247; 28, 47
InP II, 29, 121
$In_{14}P_{14}Pd_{72}$, 35A, 120
$In_{14}P_{14}Pt_{72}$, 35A, 120
In-P-Se, 27, 247
In-Pb, 29, 185; 30A, 210
In-Pb-Te, 29, 186
In_3Pr, 30A, 147
$InPt_2$, 34A, 140
$InPt_3$, 26, 225; 34A, 140
In_2Pt_3, 34A, 140
In_9Pt_{13}, 34A, 137
$In_{14}Pt_{72}Si_{14}$, 35A, 120
In-Pu, 30A, 211
InPu, 30A, 147, 148
$InPu_3$, 30A, 148

In_3Pu, 30A, 147
In_4Rb, 34A, 140
$InRu_3$, 29, 121
In_3Ru, 29, 121
In_2S_3, 26, 117; 27, 247; 30A, 62, 148
β-In_2S_3, 26, 173; 27, 245
In_6S_7, 32A, 94
In_2S_4Zn, 27, 246; 29, 121; 34A, 93
$In_2S_6Zn_3$, 32A, 95
In-Sb, 26, 42; 28, 47; 29, 186; 30A, 211
InSb, 26, 42; 27, 42, 247; 28, 47; 29, 121, 182; 30A, 148, 212
InSb II, 29, 121
$InSb_3$, 30A, 148
In_3Sb_2, 30A, 148
In-Sb-Se, 26, 42
In-Sb-Se-Sn, 29, 186
InSbSn, 30A, 148
$InSbSn_2$, 30A, 148
$InSbSn_4$, 30A, 148
In-Sb-Sn-Te, 29, 186
In-Sb-Sn-Zn, 29, 186, 187
In-Sb-Te, 26, 43; 29, 187
In_4SbTe_3, 27, 248; 29, 121
$InSc_3$, 27, 245
InSe, 26, 171
In_2Se, 30A, 62, 148
In_2Se_3, 26, 42, 170; 27, 200, 247; 31A, 44
α-In_2Se_3, 26, 170
β-In_2Se_3, 26, 171
γ-In_2Se_3, 26, 171
In_5Se_6, 30A, 148
In_2SeSn, 29, 186
In-Se-Te, 26, 43
In_3Sm, 30A, 148
In-Sn, 29, 185, 188; 30A, 212
InSn, 29, 187
In-Sn-Te, 29, 188
In_2SnTe, 29, 186
In_2Sr, 29, 3, 121
In_3Tb, 29, 121; 30A, 149
$In_{1-x}Te$, 29, 188
InTe, 28, 47; 29, 121; 30A, 149
In_xTe, 30A, 212
In_2Te_3, 26, 42, 43; 27, 200, 244, 247
α-In_2Te_3, 26, 173
β-In_2Te_3, 26, 173
In_3Te_5, 29, 121
$InTe_2Tl$, 34A, 89
In_3Th, 30A, 149
In-Ti, 29, 185
$InTi_3$, 30A, 149
In_2Ti_3, 29, 121; 30A, 149
In_4Ti_3, 27, 248; 30A, 14, 149
In-Tl, 30A, 213
In_3Tm, 29, 121; 30A, 149
In_3Y, 28, 47; 29, 121; 30A, 149
In_2Yb, 29, 3, 121
In_3Yb, 29, 121; 30A, 149
InZr, 29, 121; 30A, 149
$InZr_2$, 29, 121; 30A, 149
$InZr_3$, 27, 379
In_2Zr, 30A, 149
In_3Zr, 30A, 149
Ir, 27, 250
Ir-*Ln*-O, 30A, 452
IrLu, 30A, 149
δ-$IrMg_3$, 27, 249
$IrMg_{7.477}$, 27, 249
IrMn, 30A, 213
IrMo, 30A, 150
σ-$IrMo_2$, 29, 121
$IrMo_3$, 28, 47
Ir_3Mo, 30A, 150
$Ir_{8.5}Mo_{21.5}$, 29, 86
Ir-Nb, 29, 70, 188, 189
$α_1$-IrNb, 29, 121, 188
$α_2$-IrNb, 29, 121, 188
σ-IrNb, 29, 188
α-$IrNb_3$, 29, 121, 188, 189
Ir_2Nb_3, 29, 86, 121
α-Ir_3Nb, 29, 121
$(Ir_xRh_{1-x})Nb_3$, 30A, 213
$Ir(Nb,V)_3$, 28, 47
IrP_2, 26, 270
IrP_3, 26, 48; 34A, 117
Ir-Pd, 29, 189
Ir-Re, 27, 337
Ir-Ru, 29, 189
$IrSb_2$, 26, 35
$IrSb_3$, 26, 48, 176
IrSc, 27, 342
Ir-Si, 27, 373
IrSi, 27, 373; 35A, 120
$IrSi_{\sim 1.5}$, 27, 373; 35A, 120
$IrSi_2$, 27, 373
$IrSi_3$, 35A, 120
Ir_2Si, 27, 373
Ir_3Si, 27, 373
Ir_3Si_2, 27, 373
Ir_3Si_4, 35A, 70, 120
Ir_4Si_5, 35A, 120
Ir-Ta, 28, 47
IrTa, 28, 47
Ir_3Ta, 28, 47
Ir_2Tb, 28, 47
Ir_3Te_8, 32A, 127
IrTh, 27, 367
Ir_2Th, 26, 236
Ir_3Th_7, 26, 256
Ir_5Th, 26, 239
IrTi, 27, 250; 29, 121
δ′-IrTi, 29, 121
$IrTi_3$, 27, 250
Ir_3Ti, 27, 250
γ-Ir_3Ti, 29, 121
$\sim Ir_{40}Ti_{60}$, 29, 121
$Ir_{45}Ti_{55}$, 29, 121
$\sim Ir_{55}Ti_{45}$, 29, 121
IrTm, 30A, 150
IrU_2, 30A, 150
Ir_3U, 26, 165
Ir-V, 30A, 150
IrV, 30A, 64, 150
IrW, 30A, 150
$IrZr_3$, 29, 122
Ir_2Zr, 26, 237
$IrZr_{2+x}$(O?), 29, 121

K, 26, 3, 279
$K_2LaNb_5O_{15}$, 28, 330
$K_2LiO_9P_3.H_2O$, 27, 594
$KLiO_4S$, 27, 605
$K_2Ln_2O_6Ta_2$, 31A, 248
$KLnS_2$, 30A, 453
$K_2M'M''N_6O_{12}$, 31A, 248
KMO_9P_3, 31A, 245
$KMgO_9P_3$, 31A, 246; 32A, 379; 33A, 413
$K_2MgO_8S_2.6H_2O$, 30A, 375
$KMnO_4$, 32A, 283
K_2MnO_4, 32A, 284
$KMnO_9P_3$, 31A, 246
$K_2Mn_2O_{12}S_3$, 29, 424
$K_2MnO_8S_2.4H_2O$, 26, 447; 33A, 369
$K_2Mn_2O_{15}Si_4Zn_4$, 34A, 386
$K_{0.26}MoO_3$, 30A, 341
$K_{0.28}MoO_3$, 31A, 154
K_2MoO_4, 34A, 277; 35A, 232
$K_2Mo_2O_{11}.4H_2O$, 33A, 330
$K_2Mo_3O_{10}$, 31A, 154; 32A, 509; 33A, 333
KMo_2O_8Sc, 35A, 254
KMo_2O_8Y, 32A, 303
$KMoS_4$, 30A, 453
KNO_2, 30A, 454, 458; 31A, 181; 34A, 346
$KNO_2.5H_2O$, 28, 302
$KNO_2.8H_2O$, 28, 302
KNO_3, 27, 565; 28, 302; 34A, 338
KNO_3Os, 33A, 271
$K_{1-x}NO_3Rb_x$, 29, 439; 30A, 454
$K_2NO_7S_2$, 33A, 386
$K_3NO_9S_3.2H_2O$, 35A, 379
K-Na-Nb-O-Ta, 27, 504
$K_nNa_{2-n}O_4S$, 28, 326
$K_3NaO_8S_2$, 27, 605
$K_{1.70}Na_{1.30}O_{15}Si_6Zr$, 30A, 430
$KNbO_3$, 32A, 271
$KNbO_3.4.5H_2O$, 29, 420
K-Nb-O-Pb-Ti, 27, 504
$KNbO_5Ti$, 29, 327

$KNbO_9Ti_3$, 29, 327
$KNbO_9W_2$, 28, 334
$K_{2/3}NbS_2$, 35A, 120
$K_{2/3}NbSe_2$, 35A, 120
$K_{0.49}Nd_{1.17}O_8W_2$, 34A, 282
$K_{0.14}NiO_2$, 34A, 293
$K_{0.23}NiO_{1.8}$, 34A, 293
$K_2NiO_8S_2.6H_2O$, 26, 449
KO_2, 27, 477
KO_3, 28, 118
$K_2O.3\cdot1\text{-}6\ H_2O$, 26, 403
$K_2O_4Os.2H_2O$, 26, 400
KO_3P, 27, 592
$(KO_3P)_x$, 28, 187; 34A, 322
$KO_{12}PPb_3S_2$, 31A, 246
$KO_{12}PS_2Sr_3$, 31A, 246
KO_4PSr, 26, 469
$K_2O_{12}P_4Sr$, 31A, 245; 32A, 382
$KO_{12}P_3Th_2$, 31A, 259; 33A, 398
KO_9P_3Zn, 31A, 246
$K_2O_8PbSe_2$, 30A, 453
$K_2O_7Pb_2Si_2$, 28, 310
K-O-Pb-Ti, 30A, 457
K_2O_3Pr, 31A, 253
K_xO_3Re, 31A, 260
K_2O_4S, 26, 436; 27, 605; 28, 291; 30A, 458; 31A, 244; 33A, 367
$K_2O_{16}S_5$, 34A, 298
$K_2O_8S_2Sr$, 27, 635
$K_4O_{16}S_4Th.2H_2O$, 28, 214, 326
$K_3O_{14}Sb_5$, 31A, 135
K_2O_4Se, 27, 605; 35A, 381
$K_2O_7Si_2Zr$, 35A, 310
K_2O_3Sn, 28, 350; 34A, 258; 35A, 223
$K_2O_7Sn_3$, 29, 421
KO_4SrV, 26, 470
$K_2O_{11}Ta_4$, 29, 420
KO_9TaW_2, 28, 334
KO_2Tb, 31A, 253
K_2O_3Tb, 31A, 253
KO_4Tc, 27, 671
KO_5TeV, 34A, 318
K_2O_3Th, 31A, 245, 253
$K_{0.13}O_2Ti$, 26, 352
K_2O_3Ti, 30A, 457
$K_2O_5Ti_2$, 26, 373
$K_2O_{13}Ti$, 27, 529
$K_3O_{17}Ti_8$, 35A, 276
KO_2Tl, 26, 400
KO_3U, 27, 543
K_2O_4U, 26, 404
α-K_2O_4U, 26, 403
β-K_2O_4U, 26, 403
α-$K_2O_7U_2$, 26, 406
$K_2O_{12}U_2V_2$, 30A, 348
KO_8V_3, 31A, 141
K_3O_8V, 27, 659
$K_2O_{28}V_{10}Zn_2.16H_2O$, 31A, 142
$K_{0.30}O_3W$, 26, 356
$K_{0.45}O_3W$, 26, 357
$K_{0.52}O_3W$, 26, 357
K_2O_4W, 34A, 282; 35A, 232
$K_2O_{11}W_2.4H_2O$, 29, 332
$K_2O_{13}W_4$, 32A, 509
KO_8W_2Y, 33A, 345
$K_4O_6Xe.9H_2O$, 29, 335
K_2O_2Zn, 33A, 325
K_2O_3Zr, 28, 350; 34A, 258; 35A, 222
$K_2O_5Zr_2$, 34A, 257; 35A, 238
K_3P, 26, 53
KPb, 29, 122
K_2S_2, 27, 350
KS_2Sn, 34A, 140
$K_{2/3}S_2Ta$, 35A, 120
K_3S_4V, 29, 82, 122
KS_4W, 30A, 453
KSb, 26, 44
$K_{2/3}Se_2Ta$, 35A, 120
KSi, 26, 84, 156
KSi_6, 26, 157
$K_{6.8}Si_{45.3}$, 33A, 162
K_8Si_{46}, 32A, 117; 34A, 122
KSn, 29, 83, 122
K_8Sn_{46}, 34A, 122
Kr, 26, 286

La, 26, 234; 29, 122, 189
β-La, 26, 3
$LaLiO_2$, 30A, 452; 32A, 509
La_2Mg_{17}, 28, 47
$La_2Mg_3N_{12}O_{36}.24H_2O$, 26, 475; 27, 724
La_2MgS_4, 27, 327
$La_{1.0-0}MnO_{\sim3.0}Sr_{0-1.0}$, 26, 418
La-Mn-S, 27, 328
La_2MnS_4, 27, 327
$La_6MnS_{14}Si_2$, 35A, 77
LaN, 27, 304; 28, 47
$LaNa_2Nb_5O_{15}$, 28, 330
$LaNaS_2$, 29, 341
$LaNaSe_2$, 30A, 150
$LaNbO_4$, 27, 521; 28, 346
$LaNb_3O_9$, 29, 421; 32A, 293
$La_2Nb_2O_7$, 29, 329
$La_2Nb_4O_{13}$, 31A, 257
$LaNbO_6Ti$, 27, 537; 35A, 241
LaNi, 29, 122; 30A, 150
$LaNi_3$, 34A, 140; 35A, 120
La_2Ni_7, 34A, 140; 35A, 95
$LaNiSi_2$, 35A, 120
La_2O_3, 26, 364; 28, 344; 30A, 309
$La_2O_9Pb_6$, 29, 421
$La_4O_{19}Re_6$, 33A, 353
LaO_4Rh_2, 26, 399
LaO_4Ru_2, 26, 399
$La_4O_4Se_3$, 35A, 235
$La_2O_7Si_2$, 35A, 302
$La_2O_7Sn_2$, 33A, 286
$La_2O_7(Sn,Ti)_2$, 33A, 287
LaO_6Sr_2Ta, 28, 332
$La_{0.33}O_3Ta$, 33A, 341
LaO_4Ta, 28, 332; 31A, 248
LaO_9Ta_3, 32A, 293
$La_2O_7Ta_2$, 29, 329
La_2O_2Te, 27, 424; 32A, 308
La_2O_5Ti, 31A, 138
$LaO_{3.56}U$, 27, 545
LaO_4U, 27, 545
$LaO_{4.32}U$, 27, 545
$La_6O_{12}W$, 29, 420
La-O-Yb, 26, 364
LaO_3Yb, 26, 364
$La_2O_7Zr_2$, 27, 535; 28, 332; 35A, 245
La_4Pb_3, 33A, 152
La_5Pb_3, 30A, 150
La_5Pb_4, 34A, 137
$LaPd_3$, 26, 165; 30A, 150
La-Pt, 27, 322
LaPt, 30A, 150
$LaPt_2$, 26, 236
$LaPt_5$, 26, 239; 27, 322
La-Pu, 26, 231
LaRh, 30A, 150
$LaRh_3$, 34A, 140
La_2Rh_7, 34A, 140
La_4Rh_3, 34A, 140; 35A, 77
$LaRu_2$, 30A, 214
LaS, 29, 190
La_2S_3, 34A, 98, 294
La-S-Sr, 27, 328
La_2S_4Sr, 27, 327
La_2S_4Yb, 29, 122
$LaSb_2$, 32A, 20
La_2Sb, 35A, 12
$LaSe_2$, 32A, 20
La_2Se_3, 29, 122
La_3Se_4, 30A, 150, 215
La_2Se_4Sr, 29, 122
LaSi, 29, 122; 30A, 151
$LaSi_{2-n}$, 30A, 151
$LaSi_2$, 29, 122; 30A, 151
La_3Si_2, 30A, 150
La_5Si_3, 30A, 150
$LaSn_3$, 30A, 151
LaTe, 27, 335; 30A, 151
$LaTe_{2-x}$, 30A, 215

$LaTe_2$, 28, 47; 29, 64, 122; 30A, 64, 151; 31A, 47; 32A, 20
$LaTe_3$, 27, 336; 30A, 151, 215
La_2Te_3, 28, 47; 29, 122; 30A, 151; 31A, 47
La_2Te_5, 30A, 151
$La_{3-x}Te_4$, 30A, 215
La_3Te_4, 28, 47; 30A, 151
La_4Te_7, 29, 122; 30A, 215
La-Y, 27, 330; 29, 190
Li, 26, 3; 27, 264
$Li\mathit{Ln}S_2$, 30A, 453
$LiLuO_2$, 27, 516
$Li\mathit{M}_2O_4$, 29, 423
$LiMgO_4P$, 30A, 454
$LiMg_{0.5}O_4Ti_{1.5}$, 28, 346
$LiMgO_4V$, 28, 344
Li_2MgPb, 26, 185; 28, 28
$LiMg_2Tl$, 30A, 151
$LiMnO_2$, 34A, 291
$Li_{1.33}Mn_{1.67}O_4$, 28, 346
$Li_2MnO_7P_2$, 26, 309
$LiMnO_4Rh$, 28, 346
$Li_2MnO_8S_2$, 26, 309
$LiMnO_4Ti$, 28, 346
$Li_2Mn_3O_8Zn$, 29, 423
$Li_{6.5-4.0}Mn_{1.5-4.0}P_4$, 26, 188
$Li_{7.33-6.6}Mn_{0.67-1.4}P_4$, 26, 187
$Li_xMn_{1-x}Se$, 26, 189
$Li_{0.04}Mn_{0.96}Se_7Te_{1-y}$, 29, 190
$Li_xMn_{1-x}Te$, 29, 190
Li_9MoN_5, 26, 187
Li_2MoO_3, 31A, 245; 32A, 296
Li_2MoO_4, 27, 724
Li_2MoO_5, 31A, 245
Li_4MoO_5, 29, 316
$Li_6Mo_2O_7$, 32A, 296
LiN_3, 33A, 168
$LiNO_3$, 26, 472
$LiNO_3.3H_2O$, 26, 472
Li_9N_5W, 26, 187
$LiNaO_4S$, 32A, 328
Li-Nb-O, 30A, 452
$LiNbO_3$, 28, 354; 30A, 452; 31A, 147
$Li(Nb,Zn)O_4$, 28, 344
Li_3NbO_4, 28, 348; 29, 316; 30A, 331
$LiNbO_4Zn$, 28, 346
$Li_{2/3}NbS_2$, 35A, 120
$LiNdO_2$, 32A, 509
$LiNiO_4P$, 30A, 454
$LiNiO_4V$, 26, 409; 28, 346
$LiNi_2Si$, 34A, 140
Li_2O, 26, 187; 27, 718, 724; 34A, 238
Li_3O_4P, 32A, 355
Li_8O_6Pb, 29, 316
LiO_2Pr, 32A, 509
LiO_2Rh, 26, 390
LiO_4Rh_2, 26, 399
LiO_4RhTi, 27, 486
$LiO_4(Rh,Ti)$, 28, 346
Li_2O_4S, 27, 606
$Li_2O_4S.H_2O$, 28, 207; 30A, 363; 33A, 365
$Li_2O_6S_2.2H_2O$, 28, 306; 32A, 323
Li_3O_4Sb, 28, 348
Li_7O_6Sb, 34A, 273
$LiO_4(Sb,Zn)$, 28, 344
LiO_2Sc, 26, 370
$Li_2O_4Se.H_2O$, 27, 606
Li_2O_3Si, 33A, 452
$Li_2O_5Si_2$, 26, 506
Li_4O_4Si, 33A, 467
$Li_6O_7Si_2$, 34A, 363
LiO_2Sm, 28, 134, 308; 32A, 509; 35A, 214
Li_2O_3Sn, 28, 324; 31A, 256; 35A, 221
Li_8O_6Sn, 33A, 288; 34A, 263
Li-O-Ta, 30A, 452
LiO_3Ta, 31A, 249; 32A, 312
Li_3O_4Ta, 29, 316
Li_2O_3Tb, 31A, 253
Li_2O_3Te, 30A, 454
LiO_5TeV, 34A, 318
LiO_2Ti, 27, 527
Li_2O_3Ti, 34A, 254
$Li_xO_8Ti_{4-x/4}$, 29, 318
$LiO_4(Ti,V)$, 28, 344
$LiO_4Ti_{1.5}Zn_{0.5}$, 28, 346
$Li_2O_8Ti_3Zn$, 28, 348
α-LiO_2Tl, 26, 401
β-LiO_2Tl, 26, 401
γ-LiO_2Tl, 26, 401
Li_3O_3Tl, 26, 401
LiO_2Tm, 27, 516
LiO_3U, 27, 543; 30A, 347
α-Li_2O_4U, 26, 404
$Li_2O_{11}U_4$, 27, 543
Li_3O_4U, 29, 316
Li_4O_5U, 27, 544; 29, 340; 31A, 257
$Li_{0.27}OV_{0.73}$, 27, 515
$Li_{0.3}O_5V_2$, 29, 302
$Li_{0.48}O_5V_2$, 35A, 274
LiO_2V-OV, 27, 515
LiO_5V_2, 30A, 343
LiO_8V_5, 29, 326
Li_3O_4V, 29, 316
$LiO_{7.5}V_2W$, 31A, 247
LiO_4VZn, 26, 417; 28, 344, 346
$Li_{0.365-0.394}O_3W$, 26, 357
Li_2O_4W, 26, 391; 27, 724; 33A, 351
Li_6O_6W, 34A, 280
$Li_{14}O_{28}W_7.4H_2O$, 31A, 156
LiO_8W_2Yb, 35A, 249
$Li_2O_{12}W_3Zr$, 32A, 317
LiO_2Y, 27, 516
LiO_2Yb, 27, 516
LaO_3Yb, 34A, 253
Li_2O_3Zr, 30A, 452
Li-Pt, 27, 323
$LiPt_7$, 27, 323
LiRh, 28, 50; 30A, 65, 151
LiS_2Sn, 34A, 140
$Li_{2/3}S_2Ta$, 35A, 120
$Li_4Sb_4Sr_3$, 35A, 9
Li_2Si, 30A, 65, 152
Li_7Si_2, 30A, 66, 151
$Li_{10}Si_3$, 30A, 152
$Li_{22}Si_5$, 31A, 48
$Li_4Si_{10}Sr_7$, 34A, 140
Li_2SiZn, 34A, 140
Li_2Sn_5, 34A, 99
$Li_{22}Sn_5$, 29, 122
$LiSr_7$, 27, 265
$LiSr_8$, 27, 265
Li_2Sr_3, 27, 265
$Li_{23}Sr_6$, 27, 265; 28, 47
Ln_2MoO_6, 31A, 250
$\mathit{Ln}Mo_{0.5}O_3Ti_{0.5}$, 29, 423
$\mathit{Ln}NaO_2$, 31A, 250
Ln_2O_3, 31A, 248
$\mathit{Ln}O_4P.H_2O$, 30A, 454
$\mathit{Ln}O_{12}Re_3.4H_2O$, 30A, 458
$\mathit{Ln}O_2Se_2$, 31A, 241
Ln_2O_4Sr, 29, 421; 30A, 452
$\mathit{Ln}O_3Ti_{0.5}W_{0.5}$, 29, 423
$\mathit{Ln}_{0.1}O_3W$, 31A, 248
$\mathit{Ln}Sb$, 28, 47
$\mathit{Ln}Si_{2-x}$, 28, 48
Lu, 26, 234
β-LuMg, 29, 190
Lu_5Mg_{24}, 29, 122
Lu_2MgS_4, 29, 122
$LuMn_2$, 29, 122
$LuMn_5$, 31A, 48
Lu_6Mn_{23}, 29, 122; 30A, 152
$LuMnO_3$, 28, 138
Lu_2MnS_4, 29, 122
$LuNbO_4$, 27, 521
$LuNbO_6Ti$, 28, 326
LuNi, 29, 122; 30A, 152
$LuNi_2$, 30A, 152
$LuNiSi_2$, 35A, 121
Lu_2O_3, 27, 724
LuO_4P, 28, 304
$Lu_2O_7Sn_2$, 26, 399
LuO_4Ta, 29, 421
$Lu_6O_{12}U$, 31A, 162
LuO_4V, 28, 308
LuP, 29, 122
$LuPd_3$, 30A, 152
LuPt, 30A, 152
LuRh, 30A, 152
LuS, 29, 122

Lu_2S_3, 29, 122
$LuSb$, 28, 48; 29, 122
$LuSe$, 29, 123
Lu_2Se_3, 29, 122; 30A, 152
Lu_2Se_4Yb, 34A, 141
$LuSi$, 29, 123
$LuSi_2$, 27, 336; 28, 48
Lu_5Si_3, 30A, 152
$LuTc_2$, 29, 123
$LuTe$, 29, 123
Lu_2Te_3, 30A, 152
$LuZn_3$, 35A, 121
$Lu_{13}Zn_{58}$, 35A, 121

$M_2Mo_3O_{12}$, 30A, 452
$M'_2M''_xM'''_{8-x}O_{16}$, 31A, 253
$MO_4(P,Si)$, 32A, 359
MO_3Pb, 30A, 452
$M_5O_{12}R_3$, 30A, 324
$M_2O_{12}S_3.8H_2O$, 29, 424
MO_5TeV, 34A, 317
$M_2O_{12}W_3$, 30A, 453
σ-M_2Os, 29, 124
Mg, 27, 382
$MgMn_2O_4$, 27, 493; 30A, 457
$Mg_{0.5}MnO_4Ti_{0.5}Zn$, 26, 409
$MgMoO_4$, 28, 348
MgN_2O_6, 29, 426
$MgN_2O_6.6H_2O$, 26, 474
$MgN_6O_{18}Th.8H_2O$, 30A, 403
MgN_2Si, 35A, 136
$MgNa_6O_{16}S_4$, 29, 350
$Mg_7Na_{12}O_{52}S_{13}.15H_2O$, 35A, 441
$MgNa_4O_{28}V_{10}.24H_2O$, 31A, 242
$MgNb_2O_3$, 28, 330
$Mg_4Nb_2O_9$, 26, 389
$(Mg,Nb)O_3Sr$, 28, 330
$MgNb_{14}O_{45}Sr_9$, 28, 330
Mg-Nd, 26, 190
Mg_3Nd, 26, 190
Mg_9Nd, 26, 190
$Mg_{12}Nd$, 28, 48
$MgNd_2S_4$, 27, 327
$Mg_{0.3}Ni_{0.7}O_4S.6\frac{1}{2}H_2O$, 26, 449
MgO, 27, 545; 29, 417, 418; 30A, 450, 457
$Mg_2O_7P_2$, 26, 466; 30A, 385; 32A, 372
β-$Mg_2O_7P_2$, 26, 466
$Mg_3O_8P_2$, 33A, 394, 401
MgO_7P_2Sr, 33A, 407
$MgO_{21}Pb_8Si_6$, 29, 399
Mg-O-Pb-Ti-W, 27, 510
MgO_6Pb_2W, 27, 510,28, 310
Mg_2O_6PbW, 28, 310
MgO_4Rh_2, 26, 399
β-MgO_4S, 27, 606
$MgO_3S_2.6H_2O$, 34A, 314
$MgO_4S.4H_2O$, 27, 612; 29, 346
$MgO_4S.6H_2O$, 27, 618; 29, 347
$Mg_4O_9Sb_2$, 29, 316
$MgO_3Se.6H_2O$, 31A, 209
MgO_4Se, 29, 357
$MgO_4Se.6H_2O$, 27, 618; 29, 425
MgO_3Si, 34A, 365
γ-MgO_3Si, 26, 510
Mg_2O_4Si, 31A, 248
$(Mg,Zn)_2O_4Si$, 29, 428
$Mg_2O_6Si_2$, 26, 510
$MgO_{11}Si_2U_2.7H_2O$, 29, 410
Mg_2O_4Sn, 28, 324
$(Mg,Ta)O_3Sr$, 28, 332
MgO_6Sr_2W, 28, 308
MgO_6Ta_2, 26, 412
Mg_2O_4Tc, 29, 322
Mg_3O_6Te, 35A, 242
MgO_4Ti_2, 27, 528
MgO_6U_2, 27, 543
MgO_4V_2, 28, 286, 348
MgO_4W, 33A, 347; 34A, 285
$Mg_{14}P_{14}Pd_{72}$, 35A, 121
$Mg_{14}P_{14}Pt_{72}$, 35A, 121
MgP_2Si, 33A, 152; 34A, 141
Mg_2Pb, 30A, 152
$MgPd$, 26, 225; 27, 268
Mg_2Pd_3, 27, 268
Mg_3Pd, 27, 268
Mg_4Pd, 27, 268
Mg_5Pd_2, 33A, 110
Mg_6Pd, 27, 268
$Mg_{12}Pr$, 29, 123
$MgPr_2S_4$, 27, 327
Mg-Pt, 27, 322
$MgPt$, 26, 225
$MgPt_3$, 26, 225; 27, 322
$MgPt_7$, 27, 322
MgS, 26, 90; 27, 202; 30A, 152
MgS_4Sc_2, 29, 123
Mg_2S_4Si, 29, 123
MgS_4Sm_2, 27, 327
Mg_2S_4Sn, 29, 123
MgS_4Y_2, 27, 327
MgS_7Y_4, 26, 277
MgS_4Yb_2, 27, 327; 29, 123
$MgSc$, 30A, 152
$MgSe$, 30A, 152
Mg_2Se_4Si, 29, 123
Mg_2Se_4Sn, 29, 123
Mg_2Si II, 29, 123
Mg_2Sn II, 29, 123
Mg_4Sr, 30A, 24, 153
$Mg_{17}Sr_2$, 26, 175; 27, 64, 269; 28, 48
$Mg_{23}Sr_6$, 26, 176; 30A, 23, 153
β-Mg-Tb, 29, 191
Mg_2Tb, 29, 123
$Mg_{24}Tb_5$, 29, 123
$MgTe_2$, 34A, 140
$Mg_{23}Th_6$, 26, 176
Mg_2Tl, 35A, 62
β-$MgTm$, 29, 191
$Mg_{24}Tm_5$, 29, 123
Mg_2Y, 29, 123
$Mg_{24}Y_5$, 27, 195
Mg-Yb, 30A, 216
Mg_2Yb, 29, 123; 30A, 153, 216
$MgZn_2$, 35A, 8
Mn, 32A, 98
α-Mn, 34A, 101; 35A, 78
β-Mn, 27, 165
$MnMoO_4$, 28, 348; 30A, 327
$Mn_{47.5}Mo_{25.0}Si_{27.5}$, 26, 269
$Mn_{78}Mo_3Si_{19}$, 26, 270
Mn_2N, 33A, 111
η-Mn_3N_2, 27, 272
$Mn_4N_{0.8}$, 27, 272
$Mn_4N_{0.92}$, 27, 272
Mn_4N, 27, 272
ε-Mn_4N, 27, 271
$Mn_4N_{1.01}$, 27, 273
ζ-Mn_5N_2, 27, 271
θ-Mn_6N_5, 27, 272
MnN_2O_6, 29, 426
$Mn_{3.5-3.9}N_{0.97-1.00}Sn_{0.5-0.1}$, 27, 273
$MnNa_{12}Nb_{12}O_{38}.50H_2O$, 34A, 271
$MnNaO_4P$, 30A, 454
$Mn_2Na_2O_7Si_2$, 29, 428; 32A, 440; 34A, 384
$Mn_4Na_4O_{18}Ti_5$, 33A, 301
$Mn(Nb,Ta)_2O_6$, 31A, 244
$Mn_4Nb_2O_9$, 26, 389
$MnNb_3S_6$, 35A, 111
$Mn_{1/4}NbSe_2$, 35A, 121
$Mn_{1/3}NbSe_2$, 35A, 121
$Mn_{1/2}NbSe_2$, 35A, 121
$MnNbSi$, 32A, 140; 33A, 86; 35A, 74
$Mn_{79}Nb_5Si_{16}$, 26, 270
Mn_2Nd, 29, 123
$Mn_{23}Nd_6$, 30A, 153
$MnNdO_3$, 33A, 318
Mn-Nd-S, 27, 328
$MnNd_2S_4$, 27, 327
Mn-Ni, 27, 165; 30A, 217
$MnNi$, 30A, 153, 217
Mn_2NiO_4, 30A, 458
$Mn_{2+x}Ni_{1-x}O_4$, 27, 495
$MnNiO_4Tc$, 29, 322
$MnNi_xO_4V_{2-x}$, 34A, 265
$(Mn,Ni)_2O_4V$, 30A, 451
Mn-Ni-O-Zn, 30A, 451
Mn-Ni-Si, 27, 165
$(Mn,Ni)Si$, 27, 374
$MnNi_{1.1}Si_{0.9}$, 27, 376

$MnNi_{1.55}Si_{0.45}$, 27, 376
Mn_3Ni_2Si, 27, 376; 28, 28
η'-Mn_3Ni_2Si, 27, 166
$Mn_6Ni_{16}Si_7$, 26, 175; 27, 376
$(Mn,Ni)_7Si_3$, 27, 376
MnO, 27, 474; 29, 417; 30A, 450
MnO_2, 28, 289
β-MnO_2, 26, 357
Mn_2O_3, 27, 497; 32A, 256
Mn_3O_4, 26, 407; 27, 493
Mn_5O_8, 32A, 285
$Mn_2O_7P_2$, 26, 466
$Mn_3O_8P_2$, 33A, 401; 34A, 330
$MnO_{11}Pb_4Si_3$, 29, 428
$MnO_{21}Pb_8Si_6$, 29, 399; 31A, 220
$Mn_2O_9Pb_2Si_2$, 27, 714
MnO_6Pb_2Ta, 28, 332
MnO_3Pr, 33A, 318
$Mn_{0.75}O_4Rb_{0.75}Ti_{1.25}$, 33A, 303
MnO_4Rh_2, 26, 399
MnO_4S, 26, 436; 30A, 369
$MnO_4S.H_2O$, 31A, 199
$(Mn,Zn)_{14}O_{29}Sb_2Si_4Zn_2$, 31A, 223
$MnO_3Se.2H_2O$, 31A, 243
MnO_4Se, 29, 357; 33A, 388
Mn_2O_4Si, 26, 515
$MnO_7Si_2Sr_2$, 26, 523
Mn_2O_4Sn, 33A, 285
$Mn_2O_8Sr.3H_2O$, 31A, 159, 248
MnO_6Sr_2Ta, 28, 328
$Mn_4O_9Ta_2$, 26, 389
MnO_3Tb, 33A, 318
Mn_2O_4Tc, 29, 322
$Mn_2O_8Te_2$, 30A, 458
$(Mn,Zn)_2O_8Te_3$, 27, 716
MnO_4Ti_2, 26, 408; 27, 528
Mn_2O_4Ti, 33A, 320
α-Mn_2O_4Ti, 29, 325
β-Mn_2O_4Ti, 29, 325
Mn-O-Ti-V, 30A, 457
MnO_3Tm, 28, 138
$MnO_{3.70}U$, 26, 406
$MnO_{3.84}U$, 26, 406
MnO_4U, 26, 406; 27, 523; 31A, 159
$MnO_{10}U_3$, 28, 334
MnO_4V_2, 27, 490; 28, 286
$Mn_2O_7V_2$, 32A, 272
MnO_4W, 31A, 260; 32A, 315
MnO_3Y, 28, 138
MnO_3Yb, 28, 138
Mn_2O_4Zn, 27, 493
$Mn_{3-x}O_4Zn_x$, 31A, 250
MnP, 27, 319; 30A, 98, 153
Mn_2P, 27, 274
Mn_3P, 27, 274
ϵ-Mn_3P, 27, 102
$MnPd_3$, 27, 274; 30A, 153, 217
Mn_2Pd_3, 28, 48; 29, 123
$MnPd_2Sb$, 33A, 97
$MnPd_2Sn$, 33A, 97
Mn_2Pr, 29, 123
$Mn_{23}Pr_6$, 30A, 153
Mn-Pr-S, 27, 328
$MnPr_2S_4$, 27, 327
Mn-Pt, 29, 191
$MnPt$, 30A, 217
$MnPt_3$, 27, 254
Mn_3Pt, 30A, 153
Mn-Re, 26, 192
$Mn_{55}Re_{45}$, 33A, 152
$Mn_{70}Re_{30}$, 33A, 152
$MnRh$, 28, 48
MnS, 26, 90, 251; 27, 202; 28, 48; 30A, 217
MnS_2, 30A, 218
MnS_4Sc_2, 29, 123
Mn_2S_4Si, 29, 123; 30A, 441
Mn-S-Sm, 27, 328
MnS_4Sm_2, 27, 327
MnS_7Tm_4, 29, 123
Mn-S-Y, 26, 90
MnS_4Y_2, 27, 327
MnS_7Y_4, 26, 277; 29, 123
MnS_4Yb_2, 27, 327; 29, 123; 32A, 104
MnS_7Yb_4, 29, 123
Mn_2Sb, 28, 48
Mn_2Sc, 26, 237; 29, 123
$MnSe$, 29, 123
Mn_2Se_4Si, 29, 123
$MnSe_{1-x}Te_x$, 29, 191
$MnSi$, 27, 375
$MnSi_{1.25}$, 27, 374
$MnSi_{1.72}$, 29, 124
$MnSi_2$, 27, 375
Mn_4Si_7, 29, 124; 34A, 101
Mn_5Si_3, 27, 374; 29, 124; 32A, 102
Mn_6Si, 29, 124; 30A, 153
$Mn_{11}Si_{19}$, 29, 64, 123; 32A, 86
$Mn_{15}Si_{26}$, 32A, 99
$Mn_{79}Si_{16}Ta_5$, 26, 270
Mn_2Si_2Th, 30A, 153
$MnSi_4Ti$, 29, 124
$Mn_{27}Si_{66}Ti_7$, 30A, 153
$Mn_{50}Si_{14}Ti_{30}$, 26, 269
$Mn_{79}Si_{19}Ti_2$, 26, 270
$Mn_5Si_3U_4$, 29, 124
Mn_3SiW_2, 26, 269
$Mn_{78}Si_{19}W_3$, 26, 270
$Mn_{23}Sm_6$, 29, 124; 30A, 153
Mn-Sn, 27, 259
$MnSn_2$, 26, 192; 28, 48
$Mn_{1.5-1.8}Sn$, 26, 193; 27, 259
$Mn_{1.74}Sn$, 30A, 153
~Mn_2Sn, 27, 259
$Mn_{3.67}Sn$, 26, 192
$Mn_{12}Tb$, 31A, 48
$Mn_{23}Tb_6$, 30A, 153
Mn-Tc, 27, 361
$MnTe$, 26, 151; 29, 124
Mn_2Th, 29, 149
$Mn_{12}Th$, 33A, 132
$Mn_{0.83}Ti_{0.17}$, 26, 193
~$MnTi$, 27, 278
Mn_2Ti, 27, 278
Mn_3Ti, 26, 193
$MnTm$, 27, 196
Mn_2Tm, 26, 192; 27, 264; 29, 124
$Mn_{12}Tm$, 31A, 48
$Mn_{23}Tm_6$, 29, 124; 30A, 153, 154
MnV, 27, 279; 28, 48
Mn_4V, 27, 279
Mn_2Y, 26, 236; 27, 279
Mn_4Y, 27, 279
$Mn_{12}Y$, 27, 279
$Mn_{23}Y_6$, 30A, 154
β-$MnZn$, 29, 124
β_1-$MnZn$, 29, 124
$MnZn_3$, 27, 271
α'-$MnZn_3$, 29, 124
ϵ'-$MnZn_3$, 29, 124
$MnZn_{13}$, 27, 173
Mo, 27, 129
α-Mo, 27, 131, 281
$Mo_{0.8}N$, 28, 32
δ'-MoN, 26, 197
δ''-MoN, 26, 198
$MoNa_2O_4$, 27, 724; 31A, 250
$MoNa_2O_4.2H_2O$, 34A, 276, 281
$Mo_2Na_2O_7$, 32A, 298
$Mo_6Na_{0.9}O_{17}$, 31A, 153
$MoNaO_6P$, 27, 582
$MoNaO_6V$, 33A, 336
$Mo_xNa_xO_5V_{2-x}$, 35A, 275
Mo-Nb, 30A, 218
$MoNb$, 27, 319
$MoNbO_4$, 27, 521
$Mo_{1.5}Nb_{0.5}O_4$, 27, 521
$Mo_{1.03}Nb_{1.03}S_3$, 35A, 121
$MoNbSe_3$, 35A, 80, 121
$(Mo,Nb)Si_2$, 30A, 220
$(Mo,Nb)_5Si_3$, 30A, 220
Mo-Nb-U, 29, 191
$Mo_3Nd_2O_{12}$, 34A, 278
Mo-Ni, 26, 197; 28, 29; 29, 192
γ-$MoNi_3$, 26, 197
$MoNi_4$, 29, 124
β-$MoNi_4$, 26, 197
$MoNiO_4$, 27, 522; 28, 348

$MoNiO_6Sr_2$, 26, 420; 27, 509
$MoNiSi$, 27, 376
Mo_2Ni_3Si, 26, 269
$Mo_{58}Ni_{24}Si_{18}$, 30A, 154
MoO_2, 32A, 259
MoO_3, 26, 355, 391; 28, 128, 320
$(Mo,V)_2O_5$, 31A, 252
Mo_4O_{11}, 28, 125; 29, 309
Mo_5O_{14}, 29, 308
Mo_8O_{23}, 29, 308
$Mo_{17}O_{47}$, 28, 127
$Mo_{18}O_{52}$, 29, 308
MoO_5P, 29, 372; 35A, 337
MoO_8P_2, 27, 597
$Mo_2O_{11}P_2$, 27, 591
$Mo_{12}O_{40}P$, 27, 672
$(Mo,Sn)O_3Pb$, 30A, 452
MoO_4Pb, 27, 522; 30A, 328
MoO_4Rb_2, 35A, 233
$Mo_3O_{12}Sc_2$, 30A, 452
$Mo_3O_{12}Se_2$, 31A, 249
MoO_4Sr, 32A, 302
MoO_6Sr_2Zn, 27, 509
Mo_2O_8Th, 28, 308; 31A, 254; 33A, 338
$Mo_{0.5}O_4Ti_{1.5}$, 27, 521
MoO_4Ti, 27, 521
Mo-O-U, 30A, 453
MoO_5U, 30A, 453
MoO_6U, 29, 420; 30A, 454
$MoO_6U.2H_2O$, 28, 225
$MoO_6U.4H_2O$, 26, 391
Mo_2O_8U, 30A, 453; 31A, 254; 32A, 305
$Mo_{0.6}O_5V_{1.4}$, 32A, 273
MoO_5V, 31A, 145
MoO_8V_2, 31A, 145
$Mo_4O_{25}V_6$, 29, 419
$Mo_6O_{40}V_9$, 29, 419
$MoO_{36}W_{11}$, 26, 355
$MoO_{45}W_{14}$, 26, 355
MoO_4Zn, 28, 348; 32A, 299
$Mo_3O_8Zn_2$, 31A, 155
Mo_2O_8Zr, 31A, 256
Mo_2Os, 29, 124
Mo_3Os, 27, 282
$Mo_{19.5}Os_{10.5}$, 29, 86
Mo-Os-W, 30A, 219
$Mo_3(Os,W)$, 30A, 219
MoP, 28, 48
MoP_2, 28, 33
Mo_3P, 28, 48; 30A, 69, 154
Mo_4P_3, 30A, 67, 154
Mo-Pd, 29, 192; 30A, 219
ε-$MoPd$, 29, 124
$MoPt$, 30A, 154
γ-$MoPt_2$, 29, 124
β-$MoPt_3$, 29, 124
ε-Mo_3Pt, 29, 124
δ-Mo_3Pt_2, 29, 124
Mo-Pu-U, 26, 199
~$MoPu_3U_{12}$, 26, 199
$MoRbS_4$, 30A, 453
$MoRe$, 30A, 220
Mo-Re-Si, 29, 193
$(Mo,Re)Si_2$, 29, 193
MoS_2, 26, 199; 29, 68, 124, 192; 30A, 154
Mo_2S_3, 26, 199
$Mo_{2.06}S_3$, 35A, 79, 121
Mo_2SSe_3, 30A, 154
Mo_2S_3Se, 30A, 154
Mo_2STe_3, 30A, 154
Mo_2S_3Te, 30A, 154
Mo_3Sb_7, 30A, 154; 31A, 18
$MoSe_2$, 27, 342; 28, 48; 30A, 154
$(Mo,W)Se_2$, 30A, 220
Mo_3Se_4, 35A, 80, 121
$Mo_{x/2}Se_2Ta_{1-x}W_{x/2}$, 28, 48
$Mo(Se,Te)_2$, 30A, 220
$(Mo,W)(Se,Te)_2$, 30A, 220
Mo_2SeTe_3, 30A, 154
Mo_2Se_3Te, 30A, 154
$MoSi_2$, 27, 284; 29, 192, 193; 30A, 154
Mo_3Si, 27, 213, 284
$(Mo,V)_3Si$, 27, 284
$(Mo,V)_5Si_3$, 27, 284
$(Mo,W)_5Si_3$, 27, 284
$MoSn_{1.5}$, 30A, 155
Mo_3Sn, 29, 124
Mo-Tc, 27, 285, 361
$MoTc$, 28, 48
$MoTc_3$, 27, 285
$MoTe_2$, 26, 202; 27, 343; 30A, 155; 31A, 69
$(Mo,W)Te_2$, 30A, 220
Mo_3Te_4, 35A, 80, 121
β-Mo-Ti, 26, 202
α'-Mo-U, 26, 203
α"$_b$-Mo-U, 31A, 49
α"-Mo-U, 26, 203
γ°-Mo-U, 26, 203
γ^s-Mo-U, 26, 203
Mo-W, 30A, 221

N, 28, 296
α-N, 26, 287
N_2, 35A, 129
α-N_2, 29, 229
β-N_2, 27, 393
N_3Na, 28, 291; 33A, 168
$NNaO_2$, 26, 473; 27, 572; 30A, 458
$NNaO_3$, 27, 564; 34A, 338
$NNa_3O_7S.H_2O$, 32A, 332
δ-N-Nb, 29, 193
$N_{0.40-0.50}Nb$, 26, 215
$N_{0.75-0.79}Nb$, 26, 215
$N_{0.88-0.98}Nb$, 26, 215
$NNb_{0.987}$, 26, 216
NNb, 30A, 155
NNb_2, 30A, 155
N_3Nb_4, 30A, 155
N_xNb_2Ni, 29, 124
$N_yNb_xO_{1-y}$, 26, 293
$N_{0.75-0.82}Nb_{0.69-0.76}O_{0.25-0.19}$, 26, 294
$N_{0.81}Nb_{0.82}O_{0.19}$, 26, 294
N_xNb_2Zn, 30A, 155
NNi, 30A, 155
NNi_3, 27, 289
NNi_4 I, 27, 289
NNi_4 II, 27, 289
N_2NiO_6, 27, 571; 29, 426
$N_2NiO_6.4H_2O$, 28, 302; 32A, 348
$N_2NiO_6.6H_2O$, 27, 571
N-Ni-Zn, 26, 219
N_2O, 26, 344
N_2O_2, 26, 347
N_2O_3, 34A, 240
N_2O_4, 26, 474; 28, 53; 31A, 258
N_2O_5, 27, 718
$(N,O)_2PTh_2$, 34A, 135
$N_{3x}O_yPb$, 26, 295
$N_3O_9Pr.6H_2O$, 26, 475; 29, 379
NO_3Rb, 27, 568
$N_3O_{11}RbU$, 30A, 406
$N_2O_2S_3$, 26, 345
$N_2O_{14}S_3$, 29, 344
$N_4O_3S_4$, 26, 294; 27, 410; 29, 245; 30A, 256
$NOSbTh_2$, 35A, 113
N_2OSi_2, 29, 244, 414
$N_4O_{12}Sn$, 32A, 350
$N_{0.38-0.83}O_{0.65-0.20}Ta_2$, 26, 294
$N_{1-x}O_xTa_2$, 26, 294
$NOTa$, 34A, 168
N_2OTh_2, 31A, 68
$N_4O_{12}Th.5H_2O$, 31A, 179
$N_4O_{12}Ti$, 31A, 176
NO_3Tl, 26, 476; 27, 569
$N_2O_8U.2H_2O$, 26, 477; 27, 570
$N_2O_8U.3H_2O$, 26, 477; 28, 180
$N_2O_8U.6H_2O$, 30A, 404; 32A, 355
$N_2O_6Zn.2H_2O$, 34A, 341
$N_4O_{12}Zn.4H_2O$, 35A, 324
N_2OZr_2, 27, 415; 29, 414
$N_2O_{11}Zr_7$, 27, 415; 29, 414
$N_4O_8Zr_7$, 27, 415; 29, 414
N_2PU_2, 34A, 135
NP_3V_5, 33A, 130
N_6Pb, 28, 55; 34A, 170
N_2PdS_6, 31A, 261
NPr, 27, 304
N_3Rb, 30A, 256
N-Ru, 26, 207

N_xRuZr_2, 29, 124
N_4S_4, 28, 54
N_2STh_2, 34A, 135
N_2SU_2, 34A, 135
N_2SbTh_2, 35A, 113
N_2SbU_2, 35A, 113
$(NSe)_x$, 26, 220, 295
N_4Se_4, 31A, 82
N_2SeTh_2, 34A, 135
N_2SeU_2, 34A, 135
N_4Si_3, 34A, 166
β-N_4Si_3, 26, 294
N_6Sr, 33A, 168
N-Ta, 26, 253; 27, 307
$N_{0.98}Ta_2$, 26, 294
$N_{0.76-1.06}Ta_2$, 26, 294
NTa, 35A, 121
NTa_{27}, 27, 307
N_2Ta, 30A, 155
N_5Ta_4, 34A, 168
N_6Ta_4, 34A, 137
N_6Ta_5, 34A, 137, 168
NTb, 27, 335
N_2TeTh_2, 35A, 113
$NTeTi_3$, 29, 125
N_2TeU_2, 35A, 113
NTh, 28, 48
N_4Th_3, 31A, 68
$N_{0.62}Ti$, 27, 308
$N_{0.95}Ti$, 29, 193
N_xTi, 30A, 221
$NTi_{0.85}$, 26, 264
NTi, 27, 308
NTi_2, 34A, 115
ε-NTi_2, 27, 308
N_xTi_2Zn, 29, 125
NTm, 27, 335
NU, 28, 48
N_2U, 29, 194
N_3U_2, 29, 193
$N_{3+x}U_2$, 32A, 137
δ-N-V, 29, 194
NV_3Zn_2, 29, 125
N-W, 26, 220
$N_{0.85}W$, 27, 309
$NW_{0.5}$, 26, 222
$NW_{0.6}$, 26, 222
$NW_{0.64}$, 26, 222; 28, 32
$NW_{1.1}$, 28, 32
$NW_{1.15-1.35}$, 26, 222
$NW_{1.17}$, 26, 222
NW_2, 26, 222
N_2W_3, 26, 221
γ-N_4W_3, 31A, 54
N_xZnZr_2, 29, 125
$N_{0.93}Zr$, 29, 194
$N_{1-x}Zr$, 26, 74
NZr_x, 29, 194
Na, 26, 3, 279; 29, 125
Na-Nb-O, 30A, 452
$NaNbO_3$, 26, 381; 27, 503; 28, 330; 30A, 452; 31A, 148; 34A, 268
$NaNb_{13}O_{33}$, 30A, 448
$Na_2Nb_4O_{11}$, 35A, 257
$Na_{2/3}NbS_2$, 35A, 121
$Na_{2/3}NbSe_2$, 35A, 121
$NaNdS_2$, 29, 341
$NaNdSe_2$, 30A, 155
$NaNi_{0.5}O_2Sn_{0.5}$, 27, 518
$NaNi_{0.5}O_2Ti_{0.5}$, 27, 518
NaO_3, 27, 482; 29, 416
$(NaO_3P)_x$, 26, 461; 28, 186, 293, 304; 33A, 411
$Na_3O_9P_3$, 30A, 387
$Na_3O_9P_3.H_2O$, 30A, 387
$Na_4O_7P_2.10H_2O$, 29, 360; 32A, 370
$Na_5O_{10}P_3$, 29, 360
$Na_4O_{12}P_4.4H_2O$, 26, 455; 29, 365
$Na_5O_{10}P_3.6H_2O$, 31A, 259
$Na_5O_8P_3.14H_2O$, 34A, 321
$Na_6O_{18}P_6.6H_2O$, 30A, 393
$NaO_{12}PPb_4SSi$, 28, 324
$NaO_{25}P_5Pb_9Si$, 28, 324
$Na_5O_{13}PSi_2Ti_2$, 27, 689
NaO_4PSr, 26, 469
$NaO_{12}P_3Th_2$, 30A, 454
$NaO_{12}P_3Ti_2$, 33A, 397
$NaO_{12}P_3U_2$, 30A, 454
NaO_6PW, 27, 582
$NaO_{12}P_3Zr_2$, 33A, 397
Na_6O_5Pb, 34A, 264
Na_2O_3Pd, 31A, 253
Na_xO_3Re, 31A, 260
$Na_2O_3S.7H_2O$, 28, 306
$Na_2O_3S_2$, 26, 435
$Na_2O_3S_2.5H_2O$, 27, 630; 33A, 159
Na_2O_4S, 27, 604; 28, 291
$Na_2O_4S.10H_2O$, 26, 433; 27, 620; 28, 306
$Na_2O_6S_2$, 28, 306
$Na_2O_6S_2.2H_2O$, 28, 220, 306; 32A, 323; 33A, 159
$Na_2O_6S_4.2H_2O$, 29, 345
$Na_3O_7ScSi_2$, 34A, 376
NaO_4ScSn, 33A, 294, 295
NaO_4ScTi, 33A, 294, 295
NaO_4ScZr, 33A, 294, 295
Na_2O_4Se, 28, 222; 35A, 381
Na_2O_3Si, 32A, 448
$Na_2O_3Si.5H_2O$, 31A, 241
$Na_2O_3Si.6H_2O$, 31A, 241; 32A, 449
$Na_2O_3Si.8H_2O$, 31A, 241
$Na_2O_3Si.11H_2O$, 31A, 241
$Na_2O_5Si_2$, 33A, 461
α-$Na_2O_5Si_2$, 26, 506
$Na_2O_7Si_3$, 32A, 511
NaO_4SiSm, 32A, 430
$Na_2O_9Si_2Ti_2$, 34A, 381
$Na_2O_{11}Si_4Ti$, 27, 699
$Na_4O_{26}Si_8Ti_4.nH_2O$, 32A, 511
Na_2O_4SiZn, 29, 428; 31A, 218
$Na_2O_6Si_2Zn$, 29, 428
$Na_2O_7Si_2Zn_2$, 29, 428; 32A, 437
$Na_2O_8Si_2Zn_3$, 29, 428
$Na_2O_8Si_3Zn$, 29, 428
Na_2O_5SiZr, 35A, 464
$Na_2O_7Si_2Zr$, 35A, 309
$Na_2O_{11}Si_4Zr$, 26, 531; 32A, 465
$Na_2O_{15}Si_6Zr.3H_2O$, 28, 276; 29, 404
Na_2O_3Sn, 28, 324
NaO_4SrV, 26, 470
Na-O-Ta, 30A, 452
NaO_3Ta, 27, 503; 28, 330
Na_2O_3Tb, 27, 520
NaO_4Tc, 27, 671
NaO_5TeV, 34A, 318
Na_2O_3Th, 31A, 253
$Na_{0.2}O_2Ti$, 26, 352
NaO_2Ti, 27, 518
Na_xO_2Ti, 30A, 451
$Na_xO_8Ti_4$, 27, 532
$Na_2O_7Ti_3$, 26, 375
$Na_2O_{13}Ti_6$, 27, 528
$Na_2O_{15}Ti_7$, 33A, 297
NaO_2Tl, 26, 400, 402; 29, 339
NaO_3U, 27, 543; 35A, 220
α-Na_2O_4U, 26, 405
β-Na_2O_4U, 26, 404
$Na_2O_7U_2$, 26, 406
Na_4O_5U, 27, 544; 29, 340
$Na_4O_8U.9H_2O$, 31A, 258; 33A, 358
$Na_{11}O_{16}U_5$, 35A, 220
NaO_5V_2, 31A, 140
$Na_2O_{16}V_6.3H_2O$, 28, 352
Na_2O_4W, 27, 724; 30A, 329; 31A, 250
$Na_2O_4W.2H_2O$, 26, 392; 34A, 281
$Na_4O_6Xe.6H_2O$, 28, 228; 29, 335
$Na_4O_6Xe.8H_2O$, 28, 229; 29, 335
$Na_2O_3Zn_2$, 30A, 325
$NaPrS_2$, 29, 341
$NaPrSe_2$, 30A, 155
$Na_2S.5H_2O$, 30A, 358
$Na_2S.9H_2O$, 30A, 358
α-Na_2S_2, 27, 350
β-Na_2S_2, 27, 351
NaS_2Sm, 29, 342; 34A, 140
NaS_2Sn, 34A, 140
$Na_{2/3}S_2Ta$, 35A, 121
NaS_2Tb, 29, 342
NaS_2Y, 29, 342
$Na_2Se.5H_2O$, 30A, 358
$Na_2Se.9H_2O$, 30A, 358
Na_2Se_2, 27, 351
$NaSe_2Sm$, 30A, 155, 156

$Na_{2/3}Se_2Ta$, 35A, 121
$NaSe_2Tb$, 30A, 156
$NaSe_2Y$, 30A, 156
NaSi, 26, 156; 29, 57, 125
$NaSi_6$, 30A, 156
Na_xSi, 30A, 156
Na_xSi_{136}, 30A, 96, 156
Na_8Si_{46}, 30A, 96, 156
$Na_2Te.5H_2O$, 30A, 358
Na_2Tl, 32A, 124
Nb, 26, 217; 28, 320; 29, 194; 30A, 221; 31A, 54
α-Nb, 26, 215
β-Nb, 26, 215
γ-Nb, 26, 215
δ-Nb, 26, 215
δ'-Nb, 26, 216
ε-Nb, 26, 216
$NbNdO_4$, 26, 386; 27, 521; 28, 346
$Nb_2Nd_2O_7$, 29, 329
$NbNdO_6Sr_2$, 28, 332
$NbNdO_6Ti$, 27, 537
NbNi, 27, 287
$NbNi_2$, 30A, 156
$NbNi_3$, 26, 113; 27, 26
Nb-Ni-O, 30A, 452
Nb_2NiO_x, 29, 125
$Nb_2Ni_4O_9$, 26, 389
$(Nb,Ni)O_3Sr$, 28, 330
Nb_3NiS_6, 35A, 111
$NbNi_{1/4}Se_2$, 35A, 121
$NbNi_{1/3}Se_2$, 35A, 121
Nb_2NiSi, 30A, 44, 156
Nb_2Ni_3Si, 26, 269
Nb_3Ni_2Si, 27, 376; 28, 28
$Nb_6Ni_{16}Si_7$, 26, 175; 27, 376
$NbNi_2Sn$, 35A, 121
Nb-O, 30A, 450
NbO_x, 27, 483
NbO, 27, 475; 31A, 122
NbO_2, 27, 479; 28, 122
$NbO_{2.464}$, 34A, 241
Nb_2O_5, 29, 305; 30A, 450; 31A, 243, 251, 258; 32A, 257
$(Nb_2O_5)_m$, 35A, 209
Nb_4O, 30A, 450
$Nb_{11}O_{27}$, 31A, 258
$Nb_{12}O_{29}$, 28, 123; 31A, 122
$Nb_{25}O_{62}$, 31A, 258
NbO_5P, 31A, 147
$Nb_9O_{25}P$, 30A, 336
$Nb_{18}O_{50}P_2$, 29, 422
Nb_2O_6Pb, 27, 506, 540; 29, 439; 31A, 149, 250; 32A, 295
NbO_6Pb_2Y, 28, 332
NbO_6Pb_2Yb, 28, 332; 29, 420
NbO_4Pr, 26, 386; 27, 521
NbO_6PrSr_2, 28, 332
NbO_6PrTi, 27, 537
NbO_4Pu, 29, 315
$Nb_4O_{12}Pu$, 29, 315
NbO_9RbW_2, 28, 332
NbO_4Sb, 27, 522; 28, 354
$Nb_6O_{18}Sb_2$, 31A, 241, 258
NbO_4Sc, 27, 522; 28, 348
NbO_7Sc_3, 29, 329
NbO_4Sm, 27, 521; 28, 346
NbO_7Sm_3, 29, 329
NbO_6SmSr_2, 28, 332
NbO_6SmTi, 27, 537
$(Nb,Ta)_2O_7Sn$, 32A, 509
$(Nb,Zn)O_3Sr$, 28, 330
NbO_6Sr_2V, 31A, 247
NbO_6Sr_2Y, 28, 332
NbO_6Sr_2Yb, 28, 332
$Nb_4O_{15}Ta_2$, 27, 538; 29, 422
NbO_4Tb, 27, 521
NbO_6TbTi, 28, 326
$NbO_3Th_{0.25}$, 31A, 150, 261
NbO_4Ti, 27, 520
$Nb_{1.6}O_4Ti_{0.4}$, 27, 520
$Nb_{1.8}O_4Ti_{0.2}$, 27, 520
Nb_2O_7Ti, 26, 382
$Nb_{10}O_{29}Ti_2$, 26, 382
$Nb_{24}O_{62}Ti$, 30A, 338
NbO_6TiTm, 28, 326
$Nb_2O_{10}TiU$, 34A, 270
NbO_6TiY, 26, 387
NbO_6TiYb, 28, 326
NbO_4Tm, 27, 521
Nb-O-U, 30A, 453
$Nb_3O_{10}U$, 30A, 453; 33A, 329
Nb-O-V, 27, 475
NbO_4V, 27, 521
$Nb_{18}O_{50}V_2$, 29, 422
Nb-O-W, 30A, 340, 452
$Nb_2O_{14}W_3$, 29, 423
$Nb_4O_{31}W_7$, 33A, 349
$Nb_8O_{47}W_9$, 34A, 288
$Nb_{12}O_{33}W$, 30A, 341
$Nb_{12}O_{63}W_{11}$, 33A, 348, 350
$Nb_{14}O_{44}W_3$, 30A, 341
$Nb_{16}O_{55}W_5$, 30A, 341
$Nb_{16}O_{77}W_4$, 31A, 148
$Nb_{16}O_{94}W_{18}$, 31A, 148
$Nb_{18}O_{69}W_8$, 30A, 341
$Nb_{18}O_{93}W_{16}$, 33A, 348
$Nb_{18}O_{96}W_{17}$, 33A, 348
NbO_4Y, 26, 386; 27, 521; 28, 330, 348
β-NbO_4Y, 26, 386
$(Nb,Ta)O_4Y$, 28, 330
NbO_7Y_3, 29, 329
NbO_4Yb, 27, 521
$Nb_{0.67}O_4Zn_{2.33}$, 28, 348
Nb_2O_6Zn, 28, 350
$Nb_2O_8Zn_3$, 28, 350
Nb-O-Zr, 27, 304; 30A, 452
Nb_3Os_2, 29, 86, 125
$Nb_3(Os_xRh_{1-x})$, 30A, 221
NbP_2, 29, 125
Nb_3P, 31A, 53
Nb_7P_4, 31A, 51
Nb_3Pb, 28, 48
Nb-Pd, 30A, 221
NbPd, 27, 318
$NbPd_2$, 29, 125; 30A, 156
β-$NbPd_3$, 29, 70, 125
Nb_3Pd_2, 29, 125
$Nb_3(Pd_xRh_{1-x})$, 30A, 221
α-Nb-Pt, 29, 125
$NbPt_2$, 29, 70, 71, 125; 30A, 77, 156
$NbPt_3$, 26, 216
α-$NbPt_3$, 29, 70, 125
β-$NbPt_3$, 29, 70, 125
Nb_2Pt, 26, 217
Nb_3Pt, 26, 217
α-$Nb_{48}Pt_{52}$, 29, 70
$Nb_3(Pt_xRh_{1-x})$, 30A, 222
Nb-Re, 26, 217
σ-NbRe, 29, 125
χ-$NbRe_3$, 29, 125
$Nb_{13.5}Re_{16.5}$, 29, 86
$Nb_{14.5}Re_{43.5}$, 29, 73
α-Nb-Rh, 29, 195
$α_4$-Nb-Rh, 29, 125
$α_5$-Nb-Rh, 29, 125
$α_6$-Nb-Rh, 29, 125
σ-Nb-Rh, 29, 195
$Nb_{0.75-0.96}Rh_{1.25-1.04}$, 29, 70
$α_3$-$Nb_{0.90}Rh_{1.10}$, 29, 125
$α_2$-NbRh, 29, 125
$α_5$-NbRh, 29, 125
α-$NbRh_3$, 29, 126
$Nb_{1.3}Rh_{2.7}$, 29, 70
σ-Nb_2Rh, 29, 125
α-Nb_3Rh, 29, 125
Nb_3Rh_5, 29, 71
NbRhSn, 35A, 121
Nb-Ru, 28, 48; 29, 195
$NbRu_3$, 29, 126
$Nb_3Ru_{1-x}Th_x$, 30A, 222
Nb_3S_4, 33A, 115
$Nb_{21}S_8$, 33A, 115, 116
NbS_4Tl_3, 29, 91, 125
$NbSb_2$, 29, 19, 125; 30A, 14, 156
Nb_3Sb, 29, 126
Nb_5Sb_4, 29, 20, 125
$Nb_3Sb_2Te_5$, 32A, 17
NbSe, 27, 343; 30A, 156
$NbSe_2$, 29, 73, 74, 125; 30A, 79, 157
α-$NbSe_2$, 27, 343
β-$NbSe_2$, 27, 343
$Nb_{1.00-1.29}Se_2$, 29, 196
$Nb_{1.14}Se_2$, 30A, 157
$NbSe_3$, 30A, 157
$NbSe_4$, 30A, 157

Nb_2Se, 34A, 110
Nb_2Se_3, 33A, 113
Nb_3Se_4, 29, 75, 195
Nb_5Se_4, 28, 48
$NbSe_4Tl_3$, 29, 91, 125
$NbSe_2V_{1/4}$, 35A, 121
$NbSe_2V_{1/3}$, 35A, 121
Nb_3Si, 28, 48; 29, 126; 30A, 157; 34A, 140
Nb_5Si_3, 26, 247
Nb_5SiSn_2, 35A, 84
$NbSn_2$, 27, 302; 28, 34; 29, 126; 30A, 157
Nb_2Sn_3, 27, 303
Nb_3Sn, 27, 213; 29, 126, 196, 197; 30A, 157, 222
Nb_3Sn_2, 27, 303; 30A, 157
Nb_6Sn_5, 29, 126; 30A, 80, 157
$NbSnTa_2$, 29, 126
Nb_2SnTa, 29, 126
NbSnTaV, 29, 126
$NbSnV_2$, 29, 126
Nb_2SnV, 29, 126
NbTa, 27, 318
Nb-Tc, 26, 255; 27, 361
$NbTc_3$, 26, 255
NbTe, 27, 343; 30A, 157
$NbTe_2$, 27, 342; 30A, 157; 31A, 69
$NbTe_3$, 30A, 157
$NbTe_4$, 29, 77, 125
Nb_3Te_4, 29, 75, 197
Nb_5Te_4, 28, 48
Nb-Ti, 26, 218
$Nb_{0.20}Ti_{0.80}$, 29, 126
$(Nb_{0.03}Ti_{0.97})Ni_3$, 31A, 16
Nb-U, 26, 218; 29, 197, 198; 30A, 222
$Nb_{16.60}U_{77.76}Zr_{5.64}$, 34A, 112
$NbZn_2$, 26, 218
Nb_6Zn_7, 33A, 153
Nb-Zr, 30A, 158
β-Nb-Zr, 27, 303
NbZr, 27, 319
Nd, 26, 234; 27, 334; 29, 126
NdNi, 29, 126; 30A, 158
$NdNi_3$, 35A, 122
Nd_2Ni_7, 35A, 121
Nd_2Ni_{17}, 34A, 140
$NdNiSi_2$, 35A, 122
Nd_2O_3, 30A, 450; 31A, 125
NdO_6Pb_2Ta, 28, 332
$Nd_{0.1}O_4Pr_{0.9}Ta$, 27, 521
$Nd_4O_{11}Re_2$, 35A, 256
NdO_3Rh, 28, 133
$Nd_2O_7Si_2$, 35A, 302
$Nd_{1.8}O_3Sm_{0.2}$, 27, 546
Nd_2O_7SnTi, 33A, 287
NdO_6Sr_2Ta, 28, 332
NdO_4Ta, 27, 521; 28, 346; 29, 421
$Nd_2O_7Ta_2$, 29, 329
Nd_2O_2Te, 27, 424; 32A, 309
NdO_4V, 27, 589; 28, 306
Nd_2O_6W, 35A, 239
$Nd_4O_{15}W_3$, 35A, 261
Nd-O-Zr, 30A, 452
$Nd_2O_7Zr_2$, 27, 535
$NdPd_3$, 30A, 158
Nd-Pr, 30A, 223
NdPt, 30A, 158
$NdPt_5$, 26, 239
NdRh, 30A, 158
Nd-S, 29, 198
$Nd_4S_{12}Si_3$, 34A, 140
Nd_2S_4Sr, 27, 327
$NdSb_2$, 32A, 20
$NdSe_{1.9}$, 31A, 50
$NdSe_2$, 32A, 20
Nd_2Se_3, 29, 126, 127
Nd_3Se_4, 29, 126
Nd_xSeSm_{1-x}, 29, 198
Nd_2Se_4Sr, 29, 127
Nd_xSeYb_{1-x}, 29, 198
NdSi, 29, 127; 30A, 158
$NdSi_2$, 29, 127
Nd_5Si_3, 30A, 158
$NdSn_3$, 29, 127; 30A, 158
NdTe, 29, 127; 30A, 158
$NdTe_{1.89}$, 31A, 47
$NdTe_{2-x}$, 30A, 224
$NdTe_2$, 30A, 158; 31A, 51; 32A, 20
$NdTe_3$, 30A, 159; 31A, 50
Nd_2Te_3, 29, 127; 30A, 158
Nd_2Te_5, 30A, 158; 31A, 51; 32A, 104
Nd_3Te_4, 28, 48; 29, 127; 30A, 158, 224
NdZn, 29, 127
$NdZn_{11}$, 30A, 159
Nd_3Zn_{22}, 35A, 122
$Nd_{13}Zn_{58}$, 35A, 122
Ne, 26, 285; 27, 391; 28, 296
Ni, 26, 205, 209; 27, 285; 30A, 159
NiO, 26, 358; 28, 289; 29, 417
Ni_2O_3, 26, 358
$Ni_2O_7P_2$, 32A, 374; 33A, 409
$NiO_{21}Pb_8Si_6$, 29, 399
$NiO_2Rb_{0.10}$, 34A, 293
NiO_4Rh_2, 26, 390
$NiO_3S.6H_2O$, 27, 633, 643; 29, 353; 34A, 316
NiO_4S, 26, 436; 27, 526
$NiO_4S.H_2O$, 31A, 199
$NiO_4S.4H_2O$, 29, 424
$NiO_4S.6D_2O$, 31A, 200
$NiO_4S.7H_2O$, 33A, 160
NiO_6S_2, 27, 643
$NiO_6S_2.6H_2O$, 35A, 380
NiO_4Se, 28, 306; 33A, 388
$NiO_4Se.H_2O$, 28, 306
$NiO_4Se.6H_2O$, 29, 425; 35A, 383
Ni_2O_4Si, 28, 350
$Ni_3O_{11}Si_4.H_2O$, 28, 350
$(Ni,Ta)O_3Sr$, 28, 332
NiO_6Sr_2Te, 35A, 242
NiO_6Sr_2W, 27, 509
NiO_6Ta_2, 26, 412
NiO_4TcZn, 29, 322
Ni_3O_6Te, 32A, 306
NiO_3Ti, 27, 507
Ni_2OTi_4, 28, 20, 140
Ni_2O_4Ti, 26, 413
NiO_4U, 31A, 254
NiO_6U_2, 33A, 356
$NiO_{10}U_3$, 28, 334
$NiO_{12}U_2V_2.4H_2O$, 35A, 364
$Ni_3O_8V_2$, 35A, 362
Ni-Os, 29, 198
NiP, 30A, 72, 159
NiP_2, 26, 270; 30A, 73, 159
NiP_3, 34A, 117
Ni_2P, 27, 291
$Ni_{2.55}P$, 30A, 159
Ni_3P, 27, 290
ε-Ni_3P, 27, 102
Ni_5P_2, 29, 127
Ni_5P_4, 30A, 70, 159
$Ni_{12}P_4$, 30A, 71
$NiPS_3$, 30A, 159
$NiPSe_3$, 30A, 159
NiPb, 33A, 152
Ni-Pd, 29, 198, 199
$(Ni,Pd)_{1.1-1.4}S$, 27, 290
NiPr, 29, 127; 30A, 159
Ni_3Pr, 35A, 122
Ni_7Pr_2, 35A, 122
$NiPrSi_3$, 35A, 122
$(Ni_{0.11},Pt_{0.89})_3Ti$, 31A, 16; 34A, 105
Ni-Rh, 29, 199; 30A, 224
$NiRh_2Se_4$, 34A, 109
Ni-Ru, 29, 199, 200
Ni-S, 26, 208; 27, 293
α-$Ni_{1-x}S$, 27, 59, 294
NiS, 27, 724; 28, 48
α-NiS, 26, 209
β-NiS, 26, 209, 214; 27, 294
NiS_2, 27, 294; 34A, 73
$NiS_{2\pm x}$, 26, 209
Ni_3S_2, 27, 294
α-Ni_3S_2, 26, 209
β-Ni_3S_2, 26, 209
$Ni_{3+x}S_2$, 27, 60
Ni_3S_4, 26, 209; 27, 294
$Ni_4S_{3\pm x}$, 26, 208
α-Ni_7S_6, 27, 60
Ni_9S_8, 26, 121, 208
$Ni(S,Se)_2$, 27, 168
$Ni_3(S,Se)_4$, 29, 127

NiS_4V_2, 30A, 159
$NiSb_2$, 33A, 63; 35A, 10
Ni_5Sb_2, 29, 127
NiSbTi, 28, 48
NiSbV, 28, 48
NiSc, 27, 342
$NiSc_2$, 27, 342
Ni_2Sc, 26, 236
$Ni_{16}Sc_6Si_7$, 27, 162, 376; 28, 48
Ni-Se, 26, 208
β-NiSe, 26, 210; 29, 127
β'-NiSe, 26, 210
γ-NiSe, 29, 127
$NiSe_2$, 29, 127; 34A, 73
$NiSe_{2\pm x}$, 26, 210
α-Ni_3Se_2, 26, 210
Ni_3Se_4, 29, 127
~Ni_6Se_5, 26, 208
~$Ni_{21}Se_{20}$, 26, 208
NiSeTe, 29, 127; 30A, 159
$NiSe_4Ti_2$, 30A, 159
$NiSe_4V_2$, 30A, 159, 160; 31A, 51
Ni-Si, 27, 293
NiSi, 26, 124; 27, 27; 35A, 86
$NiSi_2$, 26, 125
Ni_3Si_2, 26, 211
Ni_5Si_2, 26, 211; 29, 127
$NiSi_2Sm$, 35A, 122
Ni_2SiTa_3, 27, 376
Ni_3SiTa_2, 26, 269
$Ni_{16}Si_7Ta_6$, 26, 175; 27, 376
$NiSi_2Tb$, 35A, 122
Ni_2Si_2Th, 30A, 160
NiSiTi, 30A, 75, 160; 33A, 78
Ni_3SiTi_2, 26, 269
$Ni_{16}Si_7Ti_6$, 26, 175; 27, 376
$NiSi_2Tm$, 35A, 122
$Ni_{1.2}Si_{0.8}V$, 26, 174; 27, 376
Ni_2SiV_3, 27, 376; 28, 28
Ni_3SiV_6, 27, 376
$Ni_5Si_3V_4$, 26, 269
$Ni_{16}Si_7V_6$, 26, 175; 27, 376
$Ni_{40}Si_{25}V_{41}$, 26, 270
NiSiW, 27, 376
Ni_3SiW_2, 26, 269
$NiSi_2Yb$, 35A, 122
Ni_2SiZn_3, 33A, 112
$Ni_{16}Si_7Zr_6$, 26, 175; 27, 376
NiSm, 29, 127; 30A, 160
Ni_3Sm, 35A, 122
Ni_7Sm_2, 35A, 122
Ni-Sn, 27, 259, 293
$Ni_{1.4}Sn$, 27, 259
$Ni_{1.5}Sn$, 32A, 107
NiSnTi, 35A, 122
Ni_2SnTi, 28, 48; 35A, 122
Ni_2SnV, 28, 48
NiSnZr, 35A, 112
Ni_2SnZr, 35A, 112
Ni-Ta, 29, 200
NiTa, 27, 287
$NiTa_2$, 27, 294
Ni_2Ta, 29, 69, 125; 30A, 160; 32A, 140
Ni_3Ta, 29, 127; 30A, 160; 32A, 140
NiTb, 29, 127, 128; 35A, 80
Ni_3Tb, 35A, 122
Ni_7Tb_2, 35A, 122
Ni-Te, 27, 295
$NiTe_{0.6}$, 30A, 160
$NiTe_{0.9}$, 30A, 160
NiTe, 27, 295; 28, 34
$NiTe_2$, 27, 295; 30A, 76, 160
$Ni_{1+x}Te_2$, 27, 295
$Ni_{2.8}Te_2$, 30A, 160
$Ni_{3\pm q}Te$, 30A, 75
$Ni_{3.3}Te_2$, 30A, 160
Ni_2Th, 26, 256
$Ni_{17}Th_2$, 33A, 132
NiTi, 30A, 160
$NiTi_2$, 27, 296; 28, 20
Ni_3Ti, 31A, 16
Ni_2TiSn, 27, 296
Ni_2TiZn, 27, 296
NiTm, 29, 128; 30A, 161
Ni_2Tm, 30A, 161
Ni_2V, 29, 128; 30A, 77, 161
NiY, 29, 128; 30A, 77, 161
Ni_2Y, 26, 236, 276
Ni_3Y, 30A, 78, 161
Ni_5Y, 26, 239
Ni_7Y_2, 34A, 140; 35A, 122
Ni_2Zn_{11}, 33A, 108
Ni_4Zn_{22}, 33A, 108
Ni-Zr, 26, 205
NiZr, 26, 206; 27, 237, 297
$NiZr_2$, 26, 205; 27, 297
Ni_5Zr, 26, 207
Ni_5Zr_2, 26, 207
Ni_7Zr_2, 26, 207
$Ni_{10}Zr_7$, 26, 206; 27, 298
$Ni_{11}Zr_9$, 26, 206
$Ni_{19}Zr_{15}$, 27, 301
Np_2O_5, 29, 418
Np-Pu, 26, 228
$Np_{57}Pu_{43}$, 29, 128

O_2, 27, 394; 31A, 259; 32A, 147
γ-O_2, 29, 229
O_2Os, 34A, 246; 35A, 211
O_4Os, 30A, 311; 31A, 239
O_5P_2, 29, 360
O_6P_4, 34A, 240
O_7P_4, 31A, 120
O_8P_4, 29, 300; 31A, 120; 34A, 240
O_9P_4, 29, 300; 32A, 252
$O_{10}P_4$, 27, 718; 29, 360
O_6P_2Pb, 29, 367
$(O_6P_2Pb)_x$, 28, 304
O_7P_2Pb, 28, 304
O_8PPb_3, 27, 587
$O_8P_2Pb_3$, 35A, 330
$O_{12}PPb_4Si_2$, 28, 324
$O_{12}P_4Pu$, 27, 598
$(O_3PRb)_x$, 29, 361
$O_{12}P_3Rb_2Sr$, 31A, 245
$O_6P_4S_4$, 32A, 220
O_4PSb, 35A, 328
O_4PSc, 28, 304
O_7P_2Si, 28, 304; 34A, 327; 35A, 341
$O_9P_2Si_2$, 28, 304, 322
O_7P_2Sn, 28, 304
$O_7P_2Sr_2$, 32A, 377; 33A, 407; 34A, 326
$O_{24}P_6Sr_9 \cdot \frac{1}{2}H_2O$, 26, 469
$O_{12}P_4SrTl_2$, 31A, 245
$O_{50}P_2Ta_{18}$, 29, 422
O_4PTb, 28, 304
$O_4PTh.0 \cdot 2H_2O$, 26, 454
O_7P_2Th, 28, 291
O_7P_2Ti, 28, 304
$O_4PTl.2H_2O$, 26, 464
O_4PTm, 28, 304
O_7P_2U, 28, 291, 304
$O_{12}P_4U$, 27, 598
$O_{11}P_2W_2$, 29, 373
O_4PY, 28, 304; 30A, 386
O_4PYb, 28, 304
$O_7P_2Zn_2$, 30A, 385; 35A, 343
$O_8P_2Zn_3$, 28, 189; 30A, 383; 32A, 357
$O_8P_2Zn_3.4H_2O$, 26, 462; 27, 586; 28, 190; 33A, 395
O_7P_2Zr, 26, 467; 28, 304
OPb, 26, 360, 388
O_2Pb, 27, 477; 30A, 450
O_3Pb_2, 34A, 240
O_4Pb_3, 27, 383; 30A, 313
O_6Pb_2PrTa, 28, 332
O_3PbRb_2, 35A, 223
$O_7Pb_2Rb_2Si_2$, 28, 310
$O_pPb_nS_m$, 26, 446
O_4PbS, 28, 78, 208
O_5Pb_2S, 26, 446; 35A, 374
O_8Pb_5S, 26, 446
$O_7Pb_2Sb_2$, 27, 534
O-Pb-Sb-Sn, 27, 534
O-Pb-Se, 30A, 453
O_3PbSe, 28, 306; 30A, 458
O_4PbSe, 27, 609
α-O_4PbSe, 27, 609

β-O_4PbSe, 27, 609
O_5Pb_3Se, 26, 432; 30A, 453
O_7Pb_5Se, 26, 432; 30A, 453
O_3PbSi, 33A, 454
$O_{21}Pb_8Si_6Sr$, 29, 399
$O_6PbSiU.H_2O$, 28, 278; 29, 405
$O_4PbSiZn$, 32A, 428
$O_{21}Pb_8Si_6Zn$, 29, 399
O_6Pb_2SmTa, 28, 332
O_4Pb_2Sn, 28, 324
O-Pb-Sn-Sr-Ti, 26, 419
$O_3Pb(Sn,W)$, 30A, 452
O-Pb-Sr-Ti-Zr, 26, 419
O-Pb-Ta, 26, 388
O_6PbTa_2, 26, 388
O_6Pb_2TaY, 28, 332
O_6Pb_2TaYb, 28, 332
O_3PbTc, 29, 322
O_8PbTe_2U, 30A, 454
O_3PbTi, 26, 419; 27, 506; 29, 326, 418; 33A, 305
$O_{4-x}PbU$, 29, 422
O_4PbU, 26, 405
O_6Pb_3U, 32A, 317
$O_7Pb_2U_2$, 27, 543
$O_{15}Pb_5U_4$, 27, 543
$O_{27}Pb_3U_8.5H_2O$, 28, 150
$O_5Pb_{0.193}V_2$, 34A, 266
$O_7Pb_2V_2$, 28, 354; 31A, 260; 32A, 275
$O_8Pb_3V_2$, 27, 587; 31A, 260
$O_{13}Pb_8V_2$, 31A, 260
O_4PbW, 35A, 228
O-Pb-W-Zr, 30A, 452
$OPd_{2x}Rh_{2-2x}$, 30A, 452
O_3PdSr_2, 35A, 224
O_3Pm_2, 28, 320; 30A, 457
$O_{1.6-1.83}Pr$, 27, 481
O_3Pr_2, 26, 493; 27, 481
$O_{12}Pr_7$, 27, 481
$O_{16}Pr_9$, 27, 481
$O_2Pr_2S_2$, 32A, 223
$O_7Pr_2Si_2$, 35A, 307
O_6PrSr_2Ta, 28, 332
O_4PrTa, 31A, 248
O_2Pr_2Te, 27, 424
$O_7Pr_2Th_2$, 26, 493
O_4PrV, 28, 306
$O_9Pr_2W_2$, 35A, 255
O_2Pt, 33A, 270
O_4Pt_3, 33A, 270
$O_7Pt_2Tl_2$, 33A, 288
O-Pu, 30A, 451
α'-O_3Pu_2, 27, 477
β-O_3Pu_2, 27, 477
O_3PuSr, 29, 422
O_4PuSr, 29, 315
$O_{7+x}Pu_2Zr_2$, 29, 315
O_2Rb_2, 30A, 450
O_3RbReS, 34A, 292
O_4Rb_2S, 27, 604; 31A, 245
$O_6Rb_2S_2$, 28, 306
$O_4Rb_xSc_xTi_{2-x}$, 33A, 304
O_4Rb_2Se, 27, 605
O_9RbTaW_2, 28, 334
O_2Rb_xTi, 29, 318
$O_{13}Rb_2Ti_6$, 27, 529
O_2RbTl, 26, 400
O_3RbU, 27, 543
O_4Rb_2U, 26, 404
$O_7Rb_2U_2$, 26, 406
$O_3Rb_{0.27}W$, 26, 356
O_4Rb_2W, 35A, 233
O_2Re, 26, 365; 29, 417
$O_2Re.2H_2O$, 26, 365
O_7Re_2, 34A, 244
$O_7Re_2.2H_2O$, 33A, 269
O_3ReSTl, 31A, 256
O_4ReTl, 31A, 250
O_2Rh, 33A, 271
O_3Rh_2, 28, 133; 35A, 211
O_4Rh_2Sr, 26, 399
O_4Rh_2Zn, 26, 399
O_2Ru, 28, 320; 31A, 124; 35A, 212
O_xRuZr_2, 29, 128
O_2S, 34A, 242
O_9S_3, 30A, 362; 32A, 253
OS_2Sb_2, 32A, 222
O_4SSn, 27, 608
O_4SSr, 26, 443; 30A, 365
O_7STe_2, 31A, 199
$O_4STh.0{\cdot}2H_2O$, 26, 454
O_4STl_2, 27, 604; 30A, 458; 31A, 198, 244
O_6SU, 30A, 454
O_5SV, 30A, 368; 35A, 373
$O_{12}S_3Y_2.8H_2O$, 29, 424
O_4SZn, 26, 437, 438, 439
$O_4SZn.H_2O$, 31A, 199
$O_4SZn.4H_2O$, 31A, 244
$OSZr$, 27, 348, 423
O_8S_2Zr, 34A, 301; 35A, 370
$O_8S_2Zr.H_2O$, 34A, 301; 35A, 368, 369
$O_8S_2Zr.4H_2O$, 30A, 453
$O_{16}S_4Zr_2.14H_2O$, 34A, 303
O_3Sb_2, 26, 362; 28, 320
O_4Sb_2, 27, 478
$O_5Sb_2.4H_2O$, 33A, 425
O_6Sb_4, 27, 719
O_4SbSc, 28, 348
O_4SbTa, 27, 522; 28, 354
O_4SbV, 28, 354
$O_4Sb_{0.67}Zn_{2.33}$, 26, 412
$O_{12}Sb_2Zn_7$, 26, 410; 27, 490, 527; 28, 348
$O_xSb_3Zr_5$, 30A, 166
OSc, 29, 128
O_3Sc_2, 26, 350; 29, 417, 418; 32A, 254; 33A, 263
$O_7Sc_2Si_2$, 27, 702
$O_7Sc_{1-x}Sn_2Yb_x$, 33A, 287
O_4Sc_2Sr, 29, 418
O_4ScTa, 27, 522; 28, 348
O_7Sc_3Ta, 29, 329
O_4ScV, 27, 539; 28, 306
$O_{12}Sc_2W_3$, 31A, 158
O_3ScY, 30A, 452
$O_3(Sc,Y)_2$, 31A, 256
$O_3(Sc,Yb)_2$, 31A, 256
$O_{12}Sc_4Zr_3$, 33A, 291
$O_{13}Sc_2Zr_5$, 33A, 291
O_3Se, 27, 482; 30A, 306
O_4SeSr, 27, 609
O_2Se_2Y, 31A, 241
$O_3SeZn.2H_2O$, 29, 358
O_4SeZn, 30A, 453; 31A, 254
$O_4SeZn.H_2O$, 30A, 458; 31A, 256
$O_4SeZn.6H_2O$, 30A, 458
OSi, 30A, 455
O_2Si, 26, 518; 27, 674; 28, 119, 120, 289, 322; 29, 427; 30A, 420; 32A, 251; 34A, 363
$O_7Si_2Sm_2$, 35A, 302, 305
O_5SiSr_3, 30A, 440
O_7Si_2SrV, 32A, 441
O-Si-Th, 26, 403
O_4SiTh, 26, 403
$O_4SiTh.0{\cdot}2H_2O$, 26, 454
$O_7Si_2Tm_2$, 35A, 300
O_xSiV_3, 30A, 224
$O_{40}SiW_{12}$, 27, 672
O_5SiY_2, 30A, 455; 32A, 434; 33A, 471; 35A, 297
$O_7Si_2Y_2$, 32A, 511; 33A, 464, 465; 35A, 300, 301
$O_7Si_2Yb_2$, 35A, 300
O_3SiZn, 30A, 458
O_4SiZn, 30A, 458
O_4SiZn_2, 29, 428; 35A, 314
α-O_4SiZn_2, 27, 699
β-O_4SiZn_2, 27, 699
O_4SiZr, 28, 304; 29, 411
O_3Sm_2, 28, 320
$O_7Sm_2Sn_2$, 33A, 286
O_4Sm_2Sr, 28, 346
O_6SmSr_2Ta, 28, 332
O_4SmTa, 27, 521; 28, 332, 348
O_7Sm_3Ta, 29, 329
$O_7Sm_2Tc_2$, 29, 322
O_2Sm_2Te, 27, 424
O_4SmV, 28, 306
$O_{12}Sm_6W$, 29, 420
$O_7Sm_2Zr_2$, 27, 535
OSn, 26, 360; 30A, 450; 34A, 239
O_2Sn, 27, 382
O_4Sn_3, 32A, 509

$O_{24}Sn_{24}.10H_2O$, 26, 360
O_3SnSr, 26, 420; 28, 324
$O_{5+x}Sn_xTa_2$, 28, 332
O_7SnTa_2, 35A, 444
$O_7Sn_2Tb_2$, 26, 399
O_8SnTe_3, 27, 527
$O_7(Sn,Ti)_2Y_2$, 33A, 287
$O_7Sn_2Tm_2$, 26, 399
$O_7Sn_2Y_2$, 33A, 286
O_4SnZn_2, 28, 324
$O_8Sn_2Zn_4$, 27, 490
OSr, 29, 422
$O_{11}SrTa_4$, 28, 346
$O_{15}Sr_5Ta_4$, 26, 422
O_6Sr_2TaV, 31A, 247
O_6Sr_2TaY, 28, 332
O_6Sr_2TaYb, 28, 332
$O_3Sr(Ta,Zn)$, 28, 332
O_4SrTb_2, 33A, 340
O_3SrTc, 29, 322
$O_4Sr_{0.5}ThV$, 30A, 452
O_3SrTi, 27, 507; 28, 292; 29, 418; 30A, 457
O_5Sr_3Ti, 28, 326
O-Sr-Ti-Zr, 27, 506
$O_{3.58}SrU$, 31A, 257
O_4SrU, 34A, 289
O_5Sr_2U, 28, 334; 34A, 289
O_6Sr_3U, 31A, 160; 34A, 289
$O_{10}SrU_3.5H_2O$, 28, 310
$O_6SrV_2.4H_2O$, 30A, 318
O_4SrW, 31A, 157
O_6Sr_3W, 28, 310
O_6Sr_2WZn, 27, 509; 28, 310
O_4SrY_2, 33A, 340
O_4SrYb_2, 28, 346; 35A, 230
O_2SrZn, 26, 396
O_3SrZr, 26, 419; 27, 507; 29, 420
OTa, 26, 294
O-Ta, 26, 252; 30A, 450
O_xTa, 27, 483
O_yTa, 27, 483
O_zTa, 27, 483
OTa_2, 30A, 312
O_2Ta, 32A, 262
O_5Ta_2, 26, 294, 354, 388, 391; 27, 482; 29, 307; 30A, 450
O_9Ta_{32}, 31A, 251
O_4TaTb, 27, 521
O_4TaTi, 27, 520
O_5TaU, 28, 332
$O_{10}Ta_3U$, 28, 332; 30A, 453
O_4TaV, 27, 521
O_6Ta_2V, 32A, 509; 33A, 312
$O_{50}Ta_{18}V_2$, 29, 422
$O_3TaY_{0.33}$, 31A, 257; 33A, 341
O_4TaY, 27, 521; 28, 348; 32A, 311
O_7TaY_3, 29, 329
O_6Ta_2Zn, 26, 412
O_xTb, 29, 422
$O_{1.5}Tb$, 26, 364; 29, 422
$O_{1.715}Tb$, 26, 364
$O_{1.809}Tb$, 26, 365
$O_{1.823}Tb$, 26, 365
$O_{1.95}Tb$, 26, 365
O_3Tb_2, 30A, 450; 33A, 269
O_4TbV, 28, 308
O_2Tc, 29, 322
$O_2Tc_{0.4}Ti_{0.6}$, 28, 322
O_4TcTl, 31A, 250
O_2Te, 26, 362; 27, 382; 31A, 257; 32A, 261; 33A, 263
O_3Te, 33A, 262
O_7Te_2, 34A, 244
O_8Te_3Ti, 27, 527
OTeU, 26, 432; 29, 342
O_3TeZn, 32A, 307
$O_8Te_3Zn_2$, 31A, 260
O_8Te_3Zr, 27, 527
O_2Th, 26, 493; 29, 422
O-Th-Ti, 26, 403
O_6ThTi_2, 26, 403; 31A, 139; 33A, 307
$O_2(Th,U)$, 30A, 453
O_8ThW_2, 28, 308
$O_{\sim 1.0}Ti$, 26, 350
O-Ti, 26, 261
OTi, 27, 484; 31A, 259; 32A, 254
OTi_2, 27, 484
OTi_3, 27, 484; 33A, 265
OTi_6, 31A, 259
O_2Ti, 26, 351, 379; 28, 289; 31A, 248; 32A, 255
$O_{2n-1}Ti_n$, 29, 301
O_3Ti_2, 27, 484, 499; 28, 121; 34A, 239
O_5Ti_3, 26, 350
O_5TiU, 31A, 255
O_5TiY_2, 33A, 306
$O_7Ti_2Y_2$, 34A, 257; 35A, 245
$O_{11}TiYb_6$, 30A, 452
O-Ti-Zn, 26, 376
O_3TiZn, 26, 377
O_4TiZn_2, 26, 376, 413; 28, 344; 31A, 244
$O_8Ti_3Zn_2$, 26, 376
OTi_2Zr, 33A, 326
O_4TiZr, 31A, 256; 32A, 267
O_3Tl_2, 31A, 254; 33A, 262; 34A, 239
O_4Tl_2W, 27, 724
O_4TmV, 28, 308
O-U, 30A, 451
O_2U, 26, 367; 28, 334; 31A, 247, 251
$O_{2+n}U$, 28, 334
$O_{2\pm x}U$, 27, 481; 30A, 451
$O_{2.13}U$, 28, 129
O_3U, 26, 368; 28, 130, 334; 30A, 451; 31A, 126, 127
$O_3U.2H_2O$, 28, 310
O_3U_2, 31A, 250
$O_4U.2H_2O$, 28, 310
$O_4U.4H_2O$, 28, 310
O_5U_2, 29, 417; 31A, 255
O_8U_3, 29, 312, 417; 31A, 251; 34A, 242; 35A, 213
$O_{8-x}U_3$, 26, 367
O_9U_4, 26, 366; 28, 334
$O_{9-x}U_4$, 27, 481; 28, 334
O_5UV, 30A, 453; 34A, 268; 35A, 236
O_6UV_2, 28, 326; 30A, 457
O_8UV_2, 32A, 276
$O_{10}UV_3$, 28, 326
O_6UW, 29, 420; 30A, 454
O_8U_2W, 31A, 249
$O_{12}UY_6$, 27, 544; 31A, 162
$O_{12}U_2Y_5$, 31A, 162
$O_{13.3}U_2Y_5$, 31A, 162
O_nV, 28, 320
OV, 27, 475, 484, 515
O_2V, 26, 352; 35A, 204
$O_{2n-1}V_n$, 29, 301
O_3V_2, 27, 499; 28, 320; 29, 417; 35A, 205
O_4V_2, 29, 417
O_5V_2, 26, 353; 30A, 457
O_9V_4, 35A, 202
$O_{13}V_6$, 30A, 450
$O_{14}V_6.6H_2O$, 28, 326
$O_{16}V_{13}$, 35A, 206
$O_5V_{1.30}W_{0.70}$, 34A, 267
O_6V_2W, 33A, 312
O_7V_2W, 33A, 314
$O_{7.5}V_2W$, 33A, 314
$O_{8-x}V_2W$, 30A, 452
$O_{20}V_5W_3$, 35A, 264
O_4VY, 27, 589; 28, 306; 33A, 310
O_4VYb, 28, 308
O_4V_2Zn, 28, 286
$O_4V_2Zn_4$, 34A, 264
O_6V_2Zn, 33A, 313
$O_{2.83}W$, 29, 417
$O_{2.90}W$, 29, 417
O_3W, 26, 391; 27, 483; 29, 417; 31A, 123, 248
$O_{3n-2}W_n$, 30A, 312
O_4WZn, 33A, 347
O_3Xe, 28, 130
$O_{1.5}Y$, 29, 422
O_3Y_2, 27, 497; 30A, 309
O_3Yb_2, 26, 364; 27, 497
O-Yb-Zr, 26, 364
$O_7Yb_2Zr_2$, 27, 535

$O_{12}Yb_4Zr_3$, 27, 542; 35A, 259
OZn, 27, 475; 34A, 238
$O_{1-x}Zr_3$, 33A, 268
O_xZr, 33A, 118
O_2Zr, 26, 364; 27, 480; 28, 320; 30A, 307, 450, 457; 31A, 240
Os, 26, 223; 27, 282; 30A, 161
OsP_2, 33A, 63
OsPS, 29, 128
OsPSe, 29, 128
OsRe, 27, 313
OsRu, 27, 314
OsS_2, 32A, 127
OsSSb, 29, 128
$OsSb_2$, 26, 44; 33A, 63
$OsSb_{0.05}Te_{1.95}$, 26, 44
$OsSb_{0.1}Te_{1.9}$, 26, 44
$OsSb_{0.2}Te_{1.8}$, 26, 44
$OsSb_{1.95}Te_{0.05}$, 26, 44
$OsSe_2$, 33A, 34
OsSi, 27, 372
$OsSi_{1.8\pm0.3}$, 27, 372
$OsSi_2$, 35A, 84
$OsSi_{2.0\pm0.5}$, 27, 373
$OsSi_{3.0\pm0.5}$, 27, 372
Os-Ta, 30A, 224
$OsTe_2$, 26, 43; 32A, 127
Os_2Th, 26, 236
Os_3Th_7(I), 26, 256
Os_3Th_7(III), 26, 256
OsU_2, 30A, 161
Os-W, 26, 223
Os_2W, 29, 86, 128

P, 28, 48; 30A, 251; 31A, 57; 34A, 149
PPd_3, 26, 226; 28, 48
$PPd_{4.8}$, 31A, 54
$P_{1\pm x}Pd_3$, 26, 226
P_2Pd, 26, 270; 28, 35
$P_{14}Pd_{72}Sn_{14}$, 35A, 122
$P_{14}Pd_{72}Tl_{14}$, 35A, 122
$P_{14}Pd_{72}Zn_{14}$, 35A, 122
PPr, 34A, 140
P_2Pt, 30A, 161
P_2Pt_5, 32A, 108
$P_{14}Pt_{72}Si_{14}$, 35A, 122
$P_7Pt_{72}Sn_{21}$, 35A, 122
$P_{14}Pt_{72}Tl_{14}$, 35A, 122
$P_{14}Pt_{72}Zn_{14}$, 35A, 122
$P_2R_4S_4Se$, 30A, 453
$P_2R_4S_4Te$, 30A, 453
PRe_2, 26, 227
P_4Re_3, 31A, 57
P_2Rh, 26, 270
P_3Rh, 26, 48; 34A, 117
PRu, 27, 319
P_2Ru, 33A, 63
PRuS, 29, 128
PRuSe, 29, 128
P_4S_5, 30A, 353
$P_4S_{6.5}$, 30A, 354
P_4S_7, 27, 401; 30A, 353
P_4S_9, 34A, 118
P_4S_{10}, 27, 401; 30A, 353
PSc, 28, 48
P_2Si, 32A, 110, 140; 33A, 36
P_2SiZn, 29, 128; 35A, 102
$PSn_{0.3}$, 35A, 122
$PSn_{0.7}$, 35A, 122
PSn, 35A, 88
$P_{2.65}Sn_4$, 33A, 146
P_3Sn_4, 33A, 146; 35A, 88
P_2SnZn, 33A, 128
PTa_2, 31A, 58; 32A, 130
P_2Ta, 29, 128
PTb, 26, 255; 27, 334
P_4Th_3, 30A, 161
$P_{0.63}Ti$, 30A, 83
PTi, 30A, 161; 32A, 111, 113
PTi_2, 30A, 161; 32A, 111
PTi_3, 28, 50; 30A, 161; 32A, 111
P_2Ti, 29, 128; 32A, 111; 33A, 128
P_2Ti_3, 30A, 161
P_3Ti_4, 30A, 161
P_3Ti_5, 29, 128; 30A, 161; 32A, 111, 112
$P_{3.15}Ti_5$, 30A, 83
PTm, 29, 128
PU, 29, 128
P_2V, 29, 128
P_3V, 28, 48
P_7V_{12}, 35A, 89
PW, 27, 319; 28, 48
P_2W, 28, 48; 29, 128
PY, 28, 48; 29, 128
PYb, 29, 128
P_2Zn, 28, 37; 30A, 83, 161
PZr_3, 31A, 53
P_2Zr, 29, 128; 33A, 128
Pa, 30A, 225
Pb, 26, 280; 27, 385; 34A, 141
$PbPu_3$, 34A, 141
Pb_2Pu, 34A, 141
Pb_3Pu, 34A, 141
Pb_3Pu_5, 34A, 141
Pb_4Pu_5, 34A, 141
PbRb, 29, 128
Pb-S, 29, 200
PbS, 26, 90
$Pb_5S_{14}Sb_2Sn_3$, 26, 432
Pb_2S_4Si, 30A, 453
PbS_2Sn, 29, 200
PbS_3Sn_2, 29, 200
PbS_3Ti, 29, 200
Pb-Sb-Te-Tl, 27, 363
Pb_3Sc_5, 30A, 161
PbSe, 26, 63; 30A, 162
Pb-Se-Sn, 26, 153
Pb-Sn, 26, 187
(Pb,Sn)Te, 27, 211
Pb_3Tb, 29, 129
PbTe, 27, 211; 30A, 225
Pb-Te-Se-Sn, 30A, 225
PbTh, 26, 186
Pb_3Th, 26, 186
Pb_3Tm, 29, 129
PbU, 26, 186
Pb_3U, 26, 186
PbV_3, 28, 48
Pb_3Y, 28, 48; 29, 129
Pb_3Y_5, 30A, 162
$PbYb_2$, 32A, 140
Pb_3Yb, 29, 129; 32A, 140
Pb_3Yb_5, 32A, 140
Pd, 26, 432; 27, 227; 29, 200
Pd_3Pr, 30A, 162
Pd-Pt, 30A, 225
PdPu, 33A, 121
Pd-Re, 27, 337
$(Pd,Rh)_3Ta$, 30A, 64, 162
PdRu, 27, 314
$(Pd,Ru)_3Ta$, 30A, 162
Pd_3S, 33A, 121
Pd_4S, 27, 315
PdSSe, 30A, 162
$PdSb_2$, 30A, 162; 34A, 10
Pd_5Sb_2, 35A, 14
PdSc, 27, 342
$PdSc_2$, 27, 342
Pd_3Sc, 26, 165
Pd_4Se, 27, 315
$Pd_{17}Se_{15}$, 27, 315
PdSeTe, 30A, 162
$Pd_{75}Se_{13}Tl_{12}$, 35A, 122
$Pd_{72}Se_{14}Zn_{14}$, 35A, 123
PdSi, 35A, 86
Pd_2Si, 31A, 56
$Pd_{4.5}Si$, 31A, 57
Pd_3Sm, 30A, 162
PdSnZr, 35A, 123
PdTa, 28, 48
$PdTa_3$, 28, 48
Pd_2Ta, 29, 70, 129; 30A, 77, 163
Pd_3Ta, 28, 48
Pd_2Tb_5, 29, 129
Pd_3Tb, 30A, 163
PdTe, 30A, 163
$PdTe_{1+x}$, 30A, 226
$PdTe_2$, 30A, 76, 163
$PdTe_{2+x}$, 30A, 226
$Pd_{3.5}Te$, 34A, 137
$Pd_{4-x}Te$, 35A, 87
Pd-Th, 29, 201
PdTh, 29, 78, 129
$PdTh_2$, 26, 226; 29, 129
Pd_3Th, 26, 165; 27, 317
Pd_4Th, 27, 317; 28, 48
Pd_5Th_3, 28, 36
PdTi, 30A, 163
$PdTi_2$, 27, 234

Pd_3Ti_2, 33A, 126
Pd_5Ti_3, 33A, 126
$PdTl_2$, 34A, 141
Pd_2Tl, 34A, 141
Pd_3Tl, 26, 225; 34A, 141
$Pd_{13}Tl_9$, 33A, 124; 34A, 137
Pd_4U, 29, 129, 201
PdV, 27, 318
Pd_2V, 29, 129; 30A, 77, 163
Pd_3V, 26, 165
Pd-W, 29, 201
Pd_3Y, 26, 165
Pd_3Yb, 30A, 163
Pd_2Zn, 26, 225
Pd_6Zn_{20}, 34A, 115
Pd-Zr, 27, 227; 30A, 226
PdZr, 27, 227
$PdZr_2$, 27, 234
Pd_2Zr, 27, 227
Pd_3Zr, 27, 227
α-Po, 31A, 61
β-Po, 31A, 61
Pr, 26, 234; 29, 129
$PrPt_5$, 26, 239
$PrRu_2$, 30A, 227
Pr-S, 29, 201
Pr_2S_4Sr, 27, 327
Pr_2S_4Yb, 29, 129
Pr_2Se_3, 29, 129
Pr_3Se_4, 30A, 227
Pr_2Se_4Sr, 29, 129
PrSi, 29, 129
$PrSi_2$, 29, 129
$PrSn_3$, 29, 129
$PrTe_3$, 27, 336
Pr_2Te_3, 28, 49
Pr_3Te_4, 28, 49
Pr_5Tl_3, 34A, 141
$Pr_{13}Zn_{58}$, 35A, 123
Pt, 26, 216
Pt-Rh, 29, 201
PtS_2, 30A, 76, 163
$PtS_{0.75}Se_{0.25}$, 30A, 163
$PtSb_2$, 30A, 163
Pt_3Sb_2, 34A, 11
$Pt_{72}Sb_{14}Si_{14}$, 35A, 123
$PtSb_{0.7}Sn_{1.3}$, 34A, 141
PtSbSn, 34A, 141
$PtSb_{1.3}Sn_{0.7}$, 34A, 141
Pt-Sc, 27, 322
PtSc, 27, 342
Pt_3Sc, 26, 165; 27, 322
$PtSe_2$, 30A, 76, 163
PtSeTe, 30A, 163
Pt_2Si, 29, 129
Pt_3Si, 29, 129
Pt_6Si_5, 29, 79, 129
Pt_7Si_3, 29, 129
$Pt_{12}Si_5$, 30A, 163; 34A, 137
$Pt_{72}Si_{14}Sn_{14}$, 35A, 123
$Pt_{72}Si_{14}Tl_{14}$, 35A, 123
PtSm, 30A, 163
Pt-Sr, 27, 323
Pt_2Sr, 27, 323
Pt_3Sr, 27, 323
Pt_5Sr, 27, 323
Pt_2Ta, 29, 70, 129; 30A, 85, 163
Pt_3Ta, 29, 70, 129; 30A, 85, 163
Pt_4Ta, 29, 129
PtTb, 30A, 164
PtTe, 34A, 120
$PtTe_2$, 30A, 76, 164
Pt_2Te_3, 34A, 138
Pt_3Te_4, 34A, 137
PtTh, 27, 325
Pt_3Th_7, 26, 256
Pt_5Th_3, 28, 49
PtTi, 29, 129; 30A, 164
Pt_3Ti, 29, 129
Pt_8Ti, 30A, 86, 164
$PtTl_2$, 34A, 141
Pt_3Tl_2, 33A, 125
PtTm, 30A, 164
PtV, 29, 129, 130; 30A, 164
α-PtV_3, 29, 129
β-PtV_3, 29, 129
α-PtV_3(O), 29, 129
Pt_2V, 30A, 77, 164
Pt_3V, 26, 165; 29, 129
Pt_2W, 34A, 141
PtY, 30A, 164
Pt-Y, 27, 322
Pt_2Y, 26, 237
Pt_3Y, 26, 165
Pt_5Y, 27, 322
Pt_3Zn_{10}, 35A, 90
Pt_7Zn_{12}, 34A, 120
PtZr, 30A, 164
$Pt_{1+x}Zr$, 29, 130
α-Pu, 26, 228; 28, 38; 29, 202
β-Pu, 26, 229; 28, 39
δ-Pu, 26, 228; 27, 325
Pu-Sc, 29, 202
Pu-Si, 30A, 228
PuSi, 30A, 164
Pu_3Si_2, 30A, 164
Pu_3Si_5, 30A, 164
Pu_5Si_3, 29, 81, 130; 30A, 164
Pu-Ti, 26, 231
PuU, 27, 326
α-Pu-U, 29, 202
ε-Pu-U, 29, 202
δ-Pu-Zn, 26, 231
$Pu_2Zn_{\sim 9}$, 32A, 114
Pu_3Zn_{22}, 33A, 131
$Pu_{26}Zn_{116}$, 32A, 114
Pu-Zr, 26, 231
ε-Pu-Zr, 29, 202
δ-Pu-Zr, 29, 202

Ra, 33A, 134
RaS, 34A, 141
RaSe, 34A, 141
Rb, 26, 3
RbS_2Sn, 34A, 141
RbS_4W, 30A, 453
Rb-Sb, 26, 45
Rb_3Sb, 26, 54
β-Rb_3Sb, 26, 45
RbSi, 26, 84, 156
$RbSi_6$, 26, 157
RbSn, 29, 130
Re, 26, 240
Re-Rh, 27, 337
ReRu, 27, 338
Re_2Sc, 28, 49; 29, 130
$Re_{24}Sc_5$, 28, 49; 29, 130
$ReSe_2$, 30A, 86, 164
$ReSi_2$, 26, 24; 29, 84, 193
(Re,Ti)Si_2, 29, 84, 130
Re-Ta, 26, 241
Re_2Tb, 30A, 164
Re-Th, 26, 241
Re_2Th, 26, 237, 241
Re_2U, 26, 241
Re_3V, 26, 242
ReW, 30A, 228
ReW_3, 30A, 165
Re_2Y, 26, 237
Rh_2S_3, 32A, 120
$Rh_{17}S_{15}$, 27, 338
$RhSb_2$, 26, 35; 27, 61
$RhSb_3$, 26, 48
RhSc, 27, 342
Rh_3Sc, 26, 165
Rh_3Se_8, 33A, 135
$RhSi_{0.7\pm 0.1}$, 27, 373
Rh-Si, 27, 373
RhSi, 27, 373; 30A, 89, 165
Rh_2Si, 27, 373; 28, 40
Rh_3Si_4, 33A, 136
Rh_4Si_5, 33A, 136, 137
Rh_5Si_3, 28, 40
$Rh_{20}Si_{13}$, 30A, 88, 165
RhSm, 30A, 165
Rh-Ta, 29, 202
α-Rh-Ta, 29, 71
σ-Rh-Ta, 29, 202
σ-$RhTa_2$, 29, 130
α_1-$Rh_{1.21}Ta_{0.79}$, 29, 130
Rh_2Ta, 29, 70
α_2-Rh_2Ta, 29, 130
Rh_3Ta, 29, 130, 202
RhTb, 30A, 165
Rh_3Te_2, 31A, 61
Rh_3Te_4, 33A, 153
RhTh, 27, 340; 38, 49
Rh_2Th, 27, 367; 28, 49
Rh_3Th, 26, 165; 28, 49

Rh_3Th_7, 26, 256; 28, 49
$RhTi$, 27, 340; 29, 130
Rh_5Ti_3, 34A, 124
$Rh_{30}Ti_{70}$, 29, 130
$Rh_{35}Ti_{65}$, 29, 130
$Rh_{55}Ti_{45}$, 29, 130
$Rh_{57}Ti_{43}$, 29, 130
$RhTm$, 30A, 165
$RhTh$, 27, 340; 28, 49
RhU_2, 30A, 165
Rh_3U, 26, 165
$(Rh,Si)V_3$, 30A, 228
Rh_3W, 30A, 165
RhY, 30A, 165
Rh_2Y, 26, 237
Rh-Zr, 30A, 165, 228
$RhZr_2$, 30A, 165
Rh_4Zr_3(?), 29, 130
Rh_5Zr_3(?), 29, 130
Ru, 27, 383; 29, 203
RuS_2, 32A, 127
$RuSSb$, 29, 130
$RuSb_2$, 33A, 63
$RuSbSe$, 29, 130
$RuSbTe$, 29, 130
$RuSc$, 27, 342
Ru_2Sc, 26, 237
$RuSe_2$, 33A, 34
Ru-Si, 27, 372
$RuSi$, 27, 372
$RuSi_{1.8\pm0.3}$, 27, 372
Ru_2Si, 26, 243; 27, 372
Ru_2Si_3, 29, 130
Ru_4Si_3, 34A, 125
Ru_5Si_3, 35A, 123
Ru_2Sm, 26, 236
$RuSn_2$, 29, 130
Ru_2Sn_3, 29, 85, 130; 32A, 86
Ru-Ta, 28, 49; 30A, 229
$RuTa$, 30A, 165
$Ru_{1-x}Ta_{1+x}$, 26, 254
Ru_3Ta, 30A, 165
$RuTe_2$, 32A, 127
$RuTh$, 27, 367; 28, 49
Ru_2Th, 26, 236; 28, 49
Ru_3Th_7, 26, 256; 28, 49
$RuTi$, 28, 49
Ru-U, 30A, 229
RuU_2, 26, 243
RuV, 28, 49
Ru-W, 29, 203
Ru_2Y, 26, 237
$RuZr$, 28, 49

S, 26, 288; 27, 397, 725; 28, 52; 32A, 508; 34A, 149
γ-S, 27, 397
S_∞, 31A, 77
S_6, 26, 288
S_8, 26, 289; 30A, 252
S_{12}, 31A, 258
S_3Sc_2, 26, 245
S_3ScY, 35A, 101
$SSeSn$, 26, 246
$(S,Se)Sn$, 27, 345
$S_{1-x}Se_xSn$, 27, 344
SSi, 30A, 165
S_2Si, 30A, 55, 165, 453
$SSiZr$, 27, 348; 29, 130
SSm, 26, 244
S_2Sm, 26, 244
S_3Sm_2, 27, 330
S_4Sm_2Sr, 27, 327
S-Sn, 29, 203
SSn, 27, 352
S_2Sn, 30A, 165; 33A, 153
S_3Sn_2, 27, 352; 29, 131; 32A, 128
S_4SrY_2, 27, 327
S_4SrYb_2, 27, 327
S_3SrZr, 28, 151
S-Ta, 27, 359
STa_2, 34A, 129, 138
STa_6, 34A, 138; 35A, 105
S_2Ta, 27, 357
S_3Ta, 27, 357; 29, 131
S_4TaTl_3, 29, 91, 131
STb, 26, 255
$STeZr$, 27, 348
$S_{12}Th_7$, 30A, 90
STi_2, 32A, 130, 140
S_2Ti, 27, 353; 28, 49; 30A, 165
$S_{2-x}Ti$, 27, 354
$S_2Ti_{1.042-1.096}$, 27, 354
$S_{2.99}Ti$, 27, 354
S_3Ti, 27, 353
S_3Ti_2, 27, 355
S_4Ti_{2+x}, 35A, 108
S_4Ti_3, 27, 353; 31A, 68; 35A, 141
S_5Ti_3, 27, 355
S_5Ti_4, 31A, 68; 35A, 142
S_8Ti_5, 31A, 67
$S_{12}Ti_8$, 27, 356
STl_2, 35A, 106
S_4Tl_3V, 29, 91, 131
STm, 29, 131
S_3Tm_2, 34A, 141
S_7Tm_5, 29, 131
S_4Tm_2Yb, 29, 131
SU, 29, 131
S_2U, 29, 88, 131
$S_{0.95}V$, 28, 41; 34A, 141
S_4V, 29, 89, 131
S_4V_3, 30A, 165; 34A, 131
S_4V_5, 34A, 141
S_8V_5, 29, 131
S_2W, 29, 131
SY, 29, 131
S_3Y_2, 26, 90
S_3Y_5, 26, 88
S_7Y_5, 26, 277; 29, 131
S_4Y_2Yb, 29, 131
SYb, 29, 131
S_3Yb_2, 29, 131; 34A, 141
S_4Yb_3, 29, 131; 32A, 132
SZn, 26, 125, 251; 27, 201, 382; 28, 49; 30A, 229; 32A, 135; 33A, 146; 34A, 133
$S_{24}Zn_{24}$, 30A, 165
$S_{0.7}Zr$, 27, 349
$S_{0.8}Zr$, 27, 349
SZr_2, 32A, 140
S_2Zr, 30A, 166
S_4Zr_3, 35A, 69
Sb, 27, 76, 385; 28, 8; 35A, 9
Sb I, 29, 203
Sb II, 29, 204
Sb III, 29, 204
$SbSc$, 27, 42, 88
Sb_2Se_3, 27, 200; 29, 132; 30A, 166
Sb_2SeTe_2, 33A, 153
Sb_2Sm, 32A, 19, 20
Sb-$SnTe$, 26, 38
Sb_2SnTe_4, 26, 47
Sb-Sn-Te-Tl, 27, 363
Sb_2SnZn, 29, 186; 33A, 153
$(Sb,Sn)_2Zr$, 30A, 166
$SbTa_3$, 30A, 166
Sb_2Ta, 29, 131; 30A, 166
Sb_4Ta_5, 29, 131; 30A, 166
$SbTb$, 26, 255; 27, 334
Sb_2Te, 27, 42; 29, 131
Sb_2Te_2, 32A, 31
Sb_2Te_3, 26, 43; 27, 42, 200; 29, 131; 30A, 166; 33A, 153
$SbTe_2Tl$, 26, 46
Sb_4Th_3, 30A, 166
$SbTi$, 27, 45
$SbTi_{1.2}$, 27, 44
$SbTi_{1.7}$, 27, 44
α-$SbTi_3$, 27, 43
β-$SbTi_3$, 27, 44
$SbTm$, 29, 131
SbU, 29, 131
Sb_2U, 29, 131; 32A, 21
Sb_4U_3, 29, 131
SbV, 27, 46; 30A, 166
~Sb_2V_3, 29, 131
Sb_4V_5, 33A, 24
Sb_5V_6, 30A, 166
SbY, 27, 46; 29, 131
Sb_2Y, 31A, 20
$SbYb$, 29, 131; 32A, 23
Sb_2Yb, 32A, 20, 23
Sb_2Yb_5, 32A, 23
Sb_3Yb_4, 32A, 23
Sb_3Yb_5, 32A, 23
Sb_4Yb_5, 32A, 23
$Sb_{10}Yb_{11}$, 35A, 17
$SbZn$, 26, 34; 29, 22
$SbZn_{1.29}$, 30A, 166
Sb_3Zn_4, 27, 41

$Sb_{4.8}Zn_{6.2}$, 30A, 16
$SbZr_2$, 30A, 166
$SbZr_3$, 29, 132; 30A, 166
Sb_2Zr, 29, 131; 33A, 24
Sb_3Zr, 30A, 166
Sb_3Zr_5, 27, 379; 28, 27
Sc, 26, 234, 245
Sc_2Se_3, 29, 132; 30A, 167
$Sc_2(Se,Te)_3$, 30A, 230
ScSi, 30A, 167
Sc_3Si_5, 28, 49
Sc_5Si_3, 27, 341
Sc_5Sn_3, 30A, 167
Sc-Tc, 26, 255; 27, 360
ScTe, 26, 245
Sc_2Te_3, 29, 132; 30A, 91, 167
Sc-Ti, 27, 342
Sc-Y, 28, 49
ScZn, 28, 49
Sc-Zr, 28, 49
Se, 29, 204; 34A, 127
SeSiZr, 27, 348; 29, 132
SeSm, 29, 132
Se_2Sm, 32A, 20
$Se_{2-x}Sm$, 33A, 140
Se_3Sm_2, 29, 132
Se_4Sm_3, 29, 204
Se_4Sm_2Sr, 29, 132
Se_2Sn, 26, 246
Se-Sn-Te, 26, 153
SeTa, 27, 343
β-$Se_{1.67}Ta$, 29, 205
γ-$Se_{1.67}Ta$, 29, 205
β-$Se_{1.98}Ta$, 29, 205
γ-$Se_{1.98}Ta$, 29, 205
Se_2Ta, 29, 74, 132; 30A, 79, 167; 32A, 140; 34A, 138
α-Se_2Ta, 27, 343
β-Se_2Ta, 27, 343
Se_3Ta, 29, 132; 30A, 94, 167; 31A, 62
Se_3Ta_2, 33A, 113
Se_4TaTl_3, 29, 91, 132
$Se_2Ta_{1-x}W_x$, 28, 49
SeTb, 26, 255
Se_3Tb_2, 30A, 167
$(Se,Te)_2W$, 30A, 230
$Se_{0.95}Ti$, 26, 264
SeTi, 26, 265
$SeTi_2$, 32A, 140
$Se_{1.05}Ti$, 26, 265
$Se_{1.20-2.50}Ti$, 26, 266
Se_2Ti, 26, 267; 27, 346; 30A, 167
Se_2Ti_3, 26, 265
Se_4Ti_3, 27, 345; 29, 132, 205
Se_8Ti_5, 29, 132
Se_4Ti_2V, 30A, 167
Se_4Tl_3V, 29, 91, 132
SeTm, 29, 132
Se_3Tm_2, 30A, 167
$Se_{0.95}V$, 29, 133
$Se_{1.96}V$, 29, 133
$Se_{0.90-2.05}V$, 29, 205
Se_3V_2, 29, 132
Se_4V_3, 27, 345; 29, 132
Se_4V_5, 29, 132
Se_8V_5, 29, 132
Se_2W, 27, 343; 30A, 167
SeY, 29, 133
Se_2Y_5, 29, 133
Se_3Y_2, 28, 50; 30A, 167
Se_4Y_2Yb, 34A, 141
SeYb, 29, 133
Se_2Yb, 32A, 20
Se_3Yb_2, 29, 133; 30A, 168
Se_4Yb_3, 34A, 141
SeZn, 26, 246; 29, 184; 30A, 168
$SeZr_2$, 32A, 140
Se_2Zr, 30A, 168; 33A, 141; 35A, 123
Se_3Zr, 30A, 168
Se_3Zr_2, 33A, 141, 142
Se_4Zr_3, 33A, 141, 142
Si, 26, 247, 287; 27, 208, 347; 28, 49, 50; 29, 205, 206; 30A, 230
Si II, 29, 55, 133
SiSm, 29, 133; 30A, 168
Si_3Sm_5, 30A, 168
$(Si,Sn)V_3$, 30A, 231
SiSr, 27, 348
Si_2Sr, 30A, 97, 168; 31A, 63
Si_3Sr_2, 30A, 168
Si_7Sr_4, 32A, 122
Si-Ta, 26, 25
$SiTa_2$, 26, 25
$SiTa_3$, 29, 133; 30A, 168; 34A, 141
$SiTa_4$, 26, 26
$SiTa_{4.5}$, 26, 25
Si_2Ta, 26, 25, 246
Si_3Ta_5, 26, 25, 247
SiTb, 29, 133
Si_2Tb, 27, 336; 28, 49; 29, 133
Si_3Tb_5, 30A, 168
Si-Tc, 30A, 168
SiTc, 30A, 168
Si_3Tc_5, 30A, 168
SiTeZr, 29, 133; 30A, 168
Si_2Th, 26, 256
SiTi, 26, 18
$SiTi_3$, 29, 133; 30A, 169
Si_2Ti, 29, 84; 32A, 86
Si_3Ti_5, 27, 31
Si_4Ti_5, 32A, 140; 34A, 141
$Si(Ti,V)_3$, 30A, 231
$Si_3(Ti,V)_5$, 30A, 231
SiTm, 29, 133
Si_2Tm, 27, 336; 28, 49
Si_3Tm_5, 30A, 169
SiU_3, 30A, 169; 33A, 29
$Si_{1.67}U$, 26, 271
$Si_{1.88}U$, 26, 271
Si_2U, 26, 271
SiV_3, 31A, 64
$SiV_{3\pm x}$, 30A, 231
Si_2V, 27, 375
Si_3V_5, 27, 284
Si_5V_6, 32A, 140
SiW_3, 27, 159, 284
Si_2W, 26, 26, 103, 156; 27, 284; 29, 207
Si_3W_5, 26, 103, 156; 27, 159
$Si_{1.6}Y$, 32A, 140
Si_2Y, 32A, 140
α-Si_2Y, 27, 336
β-Si_2Y, 27, 336
Si_5Y_3, 28, 49
SiYb, 29, 133
Si_2Yb, 27, 336
Si_3Yb_5, 30A, 169
SiZr, 30A, 97, 169; 34A, 141
$SiZr_2$, 26, 10; 27, 32
$SiZr_3$, 29, 133; 30A, 169
Si_2Zr, 26, 64; 27, 32
Si_2Zr_3, 26, 11
Si_3Zr_5, 27, 32
Si_4Zr_5, 31A, 65; 33A, 143
$SmSn_3$, 30A, 169
SmTe, 29, 133
$SmTe_2$, 32A, 20
$SmTe_3$, 30A, 169
Sm_2Te_3, 29, 133; 30A, 169
Sm_3Te_4, 28, 49
SmZn, 29, 133
$SmZn_3$, 35A, 123
$SmZn_{11}$, 35A, 98
Sm_2Zn_{17}, 30A, 169
Sm_3Zn_{11}, 35A, 123
Sm_3Zn_{22}, 35A, 123
$Sm_{13}Zn_{58}$, 35A, 123
Sn, 26, 260, 280; 27, 369; 28, 49; 30A, 169
β-Sn, 27, 382
Sn II, 29, 133; 31A, 69
$SnTa_3$, 29, 133
$SnTaV_2$, 29, 133
$SnTa_2V$, 29, 133
$Sn_{0.97}Te$, 27, 211
SnTe, 26, 40; 27, 211
SnTe II, 29, 133
Sn_3Th, 30A, 169
Sn_5Ti_6, 30A, 169
α-Sn_5Ti_6, 29, 133
β-Sn_5Ti_6, 29, 54, 133
SnV_3, 29, 133
Sn_3V_2, 30A, 169
Sn_3Y_5, 29, 133

Sn_3Yb, 30A, 169
$SnZr_{3+}$, 30A, 170
Sn_4Zr_5, 29, 133; 30A, 169
Sr, 28, 49
β-Sr, 27, 265
γ-Sr, 27, 265
δ-Sr, 27, 265
$SrTl_2$, 29, 3, 133
$SrZn_2$, 29, 87, 133

Ta, 27, 356; 30A, 170, 231
Ta-Tc, 27, 361
TaTc, 26, 255
TaTe, 27, 343
$TaTe_2$, 27, 343; 30A, 170; 31A, 69
$Ta_{1+x}Te_2$, 30A, 170
$TaTe_4$, 29, 77, 133
TaV_2, 30A, 170; 31A, 68
Ta-W-Zr, 29, 207
TaZr, 27, 360
Tb, 26, 234; 27, 334; 30A, 170, 231
$TbTc_2$, 29, 133
TbTe, 26, 255
$TbTe_3$, 30A, 170
Tb_2Te_3, 30A, 170
Tb_5Tl_3, 34A, 141
TbZn, 29, 133
$TbZn_3$, 35A, 123
$TbZn_{12}$, 30A, 170
Tb_2Zn_{17}, 30A, 170
Tb_3Zn_{11}, 35A, 123
Tb_6Zn_{23}, 30A, 170
$Tb_{13}Zn_{58}$, 35A, 123
Tc, 30A, 170
Tc_2Th, 30A, 170
Tc-Ti, 26, 255; 27, 360
Tc_2Tm, 29, 134
TcU_2, 30A, 170
Tc-V, 27, 361
Tc-W, 27, 361
Tc_2Y, 29, 134
$TcZn_6$, 29, 134
Tc-Zr, 26, 255; 27, 360
Tc_6Zr, 26, 255
Te, 28, 49; 29, 207; 30A, 231, 232; 32A, 137
Te_2Ti, 27, 366; 30A, 170
Te_2Ti_{2-x}, 27, 364
Te_4Ti_3, 27, 345
Te_4Ti_5, 26, 267; 27, 364
Te_4Ti_2V, 30A, 170
TeTl, 34A, 138
Te_3Tl_5, 35A, 108, 109
TeTm, 29, 134
Te_3Tm, 30A, 171
Te_3Tm_2, 30A, 171
Te_2U, 35A, 110
Te_3V_4, 27, 345
Te_2W, 27, 343; 31A, 69
TeY, 29, 134
Te_3Y, 30A, 171
Te_3Y_2, 30A, 171
TeYb, 29, 134
Te_2Yb, 32A, 20
TeZn, 27, 382; 29, 134
Th, 27, 368
α-Th, 26, 258
Th-U, 26, 258
ThV, 27, 368
$ThZn_2$, 26, 256, 258
$ThZn_4$, 26, 258
Th_2Zn_{17}, 33A, 132
Th-Zr, 26, 260
Ti, 26, 261; 28, 49
α-Ti, 27, 308, 370
β-Ti, 27, 249, 340
Ti-U, 26, 268
Ti-V, 26, 268
Ti-Y, 27, 381
TiZn, 27, 372
$TiZn_3$, 27, 372
Ti_2Zn, 30A, 171
Ti-Zr, 26, 168
Tl, 26, 279; 28, 49
β-Tl, 27, 367
TlY, 28, 49
Tl_3Y, 28, 49
Tm, 26, 234; 27, 334, 368
$TmZn_3$, 35A, 123
Tm_2Zn_{17}, 30A, 171
$Tm_{13}Zn_{58}$, 35A, 123

α-U, 26, 271; 27, 379, 381; 28, 41
β-U, 31A, 72

V, 27, 380; 30A, 232; 31A, 54
V-W, 27, 138
VZn_3, 28, 49
VZn_3(O?), 29, 134
V_4Zn_5, 28, 50; 29, 92, 134

W, 26, 27, 223; 27, 381; 29, 207; 30A, 232

Xe, 26, 279, 286; 27, 392

Y, 26, 234; 28, 49
YZn, 28, 49; 29, 134
YZn_5, 29, 134
YZn_{12}, 30A, 171
Y_2Zn_{17}, 30A, 171
Y_6Zn_{23}, 30A, 171
Y-Zr, 27, 381
Yb, 26, 234
$YbZn_2$, 30A, 98, 171
$YbZn_{11}$, 30A, 171
$YbZn_{12}$, 30A, 171
Yb_3Zn_{17}, 35A, 123
$Yb_{13}Zn_{58}$, 35A, 123

Zn, 26, 279; 30A, 232
$ZnZr_2$, 29, 134; 30A, 171
Zn_2Zr_3, 26, 277
Zn_3Zr, 30A, 171
$Zn_{22}Zr$, 26, 277
Zr, 28, 49; 29, 208
α-Zr, 26, 74; 27, 241, 304
β-Zr, 26, 74

INDEX OF CARBON COMPOUNDS

In this index compounds containing carbon and hydrogen are listed primarily according to the number of carbon atoms in the formula; within groups with the same number of carbon atoms the order is that of increasing number of hydrogen atoms.

A few compounds that can be regarded as organic, even though they do not contain hydrogen, are also included. They are arranged as if the formula were written C_nH_0ABC .... Polymers with indefinite suffixes have been collected at the end.

C_1
CAgNO, 27, 747; 30B, 289
CB_3NO_2, 27, 746
$CBrCl_3Hg$, 33B, 320
CCl_4, 30B, 400
CClFO, 26, 543
CCl_6N_9Sb, 35B, 468
$CCsN_3O_6$, 31B, 52
CD_4, 26, 543
CD_5N, 30B, 400
CD_3NO, 33B, 538
CD_4N_2S, 33B, 30
CF_3NO, 30B, 400
CF_3NO_2, 27, 746
CF_2O, 27, 729
CIN_3O_6, 31B, 51
CKN_3O_6, 31B, 52; 32B, 64
CN_3O_6Rb, 31B, 52; 33B, 30
CN_2Sr, 31B, 39
$CHCsN_2O_4$, 33B, 556
$CHCl_3$, 31B, 33
$CHClF_2$, 27, 729
$CHCl_2F$, 29, 449
CHFO, 26, 543
$CHKN_2O_4$, 33B, 556
$CHI_3.C_{27}H_{21}N_3$, 27, 986
CHI_3S_{24}, 27, 747
$CHN_4Na.H_2O$, 28, 453
CHN_2O_4Rb, 33B, 556
CH_2F_3P, 33B, 538
$CH_2K_2O_6S_2$, 27, 752
CH_2N_2, 26, 586
CH_2O, 28, 664
CH_2S_3, 30B, 53
CH_3, 26, 635
CH_3AsI_2, 28, 585
CH_3Br_3Si, 33B, 538
CH_3CsO, 35B, 472
CH_3CsO_3S, 32B, 64
$CH_3CuN_3O_8$, 31B, 435
CH_3F_4P, 30B, 400
CH_3K, 33B, 545; 35B, 472
CH_3KN_2O, 30B, 48
CH_3KO, 28, 595; 35B, 472
CH_3Li, 35B, 471
CH_3NaO, 29, 452
$CH_3NaO_2S_2$, 29, 484
CH_3NO, 28, 664; 33B, 538
$CH_3NO_2.AlCl_3$, 26, 704
CH_3ORb, 35B, 472
CH_4, 26, 543
CH_4N_2O, 26, 568, 574; 27, 748; 29, 473; 30B, 52
$CH_4N_2O_2$, 31B, 44; 32B, 52; 33B, 35
$CH_4N_2O_2S$, 27, 749
CH_4N_2S, 26, 572, 574; 32B, 53; 33B, 30
CH_4N_2Se, 27, 750; 28, 659; 34B, 29
CH_4O_2S, 34B, 42
CH_5AgClN_3S, 33B, 466
CH_5CaO_5P, 27, 751
$CH_5Cl_2CuN_3O$, 30B, 407
$CH_5Cl_2N_3OZn$, 30B, 407
CH_5ClSi, 27, 732
$CH_5NO.ClH$, 29, 474
CH_5NO_2, 33B, 11
$CH_5N_3O_4$, 34B, 30
CH_5N_5O, 32B, 152
$CH_5N_5O_6$, 32B, 485 35B, 18
CH_5NO_2S, 31B, 32
$CH_5N_3O_3S$, 33B, 31
CH_5N_3S, 34B, 29, 32; 35B, 28
$CH_6AlNO_8S_2.12H_2O$, 29, 537
CH_6BrN, 26, 585
CH_6ClN_3, 30B, 44
$CH_6ClN_2NaO_2$, 28, 570
CH_6Cl_3NNi, 31B, 29
CH_6ClNO, 32B, 28
CH_6ClNO_4, 33B, 21
CH_6ClN_3O, 30B, 46
$CH_6ClN_3O_4$, 30B, 407
$CH_6N_2S_2$, 32B, 486
CH_6N_4Se, 27, 751
CH_6OSi, 35B, 888
$CH_7AsN_2O_5$, 33B, 542
CH_7CaO_6P, 26, 52; 27, 751
CH_7NaO_5S, 27, 754
$CH_7N_2O_5P$, 32B, 485
CH_8BrN_3O, 28, 569
$CH_8CdN_2O_6S_2$, 32B, 407
CH_8Si_2, 35B, 888
$CH_{11}B_5$, 33B, 538
$CH_{18}AlN_3O_{14}S_2$, 27, 828; 32B, 57
$CH_{18}CrN_3O_{14}S_2$, 32B, 57
$CH_{30}AlNO_{20}S_2$, 26, 585
$CH_{30}CrNO_{20}S_2$, 33B, 519

C_2
C_2Ag_2O, 35B, 914
C_2BrCl, 35B, 903
$C_2Br_2O_2$, 26, 546; 27, 729
C_2Br_4, 33B, 237
$C_2CaO_4.H_2O$, 30B, 408
$C_2CaO_4.2.17H_2O$, 30B, 22
$C_2CdO_4.3H_2O$, 30B, 407
C_2Cl_6, 29, 449
C_2ClF_3, 26, 587
C_2Cl_3N, 26, 588
$C_2Cl_2O_2$, 27, 729
$C_2Cl_4S_2$, 34B, 178
$C_2Cl_6S_3$, 26, 592
$C_2CsO_4.H_2O$, 30B, 407
$C_2Cs_2O_2$, 29, 467
$C_2CuO_4.nH_2O$, 33B, 402
C_2D_6, 30B, 401
$C_2D_2O_4.2H_2O$, 29, 459; 30B, 395; 32B, 3
$C_2D_2O_4.2D_2O$, 34B, 4, 6
C_2D_4, 30B, 402

$C_2D_6O_6$, 32B, 3
C_2F_6N, 35B, 891
C_2HgIKN_2, 28, 104
$C_2K_2MoO_9$, 35B, 635, 907
$C_2K_2N_4O_8$, 32B, 485; 33B, 34
$C_2K_2O_4.H_2O$, 29 463: 34B, 16
$C_2K_2O_9W$, 35B, 907
$C_2Li_2O_4$, 29, 460
$C_2Li_2O_4.H_2O_2$, 34B, 15
C_2N_2, 28, 401
$C_2N_2O_4.2H_2O_2$, 29, 462
$C_2N_3O_4Rb$, 32B, 48; 33B, 30
$C_2N_4O_8Rb$, 32B 486
$C_2O_6Pu.3H_2O$, 30B, 408
$C_2O_8Pu.xC_3H_6O.2H_2O$, 30B, 409
$C_2O_2Rb_2$, 29, 465
$C_2O_4Rb_2.H_2O$, 30B, 24
$C_2O_4Sr.2.17H_2O$, 30B, 22
C_2O_6U, 29, 762
$C_2O_6U.H_2O$, 29, 762
$C_2O_6U.3H_2O$, 29, 762; 30B,408
$C_2O_8U.2H_2O$, 30B, 409
C_2HK, 33B, 296
C_2HKO_4, 33B, 11
$C_2HLiO_4.H_2O$, 35B, 9
C_2HNa, 33B, 296
$C_2HNaO_4.H_2O$, 35B, 9
C_2HRb, 33B, 296
$C_2H_2B_{10}Cl_{10}$, 30B, 413
$C_2H_2CaO_4$, 28, 669
$C_2H_2CaO_5$, 27, 819
$C_2H_2CdO_4$, 28, 669
$C_2H_2CdO_4.2H_2O$, 28, 669
$C_2H_2Cl_2$, 30B, 402
$C_2H_2Cl_2Ge$, 35B, 889
$C_2H_2Cl_2Hg$, 31B, 341
C_2H_2ClN, 26, 588
$C_2H_2CuO_4$, 26,547
$C_2H_2CuO_4.1/2C_4H_8O_2$, 30B, 415
$C_2H_2CuO_4.2H_2O$, 30B, 346
$C_2H_2DLiO_3$, 31B, 25
$C_2H_2D_2O_4$, 35B, 890
$C_2H_2F_2$, 26, 544, 545; 28, 664
C_2H_2FN, 26, 587
$C_2H_2GeI_2$, 35B, 899
$C_2H_2KMoO_7$, 28, 669
$C_2H_2K_2O_5$, 35B, 8
$C_2H_2K_2O_6$, 32B, 21
$C_2H_2N_3O_5Rb$, 35B, 19
$C_2H_2N_2OS_2$, 31B, 155
$C_2H_2N_2S$, 28, 664; 29, 733; 35B, 889; 26, 760
$C_2H_2N_2S_3$, 28, 497
$6\cdot 4C_2H_2O.46H_2O$, 30B, 267
$C_2H_2O_2$, 33B, 530
$C_2H_2O_4$, 27, 1066; 35B, 889
$C_2H_2O_4.2H_2O$, 27, 1066; 34B, 3, 4
$C_2H_2O_6Rb_2$, 32B, 21
$C_2H_2O_4Zn.2H_2O$, 28, 669
C_2H_3Br, 28, 664
C_2H_3BrCuN, 33B, 551
$C_2H_3BrN_2$, 35B, 903
$C_2H_3Cl_3$, 33B, 539
C_2H_3ClCuN, 34B, 546
$C_2H_3ClHgO_2$, 27, 816
$C_2H_3Cl_5NSb$, 32B, 266
C_2H_3ClO, 26, 546
$C_2H_3Cl_3O_2$, 27, 731; 28, 400
$C_2H_3Cl_6OSb$, 32B, 318, 33B, 293
$C_2H_3Cl_3Si$, 29, 450
C_2H_3F, 26, 544
$C_2H_3F_3N_2$, 35B, 891
$C_2H_3F_6OSb$, 33B, 38
C_2H_3HgN, 33B, 320
$C_2H_3KO_4$, 33B, 13
$C_2H_3LiO_2$, 26,546
$C_2H_3LiO_2.2H_2O$, 35B, 11
$C_2H_3LiO_3.H_2O$, 30B, 16
$C_2H_3LiO_5$, 35, 9
$C_2H_3N_3$, 30B, 127; 34B, 122
$C_2H_3N_3.Cl_2Cu$, 27, 859
$C_2H_3NaO_2.2H_2O$, 30B, 409
$C_2H_3NaO_2.3H_2O$, 30B, 409
$C_2H_3NaO_5$, 35B, 9
C_2H_4, 30B, 401
$C_2H_4B_{10}Cl_8$, 29, 529
$C_2H_4Br_3KPt.H_2O$, 30B, 291
$C_2H_4BrN_3S_2$, 30B, 157
$C_2H_4Cl_2$, 26, 545
$C_2H_4Cl_5Cs_2O_2Ti$, 27, 816
$C_2H_4Cl_3KPt.H_2O$, 30B, 291; 34B, 437
$C_2H_4ClS_2Sb$, 32B, 319
$C_2H_4Cl_4Si_2$, 33B, 529
C_2H_4FNO, 27, 756
$C_2H_4F_3NO_2$, 29, 456
$C_2H_4N_4$, 31B, 41
$C_2H_4N_2NiS_4$, 32B, 411
$C_2H_4N_2O_2$, 31B, 53
$C_2H_4N_4O_2$, 26, 589
$C_2H_4N_2S_2$, 30B, 36
$C_2H_4O_2$, 35B, 3
$C_2H_4O_3$, 32B, 497
$C_2H_4O_4$, 35B, 890
C_2H_4Si, 28, 664
C_2H_5, 26, 635
$C_2H_5Br_2N$, 32B, 28
$C_2H_5BrNNaO$, 31B, 271
C_2H_5Cl, 27, 732
$C_2H_5Cl_2N$, 33B, 23
$C_2H_5ClN_3OS_2$, 31B, 156
C_2H_5ClO, 30B, 402
$C_2H_5Cl_4OSb$, 33B, 287
$C_2H_5CoN_3O_2$, 30B, 409
C_2H_5Li, 28, 593
C_2H_5N, 26 590
$C_2H_5N_3NiO_2$, 30B, 409
C_2H_5NO, 30B, 12
$C_2H_5NO_2$, 26, 588
$C_2H_5NO_2.1/2H_2O$, 35B, 20
$C_2H_5NO_3$, 34B, 349
$C_2H_5NO_5$, 30B, 58
$C_2H_5N_5S$, 32B, 487
C_2H_5Na, 35B, 473
C_2H_6, 27, 732; 30B, 401; 33B, 529
$C_2H_6Al_2Cl_4$, 28, 596
$C_2H_6B_4$, 35, 904
$C_2H_6BCl_2N$, 35B, 890
$C_2H_6CdCl_2N_2O_2$, 32B, 444
$C_2H_6CdN_4O_4.H_2O$, 30B, 409; 33B, 436; 34B, 547
$C_2H_6Cl_2CuN_2O$, 34B, 573
$C_2H_6Cl_6N_2P_2$, 31B, 299
$C_2H_6Cl_2Sn$, 35B,485
$C_2H_6CoO_6$, 32B, 414
$C_2H_6CuN_4S_2$, 30B, 289
$C_2H_6CuO_6$, 33B, 402
$C_2H_6F_2NP$, 34B, 376
$C_2H_6F_3P$, 30B, 400
$C_2H_6F_2Sn$, 31B, 328
$C_2H_6HgS_2$, 27, 835
$C_2H_6HgS_2$, 29, 630
$C_2H_6KO_2PS_2$, 27, 755
$C_2H_6MgO_6$, 29, 455
$C_2H_6MnN_4O_4$, 30B, 409
$C_2H_6MnN_4O_4.2H_2O$, 32B 425
$C_2H_6MnO_6$, 29, 455; 33B, 402
$C_2H_6N_2$, 35B, 890, 891
$C_2H_6N_4NiO_4.2H_2O$, 35B, 720
$C_2H_6N_4NiS_4$, 28, 661
$C_2H_6N_2O$, 33B, 529
$C_2H_6N_2O_2$, 28, 404
$C_2H_6N_2O_3$, 28, 382
$C_2H_6N_2O_4$, 33B, 36
$C_2H_6N_4O_4$, 33B, 33
$C_2H_6NiO_6$, 28,628
C_2H_6O, 28, 664; 33B, 539
C_2H_6OS, 31B, 55, 56
$C_2H_6O_2S$, 28,406; 35B, 890
$C_2H_6O_6S$, 33B, 38
$C_2H_6O_6Sr$, 26, 548
$C_2H_6P_2S_4$, 27, 942; 29, 539
C_2H_6S, 26, 586; 33B, 539
C_2H_6Si, 26, 586
$C_2H_7AsO_2$, 30B, 282
C_2H_7AuO, 33B, 321
$C_2H_7Br_2NPt$, 29, 586
$C_2H_7Br_3O_2Sn$, 28, 602
$C_2H_7CaO_5P$, 27, 752
$C_2H_7ClN_2O$, 35B, 34
$C_2H_7Cl_3O_2Sn$, 28, 602
$C_2H_7LiO_4$, 35B, 11
C_2H_7N, 33B, 530, 539
$C_2H_7NO_2$, 32B, 22
$C_2H_7NO_2.C_2H_4O_2$, 34B, 16
$C_2H_7N_3O_3$, 26, 570
$C_2H_7N^+O_3^-S$, 29, 484
$C_2H_7NO_3S$, 28, 524; 31B, 187
$C_2H_7O_2P$, 32B, 304
$C_2H_8AgClN_4S_2$, 33B, 469
$C_2H_8B_4$, 29, 526
C_2H_8BN, 34B, 359
$C_2H_8Br_2CdN_4O_2$, 35B, 909
$C_2H_8Br_2N_4O_2Zn$, 35B, 909
$C_2H_8Br_3N_2Pt$, 26, 689
$C_2H_8Br_2N_4S_2Te$, 31B, 335
$C_2H_8Br_2O_2$, 29, 451
$C_2H_8Cl_2CoN_4S_2$, 34B, 521
$C_2H_8Cl_3CuN$, 31B, 29

$C_2H_8ClCuN_4S_2$, 35B, 784
$C_2H_8Cl_2HgO_2$, 30B, 368
C_2H_8ClN, 33B, 24
$C_2H_8Cl_2N_4S_2Te$, 31B, 335
$C_2H_8Cl_2N_2Se_2$, 35B, 38
$C_2H_8CrN_2O_4.2H_2O$, 30B, 328
$C_2H_8HgI_2N_4S_2$, 33B, 467
$C_2H_8ILiN_4O_2$, 35B, 474
$C_2H_8I_2N_4S_2Te$, 26, 576
$C_2H_8MoNNaO_9$, 27, 820; 28, 629
$C_2H_8N_6NiS_2$, 27, 837
$C_2H_8N_2O_4.H_2O$, 30B, 21
$C_2H_8NO_3P$, 31B, 301
$C_2H_8NO_4P$, 26, 591
$C_2H_8N_2O_8U.3H_2O$, 29, 762
$C_2H_8N_2O_5V.2H_2O$, 33B, 543
$C_2H_8N_2S$, 32B, 485
$C_2H_9B_{10}Br_3$, 31B, 283
$C_2H_9BF_4N_2$, 26, 594
$C_2H_9B_{10}I_3$, 30B, 411
$C_2H_9CaO_6P$, 26, 552; 27 752
$C_2H_9N_5O_4S.H_2O$, 28, 669
C_2H_9NSi, 35B, 891
$C_2H_{10}B_{10}Br_2$, 31B, 283; 32B, 292
$C_2H_{10}B_{10}Cl_2$, 30B, 410
$C_2H_{10}B_{10}I_2$, 30B, 410, 412; 33B, 530
$C_2H_{10}Cl_4CuN_2$, 32B, 485
$C_2H_{10}Cl_2CuN_6O_2$, 30B, 360
$C_2H_{10}Cl_2N_2$, 28, 391
$C_2H_{10}Cl_2N_6O_2Zn$, 30B, 360
$C_2H_{10}MnN_4O_6$, 32B, 425
$C_2H_{10}N_6NiO_4S_3$, 33B, 471
$C_2H_{10}N_2O_2$, 33B, 13
$C_2H_{10}N_2O_4S$, 26, 589
$C_2H_{10}N_4O_6U$, 28, 669
$C_2H_{10}O_8S_2$, 35B, 40
$C_2H_{11}B_{10}Br$, 30B, 410
$C_2H_{11}B_{10}I$, 30B, 410, 412
$C_2H_{11}ClCoN_5O_4$, 31B, 380
$C_2H_{11}CoN_6O_6$, 35B, 586
$C_2H_{11}N_2O_2$, 33B, 37
$C_2H_{12}B_{10}$, 30B, 412
$C_2H_{12}Cl_3CrN_2O_2$, 34B, 480
$C_2H_{12}N_6O_8S_2Zn$, 35B, 733
$C_2H_{13}B_5$, 31B, 287
$C_2H_{13}B_{10}I$, 32B, 491
$C_2H_{13}N_3O_2$, 33B, 15
$C_2H_{13}N_5P_2S_2$, 33B, 273
$C_2H_{14}B_4F_2NP$, 34B, 377
$C_2H_{14}N_8NiO_8S_2$, 33B, 472
$C_2H_{14}N_8O_4Zn$, 30B, 367
$C_2H_{16}B_9N$, 26, 619
$C_2H_{16}IN_7P_2$, 33B, 274
$C_2H_{18}B_{10}$, 29, 524
$C_2H_{18}Cl_2CoN_5O_6$, 30B, 344
$C_2H_{20}N_2O_4$, 32B, 269
$C_2H_{60}Al_2N_2O_{40}S_4$, 27, 813

C_3

C_3BrN, 33B, 29
C_3Cl_6, 30B, 399
C_3ClN, 27, 758; 33B, 29
C_3F_6O, 35B, 891
$C_3O_2S_3$, 34B, 179
C_3HF_6N, 35B, 891
$C_3H_2S_3$, 28, 670
$C_3H_3AsN_2$, 31B, 312
$C_3H_3BiO_6$, 34B, 395
C_3H_3Br, 26, 549
$C_3H_3Ca_{1.5}N_2O_6P_2.6H_2O$, 33B, 107
$C_3H_3CeO_6$, 27, 815
C_3H_3Cl, 26, 549
C_3H_3ClO, 26, 550
$C_3H_3F_3O_4$, 26, 828
C_3H_3K, 33B, 296
$C_3H_3KO_4$, 35B, 12
C_3H_3N, 35B, 892
$C_3H_3N_3$, 32B, 155
$C_3H_3NOS_2$, 27, 952
C_3H_3Na, 33B, 296
$C_3H_3NaO_3$, 26, 551
$C_3H_3NdO_6$, 27, 815
$C_3H_3O_2Br$, 29, 454
$C_3H_3O_6Pr$, 27, 815
$C_3H_3O_6Sc$, 33B, 402
$C_3H_3O_6Sm$, 27, 815
C_3H_4, 35B, 892
C_3H_4ClF, 33B, 540
$C_3H_4Cl_2O_3S$, 33B, 147
$C_3H_4I_2N_2$, 26, 615
$C_3H_4N_2$, 31B, 115; 34B, 123; 35B, 152, 153
$C_3H_4N_4NiOS_4$, 33B, 556
$C_3H_4N_2O$, 35B, 21
$C_3H_4N_2OS$, 34B, 124
C_3H_4O, 33B, 530; 35B, 892
$C_3H_4O_2$, 26, 550; 28, 365
$C_3H_4O_5$, 30B, 3
$C_3H_5BrN_2S_2.H_2O$, 28, 669
C_3H_5BrO, 26, 594
C_3H_5Cl, 27, 732
$C_3H_5ClN_2S_2.H_2O$, 28, 669
$C_3H_5Cl_6OSb$, 33B, 293
$C_3H_5IN_2S_2$, 28, 669
$C_3H_5KOS_2$, 28, 407
C_3H_5NO, 31B, 3
$C_3H_5NO_4$, 31B, 3
$C_3H_5NaOS_2.2H_2O$, 34B, 41
C_3H_6, 26, 549; 29, 490
$C_3H_6AgClO_4S_3.H_2O$, 33B, 473
$C_3H_6AgNO_3S_3$, 33B, 473
$C_3H_6AgNO_3S_3.H_2O$, 33B, 473
C_3H_6AsN, 28, 586
$C_3H_6Cl_2HgS_3$, 31B, 448
$C_3H_6Cl_5N_2P$, 34B, 378
$C_3H_6N_4NiO_2S_4$, 32B, 426
$C_3H_6N_2O_2$, 30B, 413; 35B, 23
$C_3H_6N_4O_3$, 30B, 41; 31B, 45
$C_3H_6N_5O_8$, 31B, 68
$C_3H_6O.17H_2O$, 26, 344
$C_3H_6O_2$, 26, 594; 27, 736
$C_3H_6O_3$, 27, 966; 28, 480; 34B, 169
$C_3H_6O_3S$, 31B, 159; 35B, 893
$C_3H_6S_3$, 30B, 152; 32B, 181; 34B, 181
$C_3H_6S_3.AgClO_4.H_2O$, 33B, 473
$C_3H_6S_3.AgNO_3$, 33B, 473
$C_3H_6S_3.AgNO_3.H_2O$, 33B, 473
$C_3H_6Se_3$, 33B, 136
C_3H_7, 26, 635
$C_3H_7ClN_2$, 27, 951
$C_3H_7Cl_5NOSb$, 31B, 324
$C_3H_7Cl_2N_2O_2Zn$, 35B, 924
$C_3H_7Cl_2N_4Pd$, 35B, 920
$C_3H_7N_5NiS_4$, 35B, 786
C_3H_7NO, 27, 759
$C_3H_7NO.ClLi$, 27, 762
$C_3H_7NO.ILi$, 27, 762
$C_3H_7NO_2$, 27, 763; 30B, 207; 31B, 187; 32B, 5
$C_3H_7NO_2S$, 33B, 172
$C_3H_7NO_5S$, 33B, 174
$C_3H_7Na_2O_{11}P$, 31B, 176
$C_3H_7O_4P$, 30B, 194; 35B, 446
$C_3H_8BrN_3O_2$, 32B, 497
$C_3H_8ClNO_2$, 27, 764; 32B, 489
$C_3H_8NO_6P.H_2O$, 35B, 299
$C_3H_8N_4S_2$, 34B, 34
C_3H_8O, 26, 550
$C_3H_8O_3$, 33B, 157
$C_3H_9AlBr_2OSi$, 33B, 326
$C_3H_9AlCl_3N$, 34B, 403
$(C_3H_9AlS)_n$, 34B, 403
$C_3H_9As_3Cl_3N_3$, 32B, 314
C_3H_9B, 30B, 400
$C_3H_9BCl_3N$, 34B, 360
$C_3H_9BF_4OS$, 28, 409
$C_3H_9B_3N_3$, 31B, 288
$C_3H_9Cl_5OPSb$, 26, 595
$C_3H_9ClO_4Se$, 29, 766
C_3H_9CsOSi, 35B, 440
C_3H_9FSn, 29, 625
$C_3H_9HgI_3S$, 28, 686; 31B, 450
C_3H_9IPt, 33B, 322
C_3H_9ISe, 31B, 58
C_3H_9KOSi, 35B, 440
C_3H_9N, 33B, 530
$C_3H_9NO_2S_2$, 32B, 38, 485
C_3H_9OP, 30B, 401
C_3H_9ORbSi, 35B, 440
$C_3H_9O_4ReSi$, 34B, 364
$C_3H_9O_4V$, 31B, 440
C_3H_9Tl, 35B, 508
$C_3H_{10}B_{10}O$, 35B, 427
$C_3H_{10}ClN$, 33B, 25
$C_3H_{10}ClNO$, 27, 762
$C_3H_{10}ClNO_3S$, 33B, 174
$C_3H_{10}IN$, 35B, 25
$C_3H_{10}OPt$, 33B, 323
$C_3H_{10}OSn$, 30B, 413
$C_3H_{11}B_{10}Cl_3$, 30B, 411
$C_3H_{11}CaO_6P$, 26, 552
$C_3H_{11}CuNO_6$, 26, 663
$C_3H_{12}B_{10}Br_2$, 30B, 411
$C_3H_{12}B_{10}Cl_2$, 30B, 411
$C_3H_{12}CdN_6O_4S_4$, 35B, 627

$C_3H_{12}ClCuN_6S_3$, 29, 593
$C_3H_{12}Cl_2HgN_6S_3$, 26, 575; 33B, 477
$C_3H_{12}Cl_2N_2$, 33B, 26
$C_3H_{12}GaN$, 28, 596
$C_3H_{12}INaO_3$, 30B, 264
$C_3H_{12}NO_9P_3$, 32B, 304
$C_3H_{12}N_6O_4S_4Zn$, 33B, 479
$C_3H_{13}BN_2O$, 27, 763
$C_3H_{14}F_4N_6S_3Te$, 26, 580; 30B, 286
$C_3H_{14}N_8NiO_6.H_2O$, 31B, 417; 32B, 427
$C_3H_{16}B_3N$, 26, 616
$C_3H_{16}CoN_8O_7$, 30B, 413
$C_3H_{16}FeN_8O_7$, 30B, 413
$C_3H_{16}N_8NiO_7$, 30B, 414; 32B, 427
$C_3H_{16}N_8O_7Zn$, 30B, 414
$C_3H_{21}AlB_3N$, 33B, 303
$C_3H_{25}As$, 27, 885
$C_3H_{40}O_{18}$, 26, 551

C_4
C_4AgN_3, 31B, 41
$C_4CaO_8Pt.3.5H_2O$, 33B, 556
C_4Cl_8, 30B, 84
$C_4Cl_3CoN_4O$, 35B, 591
$C_4Cl_4N_4S_4$, 32B, 193
$C_4CoO_4.2H_2O$, 33B, 542
$C_4D_8N_2O_3$, 35B, 304
$C_4D_3N_3O_4.D_2O$, 29, 721
C_4F_8, 33B, 531
$C_4F_{12}N_2$, 30B, 402
$C_4F_{12}P_4$, 27, 943
$C_4K_2MnO_8.2H_2O$, 33B, 554
$C_4K_2O_4.H_2O$, 29, 674
$C_4K_2O_8Pd.4H_2O$, 31B, 406
$C_4K_2O_8Pt.2H_2O$, 29, 586
C_4N_3Na, 32B, 49
$C_4NiO_4.2H_2O$, 33B, 542
$C_4O_8Pu.2H_2O$, 30B, 409
$C_4O_8Pu.6H_2O$, 30B, 408
$C_4O_8Th,2H_2O$, 30B, 409
$C_4O_8U.6H_2O$, 30B, 408
C_4HBr, 35B, 893
$C_4HCsF_6O_4$, 30B, 13
$C_4HErO_8.3H_2O$, 35B, 636
$C_4HF_6KO_4$, 30B, 13
$C_4HF_6O_4Rb$, 30B, 13
$C_4H_2BaO_8.2H_2O$, 33B, 551; 35B, 15
$C_4H_2ClCsO_4$, 26, 554
$C_4H_2ClKO_4$, 26, 554; 30B, 25
$C_4H_2ClO_4Rb$, 26, 554
$C_4H_2F_6$, 35B, 891
$C_4H_2KN_3O_4.2H_2O$, 30B, 164
$C_4H_2K_2O_4.2H_2O$, 35B, 14
$C_4H_2N_2Na_2O_3S.6H_2O$, 34B, 224
$C_4H_2N_2O_4$, 29, 715
$C_4H_2N_3O_4Rb$, 30B, 164
$C_4H_2O_3$, 27 945
$C_4H_2O_4$, 32B, 486
$C_4H_3BrN_2O_2$, 28, 529
C_4H_3BrS, 35B, 893
C_4H_3ClS, 35B, 893
$C_4H_3FN_2O_2.C_5H_7N_3O$, 34B, 224
$C_4H_3KO_4$,26, 555, 563
$C_4H_3KO_8.2H_2O$, 29, 465
$C_4H_3N_3O_5$, 28, 499
$C_4H_3N_3O_5.3H_2O$, 29, 719
$C_4H_3n_5O_2.H_2O$, 29, 771
$C_4H_3NaO_4$, 35B, 13
$C_4H_3O_4Rb$, 34B, 17
C_4H_4, 26, 553; 27, 733
$C_4H_4AgN_3O_3$, 31B, 417
$C_4H_4Ag_2N_4O_6$, 31B, 418
$C_4H_4BrNO_2$, 26, 600
$C_4H_4CaO_6.4H_2O$, 33B, 16
$C_4H_4Cl_4Ge_2$, 30B, 414; 33B, 304
$C_4H_4ClNO_2$, 26, 599
$C_4H_4Cl_2O_3$, 34B, 5
$C_4H_4Cl_4O_2$, 33B, 130
$C_4H_4CuK_2O_{10}$, 27, 817
$C_4H_4CuN_4O_6$, 35B, 721
$C_4H_4CuO_4.2H_2O$, 31B, 408
$C_4H_4CuO_6.3H_2O$, 30B, 414
C_4H_4Ge, 35B, 889
$C_4H_4Ge_2I_4$, 30B 414; 32B, 330
$C_4H_4K_2Mo_2O_{15}$, 29, 621
$C_4H_4KN_3O_5$, 33B, 39
$C_4H_4K_2O_6.2H_2O$, 30B, 26
$C_4H_4N_4$, 26, 600; 30B, 39
$C_4H_4N_2.AgNO_3$, 27, 857
$C_4H_4N_6.{}^1/_2H_2O$, 35B, 154
$C_4H_4N_2O$, 35B, 248
$C_4H_4N_2O_2$, 32B, 203
$C_4H_4N_2O_3$, 28, 501
$C_4H_4N_2O_4.H_2O$, 30B, 167
$C_4H_4N_2O_5$, 30B, 159
$C_4H_4N_2O_5.3H_2O$, 30B, 160
$C_4H_4N_6O.H_2O$, 30B, 138
$C_4H_4N_2OS$, 32B, 489
$C_4H_4N_2O_2S$, 32B, 488
$C_4H_4N_4O_2S$, 27, 954
$C_4H_4N_2OSe$, 32B, 489
$C_4H_4N_2S$, 31B, 117
$C_4H_4N_2S_2$, 32B, 204
$C_4H_4Na_2O_6.2H_2O$, 33B, 15
$C_4H_4O_3$, 30B, 4
$C_4H_4O_4$, 30B, 414; 31B, 4,5
$C_4H_4O_6Rb_2.2H_2O$, 30B, 26
$C_4H_4O_6Sr.3H_2O$, 33B, 16
C_4H_4S, 26, 760; 35B, 893
C_4H_4Se, 33B, 540
C_4H_5ClO, 30B, 403
$C_4H_5ClO_2$, 33B, 3
$C_4H_5NO_2$, 26, 597
$C_4H_5N_3O$, 29, 703; 30B, 172
$C_4H_5N_3O_5$, 32B, 205
$C_4H_5N_3O.H_2O$, 28, 503
$C_4H_5N_3S$, 35B, 244
$C_4H_5NaO_3$, 28, 384
$C_4H_5O_4S_2Sb$, 33B, 287
C_4H_6. 33B, 530; 35B, 904
$C_4H_6BF_2FeO_5S$, 35B, 516
$C_4H_6(Ba, Cd, Pb)O_4.3H_2O$, 26, 828
$C_4H_6BaO_{10}$, 35B, 15
$C_4H_6Br_2N_2$, 33B, 238
$C_4H_6Br_4N_2Nb$, 32B, 496
$C_4H_6Br_2O_2$, 28, 485
$C_4H_6Br_2S_2$, 30B, 414; 31B, 143
$C_4H_6Cl_4Cu_2N_2$, 29, 590
$C_4H_6Cl_6Cu_3N_2$, 29, 590
$C_4H_6Cl_3NO$, 30B, 122; 32B, 153
$C_4H_6Cl_2N_2Zn$, 27, 829
$C_4H_6Cl_2O_2$, 28, 482, 483, 487
C_4H_6ClOP, 35B, 894
$C_4H_6Cl_2OS$, 32B, 199
$C_4H_6Cl_2S_2$, 30B, 150
$C_4H_6CuN_4O_6$, 33B, 453
$C_4H_6CuO_5$, 30B, 415
$C_4H_6CuO_6$, 33B, 403
$C_4H_6HgO_4$, 30B, 414
$C_4H_6Hg_2O_4$, 30B, 414
$C_4H_6I_6S_2Sb_2$, 31B, 273
$C_4H_6KN_3O_6$, 28, 454
$C_4H_6K_2O_6$, 26, 563; 35B, 14
$C_4H_6N_4NiO_4$, 32B, 428
$C_4H_6N_2O_2$, 26, 767; 35B, 249
$C_4H_6N_4O_3$, 30B, 125
$C_4H_6N_4O_4$, 35B, 245
$C_4H_6N_6O_2$, 33B, 150
$C_4H_6N_4O_4Pt$, 34B, 550
$C_4H_6N_4O_3S$, 35B, 245
$C_4H_6NO_9Y$, 32B, 415
$C_4H_6N_2S_2Sn$, 35B, 486
C_4H_6O, 30B, 402
$C_4H_6O_4$, 35B, 4
$C_4H_6O_6$, 31B, 5; 32B, 7
$C_4H_6O_2S$, 32B, 65
$C_4H_6O_2S_2$, 31B, 143
$C_4H_6O_4S$, 27, 769; 32B, 6
$C_4H_7ClN_2O_3.H_2O$, 32B, 489
$C_4H_7Cl_6OSb$, 33B, 293
$C_4H_7ErO_7.H_2O$, 34B, 527
$C_4H_7KN_4O_4$, 33B, 175
$C_4H_7KO_6$, 33B, 18
$C_4H_7KO_{10}$, 32B, 25
C_4H_7NO, 33B, 531; 34B, 84
$C_4H_7NO_2$, 34B, 124
$C_4H_7NO_3$, 27, 764
$C_4H_7NO_4$, 30B, 415; 33B, 176, 177
$C_4H_7N_3O_3$, 29, 714
$C_4H_7N_3S_2$, 35B, 155
$C_4H_7NaO_4$, 26, 556
$C_4H_7O_3Rb$, 30B, 17
C_4H_8, 26, 553, 641; 28, 666; 35B, 894
$C_4H_8Br_2HgO$, 35B, 918
$C_4H_8Br_2S$, 35B, 217
$C_4H_8Br_2SSe$, 32B, 200
$C_4H_8CaO_7$, 31B, 26
$C_4H_8Cl_2GeO_2$, 35B, 407
$C_4H_8Cl_2HgS$, 29, 632
$C_4H_8ClIOSe$, 33B, 149
$C_4H_8ClLiO_2$, 27, 761; 31B, 273

C_4H_8ClNOS, 30B, 210
$C_4H_8Cl_3O_4P$, 34B, 378
$C_4H_8Cl_4O_2Ti$, 31B, 442
$C_4H_8Cl_4Se_2$, 29, 521
$C_4H_8CuN_2O_4.H_2O$, 29, 605
$C_4H_8HgN_2O_2$, 34B, 548
$C_4H_8HgN_6S_4$, 31B, 401
$C_4H_8I_2OSe$, 31B, 272
$C_4H_8I_2STe$, 35B, 495
$C_4H_8I_4Se_2$, 26, 771
$C_4H_8N_3Na_2PO_5.4\frac{1}{2}H_2O$, 33B, 275
$C_4H_8N_2NiO_2S_2$, 33B, 480; 34B, 574
$C_4H_8N_6NiS_4$, 31B, 402
$C_4H_8N_2O_2$, 26, 596
$C_4H_8N_2O_3$, 33B, 179; 35B, 304
$C_4H_8N_2O_6$, 30B, 246
$C_4H_8N_2O_5$, 26, 764
$C_4H_8N_4O_6$, 34B, 34
$C_4H_8N_8O_8$, 28, 467; 32B, 487; 35B, 179
$C_4H_8N_2O_4Pt$, 34B, 538
$C_4H_8N_2O_9V.2H_2O$, 30B, 407
$C_4H_8N_2S$, 29, 516; 30B, 42
C_4H_8O, 30B, 402
$C_4H_8O_2$, 27, 737; 28, 666
$C_4H_8O_2.GeCl_2$, 35B, 407
$C_4H_8O_2.N_2O_2$, 27, 963
$C_4H_8O_5$, 29, 518
$C_4H_8O_7$, 32B, 8
$C_4H_8O_8$, 33B, 3, 77
$8C_4H_8O.7.33H_2S.136H_2O$, 30B, 268
$C_4H_8O_2S$, 32B, 66
$C_4H_8SSe.2I_2$, 27, 964
C_4H_8Se, 35B, 894, 904
$C_4H_8Se.I_2$, 29, 522
C_4H_9Br, 28, 666
C_4H_9Cl, 26, 554; 27, 733; 28, 666; 29, 450; 33B, 28
$C_4H_9ClN_2O_3.H_2O$, 34B, 267
C_4H_9CsO, 33B, 297
C_4H_9GeN, 31B, 329
C_4H_9KO, 33B, 297
C_4H_9LiO, 30B, 415
$C_4H_9NO_2$, 33B, 180
$C_4H_9NO_3S$, 27, 765
C_4H_9NOSi, 31B, 298
C_4H_9NSSi, 31B, 298
C_4H_9NSSn, 35B, 487
C_4H_9NSn, 31B, 330
$C_4H_9N_2Sn$, 35B, 915
$C_4H_9O_2P$, 34B, 379
C_4H_9ORb, 33B, 297
$C_4H_{10}AlCl_3O$, 30B, 415
$C_4H_{10}BF_2N$, 35B, 424
$C_4H_{10}BaCl_2N_2O_4.H_2O$, 28, 670
$C_4H_{10}BrN_2NaO_2$, 27, 761
$C_4H_{10}CdN_4O_4S_2$, 27, 835; 30B, 344
$C_4H_{10}Cl_2CuN_6O_4$, 31B, 437
$C_4H_{10}ClCuS_2$, 32B, 450
$C_4H_{10}ClHgO_3P$, 35B, 834
$C_4H_{10}Cl_4Hg_2S$, 29, 632
$C_4H_{10}ClLiO_3$, 33B, 298
$C_4H_{10}Cl_2N_6O_4Zn$, 28, 645
$C_4H_{10}Cl_3NPt$, 35B, 513
$C_4H_{10}Cl_2O_2Ti$, 33B, 454
$C_4H_{10}CuN_2O_5$, 26, 661
$C_4H_{10}HgS_2$, 30B, 374
$C_4H_{10}IN_2NaO_2$, 27, 761
$C_4H_{10}Mg$, 30B, 284
$C_4H_{10}.6Mg_{0}.82O_{13}.3Pt$, 33B, 406
$C_4H_{10}N_2$, 28, 666
$C_4H_{10}N_4NiO_6$, 31B, 419
$C_4H_{10}N_4NiS_2$, 35B, 788
$C_4H_{10}N_2O_4$, 26, 602
$C_4H_{10}NiO_2S_2$, 35B, 788
$C_4H_{10}O$, 26, 554
$C_4H_{10}O_2PbS_4$, 31B, 331
$C_4H_{10}O_6S_3$, 30B, 53
$C_4H_{10}O_{11}Th$, 31B, 408
$C_4H_{10}Si$, 27, 950; 34B, 365
$C_4H_{11}BrCl_2O$, 29, 450
$C_4H_{11}BrN_2O_2$, 32B, 9; 34B, 3
$C_4H_{11}N.8^2/_3H_2O$, 32B, 288
$C_4H_{11}N.9^3/_4H_2O$, 32B, 289
$C_4H_{11}NO_4$, 34B, 16
$C_4H_{11}NO_{10}$, 32B, 23
$C_4H_{11}N_3O_7$, 33B, 183
$C_4H_{12}Ag_2I_3N$, 28, 392
$C_4H_{12}AsBr$, 28, 587
$C_4H_{12}As_2S_2$, 29, 539
$C_4H_{12}B_4$, 28, 687
$C_4H_{12}B_6$, 33B, 262
$C_4H_{12}B_{10}Br_4$, 31B, 283
$C_4H_{12}B_2Cl_4N_2$, 28, 580
$C_4H_{12}B_2F_4N_2$, 31B, 289
$C_4H_{12}BN$, 35B, 425
$C_4H_{12}BeLi_2$, 33B, 299
$C_4H_{12}Br_3HgN$, 28, 394
$C_4H_{12}Br_3NNi$, 31B, 342
$C_4H_{12}Br_4Pd_2S_2$, 33B, 481
$C_4H_{12}Cl_2CuO_2S_2$, 35B, 734
$C_4H_{12}Cl_3HgN$, 28, 393
$C_4H_{12}Cl_2^- IN^+$, 29, 477
$C_4H_{12}Cl_3MnN$, 32B, 30
$C_4H_{12}Cl_3NNi$, 33B, 520
$C_4H_{12}ClNO_4$, 29, 476
$C_4H_{12}ClN_4O_2Re$, 35B, 924
$C_4H_{12}Cl_{11}NO_5Se_5$, 32B, 29
$C_4H_{12}Cl_2O_2PdS_2$, 32B, 450
$C_4H_{12}Cl_4O_6Re_2$, 35B, 637
$C_4H_{12}CuN_2O_4$, 28, 631
$C_4H_{12}CuN_2O_6$, 26, 657
$C_4H_{12}CuN_2O_{10}$, 27, 818
$C_4H_{12}CuN_7S_4$, 28, 670
$C_4H_{12}Hg_{13}N$, 32B, 485
$C_4H_{12}I_4Te_2$, 32B, 337
$C_4H_{12}Li_2Zn$, 33B, 299
$C_4H_{12}Mo_2N_2O_{11}.2^1/_2H_2O$, 31B, 407
$C_4H_{12}N_2NiO_6$, 33B, 416
$C_4H_{12}N_2O_2S$, 27, 769
$C_4H_{12}N_2O_8V$, 32B, 416
$C_4H_{12}N_3O_{10}V.2H_2O$, 30B, 407
$C_4H_{12}N_2S$, 35B, 895
$C_4H_{12}NiO_4P_2S_4$, 34B, 586
$C_4H_{12}NiP_2S_4$, 34B, 587
$C_4H_{12}O_4P_2S_4Te$, 32B, 336
$C_4H_{12}O_2S$, 28, 399
$C_4H_{12}O_4Ti$, 33B, 455
$C_4H_{12}O_4Tl_4$, 27, 814
$C_4H_{12}P_2$, 35B, 895
$C_4H_{12}P_2S_2$, 27, 770; 32B, 305
$C_4H_{12}S_6Sn_4$, 34B, 408
$C_4H_{13}B_7$, 33B, 263
$C_4H_{13}Cl_3CrN_3$, 35B, 665
$C_4H_{13}Cl_2N$, 30B, 415
$C_4H_{13}CoN_6O_6$, 32B, 494
$C_4H_{13}NO.5H_2O$, 31B, 282
$C_4H_{13}N_4O_{10}Re_3$, 35B, 924
$C_4H_{13}NSi$, 27, 770
$C_4H_{14}B_2$, 33B, 531
$C_4H_{14}B_6$, 35B, 426
$C_4H_{14}B_{10}Br_2$, 29, 531
$C_4H_{14}Cl_2CuN_{10}.2H_2O$, 29, 768; 30B, 415
$C_4H_{14}Cl_2MnN_2O_6$, 27, 838
$C_4H_{14}Cl_2N_2$, 28, 390
$C_4H_{14}Cl_2N_{10}Ni.2H_2O$, 30B, 415
$C_4H_{14}ClN_{10}O_8Pt$, 30B, 416
$C_4H_{14}CuK_2N_6O_8$, 26, 658
$C_4H_{14}CuN_{12}O_6.2H_2O$, 30B, 416
$C_4H_{14}CuN_2O_6S.4H_2O$, 30B, 392
$C_4H_{14}N_{10}O_{12}RuS_2.7H_2O$, 30B, 416
$C_4H_{14}N_4O_4S_3$, 27, 770
$C_4H_{14}N_4O_4S_6Te$, 26, 582; 30B, 284
$C_4H_{14}N_4O_4Zn$, 30B, 365
$C_4H_{15}B_9$, 31B, 291
$C_4H_{15}B_{10}Br$, 30B, 411
$C_4H_{16}AlB_9$, 35B, 921
$C_4H_{16}BClN_2$, 33B, 264
$C_4H_{16}B_2CuF_8N_4$, 30B, 383
$C_4H_{16}B_2N_2$, 27, 771
$C_4H_{16}BrLiN_4$, 32B, 267
$C_4H_{16}Br_2N_8S_4Te$, 26, 577
$C_4H_{16}CaCl_2O_4$, 35B, 918
$C_4H_{16}CaN_{10}O_{10}$, 26, 571
$C_4H_{16}CdN_{10}O_{10}$, 26, 571
$C_4H_{16}Cl_3CoN_4.H_2O$, 30B, 331
$C_4H_{16}Cl_2CoN_5O_3$, 28, 615
$C_4H_{16}Cl_2CoN_5O_5$, 35B, 588
$C_4H_{16}CoN_2O_2S_2$, 29, 764
$C_4H_{16}CoN_2O_2Se_2$, 29, 764
$C_4H_{16}Cl_2CoO_4$, 31B, 438; 32B, 445; 33B, 239
$C_4H_{16}Cl_4CuN_2$, 35B, 880
$C_4H_{16}Cl_5CuN_3$, 32B, 485
$C_4H_{16}Cl_2CuN_4O_8$, 32B, 379
$C_4H_{16}ClLiN_4$, 31B, 378; 32B, 268
$C_4H_{16}ClN_5NiO_6$, 27, 831
$C_4H_{16}ClN_6NiO_2$, 31B, 381
$C_4H_{16}Cl_2N_8NiS_4$, 28, 623
$C_4H_{16}Cl_2N_8O_8S_4Te$, 26, 577
$C_4H_{16}ClN_8O_3S_4Tl$, 35B, 791
$C_4H_{16}ClN_8O_4S_4Tl$, 32B, 54
$C_4H_{16}Cl_2N_4Pd$, 32B, 375
$C_4H_{16}Cl_2N_8PdS_4$, 35B, 628
$C_4H_{16}Cl_4N_4Pt$, 28, 670; 35B, 587

$C_4H_{16}Cl_2N_8S_4Te$, 26, 576; 30B, 285
$C_4H_{16}CoN_{11}O_3$, 33B, 384
$C_4H_{16}CoN_{10}O_6S_4$, 35B, 911
$C_4H_{16}CoN_{10}O_6S_4.H_2O$, 35B, 911
$C_4H_{16}CuN_6O_6$, 29, 593
$C_4H_{16}F_6N_9PRu$, 35B, 592
$C_4H_{16}MnN_2O_2S_2$, 29, 764
$C_4H_{16}MnN_2O_2Se_2$, 29, 764
$C_4H_{16}N_6NiO_4$, 34B, 481
$C_4H_{16}N_2NiO_2S_2$, 29, 764
$C_4H_{16}N_2NiO_2Se_2$, 29, 764
$C_4H_{16}N_2O_2$, 32B, 269
$C_4H_{16}N_{10}O_6S_4Te$, 26, 577
$C_4H_{16}N_8O_3S_4Tl$, 32B, 54
$C_4H_{16}N_{10}O_6S_4Zn$, 26, 575
$C_4H_{17}B_7$, 32B, 293
$C_4H_{17}Cl_4CoN_4.2H_2O$, 32B, 494
$C_4H_{18}B_2P_2$, 33B, 275
$C_4H_{18}Cl_3CoN_4O$, 35B, 591
$C_4H_{18}ClCsN_8OS_4$, 33B, 239
$C_4H_{18}Cl_2CuN_4O$, 32B, 379
$C_4H_{18}Co_2N_9O_{14}.H_2O$, 32B, 494
$C_4H_{18}F_4N_8S_4Te$, 26, 579
$C_4H_{18}N_8O_4PS_4Tl$, 33B, 241
$C_4H_{18.25}BeCl_{2.25}O_5$, 29, 763
$C_4H_{19}Co_2N_9O_{14}.2H_2O$, 32B, 494
$C_4H_{20}B_{22}$, 30B, 272
$C_4H_{20}Cl_2N_6OPt$, 27, 834
$C_4H_{20}Cl_2N_4Pt$, 35B, 666
$C_4H_{20}CsFN_8O_2S_4$, 33B, 239
$C_4H_{20}FNO_4$, 32B, 291
$C_4H_{20}F_2N_8O_2S_4Te$, 26, 580
$C_4H_{22}B_{18}CoCs$, 32B, 294
$C_4H_{22}B_{18}Ni$, 35B, 429
$C_4H_{22}Cu_2N_4O_6S.H_2O$, 31B, 421
$C_4H_{22}N_2O_6$, 33B, 113
$C_4H_{24}B_8N_2$, 28, 581
$C_4H_{24}B_{10}S_2$, 27, 772
$C_4H_{24}N_2O_{16}Sb_2$, 29, 766
$C_4H_{25}B_{11}CsN$, 32B, 296
$C_4H_{26}B_{18}FeP_2$, 33B, 553

C_5

$C_5F_{15}P_5$, 26, 620
C_5HO_5Rb, 31B, 38
$C_5H_2FN_2O_4Rb.H_2O$, 29, 712
$C_5H_3Br_2Cl_2HgNO$, 31B, 439; 34B, 576
$C_5H_3N_7.H_2O$, 33B, 151
C_5H_4ClNO, 34B, 128
$C_5H_4FeO_3$, 35B, 895
$C_5H_4N_4$, 30B, 173
$C_5H_4N_2O_2$, 26, 762
$C_5H_4N_4O_3$, 30B, 416; 31B, 162
$C_5H_4N_4O_3.2H_2O$, 30B, 416
$C_5H_4O_3$, 27, 947
$C_5H_4O_2S$, 27, 949; 29, 731
$C_5H_4O_2Se$, 27, 949
$C_5H_5BrN_2O_2$, 30B, 416
$C_5H_5BrN_2O_2.C_6H_7N_5$, 30B, 416
$C_5H_5Cl_2CuNO$, 30B, 358
C_5H_5ClLiN, 31B, 274
$C_5H_5ClN_4O.H_2O$, 34B, 236
$C_5H_5Cl_3Ti$, 26, 652; 28, 609
$C_5H_5CrNO_5$, 27, 857; 29, 541
$C_5H_5I_4N$, 26, 778
C_5H_5In, 28, 598; 29, 625
$C_5H_5NO_2$, 32B, 172
$C_5H_5NO_5$, 31B, 38
$C_5H_5N_3O$, 26, 762
$C_5H_5N_3O_2$, 28, 458
$C_5H_5N_5O.HCl.2H_2O$, 30B, 175
$C_5H_5N_5S$, 35B, 251
$(C_5H_5S)^+I^-$, 27, 965
C_5H_5Tl, 28, 598
C_5H_6, 30B, 87; 35B, 905
$C_5H_6AsF_6N$, 31B, 119
$C_5H_6BrCl_2N$, 32B, 487
$C_5H_6BrN_3O$, 35B, 157
C_5H_6ClN, 27, 959
C_5H_6ClNO, 26, 703
$C_5H_6ClN_5O.2H_2O$, 30B, 175
$C_5H_6Cl_4NSb$, 32B, 487; 35B, 875
$C_5H_6Cl_4N_2Sn$, 33B, 307
$C_5H_6CuO_8$, 29, 669
$C_5H_6F_6NP$, 31B, 120
$C_5H_6F_6NSb$, 31B, 119
$C_5H_6FeN_4O_3$, 33B, 549
$C_5H_6MnO_8$, 31B, 440
$C_5H_6N_2$, 28, 670
$C_5H_6N_2O_2$, 34B, 237
$C_5H_6N_2O_3$, 30B, 129
$C_5H_6N_8O_{13}$, 35B, 36
$C_5H_6N_2O_2.BrH$, 29, 705
$C_5H_6N_2O_2.C_7H_9N_6Br$, 35B, 275
$C_5H_6N_4OS$, 34B, 234
$C_5H_6N_2S$, 34B, 27
$C_5H_6O_3$, 28, 489
$C_5H_6O_4$, 26, 558
$C_5H_6O_5$, 32B, 10
$C_5H_6O_2S$, 32B, 182
$C_5H_6O_8Zn$, 29, 669
$C_5H_7Br_4NRe$, 30B, 293
$C_5H_7Cl_4NRe$, 28, 459
$C_5H_7CuNO_4.2H_2O$, 31B, 411
$C_5H_7N_3O.BrH$, 27, 971
$C_5H_7NO_4Zn.2H_2O$, 31B, 412
$C_5H_7NaO_3$, 34B, 19
C_5H_8, 26, 558; 35B, 896
C_5H_8ClNO, 32B, 11
$C_5H_8ClNO_3$, 32B, 489; 35B, 305
$C_5H_8N_2O_2$, 33B, 154
$C_5H_8N_2O_3$, 26, 765
$C_5H_8N_2O_5$, 29, 668
$C_5H_8N_4O_{12}$, 28, 405
C_5H_8O, 30B, 403
$C_5H_8O_4$, 31B, 6
$C_5H_8O_2S$, 35B, 217
$C_5H_8O_3S$, 32B, 488
$C_5H_9BaO_8P.5H_2O$, 27, 792
$C_5H_9Br_2O_3P$, 34B, 380
$C_5H_9ClO_5S$, 28, 410
$C_5H_9I_3O_2$, 34B, 348
$C_5H_9I_3S_2$, 26, 773
C_5H_9N, 27, 735
$C_5H_9N_3$, 26, 760
$C_5H_9N_3.Cl_2H_2$, 27, 992
$C_5H_9NO_2$, 30B, 209
$C_5H_9NO_2.HCl$, 34B, 268
$C_5H_9NO_4$, 35B, 307
$C_5H_9N_3O_3S$, 35B, 164
$C_5H_9O_4P$, 32B, 220; 33B; 278
C_5H_{10}, 26, 641; 33B, 531; 35B, 897
$C_5H_{10}CuNS_2$, 28, 625
$C_5H_{10}NO_4$, 34B, 278
$C_5H_{10}N_2O_2$, 31B, 7
$C_5H_{10}N_2O_3$, 35B, 308
$C_5H_{10}N_2S$, 32B, 487
$C_5H_{10}O_2$, 27, 739
$C_5H_{10}O_5$, 26, 621, 622; 31B, 168; 32B, 211
$C_5H_{11}B_9CsO_3Re$, 31B, 294
$C_5H_{11}ClN_2O_6$, 30B, 212
$C_5H_{11}Cl_2NO_2PdS$, 32B, 496; 35B, 646
$C_5H_{11}NO_2$, 34B, 269; 35B, 310
$C_5H_{11}NO_2S$, 34B, 203
C_5H_{12}, 29, 449; 32B, 40
$C_5H_{12}BrNO_2$, 31B, 193
$C_5H_{12}Br_2N_2STe$, 26, 581
$C_5H_{12}ClNO_2$, 31B, 188; 32B, 226; 35B, 313
$C_5H_{12}ClNO_2.H_2O$, 34B, 265
$C_5H_{12}ClNO_4$, 31B, 30
$C_5H_{12}ClNS_2Sn$, 35B, 488
$C_5H_{12}Cl_2N_2STe$, 26, 580
$C_5H_{12}N_2O_3S$, 35B, 312
$C_5H_{12}O_2$, 26, 559
$C_5H_{12}O_4$, 27, 794
$C_5H_{12}O_5$, 33B, 157; 34B, 245, 246
$C_5H_{13}BrN_2O_2$, 35B, 314
$C_5H_{13}ClN_4NiS_2$, 35B, 515
$C_5H_{13}ClN_2O_2$, 32B, 227; 34B, 266
$C_5H_{13}N_2O_2$, 35B, 489
$C_5H_{14}B_{10}Hg$, 30B, 412
$C_5H_{14}ClNO$, 26, 603
$C_5H_{16}ClN_5NiS$, 33B, 385
$C_5H_{16}Cl_2N_8O$, 33B, 31
$C_5H_{16}IN_5NiS$, 34B, 481
$C_5H_{20}CoN_5O_5S_2$, 34B, 483
$C_5H_{21}B_{18}CoS_2$, 34B, 360

C_6

C_6Br_6, 28, 670
C_6BrCl_5, 31B, 77
$C_6Br_2Cl_4$, 33B, 67
C_6Br_5Cl, 31B, 77
C_6Br_5F, 29, 769
$C_6Br_4O_2$, 26, 714
C_6Cl_6, 26, 713
C_6Cl_5F, 29, 769
$C_6Cl_4O_2$, 26, 714; 27, 899
C_6F_6, 29, 642
C_6F_{12}, 34B, 532
$C_6Gd_2O_{12}.10H_2O$, 30B, 408; 35B, 906
C_6I_6, 30B, 79; 33B, 68; 35B, 77
$C_6K_3O_{12}V.3H_2O$, 32B, 496
$C_6La_2O_{12}.10H_2O$, 30B, 408

$C_6N_4O_4$, 34B, 89
$C_6N_6O_6$, 30B, 417; 31B, 152
$C_6N_6O_{12}$, 31B, 66
$C_6N_6O_6.C_{16}H_{10}$, 30B, 416
$C_6N_6O_6.C_{20}H_{12}$, 30B, 417
$C_6Nd_2O_{12}.10H_2O$, 30B, 408
$C_6Nd_2O_{12}.10.5H_2O$, 35B, 639
$C_6O_{12}Pu_2.10H_2O$, 30B, 408
$C_6HClN_3O_6$, 32B, 90
$C_6HCl_2N_3O_6$, 27, 902
C_6HCl_5O, 27, 903
C_6HF_5, 33B, 540
$C_6H_2Br_4$, 29, 642
$C_6H_2Br_4.Br_6C_6$, 29, 644
$C_6H_2Br_2Cl_2O$, 33B, 79
$C_6H_2Cl_4$, 30B, 418
$C_6H_2Cl_5N$, 34B, 69
$C_6H_2Cl_2N_2O$, 35B, 103
$C_6H_2Cl_2O_2$, 35B, 71
$C_6H_2Cl_2O_4$, 32B, 105
$C_6H_2Cl_2O_4.2H_2O$, 32B, 106
$C_6H_2Cl_4O$, 33B, 79
$C_6H_2Cl_4O_2$, 27, 901; 32B, 101
$C_6H_2D_4N_2O_2$, 34B, 69
$C_6H_2KN_3O_7$, 34B, 67
$C_6H_2K_2N_2O_{10}S_2.H_2O$, 32B, 486
$C_6H_2K_2O_8$, 30B, 31
$C_6H_2N_4Na_2O_4.4H_2O$, 31B, 131
$C_6H_2N_4O_4Rb_2.2H_2O$, 28, 671
$C_6H_3BrN_2O_2$, 27, 974
$C_6H_3ClN_2O_2$, 27, 973
$C_6H_3Cl_2NO_2$, 30B, 418
$C_6H_3ClO_2$, 35B, 71
$C_6H_3Cl_9O_3$, 27, 967
$C_6H_3MnO_5$, 35B, 897
$C_6H_3N_3$, 31B, 85
$C_6H_3N_3O_6.C_{14}H_{10}$, 29, 686
$C_6H_3N_3O_6.C_9H_8N$, 29, 689
$C_6H_4As_2Cl_2O$, 27, 984
$C_6H_4Br_3N$, 34B, 70
$C_6H_4Br_4N_2$, 33B, 237
$C_6H_4BrNO_2$, 28, 418
$C_6H_4BrNO_2.AlCl_3$, 26, 704
$C_6H_4CaO_7.4H_2O$, 34B, 18
$C_6H_4Cl_2$, 28, 424
$C_6H_4Cl_2Hg$, 29, 628
$C_6H_4ClNO_2$, 27, 892; 30B, 63
$C_6H_4ClNO_5$, 33B, 48
$C_6H_4Cl_2N_2O_2$, 34B, 78
$C_6H_4Cl_2N_4O_4$, 34B, 194
$C_6H_4ClNO_2.AlCl_3$, 26, 704
$C_6H_4ClNO_4S$, 35B, 52
$C_6H_4Cl_7N_2Sb$, 30B, 418
$C_6H_4Cl_4N_4Zn$, 33B, 52
$C_6H_4I_4N_2$, 33B, 237
$C_6H_4INO_2.AlCl_3$, 26, 704
$C_6H_4KNO_3.1/2H_2O$, 26, 704
$C_6H_4KO_6$, 28, 385
$C_6H_4N_2$, 32B, 50
$C_6H_4N_4$, 34B, 147
$C_6H_4N_2O_2$, 26, 701; 27, 889; 28, 671; 32B, 190; 33B, 145
$C_6H_4N_2O_4$, 26, 695; 31B, 67; 34B, 195
$C_6H_4N_4O_2$, 30B, 50
$C_6H_4N_2O_3S$, 35B, 78
$C_6H_4N_2O_4S.H_2O$, 34B, 36
$C_6H_4O_5$, 29, 729; 33B, 131
$C_6H_4O_6.2H_2O$, 30B, 77
$C_6H_4Se_2$, 34B, 186
C_6H_5Br, 26, 693
$C_6H_5Br_3Cu_2N_2$, 30B, 357
C_6H_5BrHg, 33B, 532
$C_6H_5Br_3N_2$, 27, 889
$C_6H_5Br_2NO$, 31B, 120
C_6H_5BrOS, 35B, 218
C_6H_5Cl, 26, 693
$C_6H_5ClN_2$, 28, 402
$C_6H_5Cl_2N$, 28, 420
$C_6H_5ClNO_2$, 27, 893
$C_6H_5ClN_2O_2$, 30B, 68
$C_6H_5Cl_3Si$, 29, 538
$C_6H_5CrN_3O_3$, 35B, 537
C_6H_5F, 26, 693; 33B, 540
$C_6H_5F_2N_3O_3S_3$, 35B, 234
$C_6H_5FeIO_3$, 30B, 418; 31B, 343; 33B, 330
$C_6H_5FeNO_6$, 30B, 419
C_6H_5I, 26, 693
$C_6H_5IN_2$, 32B, 491
$C_6H_5K_2O_4P.1^1/_2H_2O$, 32B, 220
C_6H_5NO, 27, 888
$C_6H_5NO_2$, 26, 762
$C_6H_5NO_3$, 30B, 71
$C_6H_5N_5O_6$, 32B, 91
$C_6H_5NO_2.AlCl_3$, 26, 704
C_6H_6, 29, 636; 33B, 532, 533; 34B, 79
$C_6H_6AgAlCl_4$, 31B, 364
$C_6H_6Al_2Br_6$, 26, 653
$C_6H_6AlCl_4Cu$, 28, 684; 31B, 365
$C_6H_6AsNO_6$, 28, 671
$C_6H_6Br_2N_4$, 28, 457
$C_6H_6CdCl_2N_2O_2$, 35B, 408
$C_6H_6CdCl_2N_2O_2.H_2O$, 33B, 552
$C_6H_6Cl_6FeOSi_2$, 35B, 535
$C_6H_6Cl_4Ir_2O_4$, 32B, 493
C_6H_6ClN, 31B, 69
$C_6H_6ClNO_2$, 30B, 130
$C_6H_6Cl_2O_6$, 32B, 106
$C_6H_6K_2N_8NiO_8.4H_2O$, 30B, 353
$C_6H_6MoS_6$, 30B, 399
$C_6H_6N_2O$, 26, 762; 31B, 121
$C_6H_6N_2O_2$, 26, 696; 30B, 419
$C_6H_6N_2O_4$, 30B, 243
$C_6H_6N_4O_4$, 34B, 196
$C_6H_6N_4O_7$, 34B, 68
$C_6H_6N_6O_6$, 30B, 69
$C_6H_6N_2O_5S$, 27, 981
$C_6H_6N_2S$, 31B, 121; 32B, 156
$C_6H_6N_4S.H_2O$, 33B, 156
$C_6H_6Na_{12}C_{24}P_6.33H_2O$, 35B, 273
$C_6H_6Na_{12}O_{24}P_6.35H_2O$, 35B, 273
$C_6H_6Na_{12}O_{24}P_6.38H_2O$, 33B, 543
C_6H_6O, 28, 423; 32B, 102
$C_6H_6O_2$, 31B, 71, 72
$C_6H_6O_3$, 30B, 77
$C_6H_6O_4$, 27, 734
$C_6H_6O_6$, 27, 783
$C_6H_6O_2S$, 34B, 180; 35B, 219
$C_6H_6O_3S_2$, 34B, 181
$C_6H_7Cl_3NSb$, 33B, 288
C_6H_7ClSi, 30B, 404
$C_6H_7I_2N$, 26, 763
$C_6H_7KO_7$, 27, 735; 33B, 21
$C_6H_7LiO_7$, 30B, 28
C_6H_7N, 26, 763
$C_6H_7N_5$, 29, 708
$C_6H_7N_5.2BrH$, 27, 977
C_6H_7NO, 26, 763
$C_6H_7N_3O_3$, 27, 965
$C_6H_7N_5O$, 30B, 416
$C_6H_7N_5O.BrH$, 29, 706
$C_6H_7NO_3S$, 30B, 65; 32B, 91
$C_6H_7NO_3S.H_2O$, 27, 894
$C_6H_7NaO_6$, 34B, 247
$C_6H_7NaO_7$, 30B, 28
$C_6H_7O_3$, 34B 172
C_6H_8, 33B, 533, 534
$C_6H_8AgNO_6$, 29, 483
$C_6H_8AsNO_3$, 26, 700; 27, 905
$C_6H_8BiI_4N$, 32B, 157
C_6H_8BrN, 26, 698
$C_6H_8Br_2N_2Ni$, 29, 570
$C_6H_8Br_2O_4$, 30B, 246
$C_6H_8Br_4O_4$, 32B, 173
C_6H_8ClNO, 26, 764; 29, 725; 30B, 70
$C_6H_8Cl_2N_4Zn$, 31B, 421
$C_6H_8Cl_2O_4$, 30B, 246
$C_6H_8CuN_3NaO_4.H_2O$, 30B, 348
$C_6H_8I_2S_2$, 31B, 270
$C_6H_8I_2Se_2$, 31B, 270
$C_6H_8I_4Se_2$, 30B, 249
$C_6H_8N_2$, 33B, 54
$C_6H_8N_2.2HCl$, 30B, 129
$C_6H_8NNa_3O_7$, 32B, 25
$C_6H_8N_2O_2$, 28, 505; 31B, 153
$C_6H_8N_4O_8$, 29, 648; 32B, 47
$C_6H_8N_2O_2S$, 29, 770; 30B, 66; 32B, 93
$C_6H_8N_2O_{16}U_2$, 33B, 550
$C_6H_8N_2O_{14}V_2.6H_2O$, 30B, 407
$C_6H_8N_2S$, 28, 671
C_6H_8O, 35B, 97
$C_6H_8O_2$, 29, 491
$C_6H_8O_4$, 31B, 8; 32B, 115, 486; 34B, 86
$C_6H_8O_6$, 26, 808; 32B, 212; 33B, 158
$C_6H_8O_7$, 27, 735; 34B, 7

$C_6H_8O_2.Cl_2Hg$, 29, 492
$C_6H_8O_4S_2$, 29, 520
$C_6H_8O_4Se_2$, 29, 520
$C_6H_9ClN_2$, 30B, 47
$C_6H_9Cl_2N_3O$, 29, 724
$C_6H_9GdO_9.{}^1/_2H_2O$, 34B, 528
$C_6H_9KO_7.2H_2O$, 28, 513
C_6H_9NO, 32B, 12
$C_6H_9NO_3$, 35B, 228
$C_6H_9NO_6$, 32B, 13
$C_6H_9N_3O_2.ClH.H_2O$, 29, 701
$C_6H_9N_3O_2.ClH.2H_2O$, 35B, 316
$C_6H_9N_3O_3.Cl_2Hg$, 27, 858
$C_6H_9NaO_3$, 33B, 21
$C_6H_9O_3P$, 30B, 401
$C_6H_9O_7Rb.2H_2O$, 28, 513
C_6H_{10}, 27, 735; 32B, 116; 33B, 534; 35B, 897
$C_6H_{10}Cl_2$, 29, 490
$C_6H_{10}ClCuN_3O_4.1.5H_2O$, 27, 1049; 29, 602
$C_6H_{10}Cl_3HgN_3O_2$, 35B, 647
$C_6H_{10}ClIN_4$, 32B, 270
$C_6H_{10}Cl_2N_2$, 30B, 129; 34B, 72
$C_6H_{10}ClN_3O_2.2H_2O$, 35B, 316
$C_6H_{10}Cl_2N_2O_5$, 32B, 48
$C_6H_{10}Cl_2Pd_2$, 27, 820; 29, 768; 30B, 299
$C_6H_{10}Cr$, 32B, 493
$C_6H_{10}CuN_2O_4S_2$, 27, 838
$C_6H_{10}CuO_4$, 30B, 419; 35B, 640
$C_6H_{10}CuO_4.H_2O$, 30B, 419
$C_6H_{10}CuO_6.1{}^1/_2H_2O$, 33B, 403
$C_6H_{10}I_6Se_2$, 28, 571
$C_6H_{10}Mo_2N_2Na_2O_8S_2.5H_2O$, 33B, 549; 34B, 540
$C_6H_{10}N_4NiS_4$, 33B, 483
$C_6H_{10}N_2O_2$, 33B, 80; 34B, 129; 35B, 318
$C_6H_{10}N_2O_3S$, 33B, 55
$C_6H_{10}N_3O_4Zn.{}^1/_2SO_4.4H_2O$, 35B, 648
$C_6H_{10}NiO_2S_4$, 28, 627
$C_6H_{10}O$, 33B, 540
$C_6H_{10}O_4$, 26, 559; 30B, 4; 32B, 14
$C_6H_{10}O_6$, 32B, 214
$C_6H_{10}O_4S$, 27, 773
$C_6H_{10}O_2S_4Te$, 32B, 338
$C_6H_{10}O_2S_4Zn$, 31B, 404
$C_6H_{10}S_3$, 32B, 183
$C_6H_{11}CaCl_2N_3O_4.3H_2O$, 32B, 489; 34B, 355
$C_6H_{11}CeNO_7$, 32B, 496
$C_6H_{11}Cl$, 27, 784
$C_6H_{11}ClNO_3S.H_2O$, 30B, 420
$C_6H_{11}K_2O_9P.2H_2O$, 30B, 191
$C_6H_{11}NO$, 27, 987
$C_6H_{11}NO_4S_2$, 26, 604
$C_6H_{11}NO_3S.HCl.H_2O$, 31B, 157
$C_6H_{11}O_4P$, 33B, 279
C_6H_{12}, 28, 666
$C_6H_{12}AgNO_3S_6$, 33B, 476
$C_6H_{12}Br_2$, 35B, 41
$C_6H_{12}Br_2CaN_4.10H_2O$, 30B, 399
$C_6H_{12}Br_2N_4S_2Te$, 26, 577; 30B, 287
$C_6H_{12}Cl_3Cu_3S_6$, 33B, 550
$C_6H_{12}ClNO_3S.H_2O$, 31B, 157
$C_6H_{12}ClO_2P$, 35B, 898
$C_6H_{12}CoN_3OS_4$, 27, 846
$C_6H_{12}CuN_2O_4$, 31B, 413; 34B, 537
$C_6H_{12}CuN_2O_6$, 34B, 541
$C_6H_{12}Cu_2N_4O_{10}$, 35B, 633
$C_6H_{12}CuN_6S_2$, 26, 670
$C_6H_{12}FeN_3OS_4$, 35, 792
$C_6H_{12}I_2N_4S_2Te$, 26, 578; 30B, 287
$C_6H_{12}LiNO_8$, 32B, 26
$C_6H_{12}N_2$, 29, 515
$C_6H_{12}N_4$, 26, 603; 28, 478
$C_6H_{12}N_4.6H_2O$, 30B, 269
$C_6H_{12}NNdO_9$, 31B, 409
$C_6H_{12}N_2NiO_4.2H_2O$, 29, 481
$C_6H_{12}N_6NiS_2$, 26, 670
$C_6H_{12}N_2O_2$, 30B, 413; 31B, 9; 34B, 8
$C_6H_{12}N_2O_6Zn$, 32B, 496; 35B, 651
$C_6H_{12}N_2S_4Zn$, 31B, 405
$C_6H_{12}O_4$, 29, 762
$C_6H_{12}O_5$, 26, 624; 31B, 170; 33B, 160
$C_6H_{12}O_6$, 26, 623; 28, 516; 29, 487; 30B, 179; 32B, 215, 489; 33B, 161; 34B, 242
$C_6H_{12}O_6.H_2O$, 27, 795
$C_6H_{12}O_6.2H_2O$, 28, 425
$C_6H_{12}O_4S$, 31B, 169
$C_6H_{12}O_2Se$, 32B, 488
$C_6H_{12}S_3$, 34B, 188
$C_6H_{12}S_6.AgNO_3$, 33B, 476
$C_6H_{12}Se_3$, 32B, 488
$C_6H_{13}ClN_2O_3$, 34B, 270
$C_6H_{13}CuNO_7$, 33B, 418
$C_6H_{13}MoNO_5$, 35B, 917
$C_6H_{13}NO_2$, 32B, 15
C_6H_{14}, 26, 559
$C_6H_{14}BN_3$, 34B, 362
$C_6H_{14}BrCoN_4O_4$, 34B, 539
$C_6H_{14}BrNO_2$, 32B, 229
$C_6H_{14}BrNO_5$, 30B, 180; 34B, 243
$C_6H_{14}Cl_2MoN_2O_4$, 33B, 456
$C_6H_{14}ClN$, 34B, 92
$C_6H_{14}ClNO_4$, 27, 796
$C_6H_{14}ClNO_5$, 30B, 180; 32B, 489; 34B, 243
$C_6H_{14}ClN_4O_3$, 35B, 319
$C_6H_{14}CuO_8$, 33B, 403
$C_6H_{14}N_2$, 28, 666
$C_6H_{14}N_2.2HCl$, 30B, 89
$C_6H_{14}N_2O_{14}$, 32B, 107
$C_6H_{14}N_4O_6$, 26, 571
$C_6H_{14}N_4O_2.2H_2O$, 29, 479
$C_6H_{14}N_4O_4S_2Zn$, 32B, 408
$C_6H_{14}N_2S_2$, 32B, 488
$C_6H_{14}O_6$, 33B, 162, 163
$C_6H_{14}O_2.6H_2O$, 35B, 29
$C_6H_{15}AlO_3$, 28, 671
$C_6H_{15}BN_4$, 34B, 162
$C_6H_{15}BrN_4O_2.H_2O$, 30B, 420; 31B, 189
$C_6H_{15}ClN_4O_2.H_2O$, 30B, 420; 32B, 489
$C_6H_{15}N.HCl$, 30B, 33
$C_6H_{15}NS_2Sn$, 35B, 793
$C_6H_{15}O_3P$, 30B, 401
$C_6H_{16}Br_2CoN_4O_2.6H_2O$, 30B, 419
$C_6H_{16}Br_2N_4NiO_2.6H_2O$, 30B, 419
$C_6H_{16}Br_2N_4O_2Zn$, 30B, 419
$C_6H_{16}Cl$, 26, 642
$C_6H_{16}ClCoN_6$, 35B, 592
$C_6H_{16}ClCoN_6.H_2O$, 32B, 494
$C_6H_{16}Cl_2CoN_4O_2.6H_2O$, 30B, 419
$C_6H_{16}Cl_{10}Cu_5O_2$, 29, 590
$C_6H_{16}Cl_4MnO_8P_2$, 26, 692
$C_6H_{16}ClN$, 30B, 33
$C_6H_{16}Cl_2N_2$, 30B, 89; 31B, 87
$C_6H_{16}Cl_2N_{10}Ni.H_2O$, 35B, 667
$C_6H_{16}Cl_2N_4NiO_2.6H_2O$, 30B, 419
$C_6H_{16}Cl_2N_2O_3$, 32B, 117
$C_6H_{16}ClN_3O_3S$, 30B, 208
$C_6H_{16}Cl_2N_4O_2S_2$, 32B, 489; 33B, 184
$C_6H_{16}Cl_2N_4O_2Zn$, 30B, 420
$C_6H_{16}Cl_2O_3Ti$, 28, 674
$C_6H_{16}CoN_3O_8$, 26, 675
$C_6H_{16}CuN_6S_2$, 29, 593
$C_6H_{16}I_2N_4O_2$, 32B, 489
$C_6H_{16}I_2N_4O_2Zn$, 30B, 420
$C_6H_{16}N_2NiO_8$, 34B, 542
$C_6H_{16}N_6NiS_2$, 28, 616
$C_6H_{16}N_{10}S_6Te$, 26, 577
$C_6H_{17}ClN_4O_3$, 35B, 319
$C_6H_{18}Al_2$, 32B, 327
$C_6H_{18}As_2Cl_4Pt_2$, 35B, 865
$C_6H_{18}As_4N_6$, 32B, 315
$C_6H_{18}Cl_3CrN$, 32B, 496
$C_6H_{18}Cl_2CuN_{10}O$, 35B, 670
$C_6H_{18}Cl_2N_2OV$, 33B, 437
$C_6H_{18}Cl_8N_6P_4$, 32B, 307
$C_6H_{18}Cl_3N_2V$, 27, 828; 34B, 549
$C_6H_{18}Cl_2O_2S_2Sn$, 35B, 735
$C_6H_{18}Cl_2P_2Pt$, 29, 768; 32B, 464
$C_6H_{18}CoN_2O_8P_2$, 28, 647
$C_6H_{18}HgI_4S_2$, 31B, 57
$C_6H_{18}I_2N_2$, 28, 396
$C_6H_{18}N_3O_6P_3$, 30B, 393; 33B, 280
$C_6H_{18}OSi_2$, 26, 615
$C_6H_{19}AlN_2$, 29, 535
$C_6H_{19}ClN_2O_4$, 27, 767
$C_6H_{20}B_{20}O_2$, 35B, 427
$C_6H_{20}ClCoN_6O_4$, 35B, 672
$C_6H_{20}Cl_3CoN_4O_9$, 34B, 482
$C_6H_{20}ClCu_2N_2.H_2O$, 30B, 420
$C_6H_{20}Cl_4CuN_2$, 35B, 877

$C_6H_{20}Cl_2CuN_4O_8$, 35B, 677
$C_6H_{20}CoN_7O_9$, 35B, 678
$C_6H_{20}Cr_2CuN_2O_7$, 35B, 675
$C_6H_{20}CuN_6O_4$, 35B, 676
$C_6H_{20}CuN_6O_6$, 34B, 555; 35B, 674
$C_6H_{21}AlN_2$, 28, 598
$C_6H_{21}ClCoN_7O_6$, 35B, 679
$C_6H_{21}Cl_3N_4$, 33B, 27
$C_6H_{21}I_4K_2N_6O_6$, 30B, 257
$C_6H_{22}CuN_4O_5S$, 35B, 681
$C_6H_{22}CuN_4O_5Se$, 35B, 681
$C_6H_{24}B_3N_3$, 26, 618
$C_6H_{24}BiBr_6N_3$, 33B, 293
$C_6H_{24}Br_2CaN_{12}O_6$, 35B, 909
$C_6H_{24}Br_2N_{12}NiS_6$, 34B, 523
$C_6H_{24}Cl_3CoN_6.H_2O$, 34B, 485
$C_6H_{24}Cl_2N_{12}O_8PbS_6$, 32B, 54
$C_6H_{24}Cl_2N_{12}O_8PbS_6.2H_2O$, 32B, 54
$C_6H_{24}CuN_2O_{10}$, 26, 662
$C_6H_{24}CuN_6O_4S$, 26, 605; 35B, 594
$C_6H_{24}I_3N_{12}O_6Ti$, 35B, 630
$C_6H_{24}N_6NiO_4S$, 35B, 595
$C_6H_{24}N_{14}NiO_6S_6$, 26, 576
$C_6H_{24}N_6NiO_5Si_2.8.7H_2O$, 35B, 441
$C_6H_{24}N_{14}O_6PbS_6$, 32B, 54
$C_6H_{24}N_6O_6Pd_2S_4$, 35B, 598
$C_6H_{25}Cl_4CoN_4O_2$, 27, 833
$C_6H_{26}Br_3CoN_6O$, 27, 830
$C_6H_{26}I_4N_{12}O_{12}U_2$, 33B, 556
$C_6H_{26}O_8$, 35B, 29

C_7
C_7F_5N, 33B, 541
$C_7HF_2MnO_5$, 32B, 339
$C_7HF_6O_4Rh$, 32B, 495
$C_7H_3Br_2N_3$, 32B, 60
$C_7H_3FeNO_4$, 26, 684; 27, 824
$C_7H_4BrNO_3$, 29, 652
C_7H_4BrNS, 27, 893
$C_7H_4ClNO_3$, 29, 652
$C_7H_4ClNO_4$, 27, 897
C_7H_4ClNOS, 34B, 206
$C_7H_4CrO_3S$, 30B, 318
$C_7H_4FeO_4$, 35B, 895
C_7H_4IN, 30B, 38
$C_7H_4INO_3$, 29, 652
$C_7H_5AlCl_4O$, 31B, 326
$C_7H_5Br_3FeO_2Sn$, 35B, 539
$C_7H_5Br_2NO_2$, 27, 898; 30B, 57
$C_7H_5BrO_2$, 27, 895; 32B, 76
$C_7H_5Cl_2CoHgO_2$, 32B, 493
$C_7H_5Cl_3FeO_2Sn$, 35B, 538
$C_7H_5Cl_2NO_2$, 34B, 57
$C_7H_5ClO_2$, 26, 694
$C_7H_5Cl_6OSb$, 33B, 293
$C_7H_5IO_3$, 30B, 157
$C_7H_5KN_2O_4$, 35B, 30
$C_7H_9N_5$, $C_5H_5BrN_2O_2$, 31B, 163
$C_7H_5NO_3$, 29, 651
$C_7H_5NO_5$, 28, 671; 31B, 61; 32B, 76; 35B, 41
$C_7H_5N_5O_8$, 32B, 94
C_7H_5NOS, 35B, 235
$C_7H_5NOS_2$, 34B, 189
$C_7H_5NO_3S$, 33B, 147; 34B, 207
$C_7H_6BF_2FeO_5S$, 35B, 516
C_7H_6ClNO, 29, 657; 35B, 79
$C_7H_6ClNO_2$, 26, 711
$C_7H_6ClNO_4$, 28, 671
C_7H_6FNO, 30B, 60, 61
$C_7H_6FeO_3$, 28, 605
$C_7H_6N_2S$, 30B, 64
$C_7H_6O_3$, 30B, 56; 32B, 174
$C_7H_6O_2S$, 32B, 185
$C_7H_6O_3S$, 32B, 488
C_7H_7BrHg, 29, 628
$C_7H_7BrN_2O_2$, 32B, 486
C_7H_7BrO, 31B, 73
$C_7H_7BrO_4$, 35B, 5
$C_7H_7Cl_4FeN_2O$, 30B, 51
C_7H_7ClHg, 29, 628
C_7H_7ClHgO, 29, 628
$C_7H_7ClO_2S$, 26, 695
$C_7H_7KO_4S$, 32B, 67
$C_7H_7N_5NiOS_4$, 31B, 451
$C_7H_7NO_2$, 30B, 55; 31B, 62, 63; 32B, 77; 33B, 42
$C_7H_7NO_2.AlCl_3$, 26, 703
$C_7H_7NO.HCl$, 31B, 123
$C_7H_7O_4Rh$, 32B, 495
C_7H_8, 29, 503
$C_7H_8Ag_2N_2O_6$, 31B, 367
C_7H_8ClNO, 31B, 123
$C_7H_8Cl_2Pd$, 26, 687; 30B, 324
$C_7H_8FeO_3$, 35B, 895
$C_7H_8N_2O_2$, 30B, 420
$C_7H_8N_2O_3$, 30B, 169
$C_7H_8N_2S$, 34B, 130, 136
$C_7H_8N_4S$, 35B, 166
C_7H_8O, 31B, 74; 35B, 65
$C_7H_8O_4$, 32B, 15
$C_7H_8OS_2$, 26, 761
$C_7H_8O_2S$, 30B, 83
$C_7H_8S_3$, 27, 973
$C_7H_9BrN_6.C_5H_6N_2O_2$, 35B, 275
$C_7H_9BrN_6.C_7H_{10}N_2O_2$, 35B, 275
$C_7H_9BrN_2O$, 31B, 123
$C_7H_9BrN_2O_2S_2$, 28, 669; 31B, 146
$C_7H_9BrN_2STe$, 31B, 337
$C_7H_9BrO_2.H_2O$, 30B, 145
$C_7H_9CaNO_7$, 33B, 113
$C_7H_9ClN_2O$, 31B, 123
$C_7H_9ClN_2O_2S_2$, 28, 669
$C_7H_9ClN_2STe$, 31B, 337
$C_7H_9IN_2O$, 31B, 123
$C_7H_9IN_2O_2S_2$, 28, 670
C_7H_9N, 28, 422
$C_7H_9N_5$, 31B, 163
$C_7H_9N_4O_9Rb$, 34B, 54
$C_7H_9NO_2S$, 33B, 55
$C_7H_9NO_3S$, 35B, 80
$C_7H_9NO_5S$, 32B, 81
C_7H_{10}, 31B, 93
$C_7H_{10}Cl_2$, 33B, 534; 35B, 898
$C_7H_{10}N_2$, 31B, 42
$C_7H_{10}N_2O_2$, 31B, 166
$C_7H_{10}N_2O_2.C_7H_9N_6Br$, 35B, 275
$C_7H_{10}O_2S$, 32B, 488
$C_7H_{11}Cl_3O_3$, 29, 762
$C_7H_{11}CuNO_2$, 35B, 723
$C_7H_{11}NNiO_2$, 35B, 722
$C_7H_{11}NO_9V$, 33B, 437
C_7H_{12}, 33B, 534
$C_7H_{12}Cl_2O_4$, 33B, 165; 34B, 249
$C_7H_{12}CoN_4O_5$, 35B, 759
$C_7H_{12}N_2NaO_5S_2$, 31B, 145
$C_7H_{12}N_4OS$, 32B, 487
$C_7H_{12}O_4$, 26, 561; 28, 671; 29, 468; 31B, 10; 32B, 16; 33B, 4
$C_7H_{12}O_9S$, 35B, 43
$C_7H_{13}BrN_2O_2$, 30B, 123
$C_7H_{13}BrO_5$, 30B, 177
$C_7H_{13}ClO_5$, 34B, 249
$C_7H_{13}CrN_3O_3$, 31B, 376
$C_7H_{13}HgClO_5$, 30B, 420
$C_7H_{13}I_3N_4$, 35B, 202
$C_7H_{13}MoN_3O_3$, 30B, 355
$C_7H_{14}AgNS_2$, 34B, 525
$C_7H_{14}ClN$, 33B, 127
$C_7H_{14}N_6O_8S$, 33B, 551
$C_7H_{14}NS_2Tl$, 34B, 404
$C_7H_{14}O_5$, 26, 624
$C_7H_{14}O_6$, 33B, 168; 35B, 258
$C_7H_{14}O_7$, 35B, 259
$C_7H_{15}NS_3$, 34B, 45
$C_7H_{15}O_4P$, 31B, 179
$C_7H_{15}PS_2$, 26, 606; 27, 774
$C_7H_{16}B_9Fe$, 30B, 305
$C_7H_{16}ClNO_2$, 27, 991; 35B, 321
$C_7H_{16}ClNO_5S$, 34B, 208
$C_7H_{16}O.{}^1/_2H_2O$, 28, 671
$C_7H_{16}O_6S_3$, 34B, 46
$C_7H_{18}ClN$, 28, 397
$C_7H_{20}B_3CrNO_4$, 32B, 485
$C_7H_{22}Cl_3N_3$, 31B, 260
$C_7H_{22}CoI_2N_5O_2.2H_2O$, 32B, 496
$C_7H_{23.5}N_3O_6P_{1.5}.3H_2O$, 34B, 336
$C_7H_{27}Cl_3N_{10}O$, 30B, 258

C_8
C_8Cl_8, 35B, 99
C_8F_{12}, 29, 766; 30B 112
$C_8F_6FeO_4$, 32B, 493
$C_8F_8FeO_4$, 26, 680
$C_8F_{24}N_8O_4S_4$, 35B, 917
C_8N_4S, 33B, 138
$C_8N_4S_2$, 30B, 151
$C_8HBrN_4O_5S$, 27, 982
$C_8H_2F_8FeO_4$, 32B, 341
$C_8H_3O_9Rb$, 32B, 174
$C_8H_4CdK_2$, 33B, 324
$C_8H_4CuN_6O_8.4H_2O$, 33B, 551
$C_8H_4F_3N_5O_8$, 34B, 71
$C_8H_4FeO_8$, 32B, 343; 33B, 407

$C_8H_4Hg_2O_4$, 32B, 417
$C_8H_4K_2Zn$, 33B, 324
$C_8H_4N_4$, 33B, 77
$C_8H_4N_6O_8Sr.4H_2O$, 32B, 488
$C_8H_4N_2Se_2$, 35B, 81
$C_8H_4O_2$, 35B, 121
$C_8H_4O_4$, 31B, 140
$C_8H_5BrN_2O_2$, 28, 494
$C_8H_5ClMoO_3$, 33B, 343
$C_8H_5ClN_4$, 33B, 57
$C_8H_5F_5IORh$, 30B, 313
$C_8H_5KO_4$, 30B, 62
$C_8H_5MnO_3$, 28, 608
$C_8H_5N_3$, 30B, 131
$C_8H_5NO_2$, 28, 671
$C_8H_6Br_2$, 33B, 90
$C_8H_6Br_4$, 26, 712
$C_8H_6BrNO_3$, 26, 709
$C_8H_6Br_2N_2O_5$, 30B, 106
$C_8H_6Cl_2$, 33B, 90
$C_8H_6ClNO_3$, 26, 709
$C_8H_6HfNa_4O_{19}$, 26, 691
$C_8H_6I_2$, 33B, 90
$C_8H_6IN_3$, 35B, 59
$C_8H_6K_3MoN_6O_3S_6$, 33B, 422
$C_8H_6N_2.2H_2O$, 26, 779
$C_8H_6N_2O$, 26, 776
$C_8H_6N_2O_6$, 35B, 123
$C_8H_6N_4O_8.2H_2O$, 30B, 168
$C_8H_6N_4O_4Rb_2.2H_2O$, 28, 456
$C_8H_6Na_4O_{19}Zr$, 26, 691; 28, 632
$C_8H_6O_4$, 32B, 82
C_8H_6OS, 33B, 138
$C_8H_6O_2S_2$, 34B, 182
$C_8H_6S_2$, 33B, 140; 34B, 182; 35B, 221
$C_8H_7BrN_4O_5$, 35B, 252
$C_8H_7BrO_2$, 29, 675
C_8H_7ClO, 31B, 78
$C_8H_7ClO_3$, 30B, 420
$C_8H_7CrNO_5$, 32B, 493
$C_8H_7N_3O_7$, 33B, 49
$C_8H_7N_5O$, 35B, 254
$C_8H_7NS_2$, 26, 719; 27, 980
C_8H_8, 29, 766
$C_8H_8BiBr_4N$, 32B, 157
C_8H_8BrNO, 33B, 59
$C_8H_8BrN_3O_4$, 32B, 96
C_8H_8ClCu, 29, 587
$C_8H_8ClCuN_4O_4$, 34B, 556
C_8H_8ClNO, 29, 657; 31B, 70
$C_8H_8ClNO_3$, 31B, 76
$C_8H_8Cl_2Pd$, 31B, 494
$C_8H_8FN_5O_3.H_2O$, 34B, 226
$C_8H_8I_2O_2$, 30B, 248
$C_8H_8N_{12}.H_2O$, 35B, 154
$C_8H_8N_2O_2$, 34B, 151
$C_8H_8N_2O_4$, 31B, 10
$C_8H_8N_4O_4$, 35B, 256
$C_8H_8N_2S$, 35B, 236
$C_8H_8O_2$, 28, 671; 29, 645; 32B, 83, 108, 109, 110
$C_8H_8O_3$, 32B, 84
$C_8H_8O_4$, 28, 671
$C_8H_8O_3S$, 32B, 488
$C_8H_9ClN_2O_2$, 33B, 243
$C_8H_9MnO_5Sn$, 33B, 308
C_8H_9NO, 28, 412; 31B, 70
$C_8H_9NO_2$, 26, 701; 28, 414
C_8H_9NOS, 34B, 73
C_8H_{10}, 26, 642; 35B, 899
$C_8H_{10}Br_2O_4$, 31B, 85
$C_8H_{10}CuN_2O_4$, 35B, 725
$C_8H_{10}CuN_4O_3.1\frac{1}{2}H_2O$, 32B, 420
$C_8H_{10}K_2O_{15}Sb_2$, 35B, 469
$C_8H_{10}N_2O_2$, 30B, 68
$C_8H_{10}NO_5PS$, 35B, 446
$C_8H_{10}O$, 30B, 74, 421; 31B, 73; 32B, 102
$C_8H_{10}O_4$, 28, 671; 32B, 17; 34B, 9
$C_8H_{10}OS$, 35B, 83
$C_8H_{10}Pd$, 30B, 421; 31B, 343; 33B, 331
$C_8H_{11}BrN_2S$, 31B, 125; 33B, 114
$C_8H_{11}ClO_4S$, 29, 485
$C_8H_{11}Cl_2HgN$, 32B, 429
$C_8H_{11}ClN_2S$, 31B, 126; 33B, 114
$C_8H_{11}CsO_5S$, 30B, 421
$C_8H_{11}KN_2O_3$, 27, 972, 973; 30B, 163
$C_8H_{11}LiO_5S.2H_2O$, 30B, 421
$C_8H_{11}N$, 30B, 403
$C_8H_{11}N_2NaO_3$, 27, 973
$C_8H_{11}NO_3.HCl$, 26, 808
$C_8H_{11}N_2O_6P.2H_2O$, 34B, 254
C_8H_{12}, 33B, 535; 35B, 899
$C_8H_{12}BrN$, 26, 702
$C_8H_{12}BrN_2$, 33B, 60
$C_8H_{12}BrN_5O$, 34B, 146
$C_8H_{12}Cl_2$, 28, 671
$C_8H_{12}ClCu$, 27, 850
$C_8H_{12}Cl_4Hg_2S_2$, 30B, 376
$C_8H_{12}ClN$, 26, 702
$C_8H_{12}ClNO_2$, 33B, 231
$C_8H_{12}ClNO_3$, 32B, 263; 33B, 232
$C_8H_{12}Cl_2Ni$, 27, 848
$C_8H_{12}ClRh$, 34B, 441
$C_8H_{12}Cr_2O_8.2H_2O$, 35B, 920
$C_8H_{12}Cu_2O_{10}$, 31B, 406
$C_8H_{12}KN_3O_{10}$, 33B, 50
$C_8H_{12}N_6NiS_4$, 31B, 402
$C_8H_{12}N_2O_3$, 34B, 216
$C_8H_{12}N_6S_4Te$, 26, 578
$C_8H_{12}O_2$, 27, 786; 28, 672
$C_8H_{12}O_4$, 31B, 87; 34B, 90, 91; 35B, 106
$C_8H_{12}O_8Rh_2.2H_2O$, 35B, 920
$C_8H_{12}O_8U$, 29, 458
$C_8H_{13}AgO_6S$, 28, 672
$C_8H_{13}ClN_4S_2Te$, 31B, 340
$C_8H_{13}Cl_3O_3S_2SnW$, 34B, 585
$C_8H_{13}CuNO_2$, 35B, 723
$C_8H_{13}N_3O$, 30B, 259
$C_8H_{13}NO.ClH$, 27, 956; 29, 735
$C_8H_{13}N_5O_2.2HBr.H_2O$, 35B, 366
$C_8H_{13}NaO_3$, 29, 471
C_8H_{14}, 35B, 900
$C_8H_{14}BrN_2NaO_4$, 34B, 351
$C_8H_{14}Cl_2$, 34B, 95
$C_8H_{14}ClN.H_2O$, 35B, 135
$C_8H_{14}ClNSn$, 28, 603
$C_8H_{14}Cl_2Pd_2$, 33B, 333
$C_8H_{14}CuN_4O_4$, 35B, 760
$C_8H_{14}CuN_6OS_2$, 31B, 454
$C_8H_{14}Ge$, 35B 900
$C_8H_{14}IKN_2O_4$, 34B, 352
$C_8H_{14}N_2NiS_2$, 30B, 380
$C_8H_{14}Ni$, 30B, 298
$C_8H_{14}O_4$, 29, 470; 30B, 6, 31B, 12; 32B, 18; 34B, 11
$C_8H_{14}O_2S_2$, 32B, 186
$C_8H_{14}Pt$, 34B, 418
$C_8H_{14}Si$, 35B, 900
$C_8H_{15}B_{10}Br$, 30B, 411
$C_8H_{15}N.BrH$, 29, 507
$C_8H_{15}NO$, 32B, 253
$C_8H_{15}NO_2$, 33B, 82
$C_8H_{15}NO_6$, 31B, 171
$C_8H_{15}N_2O_3.HBr$, 31B, 190
$C_8H_{16}Br_4CaO_6P_2$, 30B, 422
$C_8H_{16}Br_2HgO_4$, 35B, 918
$C_8H_{16}BrLiO_4$, 27, 761
$C_8H_{16}BrNO_2$, 30B, 90; 31B, 88
$C_8H_{16}BrN_2O_3$, 31B, 190
$C_8H_{16}Br_2N_4S_2Te$, 26, 583
$C_8H_{16}CaF_4O_8P_2$, 34B, 397
$C_8H_{16}CdN_{12}S_8Zn$, 32B, 409
$C_8H_{16}Cl_2CuN_2O_2S$, 35B, 795
$C_8H_{16}ClGaO_4$, 26, 693
$C_8H_{16}Cl_2HgO_4$, 33B, 524
$C_8H_{16}Cl_2HgO_2S_2$, 32B, 452
$C_8H_{16}ClIr$, 35B, 912
$C_8H_{16}Cl_4MnO_8P_2$, 28, 649
$C_8H_{16}Cl_2N_6NiO_3$, 34B, 561
$C_8H_{16}ClNO_2$, 31B, 88
$C_8H_{16}Cl_2N_2O$, 29, 772
$C_8H_{16}Cl_2N_4S_2Te$, 26, 583
$C_8H_{16}ClOP$, 35B, 448
$C_8H_{16}Cl_8O_4Ti_2$, 31B, 442
$C_8H_{16}Cl_4S_2Sn$, 30B, 422
$C_8H_{16}CoHgN_{12}S_8$, 30B, 342
$C_8H_{16}Cu_2O_{10}$, 32B, 496
$C_8H_{16}Ge_2$, 32B, 330; 30B, 422
$C_8H_{16}N_2O_2$, 26, 784; 30B, 413; 31B, 12
$C_8H_{16}N_6OS_2$, 31B, 454; 34B, 24
$C_8H_{16}Ni_2S_6$, 34B, 591
$C_8H_{16}O_{10}Rh_2$, 27, 816
$C_8H_{16}O_5S$, 32B, 216
$C_8H_{17}BrN_2O_3$, 34B, 267
$C_8H_{17}GeNO_3$, 35B, 484
$C_8H_{17}I.HgO$, 27, 824
$C_8H_{17}NO$, 26, 762
$C_8H_{17}NO_2Si$, 34B, 366
$C_8H_{17}O_3P$, 35B, 449
$C_8H_{17}O_2P.H_2O$, 33B, 549
C_8H_{18}, 32B, 40
$C_8H_{18}Be$, 33B, 535

$C_8H_{18}Cl_2N_6NiS_2.H_2O$, 34B, 585
$C_8H_{18}ClNO$, 28, 672
$C_8H_{18}ClNO.H_2O$, 28, 672
$C_8H_{18}Cl_2N_2O_2Zn$, 35B, 924
$C_8H_{18}Cl_2PdSe_2$, 35B, 831
$C_8H_{18}CuN_6S_2$, 32B, 381
$C_8H_{18}CuO_8$, 33B, 403
$C_8H_{18}LiO$, 30B, 415
$C_8H_{18}N_6NiS_2$, 35B, 682, 683
$C_8H_{18}N_2O_2$, 26, 608
$C_8H_{18}N_4O_6$, 26, 572
$C_8H_{18}N_4O_4S_6Te$, 26, 582
$C_8H_{18}N_2O_{15}Sb_2$, 32B, 320
$C_8H_{18}N_6S_2Zn$, 34B, 486
$C_8H_{20}AlLi$, 29, 535
$C_8H_{20}B_3CrNO_4$, 35B, 437
$C_8H_{20}BaO_8P_2$, 31B, 177
$C_8H_{20}BrCuP_2$, 35B, 835
$C_8H_{20}Br_2MgO_2$, 32B, 322
$C_8H_{20}Br_4Pt_2S_2$, 33B, 481
$C_8H_{20}CdCl_2NO_6P.3H_2O$, 30B, 192
$C_8H_{20}ClCoN_6O_8$, 26, 676
$C_8H_{20}ClCuN_4S_2$, 27, 841
$C_8H_{20}Cl_4InN$, 34B, 619
$C_8H_{20}Cl_2N_{10}NiO_4.2H_2O$, 32B, 496
$C_8H_{20}Cl_2PdSe_2$, 35B, 798
$C_8H_{20}Cl_5Re_2S_4$, 33B, 483
$C_8H_{20}CoN_7O_7$, 26, 676
$C_8H_{20}CuN_2O_6$, 26, 665
$C_8H_{20}I_3N$, 32B, 33
$C_8H_{20}I_3N_4Rh$, 34B, 550
$C_8H_{20}N_2NiS_2$, 32B, 453
$C_8H_{20}NO_6P$, 31B, 180
$C_8H_{20}NO_6P.CdCl_2.3H_2O$, 30B, 192
$C_8H_{20}NiO_4P_2S_4$, 32B, 454
$C_8H_{20}NiP_2S_4$, 34B, 592
$C_8H_{20}O_2.BrLi.H_2O$, 29, 763
$C_8H_{20}O_4P_2S_4Zn$, 34B, 592
$C_8H_{20}P_2S_2$, 26, 609
$C_8H_{20}P_2S_2Se_2$, 31B, 301
$C_8H_{20}P_2S_2Se_2Te$, 34B, 380
$C_8H_{20}P_2S_4Zn$, 35B, 796
$C_8H_{20}P_2Se_5$, 34B, 381
$C_8H_{22}B_{18}Cs_2Fe_2O_4$, 35B, 432
$C_8H_{22}Br_4NO_2Re$, 30B, 34
$C_8H_{22}ClCoN_6O_8$, 35B, 684
$C_8H_{22}Cl_2CuN_{10}O_4$, 35B, 687
$C_8H_{22}Cl_2N_{10}NiO_4.2H_2O$, 35B, 687
$C_8H_{23}Cl_3CoN_5O_4$, 35B, 686
$C_8H_{24}Ag_{13}I_{15}N_2$, 35B, 881
$C_8H_{24}Al_3Br_5O_6Si_4$, 33B, 268
$C_8H_{24}Al_2Mg$, 34B, 398
$C_8H_{24}Al_2N_2$, 35B, 479
$C_8H_{24}BeN_2Si_4$, 35B, 901
$C_8H_{24}Br_2N_4Ni$, 34B, 489
$C_8H_{24}Cl_4CoN_4$, 32B, 31
$C_8H_{24}Cl_5CoN_5Zn$, 35B, 599
$C_8H_{24}Cl_6Fe_2O_4S_4$, 32B, 446
$C_8H_{24}Cl_4N_2Ni$, 32B, 31
$C_8H_{24}ClN_4Ni_3S_4$, 35B, 789
$C_8H_{24}Cl_4N_2Zn$, 32B, 32
$C_8H_{24}Cl_2O_4S_4Zn$, 35B, 924
$C_8H_{24}CuK_2N_4O_{12}$, 33B, 418
$C_8H_{24}CuN_6O_6$, 34B, 488
$C_8H_{24}N_2NiO_8$, 27, 839
$C_8H_{24}N_4P_4$, 26, 607
$C_8H_{24}N_4Sn$, 35B, 901
$C_8H_{24}O_2Si_4$, 28, 583
$C_8H_{24}O_4Si_4$, 26, 616
$C_8H_{24}O_8Si_8$, 31B, 298
$C_8H_{25}Br_2CuN_6.H_2O$, 34B, 489
$C_8H_{25}Cl_3CuN_4P_4$, 35B, 459
$C_8H_{26}B_{20}Hg$, 30B, 411
$C_8H_{26}CuN_8O_6$, 34B, 486
$C_8H_{28}B_{18}Br_6CoN$, 33B, 265
$C_8H_{28}Br_3N_4Pt$, 31B, 423
$C_8H_{28}Cl_2N_6NiO$, 35B, 702
$C_8H_{28}N_4Si_4$, 28, 584
$C_8H_{28}Na_4O_{26}V_2$, 33B, 409
$C_8H_{30}B_6N_2$, 30B, 271
$C_8H_{30}B_{18}Ni$, 35B, 430
$C_8H_{32}B_4P_4$, 27, 989; 29, 766
$C_8H_{32}Cl_4Co_2N_8O_6S_6.H_2O$, 30B, 330
$C_8H_{32}N_2O_8S$, 33B, 27
$C_8H_{34}Co_2N_{13}O_{14}.2H_2O$, 32B, 495
$C_8H_{38}Mo_4N_4O_{27}$, 33B, 410

C_9

$C_9F_8FeO_3$, 26, 680; 32B, 371
$C_9H_4BrS_2$, 34B, 184
$C_9H_4KN_3O_2$, 32B, 51
$C_9H_4O_3$, 30B, 422
$C_9H_5BrN_4O_4$, 34B, 126
$C_9H_5BrO_3.H_2O$, 30B, 146
$C_9H_5ClN_4O_4$, 35B, 158
C_9H_5ClO, 31B, 79
$C_9H_5CoF_6S_2$, 31B, 348
$C_9H_5CoN_2S_2$, 33B, 344
$C_9H_5F_6O_4V$, 35B, 540
$C_9H_5NbO_4$, 28, 672
$C_9H_5O_4V$, 32B, 350
C_9H_6ClN, 28, 470; 30B, 422; 33B, 123, 124
$C_9H_6CrO_3$, 26, 684; 30B, 316
$C_9H_6Fe_2N_2O_7$, 33B, 423
$C_9H_6NNaO_2S$, 29, 733
$C_9H_6INO_4S$, 35B, 186
$C_9H_6N_2O_6$, 34B, 60
$C_9H_6O_3$, 31B, 141
$C_9H_6O_4$, 34B, 108
$C_9H_6O_6$, 34B, 62
$C_9H_6S_2$, 35B, 221
C_9H_7, 33B, 536
$C_9H_7BrN_2O_2$, 34B, 197, 198
$C_9H_7BrS_2$, 28, 670
$C_9H_7ClO_4S_2$, 28, 670
$C_9H_7IS_2$, 28, 670; 30B, 147; 31B, 147
$C_9H_7N.HCl$, 30B, 137
$C_9H_7N.HCl.H_2O$, 30B, 137
$C_9H_7NO_2$, 35B, 199, 201
C_9H_8BrN, 32B, 154
C_9H_8ClN, 27, 985; 30B, 137
$C_9H_8ClN.H_2O$, 30B, 137
$C_9H_8Cl_3NO_2Se$, 32B, 158
$C_9H_8Cl_6O_2S$, 27, 785
$C_9H_8FeO_4$, 34B, 418
$C_9H_8FeO_5$, 35B, 518
$C_9H_8N_2O_6$, 26, 711
$C_9H_8O_2$, 30B, 92
$C_9H_8O_3$, 32B, 497; 34B, 304; 35B, 44
$C_9H_8O_4$, 29, 653
$C_9H_8O_5$, 30B, 422
$C_9H_8O_6$, 35B, 210
$C_9H_9BrHgO_2$, 29, 628
$C_9H_9BrO_3$, 31B, 16
$C_9H_9Cl_3$, 32B, 112
$C_9H_9ClO_3$, 35B, 73
$C_9H_9Cl_3O_3$, 29, 495
$C_9H_9CuNO_4.{}^1/_2H_2O$, 32B, 391
$C_9H_9CuNO_4.4H_2O$, 34B, 513
C_9H_{10}, 30B, 404
$C_9H_{10}Br_4$, 27, 891
$C_9H_{10}BrNO$, 32B, 34; 33B, 61
$C_9H_{10}Cl_4O_2$, 35B, 99
$C_9H_{10}FN_5O_3$, 34B, 224
$C_9H_{10}MoO_2S$, 34B, 438
$C_9H_{10}N_2O_3.HBr.H_2O$, 35B, 229
$C_9H_{10}NOSe$, 34B, 33
$C_9H_{10}N_2OSe$, 28, 675
$C_9H_{10}O_2$, 28, 672; 32B, 84
$C_9H_{11}BaN_2O_9P.7H_2O$, 30B, 206
$C_9H_{11}BrN_2O_5$, 31B, 183
$C_9H_{11}BrN_2O_6$, 31B, 183
$C_9H_{11}BrN_2O_3.H_2O$, 35B, 229
$C_9H_{11}BrN_2O_2S$, 26, 733
$C_9H_{11}ClN_2O_2S$, 26, 734; 28, 573
$C_9H_{11}CuNO_4$, 26, 776
$C_9H_{11}FN_2O_5$, 29, 710
$C_9H_{11}IN_2O_5$, 30B, 204
$C_9H_{11}IN_2O_6$, 35B, 276
$C_9H_{11}IN_2O_2S$, 26, 733
$C_9H_{11}NO$, 32B, 96
$C_9H_{11}NO_3$, 29, 647
$C_9H_{11}NO_6$, 35B, 260
$C_9H_{11}N_3OS_4$, 29, 738
$C_9H_{11}O_4P$, 32B, 222
C_9H_{12}, 33B, 536
$C_9H_{12}Ag_3N_3O_9$, 32B, 373
$C_9H_{12}ClNO_2$, 28, 672; 29, 700
$C_9H_{12}Cl_2N_6O_8Se_3Te$, 26, 579
$C_9H_{12}INO_3$, 32B, 489
$C_9H_{12}N_2$, 32B, 118
$C_9H_{12}N_4O$, 30B, 171
$C_9H_{12}N_4O_3$, 28, 510
$C_9H_{12}N_2O_6.H_2O$, 35B, 294
$C_9H_{12}N_2OS_3$, 26, 785
$C_9H_{12}N_2O_2S$, 26, 733
$C_9H_{12}N_2O_5S.1{}^1/_2H_2O$, 35B, 278
$C_9H_{12}N_2S$, 34B, 137, 145; 35B, 168
$C_9H_{12}O_6$, 27, 784
$C_9H_{12}O_2S_2$, 33B, 142
$C_9H_{13}BrN_2O_2$, 27, 945
$C_9H_{13}ClO$, 32B, 144
$C_9H_{13}I_2NO_5$, 32B, 230
$C_9H_{13}KNO_3.{}^1/_3H_2O$, 35B, 159

$C_9H_{13}N_5.ClH$, 27, 978
$C_9H_{13}N_3O_5$, 30B, 198
$C_9H_{14}ClN_3O_4$, 35B, 279
$C_9H_{14}IN_2O_4$, 35B, 262
$C_9H_{14}N_2O$, 28, 461
$C_9H_{14}N_3O_8P$, 30B, 202; 32B, 223
$C_9H_{14}O_3$, 33B, 132
$C_9H_{14}O_4S_2$, 33B, 143
$C_9H_{15}CoO_3S_6$, 34B, 524
$C_9H_{15}CrO_3S_6$, 27, 844
$C_9H_{15}O_4Rb.1\frac{1}{2}H_2O$, 32B, 485
$C_9H_{15}O_3S_6Sb$, 26, 610
$C_9H_{16}CuN_4O_5$, 32B, 421
$C_9H_{16}O_4$, 26, 561; 31B, 14; 32B, 18
$C_9H_{17}ClO_5S_2$, 35B, 223
$C_9H_{17}HgNOS$, 27, 824
$C_9H_{17}O_{10}RbS$, 29, 740
$C_9H_{18}Br_2CuNS_2$, 33B, 485
$C_9H_{18}IN$, 29, 506
$C_9H_{18}INO_6$, 31B, 172
$C_9H_{18}INaO_3$, 27, 761; 28, 575
$C_9H_{18}N_4.3HCl.\frac{1}{2}H_2O$, 30B, 142
$C_9H_{18}NO_2$, 33B, 117
$C_9H_{18}N_2O_2$, 31B, 15; 34B, 272
$C_9H_{18}N_2O_3.HBr$, 34B, 272
$C_9H_{18}O_3$, 30B, 7
$C_9H_{18}O_6$, 34B, 170
$C_9H_{19}BrN_2O_3$, 34B, 272
$C_9H_{20}CoN_7O_4S$, 26, 675
$C_9H_{20}INO_2$, 35B, 322
$C_9H_{20}N_2O$, 26, 561
$C_9H_{20}N_4O_6$, 26, 572
$C_9H_{21}AlO_3$, 28, 671
$C_9H_{21}Br_2N_3O_5U$, 28, 672
$C_9H_{21}Cl_3N_4.\frac{1}{2}H_2O$, 30B, 142
$C_9H_{21}IN_3NaO_3$, 27, 760, 762
$C_9H_{21}NO_4S$, 34B, 125
$C_9H_{23}ClCoN_5O_8$, 35B, 600
$C_9H_{24}B_3N_3$, 34B, 363
$C_9H_{30}Br_3CoN_6$, 31B, 378
$C_9H_{30}Br_3CoN_6.H_2O$, 34B, 491

C_{10}
$C_{10}Cl_8$, 26, 642; 28, 429
$C_{10}Cl_{12}O_3S$, 32B, 146
$C_{10}D_8$, 34B, 97
$C_{10}H_2F_{12}Fe_2O_6P_2$, 35B, 914
$C_{10}H_2O_6$, 30B, 422
$C_{10}H_2O_6.C_6H_6$, 30B, 423
$C_{10}H_2O_6.C_{10}H_8$, 30B, 423
$C_{10}H_2O_6.C_{16}H_{10}$, 30B, 423
$C_{10}H_4Br_4$, 27, 924; 33B, 87
$C_{10}H_4Br_2Cl_2$, 27, 923; 30B, 99
$C_{10}H_4Br_2I_2$, 27, 923
$C_{10}H_4Br_5NO$, 31B, 117
$C_{10}H_4Br_2O_2$, 31B, 97; 32B, 124
$C_{10}H_4Cl_4$, 26, 745; 27, 921
$C_{10}H_4Cl_4O$(?), 26, 745
$C_{10}H_5Br_2N$, 26, 790
$C_{10}H_5BrO_2$, 26, 742
$C_{10}H_5Cl_2FN_2O_2$, 33B, 118
$C_{10}H_5ClO_2$, 26, 742
$C_{10}H_6$, 34B, 110
$C_{10}H_6Br_2$, 26, 741
$C_{10}H_6BrNO_2$, 31B, 98
$C_{10}H_6BrNO_4$, 32B, 64
$C_{10}H_6BrN_2O$, 35B, 232
$C_{10}H_6Cl_2$, 26, 740; 29, 678
$C_{10}H_6Cl_3N$, 30B, 423
$C_{10}H_6ClNO_2$, 28, 674
$C_{10}H_6Cl_2O$, 30B, 86
$C_{10}H_6FeO_4$, 29, 552
$C_{10}H_6N_2$, 31B, 91; 34B, 28
$C_{10}H_6N_2O_4$, 30B, 98
$C_{10}H_6O_2$, 26, 741; 29, 665
$C_{10}H_6O_4$, 26, 743, 744; 27, 927
$C_{10}H_7BrCl_2O_3$, 35B, 17
$C_{10}H_7BrF_2O_3$, 35B, 17
$C_{10}H_7BrO$, 33B, 88
$C_{10}H_7Cl$, 28, 672; 30B, 423
$C_{10}H_7Cl_2N$, 30B, 423
$C_{10}H_7ClO_2$, 30B, 423
$C_{10}H_7HgI$, 28, 662, 672
$C_{10}H_7N$, 26, 789
$C_{10}H_7NO_2$, 31B, 98; 34B, 101
$C_{10}H_7NO_2.\frac{1}{2}H_2O$, 35B, 118
$C_{10}H_7NS_3$, 34B, 423
$C_{10}H_7O$, 30B, 423
$C_{10}H_7O_3V$, 26, 684
$C_{10}H_8$, 26, 738, 739; 27, 925, 926
$C_{10}H_8Br_2MoN_2O_2$, 34B, 551
$C_{10}H_8Br_2N_2.B_{10}H_{12}$, 27, 883
$C_{10}H_8BrNO_2$, 26, 720; 32B, 191; 34B, 201
$C_{10}H_8Cl_2FeO_4S_2$, 29, 552
$C_{10}H_8Cl_3NO_5S$, 32B, 103
$C_{10}H_8Cl_6Sb_2$, 34B, 357
$C_{10}H_8CrN_2O_5$, 33B, 439
$C_{10}H_8CuN_4O_4$, 34B, 552
$C_{10}H_8MoO_5$, 34B, 418
$C_{10}H_8NNaO_3S.4H_2O$, 31B, 95
$C_{10}H_8N_2O_4$, 34B, 151
$C_{10}H_8O$, 29, 677; 32B, 488
$C_{10}H_8O_2$, 32B, 121
$C_{10}H_8O_4$, 31B, 140; 142
$C_{10}H_9BrNO_2$, 28, 366
$C_{10}H_9BrO$, 35B, 84
$C_{10}H_9BrO_2$, 30B, 8; 32B, 176
$C_{10}H_9ClN_4$, 33B, 124
$C_{10}H_9ClO_2$, 30B, 423
$C_{10}H_9Cl_3O$, 33B, 95
$C_{10}H_9ClO_4S_2$, 28, 670
$C_{10}H_9CrNO_3$, 32B, 368
$C_{10}H_9NO_2$, 29, 726; 34B, 203
$C_{10}H_9NO_4S.H_2O$, 35B, 185
$C_{10}H_{10}$, 32B, 487; 33B, 96
$C_{10}H_{10}AgBF_4.H_2O$, 33B, 368
$C_{10}H_{10}Be$, 28, 672; 29, 524; 33B, 536
$C_{10}H_{10}CdCl_2N_2$, 34B, 553
$C_{10}H_{10}CdN_8$, 35B, 691
$C_{10}H_{10}Cl_2$, 32B, 486; 35B; 113
$C_{10}H_{10}ClCrN_2O_2$, 31B, 349
$C_{10}H_{10}Cl_4Cu_2N_2O_2$, 32B, 446; 34B, 575
$C_{10}H_{10}Cl_4Fe_2$, 30B, 424
$C_{10}H_{10}Cl_4N_2OSe$, 35B, 493
$C_{10}H_{10}Cl_2N_2Zn$, 31B, 424
$C_{10}H_{10}Cl_2O_4$, 27, 899; 34B, 339
$C_{10}H_{10}Cl_4OTi_2$, 26, 652; 28, 610
$C_{10}H_{10}Cl_2Zr$, 35B, 901
$C_{10}H_{10}Co_2$, 29, 767
$C_{10}H_{10}CuN_9$, 34B, 554
$C_{10}H_{10}CuN_4O_8$, 34B, 580
$C_{10}H_{10}CuN_2O_4S.2H_2O$, 29, 608
$C_{10}H_{10}F_4N_2Si$, 34B, 371
$C_{10}H_{10}Fe$, 26, 650; 33B, 536
$C_{10}H_{10}FeI_3$, 33B, 345
$C_{10}H_{10}FeO_6S_2$, 28, 681
$C_{10}H_{10}Fe_2O_6S_2$, 28, 654
$C_{10}H_{10}Hg$, 30B, 424
$C_{10}H_{10}INO_3S$, 31B, 154
$C_{10}H_{10}MoN_2O_2S_4$, 34B, 524
$C_{10}H_{10}MoO_3$, 28, 612
$C_{10}H_{10}Mo_2O_2S_2$, 32B, 350
$C_{10}H_{10}N_2OS_2$, 32B, 194
$C_{10}H_{10}N_4O_4S_2$, 35B, 280
$C_{10}H_{10}N_8Zn$, 35B, 693
$C_{10}H_{10}Ni$, 35B, 902
$C_{10}H_{10}O_2$, 27, 888; 29, 667; 30B, 423
$C_{10}H_{10}O_3$, 34B, 113
$C_{10}H_{10}O_4$, 35B, 137
$C_{10}H_{10}Pb$, 31B, 332
$C_{10}H_{10}Ru$, 33B, 536
$C_{10}H_{10}Si_2$, 30B, 424
$C_{10}H_{11}BrO_3$, 33B, 133
$C_{10}H_{11}ClCuNO_2$, 35B, 617
$C_{10}H_{11}ClO_2$, 35B, 211
$C_{10}H_{11}N_3$, 35B, 187
$C_{10}H_{11}NO_4$, 26, 809
$C_{10}H_{11}NO_2S$, 34B, 211
$C_{10}H_{11}N_3S_2$, 35B, 237
$C_{10}H_{12}$, 33B, 536
$C_{10}H_{12}AgN_3O_4$, 35B, 762
$C_{10}H_{12}AlKN_2O_8.2H_2O$, 35B, 602
$C_{10}H_{12}AsNO_5$, 34B, 148
$C_{10}H_{12}BrN_5O_5.2H_2O$, 34B, 259
$C_{10}H_{12}CaN_6O_6$, 32B, 323
$C_{10}H_{12}Cl_2Cu_2N_5$, 32B, 496
$C_{10}H_{12}ClLiN_2O$, 32B, 271
$C_{10}H_{12}ClNO$, 32B, 192; 34B, 40
$C_{10}H_{12}ClN_5O.H_2O$, 30B, 124
$C_{10}H_{12}Cl_2Pt$, 26, 689
$C_{10}H_{12}Cr_2N_4O_3$, 35B, 543
$C_{10}H_{12}CsN_3O_8$, 30B, 424; 33B, 52
$C_{10}H_{12}CuKN_2O_8.3H_2O$, 32B, 495
$C_{10}H_{12}Fe_2I_2O_6P_2$, 33B, 501
$C_{10}H_{12}KN_3O_8$, 30B, 424; 33B, 52
$C_{10}H_{12}Mo$, 27, 849; 30B, 311; 31B, 352
$C_{10}H_{12}N_4NaO_8P.8H_2O$, 34B, 261
$C_{10}H_{12}N_2NaO_8Tb$, 32B, 495

$C_{10}H_{12}N_4O_4$, 34B, 239
$C_{10}H_{12}N_4O_5$, 35B, 288
$C_{10}H_{12}N_4O_4.H_2O$, 34B, 240
$C_{10}H_{12}N_4O_5.2H_2O$, 35B, 283
$C_{10}H_{12}NO_2S_3$, 27, 954
$C_{10}H_{12}N_4O_4S$, 33B, 170
$C_{10}H_{12}N_{10}O_3S.H_2O$, 32B, 489
$C_{10}H_{12}O_2$, 28, 672; 29, 766; 30B, 108; 32B, 111; 35B, 45, 46
$C_{10}H_{12}O_4$, 31B, 16
$C_{10}H_{12}O_6$, 28, 427
$C_{10}H_{12}O_2S_4$, 31B, 148
$C_{10}H_{12}S_2$, 31B, 149
$C_{10}H_{13}Br$, 30B, 79
$C_{10}H_{13}BrO_4$, 28, 560
$C_{10}H_{13}CaN_3O_8P.6H_2O$, 26, 803
$C_{10}H_{13}Cl$, 33B, 69
$C_{10}H_{13}ClFeNO_2$, 35B, 921
$C_{10}H_{13}ClIN_5$, 28, 512
$C_{10}H_{13}CuN_5Na_2O_6.4^1/_2H_2O$, 35B, 653
$C_{10}H_{13}N_5O_4$, 31B, 194
$C_{10}H_{13}NO_2.AlCl_3$, 26, 704
$C_{10}H_{13}N_5O_3.H_2O$, 30B, 197
$C_{10}H_{13}N_5O_5.2H_2O$, 35B, 283
$C_{10}H_{13}N_4O_7P.3H_2O$, 34B, 260
$C_{10}H_{14}Br_2O$, 29, 493
$C_{10}H_{14}Br_2O_2$, 35B, 114
$C_{10}H_{14}BrO_4S.C_{19}H_{26}NO_7.H_2O$, 27, 1002
$C_{10}H_{14}Cl_2CuN_8O_5$, 35B, 694
$C_{10}H_{14}Cl_4FeS_4$, 33B, 551
$C_{10}H_{14}ClKO_4Pt$, 27, 822; 34B, 495
$C_{10}H_{14}Cl_2O_3Pt$, 34B, 468
$C_{10}H_{14}Cl_2O_4Re$, 32B, 495
$C_{10}H_{14}CuO_4$, 34B, 497
$C_{10}H_{14}FeLiN_2O_9.2H_2O$, 29, 544
$C_{10}H_{14}FeN_2O_9Rb.H_2O$, 29, 544
$C_{10}H_{14}K_2N_2O_8.H_2O$, 34B, 20
$C_{10}H_{14}MnO_6$, 29, 543
$C_{10}H_{14}N_2O_5$, 34B, 262
$C_{10}H_{14}N_4O_2$, 33B, 551
$C_{10}H_{14}N_2O_6.^1/_2H_2O$, 34B, 263
$C_{10}H_{14}N_5O_7P.2H_2O$, 31B, 185
$C_{10}H_{14}N_6O_3.H_2O$, 35B, 296
$C_{10}H_{14}N_2O_8Rb_2.2H_2O$, 31B, 28; 35B, 26
$C_{10}H_{14}Ni_2$, 34B, 438
$C_{10}H_{14}NiO_4.2H_2O$, 29, 567
$C_{10}H_{14}O_4$, 31B, 33; 32B, 20; 34B, 13
$C_{10}H_{14}O_8$, 33B, 5
$C_{10}H_{14}O_4Pd$, 35B, 605
$C_{10}H_{14}O_2S$, 32B, 68
$C_{10}H_{14}O_5V$, 26, 686
$C_{10}H_{14}O_4Zn$, 26, 673
$C_{10}H_{15}Br$, 33B, 541
$C_{10}H_{15}BrClNO$, 26, 649
$C_{10}H_{15}Br_2N_3Zn$, 35B, 696
$C_{10}H_{15}BrO$, 29, 493; 34B, 305, 306
$C_{10}H_{15}Cl$, 33B, 541
$C_{10}H_{15}ClO$, 34B, 109
$C_{10}H_{15}ClCoN_5O_4$, 30B, 295
$C_{10}H_{15}CrN_2O_9$, 28, 673
$C_{10}H_{15}F$, 33B, 541
$C_{10}H_{15}FeN_2O_9$, 28, 673
$C_{10}H_{15}GaN_2O_9$, 28, 673
$C_{10}H_{16}$, 30B, 425; 32B, 148
$C_{10}H_{16}BrNO_2$, 32B, 487
$C_{10}H_{16}Cl_2N_2O_8Pd.5H_2O$, 35B, 697
$C_{10}H_{16}Cl_2Pt$, 30B, 327
$C_{10}H_{16}CoO_4$, 30B, 334
$C_{10}H_{16}CuN_2O_9$, 34B, 492
$C_{10}H_{16}CuN_2O_6.4H_2O$, 34B, 529
$C_{10}H_{16}FeN_2O_{10}Rb$, 26, 680
$C_{10}H_{16}IN_2$, 33B, 61
$C_{10}H_{16}KNO_9S_2.H_2O$, 28, 518; 35B, 263
$C_{10}H_{16}N_4NiO_6$, 35B, 762
$C_{10}H_{16}N_2O_5$, 35B, 290
$C_{10}H_{16}N_2O_8$, 27, 775
$C_{10}H_{16}N_2O_2S_4Se$, 35B, 499
$C_{10}H_{16}N_2O_2S_4Te$, 35B, 498
$C_{10}H_{16}N_6S_4Te$, 26, 584
$C_{10}H_{16}NiO_4$, 30B, 338
$C_{10}H_{16}O$, 33B, 537
$C_{10}H_{16}O_2$, 27, 997; 29, 742
$C_{10}H_{16}O_4$, 31B, 92; 33B, 85; 35B, 107
$C_{10}H_{16}O_5Zn$, 28, 618
$C_{10}H_{18}Br_2$, 29, 509
$C_{10}H_{18}CdO_2S_4$, 30B, 371
$C_{10}H_{18}Cl_2Pd_2$, 32B, 493
$C_{10}H_{18}CoN_4O_4S_2$, 35B, 632
$C_{10}H_{18}MgO_6$, 32B, 324
$C_{10}H_{18}MnO_6$, 33B, 388; 35B, 606
$C_{10}H_{18}O_4$, 31B, 17; 35B, 6
$C_{10}H_{18}O_2PbS_4$, 33B, 319
$C_{10}H_{18}O_2Si_2$, 32B, 300
$C_{10}H_{19}KO_2$, 34B, 21
$C_{10}H_{20}BrNO$, 29, 772
$C_{10}H_{20}CdN_2S_4$, 33B, 486
$C_{10}H_{20}ClFeN_2S_4$, 35B, 798
$C_{10}H_{20}Cl_2N_6NiO_9$, 35B, 655
$C_{10}H_{20}CoN_2NaO_{12}$, 26, 675
$C_{10}H_{20}CuN_2S_4$, 26, 668; 30B, 373; 31B, 404
$C_{10}H_{20}FeN_3OS_4$, 32B, 495
$C_{10}H_{20}IN$, 28, 471
$C_{10}H_{20}N_2NiS_4$, 26, 672; 28, 673; 30B, 378
$C_{10}H_{20}N_2NiSe_4$, 32B, 495
$C_{10}H_{20}N_2O_2$, 31B, 18
$C_{10}H_{20}N_2PtS_4$, 32B, 495
$C_{10}H_{20}N_3ReS_4$, 35B, 917
$C_{10}H_{20}N_2S_4$, 32B, 39
$C_{10}H_{20}N_2S_4Se$, 35B, 500
$C_{10}H_{20}N_2S_4Zn$, 30B, 381; 32B, 496
$C_{10}H_{20}P_2S_2$, 35B, 451
$C_{10}H_{21}AgNPS$, 28, 655
$C_{10}H_{21}N.ClH.1.5H_2O$, 29, 511
$C_{10}H_{22}BiBr_5N_2$, 33B, 294
$C_{10}H_{22}CuN_2O_9S_2$, 35B, 763
$C_{10}H_{22}INO_4$, 35B, 264
$C_{10}H_{22}N_2O_4$, 26, 644
$C_{10}H_{22}N_4O_6$, 26, 572
$C_{10}H_{22}N_4O_4S_6Te$, 26, 584
$C_{10}H_{23}ClN_4NiO_2.H_2O$, 33B, 424
$C_{10}H_{24}Br_4N_4S_2Te$, 26, 581; 34B, 416
$C_{10}H_{24}Cl_2N_2$, 26, 645; 31B, 93
$C_{10}H_{24}Cl_2N_4O_3V$, 35B, 737
$C_{10}H_{24}Cl_2N_4S_2Te$, 26, 581
$C_{10}H_{24}Cl_4N_4S_2Te$, 26, 580; 34B, 416
$C_{10}H_{24}CoN_7O_7S_2$, 27, 842
$C_{10}H_{24}CoN_7O_7Se_2$, 27, 842
$C_{10}H_{24}CuN_6O_6.^1/_2H_2O$, 34B, 562
$C_{10}H_{24}CuN_6S_2$, 35B, 678
$C_{10}H_{24}I_2N_4S_2Te$, 26, 581
$C_{10}H_{25}BrMgO_2$, 29, 532; 33B, 300
$C_{10}H_{26}Al_2O_2$, 32B, 328
$C_{10}H_{28}CaO_{17}$, 27, 788
$C_{10}H_{28}Cl_2CuN_4O_6$, 35B, 656
$C_{10}H_{28}Cl_2N_2O_2$, 26, 645
$C_{10}H_{28}O_{17}Sr$, 27, 788
$C_{10}H_{29}Cl_2CoN_6O_9$, 28, 673
$C_{10}H_{29}CoI_2N_6O$, 28, 673
$C_{10}H_{30}Al_4Cl_4N_6$, 35B, 480
$C_{10}H_{30}Al_2O_2Si_2$, 28, 673
$C_{10}H_{30}Cl_4N_4$, 31B, 261
$C_{10}H_{30}N_2Si_4$, 27, 942
$C_{10}H_{30}O_5Si_5$, 26, 616
$C_{10}H_{32}N_4O_8P_2.6H_2O$, 30B, 187
$C_{10}H_{33}Cl_5N_6$, 32B, 37
$C_{10}H_{45}N_5Si_5$, 32B, 301

C_{11}
$C_{11}H_3Co_3O_9$, 32B, 341
$C_{11}H_5F_7MoO_3$, 32B, 351
$C_{11}H_6Fe_3N_2O_9$, 34B, 564
$C_{11}H_7ClN_4O_2S$, 35B, 232
$C_{11}H_8Br_2O_3$, 34B, 13
$C_{11}H_8FeO_3$, 26, 681; 27, 855
$C_{11}H_8MoO_3$, 31B, 368
$C_{11}H_8O_2$, 26, 739; 27, 906
$C_{11}H_9ClN_2O_3$, 28, 495
$C_{11}H_9ClN_2O_4$, 35B, 924
$C_{11}H_9ClO$, 34B, 87
$C_{11}H_9FeNO_5$, 33B, 553; 35B, 569
$C_{11}H_9NO_2$, 31B, 100, 101
$C_{11}H_9O_3Re$, 33B, 347
$C_{11}H_{10}Cl_2CoN_4$, 31B, 425
$C_{11}H_{10}CrO_3$, 27, 853
$C_{11}H_{10}CuN_3S$, 28, 672
$C_{11}H_{10}N_2$, 31B, 104
$C_{11}H_{10}N_2O_2S$, 26, 717
$C_{11}H_{10}O_2$, 34B, 97
$C_{11}H_{10}O_2S$, 26, 775
$C_{11}H_{11}CuNO_2$, 26, 666; 30B, 361
$C_{11}H_{11}FN_6O_3$, 32B, 206
$C_{11}H_{11}N$, 27, 957
$C_{11}H_{11}NO_2$, 26, 776
$C_{11}H_{12}BrN_7O_2$, 30B, 416; 31B, 161

$C_{11}H_{12}BrNS$, 35B, 239
$C_{11}H_{12}ClNO$, 32B, 486
$C_{11}H_{12}Cl_2N_2S$, 28, 673
$C_{11}H_{12}N_2$, 34B, 149
$C_{11}H_{12}N_2O$, 35B, 161
$C_{11}H_{12}O_6$, 35B, 212
$C_{11}H_{13}BrN_2O_2$, 32B, 231
$C_{11}H_{13}BrN_4O_4$, 34B, 153; 35B, 191
$C_{11}H_{14}AgP$, 31B, 344
$C_{11}H_{14}AuN$, 31B, 343
$C_{11}H_{14}BrN_2O_2$, 31B, 191
$C_{11}H_{14}BrN_3O_7.HBr$, 35B, 367
$C_{11}H_{14}ClN_2O_2$, 31B, 191
$C_{11}H_{14}CrNO_5$, 34B, 423
$C_{11}H_{14}O_2$, 35B, 138
$C_{11}H_{15}BrN_2O_4$, 27, 790; 29, 489
$C_{11}H_{15}Br_2N_3O_7$, 28, 673; 35B, 367
$C_{11}H_{15}Cl$, 33B, 70
$C_{11}H_{15}N_3$, 31B, 137
$C_{11}H_{15}NO_3$, 35B, 325
$C_{11}H_{15}N_4O_7P.H_2O$, 28, 521
$C_{11}H_{15}O_2Ti$, 32B, 329
$C_{11}H_{16}Br_2$, 35B, 125
$C_{11}H_{16}BrN_5O_3$, 32B, 490
$C_{11}H_{16}IP$, 32B, 308
$C_{11}H_{16}N_2O_3$, 34B, 218
$C_{11}H_{17}BrN_2$, 35B, 192
$C_{11}H_{17}BrN_2O_7$, 32B, 272
$C_{11}H_{17}Cl_3GeN_2O_2.{}^1/_2H_2O$, 33B, 219
$C_{11}H_{17}Cl_2N_5$, 32B, 59
$C_{11}H_{17}N_2O_2.GeCl_3.{}^1/_2H_2O$, 33B, 219
$C_{11}H_{17}N_3O_8.HBr$, 31B, 239; 35B, 368
$C_{11}H_{17}N_3S_4Zn$, 32B, 412
$C_{11}H_{18}BrN_3O_8$, 31B, 239; 35B, 368
$C_{11}H_{18}N_2O_3$, 34B, 220
$C_{11}H_{20}CuP$, 35B, 541
$C_{11}H_{20}O_4$, 31B, 19
$C_{11}H_{21}BrO_2$, 27, 740; 28, 367, 369, 673
$C_{11}H_{21}N_8O_2S_4Tl$, 34B, 356
$C_{11}H_{22}O_2S$, 28, 371
$C_{11}H_{24}ClN$, 29, 513
$C_{11}H_{24}CoCrN_{12}O.2H_2O$, 35B, 602
$C_{11}H_{24}CrN_{11}Ni.1.5H_2O$, 33B, 386
$C_{11}H_{26}Cl_4CoN_5O_2Zn$, 35B, 660
$C_{11}H_{28}I_2N_2$, 28, 398
$C_{11}H_{30}CoI_2N_5O_4$, 35B, 658

C_{12}

$C_{12}Al_2O_{12}.6H_2O$, 30B, 425
$C_{12}Cl_{12}$, 32B, 148
$C_{12}Cl_{14}$, 33B, 86, 555
$C_{12}Co_2F_6O_6$, 33B, 325
$C_{12}F_{10}$, 33B, 537
$C_{12}F_{12}FeO_4$, 31B, 352
$C_{12}Ni_4O_{18}P_4$, 31B, 467
$C_{12}HF_{10}OTl$, 35B, 510
$C_{12}H_2Cl_6O_2$, 34B, 343
$C_{12}H_4Cl_2CuN_2$, 34B, 566
$C_{12}H_4Cl_4N_2$, 30B, 139, 141
$C_{12}H_4ClO_5Re$, 34B, 420
$C_{12}H_4KN_7O_{12}$, 27, 866
$C_{12}H_4N_4$, 30B, 40
$C_{12}H_5Cl_7O_4$, 34B, 114
$C_{12}H_6$, 30B, 425
$C_{12}H_6Br_2$, 27, 933
$C_{12}H_6CoHg_2N_6S_6$, 29, 629
$C_{12}H_6Fe_2O_6$, 29, 767; 30B, 296, 303
$C_{12}H_6N_6O_6$, 30B, 417
$C_{12}H_6O_2$, 28, 434, 674
$C_{12}H_6O_{12}$, 26, 713
$C_{12}H_8$, 27, 935; 31B, 109
$C_{12}H_8AsS$, 35B, 467
$C_{12}H_8Br_2$, 26, 747
$C_{12}H_8BrCl$, 26, 747
$C_{12}H_8Br_2N_2$, 31B, 46
$C_{12}H_8Br_2O_2S_2$, 32B, 68
$C_{12}H_8Cl_2$, 26, 746
$C_{12}H_8Cl_6$, 30B, 425
$C_{12}H_8Cl_3CoN_2.3H_2O$, 31B, 427
$C_{12}H_8Cl_2I_2Te$, 27, 864
$C_{12}H_8Cl_2N_2$, 34B, 38
$C_{12}H_8Cl_2N_2Zn$, 31B, 427
$C_{12}H_8Cl_2O_2S$, 26, 728
$C_{12}H_8Cl_4S$, 34B, 42
$C_{12}H_8CrN_2O_5$, 30B, 350
$C_{12}H_8Fe_2O_8$, 26, 682
$C_{12}H_8Hg_2N_4O_6$, 32B, 430
$C_{12}H_8I_2$, 28, 438
$C_{12}H_8I_2N_2$, 32B, 273
$C_{12}H_8N_2$, 35B, 121
$C_{12}H_8N_4$, 35B, 108
$C_{12}H_8N_2.H_2O$, 30B, 425
$C_{12}H_8N_2O$, 35B, 121
$C_{12}H_8N_2O_2$, 26, 781, 788; 27, 889; 28, 477; 35B, 169
$C_{12}H_8N_2O_4$, 27, 908; 28, 419; 34B, 155
$C_{12}H_8N_2O_6$, 28, 674
$C_{12}H_8N_4O_4$, 29, 58
$C_{12}H_8N_2O_4S_2$, 34B, 44; 35B, 58
$C_{12}H_8N_2S$, 26, 783
$C_{12}H_8O_6Os_2$, 28, 674
$C_{12}H_8OS$, 31B, 159
$C_{12}H_8O_2S$, 33B, 144
$C_{12}H_8O_4Te$, 32B, 339
$C_{12}H_8S$, 35B, 225
$C_{12}H_8S_4$, 30B, 425
$C_{12}H_8Se$, 35B, 226
$C_{12}H_9AsBrN$, 31B, 312
$C_{12}H_9AsClN$, 30B, 283; 31B, 312
$C_{12}H_9BrClNO_2S$, 29, 770; 34B, 620
$C_{12}H_9Br_2N_3$, 26, 731; 30B, 49
$C_{12}H_9BrN_2S$, 28, 675
$C_{12}H_9BrO_3$, 32B, 275
$C_{12}H_9ClO_3$, 32B, 275
$C_{12}H_9Cl_2O_4P$, 29, 663
$C_{12}H_9I$, 28, 438
$C_{12}H_9I_2N$, 26, 728
$C_{12}H_9N$, 34B, 157
$C_{12}H_9N_3O_4$, 33B, 119
$C_{12}H_9NS$, 30B, 426
$C_{12}H_{10}$, 26, 726; 27, 907; 33B, 91
$C_{12}H_{10}AsBr$, 27, 867
$C_{12}H_{10}AsCl$, 27, 867
$C_{12}H_{10}AsI$, 28, 588
$C_{12}H_{10}BBr_2P$, 26, 729
$C_{12}H_{10}BI_2P$, 26, 729
$C_{12}H_{10}Br_2$, 28, 674
$C_{12}H_{10}BrNO_2S$, 32B, 69; 33B, 553; 34B, 622
$C_{12}H_{10}ClN_3$, 33B, 63
$C_{12}H_{10}Cl_2N_2$, 33B, 64
$C_{12}H_{10}Cl_4N_3P_3$, 33B, 280
$C_{12}H_{10}Cl_2O_2Ti$, 31B, 445
$C_{12}H_{10}Cl_3Sb$, 26, 729
$C_{12}H_{10}CuN_3$, 26, 665
$C_{12}H_{10}CuN_4O_4$, 28, 648
$C_{12}H_{10}F_4N_3P_3$, 32B, 491; 34B, 383
$C_{12}H_{10}FeO_2$, 31B, 353
$C_{12}H_{10}Hg$, 27, 863; 29, 628; 33B, 537
$C_{12}H_{10}K_2O_{12}$, 26, 563; 35B, 16
$C_{12}H_{10}Mg_3O_{14}.10H_2O$, 30B, 30
$C_{12}H_{10}N_2$, 31B, 47
$C_{12}H_{10}NNaO_4S_2$, 35B, 84
$C_{12}H_{10}N_4NiO_2.2H_2O$, 32B, 496
$C_{12}H_{10}N_2O_2$, 30B, 132
$C_{12}H_{10}N_2O_5$, 32B, 160
$C_{12}H_{10}O_2$, 28, 435; 30B, 117; 31B, 102; 32B, 123
$C_{12}H_{10}O_4$, 30B, 75; 33B, 65; 35B, 409
$C_{12}H_{10}P_2S_6$, 29, 762
$C_{12}H_{10}S_2$, 34B, 48
$C_{12}H_{11}Br$, 30B, 97
$C_{12}H_{11}NO_4S_2$, 35B, 84
$C_{12}H_{11}O_2P$, 30B, 195
$C_{12}H_{12}$, 28, 674; 30B, 96
$C_{12}H_{12}Al_2Cl_8Pd_2$, 35B, 567
$C_{12}H_{12}Al_4Cl_{14}Pd_2$, 35B, 567
$C_{12}H_{12}Br_6$, 30B, 80
$C_{12}H_{12}Br_2N_2.H_2O$, 34B, 158
$C_{12}H_{12}BrNO$, 33B, 128
$C_{12}H_{12}BrN_5O_8$, 30B, 184
$C_{12}H_{12}Br_2O_5$, 35B, 213
$C_{12}H_{12}Cl_4$, 32B, 134
$C_{12}H_{12}Cl_2Na_5NdO_{23}.6H_2O$, 35B, 642
$C_{12}H_{12}Cl_2Na_5O_{23}Yb.6H_2O$, 35B, 642
$C_{12}H_{12}Cl_3OSb$, 32B, 321
$C_{12}H_{12}Co_2O_4$, 31B, 346
$C_{12}H_{12}Cr$, 28, 613; 33B, 556
$C_{12}H_{12}CuN_2O_6$, 35B, 765
$C_{12}H_{12}Fe_2O_8P_2$, 33B, 503
$C_{12}H_{12}Mo$, 26, 650
$C_{12}H_{12}MoN_2O_5$, 34B, 426

$C_{12}H_{12}N_4Ni$, 33B, 425
$C_{12}H_{12}N_2NiO_6.2H_2O$, 32B, 496
$C_{12}H_{12}N_6O_6$, 35B, 410
$C_{12}H_{12}N_2S_2$, 32B, 70; 35B, 60
$C_{12}H_{12}Na_3NdO_{15}$, 30B, 399
$C_{12}H_{12}Si_2$, 30B, 427
$C_{12}H_{12}V$, 28, 614; 674
$C_{12}H_{12}W$, 26, 650
$C_{n+11}H_{2n+13}IN_2O_2$, 26, 717
$C_{12}H_{13}Br_2N_7O_2$, 34B, 227
$C_{12}H_{13}ClCr_3O_{13}.5H_2O$, 30B, 294
$C_{12}H_{13}Cl_2N_2NbO_2$, 35B, 766
$C_{12}H_{13}NO_4$, 32B, 162
$C_{12}H_{14}BrN_7O_2$, 31B, 163; 34B, 228
$C_{12}H_{14}Cl_4CoN_2$, 34B, 131
$C_{12}H_{14}Cl_2CuN_2$, 32B, 496
$C_{12}H_{14}Cl_4CuN_2$, 34B, 138
$C_{12}H_{14}Cl_4Cu_2N_2$, 34B, 131
$C_{12}H_{14}ClNO_2$, 34B, 923
$C_{12}H_{14}Cl_4N_2Pd$, 34B, 131
$C_{12}H_{14}FN_7O_2$, 31B, 164; 32B, 207; 34B, 230
$C_{12}H_{14}IN$, 35B, 188
$C_{12}H_{14}INO$, 29, 744
$C_{12}H_{14}INO_2S$, 30B, 144
$C_{12}H_{14}K_2N_2O_{13}Zr$, 33B, 440
$C_{12}H_{14}N_4NiO_4$, 33B, 427
$C_{12}H_{15}FN_8O_2$, 32B, 210
$C_{12}H_{15}NO_3$, 34B, 165
$C_{12}H_{15}N_5O_5$, 35B, 297
$C_{12}H_{15}N_7O_2$, 28, 507
$C_{12}H_{15}NOPt$, 30B, 364
$C_{12}H_{16}$, 34B, 95
$C_{12}H_{16}BrFO_7$, 34B, 252
$C_{12}H_{16}Br_2N_8Ni$, 34B, 568
$C_{12}H_{16}BrNO_2$, 35B, 86
$C_{12}H_{16}Br_9N_2Sb$, 33B, 289
$C_{12}H_{16}CdN_6O_4.2H_2O$, 32B, 422
$C_{12}H_{16}Cl_2N_8Ni$, 32B, 432
$C_{12}H_{16}Cl_2N_4NiO_4$, 32B, 431
$C_{12}H_{16}ClNO_2$, 35B, 86
$C_{12}H_{16}Cl_2N_2O_2Sn$, 34B, 410
$C_{12}H_{16}CoN_6O_4.2H_2O$, 35B, 664
$C_{12}H_{16}Cu_2N_6Na_2O_8.2H_2O$, 27, 1049
$C_{12}H_{16}IMoNS$, 34B, 458
$C_{12}H_{16}N_2NiO_8$, 35B, 768
$C_{12}H_{16}N_{10}NiO_8$, 34B, 569
$C_{12}H_{16}N_6NiO_4.H_2O$, 32B, 424
$C_{12}H_{16}N_4O_4$, 32B, 489
$C_{12}H_{16}N_8O_2$, 32B, 209
$C_{12}H_{16}N_2O_5Si$, 34B, 367
$C_{12}H_{16}N_2O_8Zn$, 35B, 768
$C_{12}H_{16}O$, 27, 988
$C_{12}H_{16}O_8S_2Se_2$, 26, 774
$C_{12}H_{17}BrN_2O_5$, 30B, 181
$C_{12}H_{17}ClN_4OS.ClH.H_2O$, 27, 994
$C_{12}H_{17}Cl_2NPt$, 34B, 439
$C_{12}H_{17}NO_3Si$, 33B, 270
$C_{12}H_{18}$, 27, 787; 32B, 119; 35B, 100, 902
$C_{12}H_{18}.AgNO_3.H_2O$, 35B, 570
$C_{12}H_{18}Ag_2N_2O_6$, 35B, 520
$C_{12}H_{18}Br_2Ni_2O_4$, 32B, 345
$C_{12}H_{18}CaO_{14}.2H_2O$, 30B, 182
$C_{12}H_{18}CaO_{14}.3H_2O$, 30B, 427
$C_{12}H_{18}Cl_6Cu_3NO_3$, 33B, 458
$C_{12}H_{18}Cl_2O_6Pd_2$, 35B, 521
$C_{12}H_{18}Cl_2Ru$, 33B, 335
$C_{12}H_{18}CoN_6O_5$, 33B, 421
$C_{12}H_{18}CuN_2O_2$, 28, 640
$C_{12}H_{18}CuN_2O_2.CH_6NO_4Cl$, 35B, 615
$C_{12}H_{18}CuN_2O_2.{}^1/_2H_2O$, 33B, 389
$C_{12}H_{18}CuO_4$, 32B, 384
$C_{12}H_{18}CuO_6$, 30B, 336
$C_{12}H_{18}Fe_2Ge_3O_6$, 34B, 408
$C_{12}H_{18}N_6NiO_5$, 32B, 424
$C_{12}H_{18}N_6O_6.{}^1/_2H_2O$, 28, 525
$C_{12}H_{18}N_4O_7P_2S.4H_2O$, 34B, 255
$C_{12}H_{18}N_2O_3V$, 35B, 610
$C_{12}H_{18}Ni$, 30B, 394
$C_{12}H_{19}ClN_4O_7P_2S.{}^1/_2H_2O$, 31B, 181
$C_{12}H_{19}Co$, 32B, 493
$C_{12}H_{19}NO_3$, 35B, 111
$C_{12}H_{20}$, 29, 766
$C_{12}H_{20}BrNO_6$, 32B, 232
$C_{12}H_{20}Br_2O$, 31B, 84; 34B, 96
$C_{12}H_{20}CdN_6O_6$, 32B, 422
$C_{12}H_{20}Cl_2Rh_2$, 32B, 493
$C_{12}H_{20}Cr_2$, 34B, 441
$C_{12}H_{20}CuN_2O_3$, 31B, 428; 34B, 500
$C_{12}H_{20}CuN_2O_4$, 28, 642
$C_{12}H_{20}Mo_2O_7S_8$, 29, 622
$C_{12}H_{20}N_4O_4$, 35B, 181
$C_{12}H_{20}NO_4P$, 32B, 486
$C_{12}H_{20}N_6O_6Zn$, 28, 636
$C_{12}H_{20}N_2S_3$, 35B, 170
$C_{12}H_{20}O_4$, 32B, 177
$C_{12}H_{20}O_6$, 33B, 135
$C_{12}H_{20}O_2S$, 31B, 150
$C_{12}H_{20}Pt_2$, 26, 690
$C_{12}H_{20}S_6$, 35B, 911
$C_{12}H_{21}N_3O_5.Cl_2H_2$, 27, 993
$C_{12}H_{22}CaCl_2O_{11}.4H_2O$, 28, 674
$C_{12}H_{22}CaO_{12}$, 33B, 168
$C_{12}H_{22}Cl_2N_2O_2Pd$, 32B, 432
$C_{12}H_{22}CuN_8O_{12}$, 34B, 544
$C_{12}H_{22}N_2O_2$, 33B, 120; 35B, 111
$C_{12}H_{22}N_2O_4$, 26, 637; 27, 960; 31B, 127
$C_{12}H_{22}O_4$, 31B, 20
$C_{12}H_{22}O_{11}$, 26, 625; 28, 519; 31B, 174; 33B, 161
$C_{12}H_{22}O_{11}.H_2O$, 35B, 266
$C_{12}H_{22}O_{12}Sr$, 34B, 250
$C_{12}H_{23}Cl_2N_3O_5$, 28, 545
$C_{12}H_{23}OPS$, 29, 762
$C_{12}H_{24}$, 26, 752
$C_{12}H_{24}BrN$, 35B, 139
$C_{12}H_{24}Br_2N_8S_4Te$, 26, 578
$C_{12}H_{24}ClCr_3O_{16}.6H_2O$, 32B, 496; 35B, 738
$C_{12}H_{24}Cl_2GdNa_5O_{29}$, 33B, 412
$C_{12}H_{24}Cl_2N_8NiS_4$, 32B, 410
$C_{12}H_{24}Cl_2N_6O_8S_3Te$, 26, 585
$C_{12}H_{24}Cl_2N_8O_8S_4Te$, 26, 579
$C_{12}H_{24}Cl_2N_8S_4Te$, 26, 579
$C_{12}H_{24}Cl_6N_8S_4Te_2$, 26, 579
$C_{12}H_{24}Cl_2Na_5NdO_{29}$, 33B, 412
$C_{12}H_{24}Cl_2Na_5O_{29}Yb$, 33B, 412
$C_{12}H_{24}CuN_2O_8$, 26, 660
$C_{12}H_{24}N_2O_2$, 27, 776
$C_{12}H_{24}O_2$, 26, 562; 28, 373
$C_{12}H_{24}O_{12}$, 26, 625
$C_{12}H_{24}Pd_3S_9$, 34B, 594
$C_{12}H_{26}CuN_2O_5$, 34B, 545
$C_{12}H_{26}N_2.2HBr$, 30B, 136
$C_{12}H_{26}N_2O_2$, 31B, 129
$C_{12}H_{26}N_2O_4$, 31B, 27
$C_{12}H_{26}N_6O_9Zn$, 28, 636
$C_{12}H_{28}AgNO_3P_2PtS_2$, 34B, 596
$C_{12}H_{28}Br_2N_2$, 30B, 136
$C_{12}H_{28}Br_2N_8O_2S_4Te$, 26, 578
$C_{12}H_{28}Br_2O_3Ti$, 28, 674
$C_{12}H_{28}Cl_2N_8O_2S_4Te$, 26, 578
$C_{12}H_{28}Cl_2O_3Ti$, 28, 674
$C_{12}H_{28}Cr_2N_6O_{14}$, 35B, 604
$C_{12}H_{28}CuN_6O_6$, 29, 600
$C_{12}H_{28}N_6NiO_6$, 34B, 571
$C_{12}H_{28}N_4O_4.BrLi$, 27, 762
$C_{12}H_{28}N_4O_2P_2$, 34B, 386
$C_{12}H_{28}O_4P_2PbS_4$, 34B, 412
$C_{12}H_{28}O_4P_2S_4$, 35B, 452
$C_{12}H_{28}Si_4$, 34B, 373
$C_{12}H_{29}Cl_2CoN_3$, 32B, 381
$C_{12}H_{30}Al_2FK$, 26, 611; 28, 600
$C_{12}H_{30}Al_3N_3$, 35B, 481
$C_{12}H_{30}B_3N_3$, 30B, 270
$C_{12}H_{30}Br_2CoN_4$, 32B, 383
$C_{12}H_{30}Br_2CuN_4$, 33B, 427
$C_{12}H_{30}Br_2FeN_4$, 33B, 428
$C_{12}H_{30}Br_2MnN_4$, 33B, 427
$C_{12}H_{30}Br_2N_2.2H_2O$, 30B, 34
$C_{12}H_{30}Br_2N_4Ni$, 33B, 427
$C_{12}H_{30}Br_2N_4Zn$, 33B, 428
$C_{12}H_{30}Br_2P_2Pt$, 31B, 457
$C_{12}H_{30}Cl_2N_2.2H_2O$, 32B, 485
$C_{12}H_{30}Cl_2P_2Pt$, 31B, 457
$C_{12}H_{30}Cl_6P_2Re_2$, 33B, 504
$C_{12}H_{30}N_4O_8P_2S.3H_2O$, 31B, 182
$C_{12}H_{30}N_4O_4S_6Te$, 26, 583
$C_{12}H_{30}O_6P_3S_6V$, 35B, 799
$C_{12}H_{30}OSi_2$, 26, 615
$C_{12}H_{30}O_3Si_3$, 26, 616
$C_{12}H_{32}Al_2MgO_6$, 33B, 301
$C_{12}H_{32}Br_2N_8S_4Te$, 26, 582
$C_{12}H_{32}Cl_4N_2O_2U$, 35B, 911
$C_{12}H_{32}Cl_2N_7P_3.HCl$, 33B, 547
$C_{12}H_{32}Cl_2N_8S_4Te$, 26, 581

$C_{12}H_{32}Mg_2N_4$, 34B, 399
$C_{12}H_{34}Br_2Cu_2N_4O_2$, 35B, 700
$C_{12}H_{34}Cl_2N_6NiO_8$, 35B, 702
$C_{12}H_{36}Be_3N_6$, 32B, 492; 34B, 401
$C_{12}H_{36}CoCrN_{12}O_6$, 33B, 387
$C_{12}H_{36}N_4.41H_2O$, 33B, 260
$C_{12}H_{36}O_4Si_5$, 26, 616
$C_{12}H_{36}O_6Si_8$, 32B, 302
$C_{12}H_{36}Si_{13}$, 35B, 902
$C_{12}H_{40}B_{18}N_2Ni$, 35B, 431
$C_{12}H_{41}B_9P_2PtS$, 35B, 920
$C_{12}H_{42}N_{30}O_{12}Ru_2S_3.7H_2O$, 674
$C_{12}H_{44}Br_2CaN_8O_{10}$, 32B, 276
$C_{12}H_{48}Cl_{10}Co_2Cu_2N_{12}.2H_2O$, 35B, 914
$C_{12}H_{67}FO_{20}S$, 27, 781
$C_{12}H_{118}N_4O_{41}$, 33B, 260

C_{13}
$C_{13}H_5MnMoO_8$, 33B, 347
$C_{13}H_7BrN_2$, 35B, 126
$C_{13}H_7BrO$, 35B, 126
$C_{13}H_7N_7$, 35B, 31
$C_{13}H_8Br_4GeN_2O_3W$, 35B, 870
$C_{13}H_8Br_2O_4S_2$, 34B, 45
$C_{13}H_8CrO_3$, 32B, 370
$C_{13}H_8CrO_6$, 30B, 294; 33B, 336
$C_{13}H_8CuKN_3O_4S$, 35B, 412
$C_{13}H_8Fe_2O_5$, 31B, 369
$C_{13}H_8N_2$, 32B, 486
$C_{13}H_8O_6W$, 30B, 294
$C_{13}H_9BrO_2$, 30B, 427
$C_{13}H_9Br_3O$, 30B, 88
$C_{13}H_9ClO_2$, 30B, 427; 34B, 63, 64
$C_{13}H_9CrNO_3$, 33B, 361
$C_{13}H_9FO_2$, 30B, 427
$C_{13}H_9IO_2$, 30B, 427; 35B, 90
$C_{13}H_9N$, 26, 780
$C_{13}H_9NO$, 35B, 193
$C_{13}H_9NO_2$, 31B, 128
$C_{13}H_9NOS$, 35B, 241
$C_{13}H_9N_3S$, 28, 675; 33B, 63
$C_{13}H_9N_3S_2$, 27, 957
$C_{13}H_{10}BrN$, 35B, 171
$C_{13}H_{10}BrNO$, 32B, 99; 34B, 73
$C_{13}H_{10}Cl_2FeO_2Sn$, 35B, 545
$C_{13}H_{10}ClNO$, 29, 660; 32B, 99
$C_{13}H_{10}FeO_3$, 34B, 445
$C_{13}H_{10}N_2O_4$, 33B, 187
$C_{13}H_{10}N_4S$, 35B, 162
$C_{13}H_{10}O$, 32B, 113; 33B, 71
$C_{13}H_{10}O_3Rh_2$, 32B, 352
$C_{13}H_{10}O_2S$, 32B, 187
$C_{13}H_{10}O_2S.H_2O$, 27, 915; 30B, 82
$C_{13}H_{11}Cl_2IN_2O.H_2O$, 34B, 134
$C_{13}H_{11}FeN$, 35B, 544
$C_{13}H_{11}I_2IrN_2O_3$, 34B, 430
$C_{13}H_{11}N$, 35B, 61
$C_{13}H_{11}NO_3$, 31B, 54
$C_{13}H_{11}NO_2S_2$, 34B, 64
$C_{13}H_{11}NS$, 34B, 186
$C_{13}H_{12}$, 26, 727
$C_{13}H_{12}BrNO_2$, 27, 967
$C_{13}H_{12}BrNO_2S$, 34B, 623
$C_{13}H_{12}FeO$, 30B, 307
$C_{13}H_{12}N_2$, 30B, 427
$C_{13}H_{12}O_2$, 26, 775
$C_{13}H_{13}FNP$, 35B, 454
$C_{13}H_{13}IN_4O_2$, 32B, 488; 35B, 195
$C_{13}H_{13}N_3$, 33B, 63
$C_{13}H_{13}OPS$, 34B, 388
$C_{13}H_{13}OPSe$, 34B, 388
$C_{13}H_{13}P$, 35B, 454
$C_{13}H_{13}Re$, 32B, 354
$C_{13}H_{13}Ti$, 35B, 575
$C_{13}H_{14}BrNO_2$, 27, 969
$C_{13}H_{14}Cl_2N_2O_2.2H_2O$, 34B, 140
$C_{13}H_{14}D_2O$, 32B, 497
$C_{13}H_{14}INO_2$, 26, 768
$C_{13}H_{14}N_2O$, 26, 787
$C_{13}H_{14}N_2O_3$, 34B, 214
$C_{13}H_{14}N_3O_8P$, 32B, 487
$C_{13}H_{14}N_2O_4S_2$, 32B, 239
$C_{13}H_{15}NO_2.HBr$, 30B, 427
$C_{13}H_{15}NO_2.HBr.H_2O$, 30B, 427
$C_{13}H_{15}NO_2.HBr.2H_2O$, 30B, 239
$C_{13}H_{16}As_2FeO_3$, 32B, 464
$C_{13}H_{16}BrNO_2$, 30B, 427
$C_{13}H_{16}BrNO_2.H_2O$, 30B, 427
$C_{13}H_{16}BrNO_2.2H_2O$, 29, 772; 30B, 239
$C_{13}H_{16}CuN_2O_4$, 32B, 496
$C_{13}H_{16}INO_2S$, 35B, 203
$C_{13}H_{16}N_2O_2$, 35B, 326
$C_{13}H_{16}O$, 32B, 490
$C_{13}H_{16}O_3$, 27, 988
$C_{13}H_{17}BrO_7$, 34B, 171
$C_{13}H_{17}N_3O_4.\sim 1/2H_2O$, 26, 724
$C_{13}H_{18}ClHgO_8$, 27, 796
$C_{13}H_{18}Cl_2O_5$, 33B, 170
$C_{13}H_{18}N_2O_3$, 34B, 222
$C_{13}H_{18}N_2O_2S$, 34B, 204
$C_{13}H_{18}O_2Pd$, 31B, 370
$C_{13}H_{19}LiN_2$, 35B, 474
$C_{13}H_{19}NO_5.HClO_4$, 35B, 206
$C_{13}H_{19}NO_5.1/2H_2O$, 34B, 161; 35B, 204
$C_{13}H_{19}NO_2PbS_4$, 34B, 413
$C_{13}H_{20}BrN$, 33B, 109
$C_{13}H_{20}ClNO_9$, 35B, 206
$C_{13}H_{21}AlCl_4$, 33B, 41
$C_{13}H_{21}N_6OP$, 32B, 487
$C_{13}H_{22}Br_2N_2NiO$, 33B, 441
$C_{13}H_{22}BrNO_4$, 32B, 490; 34B, 283
$C_{13}H_{24}ClCuN_3O_6$, 35B, 615
$C_{13}H_{24}O_3$, 33B, 6
$C_{13}H_{24}O_4$, 33B, 7
$C_{13}H_{24}O_{11}.CH_3OH$, 35B, 268
$C_{13}H_{24}O_2Pt$, 34B, 432
$C_{13}H_{26}I_2N_2$, 33B, 125
$C_{13}H_{26}O_{12}$, 32B, 218
$C_{13}H_{28}IN$, 27, 999
$C_{13}H_{36}Si_4$, 29, 762

C_{14}
$C_{14}H_4Cl_{12}O_2S$, 27, 785
$C_{14}H_5CoF_{12}O$, 29, 556
$C_{14}H_6Br_2O_2$, 26, 751; 28, 431; 32B, 125, 486
$C_{14}H_6Cl_2O_2$, 26, 750, 751
$C_{14}H_6F_2O_2$, 26, 751; 32B, 126
$C_{14}H_6Fe_4O_{12}S_3$, 32B, 455
$C_{14}H_6I_2O_2$, 26, 751; 28, 432; 32B, 125
$C_{14}H_6I_2O_8$, 26, 751; 28, 432; 32B, 125
$C_{14}H_6N_2O_8$, 32B, 127
$C_{14}H_6O_4$, 33B, 232
$C_{14}H_7BrO_2$, 28, 675; 34, 103
$C_{14}H_7ClO_2$, 26, 750; 28, 675; 34B, 105
$C_{14}H_7FO_2$, 28, 675
$C_{14}H_7IO_2$, 28, 675
$C_{14}H_8AgN_2O_8.H_2O$, 35B, 770
$C_{14}H_8BrCl$, 26, 748
$C_{14}H_8Br_2F_2$, 32B, 41
$C_{14}H_8BrIO_4$, 35B, 91
$C_{14}H_8BrN_5O_2.1/2C_7H_8$, 31B, 138; 32B, 167
$C_{14}H_8Br_2O$, 28, 433
$C_{14}H_8Br_2O_3$, 29, 655
$C_{14}H_8Br_2O_4$, 33B, 43; 35B, 91
$C_{14}H_8ClIO_4$, 35B, 91
$C_{14}H_8Cl_2O_3$, 31B, 65
$C_{14}H_8Cl_2O_4$, 33B, 43; 35B, 91
$C_{14}H_8Fe_2O_6$, 27, 854
$C_{14}H_8Fe_2O_8$, 26, 681
$C_{14}H_8I_2O_4$, 35B, 91
$C_{14}H_8O_2$, 26, 751; 32B, 128
$C_{14}H_8O_4$, 27, 931; 28, 675; 30B, 104; 31B, 103; 32B, 129, 486
$C_{14}H_8O_6Ru_2$, 33B, 369
$C_{14}H_9BrO_2$, 34B, 173
$C_{14}H_9Cl_2KO_4$, 28, 416
$C_{14}H_9Cl_2O_4Rb$, 28, 416
$C_{14}H_9KN_2O_8$, 26, 707
$C_{14}H_9NO_2$, 27, 975
$C_{14}H_9N_2O_8Rb$, 26, 705
$C_{14}H_{10}$, 26, 739; 27, 925; 28, 439; 29, 682; 31B, 94; 33B, 72
$C_{14}H_{10}Br_2N_2NiO_2$, 26, 672
$C_{14}H_{10}Br_4N_2O_2$, 26, 733
$C_{14}H_{10}Cl_2CuN_2O_4$, 29, 598
$C_{14}H_{10}Cl_2Fe_2GeO_4$, 32B, 355
$C_{14}H_{10}Cl_2Fe_2O_4Sn$, 32B, 356
$C_{14}H_{10}Cl_2N_2NiO_4$, 27, 871
$C_{14}H_{10}ClNO_3$, 35B, 48
$C_{14}H_{10}Cl_2N_2O_4Pd$, 27, 871
$C_{14}H_{10}CrO_3$, 34B, 472
$C_{14}H_{10}CuO_2$, 30B, 369
$C_{14}H_{10}CuO_4$, 29, 599; 30B, 370
$C_{14}H_{10}Fe_2N_2O_8Sn$, 33B, 310
$C_{14}H_{10}Fe_2O_4$, 35B, 546
$C_{14}H_{10}Fe_2O_6$, 34B, 475
$C_{14}H_{10}I_4O_4Re_2$, 34B, 530

$C_{14}H_{10}N_2S$, 30B, 156
$C_{14}H_{10}O$, 27, 932; 29, 683; 35B, 121
$C_{14}H_{10}O_2$, 30B, 81; 32B, 178
$C_{14}H_{10}O_3$, 34B, 58
$C_{14}H_{10}O_4$, 32B, 85
$C_{14}H_{10}O_4Ru$, 32, 357
$C_{14}H_{10}O_3W$, 33B, 349
$C_{14}H_{11}BrO$, 30B, 428; 32B, 81
$C_{14}H_{11}Cl_3MoN_2O_3Sn$, 34B, 409
$C_{14}H_{11}ClO$, 30B, 428; 33B, 72
$C_{14}H_{11}FO$, 30B, 427; 33B, 73
$C_{14}H_{11}IO$, 30B, 428
$C_{14}H_{11}NO_2$, 35B, 61
$C_{14}H_{12}$, 26, 753; 32B, 486
$C_{14}H_{12}As_2Co_2F_4O_6$, 35B, 921
$C_{14}H_{12}As_2F_4Fe_2O_6$, 31B, 466; 32B, 374
$C_{14}H_{12}Br_2N_2O_2$, 26, 733
$C_{14}H_{12}Cl_2N_2Zn$, 32B, 496; 34B, 577
$C_{14}H_{12}Cl_2O_2$, 32B, 104
$C_{14}H_{12}Cl_6O_2S$, 27, 786
$C_{14}H_{12}Cr_2N_4O_2$, 35B, 548
$C_{14}H_{12}CuN_2O_2$, 31B, 389
$C_{14}H_{12}CuO_4$, 29, 596
$C_{14}H_{12}N_2$, 26, 731; 35B, 63
$C_{14}H_{12}N_2NiO_4$, 32B, 392
$C_{14}H_{12}N_2O_2$, 32B, 166; 35B, 61
$C_{14}H_{12}N_2O_2$, 34B, 35
$C_{14}H_{12}N_2O_4Pd$, 35B, 619
$C_{14}H_{12}N_2OSe$, 30B, 43
$C_{14}H_{12}O$, 27, 933
$C_{14}H_{12}O_3$, 35B, 369
$C_{14}H_{12}O_4$, 32B, 179
$C_{14}H_{12}O_5$, 27, 1008
$C_{14}H_{12}O_7$, 30B, 104
$C_{14}H_{12}OS$, 34B, 191
$C_{14}H_{13}BrN_4O_3.H_2O$, 30B, 161
$C_{14}H_{13}BrO_2$, 30B, 224
$C_{14}H_{13}Cl_2NO_3$, 35B, 163
$C_{14}H_{13}Cl_2NO_4$, 28, 416
$C_{14}H_{13}ClO$, 34B, 115
$C_{14}H_{13}HgNO_3$, 35B, 771
$C_{14}H_{13}IN_4O_2.H_2O$, 34B, 156; 35B, 196
$C_{14}H_{13}KO_7$, 33B, 44
$C_{14}H_{13}MnO_2$, 33B, 350
$C_{14}H_{13}NO$, 33B, 9
$C_{14}H_{14}$, 27, 787; 34B, 82
$C_{14}H_{14}Br_2N_9O$, 34B, 235
$C_{14}H_{14}BrNO_2S$, 34B, 624
$C_{14}H_{14}Br_2O_4.{}^1/_2C_4H_8O_2$, 35B, 141
$C_{14}H_{14}Cl_4$, 34B, 88
$C_{14}H_{14}Cl_2CuN_2O$, 32B, 497
$C_{14}H_{14}ClN_2O_4Tl$, 32B, 492
$C_{14}H_{14}FeO$, 32B, 358
$C_{14}H_{14}FeO_2$, 32B, 493; 35B, 550
$C_{14}H_{14}Hg$, 30B, 428; 35B, 511
$C_{14}H_{14}N_2$, 31B, 48
$C_{14}H_{14}N_6Ni$, 34B, 558
$C_{14}H_{14}N_2O_2$, 26, 732
$C_{14}H_{14}N_2O_3$, 35B, 69
$C_{14}H_{14}N_2O_2.HI_3$, 29, 656
$C_{14}H_{14}N_2S$, 35B, 913
$C_{14}H_{14}N_4S_2Zn$, 30B, 428
$C_{14}H_{14}NiO_6$, 26, 674
$C_{14}H_{14}O_2$, 32B, 104
$C_{14}H_{14}O_2Ru$, 28, 612
$C_{14}H_{14}S_2$, 32B, 71; 34B, 49
$C_{14}H_{15}BrO_3$, 32B, 248; 35B, 371
$C_{14}H_{15}BrO_3S$, 30B, 113
$C_{14}H_{15}N$, 27, 933
$C_{14}H_{15}N_3$, 29, 662
$C_{14}H_{16}As_2CrO_4$, 30B, 428
$C_{14}H_{16}As_2MoO_4$, 30B, 428
$C_{14}H_{16}As_2O_4W$, 30B, 428
$C_{14}H_{16}BF_4Fe_2O_2S_2$, 35B, 921
$C_{14}H_{16}BF_4N_5O_2$, 30B, 331
$C_{14}H_{16}BrNO_3$, 31B, 89
$C_{14}H_{16}Cd_3Cl_6O_4.2H_2O$, 30B, 429
$C_{14}H_{16}Cl_2CuO_4$, 30B, 429
$C_{14}H_{16}Cl_{12}FeN_6$, 33B, 326
$C_{14}H_{16}ClNO_4$, 34B, 141
$C_{14}H_{16}Cl_2N_8S_4Te.2H_2O$, 30B, 285
$C_{14}H_{16}Cl_2O_4Zn$, 30B, 429
$C_{14}H_{16}CrI$, 26, 679
$C_{14}H_{16}CuN_2O_2.C_6H_5NO_3$, 35B, 413
$C_{14}H_{16}CuO_7$, 28, 634
$C_{14}H_{16}F_4N_2P_2$, 32B, 491
$C_{14}H_{16}K_2N_2O_{18}Ti_2$, 35B, 740
$C_{14}H_{16}N_2$, 30B, 64
$C_{14}H_{16}Ni_2$, 30B, 428
$C_{14}H_{16}O_3$, 27, 1028
$C_{14}H_{17}Br_2N_3Ni$, 35B, 702
$C_{14}H_{17}Cl_6O_2S$, 27, 785
$C_{14}H_{17}CoO$, 26, 652; 27, 862
$C_{14}H_{18}Cl_4$, 31B, 103
$C_{14}H_{18}Cl_2CuN_2O_2$, 34B, 578
$C_{14}H_{18}Cl_2N_2O_2Zn$, 33B, 458
$C_{14}H_{18}INO_2$, 31B, 240
$C_{14}H_{18}N_2O_2.H_2SO_4$, 35B, 67
$C_{14}H_{18}N_4O_4.2H_2O$, 35B, 172
$C_{14}H_{18}O_2$, 26, 750
$C_{14}H_{18}O_6$, 30B, 429
$C_{14}H_{18}O_6.H_2O$, 30B, 429
$C_{14}H_{18}O_7$, 26, 627
$C_{14}H_{19}NO_2$, 34B, 143
$C_{14}H_{20}$, 30B, 119
$C_{14}H_{20}Br_2N_2$, 32B, 488
$C_{14}H_{20}Cl_2$, 28, 675
$C_{14}H_{20}Cl_6N_2Re$, 32B, 98
$C_{14}H_{20}N_4O_2$, 32B, 168
$C_{14}H_{20}N_4O_4$, 34B, 241; 35B, 257
$C_{14}H_{20}N_2O_3S$, 26, 719
$C_{14}H_{20}N_2O_6S$, 35B, 67
$C_{14}H_{20}N_3S_4Zn$, 32B, 412
$C_{14}H_{20}O_7S$, 29, 762
$C_{14}H_{21}BrO$, 34B, 308
$C_{14}H_{21}N_5O_6S.H_2O$, 30B, 214
$C_{14}H_{21}OP$, 33B, 549; 35B, 456
$C_{14}H_{22}$, 33B, 206
$C_{14}H_{22}ClNO_3$, 35B, 50
$C_{14}H_{22}ClNO_4$, 35B, 173
$C_{14}H_{22}F_6Mo_2PS_4$, 35B, 921
$C_{14}H_{22}Mo_2S_4$, 35B, 921
$C_{14}H_{22}N_6O_6.3H_2O$, 35B, 328
$C_{14}H_{22}N_2S_4$, 34B, 321
$C_{14}H_{23}ClO_2$, 28, 675
$C_{14}H_{23}N_6OP$, 32B, 487
$C_{14}H_{24}$, 32B, 136
$C_{14}H_{24}CuN_2S_4$, 33B, 488
$C_{14}H_{24}O_4$, 32B, 180
$C_{14}H_{25}BrMgO_2$, 29, 532
$C_{14}H_{25}Co_3O_4S_5$, 30B, 488
$C_{14}H_{25}MoO_4P_5$, 33B, 506
$C_{14}H_{26}Cl_2Pd_2$, 33B, 334
$C_{14}H_{26}N_4O_{10}$, 33B, 244
$C_{14}H_{27}BrO_4$, 30B, 429; 31B, 21
$C_{14}H_{27}KO_2$, 30B, 429
$C_{14}H_{28}Br_2P_2$, 35B, 455
$C_{14}H_{28}CoN_6O_4Se_2$, 29, 767
$C_{14}H_{28}CuN_2O_4$, 34B, 543
$C_{14}H_{28}CuN_2S_4$, 27, 844
$C_{14}H_{28}IN$, 28, 472; 32B, 163
$C_{14}H_{28}N_6NiO_4Se_2$, 29, 767; 32B, 433
$C_{14}H_{28}N_2NiS_4$, 32B, 414
$C_{14}H_{28}O_{12}$, 35B, 268
$C_{14}H_{30}I_2N_2O_4$, 35B, 239
$C_{14}H_{31}N_3O_2S_4U$, 33B, 542
$C_{14}H_{32}Cl_8N_4Ni_2$, 35B, 878

C_{15}

$C_{15}Fe_5O_{15}S_4$, 30B, 384
$C_{15}H_6F_4FeO_3$, 35B, 922
$C_{15}H_7F_5MoO_2$, 33B, 552
$C_{15}H_7F_{12}NiS_2$, 35B, 804
$C_{15}H_8BrNO_4.H_2O$, 28, 474
$C_{15}H_8Br_2O_2$, 27, 913
$C_{15}H_8F_{12}NiS_4$, 35B, 807
$C_{15}H_8Fe_2O_5$, 32B, 376
$C_{15}H_8INO_4.H_2O$, 28, 474
$C_{15}H_9BBr_2F_2O_2$, 30B, 430
$C_{15}H_{10}BrNO_3$, 26, 722; 35B, 54
$C_{15}H_{10}Br_2O_2$, 30B, 430
$C_{15}H_{10}Cl_2$, 34B, 107
$C_{15}H_{10}ClNO_3$, 35B, 54
$C_{15}H_{10}Cl_2O_2$, 29, 655; 30B, 430
$C_{15}H_{11}Br$, 26, 749
$C_{15}H_{11}Cl_2CoN_3$, 35B, 703
$C_{15}H_{11}Cl_2N_3Zn$, 32B, 431
$C_{15}H_{11}I_4NO_4.5H_2O$, 32B, 489
$C_{15}H_{11}N$, 26, 748
$C_{15}H_{11}N_5O_6$, 35B, 190
$C_{15}H_{11}N_3O_9S_2$, 34B, 50
$C_{15}H_{12}$, 34B, 106
$C_{15}H_{12}Br_2N_4O_3$, 32B, 62
$C_{15}H_{12}Fe_2O_7$, 34B, 422
$C_{15}H_{12}O$, 35B, 55
$C_{15}H_{12}O_2$, 30B, 430; 31B, 80
$C_{15}H_{12}O_5$, 27, 1005
$C_{15}H_{13}Br$, 31B, 34

$C_{15}H_{13}F_6FeO_2Sb$, 35B, 915
$C_{15}H_{14}MoO_2$, 33B, 350
$C_{15}H_{14}N_2$, 33B, 110
$C_{15}H_{14}N_2OSe$, 28, 675
$C_{15}H_{14}O_3$, 27, 909
$C_{15}H_{15}BrO_5$, 34B, 338
$C_{15}H_{15}BrO_6$, 27, 1017; 28, 675; 30B, 245; 33B, 233
$C_{15}H_{15}ClN_4O_5$, 33B, 245
$C_{15}H_{15}IN_2O$, 34B, 135
$C_{15}H_{15}Mn_3N_4O_4$, 32B, 493
$C_{15}H_{15}MoNO$, 34B, 453
$C_{15}H_{15}Ni_3S_2$, 33B, 552
$C_{15}H_{15}Sm$, 34B, 459
$C_{15}H_{16}BrO_3$, 31B, 213
$C_{15}H_{16}CrO_3$, 33B, 371
$C_{15}H_{16}N_2O_6$, 35B, 142
$C_{15}H_{17}BrN_4O_2$, 34B, 154
$C_{15}H_{17}BrO_3$, 30B, 226; 31B, 111; 32B, 249; 33B, 209; 34B, 309
$C_{15}H_{17}BrO_5.H_2O$, 27, 1019
$C_{15}H_{17}BrO_6$, 28, 675; 29, 745
$C_{15}H_{17}ClO_5.H_2O$, 27, 1019
$C_{15}H_{17}CrNO_6$, 35B, 501
$C_{15}H_{17}IN_4O_2.H_2O$, 34B, 156
$C_{15}H_{17}MoO_3P$, 29, 767
$C_{15}H_{17}O_2Rh$, 35B, 573
$C_{15}H_{18}$, 29, 693
$C_{15}H_{18}BrClCuN_4O_5$, 35B, 705
$C_{15}H_{18}BrNO_2$, 33B, 129
$C_{15}H_{18}ClU$, 30B, 315
$C_{15}H_{18}Ge_3O_9Ru_3$, 35B, 917
$C_{15}H_{18}N_2O_5$, 26, 809
$C_{15}H_{19}BrO_3$, 35B, 372
$C_{15}H_{19}BrO_3S$, 33B, 98
$C_{15}H_{19}ClO_4Zr$, 34B, 464
$C_{15}H_{19}O_3Re$, 32B, 494
$C_{15}H_{19}Ti$, 32B, 346
$C_{15}H_{20}BrN$, 35B, 197
$C_{15}H_{20}IN_3O_9.{}^1/_2CH_4O$, 31B, 241
$C_{15}H_{20}N_2O_5$, 27, 1050
$C_{15}H_{20}O_4$, 28, 675
$C_{15}H_{20}O_5$, 35B, 345
$C_{15}H_{20}O_6$, 30B, 429
$C_{15}H_{21}CoKO_6$, 27, 822
$C_{15}H_{21}CoO_6$, 27, 821
$C_{15}H_{21}CrO_6$, 30B, 332
$C_{15}H_{21}FeO_6$, 32B, 385
$C_{15}H_{21}N_9O_6$, 31B, 274
$C_{15}H_{21}O_6V$, 34B, 507; 35B, 606
$C_{15}H_{22}BrNO_2$, 26, 769; 35B, 175
$C_{15}H_{22}BrNO_3S$, 35B, 50
$C_{15}H_{22}INO_2$, 26, 769
$C_{15}H_{23}Ag_3N_2NiO_{13}$, 31B, 381
$C_{15}H_{23}NO.HBr$, 35B, 373
$C_{15}H_{23}OP$, 35B, 457
$C_{15}H_{23}O_7Yb$, 34B, 508
$C_{15}H_{23}O_7Yb.{}^1/_2C_6H_6$, 34B, 510
$C_{15}H_{24}AgNO_3$, 32B, 494; 33B, 372
$C_{15}H_{24}Ag_2NO_3$, 29, 769
$C_{15}H_{24}Ag_2N_2O_6$, 31B, 213
$C_{15}H_{24}Br_2N_4NiO$, 33B, 428
$C_{15}H_{24}BrNO$, 35B, 373
$C_{15}H_{24}BrP$, 34B, 387
$C_{15}H_{24}ClNO$, 28, 672
$C_{15}H_{24}IN_2O_3$, 33B, 210
$C_{15}H_{24}N_4O_2S_2$, 34B, 238
$C_{15}H_{25}Br$, 26, 796
$C_{15}H_{25}BrO$, 33B, 211
$C_{15}H_{25}Cl$, 26, 796; 30B, 225; 33B, 212
$C_{15}H_{25}ClN_2O_4$, 32B, 264
$C_{15}H_{25}CuO_8.H_2O$, 34B, 497
$C_{15}H_{25}LaO_8$, 33B, 390
$C_{15}H_{25}N_6OP$, 32B, 487
$C_{15}H_{25}N_6OP.{}^1/_2H_2O$, 32B, 487
$C_{15}H_{25}N_3O_7PS$, 35B, 292
$C_{15}H_{25}NdO_8.H_2O$, 32B, 495
$C_{15}H_{26}Cl_2$, 28, 539
$C_{15}H_{26}Cl_2N_4NiO_4$, 34B, 560
$C_{15}H_{26}N_2O$, 30B, 430
$C_{15}H_{26}N_2O.HBr$, 30B, 430
$C_{15}H_{26}N_3O_8P$, 34B, 256
$C_{15}H_{26}N_3O_5PS$, 34B, 258
$C_{15}H_{26}O$, 26, 800
$C_{15}H_{27}BrN_2O$, 30B, 430
$C_{15}H_{27}ClN_2O_5$, 32B, 254
$C_{15}H_{27}I_9N_9$, 26, 615
$C_{15}H_{27}O_9Y$, 32B, 386
$C_{15}H_{28}O_2$, 32B, 484, 486
$C_{15}H_{29}O_4Rh$, 35B, 574
$C_{15}H_{30}CoN_3S_6$, 33B, 491; 34B, 527
$C_{15}H_{30}O_4$, 30B, 11
$C_{15}H_{45}N_3Si_6$, 35B, 443

C_{16}
$C_{16}Co_2F_{24}S_8$, 30B, 372
$C_{16}H_5Fe_3O_{11}Rh$, 35B, 917
$C_{16}H_5N_5O_6$, 32B, 137
$C_{16}H_6N_4O_4$, 33B, 91
$C_{16}H_8Cl_4$, 31B, 106
$C_{16}H_8Cl_2N_2O_4$, 35B, 233
$C_{16}H_8Mn_2O_6$, 33B, 364
$C_{16}H_8MoO_6$, 32B, 494
$C_{16}H_8N_4$, 30B, 417; 32B, 278
$C_{16}H_8N_6O_6$, 30B, 418
$C_{16}H_8O_4$, 27, 983
$C_{16}H_8O_6$, 30B, 423
$C_{16}H_{10}$, 27, 940; 30B, 109; 35B, 131
$C_{16}H_{10}Cl_4Co_2FeGeO_6$, 35B, 920
$C_{16}H_{10}Cl_2O_2$, 35B, 57
$C_{16}H_{10}Co_4O_{10}$, 27, 826
$C_{16}H_{10}Fe_2O_6$, 35B, 920
$C_{16}H_{10}FeN_4$, 32B, 279
$C_{16}H_{10}Hg$, 27, 864
$C_{16}H_{10}N_2$, 26, 730
$C_{16}H_{10}N_2O_2$, 26, 783
$C_{16}H_{10}O$, 34B, 117
$C_{16}H_{10}O_2$, 34B, 102
$C_{16}H_{11}BrN_4O_7$, 33B, 245
$C_{16}H_{11}ClO_3$, 34B, 174
$C_{16}H_{11}CrN_3O_{10}$, 31B, 275
$C_{16}H_{11}FeO_4P$, 34B, 600
$C_{16}H_{11}N_3O_3$, 34B, 39
$C_{16}H_{11}N_3O_6$, 30B, 250
$C_{16}H_{12}$, 26, 722; 28, 452; 35B, 144
$C_{16}H_{12}CrO_3$, 33B, 373
$C_{16}H_{12}Fe_2O_8$, 35B, 575
$C_{16}H_{12}FeN_3O_7$, 32B, 493
$C_{16}H_{12}N_2O_2.HBr.2H_2O$, 35B, 233
$C_{16}H_{12}N_4OS$, 34B, 212
$C_{16}H_{12}O_2$, 35B, 57
$C_{16}H_{12}O_{12}Re_4S_4$, 35B, 917
$C_{16}H_{13}Br$, 35B, 127
$C_{16}H_{13}BrN_2O_2.2H_2O$, 35B, 233
$C_{16}H_{13}MoN_3O_2S$, 34B, 443
$C_{16}H_{14}BrNO$, 32B, 487
$C_{16}H_{14}CdN_2O_2$, 28, 676
$C_{16}H_{14}ClFeN_2O_2$, 32B, 394
$C_{16}H_{14}Cl_2O$, 35B, 145
$C_{16}H_{14}CoN_2O_2$, 34B, 520
$C_{16}H_{14}CoN_2O_2.CHCl_3$, 34B, 514
$C_{16}H_{14}CuN_2O_2.CHCl_3$, 32B, 495; 35B, 414
$C_{16}H_{14}CuN_2O_2.C_6H_5NO_3$, 32B, 495
$C_{16}H_{14}F_2$, 33B, 74
$C_{16}H_{14}MoS_2$, 35B, 553
$C_{16}H_{14}N_6Ni$, 34B, 138
$C_{16}H_{14}N_2NiO_2$, 26, 671; 28, 676; 35B, 620
$C_{16}H_{14}N_2O.HBr$, 33B, 111
$C_{16}H_{14}N_2O_2Pd$, 28, 676
$C_{16}H_{14}O$, 26, 748; 33B, 135
$C_{16}H_{14}O_2$, 34B, 26; 35B, 56
$C_{16}H_{14}O_2S$, 35B, 51
$C_{16}H_{15}Br$, 32B, 42
$C_{16}H_{15}BrN_2O$, 33B, 111
π-$C_{16}H_{15}Co$, 29, 558
$C_{16}H_{15}Co_3O_2$, 34B, 466
$C_{16}H_{15}KO_4$, 33B, 45
$C_{16}H_{15}KO_6$, 33B, 46
$C_{16}H_{15}N_2NaO_6S_2$, 28, 676
$C_{16}H_{16}$, 35B, 93, 115
$C_{16}H_{16}Cl_4N_2O_2$, 33B, 247
$C_{16}H_{16}Cl_4Pd$, 26, 689
$C_{16}H_{16}CuN_2O_2$, 26, 667; 33B, 397
$C_{16}H_{16}CuN_2O_2.C_6H_5NO_3$, 35B, 413
$C_{16}H_{16}Fe_2O_4Pb$, 33B, 310
$C_{16}H_{16}Fe_2O_4Sn$, 33B, 310
$C_{16}H_{16}HgN_2O_2S_2$, 32B, 456
$C_{16}H_{16}HgO_2$, 33B, 328
$C_{16}H_{16}IN$, 35B, 207
$C_{16}H_{16}N_2NiO_2$, 26, 672; 32B, 395
$C_{16}H_{16}N_2NiO_6$, 32B, 434
$C_{16}H_{16}N_2O_2$, 33B, 75
$C_{16}H_{16}N_2O_5S$, 35B, 384
$C_{16}H_{16}N_2O_2Zn$, 31B, 389
$C_{16}H_{16}N_2O_3Zn$, 31B, 392
$C_{16}H_{16}NiO_4$, 28, 600, 681
$C_{16}H_{16}NiO_3S_2$, 35B, 800
$C_{16}H_{16}O_3$, 27, 911

$C_{16}H_{16}O_4$, 31B, 107
$C_{16}H_{16}Ti$, 34B, 469
$C_{16}H_{16}U$, 34B, 470
$C_{16}H_{17}BCo_3NO_{10}$, 33B, 266
$C_{16}H_{17}Mo_2O_4P$, 30B, 312
$C_{16}H_{18}Br_2O_5$, 35B, 141
$C_{16}H_{18}ClCrN_2O_4$, 35B, 621
$C_{16}H_{18}CuO_8$, 33B, 403
$C_{16}H_{18}N_6Ni$, 35B, 523
$C_{16}H_{18}N_2O_2$, 35B, 70
$C_{16}H_{18}N_2O_5S$, 28, 530
$C_{16}H_{18}N_2O_6S.CH_4O$, 34B, 285
$C_{16}H_{19}NO_2.HBr$, 34B, 325
$C_{16}H_{19}NO_4.HBr$, 33B, 220
$C_{16}H_{19}N_3O_2.HBr$, 35B, 331
$C_{16}H_{20}BrN$, 32B, 149
$C_{16}H_{20}BrNO_2$, 34B, 325
$C_{16}H_{20}BrNO_5$, 33B, 188
$C_{16}H_{20}BrN_3O_2$, 35B, 331
$C_{16}H_{20}Cl_4N_6Zn$, 31B, 46
$C_{16}H_{20}Co_2O_4$, 35B, 525
$C_{16}H_{20}Fe$, 30B, 306
$C_{16}H_{20}Fe_2Ge_2O_8$, 33B, 551
$C_{16}H_{20}GeO_{10}$, 32B, 331
$C_{16}H_{20}N_2NiO_4P_2S_4$, 35B, 808
$C_{16}H_{20}Ni_2$, 30B, 431
$C_{16}H_{20}O_2$, 30B, 119
$C_{16}H_{21}ClN_3O_4$, 32B, 71
$C_{16}H_{21}IN_3$, 31B, 59
$C_{16}H_{21}O_{11}Rb.2H_2O$, 32B, 490
$C_{16}H_{22}BrNO_3$, 28, 546
$C_{16}H_{22}Br_3NiP_2.C_7H_{11}BrNi_{0.5}P.C_6H_6$, 35B, 836
$C_{16}H_{22}ClI_5N_2$, 27, 958
$C_{16}H_{22}Fe$, 27, 862
$C_{16}H_{22}INO_2S$, 35B, 190
$C_{16}H_{22}I_2P_2Pd$, 33B, 507
$C_{16}H_{22}O_6$, 30B, 429
$C_{16}H_{23}NO$, 32B, 247
$C_{16}H_{24}BrNO_2$, 28, 463
$C_{16}H_{24}Cl_2Cu_2$, 28, 614
$C_{16}H_{24}Cl_3IrSn$, 32B, 376
$C_{16}H_{24}ClNO_2$, 28, 464
$C_{16}H_{24}Cl_4Ni_2.C_6H_6$, 27, 847
$C_{16}H_{24}Cl_2Rh_2$, 27, 853
$C_{16}H_{24}CoN_5O_6$, 32B, 497
$C_{16}H_{24}N_6NiS_4$, 30B, 377
$C_{16}H_{24}Ni$, 30B, 321
$C_{16}H_{24}O_4$, 28, 676
$C_{16}H_{25}AsN_2S_4$, 32B, 316
$(C_{16}H_{25}Mn_3O_{17})_n$, 34B, 531
$C_{16}H_{26}As_2Cl_3O_2Rh$, 34B, 604
$C_{16}H_{26}Br_2O_8Sn$, 33B, 313
$C_{16}H_{27}BrN_2O_2$, 33B, 221
$C_{16}H_{28}Br_2CaN_4O_8$, 35B, 746
$C_{16}H_{28}ClO_8Ru_2$, 34B, 536
$C_{16}H_{28}O_4$, 32B, 180
$C_{16}H_{28}O_{16}Re_4$, 35B, 643
$C_{16}H_{30}Fe_2S_{14}$, 34B, 597; 35B, 802
$C_{16}H_{30}O_6Ru_2Sn_4$, 34B, 411
$C_{16}H_{31}BrO_7$, 26, 627
$C_{16}H_{31}KO_2$, 30B, 19
$C_{16}H_{32}Br_2MgO_4$, 33B, 301
$C_{16}H_{32}Cl_2N_8NiS_4$, 33B, 492
$C_{16}H_{32}Co_2N_{12}O_2$, 35B, 707
$C_{16}H_{32}MgN_2$, 35B, 709
$C_{16}H_{36}As_3ClNNiO_4P$, 32B, 465
$C_{16}H_{36}Cl_2N_8O_2S_6Te$, 26, 584
$C_{16}H_{36}Cl_2N_8O_8S_4Te$, 26, 584
$C_{16}H_{36}O_2P_2S_2Zn$, 35B, 810
$C_{16}H_{36}O_4Ti_2$, 35B, 747
$C_{16}H_{38}Br_2Mg_2O_4$, 33B, 546
$C_{16}H_{38}Br_2N_2.2H_2O$, 30B, 34
$C_{16}H_{40}Br_2Mg_2N_2$, 32B, 492
$C_{16}H_{40}Br_6Mg_4O_5$, 29, 532
$C_{16}H_{40}Br_6N_2Pt_2$, 29, 580
$C_{16}H_{40}Cl_5InN_2$, 34B, 23
$C_{16}H_{40}Cl_4N_2Ni$, 32B, 34
$C_{16}H_{40}Cl_4N_2OTi$, 33B, 460
$C_{16}H_{40}Cl_4N_2OV$, 33B, 460
$C_{16}H_{40}Mo_2O_{11}P_4S_8.2C_6H_4Cl_2$, 34B, 598
$C_{16}H_{42}Co_3N_6O_{10}$, 34B, 556
$C_{16}H_{44}B_4N_8$, 33B, 549
$C_{16}H_{48}AlO_4SbSi_4$, 28, 601
$C_{16}H_{48}Be_4O_4Si_4$, 34B, 368
$C_{16}H_{48}Cl_{10}Cu_4N_4O$, 34B, 22
$C_{16}H_{48}Cl_2CuN_8O_{14}P_4$, 35B, 744
$C_{16}H_{48}Cl_2O_{17}S_8Zr$, 30B, 410
$C_{16}H_{48}FeO_4SbSi_4$, 28, 601
$C_{16}H_{48}GaO_4SbSi_4$, 28, 601
$C_{16}H_{48}Hg_4O_4Si_4$, 34B, 368
$C_{16}H_{48}N_8O_{16}P_8$, 33B, 281
$C_{16}H_{48}N_{12}P_4$, 27, 989
$C_{16}H_{50}Cl_4CoN_8P_8$, 35B, 461
$C_{16}H_{56}Br_6N_8Pt_2$, 31B, 423
$C_{16}H_{64}Cl_6N_8O_4Pt_2$, 26, 611
$C_{16}H_{101.6}FNO_{32.8}$, 28, 578

C_{17}

$C_{17}H_5Co_3GeO_{11}$, 35B, 914
$C_{17}H_5F_{18}Rh$, 31B, 366
$C_{17}H_8Fe_2O_5$, 35B, 572
$C_{17}H_8N_2O$, 32B, 131
$C_{17}H_{10}BrNS_2$, 32B, 196
$C_{17}H_{10}CrO_3$, 29, 542; 33B, 365, 366
$C_{17}H_{11}NO_2S_2$, 34B, 193
$C_{17}H_{12}CrO_3$, 33B, 366
$C_{17}H_{12}O_2$, 35B, 121
$C_{17}H_{12}O_6$, 35B, 374
$C_{17}H_{12}O_6.CHCl_3$, 35B, 374
$C_{17}H_{12}OS_2$, 34B, 189
$C_{17}H_{12}S_3$, 31B, 151; 34B, 191
$C_{17}H_{13}N$, 26, 748
$C_{17}H_{14}$, 26, 752
$C_{17}H_{14}KNO_3S$, 35B, 116
$C_{17}H_{14}N_4S_2$, 32B, 197
$C_{17}H_{14}O_4$, 35B, 918
$C_{17}H_{14}O_6$, 35B, 374
$C_{17}H_{14}OOs$, 30B, 431
$C_{17}H_{15}BrO_2$, 34B, 117
$C_{17}H_{15}BrO_3S$, 33B, 205; 34B, 118
$C_{17}H_{15}Cl_3CoN_2O_2$, 34B, 514
$C_{17}H_{15}Cl_3CuN_2O_2$, 32B, 495
$C_{17}H_{15}Cl_3N_2O$, 32B, 164
$C_{17}H_{16}BrClO_6$, 28, 533
$C_{17}H_{16}Fe_2O_5$, 33B, 374, 375
$C_{17}H_{16}Ge$, 32B, 492
$C_{17}H_{16}O_7$, 34B, 338
$C_{17}H_{17}BBrNO_4$, 34B, 253
$C_{17}H_{17}ClFeN_3O_4$, 32B, 396
$C_{17}H_{17}N_3O_5$, 35B, 912
$C_{17}H_{19}BrN_4O_3$, 35B, 199
$C_{17}H_{19}ClN_2S$, 30B, 431; 34B, 213
$C_{17}H_{19}I_2N_9O_4$, 34B, 231
$C_{17}H_{19}NO_5$, 32B, 256
$C_{17}H_{19}NO_3S_2$, 35B, 208
$C_{17}H_{20}ClN_2OP$, 33B, 547
$C_{17}H_{20}Cl_2N_2S$, 30B, 431
$C_{17}H_{20}I_2N_{10}O_4$, 34B, 233
$C_{17}H_{20}N_4O_{11}$, 30B, 431
$C_{17}H_{21}BrN_4O_6.H_2O$, 32B, 488
$C_{17}H_{21}BrO_2$, 32B, 490; 34B, 310
$C_{17}H_{21}BrO_5$, 27, 1026; 28, 676
$C_{17}H_{22}BrNO_2$, 34B, 289
$C_{17}H_{22}BrNO_4$, 28, 548; 33B, 223
$C_{17}H_{22}ClN_3O_4$, 32B, 100
$C_{17}H_{22}N_2O_7S$, 34B, 285
$C_{17}H_{22}N_2O_6S.H_2O$, 27, 1051
$C_{17}H_{23}BrO_3$, 34B, 340
$C_{17}H_{23}BrO_5$, 30B, 227
$C_{17}H_{23}BrO_4S$, 30B, 116
$C_{17}H_{23}IN$, 32B, 485
$C_{17}H_{24}BrNO.H_2O$, 35B, 379
$C_{17}H_{24}O_2Pd$, 35B, 576
$C_{17}H_{24}OS_2$, 32B, 488
$C_{17}H_{25}Al_2N$, 34B, 406
$C_{17}H_{25}BrO_4$, 35B, 380
$C_{17}H_{25}N.HCl$, 35B, 176
$C_{17}H_{25}NO_3S$, 34B, 163
$C_{17}H_{26}BrNO_4.{}^1/_2H_2O$, 35B, 382
$C_{17}H_{26}ClN$, 35B, 176
$C_{17}H_{27}Cl_2NPt$, 32B, 493; 35B, 526
$C_{17}H_{27}NNiP_2S_4$, 35B, 807

C_{18}

$C_{18}HCo_5O_{15}$, 35B, 502
$C_{18}H_8Fe_2N_2O_6$, 35B, 699
$C_{18}H_9ClO_2$, 32B, 141
$C_{18}H_{10}Cr_2O_6$, 26, 650
$C_{18}H_{10}Fe_2O_8Rh_2$, 35B, 913
$C_{18}H_{10}N_6O_6$, 30B, 417
$C_{18}H_{11}O_2S_2$, 32B, 488
$C_{18}H_{12}$, 26, 754; 27, 925, 938; 28, 440, 444
$C_{18}H_{12}As_2F_4Fe_3O_{10}$, 35B, 866
$C_{18}H_{12}B_3NO_6$, 35B, 439
$C_{18}H_{12}Br_3N$, 28, 676
$C_{18}H_{12}Cl_3N$, 28, 676
$C_{18}H_{12}ClNO_3$, 31B, 96
$C_{18}H_{12}Cl_2N_2O_2Ti$, 33B, 445
$C_{18}H_{12}CuN_2O_2$, 28, 612; 32B, 435

$C_{18}H_{12}F_3N$, 28, 676
$C_{18}H_{12}I_3N$, 28, 676
$C_{18}H_{12}Mo_2O_6$, 34B, 471
$C_{18}H_{12}N_5O_6$, 30B, 431
$C_{18}H_{12}N_3O_{10}P$, 32B, 489; 35B, 274
$C_{18}H_{12}N_3O_6P_3.{}^1/_2(C_6H_6)$, 32B, 309
$C_{18}H_{12}N_3O_6P_3.{}^1/_2(C_6H_5Br)$, 32B, 309
$C_{18}H_{12}N_2O_2Pd$, 31B, 432
$C_{18}H_{12}N_4O_6S$, 34B, 351
$C_{18}H_{12}N_6O_6S_2$, 30B, 267
$C_{18}H_{12}N_2O_2Zn.2H_2O$, 29, 614
$C_{18}H_{12}N_2PbS_2$, 32B, 492
$C_{18}H_{12}N_2PdS_2$, 30B, 422
$C_{18}H_{12}O_6$, 29, 675
$C_{18}H_{13}As$, 28, 589
$C_{18}H_{14}$, 26, 734; 32B, 486; 34B, 83; 35B, 95
$C_{18}H_{14}Br_2N_2$, 32B, 63
$C_{18}H_{14}Br_2O_2$, 32B, 280
$C_{18}H_{14}Cl_4$, 28, 676
$C_{18}H_{14}Cl_2O_2$, 32B, 280
$C_{18}H_{14}Fe_2O_4$, 34B, 425
$C_{18}H_{14}N_4$, 32B, 485
$C_{18}H_{14}Ru$, 27, 863; 32B, 359
$C_{18}H_{15}As$, 27, 868
$C_{18}H_{15}AsO.H_2O$, 34B, 394
$C_{18}H_{15}BF_4O$, 28, 677
$C_{18}H_{15}BrO_5$, 35B, 147
$C_{18}H_{15}BrO.2H_2O$, 28, 676
$C_{18}H_{15}Bi$, 33B, 295
$C_{18}H_{15}BiCl_2$, 33B, 296
$C_{18}H_{15}ClO.2H_2O$, 28, 676
$C_{18}H_{15}Cl_2O_2PSe$, 34B, 354
$C_{18}H_{15}Cl_2Sb$, 31B, 326
$C_{18}H_{15}ClSn$, 30B, 431; 35B, 489
$C_{18}H_{15}FeN_6O_6$, 30B, 358
$C_{18}H_{15}Ga$, 35B, 482
$C_{18}H_{15}IO$, 28, 676
$C_{18}H_{15}In$, 35B, 482
$C_{18}H_{15}KO_8$, 32B, 87; 33B, 47
$C_{18}H_{15}N_4PS_3$, 35B, 917
$C_{18}H_{15}OP$, 35B, 464
$C_{18}H_{15}O_4P$, 27, 916; 30B, 193
$C_{18}H_{15}O_8Rb$, 32B, 89
$C_{18}H_{15}O_3Rh_3$, 32B, 360; 34B, 461
$C_{18}H_{15}P$, 29, 638
$C_{18}H_{16}BrNO_2$, 33B, 99
$C_{18}H_{16}Cl_2$, 28, 676
$C_{18}H_{16}I^5/_{12}N_2$, 32B, 72
$C_{18}H_{16}Mn_2O_4$, 32B, 347
$C_{18}H_{16}N_6Ni$, 35B, 177
$C_{18}H_{16}Ni_2$, 30B, 431
$C_{18}H_{16}O_4$, 33B, 65
$C_{18}H_{17}BrO_3$, 27, 1025; 31B, 262
$C_{18}H_{17}ClO_6$, 30B, 431
$C_{18}H_{17}ClO_6.{}^1/_2CHCl_3$, 30B, 431
$C_{18}H_{17}NO_3$, 32B, 122
$C_{18}H_{18}$, 30B, 94; 33B, 545
$C_{18}H_{18}Br_2$, 33B, 100
$C_{18}H_{18}BrNS$, 29, 764
$C_{18}H_{18}CdN_6Ni$, 35B, 919
$C_{18}H_{18}Cl_2$, 32B, 43
$C_{18}H_{18}ClNO$, 35B, 149
$C_{18}H_{18}ClN_2O_4PS_2$, 31B, 302
$C_{18}H_{18}ClNS$, 29, 735
$C_{18}H_{18}Cl_4O_2$, 27, 909
$C_{18}H_{18}Cl_2Pd_2$, 34B, 444
$C_{18}H_{18}Co_2$, 32B, 494
$C_{18}H_{18}Fe_2O_7$, 35B, 578
$C_{18}H_{18}N_6$, 29, 478
$C_{18}H_{18}N_2S_2$, 32B, 488
$C_{18}H_{18}N_6S_2$, 32B, 198
$C_{18}H_{18}O$, 34B, 93
$C_{18}H_{18}O_2$, 28, 442
$C_{18}H_{18}S_2$, 29, 663
$C_{18}H_{19}Br_3N_6NiO_3.H_2O$, 34B, 565
$C_{18}H_{19}BrO_3$, 34B, 120
$C_{18}H_{19}IN_2$, 31B, 133
$C_{18}H_{19}IO_8$, 34B, 342
$C_{18}H_{19}NO_4$, 28, 387
$C_{18}H_{19}N_3O_6S.C_2H_4O_2.H_2O$, 35B, 384
$C_{18}H_{19}N_7O_8U$, 29, 769
$C_{18}H_{20}$, 27, 918; 30B, 121
$C_{18}H_{20}B_2CoN_{12}$, 35B, 434
$C_{18}H_{20}Cl_2N_2$, 35B, 33
$C_{18}H_{20}ClN_3O_6S_2.(0.65\ CH_2Br_2)$, 30B, 240
$C_{18}H_{20}CuN_2O_2$, 32B, 397
$C_{18}H_{20}CuN_2O_4$, 31B, 393; 35B, 622
$C_{18}H_{20}N_2NiO_2$, 32B, 495
$C_{18}H_{20}N_2O_2$, 33B, 550
$C_{18}H_{20}N_2O_2Pd$, 29, 582
$C_{18}H_{20}O_2$, 35B, 213, 332
$C_{18}H_{20}O_2.CH_4O.H_2O$, 34B, 76
$C_{18}H_{20}O_2.H_2O$, 34B, 77
$C_{18}H_{21}BrO_2$, 28, 536
$C_{18}H_{21}BrO_7S$, 35B, 270
$C_{18}H_{21}Cr_2NO_{10}$, 31B, 31; 35B, 506
$C_{18}H_{22}BrNO_3$, 32B, 490
$C_{18}H_{22}BrNO_4.2H_2O$, 27, 1023
$C_{18}H_{22}Cu_2N_2O_8$, 26, 669; 29, 606
$C_{18}H_{22}HgN_2$, 33B, 328
$C_{18}H_{22}I_2N_6O_2S_8$, 33B, 248
$C_{18}H_{22}N_4O_{10}S_2$, 32B, 224
$C_{18}H_{22}N_4O_4S_6Te$, 26, 582
$C_{18}H_{22}N_2O_2Si_2$, 35B, 443
$C_{18}H_{22}Na_2O_{12}$, 33B, 78
$C_{18}H_{22}Ni$, 29, 768; 30B, 322
$C_{18}H_{23}ClN_2S$, 32B, 488
$C_{18}H_{24}N_{14}NiO_6$, 34B, 572; 35B, 710
$C_{18}H_{24}N_2NiO_4P_2S_4$, 35B, 808
$C_{18}H_{24}N_2O_2Pt$, 27, 886
$C_{18}H_{24}NiO_2$, 30B, 323
$C_{18}H_{24}O_3$, 26, 799; 34B, 291
$C_{18}H_{24}O_2.H_2O$, 34B, 290
$C_{18}H_{25}NO$, 34B, 326
$C_{18}H_{26}As_2IrPdS_2$, 35B, 839
$C_{18}H_{26}BrNO.H_2O$, 34B, 326
$C_{18}H_{26}BrNO_6.{}^1/_2C_2H_6O$, 28, 550
$C_{18}H_{28}IN$, 28, 677
$C_{18}H_{28}INO_3$, 35B, 383
$C_{18}H_{30}$, 29, 502; 32B, 142; 33B, 206
$C_{18}H_{30}.0.39C_6H_{12}$, 32B, 292
$C_{18}H_{30}.0.225C_7H_{16}$, 32B, 491
$C_{18}H_{30}.1/2CHCl_3$, 32B, 491
$C_{18}H_{30}Ag_2Cl_2O_8$, 33B, 376
$C_{18}H_{30}BrNO$, 34B, 93
$C_{18}H_{30}Ca_2O_{12}Sr$, 32B, 492
$C_{18}H_{30}Cl_2$, 33B, 82
$C_{18}H_{30}Cl_4O_2Sn_2$, 26, 721
$C_{18}H_{30}FeN_8O_6$, 29, 554
$C_{18}H_{30}O_2$, 35B, 128
$C_{18}H_{30}N_2NiO_4P_2S_4$, 32B, 457
$C_{18}H_{32}ClCoN_6O_4$, 35B, 712
$C_{18}H_{32}Cl_2N_{12}O_4Zn$, 32B, 436
$C_{18}H_{32}O_{16}$, 27, 798
$C_{18}H_{32}O_{16}.5H_2O$, 35B, 270
$C_{18}H_{34}Cl_2N_2O_5S.H_2O$, 32B, 490
$C_{18}H_{34}O_2$, 27, 740; 28, 375
$C_{18}H_{35}BrO_2$, 31B, 22
$C_{18}H_{36}$, 33B, 83
$C_{18}H_{36}AuBr_2CuN_2S_4$, 35B, 810
$C_{18}H_{36}AuBrN_2S_4$, 35B, 806
$C_{18}H_{36}Au_2Br_2N_2S_4$, 33B, 493
$C_{18}H_{36}Cs_2N_2S_4$, 34B, 401
$C_{18}H_{38}$, 27, 743
$C_{18}H_{39}ClN_4NiO_6$, 35B, 772
$C_{18}H_{40}N_2O_4Pt_2$, 30B, 339
$C_{18}H_{42}Cl_4P_2Pt_2$, 34B, 606
$C_{18}H_{52}Cl_3CoN_6O_5$, 35B, 713
$C_{18}H_{54}AlN_3Si_6$, 34B, 370
$C_{18}H_{54}B_2N_4Si_6$, 34B, 374
$C_{18}H_{54}Li_3N_3Si_6$, 34B, 372

C_{19}

$C_{19}H_6BrN_7O_{10}.H_2O$, 30B, 196
$C_{19}H_{10}Fe_2N_2O_7$, 32B, 437, 497
$C_{19}H_{12}Br_2N_2$, 32B, 134
$C_{19}H_{12}Br_2O$, 34B, 80
$C_{19}H_{12}N_3O_6$, 27, 915; 32B, 74
$C_{19}H_{12}O_2$, 28, 447, 449
$C_{19}H_{13}BrN_2$, 28, 476
$C_{19}H_{13}BrOS_4$, 32B, 488; 34B, 193
$C_{19}H_{13}IS_4.CH_3O$, 30B, 149
$C_{19}H_{14}Br_2O_2$, 31B, 75
$C_{19}H_{14}I_2O$, 26, 734
$C_{19}H_{15}$, 30B, 403
$C_{19}H_{15}AuNP$, 34B, 604
$C_{19}H_{15}Br$, 26, 735; 31B, 82
$C_{19}H_{15}ClO_4$, 30B, 54
$C_{19}H_{15}Fe_2NO_{16}S$, 34B, 451
$C_{19}H_{15}MoN_3O_2S$, 35B, 527
$C_{19}H_{15}N$, 34B, 111
$C_{19}H_{15}NaO_4.C_3H_8O.(H_2O)$, 30B, 432
$C_{19}H_{16}$, 28, 677; 30B, 403

$C_{19}H_{16}Cl_4FeP$, 32B, 486
$C_{19}H_{16}FeO_2$, 34B, 454
$C_{19}H_{16}FeO_3$, 30B, 377
$C_{19}H_{17}NO_2$, 35B, 120
$C_{19}H_{17}N_3O_4S_2.HCl.H_2O$, 34B, 384
$C_{19}H_{17}P$, 33B, 545; 34B, 389
$C_{19}H_{18}O_6$, 27, 1006; 28, 562
$C_{19}H_{18}OPdS$, 34B, 444
$C_{19}H_{19}BrO_5$, 28, 563
$C_{19}H_{20}BrNO_4$, 33B, 220
$C_{19}H_{20}Co_6O_{11}S_5$, 33B, 490
$C_{19}H_{20}CuN_2O_2$, 35B, 727
$C_{19}H_{21}Br_2N_3O_3Si$, 29, 538
$C_{19}H_{21}CuNO_4$, 34B, 498
$C_{19}H_{21}CuN_3O_4$, 32B, 419
$C_{19}H_{21}NO_2$, 35B, 351
$C_{19}H_{22}N_2O$, 27, 1009
$C_{19}H_{22}N_3PPdS_2$, 35B, 840
$C_{19}H_{22}O$, 34B, 176
$C_{19}H_{23}BrO_4$, 35B, 129
$C_{19}H_{23}BrO_{10}S$, 30B, 183
$C_{19}H_{23}Cl_4N_3Sn_2$, 33B, 314
$C_{19}H_{23}NO_3$, 27, 1003
$C_{19}H_{24}BrN$, 34B, 127
$C_{19}H_{24}Br_2O_6$, 35B, 386
$C_{19}H_{24}ClNO_6$, 33B, 223
$C_{19}H_{24}FeNiO_3$, 35B, 585
$C_{19}H_{24}N_2NiO_3$, 32B, 495
$C_{19}H_{24}N_7O_{12}P.4H_2O$, 34B, 265
$C_{19}H_{24}O_2$, 28, 677
$C_{19}H_{24}O_3$, 30B, 437; 35B, 924
$C_{19}H_{24}O_5$, 28, 675
$C_{19}H_{24}O_3.H_2O$, 34B, 76
$C_{19}H_{25}BrN_2$, 35B, 354
$C_{19}H_{25}BrN_8O_8$, 30B, 200
$C_{19}H_{25}BrO_2$, 35B, 333
$C_{19}H_{25}Cl_2IrO_3S_2$, 32B, 493; 35B, 811
$C_{19}H_{25}ClN_2S$, 34B, 210
$C_{19}H_{25}IN_2O_2$, 35B, 353
$C_{19}H_{25}O_2Rh$, 34B, 467
$C_{19}H_{26}NO_7.C_{10}H_{14}BrO_4S.H_2O$, 27, 1002
$C_{19}H_{26}N_{10}O_4.H_2O$, 34B, 232
$C_{19}H_{26}O_2$, 30B, 437
$C_{19}H_{26}O_2.C_6H_5BrO$, 30B, 438
$C_{19}H_{27}BrO_2$, 29, 750; 27, 1031; 30B, 439
$C_{19}H_{28}Co_2N_2O$, 32B, 493; 34B, 455
$C_{19}H_{28}O_2$, 27, 1039; 30B, 437
$C_{19}H_{28}O_2.C_6H_5BrO$, 30B, 438
$C_{19}H_{29}BrO_3$, 35B, 334
$C_{19}H_{30}INO_2$, 30B, 432
$C_{19}H_{30}O$, 27, 1038
$C_{19}H_{30}O_2$, 27, 1038; 30B, 437; 31B, 201
$C_{19}H_{30}O_2.C_6H_5BrO$, 30B, 438
$C_{19}H_{31}Cl_2CrO_3$, 31B, 444; 32B, 448
$C_{19}H_{31}NO_2.BrH.CH_3OH$, 27, 1034
$C_{19}H_{31}NO_2,ClH.CH_3OH$, 27, 1034
$C_{19}H_{31}NO_2.HI.CH_3OH$, 27, 1034
$C_{19}H_{32}O_2$, 27, 1039
$C_{19}H_{34}$, 27, 1036
$C_{19}H_{38}O_2$, 35B, 7

C_{20}

$C_{20}H_4AmCsF_{24}O_8$, 34B, 500
$C_{20}H_4CsEuF_{24}O_8$, 34B, 500
$C_{20}H_4CsF_{24}O_8Y$, 31B, 383; 33B, 391
$C_{20}H_8Br_4O_4$, 33B, 93
$C_{20}H_{10}$, 32B, 486
$C_{20}H_{10}Cl_4$, 32B, 486
$C_{20}H_{10}CrO_5$, 34B, 426
$C_{20}H_{10}CuF_{12}N_2O_4$, 34B, 503
$C_{20}H_{10}F_6$, 30B, 417
$C_{20}H_{10}FeO_8$, 33B, 550
$C_{20}H_{10}Fe_2O_6$, 28, 767; 32B, 348
$C_{20}H_{10}Fe_2O_6S_2$, 32B, 458
$C_{20}H_{10}N_4$, 32B, 281
$C_{20}H_{10}N_6O_6$, 30B, 418
$C_{20}H_{10}O_4$, 26, 748; 34B, 103
$C_{20}H_{11}Mn_2O_8P$, 32B, 466
$C_{20}H_{12}$, 28, 451; 29, 694; 30B, 404
$C_{20}H_{12}Br_2N_2O$, 33B, 146
$C_{20}H_{12}Cl_4$, 35B, 103
$C_{20}H_{12}Fe_4O_{16}Sn_3$, 32B, 332
$C_{20}H_{12}IN_3$, 32B, 138
$C_{20}H_{12}N_2$, 27, 988; 28, 677; 30B, 143
$C_{20}H_{12}O_8Re_2Si$, 35B, 871
$C_{20}H_{13}Br$, 33B, 101
$C_{20}H_{13}OSi$, 34B, 373
$C_{20}H_{14}$, 26, 747; 35B, 149
$C_{20}H_{14}BrN$, 35B, 64
$C_{20}H_{14}CuN_2O_2$, 26, 670
$C_{20}H_{14}CuN_2O_2.CHCl_3$, 35B, 624
$C_{20}H_{14}N_4$, 30B, 216
$C_{20}H_{14}N_2NiO_2$, 26, 670
$C_{20}H_{15}OP$, 31B, 303
$C_{20}H_{15}PS$, 32B, 310
$C_{20}H_{16}$, 28, 446; 29, 696; 35B, 130
$C_{20}H_{16}CaO_8P_2.3H_2O$, 30B, 188
$C_{20}H_{16}Cl_2Fe_2$, 27, 861; 30B, 310; 31B, 354
$C_{20}H_{16}CuI_2N_4$, 26, 657; 28, 643
$C_{20}H_{16}CuN_6O_5$, 34B, 569
$C_{20}H_{16}CuO_4$, 31B, 383
$C_{20}H_{16}Fe$, 34B, 473
$C_{20}H_{16}Fe_2$, 34B, 456
$C_{20}H_{16}N_6Pd$, 30B, 354
$C_{20}H_{16}O_2$, 35B, 215
$C_{20}H_{16}O_4Ru_3$, 33B, 378
$C_{20}H_{17}Cl_3O_5$, 29, 772
$C_{20}H_{17}N_4$, 34B, 56
$C_{20}H_{17}NO_5$, 35B, 356
$C_{20}H_{18}Cl_2$, 31B, 35
$C_{20}H_{18}Cu_2N_6O_8$, 35B, 728
$C_{20}H_{18}Fe_2$, 27, 861; 29, 549
$C_{20}H_{18}GeO$, 33B, 306
$C_{20}H_{18}N_6$, 32B, 282
$C_{20}H_{18}N_2O_2Sn$, 32B, 333
$C_{20}H_{18}O_7$, 35B, 132
$C_{20}H_{18}O_4Pd$, 32B, 387
$C_{20}H_{18}O_5V$, 30B, 340
$C_{20}H_{19}AsF_6N_2O_2$, 34B, 150
$C_{20}H_{19}BrN_2O$, 32B, 170
$C_{20}H_{19}I_2NiPS_2$, 34B, 388
$C_{20}H_{19}NO_5$, 33B, 224
$C_{20}H_{20}$, 32B, 144
$C_{20}H_{20}B_2CuF_8N_4O_4$, 34B, 578
$C_{20}H_{20}BrNO_3$, 35B, 357
$C_{20}H_{20}Cl_4Co_3$, 35B, 922
$C_{20}H_{20}Cl_4Co_2Zn$, 35B, 927
$C_{20}H_{20}Cl_2CuN_4O_{12}$, 34B, 580
$C_{20}H_{20}Cl_4O_4Ti_4$, 35B, 554
$C_{20}H_{20}Co_4S_6.{}^1/_2CHCl_3$, 34B, 457
$C_{20}H_{20}CuN_2O_2$, 33B, 446
$C_{20}H_{20}Fe_4S_4$, 31B, 356
$C_{20}H_{20}N_2$, 35B, 918
$C_{20}H_{20}Ti$, 35B, 920
$C_{20}H_{20}Zr$, 35B, 556
$C_{20}H_{21}BrO_6$, 29, 751
$C_{20}H_{21}Cl_2N_6O_6Rh$, 32B, 497
$C_{20}H_{21}O_2Sb$, 33B, 290
$C_{20}H_{21}Rh_3$, 33B, 351
$C_{20}H_{22}Br_2O_5$, 32B, 265
$C_{20}H_{22}CaO_{11}P_2$, 29, 770
$C_{20}H_{22}Cl_4Cu_2N_2O_8$, 35B, 774
$C_{20}H_{22}Fe_2N_2O_4$, 35B, 557
$C_{20}H_{22}N_2O_2$, 35B, 923
$C_{20}H_{22}N_2O_3$, 28, 677
$C_{20}H_{22}N_2O_5$, 26, 627
$C_{20}H_{23}BrO_6$, 27, 1017, 31B, 223
$C_{20}H_{23}NO_5.HBr.H_2O$, 35B, 199
$C_{20}H_{24}BrNO_2$, 35B, 348
$C_{20}H_{24}BrN_3O_6$, 34B, 274
$C_{20}H_{24}Cl_5NO_3$, 34B, 166
$C_{20}H_{24}CoN_2O_4$, 33B, 392
$C_{20}H_{24}CuN_2O_2$, 31B, 393; 34B, 517
$C_{20}H_{24}INO_4$, 32B, 491
$C_{20}H_{24}N_2NiO_2$, 29, 568; 30B, 432
$C_{20}H_{24}N_2NiO_4$, 33B, 393
$C_{20}H_{24}N_2NiO_6$, 33B, 394
$C_{20}H_{24}N_4O_{13}$, 30B, 432
$C_{20}H_{24}N_8S_6V$, 32B, 485
$C_{20}H_{24}O_3$, 28, 677
$C_{20}H_{25}ClO_6$, 30B, 265
$C_{20}H_{25}Co_5O_{10}S_5$, 33B, 490
$C_{20}H_{25}IO_3$, 30B, 102
$C_{20}H_{25}N_4Na_3O_{15}P_2.12H_2O$, 30B, 432
$C_{20}H_{25}P_5$, 29, 637
$C_{20}H_{26}Br_2O$, 34B, 345
$C_{20}H_{26}CuN_4O_2.2H_2O$, 32B, 438
$C_{20}H_{26}Mn_3N_4O_{16}.10H_2O$, 29, 544
$C_{20}H_{26}N_2O_4.CH_3I$, 27, 1021
$C_{20}H_{26}O_2$, 28, 677; 33B, 214

$C_{20}H_{27}BrN_2O$, 34B, 75
$C_{20}H_{27}BrO_3$, 33B, 192
$C_{20}H_{27}IN_2$, 29, 746
$C_{20}H_{27}IN_2O_2$, 27, 1008; 30B, 235
$C_{20}H_{28}BrNO_3$, 28, 552
$C_{20}H_{28}BrN_2O_{4.5}$, 26, 793
$C_{20}H_{28}CeO_8$, 28, 622; 34B, 503
$C_{20}H_{28}CuN_2O_2.{}^1/_2C_6H_{12}$, 30B, 424
$C_{20}H_{28}CuN_2O_2.{}^1/_3C_6H_{12}$, 30B, 424
$C_{20}H_{28}CuN_2O_2.{}^2/_3C_9H_{12}$, 30B, 424
$C_{20}H_{28}O_3$, 30B, 439
$C_{20}H_{28}O_4P_2Zn$, 34B, 582
$C_{20}H_{28}O_4Pt_2$, 32B, 399
$C_{20}H_{28}O_8U$, 35B, 608
$C_{20}H_{28}O_8Zr$, 28, 619
$C_{20}H_{29}Al_2NO_2$, 35B, 916
$C_{20}H_{29}IN_2O$, 34B, 328
$C_{20}H_{30}$, 33B, 206
$C_{20}H_{30}Cl_4Co_2N_{10}O_{16}$, 29, 365
$C_{20}H_{30}Cl_3OP_2Re$, 28, 656
$C_{20}H_{30}CuN_4O_4$, 32B, 438
$C_{20}H_{30}Mn_2O_8P_2$, 33B, 509
$C_{20}H_{30}NO_7Yb$, 35B, 612, 614
$C_{20}H_{30}O$, 26, 808
$C_{20}H_{30}O_2$, 28, 377
$C_{20}H_{30}O_2S$, 27, 1002
$C_{20}H_{32}$, 26, 799; 34B, 315; 35B, 349
$C_{20}H_{32}As_4AuI_3$, 34B, 606
$C_{20}H_{32}As_4Cl_3CoO_4$, 35B, 869
$C_{20}H_{32}As_4Cl_2CoO_8$, 32B, 468
$C_{20}H_{32}As_2Cl_3N_2Rh$, 35B, 868
$C_{20}H_{32}As_8Cl_2Pt$, 29, 578
$C_{20}H_{32}As_4Cl_4Ti$, 26, 687
$C_{20}H_{32}As_4I_2Ni$, 27, 869
$C_{20}H_{32}As_8I_2Ni$, 29, 576
$C_{20}H_{32}As_4I_2Pd$, 27, 869
$C_{20}H_{32}As_4I_2Pt$, 27, 869
$C_{20}H_{32}Cl_4CuN_2$, 32B, 35
$C_{20}H_{32}Co_2O_{12}$, 31B, 384
$C_{20}H_{33}N_3$, 31B, 244; 32B, 171
$C_{20}H_{36}AgNO_3$, 32B, 378
$C_{20}H_{36}BrNO_3$, 26, 798
$C_{20}H_{36}ClNO_3$, 26, 798
$C_{20}H_{36}INO_3$, 26, 798
$C_{20}H_{36}O_2Sn$, 33B, 318
$C_{20}H_{38}FeSi_4$, 33B, 353
$C_{20}H_{40}F_8N_4NiP_4$, 28, 677
$C_{20}H_{40}N_4O_3Re_2S_8$, 35B, 917
$C_{20}H_{40}N_4Pb_2S_8$, 32B, 492
$C_{20}H_{40}N_4S_8Th$, 35B, 813
$C_{20}H_{40}O_2$, 26, 744; 33B, 10
$C_{20}H_{42}$, 35B, 907
$C_{20}H_{42}Cl_2N_2P_2Pt_2S_2$, 35B, 814
$C_{20}H_{42}N_2P_2PdS_2$, 28, 677
$C_{20}H_{44}B_4S_4$, 31B, 296
$C_{20}H_{44}IP$, 35B, 465
$C_{20}H_{46}Mn_3N_4O_{26}$, 26, 692
$C_{20}H_{62}B_{18}CuN_2$, 32B, 297
$C_{20}H_{120}FNO_{38}$, 26, 613

C_{21}

$C_{21}H_6Co_6O_{19}$, 34B, 430
$C_{21}H_{10}CoMnO_9Sn$, 33B, 316
$C_{21}H_{11}N_5NiO_6.2H_2O$, 30B, 363
$C_{21}H_{12}Cl_3N_3$, 30B, 135
$C_{21}H_{12}N_4$, 34B, 159
$C_{21}H_{12}N_2O_3$, 35B, 201
$C_{21}H_{14}BrNO_3$, 32B, 114
$C_{21}H_{15}Cl_3CuN_2O_2$, 35B, 624
$C_{21}H_{15}ClO_4$, 31B, 60
$C_{21}H_{15}ClO_4S_2$, 27, 670
$C_{21}H_{15}ClO_6Sn.CHCl_3$, 32B, 493
$C_{21}H_{15}MnNO_4P$, 33B, 429
$C_{21}H_{15}N_3$, 30B, 134
$C_{21}H_{16}CrO_5$, 31B, 372
$C_{21}H_{16}O_9Ru_3$, 35B, 914
$C_{21}H_{17}N$, 35B, 64
$C_{21}H_{18}FeN_3O_6.3H_2O$, 34B, 583
$C_{21}H_{18}Fe_2O$, 31B, 359
$C_{21}H_{18}FeORu$, 29, 551
$C_{21}H_{18}Mo_2O_6$, 33B, 379
$C_{21}H_{19}CoN_3O_2$, 35B, 625
$C_{21}H_{20}BrN_3.CH_4O$, 34B, 160
$C_{21}H_{20}BrNO_7S$, 35B, 273
$C_{21}H_{20}Br_2O_8$, 29, 770; 35B, 103
$C_{21}H_{21}As$, 27, 869; 28, 588
$C_{21}H_{21}Cl_2NiO_4PS_3$, 34B, 607
$C_{21}H_{21}P$, 33B, 545
$C_{21}H_{22}BrNO_4$, 34B, 311
$C_{21}H_{22}INO_3$, 35B, 359
$C_{21}H_{23}NO_5$, 33B, 226
$C_{21}H_{23}NO_3.HBr$, 34B, 168
$C_{21}H_{24}B_3N_3$, 29, 766
$C_{21}H_{24}BrNO_3$, 34B, 168
$C_{21}H_{24}BrO_2$, 31B, 202
$C_{21}H_{24}F_3N_3S$, 30B, 432
$C_{21}H_{24}IO_2N_2$, 33B, 549
$C_{21.5}H_{24}O_{10.5}Sn$, 35B, 751
$C_{21}H_{25}IN_2O$, 34B, 330
$C_{21}H_{26}BrN.H_2O$, 29, 503
$C_{21}H_{26}BrNO_5$, 34B, 332
$C_{21}H_{26}ClNO_7$, 33B, 223
$C_{21}H_{27}BrN_2O_2$, 32B, 491
$C_{21}H_{28}BrO_3$, 33B, 193
$C_{21}H_{28}O_5$, 27, 1037
$C_{21}H_{29}BrN_2O_3S$, 31B, 216
$C_{21}H_{29}BrO_2$, 34B, 293
$C_{21}H_{29}BrO_3$, 30B, 439; 32B, 243, 490
$C_{21}H_{29}BrO_3S$, 29, 771; 35B, 386
$C_{21}H_{29}BrO_4S$, 34B, 312
$C_{21}H_{29}Cl_5O_9Ti_2$, 32B, 388
$C_{21}H_{29}IN_2O_4$, 26, 830
$C_{21}H_{29}N_3O_6$, 26, 809
$C_{21}H_{30}O_3$, 30B, 438, 439, 440
$C_{21}H_{30}O_5$, 27, 1038
$C_{21}H_{31}BrO_3$, 34B, 294
$C_{21}H_{31}BrO_4$, 33B, 194; 34B, 296
$C_{21}H_{32}BrNO_3.CH_4O$, 29, 555
$C_{21}H_{32}O_2$, 30B, 440
$C_{21}H_{32}O_3$, 27, 1038
$C_{21}H_{32}O_5$, 27, 1038
$C_{21}H_{34}INO$, 28, 553
$C_{21}H_{34}O_2$, 27, 1038
$C_{21}H_{34}O_3$, 30B, 440
$C_{21}H_{34}O_5$, 27, 1038
$C_{21}H_{35}N_3$, 28, 684; 29, 747
$C_{21}H_{36}IN_3.{}^1/_2CH_4O$, 32B, 171
$C_{21}H_{36}Mo_3O_{15}$, 32B, 390
$C_{21}H_{36}O_2$, 28, 679
$C_{21}H_{36}O_3$, 28, 490; 30B, 440
$C_{21}H_{36}O_4$, 30B, 440
$C_{21}H_{37}Na_3O_{26}Yb$, 35B, 644
$C_{21}H_{42}O_2$, 33B, 10
$C_{21}H_{42}O_4$, 30B, 12

C_{22}

$C_{22}Co_6O_{18}$, 35B, 504
$C_{22}H_{10}F_6$, 30B, 417
$C_{22}H_{10}Fe_2O_8$, 33B, 339
$C_{22}H_{10}N_4$, 30B, 417; 31B, 278; 33B, 250
$C_{22}H_{10}N_6O_6$, 30B, 416
$C_{22}H_{12}O_2$, 34B, 111
$C_{22}H_{13}N_3O_6$, 31B, 279
$C_{22}H_{14}$, 26, 756; 27, 938
$C_{22}H_{14}O_9Ru_4$, 34B, 475
$C_{22}H_{15}ClFe_2MoO_7Sn$, 32B, 494
$C_{22}H_{16}Cl_4O_6Sn$, 32B, 493; 35B, 751
$C_{22}H_{16}O_{10}Ru_4$, 35B, 916
$C_{22}H_{17}Cl$, 34B, 121
$C_{22}H_{17}IN_2O_5$, 33B, 226
$C_{22}H_{17}N_2O_3.{}^1/_2HgCl_4.H_2O$, 31B, 246
$C_{22}H_{18}Fe$, 32B, 361; 35B, 558
$C_{22}H_{19}BrO_6$, 35B, 388
$C_{22}H_{19}CuN_3O_5$, 32B, 495
$C_{22}H_{20}Cr_2N_2O_2S_2$, 33B, 354
$C_{22}H_{20}FeN_6S_2$, 32B, 440
$C_{22}H_{20}N_4$, 33B, 250
$C_{22}H_{20}N_6$, 30B, 249
$C_{22}H_{20}N_2O_8S_2$, 33B, 553
$C_{22}H_{20}O_4$, 30B, 110
$C_{22}H_{21}CuNO_4$, 34B, 512
$C_{22}H_{22}BrN_3O_8S.{}^1/_2(C_6H_6)$, 27, 1015
$C_{22}H_{22}Br_2O_3.4H_2O$, 30B, 55
$C_{22}H_{22}Cl_2O_3.4H_2O$, 30B, 55
$C_{22}H_{22}CuO_4$, 27, 844; 30B, 337
$C_{22}H_{23}NO_2$, 34B, 99
$C_{22}H_{24}BrN_3O$, 34B, 160
$C_{22}H_{24}Br_2O_6$, 33B, 551
$C_{22}H_{24}Cl_2N_2O_8$, 28, 534
$C_{22}H_{24}INO_4$, 31B, 249
$C_{22}H_{24}N_2NiO_2$, 27, 843
$C_{22}H_{24}N_2O_4.HCl$, 30B, 221
$C_{22}H_{24}O_5Zn$, 34B, 504
$C_{22}H_{25}BrO_4$, 29, 772; 31B, 218
$C_{22}H_{25}BrO_8.{}^1/_2C_6H_6$, 32B, 490

$C_{22}H_{25}Cl_4FeO_6$, 30B, 242

$C_{22}H_{25}ClN_2O_4$, 30B, 221

$C_{22}H_{25}NO_4Si$, 33B, 271

$C_{22}H_{26}N_2O_3$, 28, 679

$C_{22}H_{26}N_2O_2Pd$, 32B, 400

$C_{22}H_{26}O_4$, 28, 677

$C_{22}H_{27}BrN_2O_4.H_2O$, 30B, 233

$C_{22}H_{27}BrO_3$, 35B, 335

$C_{22}H_{27}BrO_4$, 35B, 389

$C_{22}H_{27}IN_2O_3$, 28, 558

$C_{22}H_{27}IN_2O_3.H_2O$, 28, 557

$C_{22}H_{28}$, 33B, 553

$C_{22}H_{28}Br_2N_4$, 30B, 232

$C_{22}H_{28}Br_2N_4.2H_2O$, 27, 1013

$C_{22}H_{28}Br_2O_4$, 34B, 343

$C_{22}H_{28}CoN_2O_2$, 26, 678

$C_{22}H_{28}CuN_2O_2$, 26, 678; 27, 872; 31B, 395; 34B, 518

$C_{22}H_{28}N_2NiO_2$, 26, 678; 27, 872; 31B, 397

$C_{22}H_{28}N_2O_2Pd$, 27, 872; 29, 582; 33B, 399

$C_{22}H_{28}N_2O_2Zn$, 26, 678

$C_{22}H_{29}BrN_2O_4.CH_3OH$, 27, 1010

$C_{22}H_{29}BrO_3$, 34B, 299

$C_{22}H_{29}BrO_6$, 32B, 251

$C_{22}H_{29}ClN_2O_4$, 26, 792; 27, 1012

$C_{22}H_{29}ClO_7$, 31B, 264

$C_{22}H_{29}IN_2O_4$, 26, 792; 27, 1011

$C_{22}H_{29}IO_6$, 31B, 225

$C_{22}H_{29}N_2O_2Zn$, 26, 679

$C_{22}H_{29}N_2S_2$, 30B, 432

$C_{22}H_{29}N_3S_2$, 35B, 242

$C_{22}H_{30}CuN_2O_2.2H_2O$, 33B, 447

$C_{22}H_{30}O_3$, 35B, 336

$C_{22}H_{31}BrO_3$, 35B, 337

$C_{22}H_{31}IN_2O$, 30B, 229

$C_{22}H_{32}NO_5$, 33B, 555

$C_{22}H_{32}O_3$, 30B, 439

$C_{22}H_{33}BrN_2O_6$, 26, 792

$C_{22}H_{33}BrO_6S$, 26, 800

$C_{22}H_{33}NO_5.HBr.H_2O$, 30B, 234

$C_{22}H_{33}O_3$, 30B, 441

$C_{22}H_{34}BrNO_5.H_2O$, 30B, 234

$C_{22}H_{34}N_4O_4S_6Te$, 26, 583

$C_{22}H_{35}NO_2.BrH.H_2O$, 27, 1003

$C_{22}H_{38}NiO_4$, 31B, 386

$C_{22}H_{38}O_4Zn$, 29, 616

$C_{22}H_{42}Mo_2O_6P_4$, 32B, 469

$C_{22}H_{48}N_{10}NiS_6$, 34B, 522

$C_{22}H_{49}I_3S_2Zn$, 34B, 47

$C_{22}H_{52}Fe_2N_6O_{21}$, 32B, 441

C_{23}

$C_{23}H_{10}Fe_3O_9$, 29, 767; 31B, 347

$C_{23}H_{15}MnO_5Sn$, 32B, 334

$C_{23}H_{16}AgN_3O_2$, 33B, 443

$C_{23}H_{16}CoF_4O_3P$, 32B, 470

$C_{23}H_{17}ClN_2O_2Ti$, 35B, 559

$C_{23}H_{19}IO_2$, 35B, 117

$C_{23}H_{20}Cl_3GeNiP$, 35B, 921

$C_{23}H_{20}CuP$, 35B, 561

$C_{23}H_{20}FeIO_2P$, 33B, 340

$C_{23}H_{22}N_2OS_2$, 30B, 154

$C_{23}H_{22}O_6$, 30B, 432

$C_{23}H_{23}BrO_7$, 34B, 346

$C_{23}H_{23}IO_8$, 30B, 433; 31B, 489

$C_{23}H_{23}O_2PS$, 30B, 277

$C_{23}H_{24}BrNO_6$, 33B, 213

$C_{23}H_{24}Br_2N_4O_3S$, 30B, 242

$C_{23}H_{24}O_2Sn$, 33B, 317

$C_{23}H_{25}BrO_5$, 31B, 220

$C_{23}H_{28}FeO_4$, 35B, 529

$(C_{23}H_{23}N_3O)I$, 27, 961

$C_{23}H_{29}IN_2O_3$, 30B, 432

$C_{23}H_{29}N_3O_2S_2$, 32B, 488

$C_{23}H_{29}O_5P$, 32B, 311

$C_{23}H_{31}BrN_4O_7$, 34B, 276

$C_{23}H_{31}IN_2O_4$, 29, 772; 30B, 237

$C_{23}H_{32}N_4O_7$, 26, 809; 27, 1051

$C_{23}H_{32}O_3$, 35B, 338

$C_{23}H_{32}O_6$, 30B, 440

$C_{23}H_{33}BrO_5$, 33B, 194

$C_{23}H_{33}BrO_6$, 34B, 314

$C_{23}H_{33}BrO_3.C_7H_{16}$, 30B, 439

$C_{23}H_{33}IN_2O$, 32B, 491

$C_{23}H_{34}O_3$, 30B, 439

$C_{23}H_{34}O_4$, 30B, 440; 34B, 298

$C_{23}H_{36}BrNO_4.2H_2O$, 32B, 491

$C_{23}H_{36}O_3$, 30B, 440

$C_{23}H_{37}BrO_3$, 35B, 338

$C_{23}H_{37}NO_6.HBr$, 35B, 361

$C_{23}H_{38}BrNO_6$, 34B, 330; 35B, 361

$C_{23}H_{40}B_{18}CuP$, 33B, 267

$C_{23}H_{120}NO_{41.5}$, 27, 778

C_{24}

$C_{24}H_{10}CuN_2O_2$, 30B, 433

$C_{24}H_{12}$, 26, 739; 30B, 111; 35B, 133

$C_{24}H_{12}AsN_3O_6$, 33B, 285

$C_{24}H_{12}Cl_4N_2O_4Pd$, 30B, 262

$C_{24}H_{12}CuN_8O_8$, 30B, 263

$C_{24}H_{12}O_6$, 30B, 254

$C_{24}H_{12}O_{14}Ru$, 33B, 547

$C_{24}H_{14}Cl_4N_2O_4$, 32B, 283

$C_{24}H_{15}As$, 27, 916

$C_{24}H_{15}P$, 27, 916; 32B, 314

$C_{24}H_{15}Sb$, 27, 916

$C_{24}H_{16}Br_6Ga_2N_4$, 35B, 906

$C_{24}H_{16}Cl_2HgO_2S_2$, 34B, 589

$C_{24}H_{16}N_2O_2$, 30B, 152

$C_{24}H_{17}BrO_2$, 32B, 151

$C_{24}H_{17}NO_3Si$, 33B, 270

$C_{24}H_{18}Br_2CuN_2O_2$, 30B, 433

$C_{24}H_{18}CuI_2N_2O_2$, 30B, 433

$C_{24}H_{18}N_5O_6$, 32B, 73

$C_{24}H_{18}O_6.C_6H_5Br$, 34B, 100

$C_{24}H_{18}O_2Ru$, 30B, 433

$C_{24}H_{19}P_3$, 31B, 306

$C_{24}H_{20}$, 33B, 105

$C_{24}H_{20}As_2O$, 28, 592

$C_{24}H_{20}BRb$, 26, 735; 27, 884

$C_{24}H_{20}Cl_2N_3P_3$, 31B, 307

$C_{24}H_{20}CoO_2P_2S_2$, 35B, 810

$C_{24}H_{20}F_3NiP$, 35B, 530

$C_{24}H_{20}Fe_2O_4Sn$, 33B, 549; 34B, 414

$C_{24}H_{20}Ge$, 27, 875

$C_{24}H_{20}N_4$, 28, 679

$C_{24}H_{20}N_8O_8Zr$, 35B, 748

$C_{24}H_{20}Ni_2$, 33B, 341

$C_{24}H_{20}NiP_2S_4$, 33B, 495

$C_{24}H_{20}Pb$, 27, 875

$C_{24}H_{20}Si$, 27, 875

$C_{24}H_{20}Sn$, 27, 875; 35B, 490

$C_{24}H_{21}AsN_2O_6$, 32B, 492

$C_{24}H_{21}FeS_7$, 34B, 599

$C_{24}H_{21}InS_6$, 35B, 816

$C_{24}H_{21}OSb$, 34B, 396

$C_{24}H_{22}AsBr_4MoO_2$, 32B, 317

$C_{24}H_{22}Fe_2O_2$, 27, 861

$C_{24}H_{22}Fe_2O_4$, 27, 861; 33B, 355

$C_{24}H_{22}N_4$, 35B, 415

$C_{24}H_{22}Nb_2O_{10}$, 35B, 562

$C_{24}H_{22}O_7$, 35B, 923

$C_{24}H_{22}O_4S$, 26, 775

$C_{24}H_{24}As_4F_8O_8Ru_3$, 35B, 845

$C_{24}H_{24}CoN_2O_4$, 30B, 433

$C_{24}H_{24}CuN_2O_2$, 30B, 433

$C_{24}H_{24}Fe_2$, 32B, 494

$C_{24}H_{24}IN_3O_5.C_2H_6O$, 31B, 250

$C_{24}H_{24}Ti_2$, 33B, 380

$C_{24}H_{26}AsCl_4O_3Ru$, 31B, 315

$C_{24}H_{26}Fe_2$, 27, 861; 30B, 308; 31B, 361

$C_{24}H_{26}N_2O_8S_2$, 33B, 553

$C_{24}H_{27}As$, 27, 869; 28, 591

$C_{24}H_{27}B_2F_8FeN_7$, 35B, 915

$C_{24}H_{28}Br_2O_8$, 27, 997

$C_{24}H_{29}I$, 34B, 25

$C_{24}H_{30}Cl_6Cu_4N_4O_2$, 35B, 775

$C_{24}H_{30}N_4O_3$, 35B, 423

$C_{24}H_{31}BrO_8$, 29, 771; 31B, 227

$C_{24}H_{31}IN_2O_6$, 33B, 228

$C_{24}H_{32}CuN_2O_2$, 29, 609; 35B, 730

$C_{24}H_{32}N_2NiO$, 32B, 403

$C_{24}H_{32}N_2O_2Pd$, 32B, 404

$C_{24}H_{33}B_3N_3O_3$, 35B, 923

$C_{24}H_{33}Br_4Ni_{1.5}P_3.C_6H_6$, 34B, 601

$C_{24}H_{33}BrO_8$, 27, 1029

$C_{24}H_{34}Cl_2Pt_2$, 35B, 579

$C_{24}H_{34}K_2O_{12}$, 35B, 112

$C_{24}H_{34}N_4NiO_2$, 32B, 406

$C_{24}H_{35}IO_4$, 31B, 229

$C_{24}H_{35}NO_8.HI.3H_2O$, 29, 749

$C_{24}H_{36}Cl_6Pt_3Sn_2$, 33B, 547

$C_{24}H_{36}CuN_5S_4$, 29, 560

$C_{24}H_{36}FeN_5S_4$, 32B, 459

$C_{24}H_{36}INO_4$, 31B, 254

$C_{24}H_{36}N_3NiS_2$, 35B, 817

$C_{24}H_{36}O_3$, 30B, 439
$C_{24}H_{39}ClN_2P_2Pd$, 33B, 548; 35B, 842
$C_{24}H_{39}O_5Na$, 30B, 441
$C_{24}H_{40}O_4$, 30B, 441
$C_{24}H_{40}O_4.C_2H_5OH$, 30B, 441
$C_{24}H_{40}O_5.4H_2O$, 30B, 441
$C_{24}H_{42}Cl_2Cu_2$, 35B, 581
$C_{24}H_{42}INO_7$, 26, 795
$C_{24}H_{42}INO_{11}$, 26, 793
$C_{24}H_{42}N_2O_6$, 34B, 270
$C_{24}H_{42}O$, 30B, 441
$C_{24}H_{43}NO_5$, 34B, 346
$C_{24}H_{44}N_4O_4$, 35B, 181
$C_{24}H_{44}N_4O_8$, 26, 637
$C_{24}H_{44}P_4$, 34B, 392
$C_{24}H_{44}P_2S_2$, 29, 762
$C_{24}H_{50}N_8O_{50}Zn_{13}$, 26, 738
$C_{24}H_{56}Cr_2Li_4O_4$, 35B, 507
$C_{24}H_{60}Ni_6S_{12}$, 30B, 399
$C_{24}H_{60}O_4Si_5$, 26, 616
$C_{24}H_{72}Cl_2CoN_{12}O_{17}P_6$, 35B, 742
$C_{24}H_{72}Cl_2CuN_{12}O_{17}P_6$, 35B, 742
$C_{24}H_{72}Cl_3Cu_2N_{18}P_6$, 35B, 916
$C_{24}H_{72}Cl_2MgN_{12}O_{17}P_6$, 35B, 742
$C_{24}H_{72}Cl_2O_{20}S_{12}Th$, 30B, 410
$C_{24}H_{72}N_{18}P_6$, 33B, 282

C_{25}

$C_{25}H_{10}Fe_2O_8$, 34B, 427
$C_{25}H_{15}N_6$, 31B, 136
$C_{25}H_{17}MoO_4P$, 34B, 451
$C_{25}H_{18}FeO_3$, 35B, 582, 912
$C_{25}H_{19}MoO_3P$, 32B, 494
$C_{25}H_{20}FeO_2Sn$, 32B, 335
$C_{25}H_{20}OS$, 32B, 189
$C_{25}H_{21}CrO_5P$, 34B, 433
$C_{25}H_{22}N_4O$, 30B, 253
$C_{25}H_{22}N_4O_3$, 26, 754
$C_{25}H_{23}IO_8$, 27, 1028
$C_{25}H_{23}O_2P$, 34B, 392
$C_{25}H_{23}OSb$, 33B, 290
$C_{25}H_{23}P$, 35B, 465
$C_{25}H_{25}BrN_3O_8S$, 32B, 240
$C_{25}H_{25}Co_2$, 26, 653; 30B, 302
$C_{25}H_{25}Cu_2N_9O_{12}$, 35B, 912
$C_{25}H_{26}O_3$, 28, 677
$C_{25}H_{27}BrO_3$, 33B, 195
$C_{25}H_{29}BrO_8.0.8(CHCl_3)$, 33B, 215
$C_{25}H_{29}N_4O_{10}P$, 35B, 416
$C_{25}H_{30}Br_{11}N_5Sb_2$, 35B, 876
$C_{25}H_{31}BrO_2$, 32B, 490
$C_{25}H_{31}BrO_3$, 30B, 438
$C_{25}H_{32}BrN_3O_4.H_2O$, 30B, 236; 35B, 390
$C_{25}H_{32}IN_3O_4.H_2O$, 30B, 236; 35B, 390
$C_{25}H_{33}BrO_3$, 30B, 438; 34B, 358
$C_{25}H_{33}O_2$, 34B, 55
$C_{25}H_{34}Br_2Cl_2O_8S_2$, 33B, 190
$C_{25}H_{35}BrO_3$, 30B, 438
$C_{25}H_{38}Br_2O_3$, 34B, 300, 301
$C_{25}H_{40}O_4$, 30B, 441
$C_{25}H_{42}BrO_4$, 27, 1039
$C_{25}H_{42}O_3$, 30B, 442
$C_{25}H_{42}O_5.C_2H_5OH$, 30B, 442
$C_{25}H_{43}Br_3N_2Ni$, 33B, 430
$C_{25}H_{48}O_5S_2$, 28, 379

C_{26}

$C_{26}H_{10}AuO_3PW$, 34B, 463
$C_{26}H_{12}.C_6F_4O_2$, 28, 576
$C_{26}H_{12}F_6$, 30B, 417
$C_{26}H_{12}N_4$, 30B, 417; 32B, 284; 35B, 417
$C_{26}H_{12}N_6O_6$, 30B, 416
$C_{26}H_{12}O_6$, 30B, 423
$C_{26}H_{14}N_4$, 33B, 252
$C_{26}H_{15}AuClF_{12}PS_4$, 33B, 496
$C_{26}H_{15}O_5$, 31B, 265
$C_{26}H_{16}$, 27, 938
$C_{26}H_{16}CuN_2O_4$, 34B, 572
$C_{26}H_{16}N_2O_4Pd$, 27, 883
$C_{26}H_{16}O_2$, 28, 491
$C_{26}H_{18}CuN_2O_2$, 28, 679; 31B, 398
$C_{26}H_{18}CuN_2O_2.{}^1/_4(CCl_3H)$, 27, 873
$C_{26}H_{18}CuN_2O_2.CCl_3H$, 27, 873
$C_{26}H_{18}N_2O_2$, 32B, 132, 133
$C_{26}H_{20}As_2$, 26, 736
$C_{26}H_{20}BrOP$, 30B, 275
$C_{26}H_{20}ClOP$, 30B, 275
$C_{26}H_{20}CuN_2O_2$, 29, 597
$C_{26}H_{20}IOP$, 30B, 275
$C_{26}H_{20}N_2NiO_2$, 27, 843
$C_{26}H_{20}Ni$, 35B, 921
$C_{26}H_{20}Ni_2$, 30B, 433
$C_{26}H_{20}P_2$, 26, 736; 34B, 390
$C_{26}H_{22}CuN_2O_2$, 28, 645
$C_{26}H_{22}Se_2$, 34B, 52
$C_{26}H_{23}AsBrNO$, 31B, 415
$C_{26}H_{24}CuN_8S_2$, 26, 663
$C_{26}H_{24}F_6IrO_2P_5$, 33B, 546
$C_{26}H_{25}Cl_2NP_2Pd$, 30B, 387
$C_{26}H_{26}Cu_2N_2O_8$, 33B, 412; 35B, 922
$C_{26}H_{26}O_2$, 33B, 105
$C_{26}H_{28}Br_2N_{12}O_3$, 33B, 254
$C_{26}H_{30}BrNO$, 35B, 97
$C_{26}H_{30}N_4Ni$, 35B, 715
$C_{26}H_{30}O_5$, 27, 1032
$C_{26}H_{30}O_8$, 26, 801
$C_{26}H_{31}BrO_3$, 35B, 337, 341
$C_{26}H_{31}BrO_5$, 33B, 196
$C_{26}H_{32}Hg_5N_{10}O_4$, 35B, 918
$C_{26}H_{33}BrO_4S$, 32B, 245
$C_{26}H_{33}N_2NiP_3$, 34B, 609
$C_{26}H_{34}O_3$, 30B, 438
$C_{26}H_{36}BrNO_4$, 32B, 260
$C_{26}H_{36}Cl_2N_4NiO_2$, 31B, 400
$C_{26}H_{38}B_2Cl_2N_2$, 35B, 435
$C_{26}H_{39}BrO_5$, 30B, 244; 33B, 217
$C_{26}H_{40}Al_2Fe_2O_4$, 34B, 459
$C_{26}H_{42}MnN_4O_{12}$, 28, 639
$C_{26}H_{46}O_5$, 27, 1039
$C_{26}H_{62}Si_7$, 32B, 491

C_{27}

$C_{27}H_{18}N_2$, 33B, 127
$C_{27}H_{18}N_4NiPS_4$, 31B, 452
$C_{27}H_{19}Fe_2O_6P$, 33B, 546
$C_{27}H_{19}N_3O_5U$, 32B, 497
$C_{27}H_{19}N_3O_5U.CHCl_3$, 32B, 442
$C_{27}H_{21}N_3.CHI_3$, 27, 986
$C_{27}H_{23}MoO_3P$, 33B, 356
$C_{27}H_{25}BrO_3$, 33B, 198
$C_{27}H_{27}BrO_7$, 35B, 391
$C_{27}H_{27}BrO_{10}.2C_2H_6O$, 32B, 490
$C_{27}H_{32}BrNO_5.CH_3OH$, 30B, 237
$C_{27}H_{33}BrO_6$, 34B, 317
$C_{27}H_{37}BrO_8$, 35B, 392
$C_{27}H_{37}Cl_3NOP_2Re$, 33B, 511
$C_{27}H_{38}BrNO_7$, 31B, 255
$C_{27}H_{38}O$, 35B, 923
$C_{27}H_{42}O_5$, 30B, 442
$C_{27}H_{42}O_8$, 30B, 441
$C_{27}H_{43}NO_2.HI.CH_3OH$, 33B, 228
$C_{27}H_{43}NO_3.\ CH_3Br$, 35B, 394
$C_{27}H_{44}O_4$, 30B, 442
$C_{27}H_{44}O_6$, 30B, 222
$C_{27}H_{45}Cl$, 30B, 439
$C_{27}H_{45}Cl_2NO$, 35B, 342
$C_{27}H_{46}Br_2$, 31B, 203
$C_{27}H_{46}BrNO_2$, 32B, 258
$C_{27}H_{46}Cl_2$, 31B, 203, 206
$C_{27}H_{46}INO_2$, 32B, 260
$C_{27}H_{46}N_2O$, 28, 680
$C_{27}H_{46}O.H_2O$, 28, 680

C_{28}

$C_{28}H_{12}F_{12}NiS_4$, 34B, 354
$C_{28}H_{14}N_6O_2Pd$, 31B, 280
$C_{28}H_{16}Br_2O_4$, 32B, 86
$C_{28}H_{16}ClN_3O_4$, 26, 782
$C_{28}H_{16}Cl_2O_4$, 34B, 65
$C_{28}H_{18}FeO_4$, 27, 876
$C_{28}H_{18}O_2$, 32B, 139
$C_{28}H_{18}O_4$, 27, 936
$C_{28}H_{19}BrN_2O_3$, 31B, 50
$C_{28}H_{20}$, 31B, 112
$C_{28}H_{20}Br_4Ge_2$, 30B, 434
$C_{28}H_{20}Cl_4Ge_2$, 30B, 434
$C_{28}H_{20}Cl_3N_3O_5U$, 32B, 442
$C_{28}H_{20}Ni_2O_4P_2$, 35B, 848
$C_{28}H_{20}NiS_4$, 32B, 460
$C_{28}H_{22}$, 30B, 37
$C_{28}H_{22}CoN_2O_2$, 34B, 516
$C_{28}H_{22}Mo_2O_6$, 33B, 382
$C_{28}H_{23}FeOPS$, 35B, 924
$C_{28}H_{24}$, 30B, 85
$C_{28}H_{24}Ge_2$, 30B, 434; 31B, 333; 32B, 336, 493
$C_{28}H_{24}N_2O_6Si$, 34B, 369
$C_{28}H_{24}N_2O_4Zn$, 35B, 627

$C_{28}H_{28}Cl_2NiOP_2$, 35B, 849
$C_{28}H_{28}I_2Sn_2$, 34B, 414
$C_{28}H_{29}Br_2NNiP_2$, 35B, 850
$C_{28}H_{30}INP_2$, 30B, 278
$C_{28}H_{31}BrN_2O_4$, 29, 764
$C_{28}H_{31}ClP_2Pt$, 30B, 388
$C_{28}H_{31}IO_6$, 28, 542
$C_{28}H_{32}As_4Cl_2CoO_{12}$, 30B, 398
$C_{28}H_{32}Cl_4Cu_4$, 29, 587
$C_{28}H_{32}O_4Si_4$, 30B, 434
$C_{28}H_{33}BrO_8$, 33B, 234
$C_{28}H_{33}IO_9$, 26, 801
$C_{28}H_{35}AsN_2O_3S_4U$, 35B, 819
$C_{28}H_{35}N_2O_3PS_4U$, 35B, 819
$C_{28}H_{36}Cl_2N_4NiO_8$, 33B, 447, 449
$C_{28}H_{37}Cl_3NOP_2Re$, 33B, 511
$C_{28}H_{39}BrO_5$, 34B, 318
$C_{28}H_{39}N_5O_8$, 26, 723; 27, 1051
$C_{28}H_{40}Al_2Ti_2$, 32B, 362
$C_{28}H_{40}CaFe_2N_4O_8.8H_2O$, 31B, 429
$C_{28}H_{40}NiP_2$, 32B, 474
$C_{28}H_{41}BrO_4$, 33B, 216
$C_{28}H_{42}Br_2O_9$, 33B, 555; 35B, 396
$C_{28}H_{42}O_7$, 30B, 442
$C_{28}H_{44}Br_2O$, 34B, 303
$C_{28}H_{44}Cu_4N_4O_8$, 33B, 550
$C_{28}H_{48}Cl_4Mo_2O_{11}P_4S_8$, 34B, 598
$C_{28}H_{49}BrN_2O$, 28, 680
$C_{28}H_{55}KO$, 30B, 430
$C_{28}H_{60}N_5NpS_8$, 35B, 818
$C_{28}H_{64}Cl_2N_6Ni_2O_{10}$, 35B, 777

C_{29}
$C_{29}H_{15}Fe_3O_{11}P$, 33B, 512
$C_{29}H_{18}Fe_2N_2O_6$, 35B, 917
$C_{29}H_{20}F_5NiP$, 33B, 357
$C_{29}H_{21}BrO_{11}$, 30B, 434; 33B, 235
$C_{29}H_{25}NiP$, 34B, 463
$C_{29}H_{32}BrClO_6S$, 34B, 286
$C_{29}H_{35}BCl_3F_4IrOP_2$, 35B, 921
$C_{29}H_{39}BrN_2O_5S$, 33B, 199
$C_{29}H_{39}Br_4Ni_{1.5}P_3$, 35B, 836
$C_{29}H_{42}BrNO_{12}S$, 32B, 256
$C_{29}H_{43}IO_4$, 32B, 246
$C_{29}H_{44}O_7.1/2H_2O$, 30B, 442
$C_{29}H_{47}NO_5.CH_3OH$, 35B, 362
$C_{29}H_{48}O_2S$, 33B, 201

C_{30}
$C_{30}H_{14}O_6$, 30B, 254
$C_{30}H_{16}CuN_6O_2$, 32B, 286
$C_{30}H_{16}CuN_{14}O_{14}$, 30B, 260
$C_{30}H_{16}Fe_4O_{10}.C_2H_4Cl_2$, 34B, 477
$C_{30}H_{18}Cl_2$, 27, 940; 29, 697
$C_{30}H_{18}Co_3O_8P$, 35B, 846
$C_{30}H_{18}Fe_2O_6$, 27, 877
$C_{30}H_{18}O_2$, 33B, 93
$C_{30}H_{20}$, 35B, 134
$C_{30}H_{20}K_2N_4O_4$, 30B, 434
$C_{30}H_{20}N_4Na_2O_4$, 30B, 435
$C_{30}H_{20}Ni_2O_6P_2$, 32B, 475
$C_{30}H_{20}O_2$, 33B, 106
$C_{30}H_{22}Br_2O_{10}.2CH_4O.H_2O$, 35B, 398
$C_{30}H_{22}Cl_8O_8Re_2$, 33B, 414
$C_{30}H_{22}CuO_2$, 31B, 387
$C_{30}H_{22}CuO_4$, 34B, 505
$C_{30}H_{22}O_4Pd$, 33B, 462
$C_{30}H_{23}BrO_6$, 34B, 100
$C_{30}H_{24}CrN_6$, 28, 680
$C_{30}H_{24}N_9$, 35B, 76
$C_{30}H_{24}N_6Ti$, 28, 680
$C_{30}H_{24}N_6V$, 28, 680
$C_{30}H_{25}BrNO_4P$, 30B, 278
$C_{30}H_{25}FeOP$, 30B, 434; 31B, 361; 34B, 434
$C_{30}H_{25}MoNO_4P_2$, 30B, 387
$C_{30}H_{25}P$, 27, 885; 29, 639
$C_{30}H_{25}Sb$, 27, 885; 29, 641; 33B, 292
$C_{30}H_{26}Br_2Sn$, 35B, 491
$C_{30}H_{26}Fe_3$, 27, 862
$C_{30}H_{27}IO_{14}$, 33B, 190
$C_{30}H_{28}Cl_4I_2O_4$, 32B, 75
$C_{30}H_{29}O_7Y$, 33B, 463
$C_{30}H_{30}AgBF_4$, 32B, 494
$C_{30}H_{30}Br_2N_4O_4$, 28, 564
$C_{30}H_{30}N_8Ni$, 32B, 443
$C_{30}H_{31}BrNiP_2$, 35B, 532
$C_{30}H_{32}Al_2N_2O_2$, 33B, 550
$C_{30}H_{35}Cl_2CoN_5O_{13}$, 35B, 749
$C_{30}H_{36}As_4Cl_2O_4Pd$, 32B, 472
$C_{30}H_{36}O_4$, 27, 1007; 29, 684
$C_{30}H_{37}BrO_{12}.2CH_4O$, 34B, 344; 35B, 400
$C_{30}H_{38}ClCoN_7O_5$, 32B, 233
$C_{30}H_{38}Cl_2N_2P_3Re$, 35B, 915
$C_{30}H_{39}BrO_4$, 32B, 252
$C_{30}H_{39}IO_8$, 27, 1033
$C_{30}H_{41}N_3O_7$, 27, 1050
$C_{30}H_{42}Ni_3O_{12}$, 26, 673
$C_{30}H_{42}O_{12}Zn_3$, 26, 673; 33B, 395
$C_{30}H_{44}BrN_3O_3$, 31B, 194
$C_{30}H_{44}Br_2N_2O_2S.4H_2O$, 32B, 261
$C_{30}H_{44}Br_2O_2$, 29, 751
$C_{30}H_{44}Co_3O_{13}$, 33B, 396
$C_{30}H_{45}Cl_2NP_3Re$, 32B, 477
$C_{30}H_{45}Cl_2NiO_{23}P_5$, 32B, 491; 35B, 852
$C_{30}H_{45}Cl_9P_3Re_3$, 29, 626
$C_{30}H_{45}Cl_2P_3Ru$, 34B, 619
$C_{30}H_{47}BrO$, 32B, 252
$C_{30}H_{49}BrO_3$, 30B, 439
$C_{30}H_{49}NO_{11}.HBr.xC_2H_6O$, 35B, 402
$C_{30}H_{50}BrNO_{11}.xC_2H_6O$, 32B, 490; 35B, 402
$C_{30}H_{50}CrO_4$, 28, 680
$C_{30}H_{50}O_2$, 27, 1036
$C_{30}H_{52}$, 35B, 350
$C_{30}H_{52}Br_2N_2O_6S$, 32B, 261
$C_{30}H_{52}N_4O_2.3HClO_4$, 34B, 322
$C_{30}H_{54}Co_2P_2O_6$, 33B, 513
$C_{30}H_{55}Cl_3N_4O_{14}$, 33B, 555; 34B, 322
$C_{30}H_{90}N_{10}.80H_2O$, 35B, 422

C_{31-35}
$C_{31}H_{20}FeO_3$, 30B, 320
$C_{31}H_{25}CuN_3O_2$, 33B, 401
$C_{31}H_{25}FeO_2P$, 34B, 435
$C_{31}H_{27}NO_9Th$, 35B, 751
$C_{31}H_{30}Cl_2FeN_4O_9S_2$, 35B, 823
$C_{31}H_{30}Cl_2N_2PRh.0.61CHCl_3$, 35B, 852
$C_{31}H_{35}BrO_9$, 32B, 491
$C_{31}H_{35}I_2NNiOP_2$, 35B, 853
$C_{31}H_{36}NNa_{0.45}O_9Rb_{0.55}S$, 35B, 476
$C_{31}H_{38}BrNO_6$, 34B, 334
$C_{31}H_{39}BrO_{12}$, 32B, 490
$C_{31}H_{44}Br_2N_4O_{10}.H_2O$, 32B, 490
$C_{31}H_{48}O_7$, 29, 772
$C_{31}H_{51}BrO_2$, 35B, 403
$C_{31}H_{57}CoNO_6$, 32B, 495
$C_{32}H_{12}Cs_2N_{12}$, 29, 650
$C_{32}H_{16}$, 26, 757
$C_{32}H_{16}CoN_8$, 30B, 435
$C_{32}H_{16}CuN_8$, 30B, 435; 33B, 431
$C_{32}H_{16}FeN_8$, 30B, 435
$C_{32}H_{16}MnN_8$, 30B, 435
$C_{32}H_{16}N_8Ni$, 30B, 435
$C_{32}H_{16}N_8Pt$, 33B, 432
$C_{32}H_{16}N_8Zn$, 30B, 436
$C_{32}H_{22}CuN_4O_2$, 26, 663
$C_{32}H_{22}F_{12}NO_9PrS_4$, 32B, 391
$C_{32}H_{24}$, 31B, 114
$C_{32}H_{29}ClMoOP_2$, 35B 854
$C_{32}H_{32}Ge_2$, 30B, 435
$C_{32}H_{32}Si_2$, 30B, 273, 435; 32B, 491
$C_{32}H_{34}Cl_4N_4NiO_4.4H_2O$, 29, 571
$C_{32}H_{34}Cl_4N_4NiO_4.2/3C_2H_6O.4/3H_2O$, 29, 571
$C_{32}H_{34}N_4OV.C_2H_4Cl_2$, 34B, 279
$C_{32}H_{36}CoN_7S_4$, 35B, 821
$C_{32}H_{36}N_4Ni$, 28, 528
$C_{32}H_{39}BrO_{11}$, 29, 753
$C_{32}H_{43}BrO_2$, 34B, 320
$C_{32}H_{45}N_2NiO_6P_3$, 34B, 610
$C_{32}H_{45}N_3Si_5$, 33B, 272
$C_{32}H_{50}BrNO_5$, 31B, 257
$C_{32}H_{51}IO_2$, 31B, 232
$C_{32}H_{52}Br_2O_3$, 31B, 208
$C_{32}H_{52}INO_3$, 31B, 258
$C_{32}H_{52}IO_2$, 33B, 218
$C_{32}H_{53}BrO_4$, 28, 680
$C_{32}H_{53}IO_3$, 31B, 234
$C_{32}H_{80}O_{16}Ti_4$, 28, 651

$C_{32}H_{102}B_{34}Co_{3}N_{3}$, 35B, 921
$C_{32}H_{192}O_{64}P_{2}W$, 27, 780
$C_{33}H_{3}Co_{8}O_{24}$, 35B, 505
$C_{33}H_{20}AsFN_{6}$, 31B, 317
$C_{33}H_{24}FeO_{9}$, 31B, 362
$C_{33}H_{25}N_{3}O_{3}$, 31B, 210
$C_{33}H_{30}Cl_{2}FeN_{4}O_{5}$, 35B, 924
$C_{33}H_{30}Fe_{3}$, 35B, 563
$C_{33}H_{31}IN_{2}O_{6}.C_{3}H_{6}O$, 31B, 230
$C_{33}H_{36}BiCl_{3}N_{12}O_{9}S_{3}$, 27, 881
$C_{33}H_{37}IO_{11}S.C_{3}H_{6}O$, 31B, 196
$C_{33}H_{38}N_{4}O_{2}$, 34B, 281
$C_{33}H_{40}BrN_{3}O_{7}$, 34B, 287
$C_{33}H_{40}FeN_{2}O_{5}$, 32B, 363
$C_{33}H_{40}N_{2}O_{9}$, 33B, 230
$C_{33}H_{51}BrO_{4}$, 28, 543
$C_{33}H_{52}INO_{4}.C_{3}H_{6}O$, 34B, 333
$C_{33}H_{57}IKN_{3}O_{9}$, 34B, 288
$C_{33}H_{57}O_{6}Pr$, 35B, 752
$C_{33}H_{62}O_{6}$, 31B, 24
$C_{34}H_{16}O_{2}$, 29, 698
$C_{34}H_{21}Br_{2}N_{7}O_{3}S_{4}.1/2C_{7}H_{8}O$, 31B, 198
$C_{34}H_{24}N_{10}$, 33B, 254
$C_{34}H_{25}NbO$, 34B, 447
$C_{34}H_{26}N_{4}NiO_{2}$, 33B, 450
$C_{34}H_{30}Br_{2}O_{8}$, 28, 540
$C_{34}H_{30}Co_{2}P_{2}$, 32B, 364
$C_{34}H_{30}Ni_{2}P_{2}$, 32B, 365
$C_{34}H_{32}ClFeN_{4}O_{4}$, 30B, 217
$C_{34}H_{33}AsCoO_{0.5}S_{4}$, 33B, 514
$C_{34}H_{37}IO_{9}$, 31B, 236
$C_{34}H_{38}Cl_{2}N_{4}OV$, 34B, 279
$C_{34}H_{46}Cl_{2}OP_{4}Pd$, 35B, 855
$C_{34}H_{51}IO_{6}$, 30B, 228
$C_{34}H_{63}BrO_{6}$, 33B, 236
$C_{34}H_{68}$, 33B, 87
$C_{35}H_{20}Fe_{3}N_{4}O_{9}$, 32B, 497
$C_{35}H_{24}I_{2}N_{2}$, 35B, 184
$C_{35}H_{25}O_{2}V$, 34B, 479
$C_{35}H_{34}Fe_{3}O$, 35B, 564
$C_{35}H_{42}BrNO_{10}$, 29, 772
$C_{35}H_{46}INO_{4}$, 28, 537; 30B, 223
$C_{35}H_{47}BrO_{2}$, 26, 808; 33B, 202

C_{36-40}

$C_{36}H_{12}Cs_{2}N_{12}$, 31B, 43
$C_{36}H_{20}Fe_{3}O_{8}$, 27, 880; 30B, 297
$C_{36}H_{24}$, 35B, 134
$C_{36}H_{24}Br_{2}CoN_{6}.7H_{2}O$, 30B, 426
$C_{36}H_{24}Br_{2}FeN_{6}.7H_{2}O$, 30B, 426
$C_{36}H_{24}Br_{2}N_{6}Ni.7H_{2}O$, 30B, 426
$C_{36}H_{28}Fe_{2}N_{2}O_{6}$, 32B, 497
$C_{36}H_{28}NiP_{2}Se_{2}$, 35B, 832
$C_{36}H_{30}Al_{2}$, 32B, 492
$C_{36}H_{30}As_{6}$, 26, 736
$C_{36}H_{30}As_{2}Cl_{2}HgO_{2}$, 28, 652
$C_{36}H_{30}As_{2}Cl_{4}Hg_{2}O_{2}$, 29, 632
$C_{36}H_{30}Br_{2}NiP_{2}$, 28, 680; 33B, 515
$C_{36}H_{30}ClIrN_{2}O_{6}P_{2}$, 35B, 856
$C_{36}H_{30}Cl_{2}NP_{2}Re$, 32B, 477
$C_{36}H_{30}Cl_{2}NiP_{2}$, 28, 658
$C_{36}H_{30}CoN_{9}.C_{7}H_{8}$, 32B, 497
$C_{36}H_{30}CrO_{4}Si_{2}$, 35B, 445
$C_{36}H_{30}Fe_{2}N_{4}O_{6}$, 32B, 497
$C_{36}H_{30}I_{2}NiP_{2}$, 28, 680
$C_{36}H_{30}I_{6}P_{2}S_{2}$, 33B, 259
$C_{36}H_{30}N_{3}P_{3}$, 34B, 393
$C_{36}H_{30}Ni_{4}$, 30B, 436
$C_{36}H_{30}P_{6}$, 30B, 279; 31B, 309
$C_{36}H_{31}Cl_{4}P_{2}RhSi.xSiHCl_{3}$, 35B, 873
$C_{36}H_{33}Br_{2}CoP_{3}$, 31B, 458
$C_{36}H_{33}Cl_{3}NO_{3}P_{2}$, 34B, 611
$C_{36}H_{33}I_{2}NiP_{3}$, 31B, 458
$C_{36}H_{34}BCuP_{2}$, 32B, 478
$C_{36}(H,D)_{36}N_{4}NiO_{6}.1/2C_{6}H_{6}$, 30B, 219
$C_{36}H_{40}NNaO_{7}V.2C_{3}H_{8}O$, 34B, 532
$C_{36}H_{45}NO_{17}.CCl_{3}H$, 27, 1040
$C_{36}H_{45}NO_{17}.CH_{3}COCH_{3}$, 27, 1040
$C_{36}H_{47}I_{3}N_{6}$, 32B, 491
$C_{36}H_{52}Cl_{2}N_{10}O_{8}$, 35B, 419
$C_{36}H_{54}F_{10}P_{2}PtS_{2}$, 35B, 824
$C_{36}H_{56}Cl_{6}N_{2}O_{12}Pt.2H_{2}O$, 31B, 243
$C_{36}H_{58}INO_{5}$, 34B, 333
$C_{36}H_{60}O_{30}.1.54(C_{2}H_{3}KO_{2}).9.7H_{2}O$, 30B, 186
$C_{36}H_{61}AgO_{11}.2H_{2}O$, 35B, 754
$C_{36}H_{65}Br_{3}O_{6}$, 30B, 436
$C_{36}H_{66}Cl_{2}NiP_{2}$, 28, 659
$C_{36}H_{70}MgO_{4}$, 27, 827
$C_{36}H_{84}O_{13}P_{6}S_{12}Zn_{4}$, 30B, 436
$C_{36}H_{84}S_{12}Pd_{6}$, 33B, 497
$C_{37}H_{25}MoN_{4}PS_{4}$, 35B, 566
$C_{37}H_{30}BClF_{4}IrNO_{2}P_{2}$, 33B, 430
$C_{37}H_{30}BF_{4}IIrNO_{2}P_{2}.C_{6}H_{6}$, 34B, 614
$C_{37}H_{30}ClIrO_{3}P_{2}$, 29, 770; 30B, 386;
$C_{37}H_{30}ClIrO_{3}P_{2}S$, 31B, 460
$C_{37}H_{30}ClO_{3}P_{2}RhS$, 34B, 613
$C_{37}H_{30}P_{2}PtS_{2}$, 35B, 825
$C_{37}H_{35}MoN_{2}O_{3}P_{3}$, 35B, 857
$C_{37}H_{37}BrF_{2}N_{2}O_{5}$, 35B, 344
$C_{37}H_{42}Br_{2}N_{4}O_{7}$, 35B, 365
$C_{37}H_{43}BrO_{8}.0.5(C_{4}H_{8}O_{2})$, 33B, 202
$C_{37}H_{43}FeN_{4}O_{5}$, 30B, 218
$C_{37}H_{44}O_{5}P_{4}Pt_{4}$, 34B, 614
$C_{37}H_{52}BrNO.C_{2}H_{6}O$, 35B, 404
$C_{37}H_{82}B_{10}MoN_{2}O_{5}$, 35B, 920
$C_{38}H_{16}CuN_{10}O_{2}$, 33B, 451
$C_{38}H_{24}N_{2}O_{30}Rh_{12}$, 34B, 436
$C_{38}H_{26}Cl_{5}F_{5}NiP_{2}$, 35B, 915
$C_{38}H_{28}Cl_{12}Sb_{2}$, 32B, 486
$C_{38}H_{30}MnNO_{3}P_{2}$, 32B, 480
$C_{38}H_{32}Cl_{2}IIrO_{3}P_{2}$, 32B, 479
$C_{38}H_{34}Br_{2}NiP_{2}$, 33B, 343; 35B, 858
$C_{38}H_{34}NiP$, 32B, 493
$C_{38}H_{35}NO_{2}Si_{3}$, 32B, 303
$C_{38}H_{36}As_{2}Cl_{4}Ni$, 31B, 314
$C_{38}H_{36}N_{8}O_{6}$, 31B, 111; 32B, 287
$C_{38}H_{42}Co_{2}N_{6}O_{8}$, 35B, 779
$C_{38}H_{52}CoP_{2}$, 28, 660
$C_{38}H_{56}Cl_{2}HgO_{4}$, 33B, 203
$C_{29}H_{30}CoGeO_{3}P$, 35B, 875
$C_{39}H_{30}O_{3}Os$, 34B, 617
$C_{39}H_{35}Cl_{2}O_{3}P_{2}Re_{2}$, 33B, 416
$C_{39}H_{74}O_{6}$, 30B, 10
$C_{40}H_{20}$, 27, 925
$C_{40}H_{28}Cl_{2}N_{8}$, 32B, 489
$C_{40}H_{30}AuMnO_{7}P_{2}$, 32B, 481
$C_{40}H_{30}BrN_{3}O_{3}$, 33B, 208
$C_{40}H_{30}F_{6}OP_{2}$, 33B, 284
$C_{40}H_{30}MnO_{4}PSn$, 32B, 481
$C_{40}H_{30}Nb_{2}O_{2}$, 34B, 448
$C_{40}H_{30}O_{14}P_{2}U_{2}$, 33B, 546
$C_{40}H_{31}F_{6}N_{3}P_{2}Pt$, 33B, 549
$C_{40}H_{32}Al_{2}N_{4}O_{5}$, 35B, 732
$C_{40}H_{32}N_{4}O_{4}Th$, 27, 888
$C_{40}H_{32}N_{2}P_{2}Pt$, 35B, 859
$C_{40}H_{34}Cl_{6}N_{8}O$, 33B, 185
$C_{40}H_{36}As_{2}O_{8}U$, 33B, 542
$C_{40}H_{37}As_{2}Cl_{2}Rh$, 35B, 534
$C_{40}H_{44}I_{2}N_{4}O_{2}$, 30B, 230
$C_{40}H_{50}O$, 33B, 84
$C_{40}H_{52}O$, 33B, 84
$C_{40}H_{54}$, 29, 499
$C_{40}H_{56}$, 29, 496
$C_{40}H_{56}Ni_{4}O_{16}$, 34B, 517
$C_{40}H_{58}$, 29, 497
$C_{40}H_{62}FeO_{8}P_{4}$, 35B, 920
$C_{40}H_{64}Al_{2}Na_{2}O_{4}$, 35B, 478
$C_{40}H_{66}CoN_{11}O_{9}.11H_{2}O$, 27, 1040
$C_{40}H_{67}AgO_{11}$, 35B, 755
$C_{40}H_{72}CoN_{6}S_{4}$, 29, 560
$C_{40}H_{100}Nb_{8}O_{30}$, 33B, 546

C_{41-50}

$C_{41}H_{28}BrNO.C_{2}H_{6}O$, 34B, 143
$C_{41}H_{32}Br_{2}CrN_{9}O_{6}.3H_{2}O$, 33B, 452
$C_{41}H_{33}IrN_{2}OP_{2}$, 34B, 615
$C_{41}H_{35}FeIO_{6}P_{2}$, 33B, 358
$C_{41}H_{48}I_{2}N_{4}O.H_{2}O$, 34B, 333
$C_{41}H_{48}N_{4}O_{4}$, 30B, 241
$C_{41}H_{52}O_{17}$, 35B, 916
$C_{41}H_{64}KNO_{12}S$, 32B, 242
$C_{41}H_{66}FeN_{9}O_{24}$, 31B, 414
$C_{42}H_{18}$, 26, 758
$C_{42}H_{28}$, 27, 940
$C_{42}H_{30}$, 33B, 76
$C_{42}H_{30}Be_{4}O_{13}$, 29, 763
$C_{42}H_{30}N_{4}P_{2}Pt$, 35B, 860
$C_{42}H_{30}ReS_{6}$, 31B, 455
$C_{42}H_{30}S_{6}V$, 31B, 456; 32B, 462

$C_{42}H_{35}AsI_3NiP$, 33B, 517
$C_{42}H_{35}BO$, 28, 677
$C_{42}H_{36}ClF_6N_2O_2P_3Ru$, 35B, 920
$C_{42}H_{36}CoI_3N_6O_3$, 35B, 716
$C_{42}H_{40}Cl_3O_5P_2Re_2$, 34B, 532
$C_{42}H_{42}ClCoF_6NP_4$, 35B, 920
$C_{42}H_{42}I_2NNiP_3$, 35B, 862
$C_{42}H_{47}Br_3O_8$, 28, 567
$C_{42}H_{47}I_3O_8$, 28, 566
$C_{42}H_{48}N_{12}Ni_4O_6P_4$, 32B, 482
$C_{42}H_{56}CoN_3S_4$, 35B, 826
$C_{42}H_{82}O_4$, 29, 762
$C_{42}H_{84}Ag_6N_6S_{12}$, 34B, 525
$C_{42}H_{84}Co_3O_{24}P_6$, 33B, 464
$C_{42}H_{84}Cu_6N_6O_6S_6$, 35B, 828
$C_{43}H_{30}Fe_5N_6O_{13}$, 31B, 433
$C_{43}H_{34}BrNO_2$, 34B, 143
$C_{44}H_{28}ClFeN_4$, 32B, 234
$C_{44}H_{28}CuN_4$, 29, 617
$C_{44}H_{28}N_4Pd$, 29, 617
$C_{44}H_{28}N_4Zn.2H_2O$, 29, 617
$C_{44}H_{29}FeN_4O.2H_2O$, 29, 617
$C_{44}H_{30}BrIrN_4OP_2$, 33B, 546
$C_{44}H_{30}Cu_2N_2O_4$, 35B, 781
$C_{44}H_{30}F_2$, 33B, 542
$C_{44}H_{30}F_3$, 27, 786
$C_{44}H_{30}MgN_4O$, 34B, 278
$C_{44}H_{30}N_4$, 29, 728; 32B, 235
$C_{44}H_{30}N_4OZn$, 32B, 237
$C_{44}H_{32}Cl_5FeN_4$, 33B, 185
$C_{44}H_{32}Cl_2N_4$, 32B, 490
$C_{44}H_{42}As_2O_{14}U_2$, 33B, 542
$C_{44}H_{42}Cl_2O_6P_2Rh_2$, 35B, 583
$C_{44}H_{42}O_{14}P_2U_2$, 34B, 533
$C_{44}H_{48}Cr_2O_{12}P_2$, 30B, 333
$C_{45}H_{29}N_9$Z.., 32B, 490; 35B, 717
$C_{45}H_{31}N_5O_5Th$, 27, 887
$C_{45}H_{32}Cl_3F_{12}RhSb_2$, 35B, 512
$C_{45}H_{32}N_6O_6$, 30B, 435
$C_{45}H_{35}HoO_7$, 32B, 495
$C_{45}H_{48}EuNO_8$, 34B, 508
$C_{45}H_{53}IN_2O_{13}$, 29, 772
$C_{46}H_{30}Cr_2IN_2O_{10}P_4$, 35B, 506
$C_{47}H_{59}IN_4O_{10}.2H_2O$, 31B, 258
$C_{48}H_{28}Fe_2O_4$, 35B, 584
$C_{48}H_{28}N_8P$, 33B, 254
$C_{48}H_{32}Br_2O_2.1\ 1/2C_6H_6$, 33B, 66
$C_{48}H_{35}NbO$, 34B, 450
$C_{48}H_{40}Al_4N_4$, 28, 599
$C_{48}H_{40}As_2Cl_{11}Re_3$, 31B, 319
$C_{48}H_{40}As_2CoN_4O_{12}$, 31B, 322
$C_{48}H_{40}As_2CuN_4O_{12}$, 35B, 758
$C_{48}H_{40}As_2MnN_4O_{12}$, 35B, 756
$C_{48}H_{40}As_4N_4$, 32B, 492
$C_{48}H_{40}As_2N_4NiO_{12}$, 35B, 758
$C_{48}H_{40}As_2N_4O_{12}Zn$, 35B, 758
$C_{48}H_{40}I_2O_{16}$, 31B, 267
$C_{48}H_{40}N_4P_4$, 32B, 492
$C_{48}H_{40}Si_4$, 33B, 542
$C_{48}H_{44}I_2N_8Ni_2O_8$, 32B, 491
$C_{48}H_{72}N_2O_{16}Re_4$, 33B, 329
$C_{48}H_{108}Cl_5P_4Ru_2$, 33B, 517
$C_{48}H_{108}O_{13}P_6S_{12}Zn_4$, 30B, 436
$C_{49}H_{34}IrN_5OP_2$, 35B, 920
$C_{49}H_{40}Br_4Pb_4$, 35B, 492
$C_{50}H_{33}ClN_4Rh.H_2O$, 32B, 490
$C_{50}H_{35}MnO_4Sn$, 32B, 367
$C_{50}H_{36}ClFeN_8$, 34B, 279
$C_{50}H_{38}Al_2Br_2N_2.2C_6H_6$, 34B, 406
$C_{50}H_{40}As_2NiS_6$, 32B, 492
$C_{50}H_{40}P_2Pt$, 32B, 493
$C_{50}H_{46}F_{30}Ni_6O_{24}.C_7H_8$, 34B, 511
$C_{50}H_{75}Cl_4P_5Ru_2$, 33B, 518

$C > 51$

$C_{52}H_{35}Cl$, 35B, 923
$C_{52}H_{40}As_2N_5NbOS_5$, 35B, 782
$C_{52}H_{40}As_2N_4NiS_4$, 33B, 498
$C_{52}H_{44}As_4Cl_2O_2Rh_2$, 34B, 616
$C_{52}H_{48}ClP_4O_4Rh$, 35B, 865
$C_{54}H_{42}As_4Br_2Hg.CH_2Cl_2$, 31B, 463
$C_{54}H_{42}As_4Br_2Ru$, 30B, 391
$C_{54}H_{45}Cl_2P_3Ru$, 30B, 390
$C_{54}H_{46}CoN_2P_3$, 34B, 618
$C_{54}H_{52}Cu_2N_2P_4S_2$, 35B, 829
$C_{54}H_{60}.1.19C_6H_6$, 30B, 436
$C_{54}H_{60}.0.20C_7H_8$, 30B, 436
$C_{54}H_{60}.2.92CHCl_3$, 30B, 436
$C_{54}H_{60}.1.11C_6H_5Cl$, 30B, 436
$C_{54}H_{60}.0.55C_3H_6O$, 30B, 437
$C_{54}H_{60}.1.95C_4H_8O_2$, 30B, 437
$C_{54}H_{60}.0.96CS_2$, 30B, 437
$C_{54}H_{108}ErN_9S_6$, 33B, 553
$C_{55}H_{44}As_4Br_2Cl_2Hg$, 31B, 463
$C_{55}H_{46}BrOOsP_3$, 27, 873
$C_{55}H_{46}OP_3Rh$, 28, 685; 30B, 389
$C_{55}H_{60}Cl_2CuN_{10}O_{13}$, 30B, 425
$C_{56}H_{40}$, 29, 770; 30B, 93
$C_{56}H_{40}As_2CoF_{12}O_8$, 31B, 410
$C_{56}H_{72}Cl_{16}Co_2N_2S_8$, 33B, 499
$C_{56}H_{108}MoN_{11}$, 35B, 879
$C_{57}H_{84}O_6.C_6H_5BrO$, 30B, 438
$C_{58}H_{52}Br_3NO_{16}$, 34B, 287
$C_{58}H_{56}CoN_2O$, 33B, 546
$C_{60}H_{30}F_{20}P_2Pd_2S_4$, 35B, 830
$C_{60}H_{40}Ag_2FeN_6S_6$, 32B, 492
$C_{60}H_{40}Br_2Mo_2O_4$, 34B, 479
$C_{60}H_{45}N_6O_6P_3.1/2C_7H_8$, 32B, 492
$C_{60}H_{50}Cl_2O_2Pd_2$, 29, 768; 30B, 326
$C_{60}H_{50}Si_5$, 33B, 542
$C_{60}H_{56}Cu_8N_{16}S_{12}$, 33B, 553
$C_{62}H_{57}P_4Re$, 31B, 464
$C_{63}H_{88}CoN_{14}O_{14}P.{\sim}18H_2O$, 28, 1042
$C_{63}H_{88}CoN_{14}O_{14}P.22H_2O$, 29, 754
$C_{63}H_{89}BrO_7$, 30B, 438
$C_{66}H_{60}Br_2O_{12}.2C_6H_6$, 35B, 405
$C_{66}H_{72}CaCl_2N_{12}O_{14}$, 30B, 425; 33B, 465
$C_{66}H_{72}Cl_2MgN_{12}O_{14}$, 30B, 425; 32B, 324
$C_{66}H_{72}Cl_2N_{12}O_{14}Pb$, 31B, 334
$C_{66}H_{72}Cl_2N_{12}O_{14}Zn$, 30B, 425
$C_{68}H_{72}N_{16}Ni_2O_4$, 33B, 435
$C_{68}H_{108}BaO_{17}$, 35B, 406
$C_{72}H_{60}Cl_6Cu_4O_5P_4$, 31B, 446; 32B, 449
$C_{72}H_{60}Si_6$, 33B, 542
$C_{72}H_{60}Sn_2.C_{16}H_{20}$, 29, 769
$C_{72}H_{62}As_4Br_6Hg_2O_4$, 32B, 492
$C_{72}H_{100}CoN_{18}O_{17}P$, 26, 807
$C_{72}H_{100}CoN_{18}O_{17}P.17H_2O$, 33B, 186
$C_{78}H_{62}As_4BIPt$, 26, 690
$C_{83}H_{62}Cl_2Cr_2N_2O_{10}P_4$, 35B, 506
$C_{83}H_{62}Cl_2Mo_2N_2O_{10}P_4$, 35B, 506
$C_{84}H_{52}Mn_2N_{20}O$, 32B, 238
$C_{84}H_{84}Br_4O_{16}$, 33B, 95
$C_{88}H_{56}Fe_2N_8O.C_8H_{10}$, 34B, 283
$C_{88}H_{80}Sn_6$, 28, 604
$C_{90}H_{135}Cl_6P_9Ru_3$, 34B, 619
$C_{94}H_{190}$, 27, 746
$C_{96}H_{66}Fe_2N_8O$, 34B, 283
$C_{126}H_{84}Au_{11}Cl_{21}I_3P_7$, 35B, 916
$C_{168}H_{84}Cl_4N_{36}$, 35B, 421

POLYMERS

$(C_2F_4)p$, 27, 804
$(CH_2)n$, 35B, 882
$(CH_2O)p$, 27, 809
$(CH_3As)n$, 35B, 466
$(C_2H_2F_2)n$, 35B, 882
$(C_2H_2O_2)n$, 30B, 521
$(C_2H_3Cl)p$, 26, 634
$(C_2H_3F)p$, 26, 633
$(C_2H_4)p$, 26, 631; 27, 799
$(C_2H_4O)p$, 26, 631, 635; 27, 809, 810

$(C_2H_4S)p$, 27, 811
$(C_2H_6OSi)p$, 27, 813
$(C_3H_3N)p$, 26, 634; 27, 804
$(C_3H_4O)p$, 26, 632
$(C_3H_4O_2)p$, 27, 806
$(C_3H_5ClO)n$, 26, 632; 35B, 883
$(C_3H_5NO)n$, 35B 883
$(C_3H_6)p$, 26, 631; 27, 801
$(C_3H_6O)n$, 26, 635; 33B, 521
$(C_3H_6O_3)p$, 26, 638
$(C_3H_8O_2)n$, 35B, 884
$(C_3H_9AlS)n$, 34B, 403
$(C_4H_3F_3O_2)p$, 26, 634
$(C_4H_6)n$, 35B, 884
$(C_4H_7NO)n$, 31B, 31
$(C_4H_8O)p$, 26, 635; 27, 812
$(C_5H_8)n$, 26, 636; 32B, 44
$(C_5H_{10})n$, 35B, 884
$(C_6H_5PS)n$, 30B, 418
$(C_6H_8NO_3)p$, 27, 812
$(C_6H_{10})n$, 33B, 522
$(C_6H_{10}O_2)n$, 35B, 885
$(C_6H_{10}O_5)p$, 27, 798
$(C_6H_{12})p$, 27, 801
$(C_7H_{12})n$, 27, 801; 32B, 44
$(C_8H_7Br)p$, 27, 803
$(C_8H_7Cl)p$, 27, 803
$(C_8H_7I)p$, 27, 803
$(C_8H_8)p$, 27, 802
$(C_8H_8O)p$, 27, 810
$(C_8H_{12}O_4)p$, 27, 806
$(C_8H_{13}NO_5)n$, 33B, 522
$(C_8H_{14})n$, 32B, 45; 33B, 523
$(C_8H_{15}NO_6)p$, 26, 821
$(C_{10}H_8O_4)p$, 27, 808
$(C_{10}H_{12})n$, 33B, 524
$(C_{10}H_{16}O_4)p$, 27, 806
$(C_{10}H_{18})n$, 32B, 45; 33B, 523
$(C_{12}H_{22})n$, 32B, 46; 33B, 523
$(C_{16}H_{14}O_3)p$, 27, 809
$(C_{16}H_{18}O_4)p$, 26, 639
$(C_{16}H_{25}Mn_3O_{17})n$, 34B, 604
$(C_{18}H_{12}O)n$, 35B, 886
$(C_{20}H_{44}O_4P_2Zn)n$, 33B, 525

AUTHOR INDEX

Modified letters [*å, ä, č, é, ė, š, ž, etc.*] are placed in, or immediately following the positions where the unmodified letters would occur in the normal English alphabetical order. Names beginning with *Mc* or *Mac* are collected at the beginning of the M's. Names beginning with a separated *de*, *van* or *von* are listed according to the main component, thus de Vries occurs with the V's.

Aava, U., 34B, 401; 35B, 828
Abagjan, S.A., 30A, 244
Abasolo, C., 26, 571
Abbasov, A.A., 33B, 539
Abbattista, F., 26, 523
Abdullaev, G.K., 30B, 392
Abdurakhmanova, A.A., 30A, 248
Abe, A., 26, 630
Abe, J., 35B, 402
Abe, Y., 33B, 532
Abel, E.W., 30B, 399
Abel, J.E., 26, 293
de Abeledo, M.J., 27, 462, 967
Abello, L., 28, 316; 29, 233
Aben, P.C., 27, 239
Aberg, M., 34A, 225; 35B, 905
Abernethy, L.L., 26, 415
Ablov, A.V., 27, 842; 38, 631; 31B, 380, 427; 33B, 412; 35B, 408
Abov, Ju. B., 26, 120
Abowitz, G., 34B, 122
Abračev, K.K., 26, 537
Abrahams, S.C., 26, 288, 728; 27, 250, 281, 436; 28, 121; 29, 51, 136, 236, 433; 30A, 252, 322, 327; 30B, 53; 31A, 147, 158, 187; 31B, 352; 32A, 287, 292, 299, 312; 33A, 327, 328, 335; 33B, 43; 34A, 238, 269, 278, 294; 35A, 45, 102, 164, 165
Abrahamsson, S., 27, 740; 28, 371, 530, 566; 30B, 392; 31B, 22, 180, 211, 218, 265; 32B, 96, 194; 34B, 25
Abramova, A.V., 26, 48
Abršev, K.K., 27, 704; 28, 357; 29, 398, 436
Adam, G., 33B, 228
Adams, E., 27, 198
Adams, K.M., 29, 239
Adams, L.H., 27, 425, 565; 28, 316; 29; 433
Adams, M.J., 35B, 647
Adams, W.J., 35B, 897
Adamson, G.W., 30B, 392; 35B, 443, 454
Adamson, M.G., 27, 101; 31B, 315
Addison, C.C., 26, 474
Addison, W.E., 30A, 400
Adelson, E., 27, 57, 396; 30A, 178, 246; 33A, 85
Adeoye, S.A., 30B, 392
Adiwidjaja, G., 35A, 19
Adler, I., 29, 136, 435
Adman, E., 32B, 98, 279; 33B, 78; 34B, 86; 35B, 256
Admiraal, L.J., 34B, 321
Adolphe, C., 29, 135; 33A, 98, 362; 34A, 98; 35A, 235
Adrian, H.W.W., 34A, 172
Afonskij, N.S., 33A, 341
Agaev, K.A., 30A, 28, 175; 31A, 18, 25; 33A, 42, 43; 34A, 88
Ageev, N.V., 27, 255; 35A, 124
Aggarwal, P.S., 26, 348
Agranovskaja, A.I., 28, 341
Agrawal, V.K., 33A, 517
Agre, V.M., 33B, 486
Agrell, I., 31A, 77; 32A, 232; 33A, 510; 34B, 554; 35B, 691, 693
Agrell, S.O., 26, 512
Agron, P.A., 26, 325; 28, 73, 75; 32A, 395
Ahmed, A.H.M., 34A, 252; 35A, 292, 294
Ahmed, F.R., 26, 353, 782; 27, 1023; 28, 204, 440, 462; 29, 503; 30B, 280; 31A, 142; 31B, 307; 33B, 224, 226; 34B, 75, 176, 393
Ahmed, N., 27, 534
Ahmed, N.A.K., 33B, 36
Ahmed, S.J., 32A, 510
Aia, M.A., 28, 200
Aiello, R., 28, 339
Aihara, A., 35B, 108
Aime, S., 35B, 118
Ainalem, I.-B., 33B, 439
Aitken, E.A., 29, 435
Aizlewood, J.M., 29, 760
Ajtkhozin, E.S., 30A, 248
Ajzikovič, A.N., 28, 340
Akabori, S., 27, 1050
Akamatu, H., 31B, 278
Akazuki, M., 28, 357
Akers, C., 33A, 252
Akhavan Niaki, A.-N., 26, 468
Akhmed, N.A., 35B, 490
Akhmetzjanov, K.G., 30A, 181, 247
Akhtar, M., 31A, 82
Akhundov, Ju. A., 26, 521

Akimoto, H., 34B, 100
Akimoto, S., 29, 434
Akimoto, S.I., 34A, 262
Akišin, P.A., 26, 298, 362, 486, 550, 558, 650; 27, 718, 759; 28, 362; 29, 450, 538; 30B, 405; 34A, 173
Akopjan, Z.A., 28, 683; 30B, 98, 443; 31B, 66
Akulinin, O.B., 35B, 904
d'Alagni, M., 29, 760
Albano, V., 28, 495, 659; 31B, 464; 33A, 165
Albano, V.G., 34A, 159; 34B, 430, 436
Albers, W., 27, 344; 30A, 176
Albert, R., 35B, 217
Alberti, A., 34A, 393
Alberts, L., 35A, 49
Albertsson, J., 30B, 399; 33B, 412; 35B, 642, 644
Albrecht, E.D., 26, 233
Albrecht, G., 28, 683; 30B, 443
Alcock, N.W., 30A, 86, 181; 30B, 392; 31B, 148; 32B, 354; 33A, 358; 33B, 317, 318, 450, 518
Alden, K.-I., 29, 369
Alden, R.A., 27, 1007; 28, 684; 29, 684, 33B, 206
Alderman, P. R. H., 27, 846
Aldred, A. T., 27, 342
Aleby, S., 27, 744; 33B, 10
Alejnikova, K.B., 30A, 175
Aleksandrov, G. G., 34B, 467, 479; 35B, 490, 540, 573, 574
Aleksandrov, K. S., 28, 244
Aleksandrov, S. M., 29, 437
Aleksandrov, V. B., 26, 381; 27, 535, 536; 28, 342; 33A, 334; 35A, 172
Aleksandrova, M.M., 33A, 217
Alekseev, N. V., 28, 668; 33B, 531; 35B, 900, 901
Alekseevskij, N.E., 35A, 124
Aleonard, R., 26, 389; 31A, 133
Aléonard, S., 29, 223; 31A, 87; 32A, 163; 33A, 180, 186, 443; 34A, 186; 35A, 173
Aleshin, E., 27, 534
Alexander, D.G., 33A, 153
Alexander, E., 32A, 135; 33B, 99
Alexander, L. E., 28, 119, 584; 29, 241; 31B, 42; 32B, 300; 33B, 4, 120
Alexander, R., 31B, 367
Alexanian, C., 26, 287
Ali, S. A. D., 34A, 176
Aliev. M. I., 30A, 248
Alieva, A. G., 26, 37
Alikhanov, R. A., 27, 561; 32A, 147
Alisova, S. P., 29, 137, 138
Aljamovskij, S. I., 26, 105
Allain, Y., 30A, 322
Alleaume, M., 26, 750; 30B, 66, 392; 31B, 62; 33B, 55
Allegra, G., 26, 317, 328, 611, 633, 651, 652, 684, 685; 27, 441, 787, 801; 28, 597, 600, 609, 610, 682; 29, 502, 760; 30B, 393, 398; 31B, 140; 32B, 119, 142, 292; 34B, 430, 441; 35B, 6, 106, 107, 111, 217, 567
Allen, C. W., 34B, 383
Allen, D. W., 35B, 467
Allen, F. H., 33B, 211; 35B, 121, 123, 404, 558
Allen, K. W., 27, 991; 28, 314
Allen, W. C., 29, 296
Alleyne, A. B., 28, 489
Allison, G. B., 35B, 190
Allmann, R., 27, 958; 29, 89; 31A, 174, 31B, 302; 32B, 198; 33A, 217, 277, 278, 439, 491; 33B, 66; 34B, 143; 35B, 131
Allpress, J. G., 29, 282; 30A, 447
Almenningen, A., 26, 553, 641; 28, 103, 362; 29, 449, 524, 642; 33B, 535, 537; 34A, 162, 163; 35B, 888, 890, 893, 894
Almodovar, I., 29, 331; 33B, 402
Alper, A. M., 27, 545
Alperin, H. A., 26, 358; 27, 496
Alsdorf, H., 33B, 297; 35B, 472
D'Alte de Viega, L.M., 27, 30, 166; 28, 17; 30A, 43, 176; 31A, 36
Altenburg, H., 34A, 321
Altman, C., 30A, 182
Altona, C., 28, 482, 484, 485, 487; 31B, 159; 32B, 199
Alver, E., 32B, 308; 34B, 379
Alyea, E. C., 34B, 601
Amakasu, O., 28, 687; 31B, 239, 241
Amano, Y., 29, 226, 609; 30B, 443; 33B, 447
Amanto T., 34A, 244
Ambady, G. K., 33B, 15, 16; 34B, 19
Ambats, I., 30B, 152
Amberger, E., 33A, 153
Amendola, A., 28, 682, 683
Amilius, Z., 34A, 207
Amirov, S. T., 32A, 437, 482
Amirthalingam, V., 29, 437
Amit, A., 33B, 96
Amit, A. G., 31B, 46
Amma, E. A. 33B, 469
Amma, E. L., 26, 295; 27, 858; 28, 682, 684; 31B, 364, 365, 417, 457; 32B, 327, 453, 464, 474; 34B, 523; 35B, 628, 784
Amma, E. R., 33A, 208
Ammon, H. L., 31B, 107, 133; 32B, 168, 34B, 103, 111; 35B, 201, 219
Amonenko, V. M., 26, 56; 27, 65
Amoros, J. L., 26, 297, 561, 571, 738, 780; 27, 566

Ampaio, A. O., 33A, 153
Amstutz, G. C., 26, 431; 29, 223
Amzel, L. M., 34A, 314; 34B, 522; 35B, 598
Anagnostidis, M., 29, 224
Ananthakrishnan, N., 29, 434
Anantharaman, T.R., 26, 139; 30A, 173, 180, 244
Anaškin, M. G., 33B, 532, 537
Anca, R., 32B, 84
Ancyškina, A. S., 27, 817; 30B, 399; 31B, 408, 417; 32B, 414, 433; 33B, 385, 402; 34A, 301; 34B, 481
Anderko, K., 27, 304
Andersen, B., 29, 449; 35B, 899
Andersen C. A., 29, 433
Andersen, E. K., 29, 648; 32B, 47, 48, 105, 106, 107
Andersen, P., 27, 915; 29, 450; 30B, 55, 405; 32A, 208, 234; 32B, 49, 74
Anderson, B. F., 35B, 349
Anderson, B. W., 26, 538
Anderson, C. P., 35A, 469
Anderson, E., 29, 136, 224
Anderson, J. C., 29, 322
Anderson, J. S., 35B, 666
Anderson, K., 29, 135
Anderson, K. D., 30A, 65, 180; 34A, 144
Anderson, O. P., 35B, 499
Anderson, P., 28, 107
Andersson, B., 35A, 206
Andersson, S., 26, 352, 373, 375; 27, 528, 532; 29, 137, 169, 318; 30A, 335, 336, 344, 448; 31A, 148; 32A, 257; 33A, 46, 189; 34A, 189, 217; 35A, 34, 38, 124, 153, 209
Andiranov, V. I., 32B, 153
Andjanov, V. G., 27, 665
Ando, K., 34B, 286
Ando, O., 32B, 226
Ando, T., 32B, 104
Andreassen, A. L., 35B, 891
Andreetti, G. D., 32B, 408; 33B, 59, 479; 34B, 189, 486; 35B, 28
Andreev, S. N., 26, 334; 28, 102
Andreeva, N. A., 26, 630; 28, 682
Andresen, A. F., 29, 136; 30A, 247; 33A, 106, 160; 35A, 47, 72
Andresen, O., 27, 889
Andrew, J. E., 34B, 517, 561
Andrews, A. C., 26, 760
Andrews, K. W., 29, 135, 136
Andrianov, V. G., 26, 400; 27, 665; 33B, 337, 340, 358
Andrianov, V. I., 32A, 150
Andrievskij, A. I., 27, 200
Andrievskij, R. A., 32A, 44
Andrist, A. H., 35B, 905
Andron, B., 31A, 36, 37; 32A, 54
Andruščenko, N. K., 26, 539
Andrusko, A. F., 27, 115; 30A, 182
Anfinsen, I. M., 35B, 890, 894
Angenault, J., 33A, 312, 313
Angoso, A., 29, 151
Angstadt, R. T., 27, 478
Anikina, L. I., 26, 391; 28, 225
Anisimov, K. N., 33B, 310
Annamamedov, R., 30A, 173
Annapoorni, L., 30B, 443
Anselin, E., 26, 198
Anselin, F., 31A, 32
Ansell, G. B., 30B, 393; 31A, 155; 33B, 281; 34B, 89, 587
Antenen, K., 27, 961
Anthony, J. W., 35A, 461
Antonova, L. I., 29, 180, 247; 30A, 180, 247
Anzai, S., 34A, 146
Anzenhoffer, K., 31B, 288; 32B, 419; 34A, 70; 35A, 111; 35B, 149
Aoki, I., 27, 494
Aoki, T., 34B, 441
Aota, K., 35B, 372
Aoyagi, K., 27, 243; 30A, 177
Apinitis, S., 28, 682
Apinitis, S. K., 30B, 443
Apostolov, A., 31A, 137, 241
Appel, J., 27, 63
Appelman, E., 34A, 361
Appleman, D. E., 26, 484, 505, 525; 27, 554, 577, 683, 688; 28, 339, 358; 29, 259, 375, 391; 30A, 348, 435
Aragon de la Cruz, F., 27, 396
Arai, G. J., 27, 471
Arajs, S., 27, 260; 29, 225; 30A, 246
Araki, T., 26, 439; 27, 623; 28, 211; 32A, 424; 33A, 403, 404; 34A, 363
Aramaki, S., 26, 504; 27, 707; 28, 358
Aravamudan, G., 26, 472
Arbuckle, J., 27, 341
Arbuzov, M. P., 27, 26, 124; 30A, 172
Arcovito, G., 34B, 35
Arerbuch, M.-T., 35A, 347
Arevalo, M.I., 26, 559
Argay, G., 28, 222; 30A, 308, 320; 31A, 83
Argent, B. B., 26, 95; 27, 318; 29, 225
Argos, P., 35B, 176
Arima, K., 34B, 286
Arimoto, H., 27, 811; 29, 760
Ariya, S. M., 27, 484
Ariyaratne, J. K. P., 34B, 418
Armantrout, C. E., 27, 198, 249
Armitage, C. H., 30A, 180
Armstrong, R. L., 28, 358
Armstrong V. S., 27, 851
Arndt, D., 34A, 264
Arnold, G. P., 27, 197, 475; 33A, 53, 56, 256, 258; 34A, 59
Arnold, H., 27, 674

Arnone, A., 33B, 244
Arnott, J. H., 30B, 443
Arnott, R. J., 27, 495; 28, 133, 255; 35A, 158, 373
Arnott, S., 26, 801; 31B, 193; 33B, 185
Aronsson, B., 27, 93, 386; 28, 13
Arora, S. K., 34B, 57
Arrese, F., 27, 566
Arrington, D. E., 35B, 234
Arsakuni, A. A., 29, 434
Arthur, P., 26, 354; 29, 650; 31B, 43
Aruja, E., 26, 369
Arunsingh, 34A, 144
Arutjunjan, E.G., 28, 114, 214, 338; 31A, 119, 205; 31B, 408
Asahi, Y., 34B, 238
Asai, M., 35B, 388
Asanabe, S., 29, 139
Asanuma, M., 26, 193; 27, 258
Asbrink, S., 29, 309, 373; 35A, 207
Aselius, J., 27, 93
Asher, J. D. M., 27, 1008; 28, 686; 30B, 226, 227, 235; 31B, 267
Ashida, T., 26, 644, 788; 28, 390, 391; 30B, 132, 399; 31B, 91, 106, 128, 191; 32B, 67, 76, 150, 176, 226, 227, 232, 392; 33B, 174, 218, 418; 34B, 88, 270, 276, 513; 35B, 169
Ashworth, R. S., 33B, 473, 476
Ashworth, T., 31A, 72
Asirov, A., 27, 551, 557; 28, 160, 161
Askanov, L. A., 34B, 497
Askew, W. B., 33B, 266
Aslanov, L. A., 29, 137, 225; 31B, 407; 32A, 510; 33B, 410; 34B, 508, 35B, 614
Asmussen, R. W., 34A, 211
Aso, K., 29, 435
Asprey, L. B., 27, 463; 29, 433; 33A, 195; 35A, 167
Assarsson, P., 27, 988
Astakhova, I. S., 34B, 420; 35B, 582
Astakhova, L. P., 32A, 440; 34A, 384
Aström, A., 29, 269; 31A, 123
Atabaeva, E. Ja., 29, 138 33A, 45
Atkinson, V. A., 26, 642
Atoji, M., 26, 91, 92; 30A, 172; 31A, 32; 32A, 91, 118, 119, 120; 33A, 52, 91; 34A, 60, 81, 129; 35A, 134
Atovmjan, L. O., 26, 400; 27, 665, 820; 28, 629; 30B, 122; 31B, 51, 424; 32A, 150, 347, 510; 32B, 34, 64, 153, 154; 33A, 227, 271, 450; 33B, 61; 34A, 276, 349; 35A, 423, 431; 35B, 484
Atroscenko, L. V., 26, 133; 30A, 245
Atwood, J. L., 32B, 328; 33B, 301; 34B, 398, 401; 35B, 481
Au, P. K. L., 32A, 274; 34A, 327
Au, R., 32A, 185
Aubry, A., 34A, 387
Aubry, J., 30A, 180
Auer-Welsbach, H., 26, 103; 27, 423
Auld, J. H., 26, 118
Aurivillius, B., 27, 539, 657; 29, 271, 320, 454; 31A, 135
Aurivillius, K., 26, 444; 29, 67, 283, 310; 30A, 301; 32A, 191; 33A, 221, 235
Austerman, S. B., 30A, 304
Austin, A. E., 27, 57, 396; 30A, 178, 246; 33A, 85
Auvray, P., 34B, 28
Averbach, B. L., 27, 19
Averbuch-Pouchot, M.-T., 32A, 379; 33A, 395, 396
Averjanova, L. N., 29, 435
Averkieva, G. K., 27, 50
Avertisjan, E. I., 35A, 299
Avery, J. G., 30A, 244
Avilov, A. S., 34A, 88, 127
Avitabile, G., 31B, 12; 32B, 10; 34B, 11, 96
Avojan, G. L., 26, 745, 746
Avojan, R. L., 27, 784, 860, 933, 939; 28, 438, 683, 685; 29, 549, 679, 697; 30B, 443; 31B, 343, 361; 33A, 217
Avrova, N. P., 29, 436
Axel, H., 30A, 65, 66, 180; 31A, 48; 32A, 27; 33A, 39; 34A, 53
Aylett, B. J., 34A, 155
Aynsley, E. E., 28, 315
Ayphassorho, C., 26, 799
Ayres, F. D., 29, 570
Ayyar, R., 32B, 231
Ayyar, R. R. 30B, 393; 33B, 174
Azarkh, Z. M., 29, 225
Azaroff, L. V., 28, 118
Azenhofer, K., 35B, 534
Aziz, N. E. A., 26, 550; 29, 453
Azuma, K., 28, 65

Babad-Zakhrjapin, A. A., 26, 264; 27, 672; 34A, 144
Babcock, K., 29, 434
Babcock, K. R., 27, 478
Babel, D., 27, 453; 28, 314; 30A, 271, 277, 333; 31A, 138; 32A, 508; 34A, 203; 35A, 158
Babezova, V. I., 26, 19
Babicyna, A. A., 29, 225
Babuškina, T. A., 30B, 79; 33B, 68, 320; 34B, 69
Bacchella, G. L., 31A, 33
Bach, B., 28, 65
Bachmann, H. G., 26, 353, 354, 451; 27, 589
Bachmann, M., 30A, 347; 31A, 159; 32A, 292

Bacon, G. E., 27, 220, 614; 29, 636
Badachhape, S. B., 27, 352
Badamir, D. V., 27, 395
Baenziger, N. C., 26, 235, 364, 687, 688, 690; 27, 481; 28, 684; 29 5, 587, 668; 30B, 324, 327; 31B, 38, 367; 32A, 322; 32B, 75, 270; 33A, 161; 33B, 41; 34B, 43
Baetslé, L., 26, 467
Baetsle, L. H., 33A, 425
Baffier, N., 26, 412
Bagarjackij, Ju. A., 26, 264; 27, 371
Baggage, W. D., 28, 358
Baggio, S., 34A, 314, 316; 34B, 483, 522; 35A, 408, 35B, 598
Baglio, J. A., 33A, 310; 35A, 415, 416
Bagnell, K.W., 33A, 223
Bagouin, M., 28, 65
Bahezre, C., 32A, 230
Baikie, D. E., 34A, 472
Baikie, P. E., 30B, 393; 33B, 273
Bailey, A. S., 30B, 260; 35B, 190
Bailey, D. H., 32A, 14
Bailey, D. M., 26, 128, 164; 27, 297
Bailey, M., 32B, 82, 127, 132, 133
Bailey, M. F., 30A, 443; 30B, 316, 318; 35B, 772
Bailey, N. A., 30B, 393; 31B, 94, 352; 33B, 303, 325, 507
Bailey, S. W., 27, 696; 28, 270; 29, 407; 30A, 428; 31A, 237; 32A, 474, 475; 34A, 368
Bain, J., 30B, 443
Bain, V. A., 30B, 443; 34B, 371
Baird, H. W., 28, 682, 687; 31B, 348; 32A, 399, 508; 34A, 103
Baird, J. C., 26, 345
Baird, W. C. 28, 614
Baiwir, M., 31A, 43
Bak, B., 26, 760
Bakakin, V. V., 31A, 221; 32A, 493; 33A, 440; 35A, 188, 473, 477
Baker, D. E., 30A, 182
Baker, E. N., 31B, 389; 32B, 397; 35B, 413, 414, 615
Baker, J. G., 26, 345
Baker, T. W., 27, 68
Baker, W. E., 28, 342
Baker-Hawkes, M. J., 33B, 499
Baklagina, Ju. G., 30B, 443; 31B, 161
Baklanova, K. A., 26, 454
Bakulina, V. M., 29, 433
Bakuta, S. A., 30A, 248
Balaban, A. T., 34B, 141
Balasubramanian, K., 28, 682
Balasubramanian, R., 32A, 330; 33B, 384
Balchin,A. A., 30B, 183, 443
Baldock, P., 27, 125
Baldwin, J. E., 31B, 138; 32B, 167, 170, 196; 33B, 142; 34B, 87; 35B, 905
Baldwin, N. L., 29, 135, 224; 30A, 245
Bale, M. G., 26, 231
Baliceva, T. G., 35A, 387
Ball, M. C., 26, 348
Ballentyne, D. W. G., 26, 251
Ballestracci, R., 29, 39, 341; 30A, 44, 176, 177; 31A, 65; 32A, 223, 308, 510
Ballman, A. A. 27, 562
Balluffi, R. W., 29, 224
Bally, R., 28, 687; 30B, 393; 32B, 316; 35B, 446
Balta, E. Ja., 31B, 408
Baltzer, P.K., 33A, 235
Baluskina, T.A., 31B, 77
Bälz, U., 34A, 25
Bambauer, H. U., 35A, 282
Ban, N. T., 32A, 325
Ban, Y., 34B, 328
Ban, Z., 29, 90, 137; 30A, 176; 32A, 141
Band, E., 35A, 399
Bandoli, G., 34B, 533; 35B, 464, 859, 860
Banerjee, B. K., 30B, 445
Banerjee, N. K., 28, 682
Banerjee, R. L., 26, 561
Bang, E., 26, 335; 33A, 206
Bang, W. -B., 29, 252
Bankovskij, J. A., 30B, 447
Banks, E., 26, 415; 27, 336; 28, 358; 32A, 202; 33A, 344
Banks, F., 35A, 192
Bannister, A. J., 27, 409
Banus, M. D., 27, 244, 281; 29, 136
Baraniak, E., 35B, 759
Baranova, E. N., 26, 483
Baranova, R. V., 26, 335; 29, 48; 30A, 61, 179; 32A, 61; 33A, 80, 153; 34A, 79
Barassin, J., 26, 720; 28, 366
Barbariol, I., 28, 341
Barber, P. G., 35B, 69
Barclay, G. A., 26, 547, 657, 666, 776; 27, 654, 702; 28, 175, 642, 643, 683; 30A, 406; 30B, 336, 361; 34B, 565, 591; 35B, 672
Barczak, V. J., 26, 5
Bardi, R., 26, 761
Bardos, T. J., 26, 269; 32B, 224
Bare, D. W., 29, 223
Barendregt, F., 35B, 877
Bariand, F., 28, 358
Bariand, P., 27, 602; 28, 340, 357
Barker, D. L., 29, 703
Barker, S. M., 30A, 247
Barlow, M., 28, 409; 34B, 265
Barnes, W. H., 26, 353, 354, 380, 563, 782; 27, 589, 604, 1023; 28, 198, 204, 462, 548; 30B, 280; 31A, 142; 31B, 307; 34B, 393

Barnett, B. L., 35B, 142, 144
Barnett, J. D., 29, 136, 224, 225; 31A, 69
Barnhart, D. M., 33B, 307
Bårnighausen, H., 26, 220, 293, 337, 830; 27, 476; 28, 356, 494; 29, 275; 30A, 83, 289; 31A, 82; 32A, 245; 33A, 226, 229; 34A, 181, 182, 235; 35A, 234
Barns, R. L., 29, 435
Baro, R., 29, 434
Baron, V. V., 27, 284; 30A, 244, 248
Barr, R. E., 35B, 176
Barrans, Y., 31B, 78
Barrante, J. R., 26, 420
Barrer, R. M., 26, 517, 518; 34A, 371; 35A, 458
Barrett, A. N., 34B, 254
Barrett, C. A., 29, 434
Barrett, C. S., 26, 145; 27, 75; 28, 8, 41; 29, 433; 30A, 178; 31A, 42, 64, 240; 32A, 147; 33A, 157; 34A, 13
Barrett, E. K., 34B, 151
Barrett, R., 30A, 29, 174
Barro, R., 35A, 406
Bart, J. C. J., 33B, 76, 84, 147; 34B, 389, 390, 392; 35B, 454
Bartell, L. S., 26, 543; 29, 625; 30A, 266, 303; 30B, 405; 33B, 531; 34A, 171; 35B, 890, 897, 902
Barthel, J., 29, 434
Barthelemy, P., 31A, 32
Bartholin, H., 30A, 175
Bartl, H., 31A, 167, 168; 32A, 483, 34A, 251
Bartlett, M. F., 27, 1008
Bartlett, N., 26, 301, 313; 27, 665, 666; 31A, 83; 32A, 155, 160; 33A, 174; 34A, 219
Barton, C. J., 27, 451
Barton, G. H., 29, 135
Barton, J.W., 27, 296
Barton, R., 34A, 391
Bartram, F., 26, 492
Bartram, S. F., 26, 376; 29, 435, 438; 31A, 162; 34A, 377; 35A, 219
Barzdain, P. P., 33B, 531
Basak, B. S., 34B, 157
Basinski, S. J., 26, 65
Basinski, Z. S., 26, 147
Baskin, Y., 26, 102, 494; 29, 141
Bassett, D. C., 29, 760
Bassett, W. A., 29, 140; 34A, 146
Bassi, G., 28, 133, 317; 30A, 313, 326, 331, 347; 31A, 133; 33A, 417
Bassi, I. W., 26, 633, 635; 27, 787, 801; 29, 760; 32B, 44, 45, 46; 33B, 376, 521, 522, 523
Bassi, W., 34B, 365
Basson, S. S., 34B, 138; 35A, 423, 424; 35B, 177
Bastiansen, O., 26, 553, 641, 739; 28, 362; 29, 233, 490, 524, 642; 33B, 537; 35B, 890
Basu, G., 27, 844
Bataleva, S. K., 30A, 177
Batalieva, N. G., 29, 411; 32A, 511; 33A, 464
Batalli-Cosmovici, C., 34A, 144
Batchelder, D. N., 29, 225, 533; 30A, 244
Bates, C. H., 30A, 246
Bates, R. B., 34B, 308; 35B, 191, 389
Batt, A., 27, 510
Battelle, L., 32B, 200
Battelle, L. F., 30A, 378
Batterham, T. J., 34B, 146
Batterman, B. W., 27, 208; 31A, 64
Battersby, A. R., 26, 830
Battey, M. H., 27, 520
Battistini, G. G., 30B, 360, 397, 447; 31B, 402
Baturina, E. A., 31A, 84
Bau, R., 32A, 396, 510; 33B, 329
Baudour, J., 30B, 79
Bauer, D., 30A, 264; 33A, 203
Bauer, J., 28, 314
Bauer, S. H., 28, 362; 31B, 93, 298; 33B, 532, 534, 535; 34A, 152, 153, 185; 35B, 891, 896, 899, 902
Baughman, R. H., 29, 744
Baumann, I., 29, 89
Baun, W. L., 26, 254, 564; 27, 822
Baur, H., 35A, 335
Baur, W. H., 26, 350; 27, 612; 29, 346, 348, 351, 404, 437; 32A, 367; 33A, 368; 34A, 331, 332; 35A, 199, 271, 284
Bauren, E. -B., 32A, 511
Baxter, R.D., 29, 141, 226
Baxter, R.M., 34B, 184
Bayer, G., 26, 412; 27, 526; 28, 356; 35A, 242
Bayliss, P., 32A, 26; 33A, 359
Beagley, B., 28, 382, 686; 29, 232, 460; 30A, 445; 32A, 252; 33A, 411; 33B, 530; 34A, 155, 163, 178, 240; 35B, 888, 895
Beal, H., 32B, 292
Beale, J. P., 35B, 839
Beall, R., 35B, 50, 217
Beaman, D. R., 29, 224
Bean, V. E., 31A, 69
Bear, C. A., 35B, 624
Bear, I. J., 34A, 303; 35A, 367
Beard, L. N., 33B, 107
Bearth, P., 29, 437
Beattie, I. R., 30B, 443
Beauchamp, A. L., 33B, 292; 34B, 396

Beaudet, R. A., 33B, 538; 35B, 904
Beaudry, B. J., 27, 342; 29, 224
Beaute, C., 34B, 3
Becher, H. J., 27, 3, 65; 31A, 3; 35A, 6
Becherer, G., 27, 559
Beck, B., 26, 387
Beck, F. R., 27, 551
Beck, P. A., 26, 269; 27, 46, 278
Beck, R. L., 27, 241
Beck, W., 28, 71
Becka, L. N., 26, 603; 28, 71, 478; 32A, 326; 34A, 314, 316, 317; 34B, 483, 522; 35B, 598
Becker, L. 27, 809
Becker, M. R., 32B, 144
Becker, W., 33A, 153; 34A, 93
Becker, W. -J., 34A, 257; 35A, 244, 288
Beckett, D. R., 29, 760
Beckmann, O., 26, 305; 31A, 99; 34A, 144
Beckmanon, O., 33A, 153
Bècle, C., 32A, 8
Bedford, R. G., 26, 58
Bedlivy, D., 30A, 358
Bednowitz, A. L., 31B, 5, 153; 32B, 15, 146, 33B, 93; 35A, 158
Beegle, L. C., 30A, 248
Beeman, W. W., 27, 1061
Beerntsen, D. J., 29, 31; 30A, 79, 180
Beeskow, H., 26, 161; 29, 223
Beeson, E. L., 27, 729
Beevers, C. A., 26, 31; 28, 425; 30B, 191
Beger, R. M., 32A, 511
Begum, R. J., 30A, 178, 180; 34A, 9
Behnke, J., 34A, 264
Beineke, T. A., 33A, 115, 426; 34B, 380
Beintema, J., 30B, 96
Bekebrede, W. R., 26, 217, 474; 27, 40
Bekoe, D. A., 30B, 392; 32B, 323
Belavceva, E. M., 29, 760
Belbeoch, B., 26, 366, 629
Belcher, S. C., 29, 226
Belechij, M. S., 26, 189
Belford, R. L., 27, 844; 30B, 337, 340; 31B, 383; 32B, 387; 33B, 409, 437; 34B, 504, 532
Belforti, D., 26, 291
Belikov, A. M., 27, 492
Belikov, M., 27, 123
Belinskij, I., 35B, 676
Belitskus, D., 28, 683; 30B, 7, 136
Beljaev, I. N., 28, 318; 29, 435
Beljaeva, G. F., 27, 27
Beljaeva, I. I., 29, 435
Beljaeva, K. F., 31B, 409
Belkovskij, A. A., 29, 436
Bell, J. D., 31A, 102
Bello, J., 26, 817; 27, 1059
Bellobono, I. R., 34B, 118
Bellon, B. L., 34A, 159
Bellon, P. L., 26, 540; 28, 495, 659; 31B, 464; 34B, 430, 436, 604
Belluco,U., 35B, 859
Belopolskij, M. P., 27, 536
Belov, K. P., 27, 511; 30A, 179
Belov, N. V., 26, 301, 303, 462, 476, 509, 521; 27, 496, 510, 540, 551, 557, 675, 689, 704, 705, 714; 28, 190, 247, 254, 261, 264, 268, 275, 276, 342; 29, 379, 398, 404, 30A, 270, 377, 378, 419; 31A, 156-158, 218, 229, 232, 240; 31B, 380; 32A, 278, 302, 354, 368, 430, 437, 447, 454, 482, 511; 32B, 320; 33A, 177, 337, 347, 351, 446, 452, 469, 470, 471, 473, 481, 506; 34A, 12, 275, 286, 376, 379, 381, 386; 35A, 15, 226, 228, 229, 239, 249, 261, 283, 290, 291, 297, 299, 310, 312, 314, 315, 316, 388, 391, 445, 464, 465, 470, 477, 480
Belova, E. K., 26, 133; 30A, 177, 178, 245, 248
Belova, E. N., 26, 535
Bel'skaja, N. P., 35B, 602
Bel'skij, N. K., 30A, 275
Belt, R. F., 32A, 322
Benacerraf, A., 29, 136, 224
Bénard, J., 29, 135, 223
Bencze, W. L., 35B, 117
Benda, K. V., 33A, 31
Bender, S. L., 27, 498
Benedetti, E., 32B, 16; 34B, 90, 91, 129, 440; 35B, 6, 97, 106, 107, 111, 217
Benesovsky, F.,26, 10, 11, 18, 21, 25, 58, 68, 73, 81, 93, 101, 104, 156; 27, 30, 105, 126, 133, 135, 136, 138, 159, 213; 28, 3, 12; 29, 135, 136, 137, 138, 139, 226; 30A, 172, 174, 175, 244, 245, 246; 31A, 27; 32A, 141; 33A, 153; 34A, 39, 144
Bennet, M. J., 34A, 209
Bennett, I., 35B, 316
Bennett, J., 35B, 834
Bennett, J. M., 32A, 484; 33A, 492, 497; 34A, 371
Bennett, M. J., 27, 849; 28, 612; 31B, 353; 32A, 509; 32B, 446, 450, 482; 33A, 238; 33B, 292, 371, 378, 391, 395, 414, 483, 509; 34B, 396, 536
Bennett, S. L., 30A, 244
Bennion, R. B., 32A, 29
Bensey, F. N., 27, 446
Benson, K. E., 28, 21
Benson, R. C., 35B, 905
Bentle, G. G., 26, 258

de Benyacar, M.R., 27, 462, 967
Benz, R., 31A, 68; 34A, 135; 35A, 113
Berca, D.M., 30A, 181
Berdonosov, S.S., 27, 444; 28, 315; 29, 433
Berdonosova, D.G., 27, 445; 28, 315; 29, 433
Berdot, J.-L., 32A, 270
Berg, H., 26, 640
Berg, L., 27, 916
van den Berg, J.M., 26, 320; 29, 82; 32A, 96; 35A, 111
Berger, J.A., 30A, 246
Berger, J.E., 26, 820
Berger, L. I., 30A, 173
Berger, R. M., 34A, 370
Bergerhoff, G., 27, 420, 430; 33A, 192, 284; 34A, 344; 35A, 173
de Bergevin, F., 26, 178; 27, 256, 523, 29, 435
Berghout, C. W., 27, 303
Bergin, R., 32B, 263; 33B, 231
Bergman, B. G., 29, 87
Bergman, J. G., 31B, 322, 410
Bergsma, J., 26, 27; 27, 118
Bergstein, A., 28, 285
Bergstol, S., 29, 434
Berkebile, C.A., 33B, 47
Berking, B., 34B, 93; 35B, 342
Berkowitz, A. E., 29, 140
Berlincourt, D., 28, 317
Berlincourt, T. G., 26, 202
Berliner, L. J., 33B, 117
Berman, H., 32B, 52
Berman, H. M., 33B, 162, 163, 168; 34B, 124, 125; 35B, 270
Bernal, I., 28, 316; 32B, 459; 34B, 44, 527; 35B, 58
Bernal, J., 34B, 44, 527
Bernard, J., 28, 340
Bernard, M., 26, 436
Bernard, M.-J., 34B, 546
Bernard, W.H., 35B, 700
Bernardini, G.P., 33A, 79
Bernauer, K., 34B, 333
Berndt, A.F., 26, 243; 27, 326; 28, 608; 29, 136, 225; 30A, 179, 182; 32A, 141
Berndt, F., 29, 140
Berner, R.A., 27, 257; 29, 136, 435
Bernier, J.-C., 26, 409; 27, 491; 28, 356; 31A, 161; 32A, 509; 33A, 312
Bernostat-Wulf, H., 34B, 338
Bernstein, H., 29, 224
Bernstein, J.L., 29, 51, 136; 30A, 322; 31A, 147, 158, 187; 32A, 287, 292, 312; 33A, 327, 335; 34A, 238, 269, 278, 294; 35A, 102, 164, 165
Bernstein, T., 33B, 345
Bernusset, P., 27, 345, 346; 29, 135, 223
Berodias, G., 30A, 176, 177; 31A, 37; 33A, 66; 34A, 96
Berry, L. G., 28, 340, 356
Berspächer, O. E., 33A, 323
Bersted, B. H., 33B, 437
Berta, D. A., 35B, 628
Bertaut, E. F., 26, 389, 390; 27, 256, 522, 523; 28, 133, 134, 136, 138, 151, 318; 29, 341; 30A, 42, 44, 175, 176, 177, 178, 313, 316, 347; 31A, 36, 133, 138, 159, 241; 32A, 53, 268, 283, 292, 426, 434; 33A, 153, 316, 320; 34A, 91
Bertaut, F., 26, 117, 400; 27, 163, 345, 516
Berthold, H. J., 26, 592; 29, 433
Berthold, I., 28, 223; 32A, 323
Berthold, J. H., 31A, 100
Berthon, J., 31A, 161
Berthou, J., 27, 972; 30B, 163; 32B, 63; 34B, 45; 35B, 152
Bertie, J. E., 28, 118
Bertinotti, A., 28, 422
Bertinotti, C., 35A, 201
Bertinotti, E., 28, 315; 35A, 201
Bertrand, J. A., 28, 87; 31A, 181; 31B, 179, 446, 458; 32B, 449; 34B, 22, 556; 35B, 617, 722, 723
Besançon, P., 34A, 98, 294
Best, K. J., 34A, 142
Bettinali, C., 28, 317
Bettman, M., 34A, 250
Betz, A., 32A, 142; 33A, 153
Beu, K. E., 27, 347; 29, 223
Beucher, M., 33A, 417
Beurskens, G., 28, 683; 29, 254, 297
Beurskens, P., 30B, 58
Beurskens, P. T., 33B, 485; 35B, 806, 810
Bevan, C. W. L., 30B, 233
Bevan, D. J. M., 33A, 291; 35A, 182, 259
Bevan, J. A. 28, 687
Bevan, T. H., 27, 783
Bever, H., 34B, 178
Bever, M. B., 29, 136
Beyer, H., 32A, 261; 33A, 269, 387; 35A, 279
Beyer, H. H., 34A, 244
Bezjak, A., 29, 458
Beznosikova, A. V., 33A, 121
Bezzi, S., 26, 761; 28, 424; 29, 760
Bezzubov, V. M., 35B, 898
Bhan, S., 33A, 124, 126, 153; 34A, 11, 142, 145; 35A, 14, 108
Bhatnagar, V. M., 30B, 443
Bhattacharjee, S., 28, 342
Bhattacharjee, S. B., 28, 342
Bhattacharyya, G. C., 28, 683
Bhowmik, G., 26, 449
Biagini, M. C., 30B, 443

Biagini, S., 30B, 4; 34B, 195, 199, 201; 35B, 702
Bianco, V. D., 28, 659
Bideau, J. -P., 34B, 214, 222
Bidwell, L. R., 29, 223
Bielen, H., 27, 606
Bienenstock, A., 28, 290
Bienfait, M., 27, 888
Bierrum, A. M., 34B, 418
Bierstedt, R., 27, 965
Bigliardi, G., 27, 471; 28, 89, 92, 315; 30A, 298, 409; 30B, 365, 367, 444; 31A, 159
Bigoli, F., 33B, 436; 34B, 547; 35A, 396, 397; 35B, 720
Bigolt, F., 34A, 340
Bijuskin, V. N., 31B, 380
Bijvoet, J. M., 30B, 26
Billier-Pandraud, H., 34A, 218
Billiet, Y., 28, 356; 34A, 90
Binas, H., 27, 575, 576; 31A, 194; 32B, 266
Binnie, W. P., 35A, 144
Bird, G. R., 26, 345
Bird, P. H., 30B, 393; 32A, 508; 33B, 303, 364, 379, 382; 34B, 475, 477
Birjukov, B. P., 33B, 310, 316, 348; 34B, 414
Birkhan, J., 34A, 144
Birle, J. D., 33A, 477
Birnbaum, G. I., 30B, 393; 32B, 261, 264; 33B, 215; 34B, 330; 35B, 361
Birnbaum, K. B., 34B, 317; 35B, 351
Birnstock, R., 32A, 351; 35A, 190
Birun, N. A., 26, 114; 27, 268
Bishop, J. E., 26, 391
Bissert, G., 34A, 327; 35A, 341
Biswas, A. B., 26, 551; 27, 493, 506; 28, 384; 29, 471, 481; 33B, 179; 34A, 327; 34B, 19; 35A, 341
Biswas, T., 34A, 142
Bither, T. A., 34A, 144
Bitjuckaja, L. A., 30A, 174
Bitjuckij, P. N., 34A, 293
Bitti, R. R., 33A, 153
Bittiger, H., 35B, 885
Bittner, H., 27, 121; 29, 139; 30A, 35
Bjåmer, K., 28, 684; 29, 489; 31A, 62; 32B, 201, 252; 33B, 216
Bjerkelund, E., 29, 77, 135; 30A, 94, 181; 32A, 143; 35A, 124
Bjőrnstjerna, V., 28, 7
Bjørseth, A., 35B, 903
Bjorvatten, T., 26, 546, 773; 27, 747, 986; 28, 103, 362, 571; 31B, 273, 33B, 29; 34B, 348
Bkhargava, K., 28, 356
Bkouche-Waksman, I., 31B, 438; 32B, 445; 33B, 239
Black, D. R., 30B, 446
Black, J. B., 29, 224
Black, M., 34B, 437, 606
Black, P. J., 26, 6, 7, 16
Blackedge, J. P., 27, 241
Blackman, L. C., 26, 287
Blackman,M., 26, 296, 31A; 240
Blackstone, M., 31B, 387
Bladon, P., 30B, 236
Blaisse, B. A., 32A, 65
Blake, A. B., 29, 622; 34B, 517, 561
Blake, A. R., 29, 296
Blake, C. C. F., 27, 1058, 1059
Blake, R. L., 31A, 124
Blanconnier, P., 29, 141; 30A, 178
Bland, J. A., 26, 68, 325, 370; 27, 773
Blankenburgs, G., 30A, 172
Blasse, G., 28, 285, 356; 29, 136, 435; 31A, 144
Blazko, L. P., 27, 558
Bleidelis, Ja. Ja., 35B, 484
Bless, P. W., 35A, 272
Bliesner, W. C., 27, 798
Blinova, N. I., 29, 434
Bliznakov, G., 29, 136
Bloc, F., 28, 65
Block, S., 26, 455; 28, 164, 292; 29, 389, 392; 30A, 254, 413; 30B, 272; 31A, 134, 141, 169, 171; 31B, 427; 32A, 355, 497; 32B, 185; 34B, 79
Blockhammar, I., 30A, 447
Blom, E. A., 34B, 410
Bloor, J. H., 32B, 270
Bloss, F. D., 26, 531; 28, 358
Blount, A. M., 34A, 368
Blount, J. F., 31B, 347; 32B, 420; 34B, 556; 35B, 653
Blow, D. M., 26, 811; 27, 1063
Bloxam, T. W., 29, 437
Blukis, U., 28, 668
Blum, P. -L., 30A, 40, 175, 182; 31A, 33; 33A, 54
Blumberg, A. A., 29, 759
Blundell, T. L., 30B, 396; 32B, 472
Boates, T. L., 35B, 902
Bobro, Ju. G., 29, 225
Bochynski, Z., 33B, 528
Bock, C. M., 33B, 77; 35B, 427
Bockelmann, W., 35A, 61, 124
Bocquet, J.-P., 33A, 306
Bodak, O., 34A, 63
Bodak, O. I., 31A, 35; 32A, 5; 34A, 63, 64, 143; 35A, 44, 96, 124
Bode, H., 30A, 180
Bodenseh, H. K., 34A, 233
Bodnar, R., 31A, 20
Bodnar, R. E., 32A, 22
Bodor, G., 32B, 15
Boelema, E., 32B, 182
Boer, F. B., 28, 687
Boer, F. P., 29, 230, 526; 32A, 510; 33B, 38, 235, 270, 271; 34B, 34, 366,

367, 369; 35B, 99, 443, 491
de Boer, B. G., 33A, 158; 33B, 265; 35B, 715
de Boer, J. H., 29, 298
de Boer, J.J., 35B, 149, 534
de Boer, J.L. 28, 82; 31A, 214; 33B, 61, 247
Boeters, K.E., 29, 137
Boetticher, H., 30B, 393
Boeyens, J. C. A., 30B, 254, 443; 32B, 54; 33A, 253; 33B, 93, 239, 241; 34A, 389; 34B, 351, 352, 356; 35B, 159, 160, 421, 746, 752, 791
Bogdanov, S. V., 26, 701
Boggs, J. E., 30B, 405; 33B, 538
Bogorodskij, O. V., 26, 523
Bohloff, R., 35A, 282
Bohmann, T., 26, 187
Bohn, C. R., 26, 634
Bohn, R. K., 35B, 896
Boiko, E. B., 32A, 44
Boiko, L. G. 35A, 124
Boileau, S., 27, 811
Bois, C., 31B, 74; 35B, 65
Boive, T., 35B, 887
Bojarski, Z., 30A, 38; 32A, 45
Bojesen, I., 33B, 540
Bok, L. D. C., 34B, 138; 35A, 423, 424; 35B, 177
Bokij, G. B., 26, 330; 27, 668, 669, 698, 820; 28, 629; 29, 27, 586; 30A, 16, 181, 265; 30B, 291
Bokij, N. G., 28, 686; 30B, 51, 274, 443, 447; 31B, 333; 32B, 330, 331, 336; 33B, 304, 35B, 407, 489, 556
Bokij, N. K., 32A, 511
Bokov, V. A., 29, 435; 33A, 324
Boldrini, P., 27, 835, 31A, 123; 31B, 344, 360, 447; 32A, 156, 346; 32B, 407; 35B, 628
Bolgov, I. S., 29, 225
Bolhuis, F. van, 30A, 353; 31A, 214; 32A, 149; 35B, 221
Boller, H., 26, 9; 28, 9, 27; 29 137; 30A, 174, 181; 32A, 24; 33A, 28, 130, 153; 34A, 144
Bolton, W., 28, 499, 501; 29, 698, 699, 715; 30B, 105, 167, 443
Bolz, L.H., 27, 391; 28, 314
Boman, C. -E., 35A, 211, 212
Bombieri, A., 31B, 332
Bombieri, 32B, 397; 34B, 517, 533, 604; 35B, 762, 819, 860, 868
Bommert, D., 29, 435
Bonamico, M., 26, 768; 27, 778, 967, 979, 1039; 28, 578, 683, 687; 29, 286; 31B, 373, 378, 381; 30A, 444; 32B, 35, 326; 33B, 268; 34B, 35; 35B, 816
Bonar, L. C., 26, 826
Bonart, R., 27, 800
Bond, W. L., 26, 288
Bondarenko, V. S., 26, 372
Bonefacic, A., 26, 443; 27, 610; 28, 212
Bongers, P. F., 33A, 61
Bonham, R. A., 26, 760; 28, 668; 29, 450, 733
Boniszewski, T., 26, 166
Bonnelle, C., 26, 205
Bonnemere, C., 26, 52; 30B, 242
Bonnet, J. -J., 35B, 655, 662
Bonnetain, L., 33B, 528
Bonnier, E., 29, 223
Boon, J., 34B, 177; 35B, 886
Boonstra, E.G. 26, 830, 28, 419; 33A, 319
Boorman, P.M., 33A, 220
Bootsma, G.A., 32B, 7, 8
Bordet, P., 29, 433
Bordner, J., 35B, 215, 338, 348
Borene, J., 33A, 314; 35A, 364, 375
Borensztajn, J., 31A, 146, 185
Borgeaud, P., 26, 247
Borgen, B., 27, 758
Borgen, O., 26, 292
Boriskina, N. G., 29, 225
Borisov, S. V., 27, 540; 28, 188; 29, 374; 30A, 377, 378; 31A, 201; 32A, 303; 33A, 331,345, 351, 518; 34A, 280, 282, 286, 287; 35A, 239, 247, 255, 261, 477
Borlera, M. L., 28, 356; 29, 434
Born, L., 34B, 386
Borodin, L.S., 27, 514
Borovjak, T. E., 35B, 544
Borscevskij, A. S., 27, 51
Borschutzky, J., 32A, 464
Borsini, G. C., 29, 760
Bort, D. N., 26, 633
Bortin, O., 33A, 209
Bortolozzo, G., 35B, 464
Borup, R. A., 27, 563
Borusevič, L. K., 29, 138, 225; 30A, 175, 176
Bosch-Figueroa, J. M., 30B, 444
Bosio, L., 30A, 48, 178; 34B, 82
Bosnich, B., 30B, 393
Bostsarron, A. -M., 27, 201
Bottcher, B., 34B, 368, 372
Botterweg, R., 30A, 353
Bottomley, F., 33A, 211
Bouchand, J. P., 33A, 153
Bouchard, J.-P., 29, 139; 30A, 36, 175; 32A, 102, 141
Bouchard, R.J., 30A, 181; 34A, 109
Boucahrd, R.T., 31A, 51
Boucher, B., 30A, 322
Boucher, M. L., 33A, 454
Boucher, R. R., 30A, 11, 173
Boueke, K., 27, 802
Bouissières, G., 29, 434
Bouloux, J. -C., 34A, 265
Bouska, V., 28, 342

Boutin, H., 27, 421; 29, 343
Boutin, H. P., 34A, 170; 35B, 179
Bouvaist, J., 34A, 240; 34B, 454
Bovio, B., 27, 989; 28, 683, 687; 30B, 139, 141, 143
Bowden, F. P., 26, 296
Bowman, A. L., 26, 104; 30A, 40, 175; 31A, 122; 33A, 53, 56, 256; 34A, 59
Bowman, M. G., 26, 3, 80
Boyce, C.B., 30B, 237
Boye, C.A., 26, 639
Boyer, W. H., 27, 716
Boyko, B. T., 27, 369
Boyko, E. R., 29, 693; 31B, 393
Boyle, R. W., 29, 438
Bozorth, R. M., 26, 127
Bracher, B. H., 32B, 5; 35B, 20
Brackett, E. B., 28, 77, 80, 97, 98
Brackett, T. E., 28, 77, 80, 97, 98
Bradford, S. A., 27, 380
Bradley, D. C., 27, 835; 29, 630; 30B, 374
Bradley, J. N., 32A, 216
Bradley, R. S., 29, 438
Bradley, W. F., 29, 464, 561; 30A, 376; 31A, 47, 172
Brady, G. W., 29, 760
Brafman, O., 32A, 135; 33A, 146
Braibanti, A., 26, 336, 342, 500; 27, 471, 640, 641; 28, 89, 92, 315; 30A, 298, 350, 409; 30B, 365, 367, 444, 445; 31A, 103, 159, 208; 32A, 151, 352, 509; 32B, 425, 427; 33B, 436; 34A, 340, 349; 34B, 34, 547; 35A, 397; 35B, 720
Braitsch, O., 26, 449, 494
Brajnik, I. E., 26, 166
Brajovic, V., 27, 421
Branche, G., 28, 357
Brand, J. A., 26, 110, 113; 27, 140
Brand, J.C. D., 34A, 243
Brand, P., 28, 99, 315; 31A, 105; 32A, 107
Brändén, C. I., 26, 595; 27, 468; 28, 94, 95, 652; 29, 632; 31B, 26, 324; 32B, 436, 450; 34B, 134; 35B, 887
Brandhorst, H., 29, 137
Brandl, F., 32B, 138
Brandle, C. D., 34B, 253
Brandon, J. K., 32B, 64; 33A, 316
Brandt, B. G., 32A, 259
Bratcher, L.M., 27, 451
Bratek-Wieworowska, M.D., 32B, 264
Brathovde, J. R., 29, 543
Bratschun, W. R., 28, 282
Bratton, W. K., 33B, 414; 34B, 530, 35A, 426
Brauer, D. J., 34B, 403; 35B, 478
Brauer, G., 26, 215, 293; 27, 475, 476, 483; 28, 101, 284, 338; 29, 136, 224, 225, 433
Braun, P. B., 28, 7; 35B, 138
Braun, P. J., 35B, 336
Braun, R. L., 31B, 397, 400; 32B, 403, 404
Braxton, H., 29, 759
Braye, E. H., 27, 647
Breazeale, J. D., 26, 674
Breck, D. W., 26, 517; 33A, 480
Bredig, M.A., 26, 87
Bredikhina, T. N., 26, 650
Breece, J. L., 35B, 617
Breedis, J. F., 27, 155, 251
Bregeault, J. -M., 29, 435; 35A, 372
Bregman, J., 27, 734, 787; 29, 660; 30B, 94
Brehler, B., 26, 319; 28, 86; 29, 278; 32A, 207; 33A, 233, 451; 34A, 205, 206; 35A, 149, 161
Brehm, L., 35B, 5
Breitling, G., 27, 385
Brencic, J. V., 34A, 207, 209; 35A, 418
Breneman, C. L., 32A, 203; 34A, 164, 180
Brenet, J., 26, 365
Brennan, T., 34B, 527
Brennan, T. F., 35B, 322
Brenner, S., 29, 667
Brenner, S. A., 30B, 214; 31B, 6
Breskovska, V., 26, 523; 28, 340
Breton, M., 30B, 393; 31B, 97, 102
Breton-Lacombe, M., 32B, 123, 124
Brett, R., 29, 223
Breusov, O. N., 26, 299
Brewer, F. M., 26, 693
Briancon, D., 26, 368; 27, 519
Brice, J. C., 26, 149
Brice, M. D., 35B, 846
Bridle, C., 30B, 444
Bright, D., 33B, 511; 34A, 226; 34B, 609, 611; 35B, 476
Bright, N.F. H., 26, 349, 382; 27, 537
Brill, R., 27, 382
Brimacombe, J. S., 30B, 170; 35B, 262
Brindley, G. W., 26, 346, 529
Briner, R. C., 30B, 237
Brinkel, H., 26, 736
Brisi, C., 26, 406; 523; 28, 341
Brisse, F., 30A, 317; 33A, 286;
Britt, C. O., 33B, 538
Brittain, A. H., 34A, 240
Britton, D., 27, 973; 28, 105; 29, 290 30A, 294; 30B, 38, 289; 31B, 41, 312, 329, 330; 33B, 138, 321; 35A, 190
Britts, K., 29, 667, 726; 30B, 86; 31B, 85, 137, 182; 32B, 39, 152, 186, 247
Brixner, L. H., 27, 342; 29, 434, 435
Brixner, L. J., 30A, 172
Brjukhova, E. V., 33B, 320
Broadbent, T. A., 30A, 365
Broch, N. C., 30B, 398; 31B, 326; 32A, 260
Brochu, R., 35A, 28
Brodersen, K., 26, 324; 30A, 276; 31A,

106; 33A, 236, 240; 34A, 183
Broggi, R., 29, 502
Bronger, W., 27, 321; 33A, 3
Brophy, G. P., 27, 633
Brophy, J. H., 29, 136, 224
Brosset, C., 26, 479; 27, 406, 667, 676; 28, 184, 292; 29, 244
Brostigen, G., 34A, 97; 35A, 75
Brouns, E., 29, 380
de Brouwer, J. L., 30A, 176
Brown, A., 26, 186, 256, 271; 27, 90, 216; 30A, 21, 173, 251
Brown, A. R. G., 29, 137
Brown, B. E., 27, 678, 696; 28, 270; 29, 31, 407; 30A, 79, 180
Brown, B. W., 26, 670; 28, 616; 29, 595, 598
Brown, C. J., 27, 871; 30B, 81; 31B, 4, 27, 47, 48, 70, 71, 95, 129, 174; 32B, 59, 62, 77, 82, 127, 132, 133; 33B, 431, 432
Brown, D., 29, 281; 32A, 159; 33A, 223; 34A, 180, 197; 35B, 813, 818
Brown, D. A., 27, 783
Brown, D. A. S., 34B, 23
Brown, D. H., 27, 672
Brown, D. P., 34A, 178
Brown, D. S., 29, 686; 30B, 251; 32B, 464; 33B, 383; 34B, 578, 580
Brown, F., 29, 224
Brown, G. E., 34A, 388
Brown, G. H., 33B, 528
Brown, G. L., 27, 1055
Brown, G. M., 27, 731; 28, 414, 519, 682; 30B, 179; 31A, 90; 33B, 23, 447, 449; 34A, 305
Brown, H. A., 26, 217
Brown, I., 27, 983
Brown, I. D., 26, 655; 29, 286, 343, 409; 31A, 109; 32B, 64; 33A, 316; 35A, 168, 385
Brown, J. J., 29, 436
Brown, J. N., 35B, 34, 184, 207, 455
Brown, P. J., 26, 19; 27, 173, 525; 32A, 102, 415
Brown, R., 33B, 93
Brown, R. D., 33B, 540
Brown, R. N., 26, 599; 27, 566, 568, 569; 28, 683
Brown, T. L., 29, 535
Brown, W. A. C., 26, 747; 28, 533; 30B, 116; 31B, 372
Brown, W. E., 27, 575, 577; 34A, 319; 35A, 296
Browne, J. D., 26, 218
Brownlee, L. D., 28, 135
Brozel, R., 27, 386
Bruce, J. A., 30A, 324
Bruckner, S., 34B, 520
Brueckner, D. A., 26, 830; 27, 1000
Brufani, M., 26, 769, 776, 779; 28, 456, 682, 687; 31B, 131; 33B, 191
van Bruggen, C. F., 33A, 61
Brugger, R. M., 32A, 29
Bruins, D., 35B, 610
Brukl, C., 26, 18, 21, 25, 70, 156
Brümmer, O., 26, 109; 27, 559
Brun, E., 27, 487
Brun, K., 29, 224; 30A, 246, 248
Brun, L., 31B, 324, 442
Brunel, M., 28, 357; 29, 435
Brunie, S., 28, 135, 136, 138
Bruning, H. A. C. M., 29, 54; 30A, 174
Brunton, G., 31A, 93; 32A, 179; 34A, 188, 190, 192, 193, 195, 198, 199; 35A, 168
Brunton, G. D., 33A, 176; 181; 35A, 166; 35B, 636
Brusencev, F. A., 29, 586; 30A, 377, 378
Brusilovskij, B. A., 26, 166
Brusset, H., 31A, 149; 31B, 74, 378, 438, 32A, 270; 32B, 102; 33B, 239; 34A, 218
Bruzzone, G., 26, 164; 29, 136, 140; 30A, 174; 34A, 145; 35A, 97, 124
Bryan, J. G. H., 35B, 262
Bryan, R. F., 26, 663, 665, 831; 27, 971, 977; 28, 387, 636; 32B, 84, 334, 335, 367, 458, 481; 33B, 308; 35B, 432, 538, 539, 545, 546, 800
Bryant, M. W. R., 34B, 48, 49; 35B, 60
Bryden, J. H., 26, 589; 27, 52
Brydon, J. E., 26, 529
Bryžina, M. F., 27, 508; 28, 341
Buchanan, D. R., 33A, 202
Buchanan, G. L., 27, 945
Buchardt, O., 33B, 146; 35B, 232
Büchi, G., 28, 683
Büchner, W., 29, 467
Buckel, W., 27, 340
Buckingham, D.A., 35B, 686
Buckles, R.E., 32B, 75; 34B, 43
Buckton, K.S., 34A, 215
Budberg, P.B., 26, 218; 29, 137, 138
Budzinski, H., 28, 314; 29, 433
Buehler, W.J., 30A, 181
Buerger, M.J., 26, 471, 502; 27, 528, 699, 701, 712; 28, 19; 32A, 384, 495, 511; 35A, 363
Buerskens, P.T., 31B, 3
Buessem, W.R., 27, 495
Bugg, C., 30B, 131; 32B, 51
Bugg, C.E., 32B, 223; 34B, 259; 35B, 251, 283
Bugli, G., 27, 169
Buisson, G., 31A, 133; 32A, 434; 35A, 237
Bujnov, N.N., 27, 22
Bukin, V.I., 32A, 490
Bukovska, M., 26, 478, 497
Bukowska-Strzyzewska, M., 28, 631; 30B, 346, 444; 31B, 406
Bulakh, A.G., 26, 454, 483

Bulatov, A.S., 27, 285
Bulatova, R.F., 26, 285
Bullen, G.J., 26, 673, 729; 27, 990; 30B, 339, 393, 444; 33B, 281; 34A, 155; 34B, 362; 35B, 425, 439
Bulliner, P.A., 34A, 219
Bundy, F.P., 32A, 147
Bunn, C.W., 27, 807
Burbank, R.D., 27, 446; 31A, 97; 35A, 177
Burden, F.R., 33B, 540
Burdese, A., 28, 291, 356; 29, 434
Burewicz, A., 33A, 266
Burge, R.E., 27, 1054
Burger, K.O., 26, 124; 27, 11, 375
Burgers, W.G., 27, 239
Bürgi, H.B., 33B, 85; 35B, 61
Burgner, J.W., 34B, 124, 125
Burke, M., 27, 862
Burkhardt, K., 30A, 57, 173, 174; 33A, 89; 34A, 142
Bürki, H., 26, 426, 738; 33A, 243
Burlet, P., 30A, 178
Burley, B.J., 26, 531
Burley, G., 28, 100, 290
Burn, A.J., 30B, 444
Burnašova, V.V., 30A, 179; 34A, 143; 35A, 7, 53
Burnham, C.W., 26, 502; 27, 707, 712; 28, 233, 237, 357; 31A, 228; 32A, 469
Burnham, J.B., 29, 225
Burns, J.H., 27, 451, 460, 643; 28, 74, 75; 29, 255, 262; 30A, 266, 269, 272; 31A, 86; 33A, 175, 181, 185, 195; 34B, 500; 35A, 151
Burota, T.A., 26, 381
Burov, I.V., 26, 142; 27, 161
Burova, T.A., 28, 342
Burow, N.E., 27, 735; 33B, 21
Burr, A.H., 34B, 73
Burri, G., 32A, 230
Burtlich, J.M., 32A, 510
Burzlaff, H., 26, 453; 29, 25; 32A, 504; 33B, 126, 143; 34B, 187; 35B, 221
Busch, G., 26, 245
Buschow, K.H.J., 29, 140; 30A, 172, 173; 31A, 12; 32A, 6; 33A, 153; 34A, 68, 144, 145; 35A, 53, 94, 95, 124
Busetta, B., 34B, 76, 77, 160, 290
Busetti, V., 27, 966; 28, 480; 33B, 3; 34B, 169, 181, 188; 35B, 155
Bush, M.A., 32B, 319, 355; 33B, 506; 35B, 537, 548
Bushnell, G.W., 32B, 464
Busing, W.R., 27, 422; 30A, 269, 303; 33B, 449; 34B, 30
Buslaev, J.A., 34A, 168
Busmann, E., 26, 44, 84; 27, 350; 29, 83
Butikova, I.K., 35A, 305
Butler, J.F., 26, 99
Butler, S.R., 27, 495
Butman, L.A., 29, 380; 32A, 510
Butte, W.A., 27, 810
Buttner, G., 28, 338
Buxbaum, G., 34A, 265
Buys, H.R., 35B, 897
Byckova, M.I., 30A, 248
Byk, D.S., 29, 138
Bykov, V.N., 27, 151; 34A, 142; 35A, 115
Bykova, A.V., 26, 379; 27, 514; 28, 339
Bylickina, T.I., 35A, 226

Caban, N.F., 34A, 6
Cable, J.W., 26, 141, 318; 27, 151, 274, 333, 334, 368, 434, 440; 28, 27; 29, 136, 137
Cabri, L.J., 30A, 172
Cabrini, A., 26, 672
Cadoff, I.B., 27, 181
Cady, H.H., 28, 467, 686; 30B, 69; 31B, 152; 32B, 94
Cagle, F.W., 26, 195
Caglioti, G., 27, 474, 720
Cahn, R.W., 26, 14; 27, 12, 383
Calabrese, J.C., 31B, 64; 35B, 17,50
Calas, M.-R., 34B, 210
Calder, R.S., 27, 397
Calderon, J.L., 34B, 453
Calhoun, P.A., 30A, 177, 247
Calleja, F.J.B., 29, 760
Calleri, H., 31B, 53
Calleri, M., 28, 241; 29, 663; 31A, 219; 31B, 65; 32A, 387; 32B, 428; 34B, 150, 197, 198; 35B, 213
Calligaris, M., 31B, 89; 34B, 520; 35B, 625, 779, 796, 810
Calvarin, F., 34B, 454
Calvert, D., 30B, 443
Calvert, L.D., 26, 55, 701; 27, 119, 182, 482, 828; 28, 118; 29, 18, 33, 137, 433; 30A, 173; 35A, 12, 93, 124
Calvo, C., 28, 189; 30A, 383, 385; 32A, 274, 357, 372, 376; 33A, 399, 401, 407, 410, 419; 34A, 327, 330; 34B, 29; 35A, 343, 350, 357, 359; 35B, 643
Camellini, A.T., 35A, 396
Camellini, M., 35B, 425
Camellini, M.T., 35A, 396
Camerman, A., 29, 694; 30B, 109, 184, 185, 229, 283; 34B, 141; 35B, 116, 257
Camerman, N., 28, 585, 586; 29, 539, 746; 30B, 185, 204, 229, 444; 34B, 239, 241; 35B, 257
Cameron, A.F., 30B, 393; 33B, 98; 34B, 120, 310, 340; 35B, 147
Cameron, E.N., 26, 138
Cameron, N., 26, 711
Čamova, V.N., 27, 560
van Camp, A., 26, 623
de la Camp, U., 35B, 83

de Camp, W.H., 35B, 161
Campbell, A.E., 27, 1038
Campbell, F.A., 27, 690
Campbell, J.C., 34B, 252
Campbell, R.B., 26, 756; 27, 938
Campbell, W.J., 26, 347
Campos, F.P., 28, 23
Camy, G., 30B, 393
Canali-Padovani, R., 30B, 444
Candlin, R., 30B, 395; 32B, 422; 35B, 664
Canepa, F.G., 28, 398
Canepari, R., 27, 520
Cann, J.R., 32A, 465
Cannas, M., 28, 682; 30B, 4, 393, 32B, 190; 33B, 145; 34B, 195, 199, 201; 35B, 702
Cannillo, E., 29, 608; 31A, 221, 232; 32A, 444; 33A, 489; 34A, 381
Cannon, P., 29, 138
Cano, F.H., 35B, 45
Cantrell, J.S., 26, 615; 34B, 343
Canut, M.L., 26, 559, 561, 738, 780; 27, 566, 1066
Capacchi, L., 33B, 483
Capelli, R., 26, 226; 27, 249; 29, 140, 141, 226
Capio, C.D., 30A, 248
Čapovskij, Ju.A., 31B, 361
Capparelli, M.V., 34A, 317
Captuller, J., 35A, 124
Carazzolo, G., 27, 966; 28, 480; 29, 760; 30B, 152; 33B, 136
Cardillo, M.J., 35B, 902
Carel, C., 29, 434
Carfagna, P.D., 32B, 474
Carhart, R.E., 34B, 93
Carl, W., 34A, 120
Carlisle, C.H., 27, 1059; 30B, 183, 443; 31B, 117, 234; 32B, 186; 34B, 255
Carlson, O.N., 26, 275; 27, 380; 29, 223; 30A, 177; 33A, 153
Carlsson, B., 34A, 143
Carlsson, J.-O., 35A, 38, 124
Carlström, D., 30B, 41; 31B, 123; 32B, 263; 33B, 231; 34B, 168
Carmichael, J.W., 30B, 337; 35B, 666
Caron, A., 26, 288, 289; 27, 762, 957; 29, 241, 473, 701; 30A, 252, 259
Carpenter, C., 26, 687, 688
Carpenter, D.A., 31A, 181
Carpenter, G.B., 27, 448; 29, 252, 284; 31A, 112, 175; 32B, 410; 33A, 230; 35A, 285
Carpenter, R.H., 29, 135; 30A, 178
Carrell, H.L., 33B, 275
Carrera, S.M., 35B, 606
Carrison, L.C., 29, 138; 30A, 179
Carroll, B.L., 30B, 405; 33B, 531
Carron, G.J., 28, 9
Carron, M.K., 32A, 417
Carstensen-Oeser, E., 33B, 245
Carter, F.L., 27, 332, 335; 29, 22
Carter, J.R., 29, 434
Carter, O.L., 31B, 275, 349; 32B, 363, 368, 369; 33B, 361
Cartz, L., 29, 393
Casagrande, G.T., 35B, 567
Casalone, A.L., 31B, 35
Casalone, G., 34B, 82; 35B, 93
Casalone, G.L., 30B, 397; 32B, 41, 42
Casalor, A., 35A, 274
Casalot, A., 31A, 140
Casciato, C.A., 34B, 18
Cash, A.N., 26, 271
Casini, G., 31B, 131
Casinovi, C.G., 28, 683
Cassedanne, J., 29, 435
Casselton, R.E., 29, 136, 224
Cassien, M., 31A, 194
Cassol, A., 35B, 819
Castellano, E.E., 35B, 120
Castellion, G.A., 30A, 248
Castelliz, L., 30A, 317
Castiglione, P.C., 32A, 164
Castles, J.R., 32A, 254
Catani, A., 33A, 79
Catesby, C.G.C., 26, 579; 28, 569
Caticha-Ellis, S., 33B, 43
Catterall, J.A., 30A, 247
Caughan, A.P., 35B, 829
Caughey, W.S., 30B, 218, 219
Caughlan, C.N., 30B, 188, 193, 399; 31B, 176, 213, 440, 445; 32B, 189, 220, 248, 388; 33B, 307; 35B, 274, 372, 380, 446, 449, 595
Caulton, K.G., 34B, 536
Cava, M.P., 30B, 397
Cavalca, C., 35B, 67, 235, 627
Cavalca, L., 27, 837, 881; 32A, 346; 32B, 407, 408; 33B, 59, 479; 34B, 189, 206
Cavelier, C., 27, 972
Cavin, O.B., 29, 139
Čebotarev, N.T., 27, 325; 33A, 121, 193
Čečernikov, V.I., 30A, 182
Cech, F., 28, 338
Cefola, M., 32B, 391
Čelinskij, I.V., 32B, 48; 33B, 30; 34B, 54
Čepiga, M.V., 34A, 143
Čeremisinov, V.P., 28, 317
Cerisy, B., 27, 149
Čerkašin, E.E., 26, 131; 29, 138; 30A, 174, 178; 32A, 141
Černikov, V.I., 30A, 178
Cernoplekov, N.A., 33A, 81
Černoruk, S.G., 27, 536
Černov, A.N., 35A, 310
Černov, V.I., 29, 438
Černova, N.I., 33A, 446
Cerrini, S., 35B, 379
Čertkov, A.A., 33A, 81; 35A, 133

Cervinka, L., 26, 411; 27, 496; 35A, 18
Cesari, M., 26, 809; 29, 760; 31B, 37; 35B, 883
Cesbron, F., 35A, 364
Cesur, A.F., 30B, 70, 225
Četkina, L.A., 26, 751; 28, 431; 432, 683; 32B, 41, 124, 126; 33B, 74
Chackraburtty, D.M., 28, 341; 29, 435; 30A, 175
Chadwick, D., 33B, 541
Chadwick, E.M., 35B, 870
Chadwick, J.R., 26, 693
Chaipayungpundhu, S., 35A, 429
Chaisse, F., 30A, 175
Chaiwasie, S., 33B, 343
Chalmers, R.A., 29, 437
Chamberland, B.L., 33A, 332
Champion, J.A., 30A, 180, 247
Chan, F.L., 30A, 181
Chan, J., 35A, 380
Chan, L.Y.Y., 35B, 543
Chandrasekaran, R., 31B, 193; 32B, 231; 33B, 156; 34B, 72, 243, 272
Chandrasekhar, S., 31A, 222
Chandross, R., 29, 253
Chaney, M.O., 33B, 184
Chang, C., 28, 341
Chang, C.H., 34A, 153; 35B, 891
Chang, H.-C., 31A, 230
Chang, K'un, 35B, 734
Chang, L.-C., 27, 703
Chang, L.J., 34A, 99
Chang, L.L.Y., 29, 434; 32A, 317
Chang, P.-S., 27, 537
Chang, S.-C., 28, 316; 33B, 427; 35B, 738
Chang, T.-F., 32A, 141
Chantret, F., 28, 357, 358
Chao, C.C., 29, 136, 223
Chao, C.-L., 29, 437
Chao, E.C.T., 26, 304, 482; 27, 675; 29, 136, 435
Chao, G.Y., 26, 771; 27, 864
Chapman, D.R., 27, 208
Chapovskij, J.A., 33B, 358
Chappert, J., 30A, 178; 31A, 133, 241; 32A, 268
Charbonneau, G., 31B, 79; 33B, 69
Charbonneau, G.-P., 32B, 112; 33B, 70
Charette, D.J., 29, 412
Charlton, T.L., 28, 418
Charnell, J.F., 32A, 484; 33A, 497
Charrier, J., 29, 433
Chasanov, M.G., 27, 109, 116; 29, 32, 136; 30A, 175
Chasar, D.W., 32B, 187
Chase, A.B., 27, 520
Chase, G.A., 27, 544
Chasmar, R.P., 29, 224
Chastain, R.V., 31A, 204; 32A, 343
Chasteen, N.D., 34B, 504, 532
Chatani, Y., 26, 632; 27, 805, 812; 28, 365; 29, 706; 33B, 521; 35B, 69, 884
Chateau, H., 27, 419, 427; 29, 433
Chatt, J., 34B, 614
Chaudhuri, B., 27, 1033; 28, 683; 33B, 63
Chaudhuri, D.K., 30A, 247, 248
Chawdhury, S.A., 33B, 64
Cheer, C.J., 35B, 389
Cheeseman, T.P., 28, 645; 30B, 393; 31B, 395, 398
Cheesman, G.H., 35A, 147
Chen, C.C.H., 30B, 444
Chen, C.W., 29, 226
Chen, D.-C., 28, 340
Chen, Y.C., 30B, 289
Cheng, C., 29, 437
Cheng, G.C.H., 28, 207
Cheng, J.C., 33B, 118; 34B, 285
Cheng, M.P., 28, 338
Cheng, P.-T., 35B, 86
Cherin, P., 32A, 137
Chessin, H., 27, 260, 548; 29, 225; 30A, 246
Cheung, K.K., 30B, 376, 393; 31B, 230; 32A, 509; 32B, 264; 34B, 340; 35B, 857
Chevalier, R., 32A, 132; 33A, 329; 34A, 270; 35A, 236
Chevreton, M., 26, 115; 27, 345; 29, 40, 135, 136; 30A, 176, 177; 31A, 37; 32A, 54; 33A, 66
Chevrier, B., 31A, 89; 33B, 293
Chheda, G.B., 35B, 254
Chi, K.-C., 30B, 195
Chia, P.-W., 29, 437
Chiang, J.F., 31B, 93; 33B, 534, 535; 35B, 896, 899
Chiang, J.P., 35B, 892
Chiang, R., 27, 800
Chiang, T.-I., 27, 442, 816; 29, 249
Chiari, G., 35A, 335, 354; 35B, 213
Chiba, A., 32B, 227
Chiba, T., 30B, 399
Chidambaram, R., 26, 343; 29, 463; 31A, 117; 32B, 101; 33A, 369, 448; 34A, 298, 305; 35A, 197; 35B, 8
Chieh, C., 30A, 33
Chieh, P.C., 32B, 171; 34B, 373; 35B, 21, 23, 490
Chien, T.-C., 29, 436
Ch'ih-fa, L., 29, 225
Chikalla, T.D., 27, 132, 477; 29, 225
Chikawa, J., 26, 45
Child, H.R., 26, 106; 27, 274, 333, 334
Ching-Liang, C., 30A, 92
Chini, P., 33A, 165; 34A, 159; 34B, 614
Chin-Khan, 34A, 381
Chioccola, G., 33B, 272, 284, 517; 34B, 366
Chiotti, P., 26, 258; 30A, 176; 35A, 98
Chipman, D.M., 34A, 164
Chiu, C.C., 34B, 13
Chiu, Y.H., 30B, 194; 34B, 385
Cho, G.-D., 29, 483

Choe, J.O. 26, 346
Choi, C.S., 34A, 169, 170; 35B, 179
Choisnet, J., 33A, 286
Chojnacki, J., 34A, 229
Chol, G., 33A, 324
Chopra, C.S., 28, 684
Chopra, K.L., 27, 382
Chopra, K.S., 29, 141, 226
Chou, C.L., 26, 250; 27, 344
Chou, K.-D., 28, 178, 221
Chow, J.M., 35B, 486
Chrétien, A., 26, 72, 75; 29, 433
Christ, C.L., 27, 554; 29, 391, 437; 35A, 286, 446
Christensen, A.N., 29, 310; 30A, 278, 305, 310; 32A, 244, 246, 248, 260, 458; 34A, 235, 236; 35A, 278, 320
Christensen, A.T., 33B, 198; 34B, 27, 70; 35B, 344
Christensen, D., 26, 549, 760
Christie, O.H.J., 26, 525; 27, 635
Christoph, G.G., 34A, 345
Chu, S., 30B, 58
Chu, S.-C., 35B, 41
Chu, S.S.C., 27, 899; 30B, 180, 444; 32B, 218; 33B, 161
Chujo, K., 33B, 521
Chung, C., 30A, 175
Chung, M.K., 32A, 325
Church, J.F., 27, 783
Churchill, M.R., 28, 684; 29, 558; 30B, 313, 393, 394; 31B, 366, 370, 32A, 396, 508, 510; 32B, 341, 345, 352, 371, 376; 33B, 329, 344, 356, 357, 364, 379, 382; 34A, 162; 34B, 360, 425, 445, 456, 464, 473, 475, 477; 35B, 430, 434, 505, 516, 530, 532, 566, 572
Chzhan, G., 28, 274
Chzhev, S.D., 28, 338
Cialdi, G., 32A, 501; 32B, 191
Ciampelli, F., 26, 636
Ciana, A., 35B, 810
Čičagov, A.V., 31A, 156, 158; 32A, 430; 35A, 299, 315
Cid-Dresdner, H., 27, 528; 30A, 395; 30B, 221; 33A, 346
Ciechanowicz, M., 34A, 210
Cien, J.-Y., 29, 760
Čikagov, M.I., 30A, 179
le R. Cilliers, J.J., 26, 528
Cincadze, G.V., 32B, 433
Cingi, M.B., 34B, 535; 35B, 725
Cirel'nikov, P.I., 27, 718
Cistjakov, I.G., 33B, 529
Čistjakov, Ju.D., 26, 199
Čistjakov, O.D., 26, 142
Čkalova, V.V.,26, 372
Clardy, J.C., 35B, 91
Clarebrough, L.M., 26, 139
Claringbull, G.F., 26, 538
Clark, A.H., 35B, 901
Clark, C.C., 34A, 292
Clark, D., 26, 366; 29, 137
Clark, E.S., 27, 804, 1065
Clark, G.R., 32B, 397; 33B, 389; 34B, 500; 35B, 129, 840
Clark, G.W., 28, 281
Clark, H.C., 28, 685; 29, 625
Clark, H.G., 31B, 42
Clark, H.L., 35A, 17
Clark, J.A., 29, 224
Clark, J.B., 33A, 171
Clark, J.R., 26, 489; 27, 555; 28, 163; 29, 259, 390, 391; 30A, 435; 33A, 459, 495; 34A, 358; 35A, 286, 446, 481
Clark, J.S., 26, 529
Clark, L.A., 26, 50
Clark, M.J., 34A, 189
Clark, M.J.R., 35A, 189
Clark, N.H., 34B, 362; 35B, 425
Clark, R.J.H., 26, 687; 28, 685; 34B, 604; 35B, 868
Clark, S.P., 26, 505
Clarke, A.M., 30A, 173
Clarke, P.T., 31B, 296
Clarke, R.S., 28, 340, 358
Clastre, J., 30B, 394
Clausen, A., 35B, 683
Clayton, G.T., 26, 169, 286
Clearfield, A., 28, 151; 34A, 328, 336
Clemente, D.A., 35B, 464, 859
Clerc, F., 27, 511
Clevenger, T.R., 28, 282
Clews, C.J.B., 26, 734; 35B, 95
Cline, C.F., 26, 407; 27, 480; 30A, 304
Cline, F., 26, 57
Clippard, F.B., 35B, 890, 902
Cloud, W.H., 27, 57, 478
Clougherty, E.V., 26, 101
Clunie, J.S.,26, 796
Coates, R.O., 29, 434
Cocco, A., 28, 340, 341
Cocco, G., 26, 484; 27, 620, 819; 30A, 371; 31A, 190; 32A, 164, 391
Cochran, W., 26, 555; 27, 397
Cockayne, B., 26, 129
Coda, A., 27, 584; 28, 177; 31A, 221; 32A, 362, 459, 466
Cody, G.D., 29, 140
Cody, V., 34B, 414
Coeffier, G., 26, 365
Coenen, M., 28, 289
Coetzee, E., 28, 339
Coetzer, J., 35B, 583, 737, 795
Coffey, D., 30B, 405; 33B, 538
Coggon, P., 34B, 309, 314, 342; 35B, 366 371, 383, 405, 570, 621, 834
Coghi, L., 30B, 443
Cohen, D., 29, 434
Cohen, E.A., 33B, 538
Cohen, G.H., 31B, 429; 32B, 234, 237; 35B, 887
Cohen, I., 28, 342
Cohen, J.B., 34A, 246
Cohen-Addad, C., 32A, 247, 426; 33A, 289

Cohen-Addad, P.D., 28, 151
Coing-Boyat, J., 26, 392, 441; 27, 606; 28, 317; 33A, 392; 35A, 372
Cojazzi, G., 35B, 155
Cola, M., 26, 605
Cole, F.E., 35B, 305
Cole, J., 30A, 175
Cole, J.F., 35A, 458
Cole, S.J., 28, 636
Coleman, J.M., 32B, 364, 455
Coleman, L.B., 33B, 322
Coleman, L.C., 27, 700
Colin, A., 28, 315
Colin, M.-L., 28, 340
Colleter, J.-C., 31B, 121, 125, 126; 32B, 156; 33B, 114; 34B, 136, 137; 35B, 168
Collin, G., 30A, 172; 35A, 76
Collin, R.L., 26, 603
Collins, D.M., 34B, 279; 35B, 717, 751
Collins, E., 28, 587
Collins, M.F., 27, 251; 31A, 240
Collins, R.C., 26, 640
Collomb, A., 35A, 228
Collongues, R., 26, 364, 368, 403; 27, 415, 518, 542; 33A, 306
Collot, C., 29, 225
Colman, P.M., 35B, 245
Colodny, P.C., 28, 291
Colombie, M., 29, 224
Colombo, A., 30B, 398; 31B, 140; 32B, 142
Colominas, C., 32A, 53
Colominas, C.B., 29, 39
Colominas-Broquetas, C., 32A, 53
Colton, R.,H., 30B, 16; 35B, 415
Colville, A.A., 32A, 327; 33A, 478
Colville, A.F., 35A, 243
Colvin, J.R., 33B, 522
Colvin, R.V., 29, 225; 30A, 246
Comer, F.W., 31B, 220
Comer, J.J., 29, 135
Cometto, D.J., 30A, 180
Compton, V.B., 26, 127, 254, 256; 27, 245; 29, 138, 30A, 247; 32A, 141
van Con, K., 30A, 177
Conant, J.W., 29, 262; 35A, 392
Conard, B.R., 32A, 130, 142; 33A, 115
Conlin, E.T., 29, 138
Conner, R.A., 26, 165; 30A, 176; 33A, 13
Conner, S.H., 31B, 119
Connor, J.A., 34B, 423
Conrad, B.R., 34A, 110
Conrad, C.M., 27, 799
Contag, B., 26, 378
Cook, A.J., 29, 224
Cook, D.S., 34B, 255
Cook, R.E., 35A, 130; 35B, 264, 356
Cook, R.L., 28, 341
Cook, W.R., 28, 317
Cooke, J., 35B, 566
Cooper, A., 30B, 336; 33B, 192, 193, 194, 201, 203, 204; 34B, 291, 293, 294, 296, 300, 301, 358, 545; 35B, 332
Cooper, A.S., 27, 381
Cooper, J., 26, 495
Cooper, M., 31A, 8
Cooper, M.J., 29, 239
Cooper, P.J., 31B, 269
Cooper, R.D.G., 34B, 285
Cooper, T., 27, 1049
Copeland, M., 27, 198
Copeland, M.I., 29, 226
Copeland, R.F., 31B, 30, 119
Coppens, P., 27, 755, 1063; 29, 651; 30A, 323; 30B, 72, 444; 32B, 155; 34B, 6
Coppola, F., 26, 768; 27, 967, 969
Coppola, J.C., 35B, 467
Corazza, E., 30A, 371; 31A, 172; 32A, 335, 336, 416, 501
Corbett, J.D., 26, 321; 27, 450; 28, 81; 35A, 150
Corbridge, D.E.C., 31B, 95
Corden, B.J., 35B, 879
Cordes, A.W., 26, 294; 29, 433; 30A, 256; 32B, 158; 33A, 428; 35B, 493, 666
Cordfunke, E.H.P., 31A, 127
Corenzwit, E., 26, 254, 256
Corey, E.J., 35B, 326
Corey, E.R., 27, 651; 28, 356; 32B, 356; 33B, 456; 34B, 414
Corey, R.B., 27, 1058, 1059
Corfield, P.W.R., 31B, 344, 345; 32B, 212, 343, 477; 33B, 386; 34B, 93; 35B, 612
Corliss, L.M., 26, 117, 306, 391; 27, 492; 33A, 344
Corll, J.A., 29, 137
Cornforth, J.W., 28, 537
Cornforth, R.H., 28, 537
Corradini, P., 26, 317, 328, 634, 635, 636; 27, 801; 28, 315, 682; 29, 760; 30B, 169, 394; 32B, 14, 362; 33B, 521, 522; 34B, 90, 91, 129, 440; 35B, 97, 884
Corrigan, L.I., 29, 262
Corvin, I., 28, 118
Cory, E.G., 28, 71
Cossee, P., 35A, 111
Costa, P., 31A, 35
Costain, C.C., 26, 829
Coste, J., 27, 811
Costello, W.R., 31B, 447; 33A, 163
Costin, C.R., 33B, 5
Cotrait, M., 31B, 28; 34B, 20; 35B, 26
Cotta, L.J., 26, 179
Cotton, F.A., 28, 87, 613, 636, 647; 29, 565, 616, 620, 621, 622, 626; 30A, 283; 30B, 34, 295, 334, 346, 355, 394; 31A, 124; 31B, 322, 353, 376, 383, 384, 386, 410, 415; 32A, 509; 32B, 446, 450, 482; 33A, 209, 238; 33B, 274, 292, 350, 369,

371, 375, 378, 391, 395, 396, 414, 416, 463, 464, 483, 498, 504; 34A, 207, 209; 34B, 396, 453, 475, 511, 530, 532, 536; 35A, 171, 418, 426; 35B, 85, 561, 715
Coucouvanis, D., 34B, 597, 599; 35B, 802
Couffon, M.-M., 29, 321
Coulon, M., 26, 432
Coulter, C.L., 28, 410, 452; 29, 759; 34B, 256; 35B, 329
Countryman, R., 34B, 279
Courseille, C., 34B, 102
Courtine, P., 27, 169
Cousseins, J.C., 29, 433
Coutsouradis, D., 30A, 176
Couvillion, J., 28, 397, 472
Covington, A.K., 26, 63
Cowan, D.O., 33B, 322, 323
Cowgill, U.M., 28, 358
Cowley, A.H., 33B, 538
Cowley, J.M., 26, 529; 27, 384, 396
Cox, A.P., 34A, 240
Cox, D., 31A, 31
Cox, D.E., 27, 498, 515; 29, 39; 30A, 369; 31A, 42, 241; 32A, 272
Cox, E.G., 31B, 170
Cox, G.W., 32A, 154, 325; 34B, 3
Cox, J.H., 28, 314
Cox, J.R., 31B, 179
Cox, K.G., 26, 379
Cox, N.L., 26, 354
Cox, W.L., 31A, 47
Coyle, B.A., 35A, 152; 35B, 749
Cradwick, P.D., 35A, 402; 35B, 682
Craig, D.C., 30A, 18, 173; 33A, 352; 34A, 288
Cramer, E.M., 29, 137; 34A, 146
Crangle, J., 27, 254
Cras, J.A., 27, 441; 33B, 485; 35B, 806, 810
Craven, B.M., 26, 611; 27, 1017; 28, 683; 29, 714, 719, 721, 745; 31A, 165; 31B, 423; 32B, 205; 34B, 3, 216, 218, 220, 330
Credali, L., 29, 760
Creely, J.J., 27, 799
Crevecoeur, C., 29, 46, 91
Crick, F.H.C., 26, 820, 823
Crikevich, S., 27, 396
Cristensen, A.N., 31A, 129
Critchley, J.K., 30A, 33, 175
Croatto, U., 31B, 332; 35B, 464
Croeni, J.G., 27, 249
Croft, W.J., 28, 133; 29, 434
Cromer, D.T., 26, 107, 108, 127, 498; 27, 142, 416; 28, 17, 467; 29, 36, 37, 81, 261, 266, 313; 30A, 50, 178, 245, 290; 31A, 116, 202; 31B, 152; 32A, 114, 178, 339; 33A, 245; 35A, 170
Cromer, F.W., 30B, 444
Cros, C., 30A, 96, 180
Cross, A.D., 32B, 245
Cross, C., 33A, 162
Cross, J.H., 35B, 854
Crosse, B.C., 30B, 399
Crouzet, A., 34A, 186
Crowley, F.A., 27, 716
Cruickshank, D.W.J., 26, 458, 603; 27, 702, 925; 28, 215, 478; 29, 344, 346, 360, 394, 456, 754; 31A, 207; 32A, 252, 253, 370, 425, 435, 448; 33A, 461; 34A, 163, 240; 35A, 206, 381, 382, 384, 396
Cruickshank, F.R., 29, 434
Csordás, L., 26, 435; 30B, 444
Cubero, M., 26, 575; 28, 683
Cucka, P., 27, 75; 28, 8, 458; 35B, 887
Cudby, M.E.A., 27, 773
Cueilleron, J., 29, 433
Čuenko, L.J., 29, 434
de Cugnac, A., 27, 419, 427; 29, 433
Čukhrov, F.V., 33A, 472
Cullen, D.L., 35B, 594, 687
Cullen, G.W., 29, 223
Cullen, W.R., 26, 728; 27, 984; 28, 592; 31B, 466
Cullis, A.F., 26, 812
Culot, J.P., 27, 427
Culver, K.B., 28, 356
Cummings, J.P., 35A, 240
Cummins, J.D., 35A, 66
Cunningham, B.B., 27, 39; 34A, 179
Cunningham, J.A., 32B, 386; 34B, 509, 510; 35B, 832, 879
Cuprina, V.G., 27, 26
Čurakov, M.M., 29, 139
Čurbakova, T.I., 26, 403
Curien, H., 26, 145; 30A, 178; 34A, 82, 375
Curl, R.F., 26, 345, 543
Curran, R., 35B, 832
Currie, M., 32B, 23; 34B, 148; 35B, 12
Curry, N.A., 27, 614; 29, 636; 32B, 23
Curti, R., 26, 502, 781; 27, 989; 28, 687; 30B, 139, 141
Curtin, D.Y., 31B, 50
Curtis, C.S., 27, 763
Curtis, N.F., 35B, 772
Curzon, A.E., 31A, 240
Cusatis, C., 34B, 218
Cushen, D.W., 27, 439; 29, 285
Cvankin, D.Ja., 26, 641; 27, 812; 29, 760
Cyganova, I.A., 26, 241; 27, 233, 337
Cyvin, S.J., 27, 925
Čžan, P.-S., 28, 158, 340
Czou, C.-L., 30A, 172

Daane, A.H., 26, 74, 233; 27, 279, 330, 342; 29, 224
Dachille, F., 30A, 246; 32A, 255
Dachs, H., 28, 131; 32A, 315
Dagerhamn, T., 26, 363; 31A, 13, 32A, 14
D'Agostino, S., 31B, 342

Dahl, E., 32A, 108
Dahl, L.F., 26, 332, 652; 27, 440, 443, 646, 647, 651, 814, 826, 876; 28, 67, 69, 70, 71, 606, 654; 29, 249, 669; 30A, 259, 443; 30B, 300, 312, 316, 318, 322, 323, 326, 383, 384, 399; 31B, 31, 347, 356, 362, 433, 440; 32A, 399, 404, 406, 510; 32B, 341, 350, 364, 455, 465; 33A, 163; 33B, 489, 490; 34A, 160, 183; 34B, 438, 457, 466, 614; 35A, 403; 35B, 506, 584, 585, 789
Dahl, T., 29, 490; 30B, 249; 33B, 237; 35B, 202
Dahm, D.J., 33B, 512
Dahm, S., 28, 684
Dähne, W., 26, 308; 27, 442
Dakin, T.W., 34B, 152
Dale, D., 27, 1040; 30A, 280; 32B, 443
Dale, D.H., 31A, 200
Dallavalle, F., 28, 89
Dallinga, G., 30B, 405; 33B, 537
Dal Negro, A., 32A, 444, 459
Dalton, R.F., 35B, 794
Daly, J.J., 26, 604; 28, 684; 29, 538, 539, 637, 638; 30B, 279, 444; 31B, 303, 306, 309; 32B, 25, 304, 310, 448; 33B, 272, 284, 517; 35B, 443, 454, 465, 466
Daly, J.W., 34B, 278; 35B, 362
Daly, M.C., 28, 293
Dalziel, J.A.W., 31B, 447
Damaschun, G., 27, 813
Damiani, A., 27, 1008; 30B, 134, 253; 31B, 111; 32B, 287
Damjanovic, A., 26, 6, 16, 17
Damm, E., 30B, 247
Damon, J.-L., 31A, 217
Dance, I.G., 30A, 280; 34A, 298
Danford, M.D., 26, 325; 27, 722; 34B, 500
Daniel, E.J., 26, 826
Danielsen, J., 26, 692; 28, 649; 29, 495; 33B, 233; 34B, 180; 35A, 186
Danilov, A.V., 26, 48
Danø, M., 31A, 236
Danot, M., 31A, 107; 34A, 145
Daou, J.N., 30A, 246
Daoudi, A., 35A, 228
Dapporto, P., 35B, 777, 853, 862
Darby, J.B., 26, 254; 27, 285, 360; 29, 136; 30A, 173, 182
Darby, W., 27, 467, 509; 28, 357
Dariel, M.P., 35A, 124
Darlow, S.F., 26, 555, 713; 30B, 32
Darmon, R., 31A, 21
Darnell, A.J., 30A, 179
Darriet, B., 33A, 336
Darriet, J., 34A, 266, 318; 35A, 274
Darrouy, R., 26, 750
Das, A.K., 31A, 109
Das, B.N., 27, 278
Das, I.M., 27, 1033
Das, R.K., 27, 480
Dasannacharya, B.A., 27, 257
Dasarathy, C., 29, 135; 30A, 178, 246
Dasevskij, V.G., 31B, 66; 35B, 897, 898
Dasgupta, D.R., 26, 540
Datt, I.D., 30A, 364; 34A, 450; 35A, 387
Datta, R.K., 26, 413; 28, 338
Datta, S.K., 28, 682, 683
Dauphin, G., 30B, 398; 32B, 69
Dautovich, D.P., 30A, 181
Dautreppe, D., 27, 251
Daval, R., 30A, 180
Davey, G., 35B, 774
David, J., 28, 315; 35A, 140
Davidovits, J., 26, 637
Davidson, A.G.H., 35B, 316
Davie, A.W., 26, 801
Davies, A.G., 35B, 485
Davies, D.A., 29, 225; 30A, 245, 246
Davies, D.D., 26, 127
Davies, D.R., 26, 820; 27, 1055; 29, 759; 35B, 887
Davies, G.R., 32B, 474; 33B, 501; 35B, 526, 792
Davies, R.G., 26, 134; 27, 12
Davies, W.O., 27, 916; 34A, 229
Davis, B.L., 27, 425, 565; 28, 316; 29, 294, 433, 436
Davis, B.R., 34B, 618; 35B, 592
Davis, G.L., 27, 814
Davis, M., 28, 668
Davis, M.I., 30A, 259; 30B, 405; 35B, 895, 896
Davis, P.H., 35B, 630
Davis, R.E., 27, 1025; 31B, 262; 32B, 216; 35B, 142, 144
Davis, R.J., 26, 538; 29, 436
Davison, A., 31B, 353; 34A, 158
Davison, B.E., 35B, 273
Davydova, M.A., 26, 745, 831; 27, 921, 923, 924; 30B, 99, 394; 33B, 87
Day, G.F., 26, 159
Day, P., 28, 315
Day, V.W., 32B, 65; 33A, 248; 33B, 399; 35B, 751
Dayal, B., 27, 347; 29, 433; 34A, 144
Deadmore, D.L., 27, 464
Dean, G., 29, 136
Dear, P.S., 26, 521
Deardorff, D.K., 29, 226; 34A, 146
Deb, R., 30B, 447
Debaerdemaeker, T., 35B, 325
Debets, P.C., 28, 318; 31A, 126; 33A, 221
De Boer, B.G., 34A, 191; 35A, 171
De Boer, J.J., 34A, 70
De Broin, C.E., 33A, 242
Decap, J., 30B, 66, 392; 31B, 62; 33B, 55
Decarlo, J.V., 29, 223
Decian, A., 32A, 171
DeCou, F., 26, 760

De Fontaine, D., 34A, 278
Defrain, A., 26, 145; 30A, 178; 34A, 82
Degeilh, R., 30B, 358
Degreve, Y., 32B, 348
Dehli, J., 34B, 96
Deigner, P., 30A, 277
Deitch, R.H., 34A, 47
Dejace, J., 34B, 83
De Kimpe, C., 26, 529
Delacroix, S., 31B, 378
Delaey, L., 30A, 177
Delafosse, D., 29, 434
Delamoye, P., 34A, 286
Delapalme, A., 27, 256, 522, 523; 28, 133, 151; 30A, 313; 31A, 133
Delaplane, R.G., 34A, 237; 34B, 4
Delbaere, L.T.J., 35B, 541
Delettre, J., 34B, 58
Delgaudio, J., 33A, 426
DelGrosso, E.J., 28, 357
Deliry, J., 30A, 175
Dellaca, R.J., 34B, 107; 35B, 502, 504
Delord, P., 33B, 529
Del Pra, A., 33B, 136; 34B, 169
De Luca, L.B., 26, 629
Demarco, P.V., 34B, 285
Dembinski, W., 28, 682
Dembovskij, S.A., 28, 139
Demel, J.T., 34A, 145
Demelio, F., 35A, 415
Dem'janec, L.N., 31A, 156, 157
Dempsey, E., 34A, 370
De Neui, R.J., 26, 543
D'Enghien-Peteau, M.B., 33B, 248
Denisov, A.P., 29, 437
Denny, P.J., 26, 517, 518
Dent Glasser, L.S., 31A, 215
De Pape, R., 33A, 183, 184
Depatie, D.A., 31A, 240
De Pieri, R., 30A, 320, 441
Deptula, A., 28, 682
Derbyshire, S.W., 26, 420
Dereigne, A., 33A, 374
Dereppe, J.-M., 27, 428, 435, 560; 28, 220
Derissen, J.L., 33B, 176; 35B, 4
Dernier, P.D., 33A, 282; 34A, 354, 355; 35A, 205, 218
Derry, J.E., 34B, 158
DeSando, R.J., 27, 723; 31A, 61; 33B, 528
De Santis, F., 26, 754
DeSantis, P., 32B, 276
Desautels, F.E., 28, 358
Desautels, P.E., 28, 340
Desborough, G.A., 29, 135; 30A, 178
Deschanvres, A., 33A, 286
Deshpande, V.T., 26, 260, 297, 298; 27, 369, 381; 28, 291; 29, 225, 433; 30A, 248
Desiderato, R., 30A, 291; 30B, 39; 32B, 66, 160; 34B, 152
Desjardins, M., 27, 542
Despault, J.G., 35A, 93, 124
Dessy, G., 28, 687; 30A, 444; 30B, 373, 378; 32B, 35, 326; 33B, 268; 35B, 816
Destro, R., 30B, 444; 33B, 49, 52; 34B, 113, 118, 304
De Teresa Carral, M., 27, 1028
Dethlefsen, W., 33A, 31
Deuschl, H., 29, 542; 30B, 127
Devaux, J., 31A, 225
Devere, R.H., 30B, 448
De Vries, R., 34A, 298
Dewar, R., 34B, 560
Dewar, R.B.K., 31B, 369
Dey, B.N., 27, 111; 29, 137; 30A, 175
Deyris, B., 32A, 142; 33A, 86
Dhar, J., 27, 940
Diana, G., 28, 315, 490, 682
Dianoux, A.-J., 33A, 356
Dianova, I.M., 29, 436
Dias, H.W., 29, 516
Dickens, B., 26, 681; 27, 854; 32A, 510; 34A, 359; 35A, 296; 35B, 18
Dickens, P.G., 32A, 314
Dickerson, R.E., 26, 810, 811; 27, 844, 1058, 1059; 29, 535; 33B, 113; 35B, 215, 338, 348, 888
Dickinson, C., 31B, 68; 32B, 90; 33B, 39; 34B, 71
Dickinson, J.G., 27, 508
Dickinson, T., 26, 366
Dickson, J.G., 26, 370
Dideberg, O., 32B, 188
Dierks, H., 30B, 321; 33B, 380
Dieseroth, H.-J., 35A, 244
Dietrich, H., 26, 608; 28, 593, 685; 30B, 298, 321, 394; 33B, 343, 380; 34B, 469; 35B, 100
Dietze, W., 33A, 153
Dijačenko, O.A., 35A, 423
Dijkstra, A., 31B, 413
Dik-Edixhoven, L., 34B, 135
Dillingerová, T., 27, 894; 28, 682
Dima, I., 30A, 182
Dimitrijević, Ž., 33A, 266
Dinnin, J., 29, 434
Diorio, A.F., 26, 630
Dismukes, J.P., 29, 137, 225; 30A, 91, 179, 181, 248
Distler, T., 32A, 194
Dittmar, G., 34B, 368
Dittrich, G., 34A, 291
Di Vaira, M., 31B, 389, 400; 32B, 31, 383, 406; 33B, 428, 441
Dixmier, J., 33A, 153
Dixon, D.T., 30A, 354
D'jacenko, O.A., 29, 436; 33A, 271; 34A, 276
D'Jakonov, J.S., 28, 339
Djega-Mariadasov, G., 29, 435
Djerassi, C., 27, 993; 30B, 396

Dmitrie, G.A., 33A, 217
Dmitriev, A.G., 26, 208
Dmytrijeva, G.P., 29, 225, 226
Dobler, M., 27, 843; 28, 683; 29, 507; 30B, 117; 31B, 92; 33B, 191; 34B, 288
Dobozy, O., 27, 838
Dobroljubora, M.S.,27, 482
Dobrott, R.D., 27, 399; 28, 62, 314; 29, 231
Dobryden', K.A., 29, 140
Dodd, R.A., 27, 32; 29, 224
Dodge, R.P., 26, 686; 27, 880; 28, 683; 29, 552; 30B, 297, 320; 32A, 186; 33A, 223
Dodinh, C., 30A, 316
Dodson, G.G., 31B, 243
Doedens, R.J., 30B, 312; 31B, 362, 433; 32B, 467, 477, 479; 33B, 424; 34B, 564; 35B, 633, 699
Doenitz, F.-D., 33A, 102; 35A, 71
Doherty, G.O.P., 30B, 394
Doi, H., 26, 367
Doi, K., 31A, 240
Doll, W.W., 35B, 882
Dollase, W.A., 28, 87, 613; 30A, 420; 30B, 151; 32A, 251, 384; 33A, 475; 35A, 454
Dolmetsch, H., 29, 760
Domagala, R.F., 26, 276
Domage, K., 30A, 177
Doman, R.C., 27, 545
Domange, L., 26, 88, 117, 277; 27, 327, 424; 29, 135, 136, 341; 30A, 175, 176, 245
Domenicano, A., 33B, 473, 476, 486; 34B, 35
Domiano, P., 31B, 402, 32A, 346; 32B, 407, 408; 33B, 59; 34A, 234; 34B, 32; 35B, 28
Donaldson, J.D., 26, 360, 475; 28, 315; 32A, 508
Donaruma, L.E., 33B, 93
Donika, F.G., 32A, 95
Donkersloot, H.C., 29, 54
Donnay, G., 26, 457; 28, 249; 29, 406; 30B, 444; 33A, 491; 33B, 322, 323
Donnay, J.D.H., 26, 457, 464; 27, 438; 28, 249; 29, 178; 30B, 444; 31A, 231
Donohue, J., 26, 271, 287, 288, 289, 569, 696, 722; 27, 199, 446, 762, 764, 943, 957, 1063; 28, 54, 534; 29, 230, 241, 473, 484, 701, 749; 30A, 252, 259, 288; 30B, 115; 32B, 148; 33B, 48, 275
Donohue, P.C., 30A, 329, 446; 31A, 137; 33A, 36; 35A, 20, 87
Donohue, T., 33A, 153
Donolato, C., 34A, 143
Doorakian, G.A., 35B, 491
Doraiswamy, S., 33B, 540, 541
Dorfelt, C., 34B, 408
Dorfman, M.D., 26, 535; 28, 342
Dori, Z., 32B, 381; 33B, 499; 514; 35B, 817, 829
Dorin, E., 27, 520
Dorko, E.A., 33B, 532
Dorm, E., 32A, 272, 319; 34A, 300
Dorn, F.W., 26, 53
Dornberg, M.L., 28, 385; 30B, 28
Dornberger-Schiff, K., 26, 397, 513; 27, 405, 575; 29, 363; 30A, 318
Dorofeev, Y.A., 34A, 58
Dorosinskij, A.L., 28, 682
Dorrian, J.F., 34A, 254
Dorsi, R., 27, 261
Dott, I.D., 31B, 41
Doty, P., 26, 816
Dougill, M.W., 26, 607; 28, 59
Douglas, B., 30B, 237
Douglass, D.L., 26, 268
Douglass, R.M., 27, 598
Douwes, H., 27, 755
Dove, D.B., 26, 174
Dover, S.D., 33B, 185
Dow, J., 35B, 319
Dowell, L.G., 33A, 488
Downey, J.W., 26, 165, 254; 27, 234, 360; 29, 136; 30A, 173, 176, 182; 33A, 13
Downs, G.L., 29, 669
Doyle, J.R., 26, 687, 688, 690; 29, 587; 30B, 324, 327; 31B, 367
Doyle, N.J., 26, 162, 223; 27, 281; 29, 224; 30A, 179, 247
Doyne, T.H., 31B, 31; 33B, 236
Dragonette, K.S., 30B, 37, 42, 214
Dragsdorf, R.D., 26, 760; 27, 170
Drake, J.E., 33B, 437
Draper, P.H.G., 27, 482
Drapier, J.M., 30A, 176
Dratovski, M., 29, 436
von Dreele, R.B., 35A, 272
Dreiner, R., 26, 195
Dreissig, W., 33B, 343
Drew, M.G.B., 30B, 331, 394; 32A, 161; 33B, 435; 34B, 99, 315; 35A, 155, 184; 35B, 770
Drews, K., 27, 916
Dreyer, D.L., 26, 830
Drič, V.A., 26, 529, 530
Drickamer, H.G., 30A, 245, 246
van Driel, C.A.A., 32B, 11
Driessens, F.C.M., 31A, 163
Drozdov, Ju.N., 35A, 480
Druding, L.F., 26, 321
Druelinger, M.L. 28, 668
Druminski, M., 33A, 153
Drummond, A.D., 27, 405
Drummond, J., 35B, 756
Druyvesteyn, W.F., 27, 302
Duale, M., 34A, 146
Duax, W.L., 33A, 161; 34A, 336
Dubertret, A., 33A, 268
Dubey, B.L., 34A, 211

Dubinin, M.M., 26, 519
Dubois, I., 34A, 242
Dubov, S.S., 29, 760
Dubrovskaja, L.B., 27, 156; 29, 135, 224; 34A, 58
Duc, T., 34A, 91
Duc, T.-Q., 31A, 143
Duchamp, D.J., 34B, 62
Duckworth, V.F., 34B, 512, 566, 606
Ducros, P., 30B, 394; 32A, 426
Duerst, R.W., 33B, 202
Dufek, V., 29, 135
Duffield, A.M., 28, 563
Duffin, B., 31B, 435; 33B, 110, 453; 35B, 185
Duffin, W.J., 26, 149, 529; 29, 84; 31A, 45; 32A, 94; 34A, 176
Dugcak, Ju.I., 30A, 179
Duggin, M.J., 29, 136; 31A, 31, 39
Dugue, J., 35A, 235
Dukefos, T., 30B, 182
Dukhovskaja, E.L., 35A, 259
Dulac, J., 26, 390, 410; 27, 486
Dulova, V.G., 28, 687; 32A, 511; 35B, 889
Dumara, D., 35A, 163
Dumbleton, J.H., 30B, 19, 444
Dumke, W.L., 27, 913
Dummer, G., 27, 340
Dumora, D., 33A, 262
Dunaj-Jurčo, M., 32A, 509
Dunitz, J.D., 26, 317, 645, 782; 27, 847; 28, 534, 537, 683; 29, 507, 509, 513, 629, 735; 30A, 294, 370; 30B, 89, 117, 136, 289, 394; 31B, 87, 92, 93; 32B, 178, 242, 378, 484; 33B, 85, 132; 34B, 288; 35B, 61
Dunken, H., 26, 831; 28, 314; 32A, 172, 173
Dunmyre, G.R., 30A, 246
Dunne, T.G., 29, 565
Dunsmore, H.S., 28, 316
Duplessis, J.J., 26, 97
DuPont, L., 32A, 236; 35A, 200
Dupres, J., 30A, 178
Durant, F., 31B, 273, 274; 32B, 267, 271; 33B, 298; 35B, 325
Duranti, D., 26, 769, 779; 29, 286
Durif, A., 26, 398, 410, 416, 417; 27, 486; 28, 143, 316, 357; 29, 435; 31A, 143; 32A, 379, 380, 381, 382, 442; 33A, 395, 396, 413; 34A, 329, 338; 35A, 362
Durif-Varambon, A., 27, 518; 28, 151, 318, 357; 30A, 313
Durovic, S., 27, 707; 28, 236; 33A, 513; 34A, 366
Dusausoy, Y., 33A, 393; 34A, 142, 318, 387; 35B, 15
Dutčak, Ja.I., 26, 62; 27, 80, 385
Dutta, S.N., 26, 475, 610; 27, 347; 29, 366, 433, 438
Dutta, S.N., 27, 770; 30A, 54, 179
Duwez, P., 29, 136, 223, 224
Dvořák, V., 33A, 242
Dvorina, L.A., 30A, 181, 182
Dvorjankin, V.F., 26, 572, 573; 27, 750; 29, 284
Dvorjankina, G.G., 27, 295; 28, 34
Dwek, R.A., 34B, 252
Dweltz, N.E., 26, 821; 27, 1057; 33B, 522
Dwight, A.E., 26, 165, 235; 30A, 176, 179; 33A, 13
Dwyer, M., 35B, 679
Dyer, G., 31B, 463
Dyke, M., 29, 284; 33A, 432; 33B, 34
Dyson, D.J., 29, 135, 136
Dyvik, F., 26, 739
D'Yvoire, F., 33A, 418
Dzaparidze, K.G., 27, 950
Dzjana, D.I., 29, 139; 30A, 174, 177; 34A, 145
Džmukhadze, D.F., 27, 505

Eade, R. A., 28, 683
Eanes, E. D., 27, 594
Easey, J. F., 34A, 197
Eastabrook, J., 29, 137
Eastman, J. C., 29, 137
Easy, J. F., 33A, 223
Eatwell, A. J., 26, 286
Ebel, H., 26, 11, 182; 30A, 247
Eberhard, E., 27, 694
Eberhart, J. -P., 28, 342
Eberius, E., 27, 420
Eberle, D., 33A, 24
Eberspacher, O., 27, 507
Ebert, M., 26, 469; 28, 317
Eckart, D. W., 26, 4, 30A, 349
Eckerlin, P., 26, 46, 61, 87, 300; 29, 224, 433; 32A, 28; 33A, 146
Eckersley, G. H., 34A, 178
Eckhardt, E.J., 28, 340
Eckhart, D.W., 28, 340
Eckstein, F., 34B, 258; 35B, 292
Economy, J., 29, 4; 30A, 29, 31, 172, 174, 251; 35A, 4
Eddy, L. P., 31A, 215; 35B, 687
Eden, A., 30A, 59
Edenharter, A., 35A, 11
Edgar, A. D., 29, 437
Edmonds, J., 35B, 31
Edmunds, I. G., 27, 915; 30B, 82
Edmundson, A. B., 26, 811
Edshammar, L. -E., 26, 9; 27, 38; 28, 7; 29, 10; 30A, 10, 12, 173; 31A, 10, 16; 32A, 12; 33A, 15, 16
Edstrom, V. -A., 34A, 115
Edwards, A. J., 26, 311; 27, 445, 464; 28, 314; 29, 258; 32A, 508; 33A, 170, 172, 173; 34A, 172, 198, 200, 202; 35A, 147, 174, 175, 178, 185
Edwards, I. A. S., 35B, 424
Edwards, J. O., 34B, 550
Edwards, O. S., 26, 7

Edwards, W. T., 33A, 209; 33B, 369; 34B, 475
Eeles, W. T., 28, 41, 65; 29, 296; 30A, 251
Efendiev, G. K., 29, 135, 140, 141
Effimov, Ju. V., 27, 284; 30A, 244, 248
Efimov, A.F., 28, 341
Efremova, O. A., 34A, 268
Eger, C., 30B, 444
Eggleton, R. A., 32A, 474
Egmond, J. V., 34B, 95
Egorov, Ju. P., 35B, 465
Egorov-Tišmenko, Ju. K., 32A, 302, 454; 33A, 469, 470
Ehrenberg, M., 28, 442; 30B, 4; 31B, 112, 114; 32B, 139; 33B, 106
Ehrlich, H.W. W., 26, 747; 27, 796; 28, 657
Ehrlich, P., 26, 333
Eibschutz, M., 30A, 323
Eichhorn, E. L., 27, 965
Eick, H. A., 26, 364; 29, 136; 31A, 145; 32A, 309; 34B, 464
van Eijck, B. P., 30B, 3
Eikerling, G. F., 29, 434
Einstein, F., 32A, 160
Einstein, F. W. B., 29, 332; 30A, 354; 355; 31B, 431, 466; 32A, 155; 32B, 337, 339, 374, 468; 33B, 314; 34B, 23, 603, 619; 35A, 322; 35B, 543
Einstein, J. R., 26, 820; 27, 1058
Eisenberg, R., 28, 685; 30B, 377, 388, 394; 31B, 455, 456; 32B, 381, 462; 33B, 499, 514; 34B, 550, 608; 35B, 817, 821, 826, 832, 879
Eisenhuth, W., 32B, 315
Eisenmann, B., 32A, 43; 34A, 53; 35A, 24, 124, 135
Eisler, U. G., 34A, 264
Eiss, R., 33B, 395, 396, 464, 34B, 532; 35B, 518, 716
Ejima, T., 26, 350; 27, 484
Ekstrom, L., 29, 225
Elagina, E. I., 26, 63
El-Boragy, M., 35A, 14, 22, 124
Elchinger, R., 32A, 508
Elder, M., 31A, 98, 110; 32A, 192; 34A, 213; 34B, 409, 410, 585; 35A, 427; 35B, 871
Elder, R. C., 29, 620; 30B, 334; 31B, 384; 32B, 430; 33B, 392, 393
Eldridge, J. M., 30A, 180
Elema, R. J., 28, 82
Eleonora, M., 34A, 234
Eley, D. D., 26, 653
Elfström, M., 26, 71; 30A, 70, 181
Elguero, J., 32B, 63; 35B, 152
Elias, J. L., 28, 682
Eliseev, A. A., 26, 208, 214; 29, 64, 138, 139, 225, 226; 30A, 64, 180, 247; 31A, 50; 33A, 140
Ellinger, F. H., 27, 28; 28, 38, 39; 30A, 51, 178, 181, 246, 248
Ellinghaus, I., 26, 30
Elliott, J. J., 29, 760
Elliott, G.R.B., 30A, 175
Elliott, N., 26, 117, 306, 340
Elliott, R.O., 26, 230; 27, 141, 148; 30A, 245
Elliott, R.P., 26, 102, 161; 29, 760
Ellis, I. A., 34A, 155
Ellis, K. C., 27, 798
Ellis, T. G., 29, 135; 30A, 80, 180
Ellison, R. D., 26, 325, 554; 27, 422; 28, 74; 30A, 226; 30B, 25; 32A, 395; 33A, 185
Ellner, M., 34A, 142
Elmaleh, D., 35A, 228
Elmendorf, R., 33A, 50
El Saffar, Z. M., 27, 435, 630; 33A, 159
Elsen, G., 26, 177
Elson, R. E., 32A, 186, 187; 33A, 223; 34A, 174
Elston, D.P., 27, 603
Elsworthy, G. C., 30A, 367
Emerson, K., 26, 341; 28, 105; 29, 290; 31A, 113; 35A, 412; 35B, 595
Emerson, M.T., 31B, 213; 35B, 372, 380
Emes-Misenko, E. I., 34A, 143
Ence, E., 26, 28, 138
Endeman, H. J., 33B, 176
Endo, K., 27, 15
Endo, S., 29, 760
Enemark, J.H., 30B, 372; 31B, 292; 32B, 480; 33B, 429, 496; 34A, 160; 35B, 602
Engberg, A., 32A, 190; 35A, 197
Engberg, Y., 27, 425
Engebretson, G., 28, 614
Engebretson, G. R., 28, 684; 29, 655
Engel, G., 35A, 352
Engel, P., 31A, 19; 33A, 37; 34A, 17; 35A, 143
Engels, J., 33A, 153
Engert, G., 33A, 153
England, R. E., 29, 434
Engmann, R., 28, 130; 31B, 42; 32B, 300
Engström, I., 28, 40; 30A, 88, 89, 181; 33A, 87, 136; 34A, 125; 35A, 70, 84, 86
Enrione, R. E., 29, 230
Entwistle, R. F., 26, 752; 34B, 111
Épel'baum, V. A., 26, 4
Epp, O., 35B, 888
Epstein, E. F., 35B, 584, 585
Eraker, J., 32B, 116
Erämetsä, O., 27, 397, 29, 435
Erasmus, C. S., 35B, 752
Erd, R. C., 26, 489; 29, 63, 436
Erdberg, M. V., 29, 137
Eremenko, V. N., 30A, 181

Érenburg, B. G., 26, 481
Erez, G., 35A, 124
Ergin, Ju. V., 30A, 179
Erichsen, H. H., 33B, 57
Erickson, R. A., 31A,31
Erikkila, H., 30B, 447
Eriks, K., 30B, 54; 31B, 55; 33B, 423
Ermilova, L. P., 33A, 472
Ermolenko, A. S., 27, 9
Ern, V., 26, 372; 28, 340
Ernst, D., 30A, 173
Ernst, D. W., 29, 50
Erokhin, E. V., 34A, 192
Erosenkova-Lukanina, I. G., 30A, 244
Eršova, M. M., 34A, 168
Ersson, N. -O., 33A, 92; 34A, 117
Esaulov, N. P., 35A, 27
Escobar, C., 33A, 346
Eser, H., 29, 735; 32B, 484
Eskenazi, G., 26, 523
Es'kova, Y.M., 28, 342
Espinosa, G. P., 27, 511; 29, 435
Esselborn, R., 26, 215, 293
von Essen, U., 27, 304
Estlin, J. A., 32B, 39, 162, 186, 256; 34B, 121
Etar'hk, D. W., 28, 340
Étienne, J., 30B, 445; 34A, 214
van Etten, R. L., 34B, 124, 125, 127
Etter, D. E., 29, 135
Ettmayer, P., 27, 271; 30A, 66, 179; 32A, 142
Ettmayer, P. E., 31A, 26
Eugster, H. P., 27, 688; 29, 438
Euk, H. A., 33A, 433
Eulenberger, G., 26, 398; 27, 500, 541; 35A, 69, 138
Eulenberger, G. R., 32A, 484
Euler, F., 30A, 324
Evans, D. S., 26, 260; 27, 145
Evans, E. B., 29, 434
Evans, E. H., 27, 723; 29, 141
Evans, H. T., 26, 304, 370; 27, 685; 29, 136, 375, 377, 435; 30A, 348; 31A, 141, 142; 33A, 278; 35A, 140
Evdokimenko, V. I., 27, 64, 195, 268, 269; 29, 137, 138, 140
Evdokimova, V. V., 27, 424
Evers, H., 30B, 106
Eversole, W. G., 26, 517
Evertsson, B., 34B, 544
Evseev, B. A., 30A, 97, 182; 33A, 143; 34A, 101, 145
Ewing, V., 28, 362
Ewing, V. C., 28, 362
Eymond, S., 34A, 143, 338
Eyring, L., 26, 364; 27, 481
Eysel, W., 27, 606; 28, 316, 357
Ezov, J. U.S., 34A, 173

Fadeyeva, V. I., 30A, 245
Fagherazzi, G., 32B, 45; 33B, 376, 522, 523
Fagnani,G., 31A, 221
Fahey, J. J., 26, 483, 484; 27, 675, 29, 436
Fahrat, S., 26, 117
Falgueirettes, J., 26, 548; 33B, 75, 529; 34B, 109; 126; 35B, 11, 158
Falkenburg, G., 29, 223
Faller, J. G., 28, 19
Faller, J. W., 31B, 353
Falqui, M. T., 28, 223
Farr , D. -L., 29, 434
Fan, H. -F., 28, 339; 30B, 212
Fanfani, L., 31A, 186, 190; 32A, 365, 391; 33A, 405; 33B, 133; 35A, 333, 439
Fang, J. H., 27, 690; 28, 141, 358; 29, 318, 396, 403; 30A, 418; 32A, 290; 34A, 310; 35A, 265, 437, 441, 478
Fang, P. H., 27, 539
Fankuchen, 31B, 153
Faoys, J., 33A, 441
Fares, V., 35B, 816
Farina, M., 29, 502
Farmakovskaja, G. I., 26, 446
Farmer, V. C., 29, 437
Farr, J. D., 26, 3
Farrell, E. F., 28, 141
Farrell, G., 29, 224
Farrier, D. S., 35B, 383
Farrow, R. F. C., 32A, 180
Fasiska, E. J., 30B, 37, 175
Fauckler, J. P., 30B, 445
Faught, J. B., 34B, 383; 35A, 137
Faulring, G. M., 26, 370
Fauquier, D., 28, 341; 35A, 241
Faust, R. J., 28, 358
Fava, G., 27, 630, 631, 835, 837, 881, 949, 951; 28, 645; 30B, 46
Favero, P., 26, 543
Fawcett, J. K., 30B, 111; 31B, 109, 172, 208; 35B, 103, 113, 389
Fayard, M., 26, 540
Fayek, M. K., 35B, 313
Fayos, J., 35B, 43
Fedeli, W., 28, 456, 682, 687; 31B, 131; 33B, 132; 35B, 173, 379
Feder, H. M., 27, 109
Feder, R., 27, 383
Fedorko, A. S., 27, 41; 29, 225
Fedoseeva, T.I., 29, 140
Fedorov, T. F., 30A, 175; 32A, 141
Fedorov, V. E., 30A, 180
Fedosov, V. N., 29, 433
Fedotova, K. V., 28, 318
Fedulov, S. A., 26, 419; 27, 505, 506; 29, 326, 434
Fehér, F., 27, 645, 751; 28, 316; 29, 291
Fehlmann, M., 29, 371; 30A, 397; 30B, 104, 394; 35B, 236

Feigelson, R. S., 29, 434
Feighenblatt, F., 27, 815
Feikama, Y. D., 26, 499; 31A, 212
Feil, D., 26, 613; 30B, 445; 33B, 31; 34A, 172; 34B, 30
Feitknecht, W., 26, 327; 28, 315; 29, 250
Fekličev, V. G., 27, 535
Felcher, G. P., 35A, 21
Felix, D., 34B, 47
Felsche, J., 32A, 265; 33A, 275; 35A, 307, 308
Felten, E. J., 26, 492
Fender, B. E. F., 35A, 114
Fenn, R. H., 27, 1058, 1059; 28, 686; 31B, 57, 450; 33B, 343; 34B, 443, 551; 35B, 527, 824, 830, 854
Fennessey, J. P., 32B, 352; 33B, 344, 356, 35B, 354
Ferguson, G., 26, 649, 694, 831; 27, 895, 897, 1028, 1031; 30B, 236, 393, 394; 31B, 79, 246; 32A, 509; 32B, 141, 201, 252; 33B, 98, 103, 210; 215, 216, 295, 296, 366; 34B, 120, 310, 312, 317, 340, 346, 394; 35B, 147, 390
Ferugson, G. A., 30A, 179
Ferguson, R. B., 26, 546; 27, 676; 34A, 368; 35B, 541
Fergusson, J.E., 26, 659; 28, 87; 32A, 192
Fermor, J. H., 31A, 62
Fernando, Q., 30B, 380; 32B, 454, 457; 34B, 498, 592; 35B, 162, 732, 808
Fernhold, L., 35B, 899
Ferrari, A., 26, 337, 341, 342, 500; 27, 471, 640, 641, 653; 28, 89, 92, 315, 356; 30A, 298, 350, 409; 30B, 365, 367, 445; 32A, 352
Ferraris, G., 28, 241; 31A, 219; 31B, 53, 65; 32A, 387; 32B, 428; 34A, 335; 34B, 197,198, 550; 35A, 335, 354; 35B, 233
Ferraro, J. R., 34A, 237
Ferrier, R. P., 28, 447, 449
Ferrier, W. G., 27, 795, 798; 28, 516
Ferris, J. P., 30B, 237
Ferro, R., 26, 226, 255; 27, 249; 29, 12, 140, 141, 226
Fert, A., 27,497
Feščenko, N. G., 35B, 465
Fesenko, E. G., 26, 420, 422; 27, 508; 28, 318, 341; 29, 434; 33A, 424
Feserer, J. I., 30A, 181
Fiala, J., 28, 314
Fiedler, G., 27, 711
Fiegel, L. J., 27, 538, 29, 438
Fielding, P. E., 30B, 445
Figgis, B.N., 27, 832; 29, 277; 30B, 394; 34A, 230
Fillipenko, O.S., 33A, 347; 35A, 312
Filipev, V.S., 26, 422; 28, 318, 340; 29, 434
Filippakis, S. E., 32B, 12, 17, 50, 174, 185
Filippini, G., 34B, 113, 304
Finch, E. D., 33A, 255, 256, 258
Finch, N., 35B, 117
Finger, L. W., 31A, 124
Fink, W., 29, 538; 32B, 303
Finkel', V. A., 26, 56, 110; 27, 65, 356; 29, 224, 225; 30A, 247, 248; 31A, 54
Finney, A. J. T., 35A, 147
Finney, J. J., 26, 108; 27, 602, 678; 28, 197, 211, 340, 358; 29, 371, 407; 31A, 196; 32A, 359; 33A, 395, 403; 34A, 358; 35A, 481
Finnie, L. N., 27, 372
Fischer, A. G., 27, 110
Fischer, E. O., 26, 650; 27, 110; 28, 682, 683
Fischer, E. W., 27, 800
Fischer, J., 31A, 89; 32B, 444; 32A, 170, 171, 508
Fischer, K., 28, 279; 31A, 224, 227; 34A, 380
Fischer, K. F., 32A, 483
Fischer, M. S., 32B, 115; 35B, 313
Fischer, P., 31A, 163; 35A, 404
Fischer, W., 29, 350
Fischmann, 26, 709
Fischmeister, H. F., 27, 604
Fisher, C. D., 30B, 265
Fisher, C. G., 35B, 651
Fisher, D. J., 29, 436
Fisher, D. W., 26, 254
Fisher, E. S., 26, 271; 27, 242
Fisher, F. G., 27, 598
Fitts, R. B., 27, 95
Fitzwater, D. R., 28, 83; 33B, 278, 279
Flachsbart, I.,28, 191
Flack, H. S., 35B, 121
Flahaut, J., 26, 88, 117, 277; 27, 327, 424; 29, 135, 136, 224, 341, 433; 30A, 175, 176, 177, 245; 32A, 104, 132; 33A, 4; 34A, 98, 144
Flanagan, T. B., 29, 226; 30A, 244
Flanigan, E. M., 32A, 484
Flautt, T. J., 33B, 529
Fleet, S. G., 27, 686; 30A, 430; 31A, 222; 32A, 465
Fleischer, E., 32A, 510; 34B, 560
Fleischer, E. B., 28, 528, 686; 29, 617; 30B, 216, 344, 363; 31B, 369; 32B, 358; 33B, 71, 185, 428; 34B, 283; 35B, 823
Fleischer, M., 32A, 477

Le Flem, G., 33A, 310
Fleming, J. E., 32B, 181; 33B, 443
Flengas, S. N., 26, 325; 29, 249
Fletcher, J. M., 32A, 185
Fletcher, R. O. W., 26, 585; 27, 813; 29, 537
Flieher, G., 32A, 99; 33A, 84
Flink E., 31A, 67
Flinn, P.A., 26, 129
Flinter, B.H., 28, 358
Flippen, J., 32B, 197
Flippen, J. L., 35B, 213, 238
Florencio, F., 35B, 46
Florian, L. R., 33B, 456
Flory, P. J., 27, 800
Flotow, H. E., 26, 167
Flournov, P. A., 29, 434
Flugel-Kahler, E., 28, 216
Flurry, R. L., 31B, 30
Flygare, W. H., 33B, 540; 35B, 905
Flynn, C. M., 34A, 271
Flynn, J. J., 32A, 510; 33B, 235, 271; 34B, 369, 35B, 99
Folberth, O. G., 26, 227
Foley, J. H., 26, 132
Foley, R. T., 26, 411
Folkers, J.B., 32B, 34
Folkes, F.M., 27, 488
Folkmarson, L., 33A, 221
Follner, H., 33A, 233; 35A, 127, 148, 149; 35B, 9
Folting, K., 28, 62, 686; 29, 657; 34B, 124
Fontaine, F., 33A, 517
Font-Altaba, M., 30B, 444
Fontbonne, A., 34A, 142, 168
Föppl, H., 27, 350
Forder, R. A., 35B, 486, 487
Foreman, D. W., 33A, 273
Foreman, N., 35A, 454
Foresti, E., 35B, 157
Forgeng, W. D., 27, 170
Formanek, H., 35B, 888
Fornasini, M. L., 34A, 142; 35A, 97, 124
Forrat, E. F., 28, 138
Forrat, F., 26, 389; 27, 256, 522
Forrest, J. G., 32B, 416; 33B, 403
Forrester, J. D., 27, 471; 28, 71, 75, 130, 181, 228; 29, 264, 273, 274, 288, 289, 335, 560; 31B, 414
Forsberg, H. -E., 27, 469
Forsellini, E., 34B, 517; 35B, 762, 819, 860
Forst, D., 31B, 299
Forster, R. H., 30A, 316; 33A, 97
Forsyth, J. B., 26, 7, 19; 27, 23, 166; 28, 17; 32A, 102, 415
Forsythe, G. D., 34B, 308
Forwerg, W., 28, 318
Fosheim, K., 30B, 285
Foss, O., 26, 576, 577, 578, 579, 580, 581, 582, 583; 28, 116, 686; 29, 345, 484, 520; 30A, 380; 30B, 284, 285, 286, 288, 330
Fossen, S., 26, 577, 578
Foster, L. M., 26, 346
Foster, M. D., 27, 685
Fouassier, C., 32A, 296; 35A, 288
Fourme, R., 31B, 33
Foust, A.S., 35A, 403
Fowlie, A. D., 35B, 665
Fox, A. C., 32A, 185
Fox, M. R., 28, 686; 29, 568; 32B, 395
Foxman, B. M., 32A, 509; 33A, 238; 33B, 416; 34B, 532
Foxman, B. R., 33B, 504
Fraker, A. C., 26, 420; 27, 4, 305; 34A, 144
Franceschi, E., 34A, 62, 146; 35A, 58
Francis, J. N., 34B, 360
Francotte, J., 30A, 177
Frank, G. R., 27, 18
Frank, H. S., 26, 343, 551
Frank, K., 28, 23; 35A, 7, 62
Franke, W., 28, 23; 29, 435, 438; 35A, 7, 62
Frankevič, D.P., 29, 138, 141; 30A, 176, 178, 34A, 143
Frank-Kameneckij, V.A., 26, 420
Franks, W. A., 34B, 541
Franzen, H., 34A, 142
Franzen, H. F., 27, 41; 31A, 43; 32A, 130, 142; 33A, 115, 141; 34A, 110, 129; 35A, 105, 107
Franzini, M., 27, 844; 28, 627, 682
Frasci, A., 34B, 13
Fraser, K. A., 30B, 395; 32B, 412, 420, 424
Fraser, R. D. B., 26, 825; 27, 1053, 1056, 1057; 29, 759
Frasson, E., 26, 679; 27, 872; 28, 424, 598, 655, 683; 29, 582
Fratini, A. V., 32B, 100, 247; 33B, 105; 34B, 326; 35B, 421
Frazer, B. C., 26, 473; 27, 525, 572; 29, 331; 30A, 369; 32A, 272
Frazier, A. W., 27, 569, 577; 28, 357
Fredericks, R. J., 31B, 31
Freed, R.L., 29, 437; 32A, 455; 34A, 383
Freedman, H.H., 35B, 491
Freeman, A. G., 26, 528
Freeman, E. B., 26, 531
Freeman, H. C., 26, 570, 658; 27, 1049; 29, 602, 605; 30A, 280, 286; 30B, 348, 349, 354, 395; 31B, 437; 32B, 420, 421; 33B, 416; 34A, 298; 34B, 482, 538, 556; 35B, 304, 653, 658, 660, 759
Freeman, J. M., 35B, 888
Freer, S. T., 30B, 210
Fregerslev, S., 33B, 219

Frei, E., 27, 1056
Freiburg, C., 26, 667
Freise, E. J., 27, 395
Frenz, B. A., 34A, 160; 35A, 413
Frenzel, G., 26, 378
Frere, P., 27, 444
Freund, F., 29, 434
Freundlich, W., 31A, 146; 33A, 314
Frevel, L. K., 27, 451
Frey, A. J., 31B, 251
Fricke, H., 28, 339
Fridrichsons, J., 28, 550, 554; 27, 1003, 1052; 30B, 240; 32B, 239 33B, 223
Friedman, H.A., 26, 311
Friedman, L. B., 28, 314; 29, 231; 31B, 287, 292; 35A, 130
Friedrich, G., 27, 534
Friedrichsen, H., 29, 248
Fries, F. J., 26, 76
Fripiat, J. J., 33B, 528
Frit, B., 35A, 186
Fritchie, C., 27, 786, 884
Fritchie, C. J., 26, 649; 31B, 43, 104, 136, 356, 452; 32B, 296, 332; 34B, 153, 351
Fritsch, F. N., 29, 490
Fritz, G., 34B, 374
Fritz, H., 35B, 197
Fritz, H. P., 28, 683
Fritzsche, W., 27, 214
Fröhlich, C., 26, 245
Fröhlich, H., 27, 338
Frolov, A. P., 27, 794
Frondel, C., 27, 697; 28, 288
Frorath, F. -C., 26, 172
Frorath, F. -K., 27, 350
Frost, R., 30B, 334
Frounfelker, R. E., 27, 382
Fruchart, D., 33A, 153
Fruchart, E., 26, 225; 27, 95; 29, 226
Fruchart, M. R., 29, 226
Fruchart, R., 26, 226; 27, 95; 28, 15; 29, 135, 139; 30A, 36, 175; 32A, 102, 142, 143; 33A, 86, 153; 34A, 144
Fruchart-Triquet, E., 26, 225
Frueh, A. J., 26, 250; 32A, 419; 33A, 17, 228
Fruendlich, W., 33A, 338
Fryxell, R. E., 35A, 219
Fu, K. -F., 29, 436
Fu, P. -Q., 31A, 189
Fuchikami, Y., 26, 306
Fuchs, A., 30A, 246
Fuess, H., 33A, 309, 388; 35A, 362, 383
Fuger, J., 34A, 209
Fujii, K., 27, 806; 35B, 44
Fujii, Y., 34A, 101
Fujino, Y., 29, 139
Fujisawa, H., 29, 434
Fujisawa, N., 26, 286
Fujishiga, R., 29, 760
Fujita, D. K., 34A, 179
Fukatsu, T., 26, 97
Fukazawa, Y., 34B, 115
Fukuda, N., 35B, 41
Fukui, S., 31A, 112
Fukushima, F., 29, 459; 30B, 395
Fukuyama, T., 33B, 530; 35B, 892
Fukuyo, M., 31B, 312; 32B, 76
Fuller, M. W., 28, 684
Fuller, W., 27, 1055
Funes, A. J., 26, 134
Funin, V. N., 29, 225
Funk, H., 26, 511
Funke, V. F., 26, 97; 27, 102; 30A, 248
Furberg, S., 26, 624; 27, 788, 790, 792; 28, 513, 684; 29, 489; 30B, 199, 395; 33B, 154; 35B, 224, 248
Furlani, C., 35B, 799
Furue, K., 35B, 488
Furukawa, J., 29, 760
Furukawa, K., 29, 675
Furusaki, A., 28, 682; 30B, 398; 32B, 83, 134, 148; 33B, 50, 86; 34B, 441; 35B, 44, 163, 171, 367, 368, 386, 402, 763, 765, 768
Furuseth, S., 29, 18, 19, 135; 30A, 14, 76, 174, 181, 247; 34A, 73
Futergendler, S. I., 27, 395; 28, 339
Fykin, L. E., 28, 207; 33A, 326

Gabe, E. J., 26, 585; 27, 999, 1019; 28, 548; 29, 503; 30B, 396; 31B, 25, 454; 32B, 26; 34B, 24
Gabel, A., 29, 224
Gabes, W., 35A, 151
Gabrielson, O., 27, 714
Gabrusevič, V. V., 34A, 143
Gabuda, S. P., 28, 244
Gadó, P., 30A, 312
Gadret, M., 31B, 121, 125, 126; 32B, 156; 33B, 114; 34B, 130, 136, 137, 145; 35B, 168
Gaedcke, H., 31A, 106
Gaetini, A., 35B, 235
Gafner, G., 26, 831; 27, 921; 28, 429; 29, 642, 644; 34B, 321; 35B, 583
Gagnaux, P., 26, 703
Gainsford, G. J., 32A, 192; 34A, 213
Galasso, F., 26, 420, 421; 27, 467, 509; 28, 341, 357; 33A, 328
Gale, B. G., 29, 135
Galigné, J. L., 26, 548; 33B, 75; 34B, 126; 35B, 11, 70, 158
Galinella, E., 29, 760
Gallazzi, M. C., 30B, 398
Galles, C. J., 34A, 164
Gallezot, P., 28, 316; 32A, 348

Galli, E., 31A, 235; 34A, 393
Gallmeier, J., 32A, 117; 34A, 122
Galloni, E. E., 27, 462, 967
Gallot, B., 26, 564; 27, 1061
Gal'perin, E. L., 26, 189; 27, 434; 29, 760
Galy, J., 29, 302; 30A, 343; 31A, 140; 33A, 189, 336; 34A, 217, 265, 266, 318; 35A, 186, 274
Gambino, J. R., 28, 317
Gamidov, R., 26, 462
Gamidov, R. S., 28, 190
Ganem, A., 33A, 177
Ganesan, S., 27, 1065
Ganglberger, E., 30A, 35, 172, 175, 244; 33A, 67, 77, 103; 35A, 89
Ganiev, R. M., 35A, 470
Ganis, P., 26, 625; 27, 801; 28, 315, 428, 490, 609, 610, 682; 29, 468, 760; 30B, 395; 31B, 8, 12; 32B, 14, 16, 178, 378; 33B, 4; 34B, 11, 96; 35B, 581, 884
Ganter, K. -W., 28, 356
Gantzel, P. K., 28, 410; 29, 135, 224; 30A, 245; 30B, 121; 32B, 323
Ganung, G., 33A, 328
Gapotčenko, N. I., 35B, 901
Gapsering, M., 34A, 270
Garaj, J., 28, 108; 31A, 118, 177; 32A, 239; 34A, 233, 340
Garavelli, C. L., 29, 435
Garaycochea, I., 26, 800
Garaycochea-Wittke, I., 32A, 384
Garazzolo, G., 34B, 181, 188
Garbuelio, C., 33B, 3
Garbuglio, C., 29, 760
Garcia, A., 35A, 197
de Garcia Banus, D. T., 35B, 602
Garcia-Blanco, S., 26, 490; 28, 682; 31A, 208; 32B, 84; 33A, 441; 34B, 345; 35B, 45, 606
Garcia-Fernández, H., 27, 409; 31A, 81
Gardin, A. I., 27, 123
Gardner, C. D., 31A, 176
Gardner, R. A., 34A, 295
Garnder, W. E., 32A, 185
Gariand, F., 28, 358
Garner, C. D., 32A, 350; 35A, 326
Garner, E. V., 26, 368
Garske, D., 29, 438; 30A, 365
Garton, G., 26, 693; 28, 658, 683
Gashurov, G., 33A, 310
Gašimzade, F. M., 29, 136
Gasparri, G. F., 30B, 344, 360, 397, 447; 31B, 402; 32A, 346; 32B, 407, 408, 411; 33B, 466, 483; 34A, 315; 34B, 32, 206; 35B, 28, 67
Gasperin, M., 28, 202, 356; 32A, 313, 33A, 329; 35A, 236, 241
Gassman, J., 34B, 281
Gassmann, J., 29, 759
Gasson, D. B., 27, 47
Gastuche, M. -C., 27, 423
Gatehouse, B. M., 29, 305; 32A, 509; 33A, 333; 34A, 257, 258, 274, 277; 35B, 258
Gatehouse, B. W., 35A, 22, 238, 270
Gatineau, L., 28, 240
Gatos, H. C., 27, 244; 281, 303, 382; 29, 136
Gattow, G., 27, 476, 605; 28, 316, 338, 339; 29, 311, 30A, 46, 174, 177, 411; 30B, 53
Gaudefroy, C., 28, 356
Gaudefroy, G., 33A, 381
Gaudin-Louer, M., 32A, 509
Gaultier, J., 26, 741; 30B, 101, 146, 395; 31B, 98, 100, 101, 141; 32B, 121; 34B, 98, 101, 102; 35B, 118
Gaultier, M., 31A, 198; 33A, 367
Gaunt, P., 30A, 59
Gautzel, P. K., 27, 641
Gauzzi, F., 28, 23
Gavezzotti, A., 35B, 93
Gavin, R. M., 29, 625
Gay, P., 26, 512; 27, 423
Gazo, J., 34A, 340
Gazzara, C. P., 26, 179; 32A, 98
Geballe, T. H., 30A, 247
Gebelt, R. E., 29, 136
Gebert, E., 28, 73, 75; 29, 280; 34A, 284; 35A, 226
Gebhardt, E., 28, 338; 29, 137
Gee, W., 26, 729
Geffrov, J., 29, 433
Geise, H., 27, 1037
Geise, H. J., 30B, 396; 31B, 159, 202, 203, 206; 32B, 222; 34B, 303; 35B, 897
Geise, H. J. V. H., 30B, 398
Geissman, T. A., 35B, 380
Geist, D., 35A, 127
Gel'd, P. V., 26, 105, 149; 27, 156, 208; 29, 135, 137, 141, 223, 224, 225; 30A, 178, 245; 34A, 58
Geller, S., 26, 349, 414; 27, 169, 315, 338, 511, 828; 29, 224; 32A, 216, 254; 34A, 149, 262; 35B, 881
Gellert, M., 27, 1055
Gelman, M. L., 27, 689
Gemmell, K. W., 34B, 333; 35B, 386
Geneste, J., 27, 156
Genet, F., 27, 985; 30B, 33, 129, 137; 32B, 439; 34B, 576
Genkin, A. D., 27, 290; 28, 338
Gent, P. A., 33B, 170
Gentry, M., 33A, 180
George, J. W., 34B, 416
Gerdil, R., 26, 765
Gerhardt, W., 28, 338
Gerloch, M., 27, 949; 29, 277, 556; 30B, 311; 31B, 425; 32B, 394, 396; 34A, 230

Germain, G., 26, 954; 28, 404; 29, 739; 35B, 308, 325
German, G., 26, 785
Germino, F. J., 29, 760, 761
Gert, L. M., 34A, 144
Gerteis, R. L., 29, 535
Gey, W., 27, 340
Ghilardi, C. A., 35B, 730, 850
Ghose, S., 26, 526, 764; 27, 644; 28, 192; 29, 371, 381; 30A, 174, 355, 397, 425
Ghosh, S. K., 30B, 445
Ghouse, K. M., 33A, 402
Giacommelo, G., 26, 768, 769, 776, 779; 27, 968, 969, 1039; 28, 456, 682, 687; 30B, 169; 31B, 131; 35B, 379
Giacovazzo, C., 34A, 215
Giancotti, V., 33B, 525; 35A, 376
Gibb, T. R. P., 26, 166, 168
Gibbons, C. S., 32B, 151; 34B, 333
Gibbs, G. V., 26, 531; 27, 698; 28, 246; 30A, 426; 31A, 223; 33A, 480; 34A, 366, 388; 35A, 469
Gibson, D. T., 30B, 53
Gibson, E. D., 29, 223
Gibson, J. W., 34B, 165; 35B, 328
Gibson, R. E., 35B, 201
Giese, R. F., 29, 4; 30A, 17, 29, 31, 172, 173, 174, 251
Giesecke, G., 26, 49
Giess, E. A., 27, 488
Giessen, B.C., 26, 217; 27, 234; 29, 69, 70, 135, 223; 30A, 63, 77, 85, 174, 179, 180, 181; 32A, 142; 34A, 31, 124
Gierke, T.D., 35B, 905
Giglio, E., 26, 754; 27, 412, 413; 30B, 134, 253; 31B, 111, 260, 261; 32A, 207; 32B, 287
Giglio, M., 26, 540
de Gil, D. R., 34A, 183
de Gil, E. R., 34B, 438
Gilardi, R. D., 34B, 326; 35B, 338
Gilfrich, J. V., 29, 50; 31A, 48
Gill, K. J., 26, 258
Gill, N. S., 35B, 775
Gillam, E., 28, 289; 29, 135
Gillard, R. D., 34B, 537; 35B, 600
Giller, Ja. L., 26, 34
Gilles, J. -C., 27, 415; 29, 433; 34A, 142, 168
Gilles, P. W., 26, 80
Gillespie, R. B., 27, 641
Gillier, H., 28, 454; 30B, 164
Gillier-Pandrand, M., 34B, 3
Gillier-Pandraud, H., 30B, 74, 368,445; 31A, 149; 31B, 74, 378, 438; 32A, 270; 32B, 102; 33B, 239
Gillson, J. L., 33A, 36
Gilmore, C. M., 32A, 142
Giminez-Huguet, M., 27, 656
Ginderow, D., 33B, 528
la Ginestra, A., 28, 317
Gingerich, K., 26, 493; 29, 136
Gingrich, N. S., 26, 3; 27, 720
Ginsberg, A. P., 27, 671; 29, 235, 236; 31B, 352
Ginzburg, I. V., 26, 526; 29, 437
Giordano, F., 31B, 89; 32B, 304; 33B, 525; 34B, 582
Giorgi, A. L., 26, 3; 33A, 153
Giraldi, G., 27, 631, 881, 949; 28, 645; 30B, 46
Giri, S. N., 28, 683; 30B, 447
Girling, R. L., 32B, 453
Giuseppetti, G., 26, 605; 27, 584; 28, 196; 29, 608; 32A, 336, 362, 444
Givord, F., 35A, 60
Gjertsen, L., 29, 137, 225
Gjønnes, J., 35A, 8, 206
Gkuznecov, V., 30A, 176
Gladkova, V. F., 29, 358
Gladyševskij, E. I., 26, 19, 169, 174, 175, 235; 27, 152, 161, 162, 206, 212, 233, 284, 287, 375, 376; 28, 28; 29, 50, 137, 138, 139, 140, 141, 225; 30A, 44, 172, 174, 175, 176, 177, 178, 179, 180, 181, 182, 244, 245, 248; 31A, 35, 63; 32A, 5, 57, 142, 143; 33A, 64; 34A, 63, 64; 35A, 44, 50, 96, 124
Glants, G., 26, 359
Glaser, H., 28, 356
Glasner, A., 27, 815; 29, 353, 383
Glass, G. E., 33B, 321
Glasser, F. P., 26, 512, 524; 28, 144, 339, 356, 29, 397, 434, 436; 33A, 460
Glasser, L. S. D., 26, 511, 512, 524; 28, 144, 250, 339, 356; 29, 397, 401; 31A, 215; 32A, 449; 33A, 265, 386, 460; 34A, 252; 35A, 292, 294; 35B, 604
Glatz, A.C., 33A, 44
Glazer, A.M. 35A, 121
Glazova, V. V., 33A, 326
von Glehn, M., 33B, 168; 34B, 154; 35B, 195
Gleitzer, C., 32A, 296
Gleizes, A., 35A, 124
Glemser, O., 29, 434; 33A, 269; 34A, 206; 35A, 279
Glen, G. L., 26, 486, 691; 28, 55, 632
Glick, M. D., 29, 669; 30B, 323; 31B, 440; 33A, 247; 33B, 399; 34A, 279; 35A, 130; 35B, 264, 356
Glidewell, C., 35B, 888, 891
Glimcher, M. J., 26, 826
Glinskaja, L. A., 34A, 221
Glover, W. J., 29, 434
Glowiak, T., 26, 340
Gluck, M. D., 32B, 234, 237

Gludkikh, N. T., 26, 63
Glusker, J. P., 26, 724; 27, 735; 28, 385; 31B, 454; 32B, 26; 33B, 21, 151; 34B, 7, 18, 24
Glusker, M.P., 30B, 28
Gluškova, M. A., 34A, 168
Gluskova, V. B., 28, 356
Gnativ, G. M., 26, 416
Gnutzmann, G., 26, 53
Go, K. T., 30B, 395; 34B, 208
Goaman, L.C. G., 29, 493
Gobillon, Y., 27, 673, 760, 761; 28, 575; 31B, 271, 273
Godbey, W. G., 26, 823
Gobrecht, H., 29, 137
Goddard, J., 30A, 181
Gödecke, T., 33A, 124; 34A, 11, 142
Godel, D., 30A, 172
Godfrey, P. D., 33B, 540
Godonikov, A. A., 27, 88
Godovikov, A. A., 26, 120
Godycki, L. E., 26, 449, 717
Godžaez, F. M., 34A, 52
Goebel, C. V., 32A, 322
Goedertier, R. E., 28, 668
Goedkoop, J. A., 26, 27; 28, 7
Goettlicher, S., 30B, 445
Goetz, W. K., 29, 224
Goffin, J., 26, 412
Goforth, R. G., 26, 294
Gogoi, B. N., 35B, 52
Gogorisvili, P. V., 31B, 417
Gohle, R., 29, 79
Golabi, M., 29, 136, 341
Golabi, S. -M., 30A, 175, 245
Golankiewicz, K., 30B, 446
Gold, K., 34B, 360, 445, 475; 35B, 430, 434, 505
Gold, W., 30A, 173; 34A, 142
Goldak, A. J., 26, 28
Goldak, J., 31A, 42
Goldak, G. R., 29, 223
Gol'der, G. A., 26, 701, 751; 27, 560; 28, 431, 432, 683; 30B, 447; 32B, 41, 124; 33B, 74; 34B, 103, 105
Goldfarb, T. D., 32B, 301
Goldish, E., 26, 288, 696; 29, 473
Goldmann, J., 34A, 319, 325
Goldschmidt, H. J., 26, 110, 113; 27, 140; 29, 224
Goldschmied, E., 35B, 672, 703
Goldsmith, J. R., 26, 481
Goldstein, A., 26, 558
Goldstein, P., 27, 989; 32B, 118; 30B, 256; 34B, 122
Golemba, F. J., 30B, 524
Golic, L., 30B, 14, 17
Golightly, J. P., 32A, 419
Golomb, M. L., 34B, 538
Golovačev, V. P., 28, 190; 29, 438; 35A, 391, 480
Golovastikov, N. I., 26, 456; 27, 580; 31A, 225; 32B, 320
Golovina, N. I., 31B, 51; 32A, 150, 347; 32B, 64; 33A, 450; 34A, 349
Golovkin, V. S., 27, 151; 34A, 142; 35A, 115
Golyšev, V. M., 33A, 271
Goman'kov, V. I., 27, 150, 157, 163
Gomes de Mesquita, A. H., 28, 34; 29, 54; 30A, 254; 30B, 54; 32A, 6, 506; 32B, 70; 33A, 160
Gondrand, M., 27, 516; 28, 134, 318; 32A, 509; 35A, 214
Gonzalo, J. A., 31A, 241
Good, P., 29, 224
Goodacre, G. W., 34B, 382; 35B, 451
Goodall, D. C., 31B, 463
Goodenough, J. B., 28, 285; 35A, 274
Goodgame, D. M. L., 30B, 331
Goodman, S. H., 29, 230; 30A, 252; 30B, 115; 32B, 148
Goodyear, J., 26, 529; 28, 24; 30A, 62, 179; 31A, 17; 34A, 25; 34A, 49, 76
Goost, L., 33A, 192; 35A, 173
van der Goot, A. S., 33A, 153; 34A, 68, 144, 145; 35A, 94, 95, 124
Gopalakrishna, E. M., 27, 893; 30B, 64; 33B, 203; 34B, 293, 294, 300, 301, 358
Gorbatyi, L. V., 33A, 334
Gorbunov, N. S., 26, 264; 27, 672
Gorbunova, Ju. E., 33A, 438; 35A, 417
Gordeeva, G. A., 31B, 407
Gordienko, V. V., 29, 437
Gordon, E. K., 29, 433; 30A, 256; 31A, 86; 33A, 8
Gordon, G. S., 33A, 105
Gordon, J. T., 33B, 236
Gordon, R. O, 30A, 172
Gordy, W., 34A, 151, 171
Goretzki, H., 27, 121; 29, 139; 30A, 35
Gorjunova, N. A., 26, 48; 27, 50, 51; 29, 136, 137; 30A, 173, 248; 33A, 126, 153
Gorochov, O., 30A, 176
Gorodeckij, S. M., 30A, 244
Gorogockaja, L. I., 33A, 518
Gorres, B. T., 35A, 202
Gorševskaja, S. A., 26, 386
Gorski, L., 28, 314
Gorškov, A. I., 33A, 472
Goswami, A., 26, 358, 529; 27, 352
Goswami, K. N., 28, 682, 683; 30B, 445
Goto, K., 33B, 223

Gottardi, G., 26, 610; 28, 252; 31A, 235
Gottlicher, S., 28, 686; 33B, 245; 34B, 584; 35B, 197
Gotz, A., 28, 316
Götz, W., 27, 675
Gough, T. E., 34B, 503
Gougoutas, J. A., 32B, 103
Gougoutas, J. Z., 35B, 91, 326, 357
Gould, E. S., 26, 784; 27, 401
Gould, R. O, 35B, 788
Gould, R. W., 27, 260
Goursolle, M., 34B, 130, 136, 137, 145; 35B, 168
de Graaf, R. A. G., 32A, 288
Grabowski, M., 34B, 181
Grace, J. D., 29, 438
Graddon, D. P., 28, 687
Graeber, E. J., 28, 76; 30A, 274, 372, 373; 32B, 30
Graeser, S., 28, 342
Graf, D. L., 26, 481; 27, 561; 31A, 172
Graham, A. J., 34A, 229; 34B, 443; 35B, 527
Graham, J., 26, 355; 31A, 43, 154; 33A, 291
Grain, C., 26, 347
Grainger, C. T., 34B, 39
Gramaccioli, C., 30B, 395
Gramaccioli, C. M., 26, 668; 30B, 444; 31B, 411, 412; 33B, 49, 52; 34B, 113, 118, 304
van de Grampel, J. C., 34A, 156, 177
Gran, G., 27, 23
Grandjean, D., 27, 633; 28, 314; 29, 317, 353; 31A, 91, 92, 209; 32A, 174, 177; 35A, 28
Granger, M. M., 31A, 174; 33A, 381; 34A, 313
Granoff, B., 33B, 350
Grant, D. F., 27, 999; 29, 493; 30B, 225; 34B, 60, 403
Grant, I. G., 26, 830
Grant, I. J., 28, 542; 30B, 232
Grant, N. J., 26, 217; 27, 234; 29, 69, 70, 135, 223; 30A, 63, 77, 85, 174, 179, 180, 181; 32A, 142; 34A, 31, 124
Grant, W. K., 27, 906
Gray, G. W., 30B, 445
Gray, H.B., 28, 685; 31B, 456; 32B, 381, 462; 33B, 499, 514; 35B, 817
Gray, N., 33A, 228
Gray, R. T., 34B, 305
Graziani, R., 32B, 397; 34B, 517, 533, 604; 35B, 762, 819, 859, 860, 868
Grdenić, D., 26, 321; 27, 457, 458; 28, 622; 29, 458; 30B, 395; 34B, 548; 35B, 469
Greaves, M. L., 35B, 329
Grebenščikov, R. G., 26, 302; 27, 699
Green, C. D., 32B, 454
Green, M. L. H., 34B, 418
Greenaway, D. L., 30A, 181
Greenberg, B., 27, 981; 28, 683; 33B, 77, 34B, 36
Greenblatt, M., 32A, 202
Greenblatt, W., 35A, 192
Greene, P. D., 32A, 216
Greene, P. T., 34B, 549; 35B, 432, 538, 545, 546, 800, 849
Greenwood, N. N., 33A, 220
Greer, S. C., 33A, 157
Gregg, R. A., 27, 4
Gregory, E., 30A, 178
Gregory, N. W., 26, 317, 27, 432, 433; 28, 315; 30A, 277
Gregory, U. A., 35B, 814
Grenier, J. -C., 30A, 331; 31A, 143; 32A, 377, 379, 380, 381, 442; 33A, 413, 416; 34A, 326, 329
Grenthe, I., 34B, 527, 528
Grethe, G., 34B, 211
Grey, I. E., 35A, 25
Gridnev, V. N., 27, 232
Grieb, R., 30B, 395
Grierson, R., 29, 223
Griffé, M., 33B, 298; 35B, 325
Griffith, J E., 26, 301
Griffiths, A., 35B, 126
Griffiths, D., 27, 397
Griffiths, P. J. F., 28, 682; 30B, 445
Grifone, L., 26, 543
Grigor'ev, A. T., 26, 112
Grigor'ev, N. A., 28, 340; 29, 436
Grigor'eva, N. V., 31B, 52; 32B, 48; 33B, 30; 34B, 54; 35B, 19, 30
Grigoreva, V. S., 33A, 153
Grigorovici, R., 33A, 49
Grim, K., 26, 38
Grimaldi, F. S., 29, 63
Grimes, N. W., 28, 213; 30A, 374
Grimes, W. R., 27, 451
Grimmer, A. -R., 29, 438
Grimvall, S., 32B, 87
Grisafe, D. A., 29, 436
Grišin, I. A., 35A, 195
Grison, B., 27, 46
Grochowski, E. G., 29, 136
Grochowski, J., 34A, 229
Groeneveld Meijer, W. O. J., 26, 527
Groenwolt, K., 26, 63
Gromov, A. E., 29, 760
Grønbaek Hazell, R., 27, 663; 28, 687; 29, 310,629; 33B, 471, 472
Grønvold, F., 26, 264, 267; 27, 43, 315, 364; 29, 138; 34A, 131, 146
Grosskreutz, J. C., 29, 759
Grossoni, G., 34A, 234
Groten, B., 29, 760

Groth, P., 26, 546; 27, 729, 963; 28, 53, 685, 686; 29, 451, 491, 492; 30B, 90, 246, 248, 395; 32B, 177, 179, 180; 33B, 81, 82, 160; 34B, 96, 143, 170; 35B, 114, 181, 212, 855
Gruehn, R., 34A, 241; 35A, 209
Grueninger, H. W., 34A, 235
Grum-Gržimajlo, N. V., 26, 111; 27, 156
Grün, K., 26, 153
Gründer, W., 27, 478
Grundy, H. D., 33A, 168
Grunze, H., 30B, 445; 34B, 397
Grušina, R. K., 29, 759
Grützmacher, H. -F., 33B, 297; 35B, 440
Gschneidner, K., 26, 230
Gschneidner, K. A., 27, 141, 148; 29, 136, 137; 30A, 172, 244, 245, 247; 32A, 142, 143; 33A, 153, 34A, 145; 35A, 124
Guare, C. J., 28, 317
Guarnieri, A., 26, 543
Guastini, C., 30B, 443; 34B, 535; 35B, 725
Guba, F., 27, 838
Gubin, S. P., 31B, 343; 34B, 418 35B, 544
Gubler, B., 28, 683
Gudel, H. U., 33A, 242, 243, 244; 35A, 404, 405
Guenter, J. R., 33A, 153
Guentert, O. J., 27, 40, 396, 408, 546
Guerchais, J. E., 29, 434
Guerreschi, L., 28, 338
Guggenberger, L. J., 29, 533; 32A, 510; 33A, 157, 241; 33B, 266, 300; 34B, 174; 35B, 437
Guggisberg, A., 34B, 333
Guha, S., 32B, 227; 34B, 266; 35B, 656
Guilhem, J., 26, 751; 32B, 129; 35B, 353
Guillaume, G., 34A, 318
Guillemin, C., 26, 522
Guillen, M., 31A, 138
Guillet, J. E., 33B, 524
Guillory, J. P., 30B, 405
Guillot, M., 30A, 175
Guinet, P., 30A, 40, 175
Guinier, A., 33A, 153
Guiseppetti, G., 34A, 381
Guitel, J. -C., 32A, 380, 381, 33A, 395, 417
Guittard, M., 26, 117; 29, 135, 136, 224; 30A, 177, 34A, 142, 146
Guliev, T. N., 30A, 179
Gullman, J., 35A, 41
Gum, P., 29, 726
Gundersen, G., 34A, 218
Günther, F., 31A, 54
Gupta, K. P., 26, 19, 269
Gupta, M. P., 26, 563; 33B, 3; 34B, 17; 35B, 13, 14, 16
Gupta, N. P., 33B, 3
Gurr, G. E., 28, 513; 33B, 446; 34A, 29, 166; 35A, 202
Gurskaja, A. V., 33A, 81
Gurskaja, G. V., 28, 682; 29, 700
Guseinov, G. D., 34A, 52
Guseinov, G. G., 34A, 52, 88
Guseinova, M. K., 31B, 417; 32B, 414; 33B, 402
Gusev, A. I., 34B, 447, 448, 450, 479; 35B, 513, 544, 562
Guseva, L. N., 30A, 177, 245
Gush, H. P., 27, 393
Guss, J. M., 33B, 416
Gustaffson, B., 35B, 416
Guthrie, P. V., 26, 197
Gutierrez, C. P., 27, 129
Gutiérrez Rios, E., 26, 828
Gutowsky, H. S., 27, 422
Gutsche, W., 26, 333
Guttman, L., 27, 250
Guttormson, R. J., 34A, 224
Güven, N., 32A, 469; 35A, 475
Guy, B. B., 31A, 194
Gwinn, W. D., 26, 543; 27, 717

Haaf, W. R., 29, 284
Haagenson, C. O., 28, 100, 682, 686; 29, 495, 32B, 37
Haaland, A., 29, 524; 32A, 252; 33B, 535, 536, 537, 34A, 163; 35B, 890, 894, 901
de Haan, Y. M., 27, 498
Haapala, I., 30A, 427
Haas, C., 27, 345
Haas, D. J., 29, 465, 753; 30B, 44, 208, 258; 31B, 6; 35B, 888
Haas, H., 34A, 252
Haas, W. L., 26, 637; 27, 1050
Haase, D. J., 28, 575; 30A, 178, 246
Haase, J., 30B, 405
Haase, W., 28, 314; 32A, 172, 173; 33B, 454, 460
Haasen, P., 29, 433
Habermehl, G., 26, 798; 28, 555, 686; 30B, 445; 33B, 245; 35B, 197
Hadinee, I., 27, 640
Haefner, K., 28, 8
Hafner, S., 27, 487
Hagel, W. C., 26, 190
Hagenmuller, P., 26, 518; 28, 315; 30A, 96, 180; 31A, 140; 33A, 162, 183, 262; 35A, 163, 188
Hagerström, G., 26, 354
Hagihara, H., 31B, 331, 404; 33B, 319; 34B, 413, 592
Hagihara, N., 33B, 500
Hagiwara, H., 29, 140
Hagman, L. -O., 33A, 397, 407; 34B, 154
Hahemann, R. F., 30A, 180

Hahn, H., 26, 147; 27, 348, 465; 29, 131, 251; 30A, 172, 173, 174, 176, 177; 35A, 69, 138, 155
Hahn, T., 27, 606; 28, 316, 357
Hähnert, H., 26, 393
Haight, H. L., 29, 587; 31B, 367
Haikonen, L., 27, 425
Hainsworth, F. N., 31A, 172
Haisa, M., 27, 1054; 28, 682
Hajek, B., 30B, 445
Hak, J., 26, 48
Häkli, A., 29, 136, 140
Hakli, T. A., 30A, 181
Hale, W. H., 34A, 80
Hälg, W., 31A, 163; 35A, 404
Hall, D., 28, 316, 640, 645; 29, 599; 30A, 280, 404; 30B, 52, 369, 393, 395; 31A, 102, 105; 31B, 102, 243, 389, 392, 393, 395, 398, 423, 428; 32B, 380, 397, 442; 33B, 389, 398, 401; 34A, 29, 166; 34B, 409, 410, 500, 519, 585; 35A, 402, 427; 35B, 349, 413, 414, 615, 682, 781, 870
Hall, D. D., 34B, 521
Hall, E. L., 35A, 21
Hall, E. O., 30A, 316; 32A, 98; 33A, 95, 96
Hall, G. S., 33B, 425
Hall, H. T., 29, 136, 224, 225; 31A, 69
Hall, L. H., 30B, 272; 30A, 254
Hall, M. C., 35B, 864
Hall, R. D., 32B, 380
Hall, R. O. A., 26, 230, 247
Hall, S. R., 28, 684; 30B, 65; 32B, 91; 33B, 224, 226
Hall, W. E., 26, 125
Hallais, J., 32A, 143
Halpern, J., 32B, 430
Halverson, F., 27, 477
Ham, J. T., 35B, 268
Ham, W. M., 29, 224
Hamachi, T., 26, 515
Hamilton, B., 26, 395
Hamilton, D. C., 29, 136
Hamilton, J. A., 26, 792, 830; 27, 945, 1010, 1026; 28, 542; 30B, 242; 32B, 230
Hamilton, L. D., 26, 817; 31B, 269
Hamilton, P. K., 26, 391
Hamilton, W. C., 26, 344, 596; 27, 461; 28, 75, 106, 223, 229; 29, 240, 335; 30B, 12, 395; 31A, 88, 89, 92, 147, 197; 31B, 328; 32A, 213, 216, 312; 32B, 301, 311, 391, 459; 33B, 31, 47, 93, 290; 34A, 151, 158; 35A, 366; 35B, 322
Hammand, F. H., 26, 101
Hammel, F., 29, 437
Hammer, C. F., 29, 752
Hammer, H., 26, 624; 28, 513; 33B, 160
Hamor, M. J., 28, 686; 29, 545, 728; 30B, 218
Hamor, T. A., 26, 292, 830; 27, 945, 1000, 1010, 1013, 1030; 28, 542, 564, 639, 686; 29, 545, 728; 30B, 218, 219, 232; 33A, 247; 33B, 170; 34B, 158; 35B, 262
Han, K., 28, 396
Hanak, J. J., 26, 233; 29, 140
Hanazaki, F., 31B, 88
Hancock, C., 29, 224
Hand, W. D., 27, 549
Handa, B. K., 29, 275, 433
Handlovic, M., 34A, 333
Handtmann, D., 27, 386
Handwerk, J. H., 26, 494
Handy, L. B., 31B, 31; 35B, 506
Haner, B., 28, 683; 30B, 447
Haner, B. A., 30B, 447; 33B, 194
Haneveld, A. J. K., 29, 38, 342; 34A, 92; 35A, 110
Hang, C., 35A, 314
Hanič, F., 26, 669; 27, 438, 660; 28, 176; 29, 600, 606; 31A, 88; 33B, 232, 365; 32B, 162, 571; 35B, 564
Hanicová, K., 26, 669; 29, 606
Hanisch, K., 31A, 234
Hanke, A. R., 26, 391
Hanke, J., 30A, 422
Hanke, K., 28, 245, 265; 32A, 307
Hanneman, R. E., 29, 136, 223
Hansen, D. A., 30A, 77, 78, 177, 181; 32A, 124; 34A, 99
Hansen, D. H., 33A, 6
Hansen, F., 34B, 29
Hansen, K. W., 30A, 266
Hansen-Nygaard, L., 26, 760
Hanson, A. W., 26, 789, 790; 27, 650, 918; 28, 546, 576; 29, 689; 30B, 110, 249, 250; 31B, 279; 32B, 144; 33B, 135, 187, 254; 34B, 117, 155
Hanson, H. P., 30A, 259; 30B, 405
Hanson, J. R., 35B, 333
Hanson, K.W., 30B, 405
Hansson, A., 26, 487
Hansson, E., 35B, 639
Hansson, I., 33B, 287
Haque, M.U., 28, 683
Hara, T., 27, 799
Harada, J., 26, 423; 27, 399; 29, 341
Harada, S., 34B, 287
Harada, Y., 26, 494
Haraldsen, H., 27, 364; 34A, 131, 146
Harari, A., 33A, 306, 307
Harda, H., 31B, 194
Hardcastle, K., 26, 169; 27, 241; 29, 139
Hardel, K., 26, 423; 30A, 289

Hardgrove, G. L., 26, 808; 33B, 90, 202
Harding, M. J. C., 30B, 444
Harding, M. M., 28, 636; 30B, 395, 443; 31B, 22; 32B, 412, 422, 424; 33B, 172, 421; 35B, 316, 664, 788
Harding, T. T., 27, 909
Hardy, A., 26, 408; 27, 662; 28, 315, 357; 29, 302, 325, 435; 30A, 343, 446, 33A, 363
Hardy, A. -M., 28, 315
Hardy, M. A., 26, 394
Hardy, W. K., 26, 224
Hare, B. R., 26, 202
Harel, M., 34A, 276
Hargittai, I., 34A, 224; 35B, 893, 895
Hargreaves, A., 26, 726; 27, 749, 907; 29, 677; 33B, 64; 35B, 52
Hargreaves, M. E., 26, 139
Hargreaves, R. N., 34B, 432
Harkema, S., 34B, 30
Harker, D., 26, 817; 27, 1059
Harker, R. I., 29, 437
Harlow, R. L., 35B, 619
Harrah, L. A., 27, 822
Harris, C. B., 30A, 283; 33B, 498
Harris, C. M., 26, 666
Harris, D. C., 30A, 177
Harris, D. G., 29, 225
Harris, D. R., 29, 710; 30B, 44, 233, 258
Harris, G. S., 32A, 509
Harris, I. R., 29, 223, 224; 30A, 172, 176, 244
Harris, L. A., 26, 311; 27, 546; 28, 281; 29, 437; 31A, 130; 32A, 264; 33A, 271
Harris, P. M., 26, 290; 31A, 240; 33A, 202
Harrison, D. E., 26, 467
Harrison, I. T., 34B, 320
Harrison, J. G., 34A, 146
Harrison, P. M., 27, 1054
Harrison, R. W., 28, 357; 31A, 192; 33B, 306; 34A, 324
Harrison, S., 34B, 320
Harsha, K. S. S., 35A, 98
Harshbarger, W., 34A, 152, 153
Hart, H., 33B, 262
Hart, R. K., 26, 347
Harth, R., 27, 587
Hartke, K., 35B, 221
Hartl, H., 27, 751; 33A, 229
Hartley, C., 30B, 396
Hartley, C. S., 26, 254
Hartley, J. M., 33A, 263
Hartman, A., 31B, 85
Hartman, P., 27, 1037; 29, 438
Hartmann, A. O., 33B, 537
Hartmann, G., 32B, 307
Hartmann, H., 26, 828; 33B, 538, 539
Hartsuck, J. A., 28, 540; 31B, 292
Hartung, H., 31A, 99; 35B, 835
Haruna, J., 33A, 111
Harvey, J., 26, 354
Hasan Rizvi, S., 33B, 64
Haschemeyer, A. E. V., 28, 685; 30B, 196, 201
Haschke, H., 32A, 141
Hase, W., 32A, 137
Hasegawa, K., 27, 497
Hasekamp, R. C. D. E., 33B, 147
Haser, R., 30B, 398; 33A, 242
Hashagen, J. T., 30B, 445
Hassel, O., 26, 546, 763, 773, 778; 27, 729, 758, 963, 986; 28, 103, 668, 685, 686; 29, 451, 490, 491, 492; 30B, 90, 246, 247, 248, 249; 33B, 237; 35B, 202
Hassler, E., 27, 290
Hastings, J. M., 26, 117, 306, 340; 27, 461, 492; 33A, 344
Haszko, S. E., 26, 146, 193; 27, 196, 261
Hata, T., 29, 760; 35B, 402
Hathaway, B. J., 35A, 414
Hatibarna, J., 27, 480
Hatt, B. A., 26, 341
Hauck, J., 33A, 288; 34A, 263, 264, 273, 280
Haug, H., 28, 286, 287
Haug, J., 26, 187
Hauge, S., 26, 576, 579; 28, 116; 30B, 286
Haugen, L. P., 34B, 608
Haugsten, K., 34A, 94
Hauptman, H., 34B, 291
Hausen, D. M., 27, 601
Hausen, H. -D., 33A, 250
Hausmann, J., 28, 683
Haussühl, S., 26, 446; 28, 682
Hauw, C., 30B, 101, 146, 395; 31B, 98, 100, 101, 141; 32B, 121; 34B, 98, 101, 102; 35B, 118
Hauw, M. C., 26, 748
Havell, R. F., 26, 102
Havens, W. W., 27, 421
Havinga, E., 28, 476
Havinga, E. E., 35A, 23
Hawes, L. L., 27, 472
Hawkinson, S., 32A, 510; 33B, 71
Hawkinson, S. W., 32B, 358; 33B, 428; 35B, 329
Hawley, D. M., 30B, 394; 33B, 210, 295, 296; 34B, 312
Hawthorne, M. F., 34B, 360
Hay, T. G., 34B, 63
Hayashi, M., 26, 639; 27, 810; 35B, 386
Hayashi, S., 26, 432
Hayashida, T., 27, 743
Hayek, E., 28, 338
Haynes, F. G., 27, 157
Hayter, R. G., 31B, 31

Hazekamp, P., 27, 411
Hazell, A. C., 27, 409; 28, 99, 114, 315, 316, 29, 294; 31A, 80, 100; 31B, 289; 32B, 193; 35B, 683
Hazell, R. G., 32A, 246, 458; 33B, 27, 471, 472; 34A, 231, 236; 34B, 29
Heairfield, I., 30B, 444
Healy, J. H., 27, 538
Heard, H. C., 26, 481
Heaton, L., 26, 3, 286; 32A, 99; 34A, 144
Heber, I., 29, 760
Hecht, J. K., 35B, 99
Hecktfischer, S., 35B, 137
Hedberg, K., 27, 717; 28, 362; 29, 490; 34A, 174, 218; 35B, 902
Hedberg, L., 35B, 902
Hedger, H. J., 27, 125
Heemskerk, J., 27, 722
Heeres, G. J., 33B, 140
Hegenbarth, J. J., 28, 684; 29, 5, 668
Hegyi, I. J., 28, 37
Heiberger, J. J., 34A, 144
Heide, F., 27, 550
Heidelberg, R. F., 26, 345
von Heidenstam, O., 32A, 260; 33A, 72, 343
Heidner, E. G., 34B, 42
Heijkoop, E. van, 30B, 396, 398; 31B, 202
Heikes, R. R., 27, 272
Heimbach, W., 28, 340
Heine, W., 27, 187, 296, 372
Heinrich, E. W., 27, 716
Heinrich, W., 30B, 399
Heiskanen, K., 29, 433, 438
Heitsch, C. W., 28, 598
Heitz, F., 26, 432
Helfrich, W. J., 27, 32; 29, 224
Helgorsky, J., 26, 75
Helland, S., 27, 788
Helland-Madsen, G., 34B, 381; 35B, 500
Helle, J. N., 35A, 111
Helleis, O., 28, 14
Hellner, E., 26, 453; 29, 25, 89, 250; 33A, 277; 33B, 66, 126; 34B, 368
van der Helm, D., 27, 735, 952; 30B, 28, 358; 33B, 21, 151, 372; 34B, 355, 541, 542; 35B, 648, 651
Helmholdt, R. B., 32B, 346; 35B, 575
Hemenger, P., 30A, 181
Hemmerich, P., 34B, 556
Hencher, J, L., 33B, 532
Hencken, G., 35B, 471
van den Hender, J. H., 28, 614; 30B, 399; 32B, 240, 260, 308; 34B, 339
Henderson, B., 27, 108, 111, 277, 310; 29, 224
Henderson, D. M., 27, 422
Hendrix, G. S., 26, 232
Heners, J., 29, 248
Henfrey, A. W., 35A, 114
Henkel, H., 33A, 188
Henn, D. E., 28, 658, 683; 30B, 16; 35B, 415
Henney, J., 29, 226
Hennig, U., 30A, 173
Henry, W. G., 27, 182
Henshaw, D. E., 26, 693
Hentschel, G., 29, 434
Herak, R., 34A, 242
Herak, R. M., 28, 315; 30A, 272
Heraud, J. -L., 31B, 18
Herbstein, F. H., 26, 831; 27, 921; 28, 429, 682; 29, 34, 136, 642, 644; 30B, 254, 443, 445; 32B, 54; 33B, 345
Herceg, M., 32B, 390
Herda, E., 26, 289
Herdklotz, J., 34B, 203
Herdklotz, J. K., 35B, 31, 217, 322
Herman, H., 31A, 17
Hermans, I. P. H., 26, 628
Hermodsson, Y., 26, 657; 32A, 200; 32B, 29; 34A, 222; 34B, 354
Herms, G., 27, 559
Hernádi, J., 35B, 895
Herndon, J. R., 27, 401
Herold, A., 28, 65
Herpin, A., 26, 158; 29, 435
Herpin, P., 27, 602; 28, 340, 358; 30A, 401, 402; 31A, 194; 35A, 134, 372
Herr, W., 27, 653; 28, 315
Herrin, J., 35B, 406
Herriott, J. R., 33B, 275
Herschaft, A., 27, 450
Hersh, O. L., 28, 362
Hershaft, A., 28, 81
Herz, W., 31B, 213; 35B, 372
Hesper, B., 30B, 398; 31B, 54, 202; 34B, 303
Hess, H., 28, 580; 31A, 99; 31B, 299; 34B, 360, 363, 375; 35B, 479
Hess, K., 26, 641
Hesse, M., 34B, 333
Hesse, R., 28, 625; 34B, 401, 404, 425; 35B, 828
Hessel, L. W., 34B, 531
Hessevick, R. E., 31A, 124
Hestermann, K., 33A, 280; 34A, 247, 248
Hettler, H., 34B, 204
Heubner, U., 27, 186
Heumann, T., 27, 108
Hewaidy, F. J., 33A, 170
Hewaidy, I. F., 29, 83
Hewertson, 35B, 526
Hewitt, T. G., 31B, 447; 32A, 252; 33B, 530; 34A, 155, 240; 35B, 534, 895
Hey, M. H., 29, 436

Heyding, R. D., 26, 55; 27, 48, 60; 29, 137; 30A, 244, 33A, 34
Hichter, P., 29, 223
Hickford, J. H., 32A, 192
Hidron, A., 30A, 247
Higginbotham, K. H., 30B, 405
Higginson, W. C. E., 26, 830
Higgs, M. A., 28, 365
High, D. F., 27, 1007, 1060; 28, 686; 29, 684
High, F., 31B, 201
Higuchi, T., 31B, 298; 32B, 166; 302; 33B, 353
Hilderbrandt, R. L., 34A, 185; 35B, 891
Hill, D. C., 27, 545
Hill, D. K., 26, 538
Hill, N. A., 29, 139, 226
Hilmer, W., 26, 397, 511; 27, 500, 501; 28, 233; 29, 395; 30A, 392; 31A, 216; 34A, 118
Himes, R. C., 29, 141, 226
Hinderer, A., 35B, 479
Hine, R., 27, 765; 35B, 126
Hirabayashi, H., 29, 292
Hirabayashi, M., 26, 82; 30A, 32, 174, 177, 246
Hirai, N., 26, 630
Hiramatsu, M., 35B, 402
Hirato, Y., 30B, 398; 31B, 194, 255, 257
Hirowkawa, S., 26, 644, 788; 28, 390, 391; 30B, 132; 31B, 128; 33B, 26
Hirono, K., 29, 760
Hirota, E., 26, 580, 588; 29, 339; 34A, 244
Hirotsu, K., 33B, 353
Hirowatari, F., 29, 437
Hirs, C. H. W., 26, 811
Hirschfeld, D., 27, 645, 751; 28, 316
Hirschfeld, F. L., 28, 444; 31B, 85; 32B, 109
Hirshfeld, F. S., 30B, 94
Hirthe, W. M., 27, 382
Hiscocks, S. E. R., 29, 139, 226
Hiskey, C. F., 30B, 445
Hitchman, M. A., 30B, 331; 34B, 504
Hitterman, R. L., 26, 271; 27, 379; 28, 41; 31A, 179
Hjortass, K. E., 32B, 136; 33B, 532
Hoard, J. L., 26, 486, 539, 680, 691, 692; 28, 314, 619, 632, 639, 682, 686; 29, 545, 728; 30B, 112, 218, 219, 396, 397; 31B, 429 32A, 510; 32B, 234, 237; 33A, 247, 248; 33B, 399, 440; 34A, 201, 279; 34B, 279; 35B, 717, 751
Hoard, L. G., 31B, 299
Hobson, A. D., 34B, 73
Hocart, R., 26, 436
Hoch, M., 27, 542; 29, 135
Hocheid, B., 30A, 178
Hock, A. A., 26, 682
Hockings, E. F., 26, 46
Hodenberg, R. Fri. von, 27, 587
Hodges, R., 26, 830
Hodgkin, D. C., 26, 608, 807, 830; 27, 1040, 1042; 28, 530, 537; 29, 754; 30A, 280; 33B, 204; 35B, 647
Hodgson, A., 26, 528
Hodgson, D. J., 33B, 430; 34A, 219; 34B, 16, 614
Hodgson, M., 35B, 427
Hodson, H. F., 26, 830
Hoehn, H. H., 26, 680
Hoekstra, H., 29, 340; 31A, 126
Hoekstra, H. R., 33A, 288, 357
Hoenig, C. L., 26, 57
Hoeve, C. A., 29, 759
Hofer, G., 29, 140; 30A, 174
Hoff, W. D., 29, 224
Hoffman, C. W. W., 28, 200, 316
Hoffman, K., 35B, 440
Hoffman, W., 35A, 282
Hoffmeister, H., 28, 686
Hogarth, D. D., 26, 387
Hoge, R., 30A, 285; 33B, 165; 34B, 249
Hogg, J. H. C., 31A, 45; 32A, 94
Hogg, S., 34A, 238
Höhle, T., 32A, 198
Hohmann, R., 35A, 162
Hohne, E., 26, 513; 27, 405, 607; 28, 316; 29, 388; 31A, 88, 166; 32B, 173, 260; 33B, 228; 34B, 378
Höhnert, M., 29, 438
Hohnke, D., 30A, 176; 32A, 120; 33A, 135; 34A, 34
Holah, D. G., 35B, 813, 818
Holden, J. P., 28, 289
Holden, J.R., 27, 903; 30A, 177; 31B, 68; 32B, 90, 91; 33B, 39; 34B, 71
Holian, B. L., 35B, 667
Holinski, R., 32A, 207; 35A, 161
Hollan, L., 30A, 173, 244; 31A, 37
Holland, R. V., 28, 316; 35B, 653
Holland, V. F., 27, 804; 29, 760
Holle, B., 33A, 240; 34A, 183
Holleck, H, 27, 105, 213; 29, 136, 139; 30A, 246
Holley, C. E., 28, 22
Hollingsworth, E. H., 27, 18
Holloway, J. H., 29, 260
Hollowell, C. D., 30B, 405
Holm, R., 33B, 538, 539
Holm, R. H., 34B, 558
Holmberg, B., 27, 307, 484; 30A, 312; 31A, 86; 33A, 181, 186
Holmén, S., 29, 139; 30A, 394
Holmes, P. J., 27, 47
Holmesland, O., 31B, 270
Holseth, H., 33A, 63; 35A, 10

Holst, D., 27, 916
Holst, J. P. C., 30B, 398
Holt, E. M., 34A, 231; 35B, 632
Holt, S. L., 32B, 410; 35B, 632
Holtedahl, B. H., 32B, 308
Holtzberg, F., 29, 140, 226
Hon, P. K., 30B, 340; 31B, 383; 32B, 387; 34B, 504
Honea, R. M., 27, 551, 603
Honeycombe, R. W. K., 26, 260
Honjo, G., 26, 423; 32A, 89
Honzl, J., 30B, 257
Hoogsteen, K., 28, 505, 507
Hoogzand, C., 31B, 347
Hook, A. B., 28, 316
van Hook, H. J., 26, 414
Hope, H., 26, 778; 27, 964; 29, 522; 30B, 43, 145; 31B, 46, 58; 33B, 198; 34B, 38, 140, 194, 196, 562; 35B, 83, 226, 495
Hopf, V., 35A, 64
Hopff, H., 26, 782
Hopkins, B. E., 29, 224
Hopkins, T. E., 27, 401; 30B, 305; 31A, 79, 84, 101; 31B, 294, 315; 32B, 294
Hoppe, H., 33B, 454, 460
Hoppe, R., 26, 172, 308, 312, 313, 396, 400, 402, 424; 27, 442, 520, 661; 29, 338, 339; 30A, 325; 33A, 188, 280, 325; 34A, 247, 248, 250, 291; 35A, 158, 162, 213, 214, 216, 220, 223, 224, 225
Hoppe, W., 26, 745; 28, 686; 29, 313, 759; 30B, 48, 222, 398; 32B, 138; 33B, 96; 34B, 281; 35B, 132, 137
Hordvik, A., 26, 621, 622, 830; 28, 497, 682; 29, 345, 484; 30B, 147, 148, 149, 157, 396; 31B, 143, 145, 146, 147, 151, 155, 156, 168; 33A, 60; 34B, 184, 185, 189, 191
Hori, T., 27, 270; 29, 140
Horii, Z., 30B, 445
Horii, Z. -I., 30B, 239
Horio, M., 29, 761
Horl, E. M., 26, 287; 27, 394
Horn, J., 33A, 27; 34A, 50
Horn, P., 26, 803
Hornbogen, E., 26, 178; 27, 218
Hornder, L., 30B, 398
Hornstra, J., 35B, 138, 336
Horrocks, W. D., 34B, 521
Horrocks, W. de W., 33B, 394, 430, 517; 34B, 516
Horstmann, M., 27, 384
Horton, G. R., 27, 887
Horyń, R., 31A, 136; 35A, 84
Hosemann, B., 27, 800
Hoshino, S., 26, 473; 28, 687; 30A, 177; 30B, 395
Hoskins, B. F., 26, 666; 27, 654; 28, 175, 643; 30B, 361, 396; 35B, 705, 798
Hosoya, 27, 474; 31B, 159
Hospital, M., 26, 748; 29, 470; 30B, 5, 6, 393, 396, 399, 445; 31B, 7, 9, 10, 12, 15, 17, 18, 19, 20; 32B, 18; 34B, 76, 77, 160, 290
Hossain, M. B., 31B, 117; 33B, 372; 34B, 542
Hostachy, A., 33A, 392
House, D. A., 35B, 599, 665
Houska, C. R., 27, 19; 29, 224; 30A, 247
Houston, M. D., 26, 244
Housty, J., 29, 470; 30B, 5, 6, 393, 396, 399, 445; 32B, 18; 31B, 7, 9, 10, 12, 15, 17, 18, 19, 20; 33B, 7; 34B, 214, 222
Housty, M. J., 26, 727, 750
van Houten, S., 27, 197
Houze, G. L., 30A, 180
Hovenkamp. S. G., 27, 755
Hovi, V., 29, 433
Howard, J., 35B, 270
Howatson, J., 35B, 681
Howie, R.A., 33A, 63
Howlett, B. W., 26, 63
Hoy, R. C., 32B, 435
Hoy, T. G., 33B, 72; 34B, 81
Hozumi, Y., 29, 760
Hrichova, R., 28, 314
Hrizo, J., 27, 504, 540
Hrubý, A., 35A, 18
Hrušková, J., 34A, 296
Hsieh, H. -D., 29, 435
Hsieh, H. -T., 28, 338; 29, 436
Hsu, D. -F., 29, 760
Hsu, S. L., 35B, 905
Hsueh, C. -Y., 31A, 236
Hu, S. -C., 28, 221
Huang, C. -L., 30B, 289
Huang, J. -L., 31A, 118
Huart, J., 28, 316
Hubbard, C. R., 33A, 515
Hubbard, W. M., 27, 198; 29, 50
Hubbert-Paletta, E., 33A, 268; 35A, 213
Hubel, W., 27, 647
Hübell, W., 31B, 347, 362
Huber, C. P., 33B, 209
Huber, C. S., 34B, 163; 35B, 359
Huber, M., 26, 173, 412; 33A, 306
Huber, R., 28, 686; 30B, 48, 222; 33B, 96; 35B, 888
Huber-Buser, E., 26, 645; 31B, 93
Hubková, H., 31A, 199
Huch, R., 29, 12
Hudd, R. C., 27, 8
Hudson, P., 27, 947, 948; 29, 731
Huellen, A., 30B, 331
Huffstutler, M. C., 26, 7
Hug, P., 34A, 264
Hügel, R., 33B, 464

Huggins, C. W., 26, 522
Huggins, M. L., 26, 630
Hughes, D. L., 35B, 270
Hughes, D. O., 27, 756
Hughes, E. W., 26, 271, 570, 736; 27, 786, 884; 33B, 179
Hughes, J. L., 27, 219; 29, 135
Hughes, R. E., 27, 810; 28, 52, 314, 401; 29, 478, 515; 35A, 272
Hughson, M. R., 29, 436
Huhma, A., 29, 140
Huisman, R., 29, 73; 33A, 113; 34A, 142
Hull, D., 29, 137
Hull, G. W., 29, 224; 30A, 246, 247
Hüllen, A., 31A, 156
Hulliger, F., 26, 47, 137, 245; 28, 9; 29, 137, 139; 30A, 178, 181
Hulme, R., 27, 439, 888, 909; 28, 603; 29, 285; 30B, 443; 33B, 288; 34B, 357
Hulscher, J. B., 33B, 135; 35B, 492
Humble, S. G., 29, 60
Hume-Rothery, W., 27, 87, 107; 29, 136, 139, 224, 226; 30A, 178, 246
Huml, K., 30B, 445; 32B, 71, 72; 33B, 61; 34B, 117, 155
Hummel, F. A., 29, 436
Hummelink, Th. W., 35B, 810
Humphreys, J. E., 30A, 172
Humphreys, J. G., 26, 100
Hund, F., 28, 338
Hung, W. -H., 26, 536
Hunt, C. R., 33A, 153
Hunt, D. J., 34B, 263; 35B, 279
Hunt, M. R., 30A, 173
Hunt, P. D., 27, 109
Hunter, F. D., 32A, 414; 33B, 157
Hunter, W. L., 27, 804
Hurditch, R. J., 32A, 314
Hurlbut, C. S., 26, 515; 27, 603
Hurley, G. F., 29, 136, 224
Hurst, H. J., 35A, 169
Huse, Y., 30B, 187; 34B, 336
Husebye, S., 26, 774; 28, 686; 30B, 284; 31B, 301, 336, 337; 32B, 338; 34B, 380, 381, 416; 35B, 498, 499, 500
Hussain, M. S., 34B, 562; 35B, 742, 744
Husson, F., 26, 564, 826; 27, 1063
Hutchinson, D. A., 27, 939
Hutchinson, G. E., 28, 358
Hutchinson, J. H., 26, 690
Huttner, G., 34B, 426; 35B, 502
Hutton, C. O., 26, 387
Huynen, A. M., 27, 710; 28, 278
Huys, D., 33A, 425
Hvoslef, J., 33B, 158; 34B, 247
Hybl, A., 30B, 186
Hyman, A., 32A, 497
Hynes, N. B., 30B, 394
Hyvönen, L. J., 26, 139
Iandelli, A., 29, 3, 140; 31A, 42; 34A, 143
Ibach, H., 29, 135, 223
Iball, J., 26, 752; 27, 936; 28, 447, 449, 683, 686; 29, 696; 30B, 176; 31B, 183; 32B, 272, 385, 438; 34B, 111, 159
Ibers, J. A., 26, 799; 27, 459, 853; 28, 75, 223, 229, 651, 685; 29, 122, 197, 203, 240, 254, 329, 335; 30A, 326; 30B, 377, 386, 388, 389, 390, 394; 31A, 92; 31B, 455, 460; 32A, 402; 32B, 464, 467, 477, 479, 480; 33A, 201, 207; 33B, 386, 387, 429, 430, 496, 511, 513, 514; 34A, 158, 160, 214, 219, 226, 237; 34B, 4, 16, 539, 588, 601, 609, 610, 611, 613, 614, 615, 617, 618; 35A, 78, 152, 410, 413; 35B, 523, 535, 587, 592, 749, 836, 856, 873, 875
Ibrahim, A., 26, 121
Ibrahim, M., 27, 537
Ichikawa, T., 33B, 180; 34B, 272
Ickert, L., 27, 763
Ickevič, E. S., 29, 138; 33A, 45
Ido, H., 33A, 33
Idrestedt, I., 29, 244
Ievin'š, A., 26, 735; 28, 166, 682; 29, 141, 387, 438; 33A, 445
Ievin's, A. F., 27, 50, 884; 30B, 443, 447
Igarashi, M., 26, 554, 587, 590, 594
Igarashi, T., 34B, 592
Ignat'ev, N. A., 30A, 175; 35B, 124
Iguchi, S., 28, 682
Ihara, M., 31A, 166; 32A, 499; 34A, 356; 34B, 332; 35B, 199
Iida, K., 35B, 727
Iida, Y., 28, 339
Iijima, I., 34B, 328
Iijima, K., 34B, 529; 35B, 622
Iijima, T., 27, 719
Iitaka, Y., 26, 328, 426, 588; 27, 81 470, 679; 30B, 187, 447; 31B, 88, 177, 194, 227, 254, 421; 33B, 180, 217, 235; 34A, 334; 34B, 100, 115, 268, 272, 286, 318, 320, 336; 35B, 229, 310, 398, 747, 771, 886
Iizuka, M., 33A, 253
Ikeda, T., 28, 341; 29, 434; 31B, 404
Ikemoto, I., 31B, 278; 32B, 284; 33B, 250; 35B, 417
Ikuma, S., 28, 687
Ilgner, W., 29, 433
Il'inskij, A. L., 34B, 497, 508
Iljukhin, V.V., 26, 303, 537; 28, 342; 29, 398, 436; 30A, 270; 31A, 156, 158, 218; 32A, 278, 430, 437, 482, 509; 33A, 334, 452, 471, 473, 506; 34A, 12, 379; 35A, 15, 229, 297, 310, 316, 464, 465, 470

Ilschner, B., 30A, 246
Imada, K., 27, 812
Imado, S., 26, 660; 30B, 239, 445
Imahori, K., 26, 816
Imamura, R., 29, 761
Imamura, S., 26, 45
Imanov, L.M., 30B, 539
Imamov, R.M. 26, 250; 29, 14, 30, 87, 138, 311; 30A, 92, 93, 172, 446; 31A, 67; 33A, 19, 30; 34A, 10, 32, 33, 88, 127, 142; 35A, 31, 33, 87
Imelik, B., 27, 571, 642; 29, 436
Immirizi, A., 28, 597; 29, 502; 30B, 393, 398; 32B, 119, 292; 33B, 83; 34B, 441; 35B, 111, 567
Imoto, S., 27, 806; 29, 138
Imre, L., 30A, 312
Inagaki, M., 26, 286; 27, 395
Inama, P., 28, 338
Inayama, S., 35B, 372
Indenbaum, G. V., 26, 199
Inglis, F., 32A, 509
Ingraham, J. N., 26, 354
Ingram, B., 29, 136, 435, 436
Ingram, L., 32A, 422; 33A, 265
Ingri, N., 28, 147; 32A, 511
Innes, D., 31A, 42
Innes, M., 30B, 392; 31B, 265; 32B, 96; 34B, 25
Ino, N., 30A, 246
Inoue, M., 32A, 97
Insley, H., 26, 311
Ionesku-Buzhor, S., 30A, 179
Ippolitova, E. A., 26, 403, 405
Irani, K. S., 27, 493
Ireland, P. R., 35A, 412; 35B, 599, 712
Irie, H., 35B, 44
Irngartinger, H., 35B, 133, 134
Isaacs, N. W., 32A, 510; 34A, 346, 35B, 735
Isaacs, P. B., 29, 303
Isaacs, T., 28, 339
Isaev, I. D., 30B, 539
Isakov, I. V., 27, 829; 30B, 64; 31B, 3
Iselinskij, I. V., 31B, 52
Isetti, G., 30A, 445
Ishaq, M., 34B, 418
Ishihara, H., 33B, 524
Ishihara, N., 26, 515; 35B, 768
Ishii, B., 26, 367
Iskhalov, Kh. Š., 27, 506
Islam, K. M. S., 26, 831; 31B, 79; 32B, 141; 33B, 215; 34B, 346
Ismailov, M. Z., 34A, 52
Ismailzade, I. G, 27, 503; 28, 340, 341
Isoda, T., 29, 138
Isrealsson, M., 34A, 267; 35A, 264
Isupova, E. N., 28, 341; 29, 434
Itai, A., 34B, 318
Ito, A., 26, 216
Ito, J., 26, 528, 536, 537; 28, 358
Ito, K., 26, 357; 27, 463
Ito, K. -I., 28, 28
Ito, M., 35B, 684
Ito, S., 34B, 115
Ito, T., 34A, 377; 34B, 592; 35B, 409
Ivanov, V. E., 26, 56; 27, 65
Ivanov, V. I., 29, 435; 32B, 497
Ivanova, G. V., 33B, 410
Ivanova, V. Ja., 35B, 408
Ivanova, V. V., 27, 508
Ivčen, A.I., 29, 14
Ivchinova, L., 28, 340
Iwamoto, R., 33B, 524
Iwasaki, F., 32B, 281; 35B, 108, 410
Iwasaki, F. F., 32B, 3
Iwasaki, H., 27, 228, 504, 833; 28, 341, 684; 29, 459; 30A, 32, 60, 174; 30B, 395; 31A, 37; 31B, 378; 32B, 3, 282; 33A, 75
Iwasaki, M., 27, 717
Iwata, M., 34B, 485
Iwata, T., 29, 760
Iwata, Y., 31A, 112
Iyengar, P. K., 27, 257
Iyer, P. N., 32A, 293
Izotova, O. E., 35A, 172

Jack, K. H., 29, 433; 31A, 30
Jackobs, J., 34B, 192
Jackson, R. B., 32B, 373
Jackson, R. J., 27, 359
Jacobi, H., 33A, 451; 34A, 142, 205
Jacobs, A. J., 26, 30; 27, 36
Jacobs, G. D., 27, 732
Jacobs, H., 27, 420; 30A, 258; 33A, 61
Jacobson, R. A., 26, 625; 27, 989; 29, 518; 30B, 333; 31A, 103, 108; 31B, 299; 33A, 199, 241, 515; 33B, 289, 350, 512; 34A, 164; 34B, 174; 35B, 702, 852, 875, 876
Jacquir, R., 32B, 63
Jacquot, F., 26, 205
Jaffe, H. W., 26, 527; 27, 558
Jaggi, H., 26, 358; 33B, 452
Jaggi, R., 29, 139
Jagner, S., 32A, 195; 35A, 398
Jagodzinski, H., 30A, 408
Jahanbagloo, I. C., 33A, 494
Jahnberg, L., 28, 201; 29, 301; 35A, 257
Jain, P. C., 32B, 381, 392, 400
Jain, S. C., 34B, 19
Jakel, H. L., 29, 437
Jakkal, V. S., 28, 79
Jakobsen, G. G., 34A, 162
Jakobson, G., 29, 433

Jakovenko, V. I., 33B, 488
Jakovleva, S. V., 26, 523
Jambor, J. L., 27, 634; 28, 340; 29, 438
James, J. M., 35B, 759
James, K. C., 30B, 445
James, M. N.G., 31B, 198
James, W., 27, 163
James, W. J., 26, 247, 275, 350; 27, 151; 33A, 153; 35A, 353
Jameson, M.B., 30B, 97
Jamet-Delcroix, S., 32B, 267
Jamieson, J. C., 29, 139, 225; 30A, 248
Jamieson, P. B., 31A, 215; 32A, 292, 449, 511; 33A, 327, 328, 335; 34A, 269, 278
Jamzin, I. I., 26, 301, 411; 27, 496, 510; 28, 342; 35A, 265
Jan, S. -K., 28, 340
Janat'eva, O. K., 28, 340
Jander, J., 26, 293; 31A, 82; 33A, 229
Janosi, A., 29, 46
Jansen, C., 29, 317
Jansen, P. W. J., 30A, 319
Jansing, J. A., 35B, 210
Jansson, I., 33A, 407
Jansson, S., 27, 695
Janzen, J., 34A, 171
Janzon, K., 30A, 97, 182; 31A, 25
Janzon, K. H., 28, 11; 32A, 43, 122; 33A, 39, 51
Jarchow, O., 30A, 428; 33A, 219, 260; 33B, 42
Jarchow, O. H., 29, 363; 34B, 73
Jarembas, E. I., 29, 64, 138, 139, 226; 30A, 180, 247; 33A, 140
Jaremenko, Z. A., 29, 434
Jarmoljuk, Ja. P., 30A, 176, 180; 33A, 64, 70
Jarrat, H. S., 33A, 235
Jarski, M. A., 29, 596
Jarvis, J. A. J., 26, 663; 27, 859; 32B, 437; 33B, 503, 515; 35B, 792, 814, 848
Jarzemskij, Ja. Ja., 27, 558
Jaskowsky, J., 27, 515
Jastrebinskij, A. A., 27, 808
Jaunarajs, K. L., 26, 513
Jayadevan, N. C., 28, 341; 29, 435; 30A, 175; 35B, 643
Jayaraman, A., 29, 136, 224; 30A, 246
Jeannin, Y., 27, 345, 346, 353; 35A, 124; 35B, 655, 662
Jeannin, Y. P., 35A, 430
Jedlicka, H., 33A, 153; 34A, 39
Jeevaratnam, J., 28, 144
Jefferies, P. R., 28, 563
Jeffery, J. W., 27, 408; 30A, 33, 175; 32A, 393; 33A, 253; 35A, 329
Jeffrey, G. A., 26, 295, 613, 764; 27, 778, 781, 899, 975, 988; 28, 3, 66, 314, 375, 420, 503, 578, 683; 29, 165, 254, 297, 719; 30A, 37, 54, 175, 179; 30B, 7, 136, 161, 180, 237, 238, 267, 269, 444; 31A, 5, 93, 194; 31B, 282; 32A, 211, 414; 32B, 211, 212, 214, 218, 289, 291; 33A, 27, 161, 163; 34A, 29, 166; 34B, 245, 246; 35B, 29, 41, 422, 738
Jeffreys, J. A. D., 30B, 236; 31B, 246; 35B, 234, 390
Jeffs, P. W., 35B, 383
Jehanno, G., 27, 179; 33A, 184
Jeitschko, W., 27, 54, 233; 28, 3, 12; 29, 135, 136, 137, 138; 30A, 39, 175, 176; 31A, 4; 32A, 46, 83; 33A, 47; 34A, 6, 66; 35A, 73, 112, 124
Jelenić, I., 29, 458
Jellinek, F., 26, 147, 200, 830; 27, 348, 357, 423; 28, 494, 613; 29, 38, 73, 137, 342; 30A, 75, 181; 31A, 67; 32B, 346; 33A, 61, 113, 115; 34A, 92, 142; 35A, 79, 110, 124, 141
Jeminet, G., 34B, 45
Jenkins, D. R., 26, 345; 27, 719
Jenkins, I. L., 30B, 446
Jenkins, J. M., 30B, 393
Jennings, I. C., 27, 47
Jennings, P. P., 29, 296
Jensen, B., 32B, 172; 35B, 329
Jensen, G. B., 29, 648
Jensen, K. G., 34B, 40; 35B, 79, 586
Jensen, L. H., 26, 562; 27, 776, 1005, 1006; 28, 521, 684, 686; 29, 708; 30B, 56, 192, 202, 265; 31B, 24, 60, 133; 32B, 168, 203; 33B, 150, 154; 34B, 141, 234, 236; 35B, 116, 244, 256, 319
Jensen, L. S., 26, 830
Jensen, P., 29, 19; 30A, 19, 173; 31A, 18, 22, 24; 32A, 17; 34A, 27
Jensen, S. J., 29, 275; 32A, 188, 244; 33A, 212, 213; 35A, 198
Jepson, K. S., 29, 137
Jeremenko, V. N., 29, 139
Jeremic, M., 35A, 171
Jerslev, B., 28, 686; 29, 657; 31B, 44; 32B, 28, 172, 192; 33B, 79, 233; 34B, 40
Jewett, R. P., 26, 6
Jha, N. K., 27, 665
Job, B. E., 26, 587; 32B, 437
Joebstl, J. A., 30A, 256
Joensuu, O., 28, 358
Joesten, M. D., 35B, 742, 744
Johan, K., 26, 48
Johannessen, I.-J., 26, 582
Johannessen, W., 26, 580, 581
Johansson, A., 33A, 72, 108; 35A, 90
Johansson, G., 26, 454; 27, 623, 625;

28, 217, 218, 219; 31A, 133, 211; 33A, 431
Johnsen, K., 29, 520; 30A, 380; 31B, 335
Johnson, A. P., 28, 683
Johnson, A. W. S., 32A, 508
Johnson, C. K., 27, 735; 30B, 28, 30, 396; 33B, 21
Johnson, D. J., 27, 1056
Johnson, D. R., 34A, 173, 244
Johnson, I., 27, 109; 29, 135, 136; 30A, 175; 34A, 144
Johnson, K. A., 30A, 51, 181, 246, 248
Johnson, L. N., 31B, 171
Johnson, N. P., 34A, 229
Johnson, P., 34A, 229
Johnson, P. L., 34B, 193; 35B, 170
Johnson, Q., 30A, 28, 53, 56, 174, 177, 179; 30B, 272; 32A, 47, 77, 87, 186, 187; 33A, 105, 131, 205, 223; 34A, 36, 75, 84, 86, 174; 35B, 29, 124
Johnson, Q. C., 27, 71, 74; 29, 137
Johnson, R. C., 27, 695
Johnson, R. G., 27, 248
Johnson, R. H., 26, 260
Johnson, R. W., 26, 74
Johnson, S. M., 33B, 190; 34B, 194; 35B, 115, 233, 406, 569
Johnsson, T., 30A, 89, 181; 34A, 125
Johnston, G. B., 33B, 95, 96, 97
Johnston, W. D., 26, 43, 151, 372; 29, 225
Joho, P., 29, 438
Jolley, W., 29, 137
Jones, A. T., 29, 759, 760
Jones, D. W., 26, 458; 27, 574; 28, 215; 29, 456; 30A, 367
Jones, F. T., 28, 683; 30B, 446
Jones, G. R., 32A, 508; 33A, 172, 173; 34A, 172, 198, 202; 35A, 174, 175, 177, 185
Jones, J. B., 26, 524
Jones, J W. S., 29, 226
Jones, K., 35B, 794
Jones, N. D., 27, 909; 30B, 307; 33B, 118; 34B, 285
Jones, P. E., 32A, 509; 34A, 296; 34B, 587
Jones, R. A., 33A, 112
Jones, R. O., 31B, 346
Jones, R. V., 27, 251
de Jonge, R., 35A, 79, 124
Jongejan, A., 26, 382
Jonsson, I., 27, 657
Jonsson, O., 30A, 321; 35A, 389
Jonsson, P. G., 33B, 38; 35A, 366
Jordan, L.T., 27, 103
Jordan, T., 27, 769; 28, 399
Jordan, T. H., 27, 393; 29, 229; 32B, 288, 289; 35B, 423
Jose, P., 29, 481; 30B, 207; 34B, 498
Joshi, K. K., 33B, 347
Jost, K. -H., 26, 459, 461; 27, 592, 593, 706; 28, 186, 187, 317; 29, 300, 367; 30A, 393; 30B, 445; 31A, 120, 121, 216; 32A, 463; 33A, 411, 412; 34A, 240, 322; 34B, 397
Jostsons, A., 32A, 254; 33A, 265
Joswig, W., 32A, 441
Joubert, J. -C., 27, 486, 518; 28, 143, 357, 29, 435; 30A, 313; 31A, 143; 35A, 228, 288
Jouravsky, G., 28, 356
Joustra, A. H., 35B, 331
Joynson, M. A., 35B, 888
Joys, S., 30B, 157
Judinskaja, I. V., 33A, 341
Judkovskij, S. I., 26, 97; 27, 102
Juenke, E. F., 29, 435, 438
Juilfs, J., 26, 640
Julien-Pouzol, M., 34A, 142
Jumal', E. M., 35B, 620
Jung, D., 26, 417
Jungfleisch, M. -L., 29, 434
Jungk, A. E., 35B, 54
Juščuk, S. I., 33A, 324
Juza, R., 26, 187; 27, 420; 28, 314; 29, 137, 224, 247, 248; 30A, 39, 175, 258; 32A, 233; 33A, 31

Kaatz, T., 31A, 104
Kabalkina, S. A, 27, 504, 507
Kabalkina, S. S., 26, 573; 28, 72, 356; 29, 141, 224, 226; 34A, 239; 35A, 9
Kabbani, M., 35A, 134
Kabouny, M., 35A, 327
Kačan, L. A., 29, 436
Kachi, S., 28, 338
Kacnel'son, A. A., 30A, 179
Kadijk, F., 29, 73; 33A, 113, 115
Kadomceva, A. M., 26, 511; 29, 226
Kadoya, S., 31B, 88
Kadžar, Č. O., 33B, 539
Kaesz, H. D., 32A, 510
Kafalas, J. A., 26, 101; 27, 244, 281; 29, 136
Kagle, B. J., 26, 162, 223; 27, 281
Kahler, E., 27, 621
Kahn, D., 33A, 57
Kahn, J. S., 29, 296
Kai, Y., 35B, 188
Kajiwara, S., 28, 6
Kajupova, M. M., 29, 437
Kajusina, R. L., 30B, 209
Kakamura, K., 28, 683
Kakati, K. K., 28, 683; 33B, 63
Kakida, H., 35B, 884
Kakudo, M., 26, 723, 809; 27, 1050; 29, 412, 583; 30B, 88, 399; 31B, 63, 91, 106, 121, 191; 32B, 67, 76

150, 176, 226, 227, 232, 392; 33B, 174, 218, 313, 418, 500; 34B, 58, 88, 270, 276, 455, 513; 35B, 188, 488, 521
Kalalova, E., 30B, 445
Kalff, H. T., 30B, 150, 446; 31B, 143, 149
Kalicak, Ja. M., 34A, 143
Kalinin, V. P., 32A, 44; 34A, 142; 35A, 115
Kalitin, V. I., 30A, 247
Kallen D., 26, 38, 153
Kallos, J., 29, 759
Kalman, A., 26, 497; 29, 683; 30B, 446; 32B, 38; 35A, 381, 382, 396
Kalman, Z. H., 26, 415; 34A, 133
Kal'na, G. I., 27, 109
Kalt, A., 32A, 511; 34A, 364
Kaluski, A., 30B, 446
Kaluski, Z., 31B, 354, 361; 33B, 355
Kaluski, Z. L., 27, 860; 29, 549; 30B, 308, 310, 396
Kalyani, V., 32B, 86; 34B, 65, 173
Kamb, B., 29, 294; 32A, 250; 33A, 256
Kamenar, B., 26, 321; 27, 457, 458; 30B, 262, 267, 395, 396; 31B, 280; 32B, 233; 34A, 343; 34B, 548; 35B, 469, 766, 782
Kametani, T., 34B, 332; 35B, 199
Kamhi, S. R., 26, 391; 29, 174
Kamijo, N., 31B, 258
Kamija, K., 30B, 397; 31B, 264; 33B, 223; 34B, 238, 287; 35B, 388
Kamper, M. J., 29, 754
Kamsu Kom, J., 29, 137, 139
Kanamaru, F., 26, 395; 29, 685; 35A, 461
Kanaoka, Y., 34B, 161; 35B, 206
Kanda, F. A., 27, 265; 30A, 23, 174
Kandarpa, V., 31A, 31
Kandler, H., 29, 14; 30A, 180; 31A, 24
Kändler, J., 27, 453
Kane, R. H., 30A, 85, 179
Kanematsu, K., 26, 149, 192; 27, 162; 30A, 246; 34A, 145
Kannan, K. K., 30A, 375; 31B, 56
Kannewurf, C. R., 27, 215
Kanters, J. A., 30B, 3; 31B, 3; 33B, 135; 35B, 9
Kapecki, G. A., 32B, 170; 33B, 142; 34B, 87
Kapitonova, T. A., 26, 536
Kaplan, S. F., 33B, 402
Kaplow, R., 33A, 258
Kapoor, A. L., 27, 993
Kapyšev, A. G., 27, 508
Karačenceva, Z. V., 27, 494
Karaev, Z. Š., 29, 135, 140
Karakida, Y., 29, 435
Karev, V. N., 29, 138
Kargin, V. A., 26, 630; 29, 759, 760
Karklin', A. Ju., 26, 343
Karle, I. L., 26, 787; 27, 746, 909, 911; 28, 525, 684; 29, 479, 667, 726, 749; 30B, 37, 42, 86, 112, 119 154, 214, 397; 31B, 16, 85, 137, 142, 182, 244; 32B, 39, 100, 152, 162, 164, 186, 197, 247, 256; 33B, 230; 34B, 64, 121, 161, 165, 166, 240, 270, 278, 298, 326, 335; 35B, 204, 206, 213, 238, 290, 307, 328, 338, 362, 421, 474
Karle, J., 27, 746; 28, 525, 684; 29, 479, 749; 30B, 112, 119, 154, 397; 31B, 16, 85, 142, 244; 32B, 100, 162, 164, 197, 256; 33B, 230; 34B, 64, 165, 166, 298, 326, 335; 35B, 204, 213, 290, 328
Karpinskij, O. G., 30A, 97, 182; 33A, 143; 34A, 101, 145, 146
Karpova, T. P., 34A, 10
Karrer, P., 34B, 333
Kartha, G., 26, 602, 782; 27, 1023; 28, 536; 29, 750, 753; 33B, 15, 194; 34B, 228, 230, 232, 233, 358
Kartmazov, G. N., 26, 110; 27, 356; 31A, 54
Kasaev, A. A., 26, 529, 530
Kasai, N., 28, 583; 30B, 446; 33B, 313, 500; 34B, 58, 455; 35B, 188, 488, 521
Kasai, P. H., 28, 668
Kasami, K., 28, 682
Kašev, A. A., 28, 341
Kashiwagi, M., 35B, 111
Kashiwagi, T., 33B, 500
Kashiwase, Y., 26, 423
Kaski, B. A., 35B, 357
Kasper, H., 33A, 284
Kasper, J. S., 27, 250; 29, 25, 55, 137; 30A, 96, 180; 32A, 147; 33A, 162, 239; 34A, 43; 35A, 39
Kassierer, F., 30A, 289
Kastalsky, V., 32B, 454; 34B, 586
Katada, K., 27, 803; 31B, 298
Katayama, C., 32B, 83; 35B, 386
Kato, A., 26, 536; 28, 358
Kato, H., 27, 198, 249; 29, 226; 30A, 246
Kato, K., 33A, 274; 34B, 26; 35B, 369
Kato, T., 28, 341
Kato, Y., 30B, 61, 88
Katsube, Y., 27, 1050; 31B, 63; 34B, 276
Katsura, T., 26, 379
Katsuraki, H., 27 15
Katz, G., 26, 30; 27, 36
Katz, H., 27, 828
Katz, J. L., 30A, 444
Katz, L., 26, 370, 420, 421; 30A, 329, 446; 30B, 396; 31A, 137, 155; 31B, 163; 32A, 271, 509; 32B, 207;

33A, 353; 34A, 296; 34B, 587; 35A, 261, 267, 269
Katz, O. M., 30A, 246
Katzman, H., 33A, 153
Kaufmann, G., 28, 682
Kavesh, S., 35B, 882
Kawada, I., 34B, 187; 35B, 480
Kawaguchi, H., 35B, 678
Kawaguchi, S., 35B, 678
Kawahara, A., 28, 258; 32A, 275, 511; 34A, 375
Kawai, M., 34B, 344; 35B, 400
Kawai, T., 26, 587
Kawakubo, T., 26, 420; 27, 508
Kawamura, M., 28, 687
Kawamura, S., 28, 318
Kawamura, T., 30A, 255
Kawano, S., 32A, 41
Kay, H. F., 29, 213; 30A, 374; 33B, 87
Kay, L. M., 27, 1058; 28, 684
Kay, M. I., 26, 360, 473; 27, 572; 29, 331, 369; 31A, 202; 31B, 5; 32A, 339; 33B, 402
Kaye, G., 27, 126
Kazakova, M. E., 26, 531, 537; 27, 535; 28, 342
Kazicyna, L. A., 28, 686; 31B, 46; 32B, 60
Kazineč, M. M., 34A, 144; 35A, 56
Kazmazovskaja, T. S., 26, 171
Keating, P. N., 26, 42
Keaveny, I., 35A, 129
Keenan, T. K., 29, 433; 436; 35A, 166
Kefeli, L. M., 33A, 331
Keffer, C., 32A, 355
Keib, G., 32A, 170
Keil, J.G., 32A, 484
Keil, K., 29, 433
Keissig, H., 27, 810
Kejriwal, P.K., 28, 21
Kekrasov, Ju. V., 27, 285
Keler, E. K., 27, 506; 29, 434
Keller, A., 29, 760
Keller, C., 27, 520, 545
Keller, C. E., 31B, 369
Keller, D. V., 30A, 174
Keller, E. K., 28, 356
Keller, O. L., 33A, 195
Kelley, J. A., 31B, 446; 34B, 22, 556; 35B, 617, 723
Kelly, S., 29, 437
Keloglu, Ju. P., 27, 41; 29, 225
Kelsey, C. H., 29, 436
Kemme, A. A., 35B, 484
Kemmitt, R. D. W., 28, 314
Kemmler, S., 27, 543; 30A, 347
Kemmler-Sack, S., 33A, 356
Kemp, M. K., 29, 433
Kempster, C. J. E., 27, 679
Kempter, C. P., 26, 76, 230; 27, 93, 134 137; 30A, 245
Kende, A. S., 30B, 399
Kendrew, J. C., 26, 810, 811
Kennard, C. H. L., 26, 547, 657, 776; 28, 314, 643, 682; 32A, 510; 32B, 31; 33B, 320, 323; 34A, 228, 346; 34B, 557; 35A, 379; 35B, 59, 419, 520, 697, 735
Kennard, O., 26, 732; 27, 1037; 28, 573; 30B, 396; 32B, 252, 258; 33B, 450; 35B, 103, 113, 389, 396, 467
Kennedy, G. C., 29, 438
Kennedy, J. P., 29, 760
Kennedy, S. W., 26, 476
Kenney, C. N., 26, 345
Kent, P. W., 34B, 252
Keppler, U., 35A, 330
Keppler, V., 27, 591
de Kerf, J., 26, 785; 27, 954
Kergomard, A., 30B, 398; 32B, 69
Kern, R., 27, 633, 888; 28, 317; 29, 433; 31A, 114; 33A, 242
Kern, W., 26, 637
Kerr, G. T., 26, 516
Kerr, I. S., 32A, 511; 34A, 374
Kerr, K. A., 32B, 114; 35B, 103, 113, 389
Kerr, P. F., 26, 391; 30A, 17, 173
Kesamanly, F. P., 29, 136, 137; 30A, 173
Kester, F., 29, 224
Kester, F. L., 30A, 245
Kettle, S. F. A., 32A, 159
Keve, E. T., 31A, 29; 32A, 509; 33A, 163; 33B, 18; 35A, 45, 164, 165
Kewley, R., 27, 719; 28, 362
Khacianova, T. L., 31B, 77
Khaenko, B. V., 27, 124
Khaikin, L. S., 33B, 530
Khajenko, B. V., 30A, 172
Khalezova, E. B., 28, 342
Khalilov, A. D., 31A, 240
Khamidov, O. Kh, 30A, 178
Khan, A. A., 28, 291; 35A, 335, 353
Khan, G. A., 29, 438
Khan, I. H., 26, 296
Khan, Y., 35A, 41, 56, 124
Khanna, A. Ju., 29, 140
Khanolkar, D. D., 26, 399
Khantanov, Ju. A., 35A, 297
Kharakh, E. A., 35A, 315
Kharčenko, V. A., 26, 166
Khare, G. P., 35B, 821
Kharitonov, Ju. A., 33A, 337, 452, 471; 34A, 275; 35A, 388, 391
Khar'kin, V. S., 34A, 142; 35A, 87
Khar'kiv, A. D., 29, 436
Kheiker, D. M., 35A, 290, 291, 312
Kheirov, M. B., 28, 247
Khejker, D. M., 34A, 382
Khitrova, V. I., 26, 220; 27, 309; 28, 32; 32A, 262; 34A, 293

Khlapova, A. N., 27, 674
Khocjanova, T. L., 28, 683; 29, 678; 30B, 79; 33A, 68; 34B, 69
Khodadad, P., 29, 433; 35A, 235
Khodašova, T. S., 26, 330; 27, 669; 28, 93; 29, 380; 30A, 282
Kholilov, A. D., 34B, 497
Khomjakov, A. P., 29, 433
Khotkevič, V. I., 27, 264
Khoury, F., 27, 746
Khvatkina, A. N., 26, 586; 30B, 156
Khvostova, V. A., 28, 146
Kido, Y., 35B, 771
Kieffer, B., 27, 338
Kieffer, R., 27, 159; 32A, 44, 507
Kienast, G., 26, 86
Kierkegaard, P., 26, 356; 27, 582, 591, 597; 29, 372, 373; 30A, 368, 380, 394; 31A, 147; 33A, 384, 394, 397; 34B, 154, 156, 524; 35A, 204, 337; 35B, 192, 195, 445
Kiflawi, I., 34A, 133
Kihlborg, L., 26, 354; 28, 125, 127, 128; 29, 308, 309; 31A, 145; 32A, 273; 34A, 267, 284; 35A, 226, 264
Kikkawa, S., 33B, 313
Kikot', B. S., 28, 686
Kikuchi, A., 29, 434; 31A, 112; 32A, 223
Kikuchi, M., 34A, 60; 35A, 134
Kikuchi, T., 33B, 223
Kikuchi, Y., 26, 588
Kilbourn, B. T., 30B, 396; 32B, 242, 437; 34B, 47, 600; 35B, 97, 792, 814, 858, 864
Kilian, H. G., 27, 802
Killean, R. C. G., 27, 795; 30B, 55, 443; 34B, 371, 403
Killpatrick D. H., 29, 136
Kim, C. D., 30A, 244, 246, 248
Kim, H. S., 33B, 163; 34B, 245, 246; 35B, 29, 41
Kim, S. -H., 32B, 52, 206, 211, 212, 215; 33B, 168, 254; 34B, 224
Kim, Y. S., 26, 169
Kime, N. E., 34A, 214; 34B, 459
Kimoto, K., 32A, 51
Kimura, H., 26, 102, 216
Kimura, K., 31B, 298
Kimura, R., 27, 15
Kimura, T., 35B, 488
Kinberger, B., 35A, 328
Kindler, E., 34A, 206, 293
Kindler, H., 30B, 92
King, A. J., 27, 265; 30A, 23, 174
King, E., 26, 230
King, G. J., 28, 291
King, G. S. D., 27, 245, 877; 31B, 10; 35B, 115, 154, 232
King, G. W., 34A, 218
King, H. P., 27, 392
King, H.W., 26, 248; 27, 190; 30A, 244
King, M. V., 27, 1059
King, R. B., 33A, 163
Kinloch, M. E., 31A, 393
Kinoshita, Y., 28, 503; 32B, 432, 35B, 111
Kinsey, J. L., 26, 345
Kintner, P. L., 34A, 149
Kiosse, G. A., 32A, 95
Kirby, A. J., 31B, 148
Kirchhoff, W. H., 27, 411; 34A, 242
Kirchmayr, H. R., 29, 139
Kirchner, E., 32A, 428
Kirianenko, 29, 139
Kirilenko, R. V., 26, 192
Kirillov, A. S., 29, 436
Kirillova, N. I., 35B, 562
Kiriyama, H., 29, 292
Kiriyama, R., 29, 292
Kirkinskij, V. A., 29, 435
Kirkpatrick, J. L., 30B, 237
Kirkpatrick, M. E., 26, 205; 27, 297, 298
Kirkwood, C.E., 35B, 722
Kirner, H., 29, 225
Kirov, G.K., 29, 437
Kirov, G.N., 28, 339
Kirschke, E., 32B, 173
Kiryu, S., 28, 682; 32B, 434
Kischio, W., 29, 137; 33A, 146
Kiselev, S. V., 27, 155, 506
Kiseleva, K. V., 28, 289
Kishi, T., 34B, 287
Kishida, E., 32A, 97
Kishikawa, M., 31B, 264
Kisis, B., 35B, 117
Kispert, L. D., 33B, 202
Kissing, W., 35B, 197
Kistenmacher, T. J., 32B, 34; 33A, 215; 35A, 160; 35B, 228
Kitadai, Y., 26, 386; 28, 357
Kitahara, Y., 30B, 399; 32B, 176; 34B, 88
Kitaigorodskij, A. I., 26, 641, 745; 28, 668, 685; 29, 628, 697; 30B, 98; 31B, 341; 33B, 72
Kitamura, N., 26, 423; 27, 399; 29, 341
Kitanishi, Y., 26, 639
Kitchingman, W. J., 29, 53, 224
Kittl, J., 30A, 245
Kittl, J. E., 27, 175
Kizaev, S. A., 29, 435
Kjekshus, A., 26, 176, 267; 27, 43, 364; 29, 18, 19, 73, 77, 135, 224; 30A, 14, 19, 76, 86, 94, 173, 174, 180, 181, 246, 247, 248; 31A, 18, 22, 24, 62; 32A, 17, 143; 33A, 63; 34A, 16, 27, 73, 97; 35A, 10, 75, 124, 208

Kjøge, H. M., 30B, 147, 288, 396; 31B, 146; 34B, 185, 379
Klanberg, F., 32A, 510; 33B, 266
Klatyk, K., 27, 457
Kleber, E. V., 26, 118, 142, 143
Kleber, W., 26, 457, 463, 464
Kleckovskaja, V. V., 32A, 262
Klee, W. E., 35A, 352
Klein, C., 29, 438
Kleinberg, R., 34A, 178; 35A, 402
Kleinwachter, V., 29, 759
Klem, E., 30B, 363
Klement, R., 26, 469; 27, 587
Klement, U., 34B, 573
Klement, W., 27, 81, 160, 221, 292; 30A, 172, 244, 245, 248
Klemm, D. D., 27, 164, 168
Klemm, W., 26, 53; 27, 304, 321; 29, 12, 83
Kleščev, G. V., 27, 12
Klescinskij, L. I., 33A, 126
Klett, D. S., 28, 682
Klevcov, P.V., 29, 305; 31A, 98; 32A, 199; 33A, 216, 331; 35A, 248, 253, 254
Klevcova, R. F., 28, 188; 29, 242, 305, 370, 374; 30A, 16, 181, 377, 378; 31A, 98, 201; 32A, 199,303; 33A, 216, 331, 345, 351; 34A, 221, 280, 287; 35A, 248, 249, 250, 253, 254, 255
Klewe, B., 27, 915; 28, 107; 32A, 234; 32B, 49, 74
Kličkaja, G. A., 32A, 150; 33A, 450
Klimasenko, N. L., 34B, 103, 105
Klimova, A. I., 30B, 448
Klimova, N. S., 28, 314
Kling, O., 26, 539
Klinga, M., 35B, 675
Klingen, W., 30A, 173, 176; 35A, 138
Ključarov, Ja. V., 29, 435
Kločkov, V. P., 26, 615
Klodt, D. T., 27, 381
Klose, W., 27, 420; 29, 224, 247; 30A, 258
Kloss, R., 35A, 127
Kloster-Jensen, E., 35B, 893, 903
Klug, H. P., 29, 241; 30B, 78; 31B, 405; 33B, 55; 35B, 80
Klym, N. M., 27, 80, 385
Knappe, J., 30B, 242
Knausenberger, M., 28, 284; 29, 136; 30A, 83
Knecht, H., 31A, 100
Knight, R. J., 27, 40
Knipovič, Y. N., 28, 339
Knjazeva, A. N., 33B, 462; 34B, 505; 35B, 605
Knobler, C., 28, 485, 487; 29, 593; 32A, 250; 32B, 200; 33B, 149; 34B, 276; 35B, 495
Knödler, H., 29, 139, 226; 31A, 41
Knop, O., 26, 121, 201; 30A, 317; 32A, 126; 33A, 263, 286, 291
Knopf, P. M., 26, 663
Knorring, O., 26, 379; 27, 599
Knorring, O. V., 28, 339; 29, 437
Knott, H., 33A, 13
Knott, H. W., 26, 271; 27, 187, 379; 28, 20, 140; 32A, 99
Knowles, C. R., 29, 21, 401; 30A, 424
Knowles, J. W., 27, 561
Knox, J. R., 33B, 423; 34B, 426, 458, 540, 598
Knox, K., 26, 305, 309; 27, 671; 29, 235, 236, 656; 30A, 52, 178
Knutson, C. D., 35A, 202
Kobayashi, H., 35B, 76, 883
Kobayashi, K., 27, 800
Kobayashi, M., 35B, 884
Kobayashi, N., 35B, 398
Kobayashi, T., 28, 357
Kobayashi, Y., 35B, 771
Kobcev, B. M., 33A, 337
Kobori, K., 35B, 111
Kobzenko, G. F., 29, 226
Kočemasov, G. G., 28, 339
Koch, C. W., 28, 228; 29, 335
Koch, F., 34A, 246
Koch, M. H. J., 35B, 308
Koch, R., 29, 135, 223
Kodama, H., 30B, 447; 32A, 421
Koehler, W. C., 26, 106, 141; 27, 151, 274, 333, 334, 368, 434, 440; 28, 27, 138; 29, 136, 137
Koenig, D. F., 30B, 217; 34B, 13, 179
Koetzle, T. F., 33B, 263; 35B, 426
Kofstad, P., 26, 139
Kogan, V. S., 26, 285; 27, 264, 285, 393; 28, 314
Köhl, P., 35A, 242
Kohler, K., 29, 435, 438
Kohlmuller, R., 26, 23
Kohlschmidt, R., 27, 465
Kohlstrung, G., 27, 483
Kohn, J. A., 26, 341; 27, 514; 28, 340, 358; 30A, 349
Koisse, G. A., 32B, 320
Koizumi, H., 28, 634; 29, 434
Kojima, H., 26, 630
de Kok, A. J., 34B, 5; 35B, 211
Kok, R. B., 30A, 75, 181
Kokkoros, P. A., 26, 443; 28, 210
Kokot, E., 26, 666
Kokotailo, G. T., 26, 516; 30A, 438; 32A, 484; 33A, 497; 34B, 412
Kokšin, V. M., 26, 133
Kola, G., 33A, 140
Kolakowski, B., 27, 394; 34B, 8
Kolbe-Rohde, H., 26, 655
Kolditz, L., 28, 314
Kolesnikov, S. P., 35B, 407
Kolesnikov, V. N., 26, 63
Kolesova, R. V., 32A, 388; 33A, 424
Kolinov, O. V., 28, 682
Kolobjanina, T. N., 35A, 9

Kolobnev, I. F., 32A, 141
Kolobova, N. E., 33B, 310
Kolomijčuk, V. N., 29, 284; 30A, 284
Kolpakov, D. A., 27, 536
Kolster, B. H., 30A, 177
Komarek, K. L., 30A, 180
Komissarova, L. N., 26, 245, 259; 29, 433, 435
Komiyama, Y., 28, 90; 29, 593; 32A, 193, 197; 34B, 529; 35B, 622, 678, 727
Komjathy, S., 26, 242
Komkov, A. I., 27, 536; 28, 339, 342
Komnik, Ju. F., 26, 133; 30A, 245
Komura, U., 27, 12
Komura, Y., 32A, 97
Kon, A. Ju., 31B, 427
Konakhovič, Ju. Ja., 28, 143
Kondo, R., 30A, 431
Kondrašev, Ju. D., 26, 731; 28, 682; 29, 358, 662; 30B, 49, 443; 30A, 26; 31B, 161; 33B, 64
Kondrašev, Y. D., 32B, 61
Kondrat'eva, V. V., 26, 483; 27, 558; 28, 356; 29, 436
König, B., 34A,206
König, U., 33A, 324
Koning, R.E., 35B, 128
van Koningsveld, H., 33B, 157; 35B, 175
Konishi, H., 33B, 174
Kon'kova, G. S., 30B, 448
Konnert, J., 31B, 41; 34B, 270; 35B, 290
Konnert, J. A., 35A, 286, 446, 481
Konnert, J. H., 33B, 321
Konovalov, O. M., 26, 63
Koo, C. H., 30B, 47; 35B, 86
Kools, F. X. N. M., 34A, 282; 35A, 232
Koopstra, J., 33A, 61
Kopeckij, Č. V., 26, 192
Köpf, H., 35B, 553
Kopp, O. C., 28, 281
Kopwillem, A., 34B, 524
Korczak, P., 32A, 439
Korczyński, A., 27, 660; 28, 113, 316; 30A, 299; 30B, 342; 31B, 401; 32A, 240; 32B, 409; 33B, 467, 477
Kordes, E., 34A, 252
Korenev, Ju. M., 27, 430
Koridze, A. A., 35B, 544
Korjtnji, E. F., 35B, 614
Kornblau, M. J., 28, 687; 29, 478
Korneeva, I. V., 26, 246
Kornelson, R. O., 26, 53
Kornilov, I. I., 29, 225, 226; 30A, 173, 247; 33A, 326
Korst, W. L., 27, 237, 239; 29, 62
Koršunov, V. A., 29, 137
Korvenranta, J., 35B, 678
Korytnyj, E. F., 26, 678
Kosevič, V.M., 26, 42
Koskas, R., 29, 434
Koski, W.S., 27, 438
Kost, M.E., 33A, 81
Koster, A.S., 34A, 282; 35A, 232
Koster, D.A., 35A, 348
Koster, P.B., 32A, 149; 35B, 221
Köster, W., 27, 22; 30A, 172
Kosterin, E.A., 33B, 529
Kostiner, E., 35A, 272
Kostjanovskij, R.G., 32B, 34, 153, 154; 33B, 531
Kostur, N. L., 29, 140
Kostur, T. A., 29, 140
Kosturkiewicz, A., 32B, 68
Kosturkiewicz, Z., 33B, 88
Kot, M. V., 29, 226
Kotick, M. P., 32B, 224
Koto, K., 31A, 40; 34A, 262, 365; 35A, 57
Kouvo, O., 28, 338
Kovacs, A. L., 30B, 399; 32B, 276
Kovaleva, E. S., 33A, 438
Kovalyk, D. A., 29, 138; 34A, 143
Kovba, L. M., 26, 403, 405; 27, 109, 544; 28, 318, 340, 341, 356; 29, 137, 438; 31A, 150; 33A, 341; 34A, 268
Koyama, B. G., 31B, 194
Koyama, H., 28, 461; 33B, 195, 205, 220, 480, 34B, 119, 171, 283, 299, 574; 35B, 260, 337, 373, 382, 403, 755
Koyama, Y., 30B, 394
Koyano, N., 31A, 112
Kozeeva, L. P., 32A, 199
Kozima, K., 33B, 540
Kožina, I.I., 26, 48, 334; 27, 51; 29, 138
Kozmanov, Ju. D., 26, 412
Koz'min, P. A., 28, 459; 29, 269; 30B, 293; 35B, 637
Kradinova, L. V., 33A, 153
Krahe, E., 34B, 374
Krainer, E., 35A, 49
Krajewski, J., 34B, 288
Kramer, W. E., 27, 211
van Kranendonk, J., 27, 393
Krasan, Ju. P., 35B, 465
Krašnicki, S., 33A, 266
Krause, B. H., 28, 316
Krausse, J., 26, 831; 33A, 179; 35B, 507
Kraut, J., 26, 591; 27, 1060; 28, 521; 29, 487, 684; 30B, 210; 31B, 201; 33B, 206
Krautwasser, P., 33A, 126
Kravčenko, S. M., 26, 537; 28, 341
Kravčenko, V. B., 27, 698; 28, 316; 29, 387, 409; 30A, 415; 34A, 285
Krebs, B., 30A, 411; 30B, 53; 33A, 269, 387; 34A, 174, 206, 244, 293; 34B, 13, 178, 179; 35A, 149, 279

Krebs, H., 26, 38, 153; 31A, 57; 34A, 149
Krebs, J. -P., 31A, 34
Krein, O. E., 26, 199
Krén, E., 29, 225
Kresse, P., 26, 469
Kretsinger, R. H., 28, 636
Kreutz, W., 27, 1061
Kreuzburg, G., 35A, 213, 220
Krieghoff, N. G., 33B, 322, 323
Krigbaum, W. R., 33B, 9; 35B, 69
Krikorian, N. H., 29, 223; 33A, 53, 56, 153; 34A, 59
Krikorian. O. H., 26, 58; 27, 71, 74, 92, 134; 35A, 29, 124
Krimm, S., 26, 824, 825; 27, 809
Kripjakevic, P.I., 26, 57, 58, 136, 174, 175, 235; 27, 5, 29, 64, 152, 165, 195, 268, 269, 287, 294; 28, 28; 29, 137, 138, 139, 140, 141; 30A, 3, 172, 173, 174, 176, 177, 178, 180, 245; 31A, 12, 14, 27, 35, 63; 32A, 5, 6, 57, 141, 142; 33A, 64, 70, 153; 34A, 145; 35A, 7, 44, 50, 53, 55, 124
Krischner, H., 34A, 169; 35A, 26
Krisckoja, I.I., 31B, 343
Krisement, O., 26, 268; 27, 122
Krishna, P., 28, 64; 29, 238
Krishna, R.P., 27, 407
Krishnamachari, N., 35A, 350, 359
Krishna Rao, H., 31A, 17
Krishnan, V., 29, 436; 30B, 446
Kroebel, R., 29, 137, 224
Krogh-Moe, J., 26, 488; 27, 459, 548, 559; 29, 389; 30A, 411, 412; 31A, 166; 32A, 496, 499; 33A, 444; 34A, 356
Krogmann, K., 28, 628; 29, 586; 31A, 3; 31B, 406; 33A, 250; 33B, 406
Krol, O. F., 29, 438
Kromholtz, K. L., 27, 71, 74
Krönert, W., 30A, 182
Kronfeld, L.R., 33B, 144
Kronman, A. G., 26, 633
Kroon, J., 30B, 3, 26; 31B, 3; 35B, 9, 492
Kroon, P. A., 35B, 575
Kropf, R., 33B, 93
Krstanović, I., 26, 529; 29, 411, 436; 30A, 386
Kruceanu, E., 30A, 179
Krueger, M., 34B, 87
Krueler, M., 34A, 143
Krug, D., 29, 271
Krug, M., 27, 198
Krug, M. P., 31A, 29
Kruger, C., 34B, 438; 35B, 575
Kruger, G. J., 33B, 93; 35B, 159, 160
Kruh, R. F., 26, 169, 294, 431; 27, 722; 28, 85; 29, 233, 433; 30A, 256; 34A, 178
Krukonis, A. P., 30B, 399; 32B, 137; 33B, 91
Krumpelt, M., 34A, 144
Krupička, S, 26, 411
Krupka, M. C., 27, 92; 30A, 153; 34A, 59
Kruse, E., 30A, 240
Kruse, F. H., 26, 331, 340; 27, 463; 28, 104; 29, 433, 436; 33A, 195; 34A, 193
Krylov, E. I., 26, 386; 28, 341
Kšnjakina, A. N., 31B, 41
Khan, Y. -H., 28, 268
Kubina, L., 35B, 334
Kubo, M., 27, 463
Kubota, B., 26, 354
Kubota, T., 28, 341
Kuchitsu, K., 26, 543; 30B, 405; 33B, 529, 530, 532; 34A, 172; 35B, 892, 900
Kuczkowski, R. L., 28, 362; 34A, 240
Kudielka, H., 30A, 35, 175
Kudielka-Artner, E., 27, 318
Kudrin, V. S., 26, 536
Kudrina, M. A., 26, 536
Kudzin, A. Ju. 27, 507
Kuhn, K. J., 30B, 396
Kuhn, L., 33B, 42
Kühner, G., 27, 483
Kuhr, H., 33B, 297
Kukharčik, M. V., 26, 536
Kukharenko, A. A., 26, 454
Kukhto, O. L., 26, 285
Kukina, G. A., 27, 668; 28, 586; 30A, 265; 30B, 291; 34B, 444
Kulešov, I. M., 27, 473
Kulikova, A. A., 30A, 182
Kulisov, V. I., 35B, 407, 556
Kuli-žade, T. S., 34B, 444
Kuljkova, A. O., 30A, 180
Kullerud, G., 26, 137; 27, 293; 28, 20
Kulp, B. A., 30A, 245
Kulpe, S., 30B, 397; 32B, 429; 34B, 78
Kumakura, S., 32B, 281
Kumbasar, I., 33A, 395; 34A, 358; 35A, 481
Kumra, S. K., 30B, 32, 241
Kumskova, N. M., 28, 342
Kunchur, N. R., 26, 675, 676; 27, 835; 29, 630; 30B, 374; 33B, 497; 35B, 511, 670

Kunitomi, N., 34A, 101
Kunnmann, W., 33A, 344
Kunz, V., 26, 426, 783; 32B, 370
Kunze, G., 27, 188
Kunzmann, R., 27, 352
Kuo, C. -K., 29, 435
Kuo, C. -L., 29, 237; 30A, 255
Kupcik, V., 32A, 222; 32B, 251; 33A, 41
Kupfer, D., 26, 764
Kupfer-Tsoucaris, D., 29, 724, 725
Kupletskaja, N. B., 32B, 60
Küpper, F. W., 33B, 328
Kuprijanov, M. F., 26, 420; 27, 508; 28, 340, 341
Kuprina, V. V., 30A, 177
Kuragano, M., 29, 760
Kurahashi, M., 32B, 76
Kurbanov, Kh. M., 28, 172; 30A, 419
Kurbatov, V. V., 29, 434
Kurdjumova, R. N., 26, 337
Kuribayashi, S., 32B, 104
Kuriyama, M., 27, 474
Kurka, F., 32A, 142
Kurki-Suonio, K., 27, 564
Kurlina, E. V., 26, 412
Kurnick, S. W., 27, 63
Kuroda, H., 31B, 278; 32B, 284; 33B, 250; 35B, 417
Kuroda, T., 32B, 67
Kurokawa, M., 27, 800
Kuroya, H., 28, 90, 615; 32A, 193, 197, 218; 35B, 591, 592
Kursanov, D. N., 28, 687; 32A, 511
Kusakov, M. M., 33B, 529
Kushi, Y., 32A, 193, 197, 218; 35B, 162, 732
Kushiro, I., 26, 379
Kusma, J. B., 29, 136, 138, 139, 226
Kušnerev, M. Ja., 26, 408
Küster, J., 27, 465
Kutoglu, A., 29, 89; 33A, 22; 35B, 553
Kutschabsky, L., 28, 316; 29, 388; 31A, 88, 166; 32A, 504; 34A, 358; 35B, 139
Kutzelnigg, A., 27, 271
Kuwabara, S., 27, 384
Kuwano, H., 26, 762
Kuzel, H. -J., 26, 517; 29, 436
Kuz'ma, Ju. B., 26, 175; 27, 29, 152, 161, 162, 165, 206, 212, 376; 28, 28; 29, 138, 139, 140, 141, 225; 30A, 44, 172, 174, 176, 177 178; 31A, 27; 32A, 141; 33A, 153; 34A, 6, 143; 35A, 36, 124
Kuzmenko, P. P., 27, 109
Kuz'min, E. A., 32A, 278; 33A, 506; 34A, 275; 35A, 229, 316, 388, 391, 480
Kuz'min, R. N., 26, 35, 65
Kuz'minov, Ju. S., 27, 510; 29, 438
Kuz'minov, Y.S. 28, 342
Kuznecov, E.A., 30A, 179
Kuznecov, L.M., 29, 435
Kuznecov, S. I., 34B, 69
Kuznecov, V. G., 26, 208, 214; 27, 398, 482; 28, 340, 459; 29, 64, 138, 139, 225, 433, 435; 30A, 64, 175, 180, 247; 30B, 293; 31A, 50; 33A, 438; 35A, 417; 35B, 637
Kvaša, L. G., 29, 437
Kvick, A., 34B, 128
Kvitka, S. S., 27, 511; 33A, 272
Kwan, T., 31B, 421
Kyogoku, Y., 31B, 177

Laan, J., 32B, 253
van Laar, 29, 42, 137; 32A, 55; 34A, 207; 35A, 72
Labartkaja, N. I., 26, 159
Labotz, R. J., 27, 212
Labrot, G., 26, 625
Laby, R. H., 35B, 905
Ladd, M. F. C., 28, 685; 31B, 234, 32B, 246
Ladell, J., 32B, 311; 34B, 122
Ladyzinskij, P. B., 29, 326, 434
Laffaille, A., 34A, 387
Lafferty, W. J., 27, 732
Laffitte, P., 27, 571
Lafleur, W. J., 27, 303
Lagervall, E., 35A, 256
Lahiri, B. N., 34B, 157
Lai, T. F., 32B, 79; 35B, 857
Laing, C. -K., 32A, 141
Laing, M. B., 30B, 306
Lajzérowicz, J., 28, 318; 29, 339; 30A, 326; 31A, 220
Lajzerowicz-Bonneteau, J., 33B, 117
Lakh, V. I., 30A, 177, 178
Lalancette, R. A., 32B, 391
Lallement, R., 29, 136; 31A, 33
Lam, D. J., 26, 254; 27, 360; 35A, 124
Lam, S. Y., 35B, 857
Lamarque, A., 30B, 394
Lambert, J. B., 35B, 93
Lambert, V. L., 28, 292
Lambert-Andron, B., 33A, 66; 34A, 96
Lamikhov, L. K., 34A, 142
Lamkhov, L. K., 29, 138
Lamm, B., 32B, 96
Lamnevik, S., 33A, 510
Lancucki, C. J., 28, 357
Land, C. C., 27, 28; 30A, 51, 181, 246, 248
Lander, G. H., 32A, 102; 34A, 144
Lando, J., 27, 810
Lando, J. B, 31B, 37; 35B, 882
Lanfredi, A. M., 26, 337, 344; 30A, 298, 409; 30B, 365, 367; 34A, 340
Lanfredi, A. M. M., 30B, 444; 31A, 159; 32A, 151, 352, 509; 32B, 425, 427; 33B, 436

34A, 34, 547; 35A, 396; 35B, 720
Lang, J., 33A, 50; 34A, 54, 166; 35A, 140
Lang, R.E., 29, 761
Lange, R.C. 30A, 61
Lange, S., 35B, 502
Langer, G., 30A, 247
Langer, S., 29, 224; 30A, 245
Langereis, C., 28, 34
Langlet, G., 29, 318
Langmyhr, F. J., 26, 264
Langridge, R., 26, 817
Lapage, R., 29, 224
Lapasset, J., 34B, 109
Lapham, D. M., 26, 538; 28, 317
Lapickij, A. V., 27, 444; 28, 315; 29, 433
Lapidus, E. L., 30A, 92, 172
La Placa, S., 26, 67; 27, 103; 33A, 344
LaPlaca, S. J., 29, 240; 30B, 386, 389, 390, 395; 31B, 460; 32B, 311, 391; 33B, 31, 290; 34A, 158
Lapluye, G., 28, 316; 29, 223
Lappe, F., 27, 246
Laprade, M. D., 33B, 350, 374; 34B, 377
Larking I., 35A, 421
Larking J., 34B, 480
Larose, P., 26, 640
Larsen, E., 35B, 788
Larsen, E. M., 29, 249
Larsen, F. K., 32A, 248; 34A, 231; 35A, 278; 35B, 153
Larsen, I. K., 31B, 44; 33B, 35
Larsen, T. K., 34A, 211
Larsen, W. L., 26, 205; 27, 298, 359, 368; 29, 138, 225
Larson, A. C., 26, 107, 108, 127, 498; 27, 142, 416; 38, 17, 467; 29, 36, 37, 81, 261, 267, 313; 30A, 50, 178, 290, 363; 30B, 69; 31A, 116 202; 31B, 152; 32A, 114, 339; 33A, 245; 34A, 193, 310
Larson, D. W., 28, 317
Larson, E. G., 27, 495
Larsson, E., 30A, 71, 181
Larsson, G., 34B, 349
Larsson, K., 27, 740; 28, 367, 369, 379, 683, 687; 30B, 10, 11, 12, 446; 31B, 21; 35B, 127
Larticle, M., 29, 760
Laruelle, G., 29, 433
Laruelle, P., 29, 135; 30A, 177; 32A, 132; 33A, 4, 98, 307, 362; 34A, 98, 146, 295; 35A, 76, 101
Laško, A. S., 26, 187
Laško, N. F., 26, 110
Laszlo, I., 30B, 446
Laube, E., 26, 91; 27, 70, 105; 29, 139; 30A, 172, 175
Laudy, J. H. A., 28, 289
Laught, M., 35A, 346
Laugier, J., 27, 251; 29, 223
Laurent, A., 28, 685; 29, 474; 30B, 130, 397; 31B, 123; 34B, 45, 388
Laurent, Y., 33A, 50, 393; 34A, 54, 166; 35A, 140
Laurie, V. M., 28, 668
Laurie, V. W., 26, 543, 544, 553; 27, 729; 28, 668, 33B, 539; 35B, 904
Laurier, J., 33A, 140
Laveant, P., 26, 72
Laveissière, J., 32A, 157
Lavendel, H. W., 26, 81
Laves, F., 28, 461; 29, 141, 319
Lavilla, R. E., 28, 362
Lavine, M. C., 27, 244, 281
Lavut, E. G., 28, 341
Lawley, A., 26, 14
Lawn, B. R., 28, 290
Lawrence, J., 30B, 397
Lawrence, J. L., 34B, 110, 172, 403
Lawrence, L. J., 30A, 177
Lawrence, R. M., 26, 294
Lawrence, R. V., 30A, 177, 247
Lawson, F., 32A, 509
Lawson, J., 35B, 78
Lawson, K. D., 33B, 529
Lawton, D., 30B, 397
Lawton, S. L., 31A, 108; 33B, 289; 34B, 412; 35B, 452
Layden, G., 33A, 328
Layden, G. K., 33A, 342
Lazarev, B. G., 26, 285
Lazarev, G.B, 27, 393
Lazarov, I. I., 34B, 508
Leach, J. S. L., 29,135
Leake, J. A., 29, 7
Leal, G., 33A, 393
Leary, J.K., 26, 513; 27, 519
LeBihan, M. -T., 26, 428, 432; 27, 402; 31A, 233; 32B, 322; 32A, 460; 33A, 50; 33B, 301; 34A, 54, 166; 34B, 546
Lebioda, L., 34A, 229
Le Blanc, A., 34A, 145
Le Blanc, A. H., 26, 543
LeCarpentier, J. -M., 32B, 318; 33B 293
Lecerf, A., 26, 408; 27, 518, 527; 28, 285; 29, 325
Leciejewicz, J., 26, 105, 360, 361, 362; 27, 477; 29, 35, 136; 30A, 38, 313, 328; 32A, 21, 45; 35A, 265
Lecocq, F., 27, 208; 29, 135, 140, 141, 224; 30A, 173, 178, 182, 244; 31A, 37
Lecocq, Y., 29, 140
Lecouturier, C., 29, 435
Ledent, J., 32A, 235
Ledsham, A. H. C., 33A, 372; 33B, 519; 34A, 312
Lee, B., 30B, 396, 397; 32A, 510
Lee, D. E., 27, 514; 29, 436
Lee, F. M., 29, 140
Lee, G., 34A, 152

Lee, G. H., 34A, 153
Lee, J. A., 26, 228, 230; 27, 325
Lee, J. D., 27, 1002; 30B, 259; 33B, 383; 34B, 48, 49, 382, 578, 580; 35B, 60, 451
Lee, K., 27, 455
Lee, K. S., 30B, 446; 31B, 157
Lee, K. T., 29, 225
Lee, M., 27, 838
Lee, T.-Y., 30A, 33, 175; 30B, 315
Lee, Y. T., 34B, 459
Leeman, H. G., 27, 961
Leenhouts, J. I., 28, 34; 35B, 138, 336
Léfebvre, S. G., 29, 760
Lefevre, J., 27, 542; 28, 282, 339
Lefkowitz, I., 31A, 148
Leflem, G., 35A, 228
LeFur, Y., 31A, 87; 32A, 163; 33A, 180; 35A, 173
Legon, A. C., 33B, 541; 34A, 215
Legrand, E., 27, 439, 453
Legrand, M., 26, 799
Leguen, J. -C., 31B, 439; 34B, 576
Legzdins, P., 31B, 383; 33B, 378, 391, 463; 34B, 453
Lehamm, H. W., 32A, 53
Lehmann, J., 26, 203; 29, 760
Lehmann, M.S., 33B, 124; 35A, 198, 278; 35B, 153
Lehovec, K., 29, 307
Lehr, J.R., 27, 569, 577; 28, 357
Lehr, P., 33A, 268
Lehtonen, M., 35B, 677
Leibau, F., 35A, 341
Leibler, K., 27, 784
Leicht, E., 26, 87; 28, 14
Leijonmarck, M., 34B, 156
Leijonmarck, M. L., 35B, 196
Leipoldt, J. G., 34B, 138; 35A, 423, 424; 35B, 177
Leiserowitz, L., 29, 660; 30B, 8, 446; 32B, 12, 17, 50; 35B, 133, 134
Leitnaker, J. M., 26, 80; 27, 92
Lejus, A. M., 26, 368; 27, 518
Lemaire, R., 27, 163; 30A, 42, 175, 176; 32A, 8, 106; 34A, 72, 105; 35A,60,80
Lemen, J., 30B, 445
Lemmer, F., 31A, 113, 114
Lenglet, M., 29, 325
Lenhert, P. G., 26, 807; 33B, 107, 186; 34B, 305; 35B, 742, 744
Lenne, H. -U., 26, 347, 575
Lensen, M., 29, 325
Léonard, A., 27, 583; 33B, 528
Leonard, B. F., 26, 489
Leonhardt, R., 26, 280
Leonova, T. N., 26, 378
Lepard, D. W., 27, 732
Lepicard, G., 31A, 136; 34B, 388
Lepore, U., 35B, 581
Lerman, L.S., 26, 818; 29, 759
Lerner, S., 28, 358; 32A, 92
Leroux-Hugon, P., 27, 47
Leroy, F., 34B, 222
Leroy, J. 28, 357
Leslie, D. H., 26, 202
Lessor, A. E., 27, 952
Lethbridge, J. W., 35B, 604
Leung, Y. C., 27, 1058
Leutner, H., 27, 543
Levasseur, A., 35A, 288
Levayer, C., 34A, 228
Levdik, V. A., 27, 151; 34A, 142; 35A, 115
Levdik, V. F., 27, 820
Leventov, V. S., 29, 436
Leverett, P., 32A, 509; 33A, 333; 34A, 274, 277; 35A, 270
Levesque, P., 26, 217
Levin, E. M., 26, 398, 491; 29, 438
Levine, D. B., 30B, 444
Levine, E. D., 27, 21
Levine, I. N., 28, 668
Levine, M. C., 27, 382
Levinson, A.A., 27, 563
Levinson, D.W., 26, 276
Levinstein, H.J., 31A, 147
Levinstein, J.H., 30A, 248
Levitin, R. Z., 29, 226
Levšov, P. P. 29, 436
Levy, C., 30A, 177
Levy, E., 29, 251, 353, 383
Levy, H. A., 26, 324, 554; 27, 422, 639, 722, 731; 28, 73, 74, 75, 519; 30A, 266, 303; 30B, 25, 179; 32A, 395; 33A, 185, 195
Levy, J., 30B, 445
Lew, H., 29, 255
Lewin, R., 28, 581
Lewis, B., 27, 90; 29, 135
Lewis, E. L. V., 34B, 21
Lewis, J., 26, 687; 27, 822
Li, C. -C., 29, 483
Li, C. -T., 30B, 188; 33A, 456, 457; 35A, 453
Li, H. -L., 29, 435
Li, J. -M., 31A, 118
Li, N. C., 33B, 427
Li, T. -J., 31A, 232
Liang, C. -K., 30A, 48, 52, 178
Liang, T. -T., 30B, 195
Libby, W. F., 30A, 179; 33A, 153
Libiseller, R., 27, 1039
Libowitz, G. G., 27, 237, 481; 28, 341
Lide, D. R., 26, 549; 27, 735
Lidecke, W., 26, 172; 27, 520
Liebau, F., 26, 463, 506; 27, 586; 29, 393; 34A, 327
Liebe, W., 26, 308, 312
Liebig, D. M., 30B, 353; 31B, 419
Liebling, G., 30B, 87
Liebrich, O., 35A, 9, 124
Lied, R. C. 27, 132
Lihl, F., 26, 11, 182, 27, 271
Lima-de-Faria, J., 27, 423

Liminga, R., 29, 233; 30A, 366, 387; 31A, 182; 32A, 318; 32B, 269; 33B, 36, 37
Lin, C. -C., 30B, 212
Lin, C. -T., 31B, 381
Lin, T. S., 34B, 338
Lin, W., 30A, 181, 247
Linares, R. C., 27, 490
Linck, M. H., 30B, 445
Lincke, G., 28, 314
Lind, M., 26, 680
Lind, M. D., 29, 545; 30B, 396, 397; 32A, 183; 34A, 149, 262; 35A, 181; 35B, 36, 881
Lindahl, T., 34A, 77; 35A, 66
Lindberg, M. L., 26, 381; 27, 603; 29, 463
Linde, V. R., 26, 408
Lindemann, W., 26, 397, 510, 540
van der Linden, J. G. M., 35B, 806, 810
Lindenberg, W., 28, 315; 29, 433
Lindenmeyer, P. H., 27, 804
Linder, H. J., 33B, 328
Lindgren, J., 33B, 25
Lindh, B., 32B, 417
Lindheim, A., 28, 103
Lindinja, L. F., 29, 141
Lindley, P. F., 31A, 178; 32A, 411; 32B, 319; 34B, 427, 471; 35B, 578
Lindner, H. J., 34B, 149, 584; 35B, 130, 187
Lindqvist, I., 26, 595, 657; 28, 95; 29, 369; 34B, 134
Lindqvist, O., 31A, 211; 32B, 339; 33A, 241, 263; 35A, 281, 345, 483
Lindsey, D. C., 34A, 220
Lindsey, J., 27, 1042
Linek, A., 27, 640; 32B, 251
Lingafelter, E. C., 26, 317, 562, 667, 671, 674, 753; 27, 432, 433; 28, 616, 618, 686; 29, 353, 567, 568, 593, 595, 596, 597, 598; 31A, 203, 204; 31B, 375, 397, 400; 32A, 343; 32B, 31, 57, 381, 392, 395, 400, 403, 404; 33B, 388; 34B, 486; 35B, 594, 687, 696, 730, 733
Linggoatmodjo, K., 35A, 197
Linke, K. -H., 27, 645, 751; 28, 316; 29, 291; 31A, 113, 114
Linnros, B., 35A, 356
Linqvist, B., 27, 695
Lionetti, A., 30B, 398
Liopo, V. A., 29, 433
Lipin, J. V., 35A, 259
Lipova, I. M., 27, 673
Lippard, S. J., 30B, 34, 394; 31B, 353, 383, 415, 32B, 441, 478; 33B, 391; 34B, 445, 597, 599; 35B, 563, 802
Lippens, B. C., 29, 298
Lippert, E. L., 26, 346, 545; 30B, 446
Lippert, W., 26, 38
Lippman, F., 26, 483; 27, 183, 563; 32A, 417; 33A, 436
Lipscomb, W. M., 34B, 385
Lipscomb, W. N., 26, 290, 545, 600, 618, 619, 620, 625, 681; 27, 393, 399, 446, 769, 854, 993; 28, 60, 62, 314, 399, 540, 545, 581, 686, 687; 29, 98, 99, 229, 230, 231, 526, 529, 531, 657; 30A, 252; 30B, 194, 372, 397; 31B, 258, 283, 287; 32B, 292, 293; 33B, 262, 263; 35B, 426
Lipsett, M. N., 27, 1055
Lipsitt, H. A., 31A, 240
Liquori, A. M., 26, 754; 29, 760; 30B, 253, 299; 31B, 111, 260, 261; 32B, 276, 287
Li-Shen, L., 26, 630
Lisina, A. V., 30A, 244
Lisnjak, S. S., 27, 492
Lister, D. G., 34A, 220
Lister, J. S., 32A, 475
Listovnicy, V. E., 30A, 181
Littke, W., 28, 101; 30B, 446
Littler, J., 27, 675
Litvin, B. N., 29, 436
Litvin, D. F., 26, 120; 27, 150, 157, 163
Litwin, J., 29, 225
Liu, C. F., 34B, 539; 35B, 587
Liu, C. -H., 27, 215
Liu, L. -P., 29, 436
Liu, S. J., 35B, 406
Liu, U. -H., 30A, 244
Liu, Y. H., 29, 224
Liu, Y. -L., 27, 599
L'jačenko, O.A., 32A, 510
Ljan, D. -C., 29, 483
Ljaščenko, B. G., 27, 150, 157, 163; 28, 15
Ljung, H., 33A, 108; 35A, 40
Llewellyn, F. J., 27, 831; 31B, 170
Lloyd, D. J., 34A, 257, 258; 35A, 222, 238
van Loan, P., 27, 634
Lobachev, A. N., 26, 569; 27, 748
Lobanova, G. M., 32B, 113
Lobanova, V. V., 29, 436
Lobier, G., 34A, 115
Lobkovskij, E. B., 35A, 423
Lobo, P. W., 27, 428, 560; 28, 220
Locchi, S., 26, 328, 502, 738, 781; 27, 470, 629, 989; 28, 687; 30B, 139, 141, 143
Lock, C. J. L., 35B, 643
Locke, H. O., 27, 796
Loebich, O., 29, 223
Loebner, E. E., 28, 37
Löfgren, P., 26, 500; 32B, 320
Logemann, W., 26, 719
Loguidice, F., 34B, 34

Lohberg, K., 26, 99
Lohmann, D. H., 27, 666
Lohmeyer, S., 26, 44
Löhn, J., 33A, 483; 35A, 19, 295
Lohr, L. L., 27, 446; 28, 399
Lohs, K., 34B, 378
Lohse, C., 35B, 232
Lohse, H. -H., 31A, 174; 33A, 277
Lokken, D. A., 35A, 150
Lomer, T. R., 26, 562; 28, 313, 425; 30B, 19, 444; 34B, 21
Long, G., 26, 346
Long, H. A., 30B, 395; 33B, 172, 421
Long, J. V. P., 28, 338
Long, R. E., 27, 573; 30B, 40; 34B, 84
Long, R. F., 27, 822
Longo, J., 26, 418
Longo, J. M., 30A, 368; 446; 31A, 147; 32A, 104, 309; 33A, 353; 35A, 158, 204, 337, 373
Löns, J., 32A, 490
Lonsdale, K., 26, 767, 828; 30B, 34
van Look, N., 33A, 153
Look, W., 33A, 406; 33B, 285
Looney, R. W., 29, 433
Loong, W. S., 33B, 31
Loopstra, B. O., 29, 312; 27, 118; 28, 318; 32A, 156; 31A, 123, 127; 34A, 289
Lopatin, E., 33A, 235
Lopato, L. M., 29, 434
Lopez, J. S., 33A, 393
López-Castro, 28, 623, 683; 29, 485; 34B, 33
Lopez-Vieira, A., 34A, 392
Lorbacher, G., 30B, 447
Lord, R. C., 27, 733
Lorenz, R., 34A, 144
Lorenzelli, R., 29, 136; 31A, 35
Loretto, M. H., 26, 139
Loriers, G., 27, 511
Lories, J., 30A, 172
Lorimer, G. W., 30A, 173
Losakova, G. V., 33A, 126
Lošmanov, A. A., 27, 150, 157, 163
Lothrop, K. H., 26, 101
Loub, J., 31A, 199
Louisnathan, J., 32A, 511
Louisnathan, S. J., 35A, 455, 472
Love, W. E., 28, 385; 30B, 28; 33B, 151
Lovell, F. M., 31B, 117
Low, B. W., 26, 820; 27, 1058; 30B, 444
Low, G. G. E., 27, 383
Löw, M., 30B, 91
Lowde, R. D., 27, 251, 397
Löwenhielm, A., 29, 320
Loye, O., 33A, 307
Lu, C. -S., 31A, 118
Lu, C. T., 27, 1038; 28, 536; 29, 750; 33B, 192
Lu, C.Y., 27, 775
Lu, H.-S., 30A, 178
Lu, S.S., 30A, 48, 52
Lubell, M. S., 27, 211, 362
Lublin, P., 29, 224
Lucas, B., 28, 682
Luccoborlera, M., 28, 291
Ludé, A. C., 29, 760
Ludi, A., 26, 328; 27, 470; 28, 315; 33A, 242, 243, 244; 35A, 404, 405
Luger, P., 35B, 112
Lugscheider, W., 30A, 247
Lukas, H. L., 33A, 153
Lukaszewicz, K., 26, 343, 466; 28, 195; 29, 135, 223; 31A, 136, 148, 192; 32A, 372, 374; 33A, 27, 409; 34A, 15, 63, 268; 35A, 84
Luk'janjčev, Ju. A., 31A, 84
Luksa, E. A., 30B, 447
Lumbroso, H., 26, 720
Lunc, A. J., 28, 339
Lundberg, B. K. S., 31B, 421
Lundberg, M., 29, 318; 30A, 334; 35A, 179
Lundell, S., 35B, 674
Lundgren, G., 28, 147; 29, 454; 31A, 205, 211
Lundgren, J. -O., 32A, 181; 33A, 229; 35A, 194
Lundgren, L. -O., 30A, 366
Lundin, A. G., 28, 244
Lundin, C. E., 27, 241, 381; 30A, 247
Lundquist, T, 28, 341
Lundström, T., 27, 157; 28, 33; 32A, 34, 111; 33A, 46, 92; 34A, 41; 35A, 34
Lundvik, L., 27, 290
Lungu, G., 27, 570
Luo, H. -L., 27, 221; 29, 136, 223, 224; 30A, 244
Luova, P., 27, 425
Lussier, R., 34B, 550
Luth, H., 30B, 123; 32B, 339; 33B, 492; 34B, 451; 35B, 510
Lüty, U., 29, 137
Luukkonen, E., 34B, 489, 494; 35B, 677
Luxmoore, A. R., 26, 684
Luxmore, A. R., 27, 824
Luznaja, N. P., 30A, 247
Luzzatti, V., 26, 564, 803, 809, 817, 818, 819; 27, 1063; 29, 759
Lyaščenko, B. G., 26, 120
Lydon, J. E., 30B, 364; 33B, 335
Lye, R. G., 33A, 57
Lykova, L. N., 33A, 341
Lynch, R. W. 30A, 245
Lynds, L., 27, 481
Lyng, S., 27, 610
Lynton, H., 27, 702; 32B, 181; 33B, 443; 34A, 189; 35A, 189; 35B, 391
Lysak, L. I., 30A, 245
Lysenko, L. T., 34A, 144

Lytle, F. W., 29, 434

McAdam, A., 33A, 411; 35B, 895
McAdam, R. C., 28, 145, 358
McAfee, K. B., 26, 301
McAllister, J. F., 26, 489
MacAuley, E. W., 34B, 394
McBride, D. W., 35B, 541
McBride, J. M., 35B, 33
McCabe, C. L., 26, 99
McCallum, I. S., 28, 357
McCammon, C. S., 29, 655
McCapra, F., 27, 1017, 1031, 31B, 223
McCarroll, W. H., 26, 35; 30A, 245
McCinnell, J. F., 28, 687
McCleverty, J. A., 27, 849
McConnell, J. D. C., 27, 423
McConnell, J. F., 27, 997; 29, 744; 30B, 172; 32B, 454; 34B, 39, 586; 35B, 646
McConville, G. T., 29, 140
McCorkindale, N. J., 26, 800
McCoy, W. H., 28, 668
McCullough, J. D., 26, 771; 27, 864, 964; 28, 409, 410; 29, 476, 522; 31B, 272; 32B, 200; 33B, 149; 34A, 276; 35B, 226, 495
McCune, R. A., 34A, 146
McDaniel, M. C., 32B, 196
MacDonald, A. C., 29, 549; 30B, 114, 224, 446; 31B, 249, 359; 34A, 342
MacDonald, B. D., 26, 201
McDonald, B. J., 27, 671
McDonald, R. R., 27, 141, 954; 29, 267
MacDonald, S. G. G., 28, 489; 30B, 397; 34B, 110, 172
McDonald, T. R. R., 30A, 178; 32B, 415
McDonald, W. S., 32A, 253, 370, 435, 448; 34A, 407; 35B, 81, 482, 576
McDowell, J. J. H., 30B, 445; 34B, 213; 35B, 242
McEachan, C. E., 30B, 397; 31B, 216, 225
MacEwan, A. W., 35B, 887
McEwen, R. S., 30B, 393; 32B, 452, 456; 34B, 589
McEwen, W. E., 33B, 290
McGahren, W. J., 34B, 339
McGandy, E. L., 34B, 124, 125, 127; 35B, 564
McGavin, S., 27, 1054
McGeachin, H., McD , 26, 314, 367; 29, 361
McGeachin, S. G., 26, 830
MacGillavry, C. H., 26, 455, 709; 27, 482, 755; 28, 377, 570; 30B, 54 358; 32B, 253; 33B, 84, 131; 35B, 7, 135
McGinnety, J.A., 32B, 479
McGrath, J. W., 26, 344
McGregor, D. R., 33B, 46; 34B, 46
McGuire, T. R., 29, 140, 226
McHargue, C. J., 26, 106
McInnes, A. G., 33B, 522
MacIntyre, W. M., 28, 512; 30B, 139, 370
McIver, E. J., 28, 256
McKechnie, J., 32A, 509
McKechnie, J. C., 31B, 368
McKechnie, J. S., 32A, 509; 33B, 111, 196, 368; 34B, 343; 35B, 334
MacKenzie, D. R., 28, 229; 29, 335
McKie, D., 27, 600; 28, 339, 342; 29, 436
McKillop, T. F. W., 34B, 312
McKinley, S. V., 35B, 491
McKinnon, A. J., 29, 599; 30B, 369
McKinzie, H. L., 35A, 179
McLaren, A. C., 27, 566, 568, 569
McLaren, J.R.,27, 125
McLay, D. B., 27, 729; 29, 449
McLean, W. J., 32B, 291; 33B, 27; 35A, 461
McManus, G.M., 26, 129
McMasters, O.D., 27, 368; 29, 137; 30A, 244; 32A, 142, 143; 33A, 153; 35A, 124
McMillan, J. W., 29, 434
McMullan, R. K., 27, 778, 781; 28, 578, 683; 30B, 267, 268; 31B, 282; 32B, 85, 289; 35B, 422
McNally, R. N., 27, 545
McNeal, R. J., 27, 137
McNeilly, C. E., 27, 477
McPartlin, E. M., 34B, 591, 594
McPartlin, M., 35B, 811
McPhail, A. A. T., 26, 830
McPhail, A. T., 27, 1026; 28, 558, 686; 30B, 68, 230, 397; 31B, 64, 196, 213, 216, 223, 235, 236, 251, 267, 275, 349, 372, 447; 32B, 250, 363, 368, 369; 33B, 202, 213, 226, 354, 361; 34B, 342; 35B, 17, 50, 273, 371, 383, 405, 570, 621, 834
McPherson, G., 31B, 342
McPherson, G. L., 35A, 160
McPherson, W. G., 32B, 157; 33B, 293, 294
McQuillan, A. D., 26, 167; 27, 238
MacRae, T. P., 26, 825; 27, 1053, 1056, 1057; 29, 759
McSkimin, H. J., 26, 271
McVay, T. N., 27, 451
McWhan, D. B., 26, 32; 27, 39; 30A, 179, 245, 248; 31A, 205; 32A, 141
Ma, C. -S., 28, 339
Maak, I., 26, 300; 29, 224, 433
Maartmann-Moe, K., 30B, 77; 31B, 72, 335, 34B, 67, 68
Maas, K. E., 33A, 40; 35A, 28

Mabbs, F. E., 32B, 394, 396; 35B, 621
Mabis, A. J., 27, 1059; 28, 686; 31B, 24; 33B, 529; 34B, 343
Machajdik, D., 34B, 571
Machonachie, 30B, 191
Macintyre, W. M., 29, 674, 710, 712
Mack, M., 29, 225
Mackay, K. J. H., 27, 936, 1042
Mackay, M. F., 28, 685; 30B, 245, 34B, 324
Macrosson, W. D. K., 33B, 103
Madan, O. P., 32B, 220
Madar, J., 26, 576
Madaule-Aubry, F., 30B, 368; 33B, 447, 449
Maddox, H., 31B, 272; 34B, 84
Madeja, K., 30B, 443
Maeda, K., 31B, 194
Maeland, A., 30A, 244
Maeland, A. J., 26, 166, 168; 29, 137, 225, 226; 33A, 120, 160
Magarill, S. A., 35A, 250
Magat, L. M., 27, 9
Magdesieva, N. N., 35B, 904
Magnéli, A., 26, 356; 30A, 312
Magnéli, C., 33A, 407
Magnuson, V. R., 32A, 404; 34B, 399, 406
Magnusson, B., 27, 406
Magre, E. P., 34B, 177; 35B, 886
Mague, J. T., 29, 626; 31A, 124; 34B, 616; 35B, 572
Mahanta, P. C., 27, 480
Mahe, R., 31A, 149; 32A, 295
Maier, C., 30B, 444
Maier, C. A., 35B, 223
Maier, L., 29, 637
Maier, R. G., 29, 225
Maier, W., 26, 586
Mair, G. A., 26, 690
Mair, J.A., 27, 672
Mair, W. N., 26, 366
Mais, R. H. B., 28, 687; 30B, 391; 31B, 463; 32B, 437, 469, 474, 475; 33B, 347, 501, 503, 515; 34B, 437, 606; 35B, 526, 792, 848
Maisseu, J. -J., 31B, 78
Maita, J. P., 26, 254; 30A, 247
Maitland, R., 29, 433
Majeste, R., 28, 471; 30B, 144; 31B, 30; 32B, 163; 33B, 127; 34B, 289; 35B, 455
Majeste, R. J., 35B, 728
Majumdar, A., 30A, 423
Majumdar, A. J., 29, 435; 30A, 248
Majumdar, J., 26, 511
Mak, T. C. W., 26, 830; 27, 816, 892, 935; 28, 434, 435, 683; 29, 681; 30B, 68, 268, 269, 278; 31B, 282; 32B, 288; 33A, 218; 35B, 423
Makarevič, L. G., 28, 683; 30B, 448
Makaročkin, B. A., 28, 342
Makarov, E. S., 26, 391; 28, 225; 29, 435; 31A, 240
Makatsu, K., 31B, 312
Maki, A. H., 34B, 355
Makino, D., 35B, 884
Makkonen, R. J., 27, 493
Makoveckij, G. I., 29, 135
Makovicky, E., 33A, 41
Maksimov, B. A., 33A, 452, 471, 473; 34A, 12; 35A, 15, 229, 297, 310
Maksimova, M. V., 26, 112
Maksimova, N. V., 28, 146; 32A, 509
Maksimova, V. A., 33A, 153
Malament, D. S., 35B, 33
Malčić, S. S., 26, 477; 28, 315; 30A, 247, 272
Maldonado, A., 29, 135
Maleev, E. F., 29, 437
Malin, A. S., 32A, 254; 33A, 265
Malinko, S. V., 33A, 446
Malinković, A. N., 34A, 143
Malinovskij, T. I., 28, 631; 31B, 427; 35B, 640
Malinovsky, M., 27, 438
Malissa, H., 26, 655
Malkin, T., 27, 783
Malkov, Ju. S., 27, 27
Malkova, A. S., 34A, 213
Mallikarjunan, M., 34B, 243, 269, 274
Mallinson, P. R., 35B, 439
Mallio,W. J., 35A, 144
Malm, J. G., 26, 301
Malmros, B., 26, 444
Malmros, G., 35A, 210
Mal'nikova, I. E., 29, 435
Malone, J. F., 35B, 482, 576
Mamaeva, G. I., 33B, 537; 34A, 151
Mamedov, K. N., 29, 434
Mamedov, K. S, 26, 462, 521, 28; 190, 247; 29, 434; 30B, 392
Mammi, M., 26, 761; 27, 966, 973; 28, 480; 30B, 152; 33B, 3, 136; 34B, 169, 181, 188; 35B, 155
Man, L. I., 27, 89; 30A, 62, 179; 35A, 106
Manaila, R., 33A, 49
Manassero, M., 34B, 604
Manca, P., 27, 256
Manchot, J., 28, 683
Mandarino, J. A., 26, 455; 27, 716; 28, 357
Mandel, G., 26, 558
Mandelkern, L, 26, 630
Mandy, T., 27, 838
Mandyczewsky, R, 35A, 155
Maner, R. J., 34B, 43
Manfredotti, A., 31B, 402; 34B, 189

Manfredotti, A. G., 34A, 234
Mangia, A., 33B, 466; 34A, 315; 34B, 206; 35B, 67, 235, 627
Mani, N. V., 26, 496; 29, 539; 30B, 280; 31B, 307; 32B, 86
Manley, R. St. J., 26, 628
Mann, A. W., 35A, 182
Mann, C. R., 27, 729
Mann, D. E., 27, 735
Mann, F. G., 28, 587; 35B, 467
Mannan, Kh. A. I. F., 32B, 481
Manohar, H., 26, 792; 27, 1012; 28, 119; 29, 397; 32B, 86
Manoharan, P.T., 28, 106
Manojlovic, Lj., 28, 315; 32B, 87; 33B, 44, 45
Manojlovic, Lj.M., 26, 477; 27, 915; 30A, 272; 30B, 82
Manojlovic-Muir, L., 35B, 535
Mansey, R. C., 30A, 172, 244
Mansikka, K., 27, 426; 29, 438
Mansmann, M., 28, 340; 29, 48, 258; 30A, 267, 268, 440
Manukian, B. K., 26, 782
Mao, H. K., 34A, 146
Marakov, E. S., 32A, 490
Marathe, B. R., 27, 47
Marchand, R., 34A, 166
Marchessault, R. H., 29, 761; 33B, 244; 35B, 266, 885
Marcinkowski, M. J., 27, 154
Marcon, J. -P., 33A, 58; 34A, 115
Marconi, W., 31B, 37
Marcovich, M., 31A, 104
Mardix, S., 32A, 135; 33A, 146; 34A, 133
Mardon, P. G., 26, 228, 230, 244
Marek, D., 27, 972
Mareschal, J., 28, 136; 32A, 268, 283, 510; 33A, 316
Marezio, M., 26, 322; 28, 157, 162, 166, 171; 30A, 313, 314; 31A, 120, 131, 165; 32A, 257; 33A, 282; 34A, 350, 354, 355; 35A, 218
Margolin, H., 26, 28, 138
Margolis, N. V., 29, 438; 31B, 52; 32B, 48; 33B, 30; 34B, 54; 35B, 19, 30
Margrave, J. L., 26, 316, 349
Margulis, T. N., 26, 606; 27, 619, 774 29, 752; 30B, 84, 85, 108, 397; 32B, 98, 115, 279; 33B, 78; 34B, 86
Mariani, C., 26, 672; 30B, 50, 378, 397; 31B, 35; 32B, 41, 42; 34B, 82; 35B, 93
Mariano, A. N., 27, 382; 29, 136, 223
Mariengof, L. F., 30A, 245
Marikhain, V. A., 27, 808
Marinder, B. -O., 27, 479, 520; 30A, 368; 32A, 272
Marini-Bettolo, G. B., 28, 683
Marion, L., 26, 830
Mark, W., 35B, 748
Markham, N. L., 30A, 177
Markiv, V. Ja., 27, 11, 161, 162, 260, 376; 29, 138, 139, 140, 225; 30A, 173, 176, 178, 248; 32A, 141; 34A, 143; 35A, 7, 53
Markov, E. S., 27, 673
Markov, P., 35B, 889
Markova, G.S., 29, 760
Markovskij, O.I., 27, 689
Marly, L., 34B, 214
Marongiu, G., 32B, 190; 33B, 145; 34B, 195, 199, 201
Marosi, L, 35A, 3
Marøy, K., 26, 583, 584; 28, 686; 30B, 284, 288, 330, 397; 31B, 335, 340
Marples, J. A. C., 26, 228, 230; 27, 325; 29, 224; 30A, 247
Marr, H. E., 27, 414
Marsau, M. P., 30B, 80
Marsau, P., 34B, 210
Marsh, A., 34A, 292
Marsh, E., 30B, 307
Marsh, R. E., 26, 495, 724; 27, 573, 764, 767, 863, 909, 945, 965, 1058, 1059; 28, 414, 608; 29, 665, 703, 749; 30B, 87, 152, 173; 31A, 206; 31B, 166, 187, 411; 32B, 79, 223, 359; 33A, 427, 428; 34A, 345; 34B, 62, 514; 35A, 409; 35B, 263, 283, 667
Marshak, M. L., 30B, 233
Marsokova, T. A., 30A, 175
Marstokk, K. -M., 33B, 238; 35B, 903
Marstrander, A., 31A, 110
Marta, E. S., 33A, 153
Martell, C., 26, 332
Martelli, G., 35B, 157
Martin, C., 32A, 380, 382; 34A, 329, 338
Martin, G., 35B, 563
Martin, H. A., 32B, 346
Martin, J., 30B, 116
Martin, J. B., 26, 491
Martin, J. E., 27, 42; 30A, 179, 246
Martin, L. L., 32B, 174
Martin, W., 35A, 209
Martinez-Carrera, S., 29, 719; 31B, 115; 32B, 84; 34B, 345; 35B, 45
Martin-Smith, M., 27, 972
Martjanov, I. N., 26, 537
Martjenko, L. I., 35B, 602
Marton, L., 26, 287
Martuscelli, E., 31B, 8, 12; 32B, 10, 16, 18, 20; 33B, 131; 34B, 9, 11;

35B, 581
Martynenko,L. I., 28, 356; 31B, 409
Marumo, F., 26, 532; 29, 23; 30A, 357; 32A, 25, 225, 227, 230; 34B, 491; 35B, 76, 684; 707, 713
Maruya, K., 28, 341
Marvin, D. A., 26, 817
Marvin, F., 28, 358
Marx, G., 35B, 507
Marx, R., 27, 534
Marzilli, L. G., 35B, 660
Marzin, C., 32B, 63
Masakuni, M., 33B, 26
Masamoto, J., 35B, 883
Mascarenhas, Y., 29, 455, 721
Masironi, L. Di M., 28, 462
Maskov, S. A., 33A, 45
Maslen, E.N., 26, 734; 27, 894, 997; 28, 530, 563, 684; 30A, 309; 30B, 65, 66, 371; 31A, 115; 31B, 229, 232, 346, 404, 408; 32B, 91, 93, 256; 33B, 214
Maslova, E. V., 30A, 178
Mason, B., 29, 433
Mason, D. R., 29, 136
Mason, J. T., 30A, 176; 35A, 98
Mason, R., 26, 597, 673, 752; 27, 822, 849; 28, 612, 684; 29, 277, 556, 558, 682; 30B, 311, 339, 393, 397; 31B, 94, 352, 366; 32B, 371; 33A, 167; 33B, 325, 333, 334, 507, 509; 34A, 230; 34B, 468, 495, 537; 35B, 529, 579, 811, 825, 852
Mason, S. A., 35B, 34
Massalski, T. B., 26, 248; 27, 175, 190; 30A, 245
Massaux, M., 34B, 546
Massazza, F., 27, 256; 28, 340
Masse, R., 32A, 377, 379, 442; 33A, 413, 416; 34A, 326
Masson, F., 26, 809, 818
Masters, O. D., 30A, 247
Mastin, S. H., 35A, 167
Mastjukov, V. S., 28, 362, 687; 29, 450, 538; 30B, 406; 35B, 889
Masuko, A., 32B, 431
Masuno, K., 29, 435
Mathew, M., 33B, 100, 135; 34B, 479, 585; 35B, 511, 670
Mathews, F. S., 26, 618; 31B, 269
Mathews, M., 26, 675
Mathieson, A. McL., 26, 831; 27, 997, 1003, 1052, 1063; 28, 550, 554, 560, 685; 29, 744; 30B, 240, 245, 397; 31B, 169; 32B, 239; 33B, 223, 232; 34B, 324
Mathieson, D. W., 30B, 394
Mathis, B., 32A, 142
Mathisen, H., 26, 559, 560; 29, 449; 32B, 40
Matias, V. V., 28, 342
Matiašovsky, K., 27, 438
Matjusenko, N. N., 27, 70; 29, 138, 225
Matković, B., 28, 179, 622; 31A, 178; 32B, 28; 33A, 398; 34A, 204; 34B, 580
Matkovich, V. I., 26, 79, 103; 29, 4; 30A, 29, 31, 172, 174, 251; 35A, 4
Matson, L. K., 29, 141, 226
Matsubara, I., 29, 760
Matsuda, S., 33B, 313
Matsueda, S., 35B, 380
Matsumaru, H., 34B, 238
Matsumoto, K., 35B, 591,592
Matsumoto, M., 27, 806
Matsumoto, T., 28, 358; 34A, 28
Matsumura, C., 26, 588
Matsuoka, H., 26, 286
Matsuura, T., 34B, 344; 35B, 400
Matsuura, Y., 34B, 455
Matsuzaki, T., 34A, 334
Mattace-Raso, M., 29, 12, 140, 141
Mattausch, H., 35A, 245
Mattes, R., 28, 602, 628; 29, 586; 35A, 3, 6
Matthews, B.W., 29, 733; 30B, 161; 35B, 887
Matthews, J.D., 35B, 559
Matthews, R.W., 35B, 770
Matthias, B.T., 26, 254, 256; 27, 245; 30A, 247
Matveenko, I. I., 26, 105
Matzat, E., 32A, 307; 33A, 391
Mau, C. E., 35B, 434
Mauer, F., 32A, 355, 497
Mauer, F. A., 27, 391; 28, 314; 30B, 272; 32B, 432; 34B, 569
Maupin, G., 26, 416
Maurin, M., 35A, 65
Mautner, H. G., 32B, 204
Mawby, R. J., 32B, 376
Maxwell, I. E., 34B, 482; 35B, 658, 660, 679, 686
Maxwell, S., 29, 225
May, N., 35A, 24, 124
Mayer, I., 26, 346, 545; 27, 815; 29, 353, 383; 30A, 289
Mayer, I. P., 27, 336
Mayer, R. M., 28, 474
Mayers, R. F., 33B, 18
Mayweg, V. P., 30B, 399
Maze, M., 31B, 73
Mazelsky, R., 27, 211, 362; 29, 22
Mazhar-Ul-Hague, 31B, 176, 32B, 189, 220, 248; 33B, 307; 35B, 274, 372, 380, 446, 448, 449, 456, 457, 595
Mazur, J., 29, 224
Mazumdar, S.K., 30B, 446; 31B, 189;

32B, 227, 34B, 266; 35B, 319
Mazza, F., 34B, 228, 230, 232, 233; 35B, 173
Mazzarella, L., 28, 684; 29, 760; 30B, 399; 32B, 276; 34B, 9
Mazzi, F., 26, 605; 27, 584; 28, 407; 30A, 437; 31A, 232; 34B, 41
Mazzone, G., 28, 687; 30B, 381
Mead, C., 29, 136, 435
Mead, S. W., 28, 28
Meaden, G. T., 26, 230
Meagher, E. P., 32A, 306; 33A, 480
Medhi Ali, S., 30B, 21
Medlin, E. H., 35B, 245
Medrud, R. C., 30B, 327; 34B, 108
Medvedev, A. V., 32A, 509
Medvedeva, Z. A., 30A, 179
Medvedeva, Z. S., 29, 225
Meek, D. W., 33A, 207; 34B, 588, 601; 35A, 410
van Meersche, M., 26, 785; 27, 427, 428, 435, 560, 583, 673, 710, 760, 954; 28, 220, 277, 278, 404, 575; 29, 739; 30B, 296, 303; 31B, 271, 273, 274; 32A, 400; 32B, 267, 271, 348, 437; 33B, 248; 34B, 422, 451; 35A, 195; 35B, 474
Megaw, H. D., 27, 679, 686; 29, 398; 31A, 148, 222; 32A, 271; 34A, 268, 368
Meier, W. M., 26, 515; 28, 252; 30A, 438; 31A, 218
Meinhardt, D., 27, 122
Meinnel, J., 30B, 79; 33B, 69
Meisingseth, E., 27, 925
Meissner, H. -G., 27, 377; 29, 92, 139; 30A, 49, 173, 178, 246
Meixner, J., 30A, 172
Mekata, M., 27, 273; 33A, 111
Mekhtiev, K. M., 29, 434
Melik-Adamjan, V. P., 26, 62
Mellor, I. P., 35B, 86
Mellor, J., 30A, 40, 83
Melmed, K. M., 32B, 478
Mel'nik, Ja. V, 32A, 141
Mel'nikov, O. K., 29, 436
Mel'nikov, V. V., 31B, 52; 33B, 30
Mel'nikova, V. V., 32B, 48
Melnitchenko, V., 30B, 445
Melsom, B. G. A., 33B, 383; 34B, 578, 580
Melson, G. A., 35B, 539, 800
Melville, D., 31B, 230
Melville, R. D., 33B, 216
Menchetti, S., 33A, 435; 34A, 215; 35A, 376; 35B, 762
Menczel, Gy., 30B, 444
Mendel, H., 27, 1064
Mendelovici, E., 33B, 528
Mendola, A., 28, 686
Menegus, F., 28, 598
Menges, E., 30A, 180
Menke, W., 27, 1061
Menkov, A. A., 26, 245, 349; 29, 433
Menzel, D., 26, 161, 207
Meriani, S., 35B, 810
Mériel, P., 26, 158; 29, 435; 31A, 33, 34
Merkle, A. B., 32A, 511; 33A, 484
Merlino, S., 28, 470; 30B, 446; 33A, 512; 33B, 123, 124, 491; 34A, 165, 351, 353; 34B, 524
Merlo, F., 34A, 142; 35A, 97, 124
Merrill, L., 29, 225
Merritt, L. L., 27, 952; 30B, 358, 34B, 124; 35B, 185
Merryman, R. G., 30A, 245
Mers, R. C., 29, 137
Merz, G., 28, 356 356
Messager, J. -C., 30B, 79
Messer, C. E., 29, 137, 139
Messien, P., 31A, 43
Messmer, G. G., 31B, 457; 32B, 464; 33A, 208
Mesureur, C., 30B, 264
Methfessel, S., 29, 140, 226
Methke, P. M., 26, 251
Métras, J. -C., 26, 742
Meulemans, R., 35B, 474
Meunier-Piret, J., 30B, 296, 303; 32A, 400; 32B, 348; 33B, 248; 34B, 422
Meussner, R. A., 30A, 180
Meuthen, B., 26, 667
Mewis, A., 34A, 145
Meyer, A., 29, 138
Meyer, E., 27, 963
Meyer, E. F., 30B, 136, 394
Meyer, G., 27, 384
Meyer, H. J., 28, 392; 34A, 348
Meyer, H. O. A., 35A, 258
Meyer, J. W., 26, 513
Meyer, L., 29, 433; 31A, 240; 32A, 147; 33A, 157
Meyerhoffer, A., 34B, 168; 35B, 208
Meyers, E. A., 30B, 129; 31B, 30; 119; 32B, 157; 33B, 259; 35B, 34, 728
Meyers, M.,34B, 42, 50
Meyrowitz, R., 26, 381; 27, 598, 603; 29, 437
Mez, H. C., 26, 317, 782; 27, 847
Michel, A., 26, 225, 409, 410; 27, 95, 208, 490, 491; 28, 356, 357; 29, 140, 435; 30A, 36, 173, 175, 178; 32A, 142; 33A, 86; 34A, 90, 286
Michel, C., 27, 888; 32A, 434
Michel, D. J., 30A, 98, 182; 31A, 44; 33A, 100
Michel, W., 26, 109
Michelet, A., 34A, 144
Middleton, R. M., 32A , 98
Midgley, H. G.,27, 633
Miesel, S. L., 32A, 509
Migchelsen, T., 27, 411; 30A, 389, 391;

32A, 149, 219; 32B, 33
Mighell, A., 32A, 355
Mighell, A. D., 29, 518; 31A, 171; 32B, 432; 34A, 319; 34B, 79, 568, 569, 572; 35B, 710, 721
Mihajlovic, A., 30A, 247
Mihara, T., 28, 339
Mijlhoff, F. C., 27, 482; 30A, 306 32A, 198, 220; 34A, 227, 298
Mikhailov, G. M., 28, 244
Mikolajčuk, A., 30A, 179
Mikolajčuk, A. G., 27, 80, 385
Milberg, M. E., 26, 486; 28, 293, 29, 384
Milburn, G. H. W., 32B, 329; 35B, 653
Miles, M. G., 33B, 321
Mil'kova, L. P., 27, 552
Mille, L. M., 30B, 444
Milledge, H. J., 26, 829; 30B, 34; 35B, 485
Millen, D. J., 26, 327; 33B, 541; 34A, 215, 220
Miller, A., 28, 425; 29, 759; 30A, 417
Miller, A. E., 29, 224
Miller, B. E., 34A, 143
Miller, B. S., 28, 291
Miller, C. E., 26, 414
Miller, C. K, 29, 617
Miller, D. S., 27, 154, 260
Miller, E., 30A, 180
Miller, J. F., 29, 141, 226
Miller, J. L., 27, 695, 698
Miller, J. R., 26, 539
Miller, N. E., 28, 314
Miller, P. H., 27, 63
Miller, R. C., 27, 498, 503
Miller, R. F., 26, 543
Miller, R. L., 26, 810; 29, 760
Mills, A. D., 27, 490, 542, 562
Mills, H. H., 26, 556; 28, 387, 416; 30B, 44, 233, 258
Mills, J. C., 33B, 320, 323
Mills, J. F. D., 28, 491
Mills, K. C., 26, 95
Mills, O. S., 26, 682; 27, 847; 28 605, 683; 30B, 393, 446; 32B, 352 357, 360; 33B, 336, 339, 341, 351, 365, 373; 34B, 423, 426, 427, 433, 471, 472; 35B, 578
Mills, R. L., 26, 285; 27, 391; 31A, 240; 35A, 129
Milton, C., 26, 304, 483; 29, 136, 434, 435
Milton, D. J., 26, 528; 27, 675
Milton, R. M., 26, 517
Minačeva, C. Kh., 34B, 481
Minačeva, L. Kh., 29, 380
Minasjanč, M. Kh., 30B, 443; 31B, 343; 33B, 330, 332, 340
Minc, R. S., 27, 27
Minceva-Stefanova, I., 29, 437
Mineev, D. A., 28, 314
Miner, W. N., 27, 28
Mingos, D. M. P., 35B, 856
Minichelli, D., 35B, 625
Minkin, J.A. , 27, 735; 30B, 28; 31B, 454; 32B, 26; 33B, 21, 151; 34B, 7, 18, 24
Minobe, M., 35B, 409
Minomura, S., 29, 433
Minsker, K. S., 26, 633
Miron, N.F., 35A, 115
Mironov, N.N., 35A, 195
Mirosnicenko, I.S., 30A, 244
Mirri, A. M., 26, 543
Miscenko, K. S., 27, 697; 28, 318
Miskell, C. F., 30A, 23, 174
Misovic, J. D., 32B, 98
Misra, N. K., 28, 682
Misra, S.,29,136
Mitchell, A. W., 29, 140, 226; 35A, 124
Mitchell, D., 34B, 64
Mitchell, D. J., 30B, 446; 34B, 45
Mitchell, D. W., 26, 309
Mitchell, J., 35B, 791
Mitchell, R. S., 26, 443, 455; 27, 431, 716; 28, 341, 357; 29, 437
Mitchell, S. B., 27, 804
Mitchell, T. P., 35B, 700
Mitra, G. B., 28, 682
Mitra, R. P., 34A, 281
Mitra, S. N., 27, 940
Mitrofanova, N. D., 31B, 409
Mitscher, L. A., 34B, 339
Mitschler, A., 32B, 444; 34A, 222
Mitsui, T., 26, 372; 28, 358
Mitsui, Y., 34B, 268
Mitsukata, T., 26, 630
Mittendorf, H., 26, 297; 30A, 180
Mitzlaff, M., 33B, 538, 539
Miyagawa, T., 27, 812
Miyahara, S., 26, 394; 29, 434, 435
Miyake, K., 28, 682
Miyatani, K., 33A, 235
Mizuno, J., 26, 318
Mizuno, K., 35B, 388
Mizushima, T., 32B, 432
Mjan, S. -P., 28, 338
Mjasnikova, K. P., 29, 226
Mjasnikova, K. R., 30A, 247
Mjasnikova, R. M., 29, 628; 30B, 446; 33B, 72
Mladeck, M. H., 26, 830; 29, 511
Mlenik, M. P., 29, 760
Mo, F., 33B, 28
Moats, W. L., 35B, 446
Möbius, H. -H., 29, 435
Močalov, S. S., 33B, 536
Močalova, N. I., 26, 550, 558
Mocci, G., 30B, 393
Mochizuki, T., 27, 806

Modern, E., 30A, 325
Moeller, T., 35A, 137; 35B, 234
Moeller, Y., 34B, 383
Moench, R. H., 29, 437
Moers, F., 33A, 236
Moews, P. C., 31A, 101
Moh, G. H., 29, 140
Mohanty, G. P., 27, 538; 29 438
Mohl, J. S., 27, 237, 481; 28, 341
Mokeeva, V. I., 28, 318; 29, 405, 410, 438; 31A, 225
Mokuolu, J. A. A., 30B, 278, 387
Moleva, V. A., 29, 433
Molin, J., 28, 292
Molinski, V. J., 27, 558
Mollard, P., 31A, 37
Mollendal, M., 35B, 903
Møller, C. K., 31A, 107, 110
Møllerud, R., 30A, 247
Molodjan, I. P., 29, 141, 226
Molodkin, A. K., 31A, 205
Momamy, F. A., 26, 760; 28, 668; 29, 450, 733
Momin, S. N., 33A, 448
Momiyama, K., 28, 338
Momoi, H., 29, 437
Monaghan, J. J., 34A, 155
Moncet, M. -C., 27, 427
Moncrief, J. W., 30B, 397; 31B, 258
Mondange, H., 26, 403
Mondet, S., 33A, 314
Money, R. K., 26, 3
Montagnoli, G., 33A, 512; 34A, 165
Montagu-Pollock, H. M., 26, 296
Montgomery, H., 26, 671; 28, 618; 29, 354, 567; 31A, 203, 204; 33B, 388; 34B, 507
Montgomery, P. W., 35A, 202
Montgomery, R., 32A, 343
Monti, H, 29, 224
Montmory, M. -C., 26, 400
Moody, J. W., 29, 141, 226
Mooney, R. W., 28, 200, 316
Mooney-Slater, R. C. L., 26, 464; 27, 588; 29, 88; 31A, 183
Moore, A. E., 35A, 378
Moore, F. H., 30B, 446; 31B, 392
Moore, M. J., 33A, 239
Moore, P. B., 29, 438; 30A, 395; 31A, 186, 223; 32A, 385, 511; 33A, 325, 421, 422, 423, 492, 502; 34A, 372, 385, 387; 35A, 337, 360, 443, 466
Mooser, E., 26, 147
Mootz, D., 26, 734, 736; 27, 352, 916; 30B, 125, 161; 32A, 128; 32B, 314; 33A, 406; 33B, 82, 285; 34A, 319, 321, 325; 34B, 93, 368, 372; 35A, 331; 35B, 40, 43, 342
Moras, D., 34A, 222
Morassi, R., 35B, 853
Morawetz, H., 27, 810
Morawietz, H., 27, 475
Moreau, J., 30A, 177
Moreau, R., 28, 291
Moreau-Colin, M. L., 33A, 249
Morehouse, S. M., 29, 621; 30B, 346; 31B, 353; 34B, 445
Morelle, I., 35B, 316
Morente, A., 26, 575
Moret, C., 34B, 387
Moret, C. L., 33B, 109
Morgan, C. H., 28, 683; 31B, 183; 32B, 272, 385, 438
Morgan, H. J., 31B, 428
Morgan, V., 26, 489
Morgante, S., 27, 1008
Morgenstern-Badarau, I., 28, 357; 29, 435
Mori, M., 26, 326
Mori, S., 26, 395
Mori, T., 31A, 112
Moriarty, J. L., 26, 235; 30A, 172
Moriarty, R. M., 30B, 397
Morimoto, C. N., 35B, 733
Morimoto, H., 28, 249
Morimoto, N., 26, 50, 137; 27, 182; 28, 20; 29, 43, 406; 31A, 40; 34A, 262, 365; 35A, 57, 475
Morin, F. J., 26, 254
Morino, Y., 26, 550, 588; 27, 719; 33B, 530; 34A, 172, 244; 35B, 900
Morisaki, M., 33B, 217
Morisi, E., 26, 341; 27, 653, 28, 356
Morita, H., 30B, 447
Morita, T., 30A, 179
Moritani, T., 34A, 172
Mornon, J. P., 31B, 96; 32B, 122; 35B, 48
Morosin, B., 26, 340, 667, 671; 28, 76; 29, 272, 543; 30A, 274, 332, 372; 30B, 332; 31A, 94, 95, 214; 32A, 184, 204, 328; 32B, 30, 324; 33A, 204; 34A, 307, 310; 34B, 507; 35A, 146, 322; 35B, 681
Morosoff, N., 27, 810
Morozov, A. N., 27, 492
Morris, D. G., 33B, 98
Morris, D. P., 26, 172; 27, 219; 29, 135
Morris, E. D., 34B, 376
Morris, M. C., 27, 723; 29, 141
Morris, R. C., 27, 634; 28, 340
Morris, R.H., 32B, 435
Morrow, J. C., 27, 670; 30B, 358
Morrow, N. L., 33A, 353
Morss, L. R., 34A, 209
Mortaillaro, L., 29, 760
Mortland, M. M., 27, 423
Morton, J. R., 29, 433
Moser, W., 26, 360, 475; 28, 315; 33A, 386
Mosevič, I. A., 30A, 248

Moshy, R. J., 29, 760, 761
Moskatov, K. A., 27, 812
Moss, A. A., 27, 520
Moss, K. C., 26, 304
Moss, S. C., 29, 299
Mossel, A., 29, 491
Mostad, A., 27, 792; 28, 513; 30B, 182; 33B, 52
Mostovaja, S. A., 30A, 174
Motherwell, W. D. S., 34B, 111, 159; 35B, 467
Motohasi, M., 33B, 447
Motz, J., 27, 242
Motzfeldt, T., 35B, 890
Moulin, N., 27, 513
Mouty, M. A., 34A, 200
Mouvet, M., 35B, 11
Mozzi, R. L., 26, 474; 27, 408, 546; 34A, 363
Mrose, M. E., 26, 440, 483; 27, 577, 599; 28, 339
Mšenskij, V. A., 29, 226
Mucker, K., 33A, 205; 34A, 174
Mucker, K. F., 31A, 240; 34A, 84
Muecke, T. W., 35B, 896
Mueller, H. H., 34A, 144
Mueller, H. J., 30A, 256
Mueller, M. H., 26, 271; 27, 187, 379; 28, 20, 41, 140; 30A, 404, 407; 31A, 179; 32A, 99; 33A, 13
Muenter, J. S., 28, 668
Muetterties, E. L., 28, 314
Mugnier, D., 27, 156
Mugnoli, A., 28, 687; 30B, 50, 373, 397, 444; 31B, 35, 92; 32B, 41, 42; 34B, 82; 35B, 93
von der Mühll, R., 34A, 266; 35A, 163
Mühlpfordt, W., 35A, 59
Muir, K. W., 33B, 103, 366; 34B, 613, 615; 35B, 535, 873
Muirhead, H., 26, 812
Mukherjee, A. B., 26, 525
Mukherjee, B., 34A, 142
Muldawer, L., 26, 178; 27, 107
Mulford, R. N. R., 26, 233; 30A, 246
Muljava, M. P., 27, 233
Mullen, J. G., 33A, 268
Müller, A., 33A, 387; 34A, 244
Müller, F. A., 34A, 103
Müller, G., 26, 505
Muller, H., 27, 483; 28, 315
Muller, J., 35A, 288
Muller, O., 29, 322; 33A, 270
Muller, R., 28, 338
Müller, U., 35B, 468
Müller, W., 32A, 27; 34A, 145; 35A, 24, 43, 124
Müller-Buschbaum, H., 30A, 309, 316; 31A, 125; 33A, 269, 340; 34A, 248, 253; 35A, 217, 230, 244, 245, 246
Mumme, W. G., 31A, 148; 32A, 266; 33A, 294, 306; 34A, 303; 35A, 277, 367, 444
Munch, R. H., 29, 434
Munnik, J., 26, 830; 28, 494
Munns, A. R. I., 35B, 283, 288
Munro, D. C., 29, 438
Murahahi, S., 26, 632, 27, 805, 812; 29, 760; 30B, 399
Murakami, Y., 31A, 44
Muralidharan, K. V., 33B, 384
Muramori, K., 26, 394
Muraski, A., 32A, 21
Murata, I., 30B, 399; 32B, 176
Murata, J., 34B, 88
Murata, T., 30B, 397
Murav'eva, I. V., 28, 338
Murayama, W., 34B, 260
Murdoch, J., 27, 558, 706
Murdock, C. C., 26, 271
Murmann, R. K., 32B, 428
Murray-Rust, P., 33B, 403, 451; 34B, 131
Murthy, M. R. L. N., 30A, 178, 180; 34A, 9
Murthy, N. S. S., 30A, 178, 180; 34A, 9
Murti, G. S. R. K., 30B, 447
Murty, B. V. R., 30B, 443, 447; 35A, 56, 124
Murty, C. K. R., 27, 630
Murty, T. S., 27, 408
Musatti, A., 30B, 397; 31B, 402; 33B, 59, 466, 479; 34A, 315; 34B, 189, 535; 35B, 725
Musil, F. J., 27, 347
Mustafaev, N. M., 30A, 270
Mutin, J. C., 35B, 15
Muto, A., 35B, 707
Muto, F., 34B, 529; 35B, 622
Muus, L. T., 27, 804, 1065
Muxart, R., 29, 434
Muzyka, G. M., 29, 436
Myer, G. H., 27, 257
Myers, E. A., 33B, 293, 294
Myers, J. R., 30A, 244
Myers, R. J., 28, 668
Myklebust, R. L., 27, 279
Mys'kiv, M. G., 34A, 143

Naake, H. J., 29, 226
Nabitovič, I. D., 27, 200
Nachman, J. F., 30A, 247
Nadol'skij, A. P., 29, 433
Naganna, C., 28, 342; 29, 437
Nagasaki, H., 29, 433
Nagasawa, A., 29, 140; 30A, 174; 31A, 38
Nagashima, N., 34B, 260
Nagashima, S., 26, 190
Nagler, E., 26, 466
Nagorsen, G., 27, 610

Náhlovská, Z., 35B, 889, 894
Náhlovský, B., 35B, 889, 894
Nahringbauer, I., 30B, 11, 13, 15; 32B, 22; 34B, 16, 349; 35B, 3
Naidu, S. V. N., 27, 430; 28, 290
Nakagawa, Y., 27, 270; 29, 140; 30A, 17, 173; 33A, 291
Nakahara, A., 34B, 529; 35B, 622, 727
Nakai, H., 32A, 218; 34B, 161, 171; 35B, 206, 403
Nakai, Y., 28, 682, 686; 29, 455; 34A, 101
Nakajima, S., 28, 12
Nakamura, D., 27, 463
Nakamura, K., 27, 873
Nakamura, T., 28, 412
Nakanishi, N., 29, 140
Nakano, T., 31B, 258
Nakao, A., 33B, 528
Nakatsu, K., 27, 830; 28, 683
Nakatzu, K., 34B, 485
Nalivkina, E. V., 26, 521
Namba, Y., 26, 781; 28, 477; 32B, 166; 35B, 41
Nametkin, N. S., 33B, 529
Namikawa, R., 31A, 58
Naqvi, R. R., 35B, 64
Narasimhayya, T., 35A, 399
Narath, A., 29, 272
Náray-Szabó, I., 26, 497; 28, 222, 338; 30A, 308, 320; 31A, 83
Nardelli, M., 27, 630, 631, 835, 837, 881, 949, 951; 28, 645; 30B, 46, 344, 360, 397, 447; 31B, 402; 32B, 411; 33B, 466, 483; 34A, 234, 315; 34B, 32, 186, 535; 35A, 28, 38, 725
Nardin, A., 34B, 520
Nardin, E., 35B, 625, 779, 796, 810
Narten, A.H., 33B, 528
Näsäkkälä, M., 35B, 674, 675
Näsänen, R., 26, 328; 30B, 447; 31A, 108
Nash, C. P., 34B, 543
Nasibov, I. O., 29, 135, 140, 141
Naslain, R., 35A, 39
Nasledov, D. N., 29, 136, 137
Nassau, K., 27, 542
Natarajan, R., 30B, 447
Nathans, R., 26, 15, 89; 27, 253; 29, 226
Natile, G, 35B, 868
Natsuma, N., 31B, 227
Natt, J. J., 32A, 343
Natta, G., 26, 317, 611, 633, 634, 635, 636, 651, 684; 27, 801; 29, 760; 32B, 44, 45, 46; 33B, 522
Naumov, V. A., 26, 298; 27, 589; 35B, 893, 894, 897, 898
Naumova, T. N., 28, 315
Navarro, Q. O., 35A, 197
Nawapong, P. C., 30A, 98, 176; 31A, 45, 53
Nawata, Y., 34B, 286
Naya, K., 35B, 386, 402
Nayak, V. K., 28, 339
Nazarenko, I. I., 33B, 529, 531
Neckel, A., 29, 139, 226
Nedumov, N. A., 26, 19, 112
Néel, L., 27, 251
Neelakantan, K., 35A, 357
Nefedorov, A. P., 30A, 180, 182; 31A, 68
Nefedov, E. I., 28, 339
Nefedov, O. M., 35B, 407
Negas, T., 28, 357
Negita, H., 33B, 528
Negreskul, V. V., 29, 226
Neidle, S., 35B, 312
le Neindre, B., 28, 315
Neira, E., 27, 1066
Nekasov, Ju. V., 34A, 382
Nelson, A. D., 32B, 270; 33B, 41
Nelson, B. G. A., 35A, 262
Nelson, N. J., 34B, 459
Nelson, N. R., 32B, 260
Nenno, S., 27, 156
Nereson, N. G., 27, 475; 33A, 53, 56, 255
Neronov, V. A., 34A, 142
Neronova, N. N. 26, 535; 29, 404
Neronova, N. V., 28, 276
Nespor, V. S., 29, 226
Nesterova, Ja. M., 31B, 46
Nesterova, O. P., 30A, 181, 247
Neubert, H. -S., 30A, 257
Neumann, H., 26, 483; 29, 434
Neuvonen, K. J., 28, 339
Nevitt, M. V., 27, 187, 234; 29, 140, 226
Newberg, D. W., 29, 437
Newkirk, H. W., 29, 296; 30A, 307
Newkirk, J. B., 26, 6
Newman, B. A., 33B, 87
Newman, P. C., 26, 149
Newnham, R. E., 26, 201, 372, 504; 27, 498; 28, 141, 170; 29, 239, 317, 318, 396; 30A, 418; 31A, 227; 32A, 267, 268, 306, 408; 34A, 254
Newton, M. G., 31B, 179; 32B, 170, 196; 34B, 194
Nguyen-Quy-Dao, 34A, 218
Nibbering, N., 33B, 131
Nice, J. P., 32B, 352, 357
Nicholas, A. F., 35B, 651
Nicholas, H. B., 35B, 648
Nichols, J. L., 26, 231, 244
Nickel, E. H., 27, 548; 28, 145, 339, 358; 29, 412
Nickels, W., 27, 465
Nickl, J., 32A, 143; 33A, 153

Nickl, J. J., 34A, 145
Nicklow, R. M., 28, 290
Nicolaieff, E., 26, 819
Niebsch, H. -H., 31A, 190
Nieder-Vahrenholz, H. G., 30A, 261
Niegisch, W. D., 35B, 885
Niehard, H., 35A, 3, 6
Niehues, K. -J., 30A, 261
van Niekerk, J. N., 26, 830
Nielsen, J., 31B, 362
Nielsen, R. E., 30A, 417
Niemiec, J., 26, 241, 432
Nigam, G. D., 30B, 447
Niggli, A., 26, 246; 30B, 104, 394, 395; 34B, 338
Niimura, N., 33B, 250
Niizeki, N., 28, 258; 29, 292, 434; 33A, 321
Nikitin, A. V., 27, 705; 28, 261; 29, 436; 31A, 218; 32A, 437
Nikolaeva, L. E., 35A, 445
Nikolic, P., 29, 223; 30A, 248
Nikolič, P. M., 30A, 246
Nikulesku, D., 30A, 179
Nilson, L., 34A, 404, 525
Nilsson, A., 32A, 246, 260; 33B, 36
Nilsson, B., 31B, 211, 218; 30B, 95, 208, 212, 221; 34B, 25
Nilsson, J. E., 33B, 535, 536
Nimrod, D. M., 33B, 278, 279
Nishi, M., 27, 810
Nishi, Y., 31B, 106; 33B, 218
Nishida, I., 32A, 51
Nishii, M., 26, 639
Nishikawa, M., 30B, 397; 33B, 223; 34B, 238, 287; 35B, 388
Nishikawa, T, 26, 354
Nishimura, T., 30A, 415
Nishimura, Y., 26, 632
Nishiyama, Z., 27, 156; 28, 6
Nistor, N., 30A, 179
Nitsche, R., 27, 246; 29, 138; 30A, 179, 181
Nitta, I., 26, 661, 698; 28, 365, 682, 685; 29, 605; 30B, 397, 398; 31B, 194, 264; 32B, 83; 33B, 223; 35B, 44, 171, 367, 368, 386, 388, 402, 765, 768
Nitzsche, H. G., 28, 338; 29, 437
Nixon, D. E., 31A, 240
Nixon, E. R., 29 515
Noakes, D. E., 33A, 168
Nobbs, C. L., 31B, 103
Nobler, C., 35B, 226
Nockolds, C., 27, 997
Nockolds, C. E., 35B, 759
Noda, T., 26, 286; 27, 395; 35B, 402
Noguchi, T., 27, 839; 33B, 26
Noguès, M., 33A, 285
Noland, W. E., 27, 973
Nominé, M., 33B, 528
Nomura, S., 26, 420; 27, 508; 29, 396
Nomura, T., 31B, 418; 32B, 431; 34B, 491
Noordik, J. H., 32B, 68
Nord, A. G, 33A, 394
Nordheim, R., 26, 217
Nordine, P. C., 30A, 28, 174
Nordman, C. E., 27, 637, 763; 28, 596, 598, 683; 30B, 115, 241; 34B, 376, 377
Nordström, B., 35B, 887
Norin, R., 28, 123; 31A, 122; 34A, 241
Norling, B. K., 31A, 50
Nørlund-Christensen, A., 34A, 236
Norman, N., 26, 559, 560, 642; 27, 483, 799; 28, 449; 32B, 40
Norment, H. G., 26, 616; 27, 909, 911; 30B, 119
Norrby, L. -J., 33A, 141; 34A, 110; 35A, 107, 207
Norrestam, P., 35B, 195
Norrestam, R., 32A, 256; 33A, 264; 33B, 168; 34B, 154, 156, 250
Norreys, J. J., 26, 271
North, A. C. T., 26, 812, 827; 27, 1058, 1059; 35B, 888
Northolt, M. G., 32B, 300; 33B, 4, 120; 35B, 182
Norton, C. J., 29, 136
Norton, D. A., 27, 1038; 28, 536, 683; 29, 750, 751; 30B, 444, 447, 33B, 192, 193, 194, 201, 203; 34B, 291, 293, 294, 296, 300, 301, 358, 545; 35B, 332, 335
Norton, J. L., 26, 254
Norton, J. T., 29, 84
Norton, L. J., 27, 360; 30A, 173
Nosova, G. I., 27, 371
Nosova, J. I., 26, 264
Novak, C., 32B, 251
Novak, F., 29, 436
Novak, G.A., 35A, 258
Novick, D.T., 27, 307
Novick, S., 32B, 301
Novickaja, G. N., 35A, 417
Novikov, G. V., 33A, 324
Novikov, Ju. N., 35B, 905
de Novion, C. -H., 31A, 34, 35
Novoselova, A. V., 26, 432; 27, 215; 29, 225
Novy, V. F., 26, 118, 142, 143
Nowacki, W., 26, 330, 426, 738, 783, 830; 27, 81; 29, 23, 68, 511; 30A, 174, 355, 357; 31A, 19,164; 32A, 25, 227, 229, 230; 32B, 370; 33A, 37, 148; 34A, 17, 23, 28, 341; 35A, 11, 15, 30, 143, 324; 35B, 354
Nowotny, H., 26, 9, 10, 11, 18, 21, 25, 58, 68, 81, 91, 103, 124, 156, 398, 400; 27,11, 30, 54, 70, 105, 159, 213, 233, 375, 500, 541; 28, 3, 12, 317, 339; 29, 64, 85, 135,

136, 137, 138, 139, 226; 30A, 35, 172, 174, 175, 181, 244, 245, 246; 31A, 27; 32A, 24, 44, 46, 68, 70; 99, 141, 494, 507; 33A, 84, 130, 153, 467, 508; 34A, 39, 144, 259, 368; 35A, 263, 311, 318
Nozakura, S., 30B, 399
Nozick, Ju. Z., 26, 301, 411; 27, 496
Nozik, J. Z., 35B, 259
Nozoe, S., 33B, 217; 34B,318
Nuffield, E. W., 30A, 177
Nugent, L. J.,27, 735
Nunzi, A., 35A, 333, 439
Nuttall, P. M., 27, 690; 29, 72, 403
Nyberg, B., 30A, 380; 33A, 384
Nyborg, J., 35A, 186
Nyburg, S. C., 26, 830; 28, 491; 29, 571; 33A, 211; 33B, 524; 34B, 239, 241; 35B, 86
Nygaard, L., 33B, 540
Nyholm, R. S., 26, 687; 27, 822
Nylund, A., 31A, 56, 58
Nyman, F., 33B, 347

Ober, H., 27, 345
Oberhammer, H., 35B, 890
Oberhansli, W., 30B, 322
Oberhänsli, W. E., 30B, 300, 326; 32B, 243; 35B, 172
Oberteuffer, J. A., 35A, 78
Obodovskaja, A.E., 35B, 627
O'Brien, E.J., 32B, 209; 35B, 887
O'Brien, R.J., 28, 685; 29, 625
O'Brien, T. A., 32B, 345; 33B, 357; 34B, 464; 35B, 530, 532
Obrowski, W., 26, 73; 27, 149, 314
Ocenaskova, D., 31B, 59; 32B, 71
O'Connell, A. M., 31B, 229; 32B, 93, 183; 33B, 6, 34B, 346
O'Connor, B. H., 30B, 66; 31A, 200; 31B, 404, 408
O'Connor, J. E., 32B, 356; 34B, 523
O'Connor, T. E., 27, 408
Oda, K., 35B, 373, 382, 521
Oda, T., 26, 781; 28, 477; 32B, 166; 35B,41
O'Daniel, H., 32A, 243
Oden. L. L., 29, 226
O'Donnell, E. A., 28, 685; 32B, 246
Oertel, G., 26, 62
Oesterreicher, H., 29, 138
Ogawa, I., 28, 400
Ogawa, K., 28, 685
Ogawa, S., 26, 82, 306; 27, 176, 180; 30A, 32, 174, 246; 31A, 37; 33A, 75
Ogawa, Y., 26, 809
Ogihara, Y., 30B, 447; 33B, 235
Ogniben, G., 29, 436
Ogren, H. B., 28, 683
Ogren, J. R., 30A, 80, 180
Oh, Y. L., 31B, 232; 33B, 214
Ohashi, Y., 32B, 282; 33B, 250; 35B, 76
Ohizumi, C., 35B, 883
Ohloff, G., 34B, 308
Ohmasa, M., 33A, 148; 35A, 30
Ohnishi, H., 26, 411
Ohnishi, Y., 30B, 540
Ohnsorge, H., 33A, 236
Ohoyama, T., 26, 149; 27, 15, 162; 30A, 246
Ohrt, J., 35B, 254
Ohrt, J. M., 28, 683; 29, 751; 30B, 447; 33B, 194; 34B, 296
Ohta, K., 28, 65; 31A, 77
Oijen, J.W. L., van, 31B, 76
Ok, H. N., 33A, 268
Oka, T., 28, 668
Oka, Y., 31A, 112; 32A, 223
Okada, K., 34B, 119
Okamoto, F., 33A, 235
Okamato, T., 35B, 592
Okamoto, Y., 33B, 223
Okamura, K., 31A, 37; 33A, 75
Okamura, S., 26, 639; 27, 810
Okamura, W. H., 35B, 807
Okawara, R., 30B, 446
Okaya, Y., 27, 838; 29, 593, 740; 30B, 62, 132; 31A, 133; 31B, 5, 128, 153, 187, 301; 32B, 81, 146; 34B, 36, 114, 207
Okazaki, A., 26, 179, 306
Okazaki, C., 26, 395
O'Keefe, T. J., 33A, 153
O'Keeffe, M., 26, 409
Okeksiv, G. I., 35A, 55
Okita, Y., 33B, 521
Okonnikov, E. G., 30A, 248
Okuda, K., 29, 760
Okuda, S., 34B, 318
Okuda, T., 33B, 528; 35B, 259
Olazcuaga, R., 33A, 310
Olcese, G. L., 26, 255; 29, 140, 226; 34A, 62, 146
Olcott, R. J., 34B, 558
Oldham, J. W. H., 28, 686
Olekhnovič, N. M., 27, 3
Oleksiv, G. I., 29, 138; 30A, 177
Oleksyn, B., 34A, 229
Oleś, A., 29, 139, 140; 30A, 174, 179; 31A, 163
Olf, H. G., 31B, 37
Olie, K., 35A, 151
Ollivier, G., 35A, 228
Olofsson, O, 30A, 45, 172; 35A, 41, 87, 89, 124
Olovsson, I., 26, 433; 29, 233; 30A, 260; 31A, 128; 32A, 181; 33A, 229, 364, 449; 33B, 25, 36, 38; 34B, 128; 35A, 365
Ol'šanskaja, E. Ja., 27, 285
Olsen, B., 35B, 788
Olsen, C. E., 27, 197, 475
Olsen, D. H., 32A,484
Olsen, K., 26, 701
Olson, D. H., 28, 604; 33A, 497; 34A,

370
Olthof, R., 30A, 353, 389, 391; 31A, 214; 34A, 196
Omae, I., 33B, 313
Omarov, T. G., 28, 314
Omel'čenko, J. A., 30B, 49
Omel'čenko, Y. A., 32B, 61
Omloo, W. P. F. A. M., 33A, 61; 35A, 124
Omori, K.,28, 357
Omura, S., 35B, 402
Ondik, H. E., 27, 594
Ondik, H. M., 26, 455; 29, 365; 30A, 387
Onillon, M., 27, 518
Onken, H., 29, 251; 30A, 424
Ono, K., 34B, 441
Ono, S., 29, 299; 35A, 93, 124
Ono, Y., 28, 318
Onorato, E., 28, 242
Onose, K., 27, 15
Onuma, S., 35B, 606
Onyszchuk, K., 26, 539
Ooi, S., 28, 615; 32B, 457; 34B, 498; 35B, 591
van Ooijen, D. J., 27, 302; 30A, 174
Oonishi, I., 35B, 622, 678, 727
Oonishi, J., 34B, 529
Opalovskij, A. A., 30A, 180
Oppegard, A. L., 26, 354
Oppengejm, V. D., 33B, 529
Orcutt, D., 26, 291, 594; 28, 291
Ordway, F., 26, 511; 30A, 423
O'Reilly, J. M., 26, 587
O'Reilly, W., 33A, 267
Organ, T. D., 35B, 333
Organova, N. I., 26, 536; 27, 689
Orii, S., 28, 412
Orioli, P. L., 27, 873; 28, 686; 29, 568, 598; 30B, 394, 398; 31B, 389, 393, 400; 32B, 383, 406; 33B, 428, 441; 34B, 549; 35B, 850
Orlov, Ju. L., 26, 537
Orlova, G. M., 26, 48
Orlova, V. T. 28, 340
Orr, R. L., 27, 349
Orr, R. S., 26, 629
Orsini, P. G., 28, 339
Osaki, K., 28, 634, 682, 686; 29, 455, 609, 660; 30B, 443; 31B, 258; 33B, 223; 35B, 259, 400
Osaki, M., 34B, 344
Osamura, K., 31A, 44
Osborne, D. W., 26, 167
Osborne, V., 26, 529
Oshio, H., 31B, 264
Osika, L. M., 29, 223
Osipowa, O. P., 28, 682
Osmanov, E. O., 29, 136; 30A, 173; 33A, 126
Ostapčenko, E. P., 26, 391
Ostertag, W., 30A, 177; 32A, 59
Osugi, J., 31A, 58
O'Sullivan, K., 32A, 334; 35A, 191
Oswald, H. R., 26, 317, 327, 328, 358, 478; 27, 629; 29, 250; 32A, 285; 33A, 153; 35A, 325
Ota, H., 28, 684
Oteng, R., 32A, 508
Otsubo, Y., 26, 395
Ott, H., 31B, 251
Otte, H. M., 26, 3; 31A, 240
Ottenburgs, R., 30A, 177
Otto, K., 27, 194
Otvos, W., 27, 1064
Ovčinnikova, T. L., 27, 511
Overton, K. H., 31B, 230
Owen, C. V., 26, 275
Owen, E. A., 29, 225; 30A, 245
Owen, T. B., 30B, 112
Owen, W. S., 26, 100
Owens, J. P., 32A, 130
Owston, P. G., 27, 846, 978; 28, 657, 660; 32B, 437, 469, 474, 475; 33B, 347, 501, 503, 515; 34B, 437, 606; 35B, 97, 526, 792, 814, 848
Oxley, D. P., 27, 41
van Oyen, J. W. L., 33B, 147
Ozeki, K., 34B, 237
Ozerov, R. P., 26, 285; 27, 155, 393, 506, 643; 28, 207; 30A, 364; 31A, 41; 33A, 326, 450; 35A, 387
Ozol, Ja., 27, 884; 29, 387, 438
Ozol, Ja. K., 30B, 447
Ozol's, J., 28, 166
Ozolin'š, G. V., 27, 50
Ozols, J., 33A, 445

Paalman, H. H., 27, 721, 722
Paasche, D. G., 30A, 246
Pabst, A., 26, 378, 520; 28, 340
Paccard, D., 32A, 106; 34A, 105; 35A, 80
Pachler, K., 28, 682
Paddock, N. L., 33B, 281
Padera, K., 29, 437
Paderno, Ju. B., 26, 66, 81
Padilla, J., 28, 683
Padiou, J., 35A, 28
Padlo, I., 27, 477
Padmanabhan, V. M., 28, 79; 30B, 21; 32A, 197, 325, 330; 33B, 384; 34A, 212
Padwa, A., 33B, 99
Paff, R. J., 27, 110
Paff, R. R. 29, 225
Pailleret, P., 31A, 146; 32A, 276, 305
Pajunen, A., 32B, 379; 34B, 488, 489, 494, 555; 35B, 676, 677, 678
Pakhomov, V. I., 27, 643, 824; 28, 662, 682; 29, 628; 31B, 341; 32A, 509; 33B, 320; 34A, 168

Pál, L., 29, 225
Palatnik, L. S., 26, 42, 63, 133; 27, 369; 29, 225; 30A, 177, 178, 245, 248
Palenik, G. J., 27, 943; 28, 56, 453; 29, 50, 246, 473, 536, 612, 614; 30B, 124; 31B, 317; 32A, 283, 284; 33B, 48; 34B, 479, 503, 585; 35B, 166, 550, 840
Palenzona, A., 29, 140; 31A, 42; 34A, 143, 146; 35A, 103
Palevsky, H., 27, 421
Palkina, K. K., 26, 208; 30A, 175, 176
Palm, J. H., 28, 570; 31B, 69; 32B, 117
Palmer, H. T., 34B, 52
Palmer, K. J., 28, 683; 30B, 446; 31B, 157; 33B, 101
Palmer, P. E., 27, 368
Palmer, R. A., 27, 1059; 34B, 52, 254
Palmieri, C., 34B, 186; 35B, 627
Palomar, M. L., 26, 571
Pamplin, B. R., 27, 244; 30A, 244
Pan, V. M., 26, 112
Panagiotopoulos, N. Ch., 35A, 385
Panattoni, C., 26, 679; 27, 872; 28, 424, 598, 655, 683; 29, 582; 31B, 332; 32B, 397; 34B, 517, 533, 604, 35B, 464, 859, 860, 868
Panday, P. K., 30A, 173; 32A, 80; 33A, 124; 34A, 82, 142
Pandey, S. N., 27, 747
Pandraud, H., 26, 783, 830; 27, 983
Panke, D., 33B, 260; 35B, 422
Pankevič, Z. V., 30A, 181
Pannell, J., 26, 327
Pannetier, G., 27, 169; 29, 435; 31A, 198; 33A, 367, 374
Pannhorst, W., 35A, 295
Panov, V. S., 26, 97; 30A, 248
Panson, A. J., 29, 225
Pant, A. K., 27, 898; 30B, 57; 31A, 207; 32A, 425; 33A, 461
Pant, L. M., 26, 551; 28, 384; 29, 471, 481; 30B, 34, 207, 399; 31B, 61; 32B, 76; 33B, 21; 34B, 57
Panteleimonov, L. A., 29, 140; 30A, 181, 247
Pantzer, G., 29, 137
Papageorgakis, J., 27, 688
Papamantellos, P., 26, 398; 27, 541; 29, 438; 33A, 262
de Pape, R., 27, 459; 29, 433
Papike, J. J., 30A, 436; 31A, 231; 32A, 468; 33A, 459, 495; 35A, 460
Papps, M. D., 34A, 213
Pardo, M. -P., 27, 424; 29, 135; 30A, 176, 177; 32A, 104; 34A, 144
Parenti, M. A., 26, 719
Park, D. N. Y., 33B, 427
Park, J. J., 34A, 279; 35B, 751
Park, Y. S., 30A, 181
Parker, R. L., 27, 633
Parker, W., 35B, 386
Parker, W. L. 33B, 235
Parkes, A. S., 28, 52, 401; 29, 515
Parks, A. T., 30B, 405
Paromova, M. V., 33A, 341
Parpiev, N. A., 27, 1003
Parr, J. G., 26, 28; 27, 296; 29, 223
Parrish, W., 29, 225
Parrott, J. E., 27, 47
Parry, G. S., 27, 415; 31A, 181, 240
Parry, R. W., 28, 598
Parselle, M. H., 30A, 247
Parsons, F. C., 34A, 179
Parsons, J. L., 26, 486
Parthasarathy, R., 30B, 398; 31B, 188, 193; 32B, 216; 34B, 267; 35B, 12, 149, 254
Parthé, E., 27, 141, 341; 28, 9, 15, 27; 29, 53, 61, 84, 139; 30A, 39 173, 175, 176; 31A, 17; 32A, 83, 120; 33A, 21, 135; 34A, 34, 47, 143, 145; 35A, 110
Paruz, H., 27, 326
Pascard, R., 29, 136; 30A, 362; 33A, 58
Pascard-Billy, C., 26, 744; 27, 927, 928; 28, 683; 30B, 397; 31B, 240; 30A, 360, 362; 33B, 228; 35B, 252
Pascher, I., 31B, 180
Pascher, J., 34B, 346
Pasečnik, Ju. V., 26, 615
Pašinkin, A. S., 27, 109, 215
Pasnak, M., 30A, 173
Pasternak, R. A., 26, 784; 27, 401
Pašutin, V. P., 29, 314
Patel, C. C., 28, 317; 29, 436; 30B, 446, 448
Patel, C. G., 35B, 526
Patel, M. B., 30B, 233
Patel, M. M., 34A, 180
Patel, P. D., 34A, 180
Paternotte, C., 29, 739
Paterson, A. R., 30B, 448
Paterson, R., 27, 899
Paterson, W. G., 26, 539
Paton, M., 27, 997
Paton, M. G., 30A, 309; 32B, 265
Patrie, M., 26, 88, 277; 27, 327; 29, 135, 341; 34A, 146
Patscheke, E., 33A, 309
Patterman, S. P., 35B, 474
Patterson, A. L., 27, 735; 28, 385; 30B, 28; 31B, 454; 32B, 26; 33B, 21, 151; 34B, 7, 24
Patterson, J. H., 26, 476; 27, 569
Patterson, W. R., 30A, 247
Pattiaratchi, D. B., 32A, 477
Pätzold, H., 27, 478
Paul, G. H., 29, 450
Paul, G. L., 35B, 304

Paul, I. C., 26, 830; 27, 1030, 28, 564; 30B, 243, 395, 398; 31B, 138, 368; 32A, 167, 509; 32B, 167, 170, 174, 196; 33B, 111, 142, 190, 196, 368, 409, 437; 34B, 87, 193, 194, 208, 211, 343, 383, 504, 532; 35A, 137; 35B, 115, 170, 223, 234, 334, 406, 569
Paul, S., 32B, 6
Paulevé, J., 27, 251
Pauli, G. H., 33B, 537
Pauling, L., 33A, 243; 35A, 129
Pauling, P., 26, 673, 687; 29, 570; 30A, 370; 30B, 339, 398; 33A, 243; 31B, 314
Pauling, P. J., 27, 822; 30B, 393; 34B, 495; 35B, 869
Paulson, P. L., 30B, 393
Paulus, E. F., 32B, 360; 33B, 351; 34B, 408, 461
Paulus, H., 34B, 553
Pauly, H., 27, 564; 33A, 98
Pausewang, G., 32A, 508; 34A, 216
Pautard, F. G. E., 26, 826; 30B, 443
Pauthenet, R., 26, 389; 27, 256, 522, 523; 31A, 133, 35A, 362
Pavin, J. L., 29, 317
Pavljukov, A. A., 30A, 172
Pavlovic, A. S., 27, 507
Pavolič, S., 29, 436
Pawar, R., 27, 381; 30A, 248
Pawar, R. R., 29, 225
Pawlek, F., 30A, 173
Pawley, G. S., 31A, 77; 34B, 97
Payne, D. S., 30B, 278, 387
Payne, N. C., 34B, 537, 618; 35B, 600
Paxton, H. W., 26, 99
Peacock, R. D., 26, 311; 27, 445, 464; 29, 260
Peacor, D. R., 27, 699, 701, 712; 28, 246, 258, 340; 29, 438; 30A, 365; 32A, 455; 33A, 454, 456, 458, 504; 34A, 383; 35A, 454
Pearce, J. H., 26, 228, 244
Pearse, R. J., 30A, 248
Pearson, E. F., 34A, 151, 171
Pearson, J., 29, 317
Pearson, W. B., 29, 139, 224, 226; 30A, 246, 247, 248
Pease, L. F., 29, 224
Pebler, A., 27, 235
Peča, L., 33A, 445
Pech, L. Ja., 30B, 447
Pechin, W. H., 29, 138, 225
Pedersen, B., 26, 692; 28, 685; 29, 462, 545; 30B, 398, 447; 33A, 433; 34A, 131, 146
Pedersen, B. F, 27, 523; 28, 685; 29, 462, 463; 30B, 24, 398, 447; 32B, 21, 40, 96; 33B, 11
Pedersen, G., 26, 176
Pedersen, T., 33B, 540
Pederson B.J., 34B, 15
Ped'ko, A. V., 30A, 179
Pedone, C., 28, 315, 428, 682; 29, 468; 30B, 169, 395; 32B, 16, 305, 343; 33B, 131, 407; 34B, 90, 91, 129, 440; 35B, 6, 97, 106, 217
Peerdeman, A. F., 30B, 26, 398; 33B, 176; 35B, 374, 492
Peiren, M. A., 35B, 154
Peisker, E., 31A, 3
Pej-Shan', C., 26, 484
Pelizzi, G., 33B, 483; 34B, 206; 35B, 67, 235
Pelleg, J., 30A, 177
Pellinghelli, M. A., 35A, 397
Pells, G. P., 29, 135, 223
Pelsmaekers, J., 26, 467
Pelter, A., 28, 683
Pen, C. -C., 28, 149, 158, 270, 274, 339
Pence, D. T., 27, 729
Penfold, B. R., 26, 600; 28, 87; 29, 332; 30A, 354; 355; 30B, 97; 31A, 98, 110; 31B, 319, 431; 32A, 192; 33B, 314; 34A, 213; 34B, 107, 410; 35B, 502, 504, 846, 866
Peng, C. -C., 31A, 236
Peng, C. -J., 27, 599, 703
Penneman, R. A., 33A, 195; 34A, 194; 35A, 166
Penney, G. J., 35A, 157
Penta, M., 28, 242
Pepinsky, R., 27, 838
Perales, A., 34B, 345
Percev, N. N., 29, 437
Perego, G., 26, 611, 685; 28, 597, 600; 29, 760; 31B, 37; 35B, 883
Percival, A., 26, 369
Perel, J., 27, 511
Peretti, E., 29, 437
Perez, G., 30A, 441; 34A, 146
Perez-Rodriguez, M., 34B, 33
Perez y Jorba, M., 26, 364,403; 27, 535, 542; 29, 434
Perio, P., 26, 366; 27, 179; 33A, 184
Perkins, H. K., 33A, 294
Perl'mutter, B. L., 35B, 407
Perloff, A., 28, 164; 29, 389, 392, 524; 30A, 254, 413; 30B, 272; 31A, 134, 169, 171; 31B, 427; 32A, 497; 35A, 419
Permingeat, F., 28, 356; 31A, 194, 217; 33A, 381
Perrault, G., 33A, 190
Perrin, M., 30B, 394
Perrin, R., 30B, 394
Perrotta, A. J., 28, 357; 29, 399; 30A, 425; 32A, 290, 488; 34A, 3
Persson, E., 33A, 136
Pérucaud, M. -C., 33B, 301

Perutz, M. F., 26, 812; 28, 684; 31B, 269
Peshev, P., 29, 136
Pessall, N., 26, 167
Petch, H. E., 29, 409; 31A, 172
Petcher, T. J., 34A, 180
Peterlin, A., 31B, 37
Peters, C. R., 29, 384; 34A, 250
Peters, E. H., 30A, 176
Peters, J. M., 28, 318
Peterse, W. J.A. M., 28, 316
Petersen, C. S., 27, 790; 28, 684; 29, 489
Petersen, D. R., 26, 277
Peterson, D. T., 26, 272; 29, 137
Peterson, J. R., 33A, 175; 35A, 151
Peterson, S. W., 26, 554; 29, 62; 30B, 448; 31A, 178; 32A, 197; 32B, 28; 33A, 252, 365; 398; 33B, 23
Peterson, W. S., 34A, 204
Petraccone, V., 34B, 96; 35B, 884
Petrie, M., 26, 344
Petrov, I. I., 33A, 19, 30; 34A, 32; 35A, 31, 33
Petrov, Ju. M., 26, 641; 29, 760
Petru, F., 30B, 445
Petrukhin, B. S., 35B, 886
Petrukhin, O. M., 34B, 497; 35B, 614
Petrunin, V. F., 31A, 68; 35A, 133
Petrunina, A. A., 34A, 12; 35A, 15
Petrusevič, V. A., 26, 173
Pettersen, R. C., 33B, 215; 34B, 279
Pettit, L. D., 35B, 81
Pettit, R., 31B, 369
Petuškova, S. M., 29, 435
Petz, J. I., 26, 431; 29, 233
Petzow, G., 26, 30; 27, 20; 29, 135, 223; 30A, 245; 33A, 153
Petzow, I. G., 26, 83
Peyronel, G., 27, 844; 32B, 414
Pezerat, H., 29, 434; 32A, 301
Pfeifer, H. -U., 33A, 87, 153
Pfeifer, N. -U., 31A, 65
Pfister, H., 26, 49, 227; 30A, 248
Pflug, J. L., 35B, 427
Pfluger, C. E., 26, 711; 27, 957; 30B, 340; 31B, 50, 383; 32B, 387; 34B, 504; 35B, 619
Pharr, R. F., 26, 443
Philippot, E., 35A, 65, 345, 483
von Philipsborn, H., 29, 141
Phillips, B., 29, 434
Phillips, D. C., 26, 811; 27, 1058, 1059; 28, 686
Phillips, J. H., 29, 224
Phillips, T., 33B, 390; 34B, 510
Phillips, W. L., 26, 299; 29, 223
Piatkowiak, E., 26, 463
Piccinin, A., 32A, 236
Pichat, P., 29, 433
Pickart, S. J., 26, 15, 189; 27, 253, 496; 30A, 181
Picon, M., 29, 433
Pidcock, A., 35B, 834
Pidgeon, L. M., 28, 314
Piekarski, C., 26, 366
Pierce, J. W., 35A, 179
Pierce, L., 26, 587; 28, 668
Pieret, A. F., 35B, 325
de Pieri, R., 31A, 219
Piermarini, G. J., 27, 425; 28, 292; 29, 138; 34B, 79
Pierpont, C. G., 35B, 826
Pierre, J., 33A, 153
Pierron, E. D., 29, 434; 31A, 77; 31B, 467
Pierrot, M., 28, 317; 29, 433; 31A, 114; 33A, 242
Pierrot, R., 26, 470, 522; 28, 340; 29, 436; 33A, 393
Pietraszko, A., 33A, 409; 34A, 15
Pietrokowsky, P., 26, 28; 27, 222; 30A, 86, 181
Pignataro, E. H., 27, 1059
Pignedoli, A., 27, 844; 32B, 414
Pignolet, L. H., 34B, 516
Pillai, S. P. 26, 829
Pilling, D., 29, 754
Pilström, G., 26, 211
Pinaeva-Strelina, M. M., 28, 341
Pinch, H. L., 35A, 44
Pinder, P. M., 34A, 163
Pindzola, D., 26, 558
Pinella, A., 33A, 44
Pings, C. J., 27, 721
Pinkerton, J. M. H., 32B, 254
Pinkerton, M., 27, 1059; 35B, 754
Pinot, M., 31A, 33
Pinsker, Z. G., 26, 37, 197, 220, 250, 446; 27, 295, 344; 28, 32, 34, 206; 29, 14, 30, 48, 87, 138, 311, 319; 30A, 61, 92, 93, 172, 179, 446; 31A, 67, 91; 32A, 262; 33A, 19; 34A, 79
Pioda, L. A. R., 32B, 242
Pipal, J. R., 35B, 715
Piret, P., 26, 305, 785; 27, 427, 760, 761, 954; 28, 404, 575; 29, 739; 30B, 264, 296, 303; 31B, 271, 273; 32A, 400; 32B, 267, 271, 348, 437; 34B, 422, 451; 35A, 195; 35B, 474
Piret-Meunier, J., 27, 583; 28, 277, 278
Piron, J., 32B, 437; 34B, 422
Pistroius, C. W. F. T., 26, 392, 436, 437, 439, 475, 520, 521; 27, 609; 28, 317, 357; 29, 357, 435, 438; 33A, 171, 253
Pistorius, M. C., 27, 609
Pitman, I. H., 35B, 164
Pitsch, W., 27, 124
Pizzarello, F. A., 27, 51
Pjatenko, Ju. A., 26, 379, 531; 27, 473, 699; 29, 319, 411; 32A, 166, 417, 433, 510, 511, 33A, 464; 35A, 309, 451

Pjatigorskaja, L. I., 29, 434
la Placa, S. J., 28, 685
Plakhina, A. M., 32A, 141
Plakhov, G. F., 35A, 228
Planje, M. C., 30B, 405
v. Planta, C., 26, 808
Plass, H., 33B, 296, 324
Plate, N. A., 29, 759
Pleass, C. M., 27, 48, 828
Plecko, R. L., 26, 48
Pletcher, J., 31B, 181; 35B, 239, 416
Plettinger, H. A., 26, 322, 391; 28, 157, 162, 166, 168, 171; 30A, 420
Plieth, K., 26, 727, 734; 30A, 182, 395; 30B, 331, 393, 396; 32A, 389; 32B, 43; 35B, 112
Plovnick, R. H., 33A, 153
Plumier, R., 27, 439, 453, 490; 28, 357; 30A, 177
Plyler, E. K., 27, 732
Plymaté, D. L., 31B, 458
Plyasova, L. M., 33A, 331
Pobedimskaja, E.A., 29, 27; 31A, 157; 32A, 440; 33A, 177, 337, 347; 34A, 386; 35A, 226, 228, 312
Podberezskaja, N.V., 33A, 518
Podder, A., 30B, 447
Poix, P., 26, 409, 410; 27, 490, 491; 28, 356, 357; 29, 435; 31A, 161; 32A, 509; 33A, 285, 312, 356; 34A, 286
Pokrovskij, N.V., 28, 340
Poljak, R. J., 26, 665; 27, 1058, 1059
Poljakova, V. P., 27, 313, 314, 338
Poljanskaja, T. M., 34A, 286; 35A, 239, 261
Poljnova, T. N., 35B, 602
Pollack, S. S., 27, 800; 29, 438, 759
Pollard, D. R., 30B, 394
Pollick, P. J., 32B, 455
Polunina, G. P. 26, 403, 405
Poluškina, A. P., 28, 356
Polynova, T. N., 26, 729; 28, 686; 30B, 51, 447; 31B, 326, 407; 32B, 321; 33B, 410
Pompa, F., 27, 412; 28, 495, 659; 34A, 201
Pond, J. S., 32B, 430
Ponjatovskij, E. G., 27, 366; 29, 226
Ponomarev, V. I., 33A, 347; 34A, 382; 35A, 290, 291, 312
Ponsioen, J. C. M. A., 32A, 142
Pontchour, C. -O., 34A, 143
Poole, D. M., 26, 231, 244
Poor, E. W. 28, 37
Popma, T. J. A., 35A, 79, 124
Popov, A. I., 32B, 270
Popov, Ju. G., 27, 484
Popova, E. G., 30B, 447
Popova, E. P., 29, 435
Popova, N. M., 30A, 175
Popova, S. V., 28, 72, 356; 29, 138; 33A, 45; 35A, 124
Popova, Z. V., 30A, 176; 30B, 293
Popovkin, B. A., 27, 609; 28, 317
Popp, K., 28, 338
Poppe, H., 30B, 398
Poppinger, D., 34A, 143
Poppleton, B., 31B, 169
Poppleton, B. J., 33B, 232; 35B, 258
Pops, H., 29, 223
Poquet, E., 28, 289
Porai-Koshits, A., 30A, 299
Poraj-Košic, A., 30A,299
Poraj-Košic, E. A., 28, 293, 316
Poraj-Košic, M.A., 26, 400, 478, 497, 671, 729; 27, 552, 665, 816, 820; 28, 109, 114, 214, 338, 686; 30B, 51, 343, 399, 447; 31A, 119, 205; 31B, 46, 326, 381, 407, 409, 417, 424; 32A, 509, 510; 32B, 60, 321, 414, 433; 33B, 385, 402, 410; 34A, 301; 34B, 444, 481, 508; 35A, 226; 35B, 602, 614
Porri, L., 26, 636, 652; 30B, 393, 398; 35B, 567
Porta, P., 32B, 376; 33B, 495, 35B, 799
Portas, J., 28, 150
Porter, D. W., 35B, 869
Porter, R. F., 34A, 152, 153; 35B, 891
Porter, S. K., 35B, 875, 876
Portheine, J., 32A, 220
Portheine, J. C., 35B, 341
Portier, J., 33A, 183, 184
Posner, A. S., 26, 630; 29, 369
Post, B., 26, 67, 568, 642; 27, 103, 336, 510, 674, 29, 334; 31B, 5; 32A, 149, 202; 32B, 15; 33B, 28,77; 34B, 91; 35A, 192
Potaško, K. A., 28, 340
Potenza, J. A., 29, 529; 31B, 283; 34A, 3
Potenza, M. F., 27, 520
Potter, R., 27, 960; 31B, 127
Potter, R. A., 27, 546; 29, 437
Pötzschke, M., 27, 16, 203, 377
Pouchard, M., 30A, 96, 180; 31A, 140; 33A, 162
Pouget, R., 28, 357, 358
Poulsen, F. R., 35A, 159, 278
Poulsen, S. J., 33A, 419
Pouradier, J., 27, 427; 29, 433
Pouzet, C., 33A, 186
Povarova, K. B., 26, 23, 217
Povet'eva, Z. P., 28, 113; 33B, 264
Povondra, P., 28, 338
Powell, F. X., 34A, 173, 244
Powell, G. W., 31A, 31

Powell, H. M., 26, 441, 690; 27, 669; 28, 658, 683, 687; 30B, 263, 391, 396; 31B, 296, 463; 32B, 350, 376, 470, 472; 33B, 326; 34B, 463; 35B, 858
Power, L., 35B, 239
Poyer, N., 31B, 82
Pöyhönen, J., 27, 426; 29, 433, 438
Poynter , R. L., 35B, 904
Pozdeev, N. M., 35B, 904
Prakash, A., 27, 931; 30B, 104; 32A, 250; 32B, 128; 33A, 256
Pramatus, S., 32 A, 32
Prandl, W., 31A, 228
Prat, M. -T., 26, 749
Pratt, L., 26, 680
Pratt, M. R. A., 28, 474
Predecki, P., 30A, 174
Predel, B., 27, 108; 29, 224
Preisinger, A., 27, 675, 1039; 29, 64; 30A, 358, 434; 32A, 70, 428
Preiss, H., 28, 314; 31A, 85; 33A, 196; 33B, 287
Prelesnik, B., 34B, 69
Presley, C. T., 35B, 103
Presnjakov, A. A., 30A, 248
Presnjakova, V. M., 27, 759
Preston, R. S., 33B, 320, 323; 34B, 557
Preston, R. D., 27, 1056
Prettre, M., 27, 642; 29, 436; 32A, 348
Prewitt, C. T., 26, 471; 28, 246, 357; 30A, 55, 179; 31A, 228; 31B, 356; 32A, 428, 469; 33A, 60, 82, 259, 281; 34A, 239, 260; 35A, 211
Price, C. C., 27, 810
Price, C. E., 30A, 181, 244
Priebe, E., 28, 683
Prilepskij, V. I., 30A, 246
Prince, E., 26, 409
Prince, M. Y., 27, 148
Prince, R. H., 33B, 450
Pringle, G. E., 30A, 365; 33A, 168
Prins, J. A., 28, 52
Prjakhina, L. I., 26, 213
Procter, I. M., 34B, 569
Procukhan, V. D., 33A, 153
Prodič, B., 30A, 403; 33A, 398
Projas, J., 33A, 381, 393
Prokof'ev, A.K., 35B, 901
Prokof'ev, D. I., 26, 111, 27, 156
Prokof'eva, I. G., 33A, 341
Prokoškin, D. A., 26, 110
Prosen, R. J., 27, 248
Proskurjakov, O. B., 33A, 324
Prostka, H. I , 27, 688
Protas, J., 27, 672; 29, 321, 435; 31A, 136, 174, 217; 34A, 142, 313, 318, 387; 35B, 15
Protasov, V. S., 29, 138, 140; 30A, 172, 178, 181
Prothero, J. W., 29, 759
Prout, C. K., 26, 441; 27, 655, 669, 851; 29, 555; 30B, 260, 262, 263, 267; 31B, 280, 432; 32B, 233, 283, 416; 33B, 403, 473, 476; 34B, 131, 252, 418, 426, 458, 540, 598; 35B, 120, 190, 469, 766, 782
Przedmojski, J., 27, 784
Przybylska, M., 26, 701, 793, 795, 830; 27, 1021; 28, 552, 685; 30B, 234; 34B, 322; 35B, 392
Psarov', V. I., 29, 140
Pšežackij, V. S., 29, 760
Puckett, R. T., 31B, 50; 32B, 245; 35B, 117
Pudovkina, Z. V., 26, 379; 27, 699; 29, 319, 433; 32A, 166; 35A, 451
Puff, H., 27, 465; 28, 356; 30B, 447
Pugačev, N. S., 27, 70
Puhl, H., 32A, 128
Puliti, R., 30B, 253; 31B, 111, 260, 261; 32A, 207
Pullia, A., 26, 668, 672
Puotinen, D., 26, 201
Purdy, G. R., 30A, 181
Purvinskij, O. F., 28, 102
Puščarovskij, D. Ju., 33A, 177; 34A, 386
Putkey, E., 35B, 299
Putkey, E. F., 35B, 299
Puxley, D.C., 35B, 485
Puzei, I.M., 27, 163
Pylaeva, E. N., 27, 287, 294; 30A, 173; 32A, 142
Pyle, J., 28, 341

Quader, M. A., 29, 137; 30A, 175
Quader, M. D., 27, 111
Quadling, F. M., 27, 706
Quail, J. W., 26, 313
Quareni, S., 29, 436; 30A, 320, 441; 31A, 219
Quarrell, A. G., 26, 95
Quesnel, J. C., 27, 60
Queyroux, F., 26, 364; 27, 542; 29, 435
Quezel-Ambrunaz, S., 28, 340; 33A, 318
Qui, D. T., 35A, 346, 347
van Qui, V., 32A, 53
Qui, V. V., 34A, 91
Quigley, G. J., 35B, 266
Quinby, M. S., 35B, 602
Quiring, W., 28, 14
Quist, A. S., 26, 345, 551
Qurashi, M. M., 26, 380; 28, 198

Raaum, F., 26, 267; 27, 364
Raaz, F., 32A, 439
Rabenau, A., 26, 300; 29, 224, 433
Rabideau, S. W., 33A, 255, 256, 258
Rabin'kin, A. G., 33A, 153
Rabinovich, D., 29, 645; 30B, 94;

32B, 15, 108, 109, 110, 111; 35B, 55, 56, 57
Rabinowitz, I. N., 29, 487
Raccah, P. M., 32A, 104, 309
Rachinger, W. A., 29, 136
Racine-Weisbuch, M., 34B, 80
Radaucan, S. I., 27, 247; 29, 141, 226
Radautsan, S. I., 32A, 95
Radošević, S., 26, 529
Radoslovich, E. W., 27, 679; 28, 357
Radtke, A. S, 27, 491; 28, 338
Rae, A. D., 28, 640; 30A, 404; 31A, 105; 32B, 442
Rae, A. I. M., 27, 894; 28, 563; 31A, 115; 32A, 509; 33A, 167; 35B, 825
Raeburn, U. A., 34B, 600; 35B, 647
Rafalovskij, V. A., 27, 232
Rahl, F. J., 30B, 448
Rahman, A., 35B, 276
Rajagopal, H., 33A, 448
Rall', I. V., 35A, 27
Ramachandran, G. N., 26, 823; 27, 136; 35B, 319
Ramamurthy, P., 28, 317
Raman, A, 29, 92, 135, 139; 30A, 14, 172, 173; 32A, 142; 33A, 153; 34A, 145; 35A, 77
Raman, S., 27, 136, 993; 28, 545; 29, 359
Ramanamurty, B. V., 27, 893
Rama Rao, B., 32A, 367
Ramasesham, S. M., 26, 472, 496, 792; 27, 1011, 28, 119, 290; 29, 297; 32B, 98
Rambaldi, G., 26, 255; 27, 249; 29, 12, 140, 141, 226
Rambidi, N. G., 26, 650; 27, 718; 28, 362; 34A, 173, 238
Ramirez, F., 30B, 395; 32B, 220, 311
Ramsey, T. H., 26, 415; 30A, 180
Ramsey, W. J., 26, 185; 28, 28; 29, 29; 30A, 180; 34A, 146
Rand, C., 29 33
Rand, M. H., 29, 226
Rand, R. P., 27, 119
Randaccio, L., 31B, 89; 33B, 525; 34B, 520, 582; 35B, 625, 779, 810
Rander, M. E., 28, 317
Range, K. -J., 33A, 153; 34A, 93, 143
de Rango, C., 31A, 85, 132, 225
Rankin, D. W., 34A, 155
Rankin, D. W. H., 34A, 155; 35B, 888, 891
Rannev, N. V., 28, 207; 30A, 364; 31B, 41; 33A, 450; 35A, 387
Rannev, V. I., 27, 643
Rao, B. R., 26, 438, 449; 32A, 367
Rao, C. N., 29, 225; 30A, 248
Rao, C. N. R., 26, 352
Rao, K. K. 29, 225; 30A, 248
Rao, K. V. K., 27, 430; 28, 290, 683; 30B, 447, 448
Rao, N. N. 32A, 359
Rao, P. B., 27, 151
Rao, P. R., 32A, 155
Rao, P. V., 28, 683; 30B, 447, 448
Rao, R. C., 31A, 125
Rao, S. S. 26, 139
Rao, S. T., 30B, 398; 31B, 190; 33B, 177; 34B, 92, 261, 265, 267, 269, 274; 35B, 297
Raphael, R. A., 26, 830
Rapp, G., 27, 535
Rapperport, E. J., 26, 254; 27, 21; 29, 223
Rascvetaeva, R. K., 32A, 511
Raseta, R., 32A, 141
Rasmussen, S. E., 26, 692; 27, 663; 28, 100, 649, 687; 29, 310; 30A, 278; 30B, 398; 31B, 326; 32A, 248; 32B, 440; 33B, 27, 124, 219; 34A, 231; 35A, 159, 278; 35B, 153, 798
Raspin, K. A, 33B, 518; 34B, 619
Rassaerts, H., 30A, 245; 31A, 26
Rastogi, R. P., 34A, 211
Rastorguev, L. N., 31A, 84
Rastrup-Andersen, J., 26, 760; 33B, 540
Ratner, L., 28, 682
Rau, R. C, 29, 434, 438
Raub, C. J., 29, 136, 138; 30A, 247
Raub, E., 26, 161; 27, 214; 29, 223
Rauch, R., 26, 745
Rault, M., 29, 325
Rausch, J. J., 26, 276
Rausell-Colom, J. A, 31A, 208
Rauterkus, K. J., 26, 637
Raveau, B., 33A, 286
Ravet, J., 27, 459
Ravez, J., 29, 433; 35A, 163
Ravn-Jonsen, E. J., 33B, 233
Ray, A. E., 31A, 46; 33A, 105
Raychaudhuri, B., 29, 438
Raymond, K. N., 33A, 207; 33B, 386, 387; 34B, 470
Raynal, J. M., 27, 811
Rayne, J.A., 26, 129
Raynor, G. V., 26, 129, 132, 260; 27, 108, 111, 145, 277, 310; 29, 223, 224; 30A, 172, 176, 244
Read, M. H., 30A, 182
Reau, J. -M., 32A, 296
Rebouillat, H. P., 31A, 133
Rebouillat, J. P., 32A, 268
Reck, G., 31A, 166; 32B, 9; 35B, 125, 139
Recke, I., 33A, 236
Recker, K. 28, 283
Reckin, V. N., 29, 138; 34A, 142
Redding, R. W., 34A, 244
Reddy, J., 27, 993; 28, 545

Reddy, J. M., 27, 1058, 1059; 29, 656; 30A, 52, 178, 322, 327; 31A, 147
Redhouse, A. D., 29, 238; 30B, 294; 33B, 336, 339; 34B, 433
Redman, M. J., 27, 474; 28, 170; 35A, 144
Reeber, R. R., 30A, 245
Reed, H. J., 27, 995
Reed, J. W., 26, 290; 27, 520
Reed, L. L., 35B, 602
Reed, R. P., 27, 155
Reed, T. B., 27, 244, 303
Reeke, G. N., 31B, 166
Rees, A. H., 30B, 233
Rees, B., 30B, 398; 35B, 72
Rees, M., 30B, 445
Regourd, M. ,29, 436
Reid, A. F., 32A, 310, 478, 492, 509; 33A, 280, 294, 505; 35A, 313
Reid, F. J., 29, 141, 226
Reid, K. I. G., 32A, 126; 33A, 291
Reid, S. T., 27, 982
Reiffenstein, E., 30A, 244
Reimann, C. W., 31B, 427; 32B, 432; 34B, 568,569, 572; 35B, 710, 721
Reimlinger, H., 35B, 154, 232
Reinen, D., 26, 410; 35A, 242
Reistad, T., 29, 520
Reltham, R. D., 34A, 183
Remeika, J. P., 27, 503; 31A, 131, 165; 32A, 257; 33A, 282; 34A, 354, 355; 35A, 218
Remeika, M. P., 30A, 314
van Remoortere, F. P., 35B, 99, 443
Renaud, M., 31B, 33
Renner, J., 30A, 312
Renouf, T. J., 26, 31
Rentsch, J., 33A, 179
Rentzeperis, P. J., 26, 437, 438, 443; 27, 608; 28, 210, 230; 35A, 467
Repko, V. P., 34A, 168
Repmann, H., 30B, 393; 32B, 43
Rérat, B., 27, 972; 29, 735; 30B, 163, 398; 32B, 69; 34B, 45, 620
Rérat, C., 26, 695; 27, 409, 672, 956, 959, 972, 992; 29, 474, 735; 30B, 163, 398; 31A, 81; 32B, 63, 69; 34B, 45, 388; 35B, 152
Rescikova, A. A., 29, 139; 30A, 247
Restivo, R., 35B, 166, 643
Reuter, B., 26, 423; 27, 515; 30A, 289; 34A, 264, 265
Revolinsky, E., 30A, 180
Rexer, J., 29, 137
Rez, I. S., 27, 506, 643
Rhee, S. K., 29, 135
Rhinehammer, T. B., 29, 135
Rhines, F. N., 27, 260
Riad, S. M., 30A, 244
Riaño, E., 26, 297
Ribar, B., 28, 179; 31A, 178; 34A, 23, 341; 35A, 15, 324
Ribbe, P.H., 27, 545, 676, 679; 28, 243; 33A, 478; 34A, 366, 368, 35A, 469
Ribek, G. D., 35A, 232
Ribes, M., 35A, 65
Ricci, F. P., 27, 474, 720
Ricci, J.S., 34B , 44; 35B, 58
Rich, A., 26, 820, 823, 827; 29, 759; 30B, 396; 31B, 163; 32B, 206, 207 ; 33B, 254; 34B, 224, 226
Richards, A., 35B, 621
Richards, G. F., 26, 688; 29, 587; 30B, 324
Richards, J. H., 30B, 307
Richards, J. P. G., 26, 704; 30B, 70; 34B, 60
Richards, J. R., 26, 631
Richards, R. R., 30A, 277
Richards, S., 26, 692, 29, 545
Richards, S. M., 27, 448; 29, 55; 34A, 43
Richardson, A. W., 34A, 244
Richardson, D. C., 31B, 376
Richardson, J. H., 30A, 248
Richardson, M., 32A, 82
Richardson, M. F., 34B, 509; 35B, 612
Riche, C., 30B, 398; 31B, 84
Richter, H., 26, 280; 27, 304, 385, 386
Rickard, C. E. F., 32B, 380; 34A, 197; 35B, 813, 818
Rider, P. E., 30A, 244
Ridgley, D. H., 35A, 21
Ridley, N., 29, 223; 30A, 245
Rieck, G. D., 31A, 163; 33A, 324; 34A, 282
Rieck, H., 35A, 216
Riedel, E., 34A, 264, 265
Riedel, E. F., 35B, 852
Rieger, W., 29, 136, 137, 139; 30A, 172, 174; 31A, 27; 32A, 141; 33A, 21; 34A, 143, 145; 35A, 54, 124
Riekel, C., 32A, 43; 35A, 135
Riess, J. G., 31A, 77; 31B, 467
Rietschel, E. T., 34A, 181
Rietveld, H. M., 26, 734; 30B, 371; 31A, 160; 34A, 207, 289; 35B, 95
Riganti, V., 26, 502, 781; 27, 989; 28, 687; 30B, 139, 141
Rigault, G., 26, 449
Rigby, G. R., 26, 395
Riley, J. F., 30A, 176
Rimmer, B., 32B, 250
Rimmer, B. M., 28, 537
Rimsky, A., 26, 145; 28, 357, 358; 30A, 178; 31A, 146; 33A, 312, 313, 314, 338; 34A, 82

Rinaldi, F., 28, 250; 29, 401
Rinaldi, F. F., 30A, 424
Rinehart, K. L., 33B, 190
Ring, M. A., 26, 291; 27, 438
Ring, S. A., 29, 139
Ringertz, H., 30B, 41, 448; 31B, 45, 162, 33B, 175; 34B, 359
Ringwood, A. E., 27, 488, 489, 501; 29, 438; 32A, 478, 492; 33A, 280; 35A, 313
Rinn, H. W., 26, 277; 27, 451; 34A, 153
Rinnovatore, J. V., 27, 21; 29, 6
Ripamonti, A., 26, 754; 27, 412, 413; 30B, 134; 31B, 111, 260, 261; 32B, 287, 304; 33B, 525; 34B, 582; 35B, 779, 796, 810
Ripamonti, O., 30B, 253
Ripperger, H., 32B, 260
Ritter, D. L., 29, 135, 223
Ritter, D. M., 26, 291
Ritter, W., 29, 140
Riva di Sanseverino, L., 31B, 35; 35B, 157, 389, 396
Rivet, J., 30A, 177
Rivlin, V. G., 27, 107
Rivoir, L., 26, 490
Rizvi, S. H., 26, 726; 27, 907; 33B, 64
Rjabcev, L. A., 26, 213
Rjabov, V. R., 31A, 14; 34A, 143
Robab, V. I., 30B, 446
Robbins, C., 31A, 134
Robbins, C. R., 26, 398, 491; 29, 438
Robbins, M., 28, 358; 29, 137; 32A, 53; 33A, 235; 35A, 124
Roberts, D. E., 26, 630
Roberts, F. P., 28, 338
Roberts, J .S., 35B, 386
Roberts, P. J., 35B, 845, 866
Roberts, W. M. B., 27, 706
Roberts, W. T., 27, 370
Robertson, B. E., 32A, 376; 33A, 410; 35A,343
Robertson, B. K., 30B, 129; 32B, 157
Robertson, G. B., 26, 687, 726; 27, 822; 29, 570; 30B, 294, 393, 398; 34B, 468, 495, 537; 35B, 529, 579, 600, 869
Robertson, I., 32B, 384
Robertson, J. H., 27,948; 28, 687; 29, 731; 30B, 21, 177, 353; 31B, 419
Robertson, J. M., 26, 649, 747, 754, 756, 757, 758, 792, 796, 801, 830; 27, 926, 938, 1000, 1008, 1010, 1013, 1028, 1030, 1033; 28, 542, 558, 564, 686; 30B, 232, 235, 370, 393, 394; 31B, 230, 267; 32B, 114, 243; 250, 252; 33B, 210; 34B, 310, 312, 333, 340; 35B, 386
Robertson, W. D., 26, 166; 27, 155
Robiette, A.G., 34A, 155, 163; 35B, 888, 891
Robin, M.B., 29, 656
Robinson, A. B., 35A, 129
Robinson, B., 29, 677
Robinson, B. H., 35B, 502, 504
Robinson, D. J., 35B, 419, 520, 697
Robinson, G., 27, 1049; 28, 605; 29, 602; 30A, 286
Robinson, J. M. M., 28, 461
Robinson, K., 31A, 8
Robinson, P. D., 34A, 310; 35A, 437, 441, 478
Robinson, P. L., 26, 301
Robinson, P. M., 34B, 157
Robinson, W. H., 27, 800
Robinson, W. R., 32B,410; 33B, 414
Robinson, W. T., 28, 87; 31A, 51; 31B, 319; 32A, 402; 32B, 467; 34B, 107, 410; 35A, 412; 35B, 502, 504, 599, 665, 712, 846
Robitsch, J., 35A, 49
Robson, A., 30B, 339; 33B, 377; 34B, 451
Rocher, G., 29, 321
Rocktaschel, G., 27, 348; 29, 140
Rödder, K.-M., 26, 312, 313
Roddy, J.T., 27, 303
Rodella, A., 29, 760
Rodgers, J., 35B, 702
Rodi, F., 30A, 333
Rodier, N., 35A, 101
Rodionov, A. N., 28, 113
Rodionov, K. P., 27, 794
Rod'kina, N. I., 27, 369
Rodley, G. A., 30B, 398; 32B, 468
Rodot, H., 26, 195; 30A, 177
Rodot, M., 26, 195
Rodriques, M. M., 34B, 216
Rodriguez, A., 26, 828
Rodriguez, M. P., 26, 575; 28, 683
Rodrique, L., 31B, 271; 34B, 451
Roe, G. J., 33A, 153
Roe, R. -J., 33B, 9
Rogacev, D. L., 26, 526; 34A, 301
Rogers, D., 27, 1036; 28, 683; 30B, 331, 394; 33B, 211; 35B, 312
Rogers, D. B., 28, 285; 34A, 239
Rogers, G. E., 27,1056
Rogers, H. H., 26, 103
Rogers, J. B., 32B, 103
Roginskij, S. Z., 26, 408
Rogowski, F., 26, 314; 29, 453
Rohmer, R., 28, 682; 29, 434
Röhrborn, H. -J., 26, 402; 27, 661; 29, 338, 339
Rohrer, D. C., 35B, 249, 294, 296
Rohrl, M., 35B, 132
Roldan, L. G., 30B, 448
Rollett, J. S., 32B, 252, 258
Romain, C. R., 30B, 397
Romanova, A. V., 26, 187
Romanow, W. J., 26, 9

Romans, P. A., 30A, 246; 31A, 29; 34A, 146
Romers, C., 26, 709; 27, 1037; 28, 476, 482, 484, 485, 487; 29, 46, 491, 647; 30B, 150, 396, 398, 446; 31B, 54, 76, 134, 143, 149, 159, 202, 203, 206, 387; 32A, 65, 220, 288; 32B, 11, 199; 33B, 127, 131, 147; 34B, 5, 95, 303, 531; 35B, 211, 341
Romita, R., 28, 338
Romming, C., 26, 546, 763; 27, 889; 28, 402; 29, 490; 30B, 199, 247, 357; 31B, 270; 32B, 116; 33B, 52, 57, 243
Romo, P., 32A, 254
Ronis, M., 33A, 418
Rönnquist, A., 26, 354
Rönnquist, O., 34B, 250; 35B, 195, 199
Ronova, I. A., 33B, 531; 35B, 900, 901
Roof, R. B., 26, 108; 27, 93; 28, 17; 29, 36, 37, 81, 261, 313; 30A, 50, 175, 178, 290; 31A, 116; 33A, 245; 35A, 392
Rooksby, H. P., 26, 548; 27, 4; 28, 356; 29, 135, 329
Rooymans, C. J. M., 26, 346, 370; 27, 546; 30A, 176, 178; 32A, 288
Ropars, C., 31A, 161
Roper, W. R., 27, 770
Roques, B., 30A, 180; 34A, 142
Rorem, E. S., 28, 683
Rösch, H., 28, 154; 29, 8
Röschel, E., 29, 223
Rose, H., 29, 434
Rose, H. A., 26, 763; 28, 683
Rose, I. A., 30B, 396
Rose, K. M., 33A, 253
Roseboom, E. H., 27, 48
Rosen, S., 29, 140, 226; 30A, 172
Rosenberg, A. J., 26, 43; 29, 223; 30A, 245
Rosenberg, R. C., 31B, 456
Rosenblum, M., 30B, 397; 32B, 279
Rosenstein, R. D., 26, 433, 657, 30A, 446; 30B, 238; 32B, 212, 214, 215; 33B, 157, 162, 163; 34B, 242, 245
Rosenweiz, A., 26, 108; 30A, 332, 372, 373; 31A, 214; 32A, 178; 33A, 195; 34A, 194; 35A, 170
Rosi, F. D., 29, 140
Rosolovskij, V. Ja., 30B, 448
Ross, D. R., 28, 358
Ross, F. K., 35B, 322, 878
Ross, G. J., 32A, 421
Ross, J. W., 26, 415
Ross, M., 28, 358; 29, 375, 377, 437
Ross, R. C., 27, 87
Rossi, G., 28, 339; 30A, 437; 31A, 232; 32A, 444, 459, 466; 33A, 489; 35B, 762
Rossmann, M.G., 26, 811, 812, 816; 27, 1063; 29, 759; 35B, 888
Rossotti, F. J. C., 33B, 403
Rossovskij, L. N., 28, 342
Rossteutscher, W., 26, 280; 27, 377; 29, 92, 139; 30A, 175, 179
Røst, E., 27, 315; 29, 137, 225; 33A, 121; 34A, 94
Roth, E. A., 30B, 405
Roth, R. S., 26, 388, 491; 28, 358; 29, 438; 30A, 336, 338, 340; 31A, 172
Roth, W. L., 29, 325
Rothbauer, R., 32A, 242, 243
Rothenbacher, R., 28, 338
Rothenstein, W., 28, 316
Rothwarf, F., 27, 107
Rouault, A., 34A, 144
Roult, G., 27, 256, 523; 29, 39
Roussel, J., 27, 189
Rouwelaar, L. N. A. S., 32B, 419
Roux, J. P., 34B, 351, 352; 35B, 746
Rouxel, J., 28, 315; 31A, 107; 33A, 153; 34A, 145, 228
Rovel, J., 27, 349
Rovša, V. S., 28, 339
Rowe, J. M., 27, 820, 846, 878; 28, 660
Rowland, J. F., 26, 382; 27, 537, 548; 28, 145, 339, 358; 29, 412
Roy, A.P., 27, 257
Roy, B., 26, 251
Roy, P. K., 34B, 17
Roy, R., 26, 394, 413, 540; 27, 534, 695, 707; 28, 317, 338, 358; 29, 322, 412; 30A, 246, 248; 33A, 270
Roy, S., 27, 769; 30B, 398
Royen, P., 28, 318
Roy-Montreuil, J., 32A, 142; 33A, 86
Rozkova, R. S., 30A, 178
Ruban, G., 26, 727; 30A, 395; 30B, 331; 35B, 112
Rubas, K. V., 31A, 14
Ruben, D. H., 26, 433
Ruben, H., 27, 616; 29, 347
Rubezov, A. Z., 34B, 418
Rubinstein, C. B., 29, 435
Ruble, W. D., 29, 438
Ruby, S. L., 27, 498
Ručkin, E. D., 27, 750
Rucklidge, J. C., 30A, 434
Rud', Ju. V., 29, 136, 137; 30A, 173
Rudman, P. S., 30A, 247
Rudman, R., 32B, 301; 33B, 28
Rudnickaja, E. S., 33A, 472
Rudolph, R. W., 35B, 427
Rüdorff, W., 27, 453, 543; 28, 314, 29, 371; 31A, 138; 34A, 216
Rudy, E., 26, 58, 68, 73, 81, 93, 101, 104; 27, 136, 338; 35A, 43, 68
Rudy, Elisabeth, 27, 126, 133, 135, 138
Rudy, Erwin, 27, 126, 133, 135, 138
Rudzinski, J., 26, 104; 29, 136
Rüegg, M., 35A, 405
Rüetschi, P., 27, 478

Ruff, J. K., 35B, 506
Ruggiero, A. F., 29, 136, 140, 226
Rugowski, F., 26, 550
Ruh, R., 31A, 139
Ruiz, F. S., 35B, 606
Ruland, W., 26, 631
Rule, L., 30B, 443
Rumanova, I. M., 26, 476, 534; 27, 551, 557; 28, 160, 161, 172, 275; 29, 279, 435; 30A, 419; 31A, 229
Rumball, S., 35B, 665
Rump, I., 27, 234
Rumš, M. A., 26, 446
Rundle, R. E., 26, 317, 689; 27, 913, 28, 67, 83, 85, 604, 614, 685; 29, 532, 533, 590, 655,729;30B, 186; 31A, 103; 33B, 300
Rundqvist, S., 26, 227, 270; 27, 96, 275, 290, 319, 386; 28, 33; 30A, 67, 69, 98, 176, 180, 251; 31A, 45, 51, 57; 32A, 32; 34A, 56, 117, 143; 35A, 38
Runnals, O. J. C., 30A, 11, 173
Runnells, D. D., 29, 438
Runnsjö, G., 34A, 56
Rush, J. J., 27 421; 34A, 237
Russell, C. M., 30A, 244
Russell, D. R., 26, 304; 28, 314; 35A, 430
Russell, J. H., 34B, 138
Russo, P., 30A, 181
Ruther, W.E., 26, 348
Rutherford, J.S., 34B, 29
Rutherford, K.W., 27, 841
Rutten, E.W.M., 31B, 206; 32B, 11; 33B, 131
Ruuskanen, A., 29, 433
Ruysink, A. F. J., 33A, 115
Ryan, N. E., 26, 118
Ryan, R. R., 34A, 174, 193, 194; 35A, 167
Ryan, T. D., 26, 689
Ryba, E., 28, 21; 30A, 98, 182; 31A, 44; 33A, 50, 100
Rychnovsky, V., 33B, 138
Ryder, B., 27, 107
Ryderstedt-Nahringbauer, I., 27, 740
Rylov, E. E., 26, 633
Rzany, H., 33A, 266

Saalfeld, H., 26, 347, 446, 515, 537; 28, 357; 29, 324; 30A, 432; 33A, 260, 274
Saari, E., 29, 437; 32A, 477
Sabelli, C., 27, 558, 819; 30A, 371; 31A, 172; 32A, 332, 335, 336, 416, 501; 32B, 191; 33A, 382; 34B, 147, 201, 203
Sabine, T. M., 26, 734; 30A, 406; 31A, 165; 32A, 154, 325; 34A, 238; 34B, 3, 6; 35B, 304
Sabrowsky, H., 33A, 280
Sacconi, L., 27, 872; 28, 686; 29, 568, 582; 30B, 398; 31B, 389, 393, 400; 35B, 777,849, 853, 862
Sachdev, H. S., 35B, 326
Sackman, P., 35B, 322
Sackmann, H., 28, 99, 315
Sadagopan, V., 27, 141; 28, 15; 30A, 39, 175
Sadana, V. N., 29,62
Sadanaga, R., 26, 525, 532; 27, 707; 30A, 415; 30B, 81; 31A, 238; 32A, 123
Sadashivaiah, M. S., 29, 437
Sadikov, G. G., 26, 264; 27, 473; 30A, 329
Sadler, M. S., 26, 354
Sadov, N. I., 33B, 536
Sadova, N. I., 30B, 406
Saebo, P. C., 26, 483
Saegusa, T., 29, 760
Saenger, W., 32B, 103; 34B, 204, 258; 35B, 278, 292, 326
Safford, G. J., 27, 421
Safiannikoff, A., 26, 388
Safronov, G. M., 34A, 168
Sager, R. S., 32B, 446 ; 33B, 458; 34B, 575, 576
Sagmuller, J. R., 29, 140
Saha, N. N., 32B, 227; 34B, 266; 35B, 656
Sahama, T.G., 29, 437; 30A, 181, 427; 32A, 477
Sahl, K., 26, 324; 28, 78, 80, 208; 35A, 374
Sahu, R. G., 35B, 13, 14, 16
Said, A., 27, 464
Saini, G. S., 29, 18, 137
de Saint-Giniez, D., 33A, 4
de Saint-Giniez-Liebig, D., 34B, 388
Saito, N., 34B, 253
Saito, S.,34A, 242, 244
Saito, Y., 26, 326; 27, 833; 28, 684; 29, 459; 30B, 395; 31B, 378, 418; 32B, 3, 281, 282, 431; 33B, 250, 524; 34B, 485, 491; 35B, 76, 410, 684, 707, 713
Sakabe, N., 30B, 398; 31B, 194, 255, 257; 33B, 50, 60; 34B, 237, 328
Sakai, K., 28, 687
Sakaki, Y., 29, 139
Sakakibara, S., 27, 1050
Sakamoto, Y., 33A, 377
Sakanoue, S., 35B, 188
Sakata, K., 30B, 61
Sakata, Y., 26, 632; 28, 365
Sakavov, I. E., 28, 339
Sakhariva, G. N., 30B, 443
Sakhova, K. I., 29, 137, 138
Sakore, T. D., 30B, 399; 31B, 61; 32B, 76; 34B, 182, 227, 232, 233, 235; 35B, 275
Šakova, K. I., 26, 218
Sakowski-Cowley, A. C., 34A, 268
Saksonov, Ju. G., 28, 143; 33A, 324

Sakurai, H., 31B, 255
Sakurai, K., 26, 589; 30B, 60
Sakurai, T., 27, 899, 901, 903; 28, 420; 30B, 75; 33B, 66; 35B, 409
Salamon, R., 35B, 221
Salas-Ciminago, G., 31B, 62
Sal'dau, E. P.. 28, 339
Sales, D. L., 33B, 481
Salimova, K. V., 27, 115; 30A, 182
Salkind, A. J., 26, 540
Salomons, W., 33A, 141
Salova, G. E., 30B, 405
Salovey, R., 29, 760
Salvo, G. P., 29, 223
Samaras, D., 35A, 372
Samarskaja, V. D., 33B, 72
Samojlov, A. A., 30A, 248
Sampaio, O. A., 30A, 245
Samraj. V. F., 35A, 124
Sams, D., 26, 31
Samson, C., 29, 139
Samson, S., 26, 277; 27, 113; 30A, 5, 173; 32A, 36; 33A, 8
Samsonov, G. V., 26, 66, 81; 29, 226
Samsonova, T. I., 29, 138; 34A, 142
Samus', I. D., 27, 842; 32A, 354
Sanadze, V. V., 26, 159
Sanatina, V. N., 26, 386
Sandanaga, R., 34A, 377
Sanders, W. W., 32B, 11
Sandler, R. A., 27, 434
Sandler, S., 28, 444
Sandmark, C., 32B, 436
Sandor, E., 26, 435; 28, 682; 32A, 180
Sands, D. E., 26, 57, 58, 407; 27, 71, 74, 772; 28, 28, 314, 406; 29, 29; 32B, 65; 33B, 390; 34B, 509, 510; 35A, 29, 124; 35B, 612
Sands, D. R., 32B, 386
Sänger, G., 32A, 389
Sanger, R., 30B, 48
Sano, T., 29, 138
di Sanseverino, L. R., 30B, 396; 31B, 35; 32B, 252, 258; 35B, 157, 389, 396
Sansoni, M., 34B, 430, 604
de Santis, D., 30B, 253, 399
de Santis, P., 27, 413; 29, 760
Santoro, A., 27, 474, 720; 34B, 568, 572; 35B, 710, 721
Santoro, R., 29, 317, 396
Santoro, R. P., 29, 318; 31A, 227
Sanyal, R. M., 30B, 445
Sanz, F., 35B, 466
Sapet, A., 30A, 176
Sapet'ko, N. N., 30B, 447
Šapkin, A. A., 35B, 904
Sara, R. V., 30A, 245
Sard, R.,30A, 178
Sargentini, A., 27, 858
Sargeson, A. M., 35B, 686
Sarko, A., 35B, 266
Sarma, B. B., 35A, 19
Sarma, V. R., 31B, 69, 120; 34B, 274; 35B, 888
Sartain, D., 28, 589; 32B, 460
Sartori, F., 34A, 351
Sarver, J. F., 28, 289
Sasada, Y., 26, 723, 809; 27, 1050; 28, 412; 29, 675; 30B, 88, 399; 31B, 63, 91, 106,121, 191, 239, 241; 32B, 67, 76, 150, 176, 226, 227, 232, 392; 33B, 218, 418; 34B, 88, 253, 276, 311, 332, 513; 35B, 84, 145, 199
Sasaguri, K., 35B, 883
Sasaki, Y., 26, 102
Sasisekharan, V., 26, 823
Šaškin, D. P., 32A, 368; 33A, 446; 35A, 283
Šaškov, O. D., 27, 22
Sass, J., 26, 415
Sass, R. L., 26, 117, 831; 27, 559, 736, 739, 783; 28, 77, 80, 97, 98, 365, 682; 29, 284, 30A, 291; 30B, 39, 131; 32B, 51, 66; 33A, 432; 33B, 34, 144; 34B, 203; 35B, 31, 50, 78, 103, 217, 322
Sassmannshausen, G., 32B, 314; 33A, 406
Sastri, G. G. K., 28, 342
Sastry, K. V. L. N., 28, 362
Sasvari, K., 28, 227; 30B, 91; 31A, 93
Sathyanaraya, D. N., 30B, 448
Sato, H., 26, 135; 27, 176, 180; 30A, 247; 31A, 42; 32A, 89; 33A, 89
Sato, K., 27, 180; 30A, 17, 173, 177
Sato, M. 28, 120; 29, 438; 35A, 12
Sato, R., 26, 367, 27, 115
Sato, S., 28, 687; 30B, 395; 33B, 183
Sato, T., 31B, 154; 33B, 195, 205, 220, 480; 34B, 283, 299, 574; 35B, 337
Satomi, K., 26, 407
Satoro, R. P., 28, 170
Šatrukov, L. F., 35B, 893
Satya Murthy, N. S., 28, 23; 34A, 9
Sauer, F., 26, 23
Sauermann, G., 35B, 472, 473
Sauerwald, F., 27, 280
Sauka, Ja. Ja., 26, 343
Saunders, G., 26, 287
Saunderson, C., 26, 546, 830
Saunderson, C. P., 30B, 233
Savickij, A. V., 30A, 181
Savickij, E. M., 27, 161, 233, 268, 284, 313, 314, 337, 338; 30A, 178, 244, 248
Savinskaja, A. A., 27, 123
Savitzki, J. M., 26, 217
Savolahti, A., 29, 437
Sawaguchi, E., 29,434, 31A, 112; 32A, 223
Sawaoka, A., 29, 434
Sawyer, D. L., 28, 340

Sawyer, J. O., 28, 341; 29, 438
Sax, M., 26, 831; 28, 375; 30B, 58; 31B, 181; 32B, 85, 160, 187; 34B, 152; 35B, 239, 416
Saxer, R. K., 30A, 244
Scane, J. G., 26, 762; 27, 1009, 1036, 28, 682, 32B, 317
Scaramuzza, L., 33B, 476; 35B, 816
Scarbrough, F. E., 33B, 263
Scatturin, V., 26, 306, 540; 27, 474, 720; 28, 495, 659; 31B, 464; 33A, 165; 34A, 159
Šćavničar, S., 28, 116; 30A, 403; 32B, 390; 34A, 341; 34B, 580
Scedrin, B. M., 31B, 381
Ščerbik, L. K., 35B, 889
Schablaske, R. V., 27, 116; 29, 32, 135, 136
Schadler, H. W., 29, 223
Schaefer, H., 32A, 142
Schaefer, J. P., 31B, 33; 33B, 5, 188
Schaefer, W. P., 31A, 206; 33A, 375, 427, 428; 34A, 345; 34B, 514, 543; 35A, 406
Schaeffer, R., 30B, 272
Schaefgen, J. R., 26, 634
Schäfer, A., 27, 65
Schäfer, G., 28, 8, 153
Schäfer, H., 26, 323, 328; 28, 8, 11, 153, 316, 338; 30A, 65, 66, 97, 180, 182, 261, 263, 264; 31A, 25, 48; 32A, 27, 43, 117, 122, 210; 33A, 39, 51, 153; 34A, 53, 122, 145; 35A, 9, 24, 43, 64, 124, 139
Schafer, H. L., 30B, 358
Schäfer, L., 33B, 536
Schäfer, R., 26, 156
Schaffner, T. J., 34B, 305
Schaffrin, R., 35B, 314
Schaffrin, R. M., 35B, 225
Schairer, J. F., 26, 502; 27, 685
Schaller, W. T., 32A, 477
Schaner, B. E., 28, 342
Schapiro, P. J., 27, 771
Scharpen, R. H., 28, 668
Schatz, B., 28, 338
Scheiber, M., 27, 511
Scheie, A., 30B, 285
Schein, J. B., 32B, 57
Scheit, K. H., 35B, 278
Schelle, S., 34B, 426
Schenk, F., 34A, 248
Schenk, H., 32B, 253; 35B, 135, 218, 331, 877
von Schenk, R., 35A, 230
Scherer, H., 34B, 408
Scherff, H.L., 34A, 144
Scheringer, C., 26, 317, 667; 28, 423
Scheurman, R. F., 26, 831; 27, 559, 736, 739
Schiaffino, L , 27, 844; 28, 682
Schibilla, H., 32B, 322
Schieber, M., 26, 415; 29, 322, 435
Schiferl, D., 34A, 13
Schiff, L., 30B, 397
Schindler, A. I., 30A, 179
Schipper, D. J., 29, 136
Schirber, J. E., 26, 194
Schläfer, H. L., 26, 539
Schlatti, M., 32A, 511; 33A, 447
Schlemper, E., 28. 683
Schlemper, E. O., 30B, 38; 31A, 88, 89, 197, 31B, 312, 328, 329, 330; 32B, 333, 428; 33B, 424
Schlessinger, R. H., 34B, 182
Schlueter, A. W., 31A, 103; 33A, 199
Schlupp, R. L., 35B, 804
Schmahl, N. G., 29, 434
Schmid, H., 29, 393; 34B, 333, 338
Schmid, R., 35A, 155
Schmidt, G. F., 26, 637; 27, 800
Schmidt, G. M. J., 28, 444; 29, 645, 651, 660; 30B, 72, 94, 446; 32B, 12, 15, 17, 50, 110, 111, 174, 185; 35B, 54, 56, 57, 133, 134
Schmidt, G. M. K., 27, 734
Schmidt, J., 31A, 105
Schmitkons, D. L., 30B, 115
Schmitt, G. A., 29, 136
Schmitt, R. D., 34B, 355
Schmittler, H., 33B, 402
Schmitz-Dumont, O., 26, 410; 27, 513
Schmutzler, R., 28, 683
Schneider, H. G., 31A,54
Schneider, R., 26, 650; 28, 682, 683
Schneider, S., 35A, 158
Schneider, V. L., 29, 224
Schneider, W., 26, 447; 32A, 333; 34A, 311; 35A, 285
Schneidereit, G., 28, 338
Schnell, W. D., 29, 224
von Schnering, H.G., 26, 325, 396, 424; 27, 517; 28, 315; 29, 57, 303; 30A, 261, 263, 264, 276, 309, 316; 31A, 99; 32A, 168, 310; 33A, 203, 231; 34B, 374
Schnittler, H., 29, 436
Schob, O., 26, 10, 11, 18, 70; 27, 30, 159; 29, 53, 61, 139; 30A, 173, 176
Schöbel, J. -D., 27, 103; 29, 135, 140; 34A, 38
Schodder, G.R., 27, 345
Schödl, G., 35B, 507
Schoenborn, B.P., 27, 997; 29, 744
Scholder, R., 28, 317, 340, 356
Scholer, F. R., 34A, 162; 35B, 505
Scholl, A. W., 27, 414
Scholze, H., 26, 446

Schomaker, V., 27, 880; 28, 683; 29, 665; 30B, 297, 320
Schönemann, H., 34A, 144, 145
Schönherr, F., 32A, 172, 173
Schoone, J. C., 27, 1049; 29, 602; 30B, 348; 32B, 7, 8
Schöpf, C., 26, 798
Schor, R., 26, 824
Schott, D., 28, 316
Schrader, R., 28, 338
Schrauzer, G. N., 30B, 399
Schreiber, A. H., 34B, 503
Schreiber, D. S., 27, 478
Schreiber, K., 32B, 260; 33B, 228
Schreyer, W., 26, 502; 27, 685, 694
Schröder, A., 26, 701
Schröder, F., 33A, 219
Schröder, H., 27, 476
Schroeder, L.W., 33A, 201; 35A, 152
Schubert, K., 27, 16, 161, 203, 223; 28, 23; 29, 79, 92, 135, 139; 30A, 14, 47, 49, 57, 172, 173, 174, 175, 178, 179, 246; 31A, 65; 32A, 80; 33A, 24, 89, 124, 126, 153; 34A, 11, 25, 82, 120, 142, 145; 35A, 7, 14, 22, 41, 56, 62, 108, 124
Schubert, W. M., 30B, 265
Schuch, A. F., 26, 285; 27, 391; 31A, 240; 35A, 129
Schuckmann, W., 31A, 167, 168; 33A, 444; 34A, 356
Schugar, H., 32B, 441
Schuhmann, H., 27, 280
Schuldt, H. S., 26, 364
Schulkes, J. A., 28, 285
Schulte-Elte, K. -H., 34B, 308
Schultz, J. M., 35B, 882
Schultz, N., 30A, 289; 34A, 182
Schultze-Rhonhof, E., 27, 420; 33A, 192; 34A, 344; 35A, 242
Schulz, H., 30A, 432; 32A, 490; 33A, 219, 483; 35A, 242, 452
Schulz, M., 32B, 173
Schulze, A., 30A, 172
Schulze, G. E. R., 26, 90
Schulze, H., 29, 139; 30A, 174, 177
Schulze, H. J., 34A, 322
Schumacher, D. P., 26, 168
Schumacher, H., 28, 314
Schunn, R. A., 31B, 356; 32B, 332
Schuster, H. -U., 30A, 177; 32A, 233; 34A, 144, 145; 35A, 61, 124
Schütze, D., 28, 338; 29, 311
Schuyff, A., 31A, 181
Schwab, A. M., 35A, 288
Schwab, H., 26, 364; 29, 435
Schwartz, C. M., 26, 252; 27, 481, 520, 28, 357; 29, 138, 319; 30A, 179
Schwartz, H., 29, 435, 436
Schwartz, L. H., 35A, 21
Schwartz, M., 27, 21; 29, 6
Schwarz, H., 28, 316, 317
Schwarzenbach, D., 31A, 183; 33B, 96, 113; 34A, 182; 35B, 740
Schwarzmann, S., 29, 266
Schweikert, W. W., 33B, 259
Schweizer, J., 30A, 42, 175, 176; 34A, 72
Schwendeman, R. H., 27, 732
Schwochau, K., 27, 653, 671; 28, 315
Schwomma, O., 29, 64, 85
Sclar, C. B., 28, 357; 29, 138; 30A, 179
Scordari, F., 35A, 376
Scott, A. I., 27, 1017, 1031
Scott, A. T., 31B, 223
Scott, C. C., 27, 982, 1008
Scott, E. S., 27, 633
Scott, H. G., 34B, 157
Scruton, J.C., 33B, 288
Ščukarev, S. A., 28, 102
Seabaugh, P., 27, 440
Seabaugh, P. W., 29, 249
Seabrook, M., 27, 488, 501; 29, 438
Sears, D. R., 27, 392; 29, 262; 34A, 195, 201
Šebatinov, M. P., 29, 224
Seccombe, R. C., 34A, 228; 35B, 59
Sechser, L., 30A, 172
Secrist, D. R., 27, 126
Sedivy, J., 26, 469; 28, 317
Sedlacek, P., 30A, 318
Sedlatschek, K., 26, 101
Seeger, K., 35A, 223
Seeliger, E., 26, 515
Seelkopf, C., 35B, 362
Seff, K., 32A, 479; 33B, 119, 256; 34B, 42
Šeftel', I. T., 26, 412
Segal, D. J., 31A, 227
Segmüller, A., 31A, 163
Segnit, E. R., 28, 357
Segrott, G. R., 35B, 830
Segrott, R., 35B, 824
Seidel, H., 33A, 31
Seifert, H. -J., 26, 326, 333, 625; 27, 457
Seifert, K., 27, 423
Seip, H. M., 26, 292; 29, 642; 30A, 269; 33B, 537; 34A, 162; 35B, 888, 894, 897
Seip, R., 29, 642; 35B, 888, 897
Sejna, L. P., 29, 305
Šekhtman, V. Š., 27, 255
Seki, M., 33B, 26
Seki, Y., 29, 438
Sekine, H., 26, 587
Sekine, Y., 26, 515
Selberg, B., 31A, 54
Selchow, D. H., 26, 527
Seleborg, M., 27, 520; 31A, 154; 32A, 298, 509
Šelestov, V. M., 34A, 349
Sellberg, B., 30A, 69, 180
Selle, J. E., 29, 135
Selmenov, E. I., 29, 339

Selte, K., 29, 73, 77; 30A, 76, 174, 180, 181; 34A, 16; 35A, 208
Selykh, A. I., 26, 173
Semaško, V. N., 35B, 894
Semenenko, K. N., 28, 315; 30B, 448
Semenov, E. I., 26, 484, 536; 28, 339
Semenov, V., 31A, 241
Semiletov, S. A., 26, 170, 199; 27, 89; 28, 26; 30A, 62, 175, 179; 31A, 25; 32A, 95; 33A, 42, 43; 34A, 10, 142; 35A, 87
Semin, G. K., 30B, 79, 446; 31B, 77; 33B, 68, 320; 34B, 69
Semion, V. A., 33B, 349, 358; 34B, 418, 420, 434, 435; 35B, 905
Sen, D. K., 35B, 193
Sénateur, J. P., 28, 15; 29, 135; 30A, 175; 32A, 102, 141, 143
Senderova, V. M., 29, 136, 435
Sendo, Y., 34B, 328
Sen Gupta, J. G., 29, 436
Senko, M., 26, 568
Senko, M.E., 28, 71
Šepelev, Ju. F., 32A, 436; 33A, 465; 34A, 261; 35A, 300, 305
Sepichowska, A., 26, 432; 29, 140
Sequeira, A., 29, 463; 31A, 117; 32A, 213, 216, 312; 33A, 369; 33B, 47; 35A, 197; 35B, 8
Seraphim, D.P., 27, 307
Serátor, M., 29, 600
Serebrjakova, T.I., 32A, 141
Serebrjanana, N.A., 34A, 239
Serebrjanaja, N. R., 28, 356
Serebrjanskij, V. T., 26, 4
Serewicz, A. J., 28, 684; 30B, 129
Sergeev, A. S., 29, 436
Sergeeva, V. M., 26, 173
Serment, J., 30A, 441
Sersale, R., 28, 339
Sestrich, D. E., 26, 151, 372
Seto, T., 27, 799
Settineri, W. J., 29, 761
Setton, R., 33B, 528
Setty, P. L. N., 27, 430
Setz, S., 33A, 153
Ševčík, J., 35B, 564
Sévèque, F., 34A, 286
Severov, E. A, 26, 333
Seyferth, D., 35B, 505
Seymour, R. S, 35A, 145
Sgamellotti, A., 33B, 495; 35B, 799
Sgarabotto, P., 34B, 32; 35B, 28
Sgarlata, F., 27, 707, 858; 28, 242; 30A, 440
Shafer, M. W., 27, 495
Shaffer, P. T. B., 34A, 157
Shah, J. S., 30A, 244
Shaked, Z., 35B, 57
Shalek, P. D., 29, 141
Shamraj, V. F., 34A, 146
Shan-chi Yu Chu, S. 26, 716
Shankar, J., 28, 79
Shanley, P., 26, 603
Shannon, J., 35A, 261, 267, 269
Shannon, R. D., 33A, 259, 270, 281; 34A, 239; 35A, 211
Shao, M. C., 27, 775
Sharan, B., 29, 367
Sharma, B. D., 26, 772; 27, 199; 28, 54; 29, 749; 30A, 408; 30B, 172; 33B, 179
Sharma, S. D., 33B, 540, 541
Sharp, D. W. A., 26, 304; 28, 314
Shaw, B. L., 30B, 393
Shaw, B. W., 30B, 446; 33B, 341
Shaw, D. C., 28, 684
Shaw, E. H., 29, 759
Shaw, J. A., 27, 254
Shaw, N., 29, 224; 30A, 244
Shaw, R. A., 26, 729
Shearer, H. M. M., 27, 847, 936; 30B, 392; 31B, 344, 345; 32B, 343
Sheat, S. V., 30B, 395; 33B, 398
Sheat-Rumball, S. V., 33B, 401
Shefter, E., 30B, 158, 206; 32B, 204, 224; 33B, 99, 138, 170, 245; 34B, 265; 35B, 164, 280, 322
Sheldrick, B., 26, 623; 28, 687; 30B, 177
Sheldrick, G. M., 32A, 510; 33A, 164; 34A, 155, 163; 34B, 364, 368; 35A, 157; 35B, 25, 486, 487, 508, 793, 794, 888, 891
Sheldrick, W. S., 34A, 155; 34B, 364, 368; 35B, 25, 508, 793, 794
Shell, H. R., 27, 698
Shen', C. -C., 28, 270
Shen, K. -W., 33B, 290
Shepherd, T. M., 30B, 448
Sheridan, J., 26, 587
Sherry, E., 31A, 126
Sherwood, R. C., 26, 351, 414; 27, 511; 29, 51, 136
Shetty, P. S., 34B, 592; 35B, 808
Shibata, S, 27, 716; 28, 362; 29, 625; 30A, 303; 30B, 447; 33B, 235; 35B, 398, 606
Shibuya, I., 26, 473; 31A, 112
Shi-Chein, L., 35A, 197
Shigeta, Y., 28, 90
Shih, C. -I., 29, 436
Shimacla, A., 32B, 166
Shimada, A., 29, 675; 26, 700; 27, 905; 31B, 298, 312; 32B, 76, 302; 33B, 353
Shimanouchi, H., 30B, 399; 31B, 91; 32B, 176; 34B, 88, 253, 311, 332; 35B, 84, 145, 199
Shimanouchi, T., 33B, 532

Shimao,E., 32A, 296, 510
Shimizu, K., 31B, 421
Shimizu, S., 28, 338
Shimizu, Y., 34B, 260
Shimoda, S., 30A, 442
Shinn, D. B., 33A, 433
Shinnaka, Y., 27, 565
Shinoda, D., 29, 139
Shinoda, G., 29, 226
Shiojiri, M., 26, 45
Shionoya, S., 29, 140
Shiozaki, Y., 28, 358
Shipley, G. G., 32B, 275, 279
Shirane, G., 26, 189, 467; 27, 150, 272, 498, 515; 29, 39, 226; 31A, 42, 241; 32A, 272
Shiro, A. M., 35B, 755
Shiro, M., 30B, 239, 445; 33B, 195, 205, 220, 480; 34B, 171, 283, 299, 574; 35B, 337
Shiroyama, M., 33B, 528
Shirozu, H., 30A, 428; 31A, 237
Shlichta, P. J., 29, 87
Shmueli, U., 27, 1054
Shoemaker, C. B., 26, 820; 28, 29; 29, 433, 754; 30A, 40, 75, 83, 181; 31B, 55
Shoemaker, D. P., 27, 416; 28, 29; 30A, 40, 75, 83, 181; 30B, 92, 31A, 218; 32A, 479, 484
Shokhor, I. N., 35B, 30
Shore, V. C., 26, 811
Shrivastava, G. P., 26, 829
Shrivastava, H. N., 26, 705
Shriver, D. F., 28, 596; 34B, 459
Shuto, Y., 35B, 747
Shvelashvili, A. E., 33B, 385
Siaud, E., 33A, 153
Šibaev, V. P., 29, 759; 35B, 886
Šibaeva, R. P., 28, 254; 30B, 122; 32B, 34, 153, 154; 33B, 61; 35B, 484
Šibayrva, R. I., 27, 714
Siciliano, F., 34A, 201
Siddall, J.B., 32B, 245
Siderenko, F. A., 29, 223
Sidhu, S. S., 28, 23; 30A, 65, 180
Sidorenko, E. F., 26, 416
Sidorenko, G. A., 26, 379, 386, 526, 536; 28, 356; 29, 437
Sidorov, P. M., 31A, 157
Siebert, H., 35A, 394
Siebert, J., 33A, 451
Siegel, L.A., 32B, 308
Siegel, S., 27, 544; 28, 73, 75; 29, 280, 340; 31A, 126; 33A, 288, 357; 34A, 361
Sieker, L. C., 27, 1060
Siemens, R. E., 34A, 146
Siemons, W. J., 33A, 36
Sienko, M. J., 26, 357; 33A, 294
Siepmann, R., 33A, 231
Sievers, R. E., 35B, 612
Sigler, P. B., 29, 759
Sigorin, D. N., 30B, 447
Sigwalt, P., 27, 811
Sigwart, C., 34B, 556
Sikirica, M., 29, 90, 137; 30A, 176; 32A, 141
Sikka, S. K., 29, 463; 32B, 101; 33A, 448; 34A, 298, 342, 362
Sikorski, J., 27, 1056
Sills, R. J. C., 35A, 178, 185
Šil'štein, S. Š., 35B, 133
Silver, A. H., 26, 486
Silverman, J., 28, 433; 30B, 171, 399; 31B, 103; 32B, 131, 137; 33B, 91
Silverman, J. N., 26, 330
Silverman, M. S., 32A, 141
Silvers, S., 28, 557
Silvers, S. J., 32B, 235
Silverton, E. W., 33B, 440
Silverton, J. V., 26, 486, 680, 691, 692; 28, 619, 632, 686; 29, 316; 30B, 53, 396; 31B, 267; 33B, 440
Silvestre, G., 30A, 40, 175, 182; 31A, 33; 33A, 54
Silvidi, A. A., 26, 344
Sim, G. A., 26, 649, 694, 792, 801, 830, 831; 27, 895, 897, 899, 926, 945, 982, 1000, 1008, 1010, 1017, 1026, 1028; 1030, 1031, 1033; 28, 533, 542, 558, 564, 686; 30B, 68, 116, 226, 227, 230, 232, 235, 236, 243, 376, 393, 397; 31B, 64, 196, 213, 216, 223, 225, 230, 236, 246, 251, 267, 275, 349, 372, 447; 32B, 114, 245, 250, 363, 368, 369, 452, 456; 33B, 103, 129, 199, 202, 213, 226, 354, 361, 366; 34B, 309, 314, 333, 342, 589; 35B, 17, 50, 333, 386, 390, 537, 548, 570
Simanov, Ju. P., 26, 245, 349, 403, 405; 27, 430, 609; 28, 317, 341; 29, 225, 435
Simaškevič, V. A., 29, 226
Sime, J. G., 26, 800; 30B, 348; 33B, 215; 35B, 12
Simensen, C. J., 35A, 8
Simerská, M., 27, 34
Simes, H. V., 28, 683
Simes, J. J. H., 28, 683
Simmons, G. L., 26, 667, 753
Simmons, R. O., 29, 224, 225, 433; 30A, 244
Simon, A., 28, 317; 32A, 210
Simon, B., 28, 317
Simon, G., 29, 433
Simon, M. S., 32B, 103
Simon, W., 32B, 242
Simonetta, M., 30B, 50, 444; 31B, 35; 32B, 41, 42; 33B, 49, 52; 34B, 82, 113, 118, 304; 35B, 93
Simonov, Ju. A., 28, 631; 35B, 640

Simonov, M. A., 32A, 302, 368, 454; 33A, 446, 469, 470; 34A, 381; 35A, 228, 283, 314
Simonov, V. I., 28, 254, 268; 30A, 437; 31A, 232; 32A, 440, 447, 511; 33A, 481; 34A, 376, 382, 384
Simonova, M. I., 29, 326
Simons, P.Y., 32A, 255
Simonsen, O., 35B, 232
Simonsen, S. H., 26, 711; 27, 883, 963; 30A, 407; 35A, 240; 35B, 619
Simpson, H. D., 35A, 17
Simpson, H. J., 31B, 187
Simpson, P. G., 26, 290, 619; 28, 60, 62, 581
Simpson, R. N. F., 32A, 510; 33A, 164
Simpson, T. D., 32B, 75; 34B, 43
Simpson, T. H., 30B, 394
Simpson, W. B., 26, 360; 28, 315
Simundza, G., 35B, 275
Sinclair, V. C., 26, 754
Singer, N., 35B, 559
Singh, A. K., 32A, 206
Singh, C., 30B, 159, 168
Singh, G., 29, 237
Singh, G. S. P., 32A, 80
Singh, I., 34B, 338
Singh, P., 30B, 139; 34B, 75, 176, 393
Singh, P. P., 34A, 145; 35B, 77
Singh, S. P., 27, 347
Singman, L., 35B, 887
Sinha, A. K., 34A, 105, 144
Sinha, A. P. B., 27, 493, 506
Sinha, U. C., 26, 731; 35B, 63
Sinnott, K. M., 26, 546
Sintes, A., 30B, 399; 31B, 19
Šipovalov, Ju. V., 29, 438
Sirdeshmukh, D. B., 26, 260, 298; 27, 369; 29, 433
Sirigu, A., 32B, 14, 305, 343, 362; 33B, 407; 35B, 515
Sirjatskaja, V. N., 33B, 530
Sirkar, S. C., 28, 682
Sirota, N. N., 27, 3; 29, 135
Šišakov, N. A., 26, 264, 539; 28, 339; 30A, 329
Sissman, E. E., 35B, 322
Šitov, V. A., 28, 318
Sitrin, A. A., 26, 378
Sivaramakrishnan, C. K., 28, 341
Sivardière, J., 32A, 268
Sivokon', N. V., 27, 70
Skakke, P. E., 35B, 798
Skancke, A., 33B, 533
Skancke, P. M., 26, 641
Skansen, T., 30A, 19, 173; 31A, 22; 34A, 27
Skapski, A. C., 31A, 29; 32A, 259, 509; 33A, 163; 33B, 18; 34A, 210; 35B, 554
Skavdahl, R. E., 27, 477; 29, 225
Skinner, B. J., 26, 251, 252, 304, 505; 28, 339; 29, 63; 30A, 244
Skinner, H. C. W., 29, 759
Sklar, N., 26, 568; 32A, 149
Sklar, R., 27, 1008
Šklover, V. E., 35B, 544
Skoblo, L. I., 29, 435
Skoldinov, A. P., 30B, 447
Skolnick, L. P., 28, 19
Skolnik, S., 32B, 253
Škol'nikova, L. M., 26, 676; 27, 821, 1065; 28, 648, 683; 30B, 448; 33B, 462; 34B, 505; 35B, 605, 620, 627
Skolozdra, O. E., 32A, 143
Skolozdra, R. V., 29, 140; 30A, 176; 31A, 27; 32A, 57, 142, 143
Skoulios, A., 27, 827, 1061
Skoulios, A. E., 26, 564
Škrabkina, M. M., 27, 398
Skrabs, P., 30B, 447
Skripkin, V. V., 33B,310
Skryševskij, A. F., 26, 615
Skšat, S. M., 30A, 437; 34A, 376
Slack, C. T., 27, 41
Sladky, F. O., 34A, 219
Slaughter, M., 28, 66; 32A, 511; 33A, 484; 35A, 456
Sleight, A. W., 26, 418; 29, 438; 31A, 148; 33A, 60, 82, 235, 287, 332, 353; 34A, 144, 239, 260
Slepetys, R. A., 26, 376
Sletten, E., 31B, 145, 147, 151; 33A, 60; 33B, 150; 34B, 189, 191, 234, 559; 35B, 318, 694
Sletten, J., 31B, 143, 151, 156; 33B, 150; 34B, 189, 191, 224, 234, 236
Sljukic, M., 33A, 398; 34A, 341
Slucker, A. I., 27, 808; 29, 760
Sly, W. G., 29, 499; 30B, 92
Smajkiewicz,R., 26, 466
Small, G. J., 29, 551
Small, R. W. H., 27, 445; 756; 28, 382, 686; 29, 260, 460; 32B, 5; 35B, 20
Smart, R. F., 27, 157
Smayerskij, V. Ju., 26, 34
Smaževskaja, E. G., 26, 419; 27, 506
Smeggil, J., 32A, 142
Smeggil, J.G., 34A, 129, 142; 35A, 105
Smirnov, G.I., 29, 436
Smirnov, Ju. M., 31A, 54
Smirnov, Ju. N., 29, 224, 225; 30A, 247, 248
Smirnov, I. A., 26, 173
Smirnov, V. P., 33A, 326
Smirnova, E. M., 27, 46, 86, 88, 286; 30A, 175
Smirnova, V. I., 33B, 68
Smith, A. E., 30B, 142, 299, 399

Smith, A. J., 28, 683; 32A, 159, 293; 34A, 180
Smith, B. C., 26, 729
Smith, B. L., 26, 286
Smith, C. P., 32B, 220, 311
Smith, D. K., 26, 407, 511; 27, 480, 547; 29, 296; 30A, 304, 307, 423
Smith, D. L., 26, 652; 27, 826, 862, 876; 32A, 328; 34B, 151, 212
Smith, G. C., 26, 166
Smith, G. D., 34A, 328
Smith, G. H., 30B, 393
Smith, G. S., 28, 119, 584, 682, 683; 29, 137, 303; 30A, 28, 53, 56, 174, 177, 179; 30B, 272; 32A, 47, 77, 87, 186, 187; 33A, 131, 205, 223; 34A, 36, 75, 84, 86, 174; 35A, 29, 124
Smith, G. W., 27, 520; 29, 329, 663; 30A, 326, 349; 30B, 444; 35A, 262; 35B, 350
Smith, H. E., 34B, 305
Smith, H. G., 27, 639; 32A, 197; 33A, 365
Smith, H. M., 30B, 358; 31B, 440
Smith, H. W., 28, 399; 29, 229; 30A, 252
Smith, J. A. S., 27, 574
Smith, J. F., 29, 137, 224
Smith, J. E. W., 26, 658
Smith, J. E. W. L., 31B, 437
Smith, J. F., 26, 128, 164; 27, 297, 298; 30A, 77, 78, 80, 177, 180, 181; 32A, 124; 33A, 6
Smith, J. H., 35A, 66
Smith, J. P., 27, 569, 577; 28, 357; 34A, 319
Smith, J. V., 27, 691, 694; 28, 246, 250; 29, 399, 401, 403; 30A, 424, 425, 426; 32A, 484; 33A, 486, 488, 497; 34A, 371; 35A, 472
Smith, L. E., 27, 875
Smith, M. F., 29, 223
Smith, M. J., 27, 40; 29, 137
Smith, N. O., 33B, 326
Smith, P., 26, 490; 35B, 46
Smith, P. H., 29, 136
Smith, P. J., 35B, 485
Smith, P. K., 34A, 80
Smith, P. L., 27, 42; 30A, 179, 246
Smith, R. A., 31B, 138; 32B, 167
Smith, W., 27, 769
Smolczyk, H. -G., 30A, 395
Smolin, Ju. I., 32A, 436; 33A, 509; 34A, 261, 378; 35A, 300, 305, 317; 35B, 441
Smolin, Ju. N., 33A, 465
Smoljaninov, I.V., 26, 169
Smorčkov, I.E., 26, 386
Smotrickaja, E.S., 33A, 121
Smuts, J., 29, 34; 34A, 389
Smyth, J. R., 33A, 421
Sneath, T. C., 35B, 191, 389
Sneeden, R. P. A., 32B, 448
Snell, P. -O., 32A, 111; 33A, 128
Snellgrove, R. A., 27, 633
Snetkov, A. Ja., 26, 213
Snow, M. R., 29, 605; 30B, 354; 35B, 686
Snyder, D. A., 35B, 588
Snyder, R. G., 27, 853
Snyder, R. L., 34B, 127
Snyman, H. C., 28, 317; 29, 357, 435
Snyman, J. A., 29, 136
Sobell, H.M., 28, 685; 29, 705, 706; 30B, 196, 201; 31B, 164; 34B, 182, 227, 228, 230, 231, 232, 233, 235; 35B, 275
Sobolev, A.S., 32A, 141
Sobolev, B.P., 28, 314
Sobolev, N.D., 26, 528
Sobolev, N.V., 26, 416
Soboleva, M. V., 26, 528
Soda, Y., 28, 318
Soden, R. R., 29, 224
Soderberg, B. -O., 31B, 26; 35B, 887
Soderberg, E., 35B, 192
Soderberg, R. H., 28, 647; 33B, 425
Söderquist, R., 33A, 510
Soderquist, S. D., 33A, 153
Soejima, T., 29, 140
van Soest, T. C., 35B, 374
Soffer, M. D., 31B, 153
Šokhor, I. N., 31B, 52; 32B, 48; 33B, 30
Sokhor, N. I., 34B, 54
Soklakov, A. I., 27, 61, 397, 401
Sokolora, Ju. A., 31B, 424
Sokolov, A. P., 26, 477; 27, 570; 28, 180; 32A, 355
Sokolova, G. V., 31B, 407
Sokolova, I. G., 26, 112; 29, 140; 30A, 177, 181, 247
Sokolova, M. A., 26, 208; 30A, 176
Sokolova, Z. A., 27, 473
Sokolovskaja, E. M., 26, 112; 30A, 182; 31A, 68
Solbakk, J., 35B, 248
Solbakk, J. K., 34A, 346
Sole, M. J., 29, 239
Solente, P., 29, 224, 226
Soleva, L. I., 35A, 226
Solheimsnes, S., 30B, 285
Soling, H., 32A, 190; 33A, 214; 34A, 211; 35B, 586
Solncev, V.M., 29, 434
Solodova, O.P., 33B, 316
Solov'ev, S.P., 26, 381; 33A, 326
Solov'ev, V.E., 26, 334
Solov'eva, L.N., 31A, 221
Solov'eva, L.P., 26, 509; 28, 264; 32A, 493; 33A, 440; 34A, 282; 35A, 188, 247, 473, 477

Soltwisch, M., 34D, 469
Somanathan, C. S., 30A, 178; 34A, 9
Somenkov, V. A., 31A, 68; 33A, 81; 35A, 133
Song Loong, W., 33B, 31
Soonpaa, H. H., 27, 90
Sørensen, A. M., 33B, 79
Sorokin, L. M., 26, 412; 28, 15
Sorokin, V. P., 28, 314
Sorrell, C. A., 28, 357
Sørum, H., 33B, 28
Sosedko, T. A., 26, 534
Šostackij, A. N., 28, 342
Sotnikova, M. N., 29, 436
Søtofte, I., 32B, 440; 35B, 153
Soule, F. B., 34B, 18
Souleau, C., 34A, 146
Soulen, J. R., 32A, 141
Southwick, P. L., 33B, 93
Spach, G., 26, 809
Spagin, I. V., 26, 56
Spagna, R., 35B, 513
Spak, Z. S., 26, 208
Spanswick, R. M., 26, 562
Sparks, R. A., 27, 1042; 29, 262; 30B, 40; 34B, 265
Speakman, J. C., 26, 556, 705; 27, 906; 28, 316, 387, 416; 29, 663; 30B, 14, 17, 278, 387; 32B, 23, 87; 33B, 45, 46; 34B, 46, 148, 150; 35B, 12
Spedding, F. H., 26, 245; 27, 330
Speed, C. S., 35B, 895
Spegt, P., 27, 827
Spegt, P. A., 26, 564
Speich, G. R., 29, 223
Speiser, R., 29, 223
Spencer, C. J., 26, 620
Spencer, C. W., 27, 40
Spencer, J. L., 35B, 502, 504
Spencer, M., 26, 817; 27, 1055
Spencer, P. M., 27, 244
Spencer, R. C., 33B, 450
Speranskaja, E. I., 29, 434
Speranskij, N. M., 30A, 178
Spicyn, V. I., 26, 245, 349, 403, 405; 28, 340, 356; 29, 435; 33A, 341; 34A, 268
Spinar, L. H., 27, 406
Spinat, P., 35A, 134
Spink, J. M., 32B, 415
Spinnler, M. A., 32A, 326
Spiridonov, V. P., 26, 362, 486; 27, 718; 34A, 151, 192, 213
Spiro, T. G., 33B, 323; 34A, 360
Spitsbergen, U., 30A, 319
Spofford, W. A., 32B, 474; 35B, 628, 784
Spohn, R. J., 35B, 505
Spooner, F. J., 27, 36; 29, 86; 30A, 178
Spragins, S. V., 30A, 244
Sprague, R. S., 31B, 31
Sprang, P. G., 30A, 172
Spratley, R. D., 32B, 311
Sprenger, H., 32A, 143; 33A, 153; 34A, 145
Spretnak, J. S., 31A, 31
Spriggs, R. M., 27, 498
Springthorpe, A. J., 34A, 146
Srensen, A. M., 32B, 269
Srikanta, S., 27, 506; 30B, 21; 33A, 369; 35A, 197; 35B, 8
Srinivasan, B. S., 30A, 180
Srinivasan, R., 29, 484; 30B, 393; 31B, 189; 33B, 156, 177, 190; 35B, 319
Srinivasan, S., 33B, 54
Srivastava, O. N., 27, 432; 29, 287; 30A, 288
Srivastava, P., 26, 743
Srivastava, R. C., 30B, 92; 33B, 31, 392
Srivastava, S. L., 33B, 540
Srivastava, S. N., 26, 750; 27, 932; 29, 683; 34B, 322; 35B, 392
Srivastava, T. S., 34B, 283
Stackelberg, M. v., 26, 667; 28, 682
Staddon, R. W., 30A, 172
Stadelmaier, H. H., 26, 97, 101, 102, 219, 224; 27, 4, 92, 95, 103, 305; 29, 135, 140; 30A, 174, 180, 244; 33A, 112; 34A, 38, 143, 144
Stadler, H. P., 29, 698; 35B, 424
Stadnicka, K., 34A, 229
Stadnjk, B.I., 30A, 177, 178
Stahlin, W., 35A, 325
Stainton, P., 28, 683
Stalhandske, C. -I., 34B, 395
Stalick, J. K., 34B, 601, 609, 610, 617; 35B, 523, 836, 875
Stam, C. H., 27, 770; 28, 377; 30B, 106; 31B, 35; 34B, 135
Stamatesku, I., 30A, 179
Stammler, M., 26, 290, 293, 594; 28, 291
Stamper, J. F., 28, 22
Standley, C. L., 26, 169; 27, 722
Stanfield, R. A., 33B, 93
Stanford, I. G., 26, 830
Stanford, R. H., 27, 414, 1058, 1059; 28, 497; 32B, 13; 35B, 215, 338
Stanko, J. A., 32A, 167; 33A, 429
Stanko, V. I., 30B, 399, 448
Stanley, E., 27, 916; 33A, 317; 34A, 224; 35A, 380
Stansbury, E. E., 26, 197
Stapleton, I. W., 27, 1053
Starikova, Z. A., 34B, 497
Starkweather, H. W., 27, 804
Starovskij, O. V., 26, 653, 679; 29, 552; 30B, 302

Stasova, M. M., 29, 140; 30A, 175; 32A, 30; 33A, 44
Stassen, W. N., 33A, 34; 35A, 12
Statton, W. O., 26, 634
Staudhammer, K., 32A, 327
Stchouzkoy, T., 29, 434
St. Clair, D., 35B, 429
Steadman, J. P., 35B, 880
Steadman, R., 27, 690; 29, 72, 403
Stecher, P., 29, 139, 226; 30A, 174; 32A, 141
Steeb, S., 26, 30, 278, 367; 27, 20; 29, 135, 223; 30A, 172, 312; 34A, 143
van der Steeg, M. G., 27, 164
Steele, R. M., 30A, 181
Steeple, H., 26, 585, 717; 27, 813; 29, 537; 31A, 72; 33A, 372; 33B, 519; 34A, 312
Steer, R. J., 35B, 77
Steggerda, J. J., 33B, 485
Stegherr, A., 26, 46
Stehr, H., 32A, 241
Steigmann, G. A., 30A, 62, 179; 31A, 17; 32A, 13; 34A, 49; 28, 24; 33A, 25
Steigemann, W., 35B, 137
Stein, C., 30A, 175
Stein, M. L., 26, 776
Steinberg, M., 27, 815
Steinberger, I. T., 32A, 135; 34A, 133
Steinfink, H., 26, 534; 27, 692, 696; 28, 576; 29, 255; 30A, 178, 180, 181, 246, 247, 376; 30B, 443; 31A, 20, 47, 50; 32A, 18, 22; 33A, 176; 34A, 188, 253; 35A, 17, 25; 35B, 636
Steinhauser, S., 35B, 479
Steinrauf, L. K., 26, 1058, 1059; 30B, 242, 337; 32B, 230, 254; 33B, 184; 35B, 754, 888
Steitz, T. A., 30B, 194
Stember, N. G., 26, 372
Stempelová, D., 26, 669; 27, 438; 29, 606
Stemple, N. R., 27, 307; 28, 315; 31A, 133; 31B, 5
Stenberg, E., 26, 76; 27, 93
Stensland, B., 35B, 445
Stenson, P., 34B, 572; 35B, 241
Stepa, T. D., 29, 139
Stepanov, A. V., 26, 333
Stepanov, V. I., 29, 433
Stepanova, A. A., 26, 66; 27, 85, 374; 29, 224, 226
Stephen, W. F., 35B, 322
Stephens, D. N., 35B, 191
Stephens, F., 35A, 414
Stephens, F. S., 28, 642; 30B, 275; 34A, 155; 34B, 486, 489, 492, 552, 569; 35B, 51, 412, 525, 557, 774
Stephens, J. S., 32A, 357; 33A, 399; 34A, 330; 35A, 206, 382, 384
Stephenson, D. A, 33A, 502
Stephenson, N C., 27, 834, 869; 28, 683; 29, 576. 578, 580; 30A, 18, 173, 341, 346; 30B, 445; 31A, 153, 231; 32A, 26; 33A, 348, 352, 359; 34A, 288; 34B, 512, 566, 591, 594, 606; 35B, 646, 672, 703, 839
Sterling, C., 29, 496, 497, 30B, 22; 32B, 331
Sterns, M., 29, 438; 32A, 317; 35B, 775
Sterr, G., 28, 602
Steuck, M. J., 35B, 602
Stevens, A. L., 30A, 179, 245
Stevens, K. L., 28, 543
Stevens, R. M., 29, 141; 30A, 174
Stevenson, D. L., 32A, 404; 32B, 350, 465
Stevenson, R., 29, 752
Steventon, B. R., 32A, 508; 33A, 172
Steward, E. G., 27, 474
Stewart, D., 31A, 49
Stewart, D. F., 32A, 160
Stewart, F. H., 28, 340
Stewart, J. M., 26, 674; 27, 432; 31B, 68; 32B, 57; 34B, 103; 35B, 161, 219
Stewart, O. M., 26, 252
Stewart, R. F., 29, 708; 32B, 203
Stewner, F., 35A, 220, 225
Steyn, J. G. D., 26, 301; 34A, 389
Stezowski, J. J., 34B, 464
Stiefel, E. I., 31B, 456; 35B, 817
Stigliani, W. M., 35B, 904
Stinchfield, R. P., 27, 723
Stirpe, D., 27, 721
Stišov, S. M., 27, 675
Stobo, J. J., 29, 225
Stocks, K., 35A, 69
Stockwell, A.,26, 819
Stogsdill, R. M., 29, 597
Stoicheff, B. P., 27, 732
Stojancova, Z. P., 30A, 180, 247
Stokely, P. F., 32A, 509; 35B, 85, 539
Stokes, J., 33B, 481
Stol'c, A. K., 26, 149, 386; 27, 208; 29, 141, 225; 30A, 178, 245
Stølevik, R., 35B, 889, 893
Stoljarova, T.I., 33A, 446
Stoll, E., 31A, 163; 32A, 315
Stoll, H., 26, 315
Stolz, E., 27, 161, 232, 377
Stomberg, R., 27, 857; 28, 224; 29, 302,541; 30A,292; 30B, 328, 350; 33A, 330; 33B, 439; 35A, 421; 35B, 635
Stone, A., 33B, 185
Stone, A. L., 31B, 369

Stone, P. L., 27, 129; 29, 223
Stone, R. G., 33B, 540
Stora, C., 26, 735; 31B, 75, 82
Storm, A. R., 28, 21; 30A, 52, 178
Storms, E. K., 27, 137; 30A, 40, 175, 245
Stout, E. H., 34B, 338
Stout, G. H., 26, 830; 27, 1005, 1006, 1007; 28, 543, 562; 29, 684
Stout, V. F., 27, 1006; 28, 562
Strähle, J., 33A, 226; 35A, 154
Strahs, G., 33B, 113
Strand, T. G., 35B, 889
Strandberg, B., 26, 657, 810, 811
Stratton, R. P., 29, 53, 224
Straumanis, M. E., 26, 247, 275, 352; 27, 151, 484; 29, 141, 223, 226, 30A, 244, 246, 248; 35A, 353
Strauss, A. J., 26, 43
Street, R. S., 29, 226
Streib, W. E., 27, 393; 28, 687; 29, 229, 526; 31B, 291; 32B, 373; 35B, 435
Strelcova, I. N., 26, 712, 713
Streurman, H. J., 35B, 218
Strickler, D. W., 26, 394
Strickler, P., 29, 735; 30B, 89; 31B, 87; 34B, 596
Strid, K.G., 31A, 78; 31B, 39
Strieter, F.J., 26, 831; 27, 736, 737, 891
Stroganov, E. V., 26, 334, 477; 27, 570; 28, 102, 180; 32A, 355
Stromberg, R., 34B, 480
Strømme, K. O., 33B, 238, 243, 532; 34A, 338, 346; 34B, 70; 35A, 322
Strong, S. L., 33A, 258
Stroud, R. M., 32B, 186
Struck, C. W., 27, 582; 28, 316
Stručkov, Ju. T., 26, 653, 679, 712, 713, 745, 746, 831; 27, 784, 860, 921, 923, 924, 933, 939; 28, 438, 682, 683, 685, 687; 29, 549, 552, 678, 679, 697; 30A, 275; 30B, 98, 99, 274, 302, 308, 310, 394, 396, 399, 443, 448; 31B, 66, 333, 343, 354, 361; 32A, 511; 32B, 68, 330, 331, 336; 33B, 87, 304, 310, 316, 330, 332, 337, 340, 348, 349, 355, 358; 34B, 414, 418, 420, 434, 435, 447, 448, 450, 467, 479; 35B, 407, 489, 540, 544, 556, 562, 573, 574, 900, 901, 905
Strunz, H., 26, 289, 380, 398, 438, 478 515, 537, 540; 27, 478, 488, 1036
Strydom, O. A. W., 35A, 49
Stuart, B. A., 27, 514
Stubican, V., 27, 695
Stubican, V. S., 28, 317; 29, 435
Stuckes, A. D., 29, 224
Stucky, G., 29, 532; 31B, 342
Stucky, G. D., 28, 685; 32B, 34, 328; 33A, 215; 33B, 301, 520; 34A, 271; 34B, 398, 399, 401, 403, 406; 35A, 160; 35B, 228, 474, 478, 481, 709, 878
Studd, B. F., 34B, 445
Stumpel, E. F., 30A, 173
Subasiev, V. K., 30A, 244
Subbarao, E. C., 26, 388, 402; 27, 504, 538, 540; 28, 341
Subramanian, E., 31B, 70, 111; 32B, 229; 34B, 263, 272; 35B, 23, 74, 279
Šubrtová, V., 34B, 162
Suchet, J., 30A, 172
Suchow, L., 27, 467; 28, 315; 33A, 305
Sudarsanan, K., 30A, 401, 402
Sudo, T., 33A, 253
Suehiro, K., 33B, 521; 35B, 884
Suemene, Y., 26, 306
Sueno, S., 32A, 123
Šugam, E. A., 26, 676; 27, 822, 1065; 28, 648, 683; 33B, 462, 486; 34B, 497, 505; 35B, 605, 620, 627
Sugden, T. M., 26, 347; 27, 719
Sugie, C., 26, 286
Sugihara, A., 33B, 418
Sugihara, M., 26, 8; 27, 10, 15, 243; 30A, 178
Sugino, T., 34B, 287
Sugiyama, K., 35B, 84
Suh, I.-H., 35A, 197
Suits, J.C., 29, 140, 226
Sulanin, B. M., 27, 507
Suleimanov, M. Kh., 33A, 140
Sullenger, R. D., 28, 314
Sullivan, D. G., 27, 507
Sullivan, R.A.L., 27, 749; 30A, 244; 33B, 64
Sullivan, S., 32B, 279
Sulonen, M. S., 27, 106
Sumjackaja, N. G., 27, 473; 32A, 510; 33A, 437; 34A, 379
Summer, R. H., 34B, 519
Sumner, G. G., 29, 241
Sunateur, J. P., 30A, 36
Sundaralingam, M., 26, 748; 27, 975, 988; 28, 420, 684; 30A, 397; 30B, 56, 192, 202; 31B, 60, 107, 185; 34B, 92, 103, 192, 261; 35B, 249, 294, 296, 297, 299
Sundara Rajan, A., 26, 588
Sundsfjord, J., 30B, 396
Sung, N., 33B, 71
Šur, Ja. S., 27, 9
Suražskaja, M.D., 35B, 637
Surin, A.K., 29, 225, 226; 30A, 245
Surjadi, A.J., 32A, 325
Süsse, P., 28, 86; 29, 278; 32A, 338, 419, 510; 33A, 377, 379; 35A, 363
Susz, B. -P., 26, 703
Sutarno, 30A, 317; 32A, 126; 33A, 291

Sutherland, H. H., 28, 524; 30A, 62, 179; 30B, 445; 33B, 72, 73; 34B, 63, 64, 81; 35B, 90
Sutherland, S. A., 27, 1033
Sutor, D. J., 28, 510, 550, 587; 32A, 360
Sutton, A. L., 28, 41
Sutton, D., 32A, 350
Sutton, L. E., 28, 362
Sutton, P. W., 28, 70; 32B, 341
Suzuki, K., 26, 316; 27, 15; 28, 341; 32A, 431; 33A, 276
Suzuki, T., 33A, 33
Suzuoka, T., 33A, 85
Svečnikov, V. N., 29, 226
Svedung, D. H., 33A, 246
Švejkin, G. P., 29, 135, 224
Švelašvili, A. E., 30B, 399; 31B, 381; 34B, 481
Svensson, A. -M., 34B, 603
Svetich, G., 30B, 193
Svets, G. N., 30A, 178
Swallow, A. G., 26, 831; 27, 886; 31A, 142; 33B, 445; 35B, 559
Swaminathan, S., 32A, 206; 33B, 54; 34A, 306
Swamy, N. K.A., 27, 22
Swanbeck, G., 27, 1056
Swank, D., 32B, 220
Swank, D. D., 30A, 296
Swann, P. R., 27, 12
Swanson, H., 32A, 355
Swanson, H. E., 27, 723; 29, 141
Swanwick, M. G., 34B, 418
Swaroop, B., 29, 249
Sweeney, D. M. C., 27, 732
Sweet, R. M., 30B, 173; 32B, 332
Swink, L. N., 31A, 175; 33A, 230
Switzer, G., 28, 340
Swoboda, T. J., 26, 354
Sygusch, J., 33A, 17
Syneček, V., 26, 411; 31A, 199; 34A, 296
Synorov, V. F., 30A, 179
Syrico, L.F., 29, 136, 434, 435
Szabó, P., 21, 225; 30A, 308
Szklarz, E. G., 33A, 153
Szymanski, J.T., 30B, 395; 32B, 420, 421; 34B, 357
Szytuta, A., 33A, 266

Tachikawa, R., 28, 687
Tadini, C., 27, 584; 28, 407; 32A, 362; 34A, 381
Tadokoro, H., 27, 812; 29, 760; 33B, 521, 524; 35B, 884
Taesler, I., 33A, 364; 35A, 365
Taga, T., 34A, 347; 34B, 344; 35B, 259, 400
Tagaya, M., 27, 156
Taggart, A.A., 30A, 280
Tahara, S., 29, 760
Tai, Y. -H., 35B, 896
Tain, P. D, 34B, 486
Tajukin, V. I., 29, 380
Takabatake, A., 31B, 264
Takabatke, M., 32A, 511
Takada, H., 27, 15
Takada, O., 35B, 41
Takada, Y., 29, 138
Takagi, K., 28, 668
Takagi, S., 34B, 242
Takahashi, S., 33B, 500
Takahashi, T., 29, 140; 34A, 146
Takaki, H., 33A, 111
Takaki, Y., 28, 412; 30B, 60, 61
Takanashi, S., 35B, 229
Takano, T., 28, 583; 30B, 399; 31B, 121
Takano, Y., 32A, 511
Takats, J., 33B, 371, 375; 35B, 561
Takayanagi, M., 35B, 884
Take, T., 35B, 402
Takeda, H., 28, 249; 29, 406; 31A, 231; 32A, 471
Takeda, Y., 35B, 886
Takei, W. J., 27, 150, 272, 498, 515; 29, 721
Takenaka, A., 35B, 171, 765, 768
Takéuchi, Y., 27, 707; 29, 68; 30A, 174, 355; 31A, 238, 32A, 441; 33A, 148; 34A, 377; 35A, 11
Takigawa, T., 31B, 191
Takizawa, T., 27, 805
Talhouk, S., 31A, 240
Talukdar, A. N., 27, 1033
Talybov, A. G., 27, 83; 29, 137; 30A, 179; 31A, 18, 25; 33A, 42; 34A, 33
Tamoto, Y., 29, 140
Tamura, C., 28, 529, 687; 30B, 236; 31B, 239, 241; 33B, 129, 199; 35B, 390
Tamura, G., 34B, 286
Tamura, Z., 35B, 229
Tanaka, A., 29, 760
Tanaka, J., 28, 451; 33B, 50, 60; 34B, 237
Tanaka, K., 27, 799; 34B, 441
Tanaka, N., 32B, 76, 150
Tanaka, T., 30A, 179
Tanaka, Y., 31A, 58
Tanako, I.A., 29, 225
Tananaev, I.V., 29, 435
Tang, H.-Y., 29, 237
Tangocci, P., 34B, 147
Tangri, K., 30A, 247, 248
Tani, B., 34A, 361
Tani, B.S., 27, 116; 29, 32, 135
Tani, M.E., 27, 653; 28, 356
Tani, M.E.V., 35B, 38
Tani, V., 34A, 234
Taniguchi, T., 30B, 60, 61
Tanimura, M., 35B, 111
Tanimura, M.M., 32B, 432

Tank, R., 27, 20; 29, 223
Tannenbaum, I. R., 27, 237
Tanner, L. E., 30A, 176
Tanon, A., 30A, 178
Tapscott, R. E., 33B, 409
Tapsell, Q. T., 30A, 355
Tarasenko, N.A., 35B, 901
Tardieu, A., 28, 357
Tare, V. B., 27, 506
Tarkhova, T. N., 30B, 412; 35A, 195
Tarnóczi, T., 29, 225
Tarnovskij, G. N., 29, 436
Tashiro, M., 34A, 146
Tasker, P. A, 35B, 823
Tašpulatov, Ju., 27, 1003
Tatarinova, L. I., 27, 812
Tate, E. F., 29, 225
Tate, M. E., 35B, 274
Tatekawa, M., 29, 437
Tauber, A., 28, 357
Tavale, S. S., 28, 384; 29, 471; 32B, 76; 34B, 19, 227, 231
Tavernier, B., 31A, 43
Taxer, K. J., 32A, 495; 35A, 333
Taylor, A., 27, 281; 29, 224; 30A, 179, 247
Taylor, D. M., 29, 434
Taylor, H. F. W., 27, 699; 29, 434, 437; 32A, 422, 510; 35A, 378; 35B, 604
Taylor, I. F., 33B, 469
Taylor, J., 29, 225
Taylor, J. B., 29, 18, 137; 30A, 173, 244; 35A, 93, 124
Taylor, J. C., 28, 560, 683; 30A, 404, 406; 31A, 179; 35A, 169
Taylor, K.A., 35B, 864
Taylor, M. R., 30B, 349, 396; 31B, 25, 454; 34B, 24
Taylor, P.A. M., 29, 412
Taylor, R. P., 33B, 517; 34B, 516
Taylor, S. R., 35B, 846
Taylor, T. I., 27, 421
Taylor, W. H., 27, 8, 676; 34A, 368
Taylor, W. K., 27, 1008
Tazzoli, V., 32A, 444; 34B, 41; 35B, 762
Tecotzky, M., 29, 139
Tegenfeedt, J., 31A, 128
Telegus, V. S., 34A, 143
Telešova, R. L., 29, 437
Temple, D. M., 35B, 419
Templeton, D. H., 27, 455, 571, 616, 619, 736, 737, 774, 891; 28, 71, 75, 130, 181, 226, 228; 29, 264, 273, 274, 288, 289, 331, 335, 347, 356, 560; 30A, 311; 30B, 305; 31A, 79, 84, 101, 179; 31B, 294, 315, 414; 32A, 161; 32B, 238, 294; 33A, 158; 33B, 90, 101, 265, 323, 394; 430, 435, 517; 34A, 191, 360; 34B, 316; 35B, 313, 429
Templeton, L. K., 29, 356; 33B, 90
Temussi, P. A., 28, 428; 29, 468; 32B, 14; 33B, 4
Tennissen, A. C., 33A, 181
Terao, N., 27, 252, 288; 28, 122; 29, 139; 30A, 180
Terao, Y., 31B, 258
Terekhova, V. F., 27, 161, 268; 30A, 178
Ternay, A. L., 32B, 187
Teske, C., 34A, 253
Teske, C. L., 34A, 248; 35A, 217, 246
Tesljuk, M. Ju., 27, 11, 212, 260; 29, 140, 225; 30A, 172, 176, 177
Tettenhorst, R., 33A, 477
Teufer, G., 27, 480; 29, 257
Tharp, A. G., 27, 336; 30A, 56, 179; 32A, 77, 87; 34A, 36
Thayer, T. P., 29, 434
Theobald, F., 28, 340
Thery, J., 27, 519
Thewalt, U., 31B, 451; 32B, 426; 33A, 361; 34B, 259, 392, 35A, 409; 35B, 251, 283, 480,
786
Thieberg, K. J. L.,28, 684
Thiedemann, D., 34A, 144
Thiel, R., 28, 283
Thiele, G., 30A, 276; 31A, 106; 33A, 236, 240; 33B, 183, 246
Thiessen, W. E., 34B, 194, 196; 35B, 345
Thijssen, J. J. M., 33A, 324
Thilo, E., 28, 314, 317; 29, 433, 438; 31A, 121; 32A, 463
Thoai, N., 34B, 3
Thom, E., 32A, 234; 32B, 49
Thoma, R. E., 35A, 166
Thomas, D., 29, 135, 139, 435
Thomas, G., 30A, 41
Thomas, J., 27, 408
Thomas, P. C., 34B, 211
Thomas, R., 31B, 55
Thompson, B. P., 34B, 111
Thompson, D. T., 32B, 469, 475; 33B, 501, 503; 34B, 600; 35B, 848
Thompson, J. R., 28, 36
Thompson, M. C., 34A, 80
Thompson, M. E., 27, 603
Thompson, R. C., 31A, 192; 32A, 334; 35A, 191
Thompson, R. M., 27, 405
Thompson, J. R., 27, 317, 325, 367; 29, 78, 136, 224
Thorbjornsen, T., 27, 43
Thoret, J., 33A, 338
Thornber, M. R., 33A, 291; 35A, 259
Thornley, A. S., 35B, 621
Thorup, N., 35B, 586
Thozet, A., 30B, 394
Threadgold, I. M., 28, 357; 34A, 368

van Thuijl, J., 28, 476; 31B, 134, 387; 33B, 127
Thurmann-Moe, T., 29, 450
Thurn, H., 31A, 57; 34A, 149
Thyagaraja Rao, S., 33B, 177
Tichy, K., 27, 1066; 32B, 14; 34B, 69
Tien, T. Y., 27, 504; 28, 341
van Tiggelen, P., 27, 435
Tikhmskij, G. F., 27, 65
Tillack, J. V., 35A, 379
Tillmanns, E., 34A, 255; 35A, 199, 271, 284
Tilloca, G., 29, 434
Timaseva, T. P., 30B, 406
Timasheff, S. N., 29, 759
Timkovich, R., 34B, 278
Timmons, C. J., 27, 1002
Timms, R. E., 33B, 317, 318
Ting, K. -S., 27, 703
Tippe, A., 34A, 151
Tiripicchio, A., 27, 640, 641; 30A, 350; 30B, 445; 31A, 103, 208; 32A, 151, 352, 509; 32B, 425, 427; 33B, 436; 34A, 340, 350; 34B, 34, 547; 35A, 397; 35B, 720
Tiripicchio Camellini, M., 34A, 350
Tiščenko, G. N., 28, 684
Titora, V. M., 27, 395
Titova, K. V., 30B, 448
Titze, H., 34B, 503; 35B, 608
Tjapkin, Ja. D., 30A, 244
Tjørnhom, T., 33B, 243
Tkačenko, E. V., 28, 341
Tkacev, S. P., 34A, 378
Tkačev, V. V., 33A, 227; 35A, 431
Tobias, R. S, 33B, 321
Tobisch, J., 32A, 137
Todd, P. F., 32B, 437
Todorovič, J., 33A, 266
Toft, L., 32B, 28
Tokareva, S. A., 27, 482; 29, 433
Tokonami, M., 27, 707; 31A, 77; 34A, 262, 377
Tokuma, Y., 34B, 270
Tolkačev, S. S., 27, 51; 29, 138
Tolles, W. M., 27, 717
Tollin, P., 27, 1063; 30B, 198; 33B, 73; 34B, 262; 35B, 288
Tolmačev, S. M., 34A, 238
Toman, K., 27, 1066; 28, 315; 30B, 257; 31B, 59; 32B, 71
Tomašpol'skij, Ju. Ja., 28, 135; 29, 323, 435, 760; 32A, 269; 33A, 305
Tombs, N. C., 29, 434
Tomii, Y., 31B, 194
Tomiie, Y., 28, 682; 30B, 397, 398; 31A, 44; 33B, 223; 34B, 441; 35B, 163, 367, 368, 388, 402, 763
Tomita, K., 28, 279; 29, 605, 705, 706, 759; 30B, 396; 32B, 207
Tomita, K. -I., 27, 971, 977; 31B, 163
Tomita, M., 33B, 223
Tomita, T., 28, 65; 31A, 77
Tomkins, I. B., 35A, 184
Tomko, J., 33B, 228
Tomlinson, A. A. G., 35B, 799
Tomlinson, D., 34A, 178
Tompson, C. W., 27, 720, 721
Tomue, Y., 31B, 264
Tondello, E., 35B, 762, 819
Toneman, L. H., 30B, 405; 33B, 537
Toney, J., 35B, 709
Topping, G., 32A, 185
Torbjornsson, L., 34B, 156
Tordjman, I., 32A, 382; 33A, 417; 35A, 346, 347
Torelli, L., 33B, 486
Torii, K., 35B, 310
Toropov, N. A., 27, 699; 29, 438
Torrie, B. H., 29, 409
Torti, E., 32B, 142; 33B, 83
Tóth, J., 29, 225
Toth, R. S., 27, 176, 180; 30A, 247; 31A, 42; 32A, 89; 33A, 89
Tou, J. C., 32A, 510
Tournoux, M., 28, 357; 29, 435
Toussaint, J., 32B, 188
Toussaint, N., 34A, 144
Touw, T. R., 29, 334
Tovbis, A. B., 30A, 364
Tovstjuk, K.D., 30A, 181
Towl, A. D. C., 35B, 852
Townes, W. D., 32A, 290; 34A, 3; 35A, 265; 35B, 852
Towns, R., 29, 506; 32B, 163
Towns, R. L., 34B, 127; 35B, 184
Towns, R. L. R., 35B, 207, 455
Toyoda, K., 28, 687
Tracy, J. W., 27, 432, 433
Traetteberg, M., 28, 362; 29, 449, 503; 30B, 405; 33B, 533, 534; 35B, 892, 899, 900
Trail, R. J., 28, 340; 34A, 368
Tranqui, D., 32A, 379, 380, 381; 33A, 395
Tran-Qui Duc, 31A, 143
Tranter, T. C., 29, 760
Traub, W., 27, 1054; 31B, 193
Traylor, T. G., 33B, 206
Trebble, R. S., 27, 41; 33A, 97
Trefilov, V. I., 27, 232
Trefonas, L. M., 28, 397, 471, 472; 29, 506; 30B, 144; 31B, 163; 32B, 30, 154; 33B, 109, 127, 128; 34B 127, 289, 387; 35B, 184, 190, 203, 207, 455
Treichel, P. M., 31B, 31, 356
Tressaud, A., 33A, 183, 184
Tresvjatskij, S. G., 28, 434
Tret'jakov, D. N., 29, 136
Treušnikov, E. N., 35A, 464, 465
Tribalat, S., 29, 434
Trichet, L., 33A, 153
Tridot, G., 29, 135, 139, 435
Trigunayat, G. C., 27, 431; 33A, 517

Trinderup, P., 35B, 788
Trinkl, A., 33A, 134; 34A, 146
Triquet, A. M., 27, 95; 29, 226
Troć, R., 29, 226; 32A, 21
Troickaja, N. F., 30A, 180
Troickaja, N. V., 28, 32
Troickaja, Z. V., 29, 141, 226
Trojan, A. O., 29, 140
Trojer, F., 30A, 433
Trojer, F. J., 31A, 41; 32A, 511; 33A, 453
Trömel, M., 32A, 305; 33A, 288; 34A, 263, 264
Trotter, J., 27, 405, 665, 764, 816, 864, 866, 867, 868, 892, 935, 938, 884; 28, 405, 418, 434, 435, 439, 440, 585, 586, 588, 591, 592, 612, 683, 685; 29, 539, 549, 625, 655, 681, 694, 746; 30A, 285, 287; 30B, 68, 109, 111, 114, 123, 184, 185, 204, 224, 229, 278, 282, 283, 444, 446, 448; 31A, 82, 111, 192; 31B, 67, 69, 109, 172, 208, 220, 249, 359, 466; 32A, 155, 160, 334,369; 32B, 112, 151, 171, 337, 339, 361, 374 33A, 174; 33B, 70, 165, 281, 306; 34A, 324; 34B, 249, 333, 373, 619; 35A, 191; 35B, 21, 23, 74, 121, 123, 225, 270, 314, 404, 459, 461, 490, 558, 845, 866
Troughton, P. G. H., 35B, 554
Trueblood, K. N., 27, 641, 1042; 28, 452, 534, 537; 30A, 378; 30B, 40, 121, 206, 306; 32B, 118, 323; 33B, 48, 119, 256; 34B, 42, 50, 84, 265
Truesdell, A. H., 26, 381
Trukhtanov, V. A., 33A, 324
Trunov, V. K., 28, 318, 341; 29, 436, 438; 31A, 150; 33A, 341
Truong, T. B. A., 26, 357
Trus, B. L., 34B, 153
Truter, M. R., 26, 496, 684, 831; 27, 752, 754, 824, 841, 886; 28, 589, 623; 29, 485, 516; 30B, 339, 364; 31B, 419; 32B, 53, 329, 384, 399, 460; 33B, 335, 377, 492; 34B, 432, 451; 35B, 274, 476, 510
Trykozko, R., 33A, 153; 35B, 274, 476, 510
Trzebiatowski, W., 26, 432; 29, 135, 136, 140, 223, 226
Tsai, C. -C., 31B, 291; 35B, 435
Tsai, J. H., 32A, 510
Tsao, R. -L., 27, 703
Tsao, Y. -L., 27, 703; 29, 437
Tscherry, V., 35A, 452
Tschirnich, J., 30B, 443
Tschöpp, R., 31A, 138
Tselinskij, I. V., 35B, 19, 30
Tseng, J. -K., 31A, 236
Tskhai, V. A., 34A, 58
Tsoucaris, G., 26, 695, 702, 720; 31A, 85, 87, 132, 225, 31B, 439
Tsu, I., 26, 449
Tsuboi, M., 33B, 180; 34B, 268; 35B, 886
Tsuboya, I., 26, 8; 27, 10, 15; 30A, 178
Tsuda, K., 28, 687; 31B, 239, 241
Tsu-Hsiang, L., 26, 523
Tsukiyama, K., 30A, 421
Tsukuda, Y., 33B, 195; 34B, 283, 299, 574; 35B, 260, 337
Tu, K. -C., 29, 435
Tube, M. L., 30B, 393
Tuck, D. G., 34B, 23, 619; 35, 322
Tucker, P. A., 29, 135
Tufte, T., 26, 763; 34A, 131, 146
Tuinstra, F., 28, 52; 31A, 77; 32A, 508
Tulinsky, A., 27, 1015, 1025; 28, 557; 31B, 262; 32B, 235, 240, 270; 34B, 278
Tullberg, A., 32A, 237
Tumanov, V. I., 30A, 248
Turbil, J. -P., 34A, 90
Turco, A., 28, 683
Turco, G., 26, 535
Turkdogan, E. T., 27, 548
Turley, J. W., 33B, 33, 270, 271; 34A, 153; 34B, 34, 366, 367
Turnbull, J. A., 28, 65; 30A, 251
Turner, R. W., 28, 684; 31B, 364, 365
Turner-Jones, A., 27, 799, 807
Turova, N. Ja., 30B, 448
Tustin, G. C., 35B, 807
Tutov, A. G., 29, 435
Tverdovskij, I. P., 30A, 248
Tydings, J., 30A, 173
Tylkina, M. A., 26, 23, 58, 192, 217, 241; 27, 233, 313, 314, 337, 338; 30A, 178
Tyrina, L. V., 26, 42
Tyrrell, J., 30B, 394
Tyson, G. J., 27, 671

U, C. -J., 28, 158
Ubbelohde, A. R., 26, 287; 31A, 240
Ubell, E., 27, 945; 32B, 111
Uchida, N., 30A, 177
Uchida, T., 32B, 273
Uchikawa, H., 30A, 421
Uchikoshi, H., 26, 367; 29, 138
Uchtman, V. A., 34B, 457, 466
Uda, M., 33A, 107
Udalov, Ju. P., 29, 438
Udalova, V. V., 26, 303; 28, 206
Ueda, H., 33B, 50
Ueda, I., 26, 714; 32A, 41
Ueda, R., 27, 572
Ueda, T., 27, 437; 28, 279; 32A, 358

Ueki, T., 30A, 311; 31A, 179; 32B, 227, 392; 33B, 313, 418, 500; 34B, 276, 455, 513; 35B, 521
Uesu, T., 34B, 45
Ugaj, Ja. A., 30A, 174, 175
Ugaj, V. A., 30A, 174
Uggla, R., 35B, 674, 675
Ugol'nikova, T. A., 29, 326
Ugryn, N. S., 30A, 176
Ukida, J., 27, 806
Üklü, D., 32A, 315
Ukraiskij, Ju. M., 29, 225
Ulicky, L., 27, 894; 28, 682, 683
Ullner, N. -A., 33A, 153
Ulmer, L., 29, 141
Ulrich, L., 34B, 338
Uma, D., 30A, 248
Umanskij, Ja. S., 27, 285; 30A, 245, 246
Umezawa, H., 31B, 194
Ungaretti, L., 32A, 444, 466; 33A, 489; 34B, 41
Ungelenk, J., 27, 426
Unger, P., 32A, 137
Uno, T., 30B, 443; 33B, 447
Upadhyaya, G. S., 27, 238
Upadyseva, A.V., 31B, 46
Upatova, T.V., 34A, 261
Urantani, Y., 32A, 511
Uraz, A. A., 35A, 38
Urdy, C. E., 27, 883
Urlau, R., 27, 550
Urnes, P. J., 26, 816
Uskokovic, M., 34B, 211
Ustinov, V. M., 29, 436
Ustjnjuk, Ju. A., 35B, 900
Usubillaga, A., 35B, 362
Utsumi, H., 35B, 44, 171, 765, 768
Utsumi, Y., 35B, 163, 713
Uttech, R., 28, 685; 30B, 298

Vaciago, A., 26, 668,672; 27, 968; 28, 456; 682, 683, 687; 29, 286; 30A, 444; 30B, 373, 378, 381; 31B, 131; 32B, 35; 33B, 473, 476, 486; 34B, 35; 35B, 173, 379, 759
Vadec, D. I., 26, 34
Vagg, R. S., 34B, 565
Vagliasindi, G., 32A, 164
Vaidya, S. N., 30A, 682; 30B, 270
Vaipolin, A. A., 28, 293, 316; 33A, 49, 126
de Vaira, M., 30B, 394, 398
Vajnštejn, B. K., 26, 63, 569, 572, 573, 810; 27, 83, 748, 812; 28, 682; 29, 383; 30B, 209; 34A, 33
Vajpolin, A. A., 29, 136, 139; 30A, 22, 174; 35A, 19
Valambal, V., 33B, 177
Valcha, Z., 29, 436
Valentine, J., 30B, 393
Valigi, M., 28, 317
Valjavko, L. R., 34A, 144
Valle, G., 30B, 152; 33B, 136; 34B, 181, 188; 35B, 155
Valle, S., 34B, 430
Vallet, P., 29, 434
Valletta, R. M., 27, 330; 29, 760, 761
Valov, Ju. A., 35A, 19
Van, K. -Z., 28, 149
Vancleave, A. B., 28, 317
Vand, V., 35A, 461
Van den Hende, J. H., 27, 850; 29, 343
Van Domelen, G. H., 27, 1061
Vannerberg, N. -G., 26, 479; 27, 117, 419, 667; 28, 91, 183, 184; 29, 293; 30A, 297, 447; 30B, 399; 31A, 78, 116, 215; 31B, 39; 32A, 195, 237; 33A, 246; 35A, 398, 401
Van-Wazer, J. R., 26, 467
Varadarajan, S., 26, 511
Varteva, M., 29, 433
Vasil'ev, E. K., 28, 341; 29, 436
Vasil'ev, V. A., 26, 534
Vasilevskaja, M. M., 27, 12
Vasjutinskij, B. M., 26, 110; 27, 356; 31A, 54
Vaska, L., 27, 873
Vassian, E. G., 34B, 556
Vaughan, D. A., 26, 252; 27, 481
Vaughan, P. A., 29, 693; 31A, 226; 32A, 194; 35A, 415, 416
Vaughan, R. W., 30A, 246
Vaugoyeau, H., 30A, 182
Vazquez, J. C., 26, 575
Vdovenko, V. M., 26, 477; 27, 570; 28, 180; 32A, 355
Veda, R., 26, 575; 27, 497
Veidis, M. V., 34B, 503
Vekris, J., 33B, 437
Velge, W. A. J. J., 26, 164
Velicko, F. K., 33B, 320
Venables, J. D., 33A, 57
Venanzi, L. M., 26, 690; 28, 658, 683, 687; 31B, 463; 32B, 376;35B, 513
Venevcev, Ju. N., 26, 372, 381, 419; 27, 505, 506, 508; 28, 135; 29, 323, 326, 434, 435; 32A, 269; 33A, 305
Veniaminov, N.N., 35B, 900
Venkatesan, K., 26, 645; 27, 1040, 31B, 111: 34B, 274
Venkateswarlu, K., 26, 829
Venteicher, F. R, 35A, 124
Venuto, C. J., 26, 478
Verbist, J., 31B, 274; 35A, 195; 35B, 474
Verescagin, L.F., 27, 424, 504, 507, 794; 28, 356; 29, 138, 141, 224, 226; 33A, 45; 34A, 239; 35A, 9
Verhoef, L. H. W., 33B, 241; 34B, 356

Verkade, J. G., 33B, 278, 279
Verkhogljadova, T. S., 30A, 182
Verkhorobin, L. F., 27, 70; 29, 138
Verkhovskij, V. Ja., 35A, 316
Verma, A. R., 27, 407, 431, 432; 28, 64; 29, 237, 238, 287; 30A, 288
Verma, H. K. L., 34A, 281; 34B, 106
Verma, J., 26, 86
Verschoor, G. C., 33B, 147
Vert, Z. L., 30A, 248
Vesela, L., 32B, 251
Vesel'skij, I., 26, 420
Vesnina, B. I., 28, 314
Vestersjo, E., 33A, 121
Viallard, R., 30A, 246
Vicat, J., 33A, 443
Vickery, R. C., 26, 118, 142, 143
Victor, D., 34B, 38
Vidavskij, L. M., 28, 341
Viebahn, W., 32A, 508
Vielhaber, E., 30A, 325; 33A, 325; 34A, 250
Viers, J. W., 32A, 508
Vigileva, E. S., 29, 139, 226; 30A, 247
Vijajan, M., 31B, 334
Vijayan, M., 30B, 399, 448; 32B, 324; 33A, 389; 33B, 465; 34B, 65, 173
Vijayaraghavan, P. R., 27, 257
Vildervanck, J. C., 29, 137
Vilkov, L. V., 26, 550, 558; 27, 718, 759; 28, 362, 687; 29, 450, 538; 30B, 405, 406; 33B, 529, 530, 531, 532, 536, 537; 35B, 889, 901
Villa, A., 32B, 411
Villa, A.C., 34B, 186; 35B, 38, 627
Villagrana, R.E., 30A, 41
Villarroel, H.S., 29, 436; 30B, 448
Villers, G., 27, 511; 28, 285; 29, 325
Villiger, H., 34A, 371; 35A, 458
Vimala, T. M., 34A, 306
Vimba, S., 26, 735; 27, 884; 28, 166; 29, 387
Vincent, H., 33A, 320; 34A, 91
Vincent, J. M., 26, 809
Vinciguerra, N., 33B, 495
Vinek, G., 31A, 26
Vinek, H., 35A, 263
Vinogradov, S.I., 27, 151
Viossat, C., 32B, 102
Virkar, A.V., 34A, 145; 35A, 77
Virtamo, I., 30B, 447
Visser, G.J., 29, 477; 32B, 182; 33B, 140; 35B, 128, 221
Visser, J.W., 29, 380
Visti, J., 35B, 674, 675
Viswamitra, M.A., 27, 817, 818; 28, 290; 30A, 375; 30B, 270, 399, 448; 31B, 56, 334; 32B, 324; 33B, 465
Viswanathan, K.S., 26, 675, 676
Viterbo, D., 31B, 53, 65; 32B, 428; 34B, 197, 198, 550; 35B, 213, 233
Vitulli, G., 35B, 567
Vivčar, O. I., 34A, 143
Vizzini, E. A., 33B, 469; 34B, 216, 220
Vlasov, L. B., 27, 444
Vlasov, L. G., 28, 315
Vlasov, V. A., 28, 26
Vlasov, V. G., 28, 341
Vlasova, E. V., 26, 537; 28, 341
Vlasse, M., 32A, 211; 32B, 214; 34A, 295; 35A, 179
Vliegenthart, J. A., 31B, 3
Vlisidis, A. C., 26, 489
Voblikova, V. A., 35B, 620
Voet, D., 29, 531; 32B, 293; 34B, 226
Vogt, L. H., 30A, 444; 32B, 238
Vogt, O., 30A, 178
Voicekhovsky, A. V., 29, 223
Voigtlaender-Tetzner, G., 26, 194
Vokkova, M. A., 30A, 173
Volavšek, B., 28, 682
Vold, C. L., 26, 218, 272
Vol'kenstejn, M. V., 30B, 443
Vol'kenstein, M. W., 31B, 161
von Volkmann, T., 26, 220, 293; 31A, 82
Völlenkle, H., 28, 317, 339; 29, 139; 32A, 68, 70, 99, 491, 494; 33A, 84, 467, 503, 507, 508; 34A, 258, 259, 363; 35A, 263, 311, 318
Volodina, G. F., 26, 476, 534; 28, 275; 29, 379; 31A, 229; 35B, 408
Vološčuk, Ja. V., 27, 200
Vol'pin, M. E., 28, 687; 32A, 511; 35B, 905
Volponi, L., 34B, 604; 35B, 868
Vooren, M., 35B, 135
Voorhoeve, J. M., 35A, 124
Vorbruggen, H., 30B, 396
Vorma, A., 26, 462; 28, 262; 30A, 427
Vorob'ev, V. V., 30A, 248
Voroncova, L. G., 30B, 84; 31B, 32, 150; 32B, 71
Voronkov, A. A., 26, 531; 27, 473; 28, 342; 29, 411; 32A, 417, 433, 510; 33A, 437; 34A, 379; 35A, 309
Voronova, A.A., 29, 383
Voronova, M.L., 27, 635
Vorosilov, Ju. V., 29, 138; 30A, 174, 176, 177, 178
Vos, A., 26, 315, 830; 27, 409, 411; 28, 58, 82, 494; 29, 477; 30A, 353, 389, 391; 30B, 399; 31A, 79, 80, 212, 214; 32A, 219; 32B, 33, 68, 182, 346; 33A, 198; 33B, 61, 140, 247, 282; 34A, 156, 177; 35B, 128
de Vos, K.J., 27, 164
Vossos, P.H., 28, 83
Vovk, Ja. N., 30A, 245
Vranka, R.G., 27, 858; 31B, 417; 32B, 327; 34B, 614

de Vries, A., 26, 602
de Vries, R., 29, 82
van Vucht, J. H. N., 26, 5, 27; 27, 302; 28, 6; 29, 54, 140; 30A, 173, 174; 31A, 12, 16; 32A, 142; 35A, 23
Vul, E. B., 32B, 113
Vuorelainen, Y., 28, 338; 29, 136, 140; 30A, 181

Waber, J. T., 29, 136, 223
Wachi, M., 33A, 321
Wada, K., 26, 588
Wada, T., 29, 292
Wada, Y., 33A, 235; 34B, 287
Wade, R. J., 34A, 155
Wadsley, A. D., 26, 352, 355, 373, 375, 382; 27, 528, 532; 29, 282, 305, 327; 30A, 336, 338, 340, 341, 31A, 139, 148, 154; 32A, 266, 478, 492; 33A, 294, 306, 505
Wadsten, T., 30A, 22, 174; 32A, 110, 142
Waerstad, K., 30B, 357
Wagner, A. J., 30B, 399; 33A, 115, 198; 30B, 282; 35A, 145, 348
Wagner, C. N. J., 26, 166
Wagner, E., 26, 83
Wagner, G., 35A, 279
Wagner, W. G., 32B, 386; 33B, 390; 34B, 509, 510
Wahl, K., 34A, 163
Waintal, A., 28, 357
Wainwright, C., 29, 224
Wakisaka, K., 35B, 199
Wald, A., 30A, 245
Wald, F., 26, 46
Waldrop, L., 35A, 432
Walford, L. K., 29, 9; 30A, 4, 173
Walitzi, E. M., 27, 601; 28, 194; 30A, 382, 433; 32A, 464; 34A, 169; 35A, 26
Walker, A., 26, 474
Walker, G., 29, 456
Walker, G. F., 35B, 905
Walker, J., 26, 732; 28, 573
Walker, L. A., 27, 414; 31A, 90; 34B, 124
Wallace, J. E., 35B, 781
Wallace, P. L., 34A, 146
Wallace, T. C., 27, 129; 29, 223; 30A, 40, 175; 31A, 122; 33A, 53, 56, 153
Wallace, W. E., 26, 139; 27, 235, 240; 29, 48 50, 137; 30A, 175
Wallach, D., 27, 651
Wallbridge, M. G. H., 30B, 393; 33B, 303
Wallen, J., 27, 667
Wallenfels, K., 30B, 446
Waller, J. M., 26, 100
Walley, W. B., 30B, 394
Walline, R. E., 29, 137; 30A, 175
Walling, C., 32B, 441
Wallmann, J. C., 27, 39; 29, 274, 434
Wallwork, S., 30B, 251, 259
Wallwork, S. C., 26, 653, 730; 27, 909, 1002; 29, 686; 30A, 274, 400; 31A, 176; 31B, 274, 435; 32A, 350; 32B, 275, 278, 279, 286; 33B, 252; 34B, 138; 35A, 326; 35B, 405
Walter, A. J., 29, 434
Walter, G., 27, 550
Walter, R., 34B, 13
Waltersson, K., 35A, 202
Walthers, K. K., 35B, 602
Walton, R. A., 33B, 483; 35B, 770
van Wambeke, L., 26, 388
Wampetich, M. J., 32A, 285
Wampler, D. L., 26, 316, 332; 27, 443, 646, 647, 814
Wandji, R., 34A, 142
Wang, A. -N., 27, 669
Wang, B. I. -C., 35B, 493
Wang, C. C., 28, 318
Wang, C. -H., 32B, 420
Wang, C. K., 28, 341
Wang, C. W., 27, 406
Wang, F. E., 26, 346, 545, 619; 27, 265; 29, 50; 30A, 23, 174, 177, 181; 31A, 46; 32A, 62, 142
Wang, H. K., 30B, 406
Wang, I. Y. M., 33B, 538
Wang, K. -S., 29, 435
Wang, R., 30A, 376; 31A, 20, 47; 32A, 18; 34A, 31, 124
Wanič, A., 33A, 266
Ward, R., 26, 370, 418; 29, 438; 30A, 329; 31A, 137
Warekois, E. P., 27, 382
Warf, J. C., 26, 169; 27, 241
Waring, J. L., 26, 491; 29, 438
Waring, L. J., 28, 358
Warlimont, H., 27, 12; 29, 135; 30A, 177
Warner, J., 29, 437
Warr, W. A., 35B, 190
Warren, B. E., 34A, 363
Warren, C. H., 34A, 218
Warren, H., 29, 229
Warren, I. H., 30A, 181, 244
Warren, R. C., 35B, 646
Warshaw, I., 26, 540
Wartenpfuhl, F., 26, 329
Warwicker, J. O., 26, 627, 764, 827; 27, 798
Wasai, G., 26, 627, 764, 827; 27, 798; 29, 760
Wasel-Nielsen, H. -D., 33A, 280; 35A, 214, 224
Waser, J., 26, 736; 28, 518; 35B, 263
Wasielewski, G. E., 29, 226
Wasilewski, R. J., 26, 240, 261, 294
Wasscher, J. D., 27, 345
Wasserman, J., 31A, 240; 32A, 147
Wassermann, G., 26, 177; 27, 186

Wasson, J. R., 35B, 700
Watanabé, A., 26, 563
Watanabe, D., 27, 176, 180, 274; 32A, 254
Watanabe, H., 28, 28; 30A, 17, 173
Watanabe, I., 34B, 320
Watanabe, K., 28, 356; 29, 138
Watanabé, T., 26, 326, 536, 537, 698, 781; 28, 108, 358, 477, 634, 682, 686; 29, 455, 609, 675; 30A, 415; 35B, 402
Watanabe, Y., 33B, 319, 34B, 413
Watelle, G., 35B, 15
Watenpaugh, K., 31B, 440, 445; 32B, 388
Waterhouse, C. R., 35B, 834
Waterman, M. J., 30B, 446
Waters, J. M., 27, 831; 32B, 233; 35B, 624, 781
Waters, T. N., 28, 640, 645; 29, 599; 30A, 404; 30B, 369, 393, 395; 31A, 102, 105; 31B, 389, 393, 395, 398, 428; 32B, 233, 380, 397, 442; 33B, 389, 398, 401; 34B, 500, 519; 35B, 129, 349, 413, 414, 615, 624, 781
Waterstrat, R. M., 26, 193; 27, 278, 279
Watkins, E. D., 34B, 510
Watkins, S. F., 34B, 411; 35B, 77, 865
Watling, R. C., 32B, 399
Watson, D. G., 26, 801; 27, 926; 30B, 173, 198; 35B, 103, 113, 389, 396, 467
Watson, H. C., 26, 811
Watson, J. D., 26, 820
Watson, K.J., 31B, 198; 32B, 265; 33B, 209; 34A, 231; 35B, 632, 788
Watson, W.H., 28, 518; 31B, 381; 32B, 446, 450; 33B, 458; 34B, 575, 578
Watt, G. W., 28, 682
Watterson, K. F., 26, 680
Watton, E. C., 34B, 565
Watts, J.A., 31A, 96; 35A, 276, 277
Watts, P. H., 35B, 219
Wawner, F., 28, 316
Wayman, C. M., 27, 251
Weakliem, H. A., 28, 314; 35A, 415
Weatherly, T. L., 27, 729
Weaver, C. F., 26, 311
Weaver, D. L., 30B, 394; 32B, 446, 450; 35B, 588, 610, 842
Weaver, E. E., 26, 301
Webb, L. E., 29, 617; 30B, 216
Webb, M. W., 28, 213; 30A, 374
Webb, N. C., 27, 863; 31A, 191, 32B, 144, 359; 34B, 343
Webb, W. W., 27, 185
Weber, A. H., 35B, 176
Weber, F., 27, 606; 33A, 176
Weber, H. P., 28, 683; 29, 509, 513; 32B, 334, 367, 458
Weber, R., 27, 515
Webster, M., 34B, 371
Webster, M. S., 28, 534
Webster, P. J., 33A, 97
Wechsberg, M., 34A, 219
Weeks, A. D., 26, 381; 27, 603; 28, 358
Weeks, C. M., 34B, 545; 35B, 332, 335
Wegener, J., 34A, 206
Weglowski, S., 30A, 174
Wehle, V., 30A, 39, 175; 32A, 233
Wehmeier, F. H., 35A, 45
Wei, C. H., 28, 69, 314, 606, 654, 684; 30A, 259; 30B, 383, 384; 31B, 356; 32A, 406, 510; 33B, 489, 490; 34A, 160; 35B, 789
Wei, L., 29, 597
Weichselgartner, H., 34B, 281
Weidelt, J., 29, 434
Weidenborner, J. E., 26, 413, 449, 717; 29, 434; 31A, 133; 35A, 158
Weidinger, A., 26, 628
Weidlein, J., 32A, 245
Weigel, D., 27, 571, 642; 28, 315, 316; 29, 434, 436; 32A, 348, 509; 34A, 240; 34B, 454
Weigel, F., 28, 286, 287; 33A, 134; 34A, 146
Weigers, G. A., 35A, 79, 124, 141
Weiher, J. F., 33A, 332
Weik, H., 28, 292; 30A, 181
Weill, G., 30A, 172
Weimann, J., 34B, 3
Weinblum, D., 34B, 241
Weininger, M.S., 34B, 523
Weinstock, B., 26, 301
Weir, C. E., 27, 425; 29, 138, 392; 34B, 79
Weisbach, J. A., 30B, 237
Weiss, A., 26, 553; 27, 610, 751; 28, 8, 11, 153, 223, 316; 30A, 65, 66, 97, 180, 182; 31A, 25, 48; 32A, 27, 43, 117, 122, 142, 158, 323; 33A, 39, 51, 98, 153; 34A, 53, 93, 122, 143, 145, 35A, 9, 24, 43, 64, 124, 139
Weiss, Armin, 27, 348, 610; 29, 140
Weiss, E., 28, 595; 29, 452, 467; 30A, 287; 30B, 284; 33B, 296, 297, 299, 324; 35B, 440, 471, 472, 473
Weiss, E. J., 28, 576; 20A, 178, 180, 181, 246, 247
Weiss, G., 34A, 283
Weiss, G. S., 29, 515
Weiss, J., 26, 345; 27, 410; 28, 661; 29, 245; 30A, 257; 31A, 81; 31B, 451; 32B, 307, 315, 318, 426; 33A, 361;

33B, 273; 34B, 378; 35A, 394
Weiss, R., 27, 633; 28, 314, 682; 29, 317, 353; 30B, 398; 31A, 89, 91, 92, 114, 209; 32A, 170, 171, 174, 177, 508; 32B, 444; 33B, 293; 34A, 222
Weiss, R. J., 32A, 98
Weissberg, B. G.. 30A, 174
Weitz, G., 26, 798; 27, 1034
Weitzel, H., 32A, 315
Wells, P., 35A, 66
Welsh, H. K., 30B, 397
Welsh, H. L., 27, 732
Welsh, M. J., 27, 1006; 28, 562
Wen, P. -Y., 29, 436
Wendlandt, W. W., 27, 887
Wendling, J.-P., 29, 353; 31A, 209
Wengelin, R. F., 32B, 87
Wenglowski, S., 29, 27
Wenzel, R. G., 30A, 40, 175; 31A, 122; 33A, 255, 258
Werding, G., 26, 400
Werkema, M. S., 29, 674; 30B, 139
Werner, P. -E., 26, 168; 33B, 168; 34B, 156, 250; 35B, 195, 196, 199
Werner, J. H.,26, 9, 127, 192; 27, 196, 261; 29, 51, 136, 224; 30A, 248
Wertz, D. L., 34A, 178
West, R., 27, 814
Westbrook, J. H., 26, 190; 27, 269
Westerdahl, A., 28, 371; 32B, 194
Westerlund, M., 29, 372
Westin, L., 33A, 110
Westlake, D. G., 27, 242
Westman, S., 26, 352; 27, 13; 28, 41, 30A, 3, 172, 244; 33A, 72, 108; 34A, 77, 115; 35A, 40, 90
Weston, N.E., 27, 408
Westphal, W.B., 26, 372
Westrum, E.F., 29, 138
de Wet, J. F., 26, 191
Wey, R., 32A, 511
Whalley, E., 28, 118
Wheatley, P. J., 26, 643, 719, 721; 27, 885, 919, 941, 942, 980; 28, 601, 682; 29, 639, 641, 653; 30B, 36, 93, 277, 380; 31A, 77; 31B, 33, 303, 467; 32B, 303, 304; 33B, 188; 35B, 64
Wheeler, A. G., 31B, 432; 32B, 283; 33B, 333, 334
Wheeler, D. A., 32A, 510; 33A, 97
Wheeler, M. J., 30A, 172
Wheelwright, E. J., 28, 338
Whillans, F. D., 35B, 705
Whimp, P. O., 35B, 579, 772
Whinfrey, C. G., 26, 399, 415; 28, 358
Whitaker, A., 32A, 393; 35A, 329
White, A. H., 35B, 798
White, B. M., 31B, 348
White, D., 31A, 240
White, D. E., 28, 684; 29, 436
White, D. N. J., 35B, 333
White, E. A. D., 28, 356; 29, 329
White, G. D., 27, 132
White, J., 28, 683
White, J. C. B., 35B, 34
White, J. G., 26, 46; 27, 246, 582, 1042; 28, 37, 316, 393, 29, 137; 30A, 83, 91, 179, 181, 248, 417; 32A, 53, 92; 33A, 235; 35A, 44; 35B, 210
White, W.B., 29, 322, 436
Whitfield, H.J., 33A, 220; 35A, 140; 35B, 831
Whitla, A., 32B, 350
Whitlow, S. H., 30B, 448; 31B, 69; 32A, 369; 33B, 281; 35B, 459, 461
Whitney, D. R., 27, 347
Whittaker, A. G., 34A, 149
Whittaker, E. J. W., 26, 526; 34A, 391
Whyman, P. E., 29, 663
Wiberley, S. E., 30A, 444
Wicke, E., 27, 194
Wickham, D. G., 27, 495
Wiegers, G. A., 26, 315; 27, 409; 28, 58; 30A, 75, 181; 31A, 67, 79, 80; 33A, 61, 141
Wieghardt, K., 35A, 394
Wieker, W., 29, 438
Wiesner, J. R., 31B, 375; 32B, 31
Wiewandt, T. A., 30A, 246
Wiewiorowski, M., 32B, 264
Wiggins, C. D., 28, 85
Wilchinsky, Z. W., 29, 433
Wilcox, C. F., 33B, 534
Wilcox, D. L., 28, 341
Wilcox, H. E., 27, 945
Wilcox, R. C., 30A, 177, 247
Wild, S., 31A, 30
Wilford, J. B., 32B, 350, 470; 33B, 326; 34B, 463
Wilhelm, G., 29, 225
Wilhelm, H. A., 29, 135
Wilhelmi, K. -A., 26, 500; 30A, 304, 321; 31A, 150; 32A, 277, 279, 280, 281; 34A, 236; 35A, 202, 256
Wilkens, M., 29, 135
Wilkes, C. E., 30B, 333
Wilkes, G. R., 31B, 356; 32A, 510
Wilkins, C. J., 27, 659, 770
Wilkins,M. H. F., 26, 817; 27, 1055
Wilkinson, C., 32A, 11
Wilkinson, G., 26, 680
Wilkinson, M. K., 26, 106, 141, 316,

318; 27, 151, 333, 334, 368, 434, 440
Will, G., 28, 687; 30A, 369; 31A, 3; 32A, 3; 33A, 309, 323, 388; 34A, 3, 257; 34B, 123, 281; 35A, 244, 288
Willcox, R. J. M., 29, 224
Willens, R. H., 27, 370
Willett, R. D., 28, 85; 29, 378, 590; 30A, 296; 31B, 29; 32A, 203; 33A, 252; 34A, 180, 204; 35B, 734, 880
Willett, R. E., 27, 18
Williams, D. A., 33B, 455
Williams, D.E., 26, 92; 27, 359, 913; 29, 138, 225, 729; 30B, 186, 448; 31B, 80; 32B, 73; 34B, 55, 56
Williams, D. G., 27, 798; 28, 684; 31B, 38; 35B, 268
Williams, D. J., 32A, 511; 34A, 374
Williams, E. W., 29, 223
Williams, G. I., 26, 203; 31A, 49
Williams, H. J., 26, 351, 414; 27, 511; 29, 51, 136
Williams, I., 26, 172
Williams, J. M., 30B, 448; 32B, 28; 33B, 23
Williams, K. C., 27, 41
Williams, P. P., 29, 398; 34B, 306
Williams, Q., 27, 729
Williams, R. J., 32B, 446, 450; 34B, 575
Williams, R. M., 31B, 274; 32B, 278, 286; 33B, 252
Williams, S. J., 26, 455; 27, 716; 28, 357
Williams, T., 34B, 211
Williamson, G. K., 26, 218
Williamson, S. M., 28, 75, 130, 228; 29, 335
Willis, B. T. M., 28, 129; 30A, 266
Williston, C. S., 31B, 67; 32B, 337, 361; 35B, 558
Willoughby, T. V., 34B, 355
Willstadter, E., 28, 639; 30B, 396
Wilman, H., 26, 354
Wilson, A., 26, 513; 29, 686
Wilson, A. S., 28, 338
Wilson, C. G., 26, 31; 27, 36, 199; 28, 27; 29, 86; 30A, 247
Wilson, E. B., 27, 411; 28, 362
Wilson, G. E., 35B, 217
Wilson, H. R., 28, 686; 30B, 176; 31B, 183; 32B, 272; 34B, 262; 35B, 276, 283
Wilson, J. N., 33B, 179
Wilson, R. H., 29, 25
Wilson, S. A., 29, 636
Wincierz, P., 27, 304
Wing, R. M., 30B, 355; 32B, 297; 33B, 267; 34B, 355; 35B, 431, 716, 804, 807
Winkler, H., 30B, 398
Winnewisser, M., 28, 362; 34A, 233
Winquist, B. H. C., 32B, 482; 34B, 511
Wintenberger, M., 26, 52; 27, 55; 31A, 21
Winter, G., 28, 683
Winter, J., 35A, 24, 124
Winterhage, H., 26, 195
Wise, J. J., 31B, 386
Wise, W. S., 29, 438
Wiseman, T. J., 29, 555
Wisnyi, L. G., 27, 126
Witkop, B., 34B, 161, 278; 35B, 206, 362
Witmore, W. G., 34B, 27
Witowska, A. M., 32A, 343
Witte, H., 33A, 98
Witte, J., 29, 57
Witte, W., 29, 435
Witters, R. D., 30B, 399
Witters, R. W., 31B, 213
Wittig, G., 27, 885
Wittke, O., 26, 800
Wittmann, A., 26, 9, 124, 398; 27, 11, 375, 500, 541; 28, 317, 339; 29, 64, 85, 139; 30A, 325; 32A, 68, 70, 491, 494; 33A, 467, 503, 507, 508; 34A, 258 259, 363; 35A, 311, 318
Witz, J., 26, 817, 819; 29, 759
Witzen, M. B., 26, 817, 819; 27, 467 29, 759
Witzmann, H., 29, 435
Wnuk, R. C., 29, 334
Woditsch, P., 34A, 246
Wodtcke, F., 27, 706
Woensdregt, C. F., 29, 438
Wöhrle, H., 26, 323; 28, 315; 30A, 263
Wojcicki, A., 32 B, 455
Wolberg, A., 34A, 231
Wold, A., 28, 133, 285; 30A, 181; 31A, 51; 33A, 153; 34A, 295; 35A, 179
Wolf, A. P., 33B, 290
Wolf, G. -U., 34B, 397
Wolf, H., 32A, 463
Wolf, L, 26, 344; 29, 435
Wolf, N. de, 32B, 199
Wolf, W., 30B, 158
Wolfe, G. A., 34B, 308
Wölfel, E., 26, 87, 798; 27, 1034; 28, 14
Wolff, P.A., 26, 127
de Wolff, P. M., 26, 368; 28, 130, 289, 316; 29, 380; 30A, 319; 31A, 130
Wolfrum, R., 33B, 299
Wollan, E. O., 26, 106, 141, 318; 27, 151, 274, 333, 334, 368, 434; 440; 28, 27; 29, 136, 137
Wollrab, J. E., 33B, 539; 35B, 903
Wolski, W., 33A, 266
Wolten, G. M., 32A, 311
Wolters, J., 33B, 140
Wolthius-Spuy, M., 35B, 7
Wonacott, A. J., 31B, 193
Wondratschek, H., 28, 339
Wones, D. R., 26, 525; 27, 683; 28, 288

Wong, C. H., 29, 665; 30A, 33, 175; 30B, 315; 34B, 459
Wood, A. M., 32B, 475; 33B, 347
Wood, D., 30A, 53, 179
Wood, D. H., 29, 29, 137; 33A, 105, 131; 34A, 75, 84, 86, 146
Wood, D. L., 27, 269
Wood, E. A., 27, 503
Wood, J. S., 26, 830; 28, 613; 29, 565, 571, 616, 621, 622; 30B, 295; 33B, 437; 35B, 630, 756
Wood, M. M., 35A, 436
Wood, R. M., 27, 370
Woodrow, J. P., 32A, 473
Woods, J. D., 33B, 9
Woodward, C. L., 29, 226
Woodward, I., 30B, 448
Woodward, P., 29, 238; 30B, 399; 31A, 178; 32A, 411; 32B, 319, 355; 33B, 481, 506; 35B, 77
Woolfson, M.M., 26, 610
Woolley, J.C., 26, 42; 29, 223, 224; 30A, 246, 248
Worlton, T. G., 32A, 29
Wormald, J., 34B, 425, 456, 473; 35B, 516, 572
Worrall, I. J., 30A, 274
Worsham, J. E., 34B, 30
Worzala, H., 28, 317; 31A, 121; 32A, 383; 33A,418
Wriedt, H. A., 27, 253; 29, 223
Wright, D. A., 27, 244, 767; 33B, 455
Wright, H.C., 26, 149
Wright, J. D., 30B, 262, 396; 31B, 280, 369; 33B, 451
Wright, J. G., 30A, 181
Wu, V. Y., 28, 3; 31A, 5
Wuensch, B. J., 28, 19; 29, 15; 31A, 164; 32A, 229
Wulf, R., 33A, 153
Wulfsgruber, H., 33A, 153
Wunderlich, H., 35A, 331; 35B, 40
Wunderlich, J.A., 26, 625; 27, 1002; 32B, 256; 34B, 146, 325
Wyckoff, H. W., 27, 746, 801
Wyckoff, R. W. G., 26, 340
Wydeven, T., 28, 315

Yadava, V.S., 34A, 212
Yaji, H., 32B, 66
Yakel, H.A., 26, 311
Yakel, H.L., 26, 317, 570; 27, 177, 546; 28, 138, 281; 31A, 130; 32A, 264; 33A, 271; 34A, 112, 249
Yakinthos, J., 34A, 72
Yakushi, K., 35B, 417
Yamada, A., 27, 802
Yamada, T., 34A, 101
Yamada, Y., 28, 108
Yamaguchi, G., 28, 318; 29, 299, 433; 32A, 431; 33A, 276
Yamaguchi, K., 26, 395
Yamaguchi, M., 26, 512
Yamaguchi, S., 27, 176; 33A, 118; 34A, 145
Yamamato, T., 35B, 765
Yamamoto, A.S., 30A, 247
Yamamoto, H., 28, 28
Yamamoto, K., 29, 138
Yamamuro, Y., 27, 802
Yamara, M., 26, 554
Yamashita, S., 31B, 331; 33B, 319
Yamashita, Y., 26, 6, 30
Yamauchi, H., 29, 136
Yanagida, H., 29, 299, 433
Yanagisawa, S., 34A, 146
Yanagita, M., 27, 802
Yanase, A., 26, 354
Yang, K.T., 27, 396
Yannoni, N.F., 26, 4, 393; 28, 433; 30B, 171, 399; 31B, 103; 32B, 131, 137; 33B, 91
Yarnell, J.L., 30A, 40, 175; 31A, 122
Yasaki, E., 35B, 678
Yasuda, K., 30B, 446
Yasukōchi, K., 26, 149, 192; 27, 162
Yasuoka, N., 28, 682; 32B, 104; 33B, 313, 500; 34B, 58, 455; 35B, 188, 488, 521
Yawney, D.B.W., 35B, 633
Ydo, D.J.W., 32A, 65
Yeats, E.A., 34B, 97
Yen, M.-C., 29, 436
Yen, T.-M., 30B, 315
Yen, T.-S., 29, 435
Yeowell, D.A., 26, 830
Yin, S.S., 27, 703; 29, 435
Yocom, P.N., 32A, 92
Yoda, N., 29, 760
Yoganarasimhan, S.R., 26, 352
Yokozeki, A., 35B, 900
Yonath, A., 31B, 193
Yonco, R.M., 30A, 175
Yonemitsu, O., 34B, 161; 35B, 206
Yorks, K.P., 26, 484
Yoshida, M., 33B, 313
Yoshida, N., 34B, 413
Yoshihara, T., 29, 760
Yoshii, C., 28, 356
Yoshimura, T., 29, 437
Yoshinaga, M., 28, 358
Yoshino, A., 31B, 254; 35B, 229, 747
Youdelis, W., 31A, 42
Young, A.P., 27, 520; 28, 357; 29, 319
Young, D.W., 27, 795, 1017, 1031; 28,524; 30B, 445; 31B, 223; 33B, 73; 34B, 262; 35B, 283
Young, H.S., 30A, 55, 179; 35A, 20
Young, J.E., 30B, 405
Young, J.F., 27, 659
Young, J.M., 35B, 86
Young, R.A., 27, 674; 28, 290; 29, 369

Young, W.A., 27, 481; 28, 341
Ysker, J.St., 35A, 282
Yu, D.-J., 28, 221
Yu, T.-H., 29, 436
Yuasa, T., 31A, 77
Yun, T.S., 26, 219; 27, 92
Yund, R.A., 27, 58, 293; 29, 223
Yu-p'u, Y., 26, 112
Yurko, G.A., 27, 296
Yvon, K., 32A, 44, 507; 33A, 153; 34A, 47; 35A, 110

Zabin, A. G., 26, 381; 27, 535; 28, 342
Zabšky, H. J., 28, 682
Začackaja, A. V., 29, 226
Zacharias, D. E., 30B, 237, 238; 35B, 365
Zachariasen, W. H., 26, 322, 391; 28, 22, 35, 38, 39, 155, 156, 157, 162, 166, 168, 171, 29, 386, 437; 30A, 51, 178, 420; 31A, 61, 68; 34A, 135; 35A, 113
Zacharis, H., 35B, 190
Zacharis, H. M., 33B, 128; 35B, 203
Zackrisson, F., 35A, 70, 124
Zaeske, P., 30B, 396
Zagrjazskij, V.L., 29, 225
Zagrjazsky, V.L., 30A, 178, 245
Zahrobsky, R. F., 28, 512; 30B, 370; 33A, 368
Zajceva, M. A., 27, 511
Zakharov, A. I., 27, 366; 29, 226; 33A, 153
Zakharova, G. N., 28, 438; 30B, 443; 32B, 331; 35B, 489
Zakrzewski, G., 35B, 696
Zalar, S. M., 27, 181
Zalkin, A., 26, 57, 58, 686; 27, 7, 455, 571, 616 ; 28, 71, 75, 130, 181, 226, 228; 29, 203, 264, 273, 274, 288, 289, 331, 335, 347, 356, 560; 30A, 311; 30B, 305; 31A, 79, 84, 101 179; 31B, 294, 315, 414; 32A, 161; 32B, 238, 294; 33A, 158; 33B, 265, 323, 394, 430, 435, 517; 34A, 191, 360; 34B, 316, 470; 35B, 313, 429
Zaluc'kij, I. I., 29, 140, 141; 30A, 3, 172, 173; 31A, 12, 14; 32A, 6
Zambelli, A., 26, 611
Zambonelli, L, 26, 769, 779; 27, 968, 969; 28,687; 29, 286; 30B, 373, 378, 381; 33B, 486; 35B, 513, 759
Zanazzi, P. F., 31A, 186, 190; 32A, 365, 391; 33A, 382, 405; 33B, 21, 133; 34B, 147, 201, 203; 35A, 333, 439
Zanini, G., 26, 636
Zannetti,R., 26, 559; 27, 973; 28, 682
Zanowick, R. L., 27, 240
Zaraev, Z. S., 29, 141
Zarečnjuk, O. S., 27, 5; 29, 139; 30A, 172; 32A, 141; 33A, 153; 34A, 143
Žarinova, A. I., 27, 494
Zaripov, N. M., 35B, 893, 897, 898
Zarli, B., 34B, 533; 35B, 819
Zaslavskij, A. I., 26, 173, 412; 27, 494, 508; 29, 62; 30A, 244; 35A, 32
Zaslow, B., 26, 810
Zasorin, E. Z., 27, 718; 28, 362; 34A, 238
Zauberis, D. D., 28, 23; 30A, 65, 180
Zavališin, E. I., 35B, 124
Zaval'skij, Ju. P., 30A, 174
Zav'jalova, A. A., 29, 311; 30A, 446
Zčerbak, V. I., 34A, 142
Ždanov, G. S., 26, 35, 285, 372, 381, 419 751; 27, 61, 86, 155, 286, 393, 401, 505, 506, 508, 643; 28, 135, 207, 431, 682; 29, 323; 32A, 269; 32B, 71; 33A, 305; 34B, 103
Ždanova, T. A., 34A, 168
Ždanowicz, W., 29, 135, 223
Zdorik, T. B., 26, 379
Zegler, S. T., 27, 285; 30A, 244, 247
Zeigan, D., 32A, 502
Zeil, W., 30B, 405; 35B, 890
Zelencov, V. V., 30B, 448
Zelenin, L. P., 29, 223
Zelenko, B., 31A, 178
Zelikman, A. N., 26, 199
Željaskova-Panaiotova, M., 26, 410
Zelwer, C., 31A, 85, 132, 225
Zemann, A., 26, 416; 27, 635
Zemann, J., 26, 324, 416, 451; 27, 635; 28, 245
Zemljanov, M. G., 33A, 81
Zen, E-an, 29, 437
Zeppezauer, E., 35B, 887
Zeppezaur, M., 34B, 134
Zhang, R. -Y., 28, 339
Zhao, G. -G., 28, 339
Zidkov, N. P., 31B, 381
Ziegleder, G., 35A, 24, 124
Ziegler, M., 28, 661; 32B, 347
Ziegler, M. L., 33B, 273, 274; 34B, 378
Žigach, A. F., 33B, 530
Zigan, F., 32A, 242, 243
Zijlstra, H., 27, 164
Zimina, A.N., 29, 433
Zimkina, T.M., 26, 446
Zimmerman, I. C., 28, 409
Zimmins, A., 34B, 368, 372
Ziolkowska, B., 27, 863
Ziolo, R. F., 35B, 829
Zirakzadeh, M., 29, 712
Ziser, M., 35B, 164
Žitar, V. F., 32A, 95
Zjulkovska, B., 29, 628
Zlomanov, V. P., 26, 432
Žmud, S., 26, 391
Zobel, T., 30B, 282; 31A, 111; 31B, 69
Zocchi, M., 27, 422; 33A, 282; 34B, 572

Zoer, H., 35A, 145
Zohner, K., 32A, 158
Zolotov, S., 30A, 289
Zoltai, T., 30A, 436; 31A, 124; 32A, 424, 468; 33A, 403, 404, 494; 34A, 363
Zorani, N. G., 33A, 272
Zorkij, P. M., 26, 671, 678
Zou, Z. -R., 27, 703
Zou, Z. -Y., 27, 703
Zsoldos, E., 27, 581
Zubieta, J. A., 34B, 597; 35B, 802
Zubkov, V., G., 34A, 58
Žubov, Ju. A., 29, 760; 35B, 886
Zuev, V. N., 26, 523
Žukov, A. N., 34A, 168
Žukova, L. A., 29, 319
Zukova, T. B., 28, 62; 30A, 244; 35A, 32
Zukovsky, V. M., 28, 341
Zulakaev, L. P., 35B, 124
Žuravlev, N. N., 26, 62, 65, 66; 27, 46, 85, 86, 88, 286; 374; 29, 224, 226; 30A, 175; 35A, 27
Zussman, J., 26, 526; 28, 207; 30A, 434; 34A, 392
Zuze, V. P., 26, 173
Zvjagin, B. B., 27, 697; 28, 318; 33A, 472
Zvonkova, Z. V., 26, 586; 27, 829; 28, 113, 682; 29, 437; 30B, 64, 156; 32B, 71; 33B, 264, 488
Zvyzgincev, O. E., 27, 290
Zwell, L., 27, 253; 29, 223; 31A, 31
Zwicker, U., 27, 35, 187, 296, 304, 372
Zwingmann, G., 27, 314
Zygmunt, A., 29, 140; 32A, 21
Zykalova, K. A., 28, 684